Handbook of Meat, Poultry and Seafood Quality

Second Edition

Handbook of Meat, Poultry and Seafood Quality

Second Edition

Edited by

Leo M. L. Nollet
Professor of Biotechnology
Department of Engineering Sciences
Hogeschool Gent
Ghent, Belgium

WILEY-BLACKWELL
A John Wiley & Sons, Inc., Publication

Wiley-Blackwell is an imprint of John Wiley & Sons, formed by the merger of Wiley's global Scientific, Technical and Medical business with Blackwell Publishing.

Editorial Offices
2121 State Avenue, Ames, Iowa 50014-8300, USA
The Atrium, Southern Gate, Chichester, West Sussex, PO19 8SQ, UK
9600 Garsington Road, Oxford, OX4 2DQ, UK

For details of our global editorial offices, for customer services and for information about how to apply for permission to reuse the copyright material in this book please see our website at www.wiley.com/wiley-blackwell.

Library of Congress Cataloging-in-Publication Data

Handbook of meat, poultry and seafood quality / [edited by] Leo M. L. Nollet. – 2nd ed.
p. cm.
Includes bibliographical references and index.
ISBN 978-0-470-95832-2 (hardcover : alk. paper)
1. Meat–Quality–Handbooks, manuals, etc. 2. Poultry–Quality–Handbooks, manuals, etc.
3. Seafood–Quality–Handbooks, manuals, etc. I. Nollet, Leo M. L., 1948–
TX556.M4H36 2012
363.19′29–dc23

2012004867

A catalogue record for this book is available from the British Library.

Wiley also publishes its books in a variety of electronic formats. Some content that appears in print may not be available in electronic books.

Set in 10/12pt Times by SPi Publisher Services, Pondicherry, India
Printed and bound in Singapore by Markono Print Media Pte Ltd

1 2012

Contents

Preface

The quality of a food is defined from two perspectives: scientific status and consumer preferences. Scientific factors affecting the quality of a food include composition, spoilage, colorants, additives, nutrients, flavorants, functional ingredients (affecting health), contamination, general safety, etc. Consumer preferences are linked directly to the human senses: sight, touch, smell, taste, and mouthfeel. Visual factors refer to color, moisture, overall appearance, etc. Tactile factors refer to sliminess, elasticity, softness, hardness, etc. Factors responsible for taste and smell cover many specific chemicals. Mouthfeel refers to texture, softness, tenderness, chewy sensation, etc. In the last 10 years or so, food quality has been defined by most professionals to include "health" and "safety." The nutrition and safety of foods has always been important, especially since the 1970s. The word "health" now includes manipulating certain chemical components in food to increase food's positive impact on our health. "Safety" now refers to a whole spectrum of new legal or recommended requirements for both fresh and processed foods. These requirements are designed to exclude or prevent undesirable agents (biological, chemical, physical, environmental, and extraneous) in our foods.

For ease of reference, we can consider that the quality of a food is the composite picture of many factors. In the last 5 to 10 years, many professional reference books have become available that explore the relationship between such factors and food quality.

The second edition of *Handbook of Meat, Poultry, and Seafood Quality* focuses especially on the different quality factors of muscle foods (beef, pork, poultry, and seafood).

The book consists of six parts. Part I, Quality Aspects of Products of Animal Origin, deals with general sensory aspects of muscle foods such as color, texture, and flavor and how to determine these parameters in muscle foods. Part II, Flavor, covers the sensory and chemical characterization of food; describes the process, savory, and off flavors and rancidity in foods; and deals with land animal products and marine animal and plant products.

Parts III to VI explore in depth quality parameters of beef (Part III), of pork (Part IV), of poultry (Part V), and seafood (Part VI). Parameters discussed in the different chapters of these four parts include shelf life, microbiological and sensory properties, packaging, and others, for fresh and for frozen muscle foods.

This work is the result of the combined efforts of more than 30 professionals from industry, government, and academia worldwide. They represent more than 11 countries with diverse expertise and background in the quality of muscle foods. An international editorial team of seven members from three countries led these experts. Each contributor or editor was responsible for researching and reviewing subjects of immense depth, breadth, and complexity. Care and attention were paramount to ensure technical accuracy for each topic. It is our sincere hope and expectation that it will serve as an essential reference on the quality of muscle foods for all professionals in government, industry, and academia.

The editorial team wishes to thank all the contributors for sharing their expertise throughout our journey. We also thank the reviewers for giving their valuable comments on how to improve the contents of each chapter. All these professionals are the ones who made this book possible. We trust that you will benefit from the fruits of their labor.

Leo M.L. Nollet
Terri D. Boylston
A.E. Feng Chen
A.E.P.C. Coggins
A.E. G. Hyldig
C.R. Kerth
Lisa McKee

Contributors

José Ángel Pérez-Alvarez
Departamento de Tecnología Agroalimentaria
Escuela Politécnica Superior de Orihuela
Universidad Miguel Hernández
Orihuela, Alicante, España

Hyung Hee Baek
Associate Professor
Department of Food Engineering
Dankook University
Cheonan, Korea

Sancho Bañon Arias
Profesor Titular de Tecnología de Alimentos
Grupo de Investigación en Tecnología Alimentaria
Facultad de Veterinaria
Departamento de Tecnología de Alimentos, Nutrición y Bromatología
Universidad de Murcia
Murcia, España

Karen L. Bett-Garber
Research Food Technologist
Southern Regional Research Center
Agricultural Research Service
U.S. Department of Agriculture
New Orleans, LA, USA

Terri Boylston
Associate Editor
Associate Professor
Department of Food Science and Human Nutrition
Iowa State University
Ames, IA, USA

Allan Bremner
Professor
Allan Bremner and Associates
Queensland, Brisbane, Australia

Susan Brewer
Food Science and Human Nutrition
Unversity of Illinois at Urbana-Champaign
Urbana, IL, USA

Christoph Cerny
Firmenich SA
Geneva, Switzerland

Feng Chen
Associate Editor
Assistant Professor
Department of Food Science and Human Nutrition
Clemson University
Clemson, SC, USA

Elizabeth Cobb
Department of Family and Consumer Sciences
College of Agricultrual, Consumer and Environmental Sciences
New Mexico State University
Las Cruces, NM, USA

Patti C. Coggins
Associate Editor
Assistant Research Professor
Department of Food Science and Technology
Director, Garrison Sensory Evaluation Laboratory
Mississippi State University
Mississippi State, MS, USA

Neil Da Costa
International Flavors and Fragrances, Inc
Union Beach, NJ, USA

Paul L. Dawson
Professor of Food Science and Human Nutrition
Food Science and Human Nutrition Department
Clemson University
Clemson, SC, USA

Sanja Eri
Mastertaste
Teterboro, NJ, USA

Alex Augusto Gonçalves
Professor of Seafood Technology
Universidade Federal Rural do Semi-Árido (UFERSA)
Mossoró, Brazil

Grethe Hyldig
Associate Editor
Senior Research Scientist
Danish Institute for Fisheries Research
Department of Seafood Research
Technical University of Denmark
Kongens Lyngby, Denmark

Flemming Jessen
Senior Research Scientist
Danish Institute for Fisheries Research
Department of Seafood Research
Technical University of Denmark
Kongens Lyngby, Denmark

Joseph P. Kerry
Packaging Group
School of Food & Nutritional Sciences
University College Cork
Cork, Ireland

Chris R. Kerth
Associate Professor, Meat Science
Department of Animal Science
Texas A&M University
College Station, TX, USA

V. Kuri
Food and Nutrition
School of Biological Sciences
University of Plymouth
Plymouth, UK

Jean-Luc Le Quéré
Institut National de la Recherche Agronomique (INRA)
Unité Mixte de Recherche sur les Arômes (UMRA)
Dijon, France

Juana Fernández-López
Departamento de Tecnología Agroalimentaria
Escuela Politécnica Superior de Orihuela
Universidad Miguel Hernández
Orihuela (Alicante), España

Emilia Martinsdóttir
Head of Department
R&D Division/Consumer and Food Safety
Icelandic Fisheries Laboratories
Reykjavík, Iceland

Lisa McKee
Associate Editor
Professor, Human Nutrition and Food Science
Department of Family and Consumer Sciences
New Mexico State University
Las Cruces, NM, USA

William Benjy Mikel
Professor and Head
Food Science, Nutrition, and Health Promotion
Director, Food Science Institute
Mississippi State University
Mississippi State, MS, USA

Rhonda K. Miller
Professor
Meat Science Section
Department of Animal Science
Texas A&M University
College Station, TX, USA

Narendra Narain
Departamento de Engenharia Química
Centro de Ciências Exatas e Tecnologia
Universidade Federal do Sergipe, Cidade Universitária, Jardim Rosa Elze
São Cristóvão, SE, Brazil

Jette Nielsen
Research Coordinator
Danish Institute for Fisheries Research
Department of Seafood Research
Technical University of Denmark
Kongens Lyngby, Denmark

Leo M.L. Nollet
Editor
Hogeschool Gent
Department of Engineering Sciences
Gent, Belgium

Maria Lúcia Nunes
Fundação Núcleo de Tecnologia Industrial do Ceará
Rua Prof. Rômulo Proença, S/N
Fortaleza, CE, Brazil

Maurice O'Sullivan
Packaging Group
School of Food & Nutritional Sciences
University College Cork
Cork, Ireland

Sarah Padilla
Department of Family and Consumer Sciences
College of Agricultural, Consumer and Environmental Sciences
New Mexico State University
Las Cruces, NM, USA

Ronald B. Pegg
Department of Food Science and Technology
The University of Georgia
Athens, GA, USA

María de Lourdes Pérez-Chabela
Biotechnology Department
Universidad Autonoma Metropolitana Iztapalapa
Mexico City, Distrito Federal, México

Rian Schelvis
Netherlands Institute for Fisheries Research
IJmuiden, The Netherlands

Fereidoon Shahidi
Department of Biochemistry
Memorial University of Newfoundland
Newfoundland and Labrador, Canada

Nick Spinelli
Chief Executive Chef
Kraft Foods
Chicago, IL, USA

Jack Thomas
Microbiology of Fresh and Frozen Beef
Department of Animal and Range Sciences
New Mexico State University
Las Cruces, NM, USA

Alfonso Totosaus
Food Science Lab
Tecnológico de Estudios Superiores de Ecatepec
Estado de México, México

Veronika Válková
Faculty of Veterinary Hygiene and Ecology
Department of Meat Hygiene and Technology
University of Veterinary and Pharmaceutical Sciences Brno
Brno, Czech Republic

Part One
Quality Aspects of Products of Animal Origin

1 Chemical and Biochemical Aspects of Color in Muscle Foods

José Ángel Pérez-Alvarez and Juana Fernández-López

Abstract: Color is the main aspect that defines a food's quality and that most influences consumer choice. Sensorial quality, especially color and appearance of meat, can be affected by both internal and external factors.

Chromoproteins, carotenes, and carotenoproteins are important in meat color. Carotenes are responsible for the color of beef fat, poultry meat and skin, fish, and shellfish. Of the hemoglobins present postmortem in the muscle, myoglobin is the one mainly responsible for color. Cytochromes are metalloproteins with a heme group with a role in meat coloration. The principal role of fat is in the brightness of meat products.

The color of meat may be altered by several factors, including exposure to light, microbial growth, rancidity, and exposure to oxygen.

Keywords: Muscle-based food color, carotenes, hemoproteins, myoglobin, cytochromes, fat color, melanosis, discoloration, premature browning, shelf life

1.1 GENERAL ASPECTS OF MUSCLE-BASED FOOD COLOR

The first impression that a consumer receives concerning a food product is established visually, and among the properties observed are color, form, and surface characteristics. Color is the main aspect that defines a food's quality, and a product may be rejected simply because of its color, even before other properties, such as aroma, texture, and taste, can be evaluated. This is why the appearance (optical properties, physical form, and presentation) of muscle-based products at the sales point is of such importance (Lanari *et al.* 2002).

Regarding the specific characteristics that contribute to the physical appearance of meat, color is the quality that most influences consumer choice (Krammer 1994). The relationship between meat color and quality has been the subject of study since the 1950s, indeed, since Urbain (1952) described how consumers had learned through experience that the color of fresh meat is bright red, and any deviation from this color (nonuniform or anomalous coloring) is unacceptable (Diestre 1992). The color of fresh meat and associated adipose tissue is, then, of great importance for its commercial acceptability, especially in the cases of beef and lamb (Cornforth 1994), and in certain countries, for example, the USA and Canada, there have been many studies to identify the factors controlling its stability. Adams and Huffman (1972) affirmed that consumers relate the color of meat to its freshness. In poultry, the consumers of

Handbook of Meat, Poultry and Seafood Quality, Second Edition. Edited by Leo M. L. Nollet.

many countries also associate meat color with the way in which the animal was raised (intensive or extensive) and fed (cereals, animal feed, etc.).

Color as a quality factor for meat can be appreciated in different ways in different countries; for example, in Denmark, pork meat color holds fifth place among qualities that affect consumers' purchase decisions (Bryhni *et al.* 2002). The sensorial quality, especially color and appearance (Brewer & Mckeith 1999), of meat can be affected by both internal and external factors. In the case of internal factors, in fish, for example, a particular problem that has been encountered in rearing some Pargus species is the darkening of the body after the capture of wild fish and during farming. During farming and marketing, the skin color (silver-red) turns dark gray (especially the tail and fins) (Kentouri *et al.* 1995; Lin *et al.* 1998). In the case of farmed salmon, too, feeding fish with carotenoid pigments is regarded as the most important management practice for marketing (Moe 1990), because without them, flesh and skin color would be less visually attractive, and therefore would be less valued as a food (Baker 2002).

Food technologists, especially those concerned with the meat industry, have a special interest in the color of food for several reasons—first, because of the need to maintain a uniform color throughout processing; second, to prevent any external or internal agent from acting on the product during its processing, storage, and display; third, to improve or optimize a product's color and appearance; and, lastly, to attempt to bring the product's color into line with what the consumer expects. Put simply, the color of meat is determined by the pigments present in it. These can be classified into four types: (1) biological pigments (carotenes and hemopigments), which are accumulated or synthesized in the organism antemortem (Lanari *et al.* 2002); (2) pigments produced as a result of damage during manipulation or inadequate processing conditions; (3) pigments produced postmortem (through enzymatic or nonenzymatic reactions) (Montero *et al.* 2001; Klomklao *et al.* 2006); and (4) pigments resulting from the addition of natural or artificial colorants (Fernández-López *et al.* 2002).

As a quality parameter, color has been widely studied in fresh meat (MacDougall 1982; Cassens *et al.* 1995; Faustman *et al.* 1996) and cooked products (Anderson *et al.* 1990; Fernández-Ginés *et al.* 2003; Fernández-López *et al.* 2003). However, dry-cured meat products have received less attention (Pérez-Alvarez 1996; Pagán-Moreno *et al.* 1998; Aleson *et al.* 2003) because in this type of product, color formation takes place during different processing stages (Pérez-Alvarez *et al.* 1997; Fernández-López *et al.* 2000); recently, a new heme pigment has been identified in this type of product (Parolari *et al.* 2003; Wakamatsu *et al.* 2004a, b). From a practical point of view, color plays a fundamental role in the animal production sector, especially in meat production (beef and poultry, basically) (Zhou *et al.* 1993; Esteve 1994; Verdoes *et al.* 1999; Irie 2001), since in many countries of the European Union (e.g., Spain and Holland) paleness receives a wholesale premium.

For fish, skin and flesh discoloration is a very important problem, especially in highly appreciated species. Since the skin and flesh color must be very vivid, many efforts have been directed at improving color, mainly through dietary control (carotene-enriched diets) (Fujita *et al.* 1983; Mori 1993). Without these pigments, the aquaculture industry would find it hard to undertake the production of some species because fish demand is driven through consumer demand for quality products (Baker 2002). In fish, consumer preference is often influenced by body pigmentation. Fish flesh color is an important quality parameter for most farmed fish, especially with salmonids (salmon, rainbow trout), (Francis 1995; Hyun *et al.* 1999), in which the pink or red color of fillets is an important feature (Sigurgisladottir *et al.* 1994; Sigurgisladottir *et al.* 1997). For example, a uniform red color in rainbow trout is considered to indicate a high-quality product and is a reason for its acceptability, while for the tuna fish industry, it is very important to avoid discoloration in fresh and processed meat and to increase its shelf life

(Goodrick *et al.* 1991; Tze *et al.* 2001). Fish nutrition has an important impact on several parameters that directly influence the quality of fish, some of which are color and appearance. The color of salmonid flesh is one of the most important quality parameters, because consumers have a preference for red- or pink-colored products in the case of salmonids. This is the reason for using carotenoids in aquaculture.

1.2 CHEMICAL AND BIOCHEMICAL ASPECTS OF MUSCLE-BASED FOOD COLOR

Of the major components of meat, proteins are the most important since they are only provided by essential amino acids, which are very important for the organism's correct functioning; proteins also make a technological contribution during processing, and some are responsible for such important attributes as color. These are the so-called chromoproteins, and they are mainly composed of a porphyrinic group conjugated with a transition metal, principally iron metalloporphyrin, which forms conjugation complexes (heme groups) (Whitaker 1972) that are responsible for color. However, carotenes and carotenoproteins (organic compounds with isoprenoid-type conjugated systems) exist alongside chromoproteins and also play an important part in meat color. There are also some enzymatic systems whose coenzymes or prosthetic groups possess chromophoric properties (peroxidases, cytochromes, and flavins) (Faust-man *et al.* 1996). However, their contribution to meat color is slight. Below, the principal characteristics of the major compounds that impart color to meat are described.

1.3 CAROTENES

Carotenes are responsible for the color of beef fat, poultry meat and skin, fish, and shellfish; in the last two cases, these are of great economic importance. The color of the fat is also important in carcass grading. Furthermore, carotenoids can be used as muscle-based food coloring agents (Verdoes *et al.* 1999). An important factor to be taken into account with these compounds is that they are not synthesized by the live animal but are obtained by assimilation (Pérez-Alvarcz *et al.* 2000), for instance, in the diet. Salmonids, for example, obtain carotenes in the wild in their preys, but in intensive fish culture, carotenoids must be added to the diet. Farmed fish, especially colored fish (e.g., salmon and rainbow trout), are now a major industry. For example, Norway exports a great part of its salmon production. Carotenoid pigments have been used in aquafeed for many years in order to impart the desired flesh color in farmed salmonids (Baker 2002). Astaxanthin has been the main flesh-coloring pigment of choice in most trout- and salmon-farming industries. The type of carotene used in animal feed is very important because the fish farmer may find that pigmentation takes on a heterogeneous appearance, which is contrary to general consumer acceptance (Yanar *et al.* 2006). The preferred pigments used in the Canadian aquaculture industry are synthetic canthaxanthin (Cx) and synthetic astaxanthin (Ax) (Higgs *et al.* 1995). In fats, the fatty acid composition can affect their color. When the ratio of cis-monounsaturated to saturated fatty acids is high, the fat exhibits a greater yellow color (Zhou *et al.* 1993). In the case of the carotenes present in fish tissues, these come from the ingestion of zooplankton, algae, and crustacean wastes (Ostermeyer & Schmidt 2004), and the levels are sometimes very high. This is possible because fish have the capacity to transport and deposit this pigment to specific sites in their muscles (Baker 2002). The deposition of Ax is higher in dark muscle than in light muscle (Ingemansson *et al.* 1993). The shells of many crustaceans, for example, lobster (*Panulirus argus*), also contain these compounds. Carotenoids have been extracted from crustacean wastes with organic solvents, but in many of the methods pigment degradation occurs (Charest *et al.* 2001).

The pigments responsible for color in fish, particularly salmonids (trout and salmon, among others), are Ax and Cx, although they are also present in tunids and are one of the most important natural pigments of marine origin. In the case of shellfish, their color depends on the so-called carotenoproteins, which are proteins with a prosthetic group that may contain various types of carotene (Minguez-Mosquera 1997), which are themselves water soluble (Shahidi & Matusalach-Brown 1998). Henmi *et al.* (1990a) reported that carotenoid–protein interaction in the salmon muscle is weak, and that Ax and Cx have a trans configuration in vivo. Henmi *et al.* (1990b) also reported that the actomyosins from salmonids showed a higher affinity for ketocarotenoids than those of other fish, except common mackerel. These authors also described correlations between the surface hydrophobicity of actomyosins and the combination of Ax and/or canthaxanthin with actomyosins. From a chemical point of view, astaxanthin or canthaxanthin bind via a beta-ionone ring to a hydrophobic binding site on actomyosin; the hydroxyl and keto end groups of the beta-end group of carotenoids intensify binding to actomyosin. Salmon actomyosin forms complexes with free Ax, astaxanthin monoester, canthaxanthin, echinenone, zeaxanthin, and beta-carotene, but not astaxanthin diester (in which a long-chain fatty acid residue may cause steric hindrance). The lipids in the actomyosin complex have no effect on the binding of carotenoids (Henmi *et al.* 1989). They are distributed in different amounts in the flesh, head, and carapace of crustaceans; for example, astaxanthin and its esters are the major carotenoids found in the extracts from different species of shrimp (*Penaeus monodon*, *Penaeus indicus*, *Metapenaeus dobsonii*, *Parapenaeopsis stylifera*) (Sachindra *et al.* 2005a), but there are different types of carotenes, depending on whether the crustaceans are marine or fresh water (Sachindra *et al.* 2005b). Another difference is that the concentration of unsaturated fatty acids in its carotenoid extracts was found to be higher than that of saturated fatty acids. In raw muscle, the main carotenoid concentration was strongly correlated with some color attributes (hue, chroma, and lightness) (Choubert *et al.* 1992). Torrisen *et al.* (1989) reported that a level of 4 mg/kg in fish fillets is regarded as a minimum acceptable carotenoid concentration in marketable-farmed salmon. Sex also affects carotene concentration: female muscles, which contain much more carotenoid, are more strongly colored than male muscles (Norris & Cunningham 2004).

As suggested by Torrisen *et al.* (1989), the rate of carotenoid deposition in salmonids is curvilinear throughout the life of the fish. As the growth rate is obviously under strong genetic control, the genetic correlation between the growth rate and color is high. It must be taken into account that carotenoids migrate from the muscle to the gonads. Carotene-type deposition in salmonid species differs; for example, Ax is more efficiently deposited than Cx in rainbow trout (Storebakken & Choubert 1991; Torrissen 1986), but this pattern is not the same for Atlantic salmon. These differences may be due to genetic background and/or environment (Baker 2002). Choubert *et al.* (1997) reported that in rainbow trout there is an unequal distribution of carotenoids so that the color of the muscle lightens from the head toward the tail and from the midline of the fish toward the dorsal and ventral external area of the fish.

From a chemical point of view, carotenoids are organic molecules that contain a conjugated carbon–carbon double bond system, which is responsible for their color. But this can be a problem during processing, because a high number of conjugated double bonds may be subject to oxidation, which can lead to discoloration of the carotenoids (Liaaen-Jensen 1971; Choubert & Baccaunaud 2006). As carotenoids are lipid soluble compounds, it might be thought that increasing dietary fat would increase carotene absorption and deposition, but this is not necessarily the case for all salmonids. The retention of carotenoids in the flesh is relatively poor, with only 10%–18% of pigment obtained from the diet being retained (Nickell & Bromage 1998). Astaxanthin can be found in its free, mono-, or diesterified forms. In processed shrimps, eicosapentaenoic acid (EPA) and docosahexaenoic acid (DHA) are the principal fatty acids esterified with the portion of astaxanthin linked to chitin in the carapace (Guillou *et al.* 1995). b-carotene and Ax are fat-soluble pigments found in squid oil. However, technological processes, such as refining, can remove Ax completely (Hwei & Min 1994).

In fish-derived products, the carotene content has previously been used as a quality parameter on its own; however, it has been demonstrated that this is not appropriate, and that other characteristics may influence color (Little *et al.* 1979). The carotene content and its influence on color is perhaps one of the characteristics that has received most attention (Swatland 1995). In the case of meat, especially beef, an excess of carotenes may actually lower the quality (Irie 2001), as occurs sometimes when classifying carcasses. The Japanese system for beef carcass classification identifies acceptable fats as white, slightly off-white, or slightly reddish white in color, while pink-yellowish and dark yellow are unacceptable (Irie 2001). It is precisely the carotenes that are responsible for these last two colorations. However, in other animal species, such as chicken (Castaneda *et al.* 2005), the opposite effect is observed, since a high carotene (xanthophyll) concentration is much appreciated by consumers (Esteve 1994), yellow being associated with traditional or "home-reared" feeding (Pérez-Alvarez *et al.* 2000). The use of the carotenoid canthaxanthin as a coloring agent in poultry feeds is designed to result in the desired coloration of poultry meat skins. The carotenoids used include citranaxanthin, capsanthin, and capsorubin, but Cx shows superior pigmenting properties and stability during processing and storage (Blanch 1999). To improve its color and brilliance, 0.004–0.04 wt% proanthocyanidin is added to fish feed containing carotenoids (Sakiura 2001). For rainbow trout, carotenoid concentrations could be 10.7 or 73 ppm Cx, or 47 or 53 ppm Ax.

1.4 HEMOPROTEINS

Of the hemoproteins present, postmortem in the muscle, myoglobin (Mb) is the one mainly responsible for color, since hemoglobin (Hb) arises from the red cells that are not eliminated during the bleeding process and are retained in the vascular system, basically in the capillaries (incomplete exsanguination; the average amount of blood remaining in meat joints is 0.3%) (Warris & Rodes 1977). However, the contribution of red cells to color does not usually exceed 5% (Swatland 1995). There is wide variation in the amounts of Hb from muscle tissue of bled and unbled fish. Mb content is minimal compared with the Hb content in fish light muscle and white fish whole muscle. Hb made up 65% and 56% by weight of the total heme protein in dark muscle from unbled and bled fish, respectively (Richards & Hultin 2002). Mb, on average, represents 1.5% by weight of the proteins of the skeletal muscle, while Hb represents about 0.5%, the same as the cytochromes and flavoproteins combined. Mb is an intracellular (sarcoplasmic) pigment apparently distributed uniformly within muscles (Ledward 1992; Kanner 1994). It is red in color and water soluble, and it is found in the red fibers of both vertebrates and invertebrates (Knipe 1993; Park & Morrisey 1994), where it fulfills the physiological role of intervening in the oxidative phosphorylation chain in the muscle (Moss 1992).

1.4.1 Structure of myoglobin

Structurally, Mb can be described as a monomeric globular protein with a very compact, well-ordered structure that is specifically, almost triangularly, folded and bound to a heme group (Whitaker 1972). It is structurally composed of two groups: a proteinaceous group and a heme group. The protein group has only one polypeptidic chain composed of 140–160 amino acid residues, measuring 3.6 nm and weighing 16 900 Da in vertebrates (Lehninger 1981). It is composed of eight relatively straight segments (where 70% of the amino acids are found), separated by curvatures caused by the incorporation into the chain of proline and other amino acids that do not form alpha-helices (such as serine and isoleucine). Each segment is composed of a portion of alpha-helix, the largest of 23 amino acids and the shortest of 7 amino acids, all dextrogyrating. Mb's high helicoidal content (forming an ellipsoid of 44 × 44 × 25 Å) and lack of disulfide bonds (there is no cysteine) make it an atypical globular protein. The absence of these groups makes the molecule highly stable (Whitaker 1972). Although the three-dimensional structure seems irregular and asymmetric, it is not totally anarchic, and all the molecules of Mb have the same conformation. One very important aspect of the protein part of Mb is its lack of color. However, the variations presented by its primary structure and the amino acid composition of the different animal and fish species destined for human consumption are the cause of the different colorations of meat and their stability when the meats are displayed in the same retail illumination conditions (Lorient 1982; Lee *et al.* 2003a). The heme group of Mb (as in Hb and other proteins) is, as mentioned above, a metalloporphyrin. These molecules are characterized by their high degree of coloration as a result of their conjugated cyclic tetrapyrrolic structure (Kalyanasundaram 1992). The heme group is composed of a complex, organic annular structure, protoporphyrin, to which an iron atom in ferrous state is united (Fe II). This atom has six coordination bonds, four with the flat protoporphyrin molecule (forming a flat square complex) and two perpendicular to it. The sixth bond is open and acts as a binding site for the oxygen molecule.

Protoporphyrin is a system with a voluminous flat ring composed of four pyrrolic units connected by methyl bridges (=C–). The Fe atom, with a coordination number of 6, lies at the center of the tetrapyrrole ring and is complexed to four pyrrolic nitrogens. The heme group is complexed to the polypeptidic chain (globin) through a specific histidine residue (imadazolic ring) occupying the fifth position of the Fe atom (Davidsson & Henry 1978). The heme group is bound to the molecule by hydrogen bridges, which are formed between the propionic acid side chains and other side chains. Other aromatic rings exist near and almost parallel to the heme group, which may also form pi (π) bonds (Stauton-West *et al.* 1969).

The Hb contains a porphyrinic heme group identical to that of Mb and equally capable of undergoing reversible oxygenation and deoxygenation. Indeed, it is functionally and structurally paired with Mb, and its molecular weight is four times greater since it contains four peptidic chains and four heme groups. The Hb, like Mb, has its fifth ligand occupied by the imidazole group of a histidine residue, while the sixth ligand may or may not be occupied. It should be mentioned that positions 5 and 6 of other hemoproteins (cytochromes) are occupied by R groups of specific amino acid residues of the proteins and therefore cannot bind to oxygen (O_2), carbon monoxide (CO), or cyanide (CN–), except a_3, which, in its biological role, usually binds to oxygen.

One of the main differences between fish and mammalian Mb is that fish Mb have two distinct endothermic peaks, indicating multiple states of structural unfolding, whereas mammalian Mb followed a two-state unfolding process. Changes in alpha-helix content and tryptophan fluorescence intensity with temperature are greater for fish Mb than for mammalian Mb. Fish Mb shows labile structural folding, suggesting greater susceptibility to heat denaturation than that of mammalian Mb (Saksit *et al.* 1996).

The helical contents of frozen-thawed Mb were practically the same as those of unfrozen Mb, regardless of pH. Frozen-thawed Mb showed a higher autoxidation rate than unfrozen Mb. During freezing and thawing, Mb suffered some conformational changes in the nonhelical region, resulting in a higher susceptibility to both unfolding and autoxidation (Chow *et al.*

1989). In tuna fish, Mb stability followed the order bluefin tuna (*Thunnus thynnus*), yellowfin tuna (*Thunnus albacares*), and bigeye tuna (*Thunnus obesus*); autoxidation rates were in the reverse order. The pH dependency of Mb from skipjack tuna (*Katsuwonus pelamis*) and mackerel (*Scomber scombrus*) were similar. Lower Mb stability was associated with higher autoxidation rates (Chow 1991).

1.4.2 Chemical properties of myoglobin

The chemical properties of Mb center on its ability to form ionic and covalent groups with other molecules. Its interaction with several gases and water depends on the oxidation state of the Fe of the heme group (Fox 1966), since this may be in either its ferrous (Fe II) or its ferric (Fe III) state. Upon oxidation, the Fe of the heme group takes on a positive charge (Kanner 1994) and, typically, binds with negatively charged ligands, such as nitrites, the agents responsible for the nitrosation reactions in cured meat products.

When the sixth coordination ligand is free, Mb is usually denominated deoxymyoglobin (DMb), which is purple in color. However, when this site is occupied by oxygen, the oxygen and the Mb form a noncovalent complex, denominated oxymyoglobin (OMb), which is cherry or bright red (Lanari & Cassens 1991). When the oxidation state of the iron atom is modified to the ferric state and the sixth position is occupied by a molecule of water, the Mb is denominated metmyoglobin (MMb), which is brown. There are several possible causes for MMb generation, and these may include the ways in which tunids, meat, and meat products are obtained, transformed, or stored (MacDougall 1982; Lee *et al.* 2003b; Mancini *et al.* 2003). Among the most important factors are low pH, the presence of ions, and high temperatures during processing (Osborn *et al.* 2003); the growth and/or formation of metabolites from the microbiota (Renerre 1990); the activity of endogenous reducing enzymes (Arihara *et al.* 1995; Osborn *et al.* 2003); and the levels of endogenous (Lanari *et al.* 2002) or exogenous antioxidants, such as ascorbic acid or its salts, tocopherols (Irie *et al.* 1999), or plant extracts (Xin & Shun 1993; Fernández-López *et al.* 2003; Sánchez-Escalante *et al.* 2003). The pH, which may be altered depending on postslaughter metabolism and on ingredient addition, can affect the stability of the central iron atom in Mb and Hb. At high pH, the heme iron is predominantly in the Fe21 state; low pH accelerates Fe21 conversion to Fe31 (Zhu & Brewer 2002, 2003). While oxygen can bind to Fe21 only, many other ligands (CN, nitric oxide [NO], CO) can bind to either Fe21 or Fe31 so producing a variety of colors. This change in the oxidation state of the heme group will result in the group being unable to bind with the oxygen molecule (Arihara *et al.* 1995). DMb is able to react with other molecules to form colored complexes, many of which are of great economic relevance for the meat industry. The most characteristic example is the reaction of DMb with nitrite, since its incorporation generates a series of compounds with distinctive colors: red in dry-cured meat products or pink in heat-treated products. The products resulting from the incorporation of nitrite are denominated cured, and such products are of enormous economic importance worldwide (Pérez-Alvarez 1996). The reaction mechanism is based on the propensity of nitric oxide (NO, generated in the reaction of nitrite in acid medium, readily gives up electrons) to form strong coordinated covalent bonds; it forms an iron complex with the DMb heme group independent of the oxidation state of the heme structure. The compound formed after the nitrification reaction is denominated nitrosomyoglobin (NOMb). As mentioned above, the presence of reducing agents such as hydrogen sulfide acid (H_2S) and ascorbates leads to the formation of undesirable pigments in both meat and meat products. These green pigments are called sulfomyoglobin (SMb) and colemyoglobin (ColeMb), respectively, and are formed as a result of bacterial activity and an excess of reducing agents in the medium. The formation of SMb is reversible, but that of

ColeMb is an irreversible mechanism, since it is rapidly oxidized between pH 5 and 7, releasing the different parts of the Mb (globin, iron, and the tetrapyrrolic ring).

From a chemical point of view, it should be borne in mind that the color of Mb, and therefore of the meat or meat products, depends not only on the molecule that occupies the sixth coordination site, but also on the oxidation state of the iron atom (ferrous or ferric), the type of bond formed between the ligand and the heme group (coordinated covalent, ionic, or none), and the state of the protein (native or denatured form), not to mention the state of the porphyrin of the heme group (intact, substituted, or degraded) (Pérez-Alvarez 1996).

During the heat treatment of fish flesh, the aggregation of denatured fish proteins is generally accompanied by changes in light-scattering intensity. Results demonstrate that changes in relative light-scattering intensity can be used for studying structural unfolding and aggregation of proteins under thermal denaturation (Saksit *et al.* 1998). When fatty fish meat like *Trachurus japonicus* was heat treated, the MMb content increased linearly, and the percentages of denatured Mb and apomyoglobin increased rapidly when mince was exposed to heat; however, when the temperature reached 60°C, the linearity was broken. The results indicated that MMb color stability was higher than that of Mb and that the thermal stability of heme was higher than that of apomyoglobin (Hui *et al.* 1998). Both Mb and ferrous iron accelerated the lipid oxidation of cooked, water-extracted fish meat. Ethylenediaminetetraacetic acid (EDTA) inhibited the lipid oxidation accelerated by ferrous iron, but not that accelerated by Mb. Also, with cooked, nonextracted mackerel meat, EDTA noticeably inhibited lipid oxidation. Nonheme iron catalysis seemed to be related in part to lipid oxidation in cooked mackerel meat. The addition of nitrite in combination with ascorbate resulted in a marked inhibition of lipid oxidation in the cooked mackerel meat. From these results, it was postulated that nitric oxide ferrohemochromogen, formed from added nitrite and Mb (present in the mackerel meat) in the presence of a reducing agent, possesses an antioxidant activity, which is attributable in part to its function as a metal chelator (Ohshima *et al.* 1988).

Tuna fish meat color can be improved when the flesh is treated or packaged with a modified atmosphere in which CO is included. Normally, the rate of penetration of CO or carbon dioxide (CO_2) in fish meat such as tuna, cod, or salmon, under different packaging conditions, is measured by monitoring pressure changes in a closed constant volume chamber with constant volume and temperature. Alternatively, however, the specific absorption spectrum of carboxymyoglobin (MbCO), within the visible range, can be obtained and used as an indicator of MbCO formation. Mb extracts from tuna muscle treated with CO exhibited higher absorbance at 570 than at 580 nm. Therefore, the relationship between absorbance at 570 nm and absorbance at 580 nm could be used to determine the extent of CO penetration of tuna steaks placed in a modified atmosphere in which CO was included. The penetration of CO into tuna muscle was very slow. After approximately 1–4 h, CO had penetrated 2–4 mm under the surface, and after 8 h, CO had penetrated 4–6 mm (Chau *et al.* 1997).

In products with added nitrite or nitrate, the complex nitrosylmyoglobin (MbFe[II]NO) is the main contributor to the characteristic color of cooked cured ham, and brine-cured and dry-cured meat products. Meat and meat products without nitrite/nitrate addition will normally attain a dull brown color or a gray color in heated products, which influences consumer acceptance negatively (Adamsen *et al.* 2005). In dry-cured meat products such as Parma ham produced without nitrite or nitrate addition, the characteristic bright red color (Wakamatsu *et al.* 2004a) is caused by a Zn-protoporphyrin IX (Zn-pp) complex, a heme derivative. Adamsen *et al.* (2005) showed that the use of nitrite as a curing ingredient inhibits the formation of Zn-pp. In the same work, the author described that this color compound is present in other meat products like Iberian ham, although in a lower concentration.

Virgili *et al.* (1999) reported that this color may be due to the action of low-molecular weight compounds containing electron-donating atoms, formed during maturation, in particular basic peptides or amino acids resulting from an external proteolysis, which may play a role as Fe ligands in Mb. Wakamatsu *et al.* (2004b) reported that anaerobic conditions favor the formation of Zn-pp and that endogenous enzymes as well as microorganisms may also be involved. There are several hypotheses that try to explain the formation of this compound. Wakamatsu *et al.* (2004b) described three possible substitution patterns: (1) a nonenzymatic reaction in which Zn(II) substitutes Fe(II) under anaerobic conditions, with concomitant dissociation of the heme; (2) a bacterial enzymatic reaction, whereby bacterial growths naturally degrade the meat proteins including the pigment; and (3) an enzymatic reaction where an endogenous ferrochelatase interchanges the two metals. However, Adamsen *et al.* (2005) described this process as having the three following mechanisms to explain the metal substation: (1) a nonenzymatic enzymatic reaction driven by the binding of iron in the high chloride meat matrix; (2) a bacterial enzymatic reaction; and (3) an endogenous enzymatic reaction.

Also, spectroscopic studies of Parma ham during processing revealed a gradual transformation of muscle Mb, initiated by salting and continuing during aging. Using electron spin resonance spectroscopy, Moller *et al.* (2003) have shown that the Parma ham pigment is different from MbFe(II)NO and is not a nitric oxide complex such as that found in brine-cured ham and Spanish Serrano hams. These authors also establish that the heme moiety is present in the acetone-water extract and that Parma ham pigment is gradually transformed from a Mb derivative into a nonprotein heme complex, which is thermally stable in an acetone-water solution. Adamsen *et al.* (2003) also demonstrated that the heme moieties of Parma ham pigments have antioxidative properties. Pigments became increasingly lipophilic during processing, suggesting that a combination of drying and maturing yields a stable red color (Parolari *et al.* 2003).

1.4.3 Cytochromes

Cytochromes are metalloproteins with a prosthetic heme group, whose putative role in meat coloration is undergoing revision (Boyle *et al.* 1994; Faustman *et al.* 1996). Initially, they were not thought to play a very important role (Ledward 1984). These compounds are found in low concentrations in the skeletal muscle, and in poultry, they do not represent more than 4.23% of the total hemoproteins present (Pikul *et al.* 1986). It has now been shown that the role of cytochrome (especially its concentration) in poultry meat color is fundamental when the animal has been previously exposed to stress (Ngoka & Froning 1982; Pikul *et al.* 1986). Cytochromes are most concentrated in cardiac muscle so that when this organ is included in meat products, the heart's contribution to color, not to mention the reactions that take place during elaboration processes, must be taken into consideration (Pérez-Alvarez *et al.* 2000).

1.5 COLOR CHARACTERISTICS OF BLOOD

Animal blood is little used in the food industry because of the dark color it imparts to the products to which it is added. For solving the negative aspects of blood incorporation, specifically food-color-related problems, several different processes and means have been employed, but they are not always completely satisfactory. The addition of 12% blood plasma to meat sausages leads to pale-colored products. The addition of discolored whole blood or globin (from which the Hb's heme group has been eliminated) has also been used to address color problems. Natural red pigments can be obtained from blood without using coloring agents such as nitrous acid salts; these pigments have Zn-pp as the metalloporphyrin moiety and can be used to produce favorably colored beef products, whale meat products, and fish products (including fish pastes) (Numata & Wakamatsu 2003). There was wide variation in amounts of

Hb extracted from the muscle tissue of bled and unbled fish, and the residual level in the muscle of bled fish was substantial. Mb content was minimal as compared with Hb content in mackerel light muscle and trout whole muscle. Hb made up 65% and 56% by weight of the total heme protein in dark muscle from unbled and bled mackerel, respectively. The blood-mediated lipid oxidation in fish muscle depends on various factors, including Hb concentration, Hb type, plasma volume, and erythrocyte integrity (Richards & Hultin 2002). The presence of blood, Hb, Mb, Fe12, Fe13, or Cu12 can stimulate lipid oxidation in the fillets of icefish (Rehbein & Orlick 1990; Richards & Li 2004). Kanner *et al.* (1987) reported that Hb, Mb, copper, and iron have the potential to promote lipid oxidation in muscle foods. Since iron can be released from Hb during storage, it is difficult to ascertain whether the intact heme protein, dissociated heme, or released iron is responsible for the bulk of lipid oxidation that occurs during storage. For this reason, Svingen *et al.* (1979) used the term "low molecular weight iron" instead of "free iron" since iron binds to other low molecular weight compounds to gain solubility and hence potential reactivity. Ferrous and ferric forms of iron can promote lipid oxidation processes (Gutteridge 1986; Tadolini & Hakim 1996). Iron shows a high reactivity with reactants such as hydrogen peroxide and lipid peroxides (Kanner & Harel 1987).

Mitochondria are a source of reactive oxygen species that could confound lipid oxidation reactions due to added Hb. During fish processing (e.g., tuna fish), the loss of redness can be a good indicator that lipid oxidation processes mediated by Hb are progressing. Just after death, Hb in muscle tissue is primarily in the reduced state (i.e., oxyhemoglobin [OHb] and deoxyhemoglobin [deoxyHb]).

This mixture of OHb and deoxyHb has a red color. With increased postmortem aging, Hb autoxidizes to methemoglobin (MHb), a brown pigment. MHb is considered more prooxidative than reduced Hb due to its less tightly bound heme group and its reactivity with hydrogen peroxide and lipid peroxides to form hypervalent Hb catalysts (Everse & Hsia 1997).

From a technological point of view, during meat or fish processing, rapid chilling may alter oxygen solubility in tissues resulting in less available oxygen to oxygenate either OMb or Hb. The conversion of OMb to MMb, which is brown and unattractive, occurs under conditions of very low oxygen tension as well (Nicolalde *et al.* 2005).

Field *et al.* (1978) describe how bone marrow is high in Hb, while muscle has a high Mb content. As with other meats, its color and Hb stability depend on packaging and storage conditions. Good temperature control and modified atmosphere packaging (MAP) with high oxygen atmospheres (80%) are often used to extend both microbiological and color shelf life (Nicolalde *et al.* 2005).

1.6 FAT COLOR

From a technological point of view, fat fulfills several functions, although, regarding color, its principal role is in the brightness of meat products. Processes such as "afinado" during the elaboration of dry-cured ham involve temperatures at which fat melts so that it infiltrates the muscle mass and increases its brilliance (Sayas 1997). When the fat is finely chopped, it "dilutes" the red components of the color, thus decreasing the color intensity of the finished product (Pérez-Alvarez *et al.* 2000). However, fats do not play such an important role in fine pastes since, after emulsification, the fat is masked by the matrix effect of the emulsion so that it contributes very little to the final color. The color of fat basically depends on the feed that the live animal received (Esteve 1994; Irie 2001). In the case of chicken and ostrich, the fat has a "white" appearance (common in Europe) when the animal has been fed with "white" cereals or other ingredients not containing xanthophylls, since these are accumulated in subcutaneous fat and other fatty deposits. However, when the same species are fed maize (rich in xanthophylls), the fatty deposits take on a yellow color. Beef or veal fat that is dark, hard (or soft), excessively bright, or shiny lowers the carcass and cut price. Fat with a yellowish color in healthy animals reflects a diet containing beta-carotene (Swatland 1988). While fat color evaluation has traditionally been a subjective process, modern methods include such techniques as optical fiber spectrophotometry (Irie 2001). Another factor influencing fat color is the concentration of the Hb retained in the capillaries of the adipose tissues (Swatland 1995). As in meat, the different states of Hb may influence the color of the meat cut. OMb is responsible for the yellowish appearance of fat, since it affects different color components (yellow-blue and red-green).

The different states of Hb present in adipose tissue may react in a similar way to those in meat so that fat color should be measured as soon as possible to avoid possible color alterations. When the Hb in the adipose tissue reacts with nitrite incorporated in the form of salt, nitrosohemoglobin (NOHb), a pigment that imparts a pink color to fat, is generated. This phenomenon occurs principally in dry-cured meat products with a degree of anatomical integrity, such as dry-cured ham or shoulder (Sayas 1997). When fat color is measured, its composition should be borne in mind since its relation with fatty acids modifies its characteristics, making it more brilliant or duller in appearance. The fat content of the conjunctive tissue must also be borne in mind—collagen may present a glassy appearance because, at acidic pH, it is "swollen," imparting a transparent aspect to the product.

1.7 ALTERATIONS IN MUSCLE-BASED FOOD COLOR

The color of meat and meat products may be altered by several factors, including exposure to light (source and intensity), microbial growth, rancidity, and exposure to oxygen. Despite the different alterations in color that may take place, only a few have been studied; these include the pink color of boiled uncured products, premature browning (PMB), fish skin discoloration, and melanosis in crustaceans.

1.8 PINK COLOR OF UNCURED MEAT PRODUCTS

The normal color of a meat product that has been heat treated but not cured is "brown," although it has recently been observed that these products show an anomalous coloration (red or pink) (Hunt & Kropf 1987). This problem is of great economic importance in "grilled" products since this type of color is not considered desirable. This defect may occur both in meats with a high hemoprotein content, such as beef and lamb (red), and in those with a low hemoprotein concentration, including chicken and turkey (pink) (Conforth *et al.* 1986). One of the principal causes of this defect is the use of water rich in nitrates, which are reduced to nitrites by nitrate-reducing bacteria, which react with the Mb in meat to form NOMb (Nash *et al.* 1985). The same defect may occur in meat products containing paprika, which according to Fernández-López (1998) contains nitrates that, once incorporated in the product, may be similarly reduced by microorganisms. Conforth *et al.* (1991) mention that several nitrogen oxides may be generated in gas and electric ovens used for cooking ham and that these nitrogen oxides will react with the Mb to generate nitrosohemopigments. CO is also produced in ovens, which reacts with Mb during thermal treatment to form a pink-colored pigment, carboxyhemochrome. It has also been described how the use of adhesives formed from starchy substances produces the same undesirable pink color in cooked products (Scriven *et al.* 1987). The same anomalous pink color may be generated when the pH of the meat is high (because of the addition of egg albumin to the ingredients) (Froning *et al.* 1968) and when the cooking temperature during processing is too low. These conditions favor the development of a reducing environment that maintains the iron of the Mb in its ferrous form, imparting a reddish/pink color (as a function of the concentration of hemopigments) instead of the typical grayish brown color of heat-treated, uncured meat products.

Cooking uncured meat products, such as roast beef, at low temperatures (less than 60°C) may produce a reddish color inside the product, which some consumers may like. This internal coloring is not related to the formation of nitrosopigments, but results from the formation of OMb, a phenomenon that occurs because there exist in the muscle MMb-reducing enzymatic systems that are activated at temperatures below 60°C (Osborn *et al.* 2003). Microbial growth may also cause the formation of a pink color in cooked meats since these reduce the oxidoreduction potential of the product during their growth. This is important when the microorganisms that develop in the medium are anaerobes, since they may generate reducing substances that decrease the heme iron. When extracts of *Pseudomonas* cultures are applied, the MMb may be reduced to Mb (Faustman *et al.* 1990).

1.9 MELANOSIS

Melanosis, or blackspot, involving the appearance of a dark, even black, color, may develop postmortem in certain shellfish during chilled and frozen storage (Slattery *et al.* 1995). Melanosis is of huge economic importance since the coloration may suggest a priori in the eyes of the consumer that the product is in bad condition, despite the fact that the formation of the pigments responsible involves no health risk. Melanosis is an undesirable surface discoloration of such high-value shellfish as lobsters that takes place immediately after harvesting since it starts with oxygen contact (López-Caballero *et al.* 2006). Blackspot is caused by enzymic formation of the precursors of phenolic pigments (Williams *et al.* 2003). Blackspot is a process regulated by a complex biochemical mechanism, whereby the phenols present in a food are oxidized to quinones in a series of enzymatic reactions caused by polyphenol oxidase (PPO) (Ogawa *et al.* 1984). This is followed by a polymerization reaction, which produces pigments of a high molecular weight and dark color. Melanosis is produced in the exoskeleton of crustaceans, first in the head and gradually spreading toward the tail. Melanosis of shell and hyperdermal tissue in some shellfish, such as lobsters, has been related to stage of molt, since the molting fluid is considered to be the source of the natural activator(s) of pro-PPO.

PPO (catechol oxidase) can be isolated from shellfish cuticle (Ali *et al.* 1994) and is still active during iced or refrigerated storage. Some authors have found a connection between melanosis and microbial growth in crustaceans. Thus, color formation (melanin) due to strains of *P. fragi* may occur if prawns are not properly chilled (Chinivasagam *et al.* 1998). In this respect, López-Caballero *et al.* (2006) reported that the presence of microorganisms (e.g., *Proteus* spp., *Pseudomonas*, etc.) and the H_2S produced reacted with metals of the lobster shell resulting in melanosis. Sulfites can be used to control the process (Ferrer *et al.* 1989; Gomez-Guillen *et al.* 2005), although their use is prohibited in many countries. It is well known that the inhibitory effect of blackspot is specific for each species, requiring adequate doses and formulations (Montero *et al.* 2001). The effective dose of 4-hexylresorcinol differs depending on the physiological state, season, method of application, etc., although the species being treated is one of the most important of these factors. Montero *et al.* (2004) and López-Caballero *et al.* (2006) found that, regardless of the season, a concentration of 0.25% 4-hexylresorcinol was effective in extending the shelf life of pink shrimp. Ficin (Taoukis *et al.* 1990) and 4-hexylresorcinol also functioned as a blackspot inhibitor, alone and in combination with L-lactic acid (Benner *et al.* 1994).

1.10 FISH SKIN DISCOLORATION

In fish and other vertebrates, in which the pigmentation of the skin can be changed by hormonal stimulation, the color of the background and illumination are determining factors for the intensity and/or the pattern of skin fish pigmentation (Duray *et al.* 1996; Crook 1997; Sugimoto 1997; Healey 1999; Papoutsoglou *et al.* 2000; Rotllant *et al.* 2003). In addition, temperature may also have an impact on color (Fernandez & Bagnara 1991). In some types of fish, especially those with a red skin, the color tone becomes dark immediately after killing, reducing the commercial value of the fish. Most of the color changes in fish are often related to stress. It is generally accepted that melanophores play an important role in the rapid color change of certain fish (Fujii 1969). These changes are related to hormonal (a-melanocyte-stimulating hormones) responses causing dispersion of the melanin granules in melanophores and are responsible for skin darkening (Green & Baker 1991; Lamers *et al.* 1992; Gröneveld *et al.* 1995; Arends *et al.* 2000; Burton & Vokey 2000).

In the case of Red Sea Bream, the rapid skin-color changes after killing of cultured fish is thought to be mainly due to the rapid dispersion of chromatosomes in melanophores elicited by handling and killing stresses. Potassium ions through the noradrenaline pathway can induce aggregation of chromatosomes in melanophores (Kumazawa & Fujii 1984). In fish skin, besides melanophores there are other chromatophores, such as xanthophores and erythrophores, the latter mainly contributing to the red color of the fish skin.

1.11 PREMATURE BROWNING

Hard-to-cook patties show persistent internal red color and are associated with high pH (>6) raw meat. Pigment concentration affects red color intensity after cooking (residual undenatured Mb), so this phenomenon is often linked to high pH dark cutting meat from older animals. Premature browning (PMB) is a condition in which ground beef (mince) looks well done at a lower than expected temperature (Warren *et al.* 1996). PMB of ground beef is a condition in which Mb denaturation appears to occur on cooking at a temperature lower than expected; it may indicate falsely that an appropriate internal core temperature of 71°C has been achieved (Suman *et al.* 2004). The relationship between cooked color and internal temperature of beef muscle is inconsistent and depends on pH and animal maturity. Increasing the pH may be of benefit in preventing PMB, but it may increase the incidence of red color in well-cooked meat (cooked over an internal temperature of 71.1°C) (Berry 1997). When pale, soft, exudative (PSE) meat was used in patty processing, patties containing OMb easily exhibited PMB. One reason for this behavior is that the percentage of Mb denaturation increased as cooking temperature rose (Lien *et al.* 2002).

1.12 COLOR AND SHELF LIFE OF MUSCLE-BASED FOODS

Meat and meat products are susceptible to degradation during storage and throughout the retail process. In this respect, color is one of the most important quality attributes for indicating the state of preservation in meat. Any energy received by food can initiate its degradation, but the rate of any reaction depends on the exact composition of the product (Jensen *et al.* 1998), environmental factors (light, temperature, presence of oxygen), and the presence of additives. Transition metals such as copper and iron are very important in the oxidative/antioxidative balance of meat. When the free ions of these two metals interact, they reduce the action of certain agents, such as cysteine, ascorbate, and alpha-tocopherol, oxidizing them and significantly reducing the antioxidant capacity in muscle (Zanardi *et al.* 1998). Traditionally, researchers have determined the discoloration of meat using as criterion the brown color of the product, calculated as percent MMb (Mancini *et al.* 2003). These authors demonstrated that in the estimation of the shelf life of beef or veal (considered as discoloration of the product), the diminution in the percent of OMb is a better tool than the increase in percentage of MMb. Occasionally, when the meat cut contains bone (especially in pork and beef), the hemopigments (mainly Hb) present in the medulla lose color because the erythrocytes are broken during cutting and accumulate on the surface of the bone Hb. When exposed to light and air, the color of the Hb changes from bright red (oxyhemoglobin [OHb]), the characteristic of blood, to brown (methemoglobin [MHb]) or even black (Gill 1996). This discoloration basically takes place during long periods of storage, especially during shelf-life display (Mancini *et al.* 2004). This characteristic is aggravated if the product is kept in a modified atmosphere rich in oxygen (Lanari *et al.* 1995). These authors also point out that the effect of bone marrow discoloration is minimized by the effect of bacterial growth in MAP. As in the case of fresh meat, the shelf life of meat products is limited by discoloration (Mancini *et al.* 2004). This phenomenon is important in this type of product because they are normally displayed in illuminated cabinets. Consequently, the possibility of photooxidation of NOMb needs to be taken into account.

During this process, the molecule is activated because it absorbs light; this may subsequently deactivate the NOMb and give the free electrons to the oxygen to generate MMb and free nitrite. In model systems of NOMb photooxidation, the addition of solutions of dextrose, an important component of the salts used for curing cooked products and in meat emulsions, can diminish the effect of NOMb photooxidation. When a meat product is exposed to light or is stored in darkness, the use of ascorbic acid or its salts may help stabilize the product's color. Such behavior has been described both in model systems of NOMb (Walsh & Rose 1956) and in dry-cured meat products (e.g., longanizas, Spanish dry-fermented sausage). However, when sodium isoascorbate or erythorbate is used in longanizas production, color stability is much reduced during the retail process (Ruíz-Peluffo *et al.* 1994).

The discoloration of white meats such as turkey is characterized by color changes that go from pink-yellow to yellow-brown, while in veal and beef, the changes go from purple to grayish brown. In turkey, it has been demonstrated that the presence or absence of lipid oxidation depends on, among other things, the concentration of vitamin E in the tissues. The color and lipid oxidation are interrelated since it has been seen that lipid oxidation in red and white muscle depends on the predominant form of catalyzing iron, Mb, or free iron (Mercier *et al.* 1998). Compared with red meat, tuna flesh tends to undergo more rapid discoloration during the refrigerated storage. Discoloration due to the oxidation of Mb in red fish presented a problem, even at low temperatures. This low color stability might be related to the lower activity or poorer stability of MMb reductase in tuna flesh (Ching *et al.* 2000). Another reason for the low color stability is that aldehydes produced during lipid oxidation can accelerate tuna OMb oxidation in vitro (Lee *et al.* 2003a). Tuna flesh could be immersed in an MMb reductase solution to extend the color stability of tuna fish. Also, the use of this enzyme can reduce MMb formation during the refrigerated storage of tuna (Tze *et al.* 2001). Yellowtail (*Seriola quinqueradiata*) fillets stored in gas barrier film packs filled with nitrogen (N_2) and placed in cold storage at 0–5°C stayed fresh for 4–7 days. N_2 or CO_2 packaging did not prevent discoloration in frozen tuna fillets; better results were achieved by thawing the frozen tuna meat in an O_2 atmosphere (Oka 1989). Packaging in atmospheres containing 4% or 9% O_2 was inferior to packaging in air, as these atmospheres promoted MMb formation. Packaging in 70% O_2 maintained the fresh red color of tuna dorsal muscle for storage periods less than 3 days

(Tanaka *et al.* 1996). To change the dark brown color to a bright red color, processors sometimes treat tuna with 100% carbon monoxide (CO) during MAP. Since Mb can react with CO rapidly even at low CO concentrations (Chi *et al.* 2001), MAP with 100% CO may result in high CO residues in the flesh, which may cause health problems.

1.13 MICROORGANISMS AND MUSCLE-BASED FOOD COLOR

Although the real limiting factor in the shelf life of fresh meat is the microbial load, consumers choose fresh meat according to its color. The bacterial load is usually the most important cause of discoloration in fresh meat and meat products (sausages and other cooked products), and slaughter, cutting, and packaging must be strictly controlled. Bacterial contamination decisively affects the biochemical mechanisms responsible for the deterioration of meat (Renerre 1990). Is it important to take into account that, just as with the bacterial load, the effect of discoloration on meat is more pronounced in meats that are more strongly pigmented (beef) than in less pigmented meats such as pork and chicken (Gobantes & Oliver 2000). Another variable affecting color stability in meat is the quantity of microorganisms present (Houben *et al.* 1998); concentrations in excess of 106/g have a strong effect. Although antioxidants, such as ascorbic acid, slow lipid oxidation and consequently improve color stability, these substances have little effect when bacterial growth is a problem (Zerby *et al.* 1999).

REFERENCES

Adams, D.C. & Huffman, R.T. (1972) Effect of controlled gas atmospheres and temperature on quality of packaged pork. *Journal of Food Science*, **37**, 869–872.

Adamsen, C.E., Hansen, M.L., Moller, J.K.S. & Skibsted, L.H. (2003) Studies on the antioxidative activity of red pigments in Italian-type dry-cured ham. *European Food Research Technology*, **217**(3), 201–206.

Adamsen, C.E., Moller, J.K.S., Laursen, K., Olsen, K. & Skibsted, L.H. (2005) Zn-porphyrin formation in cured meat products: Effect of added salt and nitrite. *Meat Science, Meat Science*, **72** (4), 672–679.

Aleson, L., Fernández-López, J., Sayas-Barberá, E., Sendra, E. & Pérez-Alvarez, J.A. (2003) Utilization of lemon albedo in dry-cured sausages. *Journal of Food Science*, **68**, 1826–1830.

Ali, M.T., Gleeson, R.A., Wei, C.I. & Marshall, M.R. (1994) Activation mechanisms of pro-phenoloxidase on melanosis development in Florida spiny lobster (*Panulirus argus*) cuticle. *Journal of Food Science*, **59**(5), 1024–1030.

Anderson, H.J., Bertelsen, G. & Skibsled, L.H. (1990) Colour and colour stability of hot processed frozen minced beef. Result from chemical model experiments tested under storage conditions. *Meat Science*, **28**(2), 87–97.

Arends, R.J., Rotllant, J., Metz, J.R., Mancera, J.M., Wendelaar-Bonga, S.E. & Flik, G. (2000) Alpha-MSH acetylation in the pituitary gland of the sea bream (*Sparus aurata* L.) in response to different backgrounds, confinement and air exposure. *Journal of Endocrinology*, **166**(2), 427–435.

Arihara, K., Itoh, M. & Kondo, Y. (1995) Significance of metmyoglobin reducing enzyme system in myocytes. In: *Proceedings of the 41st International Congress of Meat Science and Technology*, San Antonio, Texas, C70, pp. 378–379.

Baker, R.T.M. (2002) Canthaxanthin in aquafeed applications: Is there any risk? *Trends in Food Science and Technology*, **12**, 240–243.

Benner, R.A., Miget, R., Finne, G. & Acuff, G.R. (1994) Lactic acid/melanosis inhibitors to improve shelf life of brown shrimp (*Penaeus aztecus*). *Journal of Food Science*, **59**(2), 242–245.

Berry, B.W. (1997) Color of cooked beef patties as influenced by formulation and final internal temperature. *Food Research International*, **30**(7), 473–478.

Blanch, A. (1999) Getting the colour of yolk and skin right. *World Poultry*, **15**(9), 32–33.

Boyle, R.C., Tappel, A.L., Tappel, A.A., Chen, H. & Andersen, H.J. (1994) Quantitation of haem proteins from spectra of mixtures. *Journal of Agricultural and Food Chemistry*, **42**, 100–104.

Brewer, M.S. & Mckeith, F.K. (1999) Consumer-rated quality characteristics as related to purchase intent of fresh pork. *Journal of Food Science*, **64**, 171–174.

Bryhni, E.A., Byrne, D.V., Rødbotten, M., *et al.* (2002) Consumer perceptions of pork in Denmark, Norway and Sweden. *Food Quality and Preference*, **13**, 257–266.

Burton, D. & Vokey, J.E. (2000) The relative in vitro responsiveness of melanophores of winter flounder to a-MSH and MCH. *Journal of Fish Biology*, **56**, 1192–1200.

Cassens, R.G., Demeyer, D., Eikelenboom, G., *et al.* (1995) Recommendation of reference method for assessment of meat color. *Proceedings of the 41st International Congress of Meat Sci and Technology*, San Antonio, Texas, C86, pp. 410–411.

Castaneda, M.P., Hirschler, E.M. & Sams, A.R. (2005) Skin pigmentation evaluation in broilers fed natural and synthetic pigments. *Poultry Science*, **84**(1), 143–147.

Charest, D.J., Balaban, M.O., Marshall, M.R. & Cornell, J.A. (2001) Astaxanthin extraction from crawfish shells by supercritical CO_2 with ethanol as cosolvent. *Journal of Aquatic Food Product Technology*, **10**(3), 79–93.

Chau, J.C., Sue, M.L. & Mei, L.T. (1997) Characteristics of reaction between carbon monoxide gas and myoglobin in tuna flesh. *Journal of Food and Drug Analysis*, **5**(3), 199–206.

Chi, C.Y., Kuang, H.L. & Chau, J.C. (2001) Effect of carbon monoxide treatment applied to tilapia fillet. *Taiwanese Journal of Agricultural Chemistry and Food Science*, **39**(2), 117–121.

Ching, Y.P., Tze, K.C., Ming, L.H. & Shan, T.J. (2000) Effect of polyethylene package on the metmyoglobin reductase activity and color of tuna muscle during low temperature storage. *Fish Science*, **66**(2), 84–89.

Chinivasagam, H.N., Bremner, H.A. & Reeves, R. (1998) Can Spoilage cause blackspot (melanosis) in stored prawns? *Letters in Applied Microbiology*, **27**, 5–8.

Choubert, G. & Baccaunaud, M. (2006) Colour changes of fillets of rainbow trout (*Oncorhynchus mykiss* W.) fed astaxanthin or canthaxanthin during storage under controlled or modified atmosphere. *LWT—Food Science and Technology*, **39** (10), 1203–1213.

Choubert, G., Blanc, J.M. & Courvalin, C. (1992) Muscle carotenoid content and colour of farmed rainbow trout fed astaxanthin or canthaxanthin as affected by cooking and smoke-curing procedures. *International Journal of Food Science and Technology*, **27**(3), 277–284.

Choubert, G., Blanc, J.M. & Vallee, F. (1997) Colour measurement, using the CIELCH colour space, of muscle of rainbow trout, *Oncorhynchus mykiss* (Walbaum), fed astaxanthin: Effects of family, ploidy, sex, and location of reading. *Aquaculture Research*, **28**(1), 15–22.

Chow, C.J. (1991). Relationship between the stability and autoxidation of myoglobin. *Journal of Agricultural and Food Chemistry*, **39**(1), 22–26.

Chow, C.J., Ochiai, Y., Watabe, S. & Hashimoto, K. (1989) Reduced stability and accelerated autoxidation of tuna myoglobin in association with freezing and thawing. *Journal of Agricultural and Food Chemistry*, **37**(5), 1391–1395.

Cornforth, D. (1994). Colour—Its basis and importance. In: *Advances in Meat Research* (eds A.M. Pearson & T.R. Dutson), Vol. 9, pp. 34–78. Chapman & Hall, London, U.K.

Cornforth, D.P., Vahabzadhe, C.E. & Bartholomew, D.T. (1986) Role of reduced haemochromes in pink colour defect of cooked turkey. *Journal of Food Science*, **51**, 1132–1135.

Cornforth, D.P., Calkins, C.R. & Faustman, C.T. (1991) Methods for identification and prevention of pink color in cooked meats. *Proceedings of Reciprocal Meat Conference*, **44**, 53–58.

Crook, A.C. (1997) Colour patterns in a coral reef fish-is background complexity important? *Journal of Experimental Marine Biology and Ecology*, **217**, 237–252.

Davidsson, I. & Henry, J.B. (1978) *Diagnóstico Clínico por el laboratorio*, pp. 310–316. Salvat, Barcelona, Spain.

Diestre, A. (1992) Principales problemas de la calidad de la carne en el porcino. *Alim Eq Tecnol*, **98**, 73–78.

Duray, M.N., Estudillo, C.B. & Alpasan, L.G. (1996) The effect of background color and rotifer density on rotifer intake, growth and survival of the grouper (*Epinephelus suillus*) larvae. *Aquaculture*, **146**, 217–224.

Esteve, E. (1994) Alimentación animal y calidad de la carne. *Eurocarne*, **31**, 71–77.

Everse, J. and Hsia, N. (1997) The toxicities of native and modified hemoglobins. *Free Radical Biology and Medicine*, **22**, 1075–1099.

Faustman, C., Johnson, J.L., Cassens, R.G. & Doyle, M.P. (1990) Color reversion in beef: Influence of psychotrophic bacteria. *Fleischwirt*, **70**, 676.

Faustman, C., Chan, W.K.M., Lynch, M.P. & Joo, S.T. (1996) Strategies for increasing oxidative stability of (fresh) meat color. *Proceedings of the 49th Annual Reciprocal Meat Conference*, Provo, Utah, pp. 73–78.

Fernandez, P.J. & Bagnara, J.T. (1991) Effect of background colour and low temperature on skin color and circulating a-MSH in two species of leopard frog. *General and Comparative Endocrinology*, **83**, 132–141.

Fernández-Ginés, J.M., Fernández-López, J., Sayas-Barberá, E., Sendra, E. & Pérez-Alvarez, J.A. (2003) Physical, chemical and sensory properties of bologna sausage made with ostrich meat. *Journal of Food Science*, **68**, 1511–1515.

Férnández-López, J. (1998) Estudio del color por métodos objetivos en sistemas modelo de pastas de embutidos crudo-curados. PhD thesis. *Universidad de Murcia*, Murcia, Spain, p. 310.

Fernández-López, J., Pérez-Alvarez, J.A. & Aranda-Catalá, V. (2000) Effect on mincing degree on color properties in pork meat. *Color Research and Application*, **25**, 376–380.

Fernández-López, J., Pérez-Alvarez, J.A., Sayas-Barberá, M.E. & López-Santoveña, F. (2002) Effect of paprika (*Capsicum annum*) on color of Spanish-type sausages during the resting stage. *Journal of Food Science*, **67**, 2410–2414.

Fernández-López, J., Sevilla, L., Sayas-Barberá, E., Navarro, C., Marín, F. & Pérez-Alvarez, J.A. (2003) Evaluation of antioxidant potential of hyssop (*Hyssopus officinalis* L.) and rosemary (*Rosmarinus officinalis* L.) extract in cooked pork meat. *Journal of Food Science*, **68**, 660–664.

Ferrer, O.J., Otwell, W.S. & Marshall, M.R. (1989) Effect of bisulfite on lobster shell phenoloxidase. *Journal of Food Science*, **54**(2), 478–480.

Field, R.A., Sanchez, H.M., Ji, T.H., Chang, Y.O. & Smith, F.C. (1978) Amino acid analysis and acrylamide gel electrophoresis patterns of bovine hemopoietic marrow. *Journal of Agricultural and Food Chemistry*, **26**, 851.

Fox, J.B. (1966) The chemistry of meat pigments. *Journal of Agricultural and Food Chemistry*, **14**, 207.

Francis, F.J. (1995) Quality as influenced by color. *Food Quality and Preference*, **6**, 149–155.

Froning, G.W., Hargus, G. & Hartung, T.E. (1968) Color and texture of ground turkey meat products as affected by dry egg white solids. *Poultry Science*, **47**, 1187.

Fujii, R. (1969) Chromatophores and pigments. In: *Fish Physiology* (eds W.S. Hoar and D.J. Randall), Vol. III, pp. 307–353. Academic Press, New York.

Fujita, T., Satake, M., Watanabe, T., *et al.* (1983) Pigmentation of cultured red sea bream with astaxanthin diester purified from krill oil. *Nippon Suisan Gakkaishi*, **49**, 1855–1861.

Gill, C.O. (1996) Extending the storage life of raw chilled meats. *Meat Science*, **43**(Suppl), S99–S109.

Gobantes, I.Y. & Oliver, M.A. (2000) Problemática de la estabilidad del color en carne de vacuno: La adición de vitamina E y el envasado bajo atmósfera protectora. *Eurocarne*, **85**, 57–62.

Gomez-Guillen, M.C., Martinez-Alvarez, O., Llamas, A. & Montero, P. (2005) Melanosis inhibition and SO_2 residual levels in shrimps (*Parapenaeus longirostris*) after different sulfite-based treatments. *Journal of the Science of Food and Agriculture*, **85**(7), 1143–1148.

Goodrick, B., Milne, D., Grauf, S. & Dingle, V. (1991) When keeping cool means keeping colour. *Australian Fisheries*, **50**(11), 17–19.

Green, J.A. & Baker, B.I. (1991) The influence of repeated stress on the release of melanin concentrating hormone in the rainbow trout. *Journal of Endocrinology*, **128**, 261–266.

Gröneveld, D., Balm, P.H.M. & Wendelaar Bonga, S.E. (1995) Biphasic effect of MCH on alpha-MSH release from the tilapia (*Oreochromis mossambicus*) pituitary. *Peptides*, **16**, 945–949.

Guillou, A., Khalil, M. & Adambounou, V. (1995) Effects of silage preservation on astaxanthin forms and fatty acid profiles of processed shrimp (*Pandalus borealis*) waste. *Aquaculture*, **130**(4), 351–360.

Gutteridge, J.M.C. (1986) Iron promoters of the Fenton reaction and lipid peroxidation can be released from hemoglobin by peroxides. *FEBS Letters*, **201**, 291–295.

Healey, E.G. (1999) The skin pattern of young plaice and its rapid modification in response to graded changes in background tint and pattern. *Journal of Fish Biology*, **55**, 937–971.

Henmi, H., Hata, M. & Hata, V. (1989) Astaxanthin and/or canthaxanthin-actomyosin complex in salmon muscle. *Bulletin of the Japanese Society for the Science of Fish*, **55**(9), 1583–1589.

Henmi, H., Hata, M. & Takeuchi, M. (1990a). Studies on the carotenoids in the muscle of salmon. IV. Resonance Raman and circular dichroism studies of astaxanthin and/or canthaxanthin in salmon muscle. *Bulletin of the Japanese Society for the Science of Fish*, **56**(11), 1825–1828.

Henmi, H., Hata, M. & Hata, M. (1990b) Studies on the carotenoids in the muscle of salmon. III. Combination of astaxanthin and canthaxanthin with fish muscle actomyosins associated with their surface hydrophobicity. *Bulletin of the Japanese Society for the Science of Fish*, **56**(11), 1821–1823.

Higgs, D., Donaldson, E., Dosanjh, B., *et al.* (1995) The case for Phaffia. A yeast-based pigmenting agent shows promise for adding color to salmon diets. *North Aquaculture*, **11**(2), 20–24.

Houben, J.H., Eikelenboom, G. & Hoving-Bolink, A.H. (1998) Effect of the dietary supplementation with vitamin on colour stability and lipid oxidation in packaged, minced pork. *Meat Science*, **48**(3/4), 265–273.

Hui, H.C., Wen, H.H. & Tzu, H.H. (1998) Thermal stability of myoglobin and color in horse mackerel mince. *Food Science Taiwan*, **25**(4), 419–427.

Hunt, M.C. & Kropf, D.H. (1987) Colour and appearance. In: *Restructured Meat and Poultry Products, Advances in Meat Research Series* (eds A.M. Pearson & T.R. Dutson), Vol. 3, pp. 125–159. Van Nostran Reinhold, New York.

Hwei, M.W. & Min, H.L. (1994) Identification of beta-carotene and astaxanthin in squid visceral oil. *Journal of the Chinese Agricultural Chemical Society*, **31**(4), 431–443.

Hyun, Y.K., Seung, H.B., Soo, Y.K., *et al.* (1999) Metabolism of dietary carotenoids and effects to improve the body color of oily bittering, *Acheilognathus koreensis*. *Journal of the Korean Society of Food Science and Nutrition*, **28**(5), 1099–1106.

Ingemansson, T., Pettersson, A. & Kaufmann, P. (1993) Lipid hydrolysis and oxidation related to astaxanthin content in light and dark muscle of frozen stored rainbow trout (*Oncorhynchus mykiss*). *Journal of Food Science*, **58**(3), 513–518.

Irie, M. (2001) Optical evaluation of factors affecting appearance of bovine fat. *Meat Science*, **57**, 19–22.

Irie, M., Fujita, K. & Suduo, K. (1999) Changes in meat color and alpha-tocopherol concentrations in plasma and tissues from Japanese beef cattle fed by two methods of vitamin E supplementation. *Australasian Journal of Animal Sciences*, **11**, 810–814.

Jensen, C., Laurisen, C. & Bertelsen, G. (1998) Dietary vitamin E: Quality and storage stability of pork and poultry. *Trends in Food Science and Technology*, **9**, 62–72.

Kalyanasundaram, K. (1992) *Photochemistry of Poly-pyridine and Porphyrin Complexes*, pp. 25–56. Academic Press, New York.

Kanner, J. (1994) Oxidative processes in meat products: Quality implications. *Meat Science*, **36**, 169–189.

Kanner, J. & Harel, S. (1987) Desferrioxamine as an electron donor. Inhibition of membranal lipid peroxidation initiated by H2O2-activated metmyoglobin and other peroxidizing systems. *Free Radical Research Communications*, **3**, 1–5.

Kanner, J., German, J.B. & Kinsella, J.E. (1987) Initiation of lipid peroxidation in biological systems. *CRC Critical Reviews in Food Science and Nutrition*, **25**, 317–364.

Kentouri, M., Pavlidis, M., Papandroulakis, N. & Divanach, P. (1995) Culture of the red porgy, *Pagrus pagrus*, in Crete. Present knowledge, problems and perspectives. *Cahiers Options Mediterraneennes*, **16**, 65–78.

Klomklao, S., Benjakul, S., Visessanguan, W., Kishimura, H. & Simpson, B.K. (2006) Effects of the addition of spleen of skipjack tuna (*Katsuwonus pelamis*) on the liquefaction and characteristics of fish sauce made from sardine (*Sardinella gibbosa*). *Food Chemistry*, **98**, 440–452.

Knipe, L. (1993) *Basic Science of Meat Processing Cured Meat Short Course*. Meat Laboratory, Iowa State University Press, Ames, Iowa.

Krammer, A. (1994) Use of color measurements in quality control of food. *Food Technology*, **48**(10), 62–71.

Kumazawa, T. and Fujii, R. (1984) Concurrent releases of norepinephrine and purines by potassium from adrenergic melanosome-aggregating nerve in tilapia. *Comparative Biochemistry and Physiology*, **78C**, 263–266.

Lamers, A.E., Flik, G., Atsma, W. & Wendelaar Bonga, S.E. (1992) A role for di-acetyl alpha-melanocyte stimulating-hormone in the control of cortisol release in the teleost *Oreochromis mossambicus*. *Journal of Endocrinology*, **135**, 285–292.

Lanari, M.C. & Cassens, R.G. (1991) Mitochondrial activity and beef muscle color stability. *Journal of Food Science*, **56**, 1476–1479.

Lanari, M.C., Schaefer, D.M. & Scheller, K.K. (1995) Dietary vitamin E supplementation and discoloration of pork bone and muscle following modified atmosphere packaging. *Meat Science*, **41**(3), 237–250.

Lanari, M.C., Brewster, M., Yang, A. & Tume, R.K. (2002) Pasture and grain finishing affect the color stability of beef. *Journal of Food Science*, **67**, 2467–2473.

Ledward, D.A. (1984) Haemoproteins in meat and meat products. In: *Developments in Food Proteins* (ed. D.A. Ledward), pp. 33–68. Applied Science, London, U.K.

Ledward, D.A. (1992) Colour of raw and cooked meat. In: *The Chemistry of Muscle Based Foods* (eds D.A. Ledward, D.E. Johnston & M.K. Knight), pp. 128–144. The Royal Society of Chemistry, Cambridge, U.K.

Lee, E.J., Love, J. & Ahn, D.U. (2003a) Effects of antioxidants on consumer acceptance of irradiated turkey meat. *Journal of Food Science*, **68**, 1659–1663.

Lee, S., Joo, S.T., Alderton, A.L., Hill, D.W. and Faustman, C. (2003b) Oxymyoglobin and lipid oxidation in yellowfin tuna (*Thunnus albacares*) loins. *Journal of Food Science*, **68**(5), 1164–1168.

Lehninger, A.L. (1981) *Bioquímica. Las bases moleculares de la estructura y función celular*, pp. 230–233. Omega, Barcelona, Spain.

Liaaen-Jensen, L. (1971) Isolation, reaction. In: *Carotenoids* (ed. O. Isler), pp. 61–188. Birkhä user Verlag, Basel, Switzerland.

Lien, R., Hunt, M.C., Anderson, S., *et al.* (2002) Effects of endpoint temperature on the internal color of pork patties of different myoglobin form, initial cooking state, and quality. *Journal of Food Science*, **67**(3), 1011–1015.

Lin, M.Q., Ushio, H., Ohshima, T., Yamanaka, H. & Koizumi, C. (1998) Skin color control of the red sea bream (*Pagrus major*). *Lebensmittel-Wissenschaft und-Technologie*, **31**, 27–32.

Little, A.C., Martinse, C. & Sceurman, L. (1979) Color assessment of experimentally pigmented rainbow trout. *Color Research & Application*, **4**(2), 92–95.

López-Caballero, M.E., Martínez-Alvarez, O., Gómez-Guillén, M.C. & Montero, P. (2006) Quality of Norway lobster (*Nephrops norwegicus*) treated with a 4-hexylresorcinol-based formulation. *European Food Research and Technology*, **222**, 425–431.

Lorient, D. (1982) Propiedades funcionales de las proteínas de origen animal. In: *Proteínas Animales* (eds C.M. Bourgeois and P. Le Roux), pp. 215–225. El Manual Moderno, México.

MacDougall, D.B. (1982) Changes in the colour and opacity of meat. *Food Chemistry*, **9**(1/2), 75–88.

Mancini, R.A., Hunt, M.C. & Kropf, D.H. (2003) Reflectance at 610 nm estimates oxymyoglobin content on the surface of ground beef. *Meat Science*, **64**, 157–162.

Mancini, R.A., Hunt, M.C., Hachmeister, K.A., Kropf, D.H. & Johnson, D.E. (2004) Ascorbic acid minimizes lumbar vertebrae discoloration. *Meat Science*, **68**(3), 339–345.

Mercier, Y., Gatellier, P., Viau, M., Remington, H. & Renerre, M. (1998) Effect of dietary fat and vitamin E on colour stability and on lipid and protein oxidation in turkey meat during storage. *Meat Science*, **48**(3/4), 301–318.

Minguéz-Mosquera, M.I. (1997) *Clorofilas y caroten-oides en tecnología de alimentos*, pp. 15–17. Uni-versidad de Sevilla, Sevilla, Spain.

Moe, N.H. (1990) Key factors in marketing fanned salmon. *Proceedings of the Nutrition Society of New Zealand*, **15**, 16–22.

Moller, J.K.S., Adamsen, C.E. & Skibsted, L.H. (2003) Spectral characterisation of red pigment in Italian-type dry-cured ham. Increasing lipophilicity during processing and maturation. *European Food Research and Technology*, **216**(4), 290–296.

Montero, P., Ávalos, A. & Pérez-Mateos, M. (2001) Characterization of polyphenoloxidase of prawns (*Penaeus japonicus*). Alternatives to inhibition: Additives and high pressure treatment. *Food Chemistry*, **75**, 317–324.

Montero, P., Martínez-Alvarez, O. & Gómez-Guillén, M.C. (2004) Effectiveness of onboard application of 4-hexylresorcinol in inhibiting melanosis in shrimp (*Parapenaeus longirostris*). *Journal of Food Science*, **68**(8), 643–647.

Mori, T. (1993) Pigmentation of cultured fishes. In: *Carotenoids in Marine Organisms* (ed. W. Miki), pp. 59–68. Kouseisya, Kouseikaku, Tokyo.

Moss, B.W. (1992) Lean meat, animal welfare and meat quality. In: *The Chemistry of Muscle Based Foods* (eds D.A. Ledward, D.E. Johnston & M.K. Knight), pp. 62–76. The Royal Society of Chemistry, Cambridge, U.K.

Nash, D.M., Proudfoot, F.G. & Hulan, H.W. (1985) Pink discoloration in broiler chicken. *Poultry Science*, **64**, 917–919.

Ngoka, D.A. & G.W. Froning. (1982) Effect of free struggle and preslaughter excitement on color of turkey breast muscles. *Poultry Science*, **61**, 2291–2293.

Nickell, D.C. & Bromage, N.R. (1998) The effect of timing and duration of feeding astaxanthin on the development and variation of fillet colour and efficiency of pigmentation in rainbow trout (*Orcorhy-nchus mykiss*). *Aquaculture*, **169**, 233–246.

Nicolalde, C., Stetzer, A.J., Tucker, E.M., McKeith, F.K. & Brewer, M.S. (2005) Antioxidant and modified atmosphere packaging prevention of discoloration in pork bones during retail display. *Meat Science*, **72**(4), 713–718.

Norris, A.T. & Cunningham, E.P. (2004) Estimates of phenotypic and genetic parameters for flesh colour traits in farmed Atlantic salmon based on multiple trait animal model. *Livestock Production Science*, **89**, 209–222.

Numata, M. & Wakamatsu, J. (2003) Natural red pigments and foods and food materials containing the pigments. Patent WO 03/063615 A1.

Ogawa, M., Perdiago, N.B., Santiago, M.E. & Kozima, T.T. (1984) On physiological aspects of black spot appearance in shrimp. *Bulletin of the Japanese Society for the Science of Fish*, **50**(10), 1763–1769.

Ohshima, T., Wada, S. & Koizumi, C. (1988) Influences of haeme pigment, non-haeme iron, and nitrite on lipid oxidation in cooked mackerel meat. *Bulletin of the Japanese Society for the Science of Fish*, **54**(12), 2165–2171.

Oka, H. (1989) Packaging for freshness and the prevention of discoloration of fish fillets. *Packaging Technology and Science*, **2**(4), 201–213.

Osborn, H.M., Brown, H., Adams, J.B. & Ledward, D.A. (2003) High temperature reduction of metmyoglobin in aqueous muscle extracts. *Meat Science*, **65**, 631–637.

Ostermeyer, U. & Schmidt, T. (2004) Differentiation of wild salmon, conventionally and organically farmed salmon. *Deutschland Lebens Rund*, **100**(11), 437–444.

Pagán-Moreno, M.J., Gago-Gago, M.A., Pérez-Alvarez, J.A., *et al.* (1998) The evolution of colour parameters during "chorizo" processing. *Fleischwt*, **78**(9), 987–989.

Papoutsoglou, S.E., Mylonakis, G., Miliou, H., Karakatsouli, N.P. & Chadio, S. (2000) Effects of background color on growth performances and physiological responses of scaled carp (*Cyprinus carpio* L.) reared in a closed circulation system. *Aquacultural Engineering*, **22**, 300–318.

Park, J.W. & Morrisey, M.T. (1994) The need for developing the surimi standar. In: *Quality Control and Quality Assurance Seafood* (eds G. Sylvia, A. Shriver, M.T. Morrisey), p. 265. Oregon Sea Grant, Corvallis, Oregon.

Parolari, G., Gabba, L. & Saccani, G. (2003) Extraction properties and absorption spectra of dry cured hams made with and without nitrate. *Meat Science*, **64**(4), 483–490.

Pérez-Alvarez, J.A. (1996) *Contribución al estudio objetivo del color en productos cárnicos crudo-curados*. PhD thesis. Universidad Politécnica de Valencia, Valencia, Spain, p. 210.

Pérez-Alvarez, J.A., Sánchez-Rodriguez, M.E., Fernández-López, J., Gago-Gago, M.A., Ruíz-Peluffo, M.C., Rosmini, M.R., Pagán-Moreno, M.J., Lopez-Santoveña, F. & Aranda-Catalá, V. (1997) Chemical and color characteristics of "Lomo embuchado" during salting seasoning. *Journal of Muscle Foods*, **8**(4), 395–411.

Pérez-Alvarez, J.A., Fernández-López, J. and Sayas-Barberá, M.E. (2000) Fundamentos físicos, químicos, ultraestructurales y tecnológicos en el color de la carne. In: *Nuevas tendencias en la tecnología e higiene de la industria cárnica* (eds M. Rosmini, J.A. Pérez-Alvarez, J. Fernández-López), pp. 51–71. Miguel Hernández de Elche, Elche, Spain.

Pikul, J., Niewiarowicz, A. & Kupijaj, H. (1986) The cytochrome c content of various poultry meats. *Journal of the Science of Food and Agriculture*, **37**, 1236–1240.

Rehbein, H. & Orlick, B. (1990) Comparison of the contribution of formaldehyde and lipid oxidation products to protein denaturation and texture deterioration during frozen storage of minced ice-fish fillet (*Champsocephalus gunnari* and *Pseudochaenichthys georgianus*). *International Journal of Refrigeration*, **13**(5), 336–341.

Renerre, M. (1990) Review: Factors involved in discoloration of beef meat. *International Journal of Food Science and Technology*, **25**, 613–630.

Richards, M.P. and Hultin, H.O. (2002) Contributions of blood and blood components to lipid oxidation in fish muscle. *Journal of Agricultural and Food Chemistry*, **50**(3), 555–564.

Richards, M.P. and Li, R. (2004) Effects of released iron, lipid peroxides, and ascorbate in trout hemoglobin-mediated lipid oxidation of washed cod muscle. *Journal of Agricultural and Food Chemistry*, **52**, 4323–4329.

Rotllant, J., Tort, L., Montero, D., Pavlidis, M., Martinez, M., Wendelaar-Bonga, S.E. & Balm, P.H.M. (2003) Background colour influence on the stress response in cultured red porgy *Pagrus pagrus*. *Aquaculture*, **223**, 129–139.

Ruíz-Peluffo, M.C., Pérez-Alvarez, J.A. & Aranda-Catalá, V. (1994) Longaniza de pascua: Influencia del ascorbato e isoascorbato de sodio sobre las propiedades fisicoquímicas y de color. In: *Anales de Investigación del Master en Ciencia e Ingeniería de Alimentos* (eds P. Fito, J. Serra, E. Hernández & D. Vidal), Vol. IV, pp. 675–688. Reproval, Valencia, Spain.

Sachindra, N.M., Bhaskar, N. & Mahendrakar, N.S. (2005a) Carotenoids in different body components of Indian shrimps. *Journal of the Science of Food and Agriculture*, **85**(1), 167–172.

Sachindra, N.M., Bhaskar, N. & Mahendrakar, N.S. (2005b) Carotenoids in crabs from marine and fresh waters of India. *Food Science and Technology*, **38**(3), 221–225.

Sakiura, T. (2001) Cultured fish carotenoid and polyphenol added feed for improving fish body color tone and fish meat brilliance. U.S. Patent 2001/0043982 A1.

Saksit, C., Ogawa, M., Tamiya, T. & Tsuchiya, T. (1996) Studies on thermal denaturation of fish myoglobins using differential scanning calorimetry, circular dichroism, and tryptophan fluorescence. *Fish Science*, **62**(6), 927–932.

Saksit, C., Nieda, H., Ogawa, M., Tamiya, T. & Tsuchiya, T. (1998) Changes in light scattering intensity of fish holo-, and apo- and reconstituted myoglobins under thermal denaturation. *Fish Science*, **64**(5), 846–887.

Sánchez-Escalante, A., Djenane, D., Torrescano, G., Beltran, J.A. & Roncales, P. (2003) Antioxidant action of borage, rosemary, oregano and ascorbic acid in beef patties packaged in modified atmospheres. *Journal of Food Science*, **68**, 339–344.

Sayas, M.E. (1997) *Contribuciones al proceso tecnológico de elaboración del jamón curado: aspectos físicos, fisicoquímicos y ultraestructurales en los procesos de curado tradicional y rápido*. PhD thesis. Universidad Politécnica de Valen-cia, Valencia, Spain, p. 175.

Scriven, F., Sporns, P. & Wolfe, F. (1987) Investigations of nitrite and nitrate levels in paper materials used to package fresh meats. *Journal of Agricultural and Food Chemistry*, **35**, 188–192.

Shahidi, F. and Matusalach-Brown, J.A. (1998) Carotenoid pigments in seafoods and aquaculture. *Critical Reviews in Food Science*, **38**(1), 1–67.

Sigurgisladottir, S., Parrish, C.C., Lall, S.P. & Ackman, R.G. (1994) Effects of feeding natural tocopherols and astaxanthin on Atlantic salmon (*Salmo salar*) fillet quality. *Food Research International*, **27**(1), 23–32.

Sigurgisladottir, S., Torrissen, O., Lie, Ø., Thomassen, M. & Hafsteinsson, H. (1997) Salmon quality: Methods to determine the quality parameters. *Reviews in Fisheries Science*, **5**(3), 223–252.

Slattery, S.L., Williams, D.J. & Cusack, A. (1995) A sulphite free treatment inhibits blackspot formation in prawns. *Food-Australia*, **47**(11), 509–514.

Stauton-West, E., Todd, W.R., Mason, H.S. and Van Bruggen, J.T. (1969) *Bioquímica Médica*, pp. 616–623. Interamericana, México.

Storebakken, T. and Choubert, G. (1991) Flesh pigmentation of rainbow trout fed astaxanthin or canthaxanthin at different feeding rates in freshwater and saltwater. *Aquaculture*, **95**, 289–295.

Sugimoto, M. (1997) Regulation of melanophore responsiveness in the background-adapted medaka, *Oryzias latipes*: Change in the intracellular signalling system. *Comparative Biochemistry and Physiology, C* **117**, 259–265.

Suman, S.P., Faustman, C., Lee, S., Tang, J., Sepe, H.A., Vasudevan, P., Annamalai, T., Manojkumar, M., Marek, P., DeCesare, M. & Venkitanarayanan, K.S. (2004) Effect of muscle source on premature browning in ground beef. *Meat Science*, **68**(3), 457–461.

Svingen, B.A., Buege, J.A., O'Neal, F.O. & Aust, S.D. (1979) The mechanism of NADPH-dependent lipid peroxidation. *Journal Biological Chemistry*, **254**, 5892–5899.

Swatland, H.J. (1988) Carotene reflectance and the yellows of bovine adipose tissue measured with a portable fiber-optic spectrophotometer. *Journal of the Science of Food and Agriculture*, **46**, 195–200.

Swatland, H.J. (1995) Reversible pH effect on pork paleness in a model system. *Journal of Food Science*, **60**(5), 988–991.

Tadolini, B. and Hakim, G. (1996). The mechanism of iron (III) stimulation of lipid peroxidation. *Free Radical Research*, **25**, 221–227.

Tanaka, M., Nishino, H., Satomi, K., Yokoyama, M. & Ishida, Y. (1996) Gas-exchanged packaging of tuna fillets. *Nippon Suisasn Gakkaishi*, **62**(5), 800–805.

Taoukis, P.S., Labuza, T.P., Lillemo, J.H. & Lin, S.W. (1990) Inhibition of shrimp melanosis (black spot) by ficin. *Lebensmittel-Wissenschaft und-Technologie*, **23**(1), 52–54.

Torrissen, O.J. (1986) Pigmentation of salmonids-a comparison of astaxanthin and canthaxanthin as pigment sources for rainbow trout. *Aquaculture*, **53**, 271–278.

Torrisen, O.J., Hardy, R.W. & Shearer, K.D. (1989) Pigmentation of salmonids carotenoids deposition and metabolism. *Aquatic Science*, **1**, 209–225.

Tze, K.C., Ching, Y.P., Fang, P.N. & Shann, T.J. (2001) Effect of met-myoglobin reductase on the color stability of blue fin tuna during refrigerated storage. *Fish Science*, **67**(4), 694–702.

Urbain, M.W. (1952) Oxygen is key to the color of meat. *NatProv*, **127**, 140–141.

Verdoes, J.C., Krubasik, P., Sandmann, G. & van Ooyen, A.J.J. (1999) Isolation and functional characterisation of a novel type of carotenoid biosynthetic gene from *Xanthophyllomyces dendrorhous*. *Molecular Genetics and Genomics*, **262**(3), 453–461.

Virgili, R., Parolari, G., Bordini, C.S., Schivazappa, C., Cornet, M. & Monin, G. (1999) Free amino acid dipeptides in dry-cured ham. *Journal of Muscle Foods*, **10**, 119–130.

Wakamatsu, J., Nishimura, T. & Hattori, A. (2004a) A Zn-porphyrin complex contributes to bright red colour in Parma ham. *Meat Science*, **67**(1), 95–100.

Wakamatsu, J., Okui, J., Ikeda, Y., Nishimura, T. & Hattori, A. (2004b) Establishment of a model experiment system to elucidate the mechanism by which Zn-protoporphyrin IX is formed in nitrite-free dry-cured ham. *Meat Science*, **68**(2), 313–317.

Walsh, K.A. & Rose, D. (1956) Factors affecting the oxidation of nitric oxide myoglobin. *Journal of Agricultural and Food Chemistry*, **4**(4), 352–355.

Warren, K.E., Hunt, M.C., Kropf, D.H., *et al.* (1996) Chemical properties of ground beef patties exhibiting normal and premature brown internal cooked color. *Journal of Muscle Foods*, **7**(3), 303–314.

Warris, P.D. & Rhodes, D.N. (1977) Haemoglobin concentrations in beef. *Journal of Science and Food Agriculture*, **28**, 931–934.

Whitaker, J.R. (1972) *Principles of Enzymology for the Food Sciences*, pp. 225–250. Marcel Dekker, New York.

Williams, H.G., Davidson, G.W. & Mamo, JC. (2003) Heat induced activation of polyphenoloxidase in wester rock lobster (*Panulinus cygnus*) hemolymph: Implications for heat processing. *Journal of Food Science*, **68**, 1928–1932.

Xin, F. & Shun, W. (1993) Enhancing the antioxidant effect of alpha-tocopherol with rosemary in inhibiting catalyzed oxidation caused by Fe21 and hemoprotein. *Food Research International*, **26**(6), 405–411.

Yanar, Y., Büyükçapar, H.B. & Yanar, M. & Göcer, M. (2006) Effect of carotenoids from red pepper and marigold flower on pigmentation, sensory properties and fatty acid composition of rainbow trout. *Food Chemistry*, In Press.

Zanardi, E., Novelli, E., Nanni, N., *et al.* (1998) Oxidative stability and dietary treatment with vitamin E, oleic acid and copper of fresh and cooked pork chops. *Meat Science*, **49**(3), 309–320.

Zerby, H.N., Belk, K.E., Sofos, J.N., Mcdowell, L.R., Williams, S.N. & Smith, G.C. (1999) Display life of fresh beef containing different levels of vitamin E and initial microbial contamination. *Journal of Muscle Foods*, **10**, 345–355.

Zhou, G.H., Yang, A. & Tume, R.K. (1993) A relationship between bovine fat colour and fatty acid composition. *Meat Science*, **35**(2), 205–212.

Zhu, L.G. & Brewer, M.S. (2002) Combined effects of pH and temperature on myoglobin in a model system. *Meat Science*, **61**, 419–424.

Zhu, L.G. & Brewer, M.S. (2003) Effects of urea denaturation and pH on the ability of porcine myoglobin to undergo reduction. *Meat Science*, **63**, 427–432.

2 Sensory Methodology of Muscle Foods

Patti C. Coggins

Abstract: Sensory testing is predominantly evaluating the attributes of any consumed product. Sensory attributes are appearance, odor (aroma and fragrance), consistency and texture, and flavor.

In a sensory evaluation test, a number of steps may be distinguished: project objective, test objective, sample screening, test design, test conducting, data analysis, and interpretation of results. Different types of tests are available: triangle test, paired comparison test, duo-trio test, A-not-A test, preference tests, acceptance tests, and attribute assessment. Descriptive analysis techniques are the most complex methods in sensory evaluation.

Keywords: Sensory evaluation, appearance, odor, aroma, texture, flavor, meat tests, descriptive analysis

2.1 INTRODUCTION TO SENSORY EVALUATION

Sensory testing has been defined as predominantly evaluating the attributes of any consumed product (Meilgaard *et al.* 1991). Measuring the sensory properties of products and being able to determine the importance of them is a major accomplishment for sensory evaluation. Food companies exploiting this knowledge and applying it to their in-house quality maintenance and/or marketing program are utilizing a valuable research tool that will ultimately enhance overall perception of the products produced. The information that applied sensory evaluation provides is invaluable to a company. Without complete product acceptance from the end consumer, a company could encounter financial difficulties that could otherwise be avoided.

There has long been a need for a more detailed approach to sensory evaluation practices within the corporate environment of muscle foods companies. Muscle foods often fall fate to a detrimental perception by the American consumer who demands fresh, high-quality products for consumption. It can be said that the muscle foods sector of the food industry has a greatly reduced quality acceptance range than other natural foods. The typical American consumer will not accept a muscle food that is not "perfect" in their opinion. This aspect makes determination of quality and acceptance sometimes difficult to decipher due to biases of the consumer as well as inherent differences within the muscle food product.

2.2 SENSORY ATTRIBUTES

According to Civille, we are apt to perceive sensory characteristics of a food item in the following order: appearance, odor/aroma/fragrance, consistency and texture, and flavor. However, most of these overlap, and the subject will have a simultaneous sensory impression.

Handbook of Meat, Poultry and Seafood Quality, Second Edition. Edited by Leo M. L. Nollet.

Without proper training and experience in sensory methodology aspects, the subject will find it more difficult to distinguish between each attribute (Meilgaard *et al.* 1991). Sensory evaluation, as well as the consumer, has matured in the last decade. With this in mind, it is important for the muscle foods industry to realize that consumers are much more sophisticated in their tastes, and ability to detect differences overall, as well as what they are willing to accept as a quality muscle food product. Sensory attributes are defined as appearance, odor/aroma/fragrance, flavor, texture, and specific feeling/chemical factors.

2.2.1 Appearance

Appearance is the visual property of a product, including size, shape, color, texture, gloss, transparency, cloudiness, and so on (Lawless & Heymann 1998). It is often the only attribute on which one can base a decision to purchase or consume a product. One becomes comfortable and trained in doing this; panelists will do the same in a test booth. General appearance characteristics include the following:

- Color involves both physical and psychological components and is commonly expressed in terms of hue, value, and chroma of the Munsell color system. The evenness of color as opposed to uneven or blotchy appearance is important. Deterioration of food is often accompanied by a color change.
- Size and shape include length, thickness, width, particle size, geometric shape, and distribution of pieces, which are indicators of defects.
- Surface texture includes the dullness or shininess of a surface, and the roughness versus evenness. Does the surface appear wet or dry, soft or hard, crisp or tough?
- Clarity includes the haze or opacity of transparent liquids or solids, and the presence or absence of particles of visible size (Seibert *et al.* 1981, McDougall 1988).
- Carbonation, for carbonated beverages, involves the degree of effervescence observed upon pouring (Meilgaard *et al.* 1991).

The effects of appearance on flavor perception are sometimes overruling. Human beings are a visually driven species. In many societies with mature culinary arts, the visual presentation of a food is as important as its flavor and texture characteristics (Lawless & Heymann 1998).

Visual appearance is a key factor that influences consumers when they are assessing the quality and palatability of meat and meat products. Certain colors influence food acceptance, although the color of the meat itself may be influenced by its moisture and fat content and also by the content of hemoprotein, particularly myoglobin, and its relationship with the surrounding environment. The development of the characteristic color of fermented products is the result of the action of nitrite with myoglobin, producing the red color (Hui *et al.* 2001).

2.2.2 Odor/aroma/fragrance

Odor is the characteristic smell of a substance (Lawless & Heymann 1998). As defined by Civille (Meilgaard *et al.* 1991), the odor of a product is detected when its volatiles enter the nasal passage and are perceived by the olfactory system. Odor is sniffing of volatiles through the nose; aroma is the odor of a food product most often associated with applied heat or thermal property; and fragrance is the odor of a perfume, a cosmetic, or natural plant extractives. The amount of volatiles that escape from a product is affected by the temperature and by the nature of the compounds. Volatility is also influenced by the condition of a surface (Meilgaard *et al.* 1991).

Many volatile compounds have been identified in fermented products belonging to the following classes: alkanes, alkenes, aldehydes, ketones, alcohols, aromatic hydrocarbons, carboxylic acids, esters, terpenes, sulfur compounds, furans, pyrazines, amines, and chloride compounds. Different pathways are responsible for the formation of these volatile compounds. However, the impact of an odor component on the total aroma depends on a number of factors, such as odor threshold, concentration, solubility in water or fat, and temperature as reported for dry-cured ham flavor (Hui *et al.* 2001).

2.2.3 Consistency and texture

Consistency, texture, and viscosity are attributes that are perceived by sensors in the mouth other than taste and chemical feelings. Consistency can also be measured by sensory evaluation; in practice, some standardization is possible by the aid of consistometers. Civille defines texture as the sensory manifestation of the structure or inner makeup of products in terms of their reaction to stress and tactile feel properties. Viscosity refers to the rate of flow of liquids under some force, such as gravity (Meilgaard *et al.* 1991).

The International Organization for Standardization (ISO) defines the texture of a food product as all the rheological and structural (geometric and surface) attributes of the product perceptible by means of mechanical, tactile, and where appropriate, visual, and auditory receptors (ISO 1981). The texture of a product is perceived by the senses of sight (visual texture), touch (tactile texture), and sound (auditory texture) (Bourne 1982). In many products, one of these texture definitions may be used to describe a product's texture, whereas in others, all definitions could be applied. Ball *et al.* (1957) were pioneers in distinguishing between sight/visual and feel/tactile definitions of texture. Their studies on muscle foods greatly influenced how quality of meat is scientifically perceived today. One important discovery is how visual texture is a determination of product quality and freshness. Also, visual texture creates expectations of mouthfeel, which to the consumer is also an indicator of quality and/or freshness. Many surface characteristics of a food product not only affect the perceived appearance of the product but also the perception of the texture (Lawless & Heymann 1998). For example, the surface roughness of a biscuit can be assessed visually as well as orally; the viscosity of gravy can be assessed visually by pouring as well as orally. Tactile texture can be divided into oral tactile texture, mouthfeel characteristics, phase changes in the oral cavity, and the tactile texture perceived when manipulating by hand (Lawless & Heymann 1998). Oral tactile texture is often a measurement of size and shape, mouthfeel, phase change, as well as tensile strength. Oral tactile texture is an important measurement of the texture of a muscle food. For instance, if a piece of meat is mushy or soft, a negative perception could possibly form. Handheld tactile is most often used in evaluating certain consumer and/or personal-care products. However, it is still of importance in the texture measurement of muscle foods, especially luncheon meats and meat sticks.

2.2.4 Flavor

Flavor, as an attribute of foods, has been defined as the perceptions resulting from stimulation of the sense ends that are grouped together at the entrance of the alimentary and respiratory tracts (Meilgaard *et al.* 1991). It is a complex group of sensations comprising of olfactory, taste, and other chemical sensations, such as irritation or chemical heat (Lawless & Heymann 1998). Flavor includes the following:

- The aromatics caused by volatile substances released in the mouth via the posterior nares
- The tastes caused by soluble substances in the mouth
- The chemical feeling factors, which stimulate nerve ends in the soft membranes of the buccal and nasal cavities (Meilgaard *et al.* 1991)

Meat flavor is derived from many different sources. Fresh uncooked meat has a metallic and blood-like flavor. Cooked meat, dependent upon the length of cooking cycle, and process, can vary in flavor as well. Fermented meats have yet another interesting flavor. The characteristic flavor of fermented sausages mainly originates from the breakdown of carbohydrates, lipids, and proteins through the action of microbial and endogenous meat enzymes. But other substances added to the sausage, such as salt and spices, should be taken into account because of their important contribution to flavor. Additionally, there are other pathways, such as autoxidation, that form flavor compounds without direct enzymatic participation (Hui *et al.* 2001).

The carbohydrate fermentation is responsible for the typical tangy or sour taste. The interactions between carbohydrate and protein metabolism during meat fermentation determine the rate of pH decline and flavor development. During carbohydrate fermentation, significant amounts of acetic acid, besides lactic acid, are generated. On the other hand, pH is partially neutralized during drying as the result of further ammonia and free amino acids generation. All of these compounds have an impact on flavor (Hui *et al.* 2001).

There are internal and external parameters that influence flavor. The internal parameters are chemical (added sugars or spices) or microbiological (starter cultures); external parameters are physical, such as the temperature and humidity during the process (Hui *et al.* 2001).

2.3 SENSORY EVALUATION PERFORMANCE

A sensory evaluation study involves several steps. Sensory studies are difficult and time consuming to perform. Following is a guide for sensory evaluation studies.

2.3.1 Project objective

As with all research and any project, an objective must first be defined. Often, many questions must be answered prior to the start of a project. If the project objective is not clearly defined, the appropriate test is unlikely to be used and the data gathered could possibly not answer the questions at hand about the product.

2.3.2 Test objective

The test objective is defined once the project has been determined and an objective established. The type of sensory study is then defined at this point.

For example, the project objective may be to identify a less costly flavor-masking agent for the soy additive in a meat binder formulation that a meat company purchases from a seasoning supplier. Knowing this information and how it will be ultimately used is beneficial because it not only identifies the specific test objective but also alerts sensory evaluation to the potential of additional testing and to the fact that current production is the target (Stone & Sidel 1993).

2.3.3 Sample screening

While determining the project and test objectives, all sensory properties of the samples to be tested should be examined. Biases inherent or otherwise can be determined at this point.

This issue precedes a decision that the test is warranted and most often arises during the initial stages of sensory evaluation. At this time, any samples with significant differences may be weeded out, because it would be redundant to test such samples.

Pretest screening also will enable sensory evaluation to clarify any special requirements related to preparation and serving and to identify any additional background material required as it relates to the product or the test (Stone & Sidel 1993).

2.3.4 Test design

Designing the sensory methodology that will be utilized and the correct number of panelists, trained or untrained, must then be decided. The use of a properly designed score sheet is of extreme importance as to the outcome of the test results and data obtained.

Selection of the specific test method can be challenging. Industry personnel often are familiar with one particular sensory method and utilize that method for all of the work with which they are involved. This can be a "death sentence" to an internal sensory program or any other area of the company. In cases such as this, outsourcing is often necessary for objective opinions.

Sensory evaluation must first consider product and objective before reaching a decision about the most appropriate method for that problem. All discrimination methods have the same basic goal and all are essentially different forms of the two-sample choice situation. Consumer studies likewise have an end goal—for example, to define what the consumer likes or wants.

Test method selection is the result of a series of probing questions and answers by the sensory professional. The focus is on test objective, type and availability of product, preparation procedures, product sensory characteristics, test method capabilities, availability of qualified subjects, and the use of the results. Such an approach leads to a much higher quality database, and also provides more reliable and valid product information (Stone & Sidel 1993).

2.3.5 Test conducting

The evaluation should be performed in a controlled setting and a professional environment maintained at all times. Biases should always be kept to a minimum and control implemented at all times. Professionalism is a must in sensory evaluation studies.

2.3.6 Data analysis

The necessary statistical programs should be used when the study is complete. The data should be analyzed according to the procedure decided upon at the test design stage. The data should be examined for the main treatment effect as well as other test variables.

2.3.7 Interpretation of results

The original objectives of the project and test should enable the sensory analyst to review the analyzed results of the data and come to a conclusion about a range of variables. The analyst can then express the results in terms of the stated objectives and make recommendations for the product (Meilgaard *et al.* 1991).

2.4 STUDY CONTROL

When performing a sensory test, many variables must be controlled. These variables include test controls, product controls, and panel controls, to mention a few (Meilgaard *et al.* 1991).

2.4.1 Test controls

2.4.1.1 The booth or divided table setting

Sensory testing booths are desirable, but not all companies have an internal sensory program or testing booths. If formal booths are not available, divided areas of a table or other secluded areas have been used with few problems. Booths may be arranged side by side, in an L shape, or with two sets of three to four booths facing each other across the serving area. Dividers between the booths should extend 18 inches above the countertops to provide privacy between panelists. A small stainless steel sink and water faucet can be included for rinsing. However, many scientists opt to exclude the sink and provide water in cups for drinking and an additional set of cups for rinsing and expectoration if necessary. Sinks can harbor odors when not properly maintained; hence, whether or not the utilization of sinks is important must be well thought out. A signal system is frequently used to let the technicians know when the panelist is ready or has a question. This is usually accomplished using a form of a switch in each booth that will trigger a signal light (Meilgaard *et al.* 1991). However, there are other ways of administering contact between panelists and technicians if necessary or if there are questions.

The booths and the surrounding area should be odor-free and easy to clean. The ventilation for this room and especially for the booths is critical. Meat companies tend to have a great deal of different odors and aromas filtering through the research and development area. Meat samples are cooked for various studies and the smoking of meats can infiltrate a facility. For these reasons, the entire booth, or testing area, should have a slight positive pressure relative to other areas.

Facing each panelist seated at a booth will be a small door, referred to as a "sample pass-through door" allowing samples to be passed to the subject. The most commonly used door type is the "bread-box" design. The bread box is so constructed that when one side is open, the other is closed. This minimizes the likelihood of the subject having any view of the preparation area. Major disadvantages of the bread box include the increased amount of space required on the subject's side when the door is down, the height of the box relative to serving containers, and the inability to easily communicate with subjects (Stone & Sidel 1993).

2.4.1.2 Descriptive evaluation and training area

A large conference style room is recommended for expert panels and descriptive evaluation. Several tables work well because they can be arranged as necessary by the size and objective of the group. Audiovisual equipment may be useful. Separate preparation facilities may also be used for reference samples. When working with muscle foods or meats in general, use of a descriptive panel is of most importance.

2.4.1.3 Preparation area

The preparation area should permit preparation of all test samples of any kind that is anticipated. The laboratory should be equipped with the necessary equipment that will meet the needs of the evaluations performed.

2.4.1.4 General design factors

Color and lighting should be provided to give adequate viewing of the sample, but should also minimize distractions. Walls should be an off-white color; the absence of hues of any color will prevent unwanted difference in appearance. Many panel booths have colored lighting (red, green, and/or blue) at low intensity obtained through the use of colored bulbs or special filters. These lights are used to hide individual differences between samples in difference tests to determine by taste the samples that are identical (Meilgaard *et al.* 1991).

When performing studies with meats, the red lights are very useful at hiding small color differences that could be a result of smoking or fermentation procedure(s).

2.4.2 Product controls

Equipment that is used to handle samples should be controlled in order to reduce biases being introduced to the product. An example is the limitation of the use of plastics as food storage containers and, in some cases, food-serving containers. Controlled preparation of products requires careful regulation and monitoring of procedures used. All aspects of the test must be kept "constant" to ensure that the responses reflect only the variable being evaluated. Always keep in mind that the subjects are "looking" for something to be different with the test products.

2.5 TYPES OF TESTS

The tests discussed are for use with any food. However, the focus in this chapter is muscle foods and/or meat. It is important for the muscle foods expert to realize when a particular type of test should be utilized. The following sections provide a brief overview of the different types of tests available for use in sensory evaluation.

2.5.1 Discrimination testing

There are many different methods for consumer testing. The different tests available are divided into two categories: discrimination testing and affective testing. Discrimination testing represents one of the most useful tools available for sensory analysis (Stone & Sidel 1993). This class of sensory testing compares two products or stimuli ultimately looking for differentiation on a sensory basis (Lawless & Heymann 1998). Methods include triangle tests, paired comparison, duo-trio, and other lesser-known tests. However, all the methods are intended to answer a seemingly simple question: "Are these products perceived as different?" (Stone & Sidel 1993).

2.5.1.1 Triangle test

The triangle test is a discrimination test in which three samples are presented, two being the same and a third that is a different version of the variable under investigation. The judge's task is to choose the item that is most different from the other two (Lawless & Heymann 1998). Larmond (1982) recommended a simple but accurate method utilizing a statistical chart in which the task is to count the number of correct replies and refer to a chart for interpretation. This is probably the most widely used application for determination of statistically significant differences when using a difference test method.

This method should be used when the test objective is to determine whether a sensory difference exists between two products. It is especially useful in situations where treatment effects may have produced product changes, which cannot be characterized simply by one or two attributes. The triangle test is statistically more efficient than the paired comparison and duo-trio methods. However, it has limited use with products that involve sensory fatigue, carryover, or adaptation, and with subjects who find that testing three samples is too confusing. This method is effective when determining whether product differences result from a change in ingredients, processing, packaging, or storage and whether an overall difference exists, where no specific attribute can be identified as having been affected. It is also useful when selecting and monitoring panelists for the ability to discriminate given differences (Meilgaard *et al.* 1991).

2.5.1.2 Paired comparison tests

There are two sensory tests of this type. One is the directional paired comparison method (the two-alternative forced choice method) and the other is the difference paired comparison (simple difference test). With the former method, the panelist is presented with two samples and asked to determine whether they differ in a specific attribute, such as sweetness, crispness, etc. With the latter, the panelist is presented with two samples and asked simply whether or not the samples differ (Lawless & Heymann 1998). If the sensory analyst knows that the two samples differ in a specific sensory characteristic, then the two-alternative forced choice method is used. It is always more efficient and powerful to use a directional paired comparison test specifying the sensory attribute in which the samples differ than to ask the panelists to indicate the different sample. However, if the sensory analyst does not know if or how the sensory characteristics of the product differ, then the simple difference test may be more useful (Lawless & Heymann 1998).

2.5.1.3 Duo-trio test

In this test, the panelist receives three samples simultaneously. Two of these are the same and the other is different. On the duplicate samples is marked "reference." The panelist is then asked to identify which sample corresponds or matches the reference sample. These tests allow the sensory analyst to determine if two samples are perceptibly different, but the analyst will not know in which attributes the samples differed.

2.5.1.4 A-not-A test

This test is essentially a sequential paired difference test or simple difference test. The panelist evaluates the first sample and then it is removed. The panelist does the same with the second sample. The panelist is then asked to determine whether the two samples were perceived to be the same or different. The results of this test, again, will tell the analyst whether the panelists could significantly discriminate between the samples when they are not presented simultaneously. This test is useful when the experimenter cannot make the two formulations have exactly the same color, shape, or size; yet, these characteristics are not relevant to the objective of the study.

2.5.2 Affective testing

Affective tests assess the acceptability of products or the relative preference among a set of products (Lawless & Heymann 1998). These tests usually follow discrimination and descriptive tests, which have reduced the number of product alternatives to some limited subset, and precede larger scale testing done outside of research and development by others, such as marketing research (Stone & Sidel 1993). Affective testing includes preference tests, acceptance tests, and assessing individual attributes.

2.5.2.1 Preference tests

If the project is specifically designed to compare one product directly against another in situations such as product improvement or parity with competition, then a preference test is indicated. The preference test forces a choice of one item over another or others. However, it does not indicate whether any of the products are liked or disliked. Preference tests can include the (1) paired preference test with two samples (one sample is chosen over another, A-B), (2) ranked preference test with three or more samples (samples are ranked in a relative order, A-B-C-D), (3) multiple paired preference test with three or more samples (a series of paired samples with all samples paired with all others, A-B, A-C, A-D, B-C, B-D, C-D), and (4) multiple paired preference test with three or more samples (a series of paired samples with one or two select samples [e.g., control] paired with two or more [not paired with each other], A-C, A-D, A-E, B-C, B-D, B-E) (Meilgaard *et al.* 1991).

2.5.2.2 Acceptance tests

When a product researcher needs to determine how well a product is liked by consumers, an acceptance test is the correct choice. The product is compared to a well-liked company product or that of a competitor, and a hedonic scale is used to indicate degrees of unacceptable to acceptable, or dislike to like (Meilgaard *et al.* 1991).

2.5.2.3 Attribute assessment

As part of a consumer test, researchers may want to determine the reasons for any preference or rejection by asking additional questions about the sensory attributes (Meilgaard *et al.* 1991). Questions may be asked regarding how intense an attribute is or how well they like a specific characteristic.

2.5.3 Descriptive analysis techniques

Descriptive analyses are the most complex methods in sensory evaluation. These tests are useful when a detailed description of the innate and added attributes of any given product is needed.

Descriptive analyses utilize a developed language for determination of a product attributes' description. There are three types of languages: commonly used or "everyday," lexical, and scientific. Commonly used or "everyday" language is applied in daily conversations and varies in cultural aspects such as regions or countries. Lexical language is found in the dictionary and may also be used in common or "everyday" language conversations (Lawless & Heymann 1998). Scientific language is for technical work and is very specific and formally applied (Civille & Lyon 1996).

Descriptive analyses require a training phase for the development of a descriptive language by the panelists. A unique scientific language is thereby created for the product of study. This scientific language can often be formed into a lexical language if the situation allows. The goal of descriptive analyses is to have all trained panelists use the same concepts developed to describe a given attribute of a product.

Descriptors developed for a product should be singular rather than combinations of several terms. Combination or holistic terms such as "thick," "slimy," "hard," and "dirty" are complex and should be avoided. Terms such as these are impossible to describe and mean different things to different panelists. Terms such as these should be assimilated into their primary, analytical/technical, or elemental components. Panelists can use suitable descriptors with reliability and repeatability.

There are several descriptive analyses techniques. These include the Flavor Profile® method, trademarked to Arthur D. Little company; Quantitative Descriptive Analysis®, developed during the 1970s by Stone and Sidel; Texture Profile® method, created by scientists employed by General Foods during the 1960s; Sensory Spectrum® method, created by Civille; Generic Descriptive Analysis, in which techniques have been adapted to meet the needs of a particular situation; and Free Choice Profiling, created in the 1980s by British scientists.

Regardless of the method of descriptive analysis chosen for sensory evaluation studies of a given product, it is of greatest importance to ensure the development of a descriptive language to be as complete as possible. This ensures an excellent instrument available for description of the product in question and builds quality and reliability into the trained panel.

2.6 CONCLUSION

Sensory evaluation has proven to be invaluable to the muscle foods industry. All of the mentioned applications, procedures, and methods are exceptionally useful in the muscle foods area and should be utilized on a broader scale. Consumer demand drives this need, and the industry as a whole thrives on consumer perception. With this in mind, any company that does not utilize sensory evaluation techniques to their fullest extent in the evaluation of the food product(s) that it produces should expect failure of their product(s) in the marketplace sooner than later. Sensory evaluation is a tool that if used properly can solve problems before they arise. There is a depth of information that can be obtained from the proper utilization of sensory methodology.

The meat industry and muscle foods in general often have it somewhat more difficult than other industries and products in maintaining certain key attributes in the food (muscle). Sensory evaluation should not be overlooked because it is an invaluable tool in assisting in the control and monitoring of many of these key attributes.

REFERENCES

Ball, C.O., Clauss, H.E. & Stier, E.F. (1957) Factors affecting quality of prepackaged meat: Loss of weight and study of texture. *Food Technology*, **11**, 281–284.

Bourne, M.C. (1982) *Food Texture and Viscosity: Concept and Measurement*. Academic Press, New York.

Civille, G.V. *& Lyon, B.* (1996) *ASTM Lexicon Vocabulary for Descriptive Analysis*. American Society for Testing and Materials, Philadelphia, Pennsylvania.

Hui, Y.H., Nip, W.K. Rogers, R.W. & Young, O.A. (2001) *Meat Science and Application*. Marcel Dekker, New York.

ISO. (1981) *Sensory Analysis Vocabulary. Part 4*. International Organization for Standardization, Geneva, Switzerland.

Larmond, E. (1982) *Laboratory Methods for Sensory Evaluation of Food. Publication 1637*. Food Re-search Institute. Research Branch Canada Department of Agriculture, Ottawa, Ontario K1A 0C7. Cat. No. A73-1637/1977; 5M-7:82. ISBN: 0-662-01271-2.

Lawless, H. & Heymann, H. (1998) *Sensory Evaluation of Food, Principles and Practices*. Chapman & Hall, New York.

McDougall, D.B. (1988) Color, vision and appearance measurement. In: *Sensory Analysis of Foods* (ed. J.R. Piggot), 2nd edn., Elsevier, London, U.K.

Meilgaard, M., Civille, G.V. & Carr, B.T. (1991) *Sensory Evaluation Techniques*, 2nd edn., CRC Press, Boca Raton, Florida.

Seibert, K.J., Stenroos, L.E. & Reid, D.S. (1981) Characterization of amorphous-particle haze. *Journal of the American Society of Brewing Chemists*, **39**, 1–11.

Stone, H. & Sidel, J.L. (1993) *Sensory Evaluation Practices*, 2nd edn., Academic Press, San Diego, California.

3 Attributes of Muscle Foods: Color, Texture, Flavor

Patti C. Coggins

Abstract: The sensory attributes of muscle food products provide great satisfaction in eating and satiety to the end consumer. Acceptance is the predominant factor applied to all of the sensory attributes of muscle foods. The attributes of appearance, aroma, flavor, juiciness, tenderness, and texture must all be found acceptable when a muscle food is consumed. Sensory evaluation of muscle foods can be one of the most difficult analyses to perform due to inherent differences of the muscle(s).

Keywords: Muscle foods attributes, appearance, color, aroma, flavor, Maillard reaction, tenderness, juiciness, texture

3.1 INTRODUCTION

The sensory attributes of muscle foods are related to the senses of taste, smell, sight, feel, and sound. Of all the foods consumed, muscle foods have the lowest tolerance for complete sensorial acceptability. A muscle food is either acceptable or unacceptable with little in between. Predominately, the consumer assesses visually the color and surface texture of the muscle. The preparation technique of consumer choice is utilized, thereby altering the sensory attributes, usually completely. The consumer cooks or prepares the muscle food as they prefer, changing the surface color, appearance, and texture. The internal altering of texture and flavor is a result of the preparation or cooking process as well. This of course varies depending on the many methods applied. For instance, the muscle may be grilled, baked, broiled, or otherwise prepared, all with different fluctuating end results. Consumption of muscle foods is one of the most pleasurable eating experiences. The satiety value applied by the consumption of a muscle food is great when comparing food(s)-satisfying effect in general. This is why the sensorial properties of muscle foods can be viewed as often more important than that of other foods.

There are many different flavors associated with the sensory evaluation of muscle foods in general. Each, of course, is dependent on the type of muscle food evaluated. To evaluate the basic flavor scenario, we must first begin with a discussion of the innate taste parameters in most of us.

The basic tastes of salty, sweet, bitter, sour, and umami can be identified in all muscle foods. These basic tastes can vary depending on the cut of a particular muscle, the muscle in general, or the source of the muscle food. For instance, the basic tastes are slightly different when comparing fresh beef to that of fresh mutton.

Handbook of Meat, Poultry and Seafood Quality, Second Edition. Edited by Leo M. L. Nollet.

Smell is a very important sensory property. It is typically the second sensory attribute that is depicted. Fresh uncooked meat is evaluated by smell threshold, whereas cooked meat is evaluated by aroma threshold. Meat that is old, unfresh, or borderline, whether cooked or uncooked may be evaluated by odor threshold. The detection of aromatics and/or odors by the olfactory nerve comprises what is acceptable or unacceptable in all types of muscle food aroma.

Texture of food influences perceptions and acceptability. Tenderness of muscle foods affects consumer acceptability. Tenderness is the third sensory attribute that must be acceptable to the consumer. This attribute is of course dependent on the type of meat and the cut of that particular meat. For instance, a rib eye steak is anticipated and expected to be much more tender than a flank steak, even though taken from the same meat source. Of course, preparation and processing and/or treatment technique have a great deal to do with the end result and acceptability of a muscle food. A particular cut or source of meat has a perception acceptance value. Another influence on perception and acceptability of a muscle food is mouth coating, which can be residual, fat, or substance in the mouth and throat after the consumption of high fat meat products. If one eats a rib eye steak, there will be little mouth coating or fat residual in the mouth and throat. However, if lamb is considered, there is a great deal of mouth and throat coating (Miller 1994a).

Quality characteristics of muscle foods are influenced by muscle appearance, color, surface texture, internal texture, tenderness, juiciness, mouthfeel characteristics, fat content, connective tissue, muscle fiber characteristics, and storage temperature to mention a few. These innate quality characteristics are of course dependent on many predetermining factors that affected the animal prior to its function as a muscle food.

The following sections will depict a brief summary of the "generic" quality characteristics of muscle foods in general. Muscle foods will be discussed as a category and not singled out by breed, unless otherwise noted.

3.2 APPEARANCE

The general appearance of a muscle food, regardless of originating species, is the first sensory attribute viewed by a consumer. Appearance is therefore perhaps the most important attribute. Entailed in appearance is muscle color. The consumer is looking for that "bright cherry red" color that industry has deemed and trained the consumer to appreciate and demand. Any skewing from what is typically expected can result in a nonacceptance of the muscle food in question.

Appearance also intakes inherent parameters such as marbling, connective tissue visibility, fat pockets, in some cases "stringiness," moistness of surface, and a perception of moisture level within the muscle. There are many other factors that can affect perception and the general view of the appearance of the muscle food. The attributes mentioned, with the exception of the "bright cherry red" color, would pertain to all muscle foods.

3.3 COLOR

The color of a muscle food is dependent on several factors. The species of the muscle food, the inherent characteristics of various treatments (i.e., vitamin A) applied to the animal from which the muscle food was derived, the food source of the animal, the postslaughter processing

Table 3.1 The colors of meat.

Pigment	Color	Source	Comments
Hemes	Purple	Reducing conditions	Interior of fresh meats, exclusion of air
Myoglobin			
Hemoglobin			
Oxymyoglobin	Red	Oxygenation	Surface of fresh meat, exposed to air, "bloom"
Metmyoglobin	Brown	Oxidation	Old meat, low oxygen pressures
Cytochromes	Red		Very low concentrations

Source: From Zimmerman and Snyder (1969).

and/or treatment methods, the length of time the muscle is kept either frozen or refrigerated prior to packaging and sale, the type of packaging the food is placed in for sale, the environment the package endures (i.e., light source), as well as the length of time on the meat market shelf. There are of course inherent variables that influence color as well. These will be discussed later in this section.

Color reflects or emits a specific amount of energy at wavelengths able to stimulate the retina in the eye. The characteristic red color of meat is due to the heme pigments contained in the muscle. Colors of muscle foods vary depending on species, muscle function within the animal, age of the animal, and storage conditions to name a few variables that influence color. The range of color for muscle foods is influenced predominately by the content of myoglobin. Myoglobin is one of the major proteins in the sacroplasm, and is the main pigment in meat. It accounts for 90–95% of the heme proteins. At increased levels of myoglobin, the color intensity of the muscle ranges from white to pink to very dark red.

The heme pigments take up oxygen to produce "bloom" of fresh meats, but other characteristics are important because of the more widely varied conditions to which muscle tissue is exposed. The protein and heme structures are affected by oxidation, heat, acidity, and chemicals thereby affecting color. Not only do these pigments serve as indicators of freshness, since they turn brown with age, but they also indicate physical and chemical contaminations (Fox 1987). Therefore, this adds importance to the weight that color and appearance provides as a sensory attribute (Table 3.1).

As has been previously mentioned, beef color is expected to be "bright cherry red" as compared to other muscle foods' ranges of color. Higher myoglobin content in beef muscle is the major factor that differentiates the bright cherry red color of beef when compared to the lighter color of pork or poultry meat. As the animals within the species increase in age, myoglobin content increases within the muscle tissue, and thus muscle from mature animals is darker than muscle from younger animals (Tarladgis 1962a, b).

Meat color, although strongly influenced by myoglobin concentration, is also affected by handling and storage. The anaerobic environment created during vacuum-packaged storage versus the use of modified atmosphere packaging where oxygen, nitrogen, and carbon dioxide can be used as the storage environment affects the meat color. The lack of oxygen in vacuum-packaged meat systems causes beef exposed to the atmosphere to change from a bright cherry red color to a purplish color in the vacuum package (Castro 1971).

As previously mentioned, the sensory attributes of muscle foods change drastically during the cooking or heating process. This process is one that the end consumer will indeed perform and the end result color greatly affects overall perceptions and acceptability.

The brown pigments produced from proteins and sugars at cooking temperatures also alter color in muscle foods. These dark brown pigments are formed through interactions of deposits from smoke on the meat surface as well as endogenous compounds. Other brown hues are observed in fat, which contains a number of endogenous compounds with color hues ranging from yellow to bronze red (Fox *et al.* 1967).

3.4 COLOR OF CURED MEAT

Sensory acceptance of cured meat is somewhat different than that of fresh meat. The overall sensory profile of a cured meat product is changed from that of a profile for fresh meat. The curing process alters the appearance, color, texture, flavor, basic taste(s), and possibly acceptance. For instance, a cured meat product can pass for acceptable, regardless of age, cooking, packaging, or other pertinent variables. But, a fresh meat product might be found unacceptable due to one or more of these variables.

The color of cured meats is dependent on three factors: (1) the concentration of pigment in the tissues, (2) the degree of conversion into nitrosyl pigment, and (3) the proteins in the meat. In fermented beef products, the color is a deep dark red, partly because the tissue proteins are largely denatured (George & Stratmann 1952).

The color of the surface of cured meats is the result of browning reactions or possibly smoking. Smoked meats develop surface colors due to interaction of phenols, aldehydes, and acids (Zimmerman & Snyder 1969).

3.5 COLOR OF FROZEN MEATS

The color of fresh frozen meats is similar to that of chilled meats. However, over time, the muscle will change to more of a brown color. This is accelerated if the type of packaging used is not suitable for muscle food frozen storage.

In general, the quality of frozen muscle foods decreases with storage and age in the freezer. Appearance, predominately color, is the first thing the consumer views upon making the decision on whether to purchase a muscle food. However, the consumer's acceptance of frozen meat products is not as dependent upon color as that for fresh muscle food.

3.6 AROMA (VOLATILES)

Aroma is the sensory attribute that develops the desire for one to want to consume the cooked or prepared muscle food.

Nearly 1000 compounds have so far been identified in the volatile constituents of meat from beef, chicken, pork, and sheep. Shahidi (1998) provides a list of volatiles responsible for flavor and believes that sulfurous and carbonyl-containing volatiles are the predominant contribution to aroma.

The chemical nature of many flavor volatiles of meat from different species is similar qualitatively; however, there are quantitative differences. While most of the sulfurous volatiles of meat exhibit a pleasant meaty aroma at concentrations present in meat, at high levels, their odor is objectionable. Therefore, both qualitative and quantitative aspects of volatiles have to be considered when assessing the flavor quality of muscle foods (Shahidi 1998).

Nonvolatile precursors of meat flavor include amino acids, peptides, reducing sugars, vitamins, and nucleotides.

In the evaluation of flavor quality of meat, the contribution by amino acids, peptides, and nucleotides are important. These compounds not only interact with other compounds to produce flavor volatiles but also contribute to sweet, salty, bitter, sour, and umami sensations of muscle foods.

3.7 REACTIONS LEADING TO MEAT AROMA

Many different temperature conditions exist during meat cookery and preparation. Based on these different temperature conditions, a wide range of different flavor sensations is perceived in cooked meats.

The primary reactions occurring on heating that can lead to meat flavor include pyrolysis of amino acids and peptides, caramelization of carbohydrates, degradation of ribonucleotides, thiamine degradation, interaction of sugars with amino acids or peptides, and thermal degradation of lipid. The thermal decomposition of amino acids and peptides and the caramelization of sugars normally require temperatures over 150°C before aroma compounds are formed. Such temperatures are higher than those that are normally encountered during the cooking of the meat (Mottram 1998).

3.8 FLAVOR

Flavor is one of the most important components of the eating quality of muscle foods. A tremendous amount of research has been applied to the understanding of the analysis and chemistry of meat/muscle food flavor. In general, meat flavor depends on the stimulation of the senses of taste and smell. While most of the sensory research and chemistry applied in this area has focused on the volatile compounds that contribute to aroma, there is ongoing research on the contribution of nonvolatile compounds and their relationship with taste sensations.

There are many different compounds that contribute to and provide meat flavor. However, the desirable quality attribute that is perceived as meat flavor cannot be attributed to a single compound or a group of compounds (see Table 3.2).

Table 3.2 Summary of commonly used product quality attributes used in the evaluation of muscle foods.

Quality attribute	Description of the quality attribute
Appearance/visual	
Color	The perception, by the eye, of light waves that include blue (400–500 nm), green, and yellow (500–600 nm) can be defined by hue, value, and chroma. Color can be defined by the overall color or an estimation of positive and negative color attributes
Texture	A visual evaluation of the texture attributes of a product such as surface smoothness, consistency of texture
Fat content	An estimation of the amount of internal or external fat content of a muscle food, such as marbling, subcutaneous fat trim level, or amount of steam fat
Aroma	The volatile compounds detected within the nasal passage by the olfactory system, perceived by smelling (drawing air into the nasal and oral cavities)
Taste	The basic tastes of salty, sweet, sour, and bitter that are perceived by gustation through stimulation of taste buds mainly located on the surface of the tongue by water-soluble compounds in the mouth
Mouthfeel/feeling factors	The perception of senses in the mucosa of the eyes, nose, and mouth through the stimulation of the trigeminal nerves such as perception of heat, burn, cold, metallic, and astringency

Sources: From Miller (1994a, b).

The flavor of uncooked meat is "bloodlike" and metallic. The flavor that cooking imparts is unique and dependent upon the length and procedure used. It can vary, and each individual has a preferred way of consuming prepared cooked meat.

Flavor is the result of compounds stimulating the olfactory and taste receptors in the oral and nasal cavity. The muscle system can be divided into the lean portion and the lipid or fat portion, with each component contributing to meat flavor. According to Miller (1994a, b),

Meat flavor is composed of: (1) meat-like flavor derived from water-soluble reducing sugars and amino acids; (2) species-specific flavors which are due to differences in fatty acid composition and aromatic, water-soluble compounds that are stored in lipid depots of the animal; and (3) off-flavor development as the result of oxidation of lipid double bonds, defined as lipid oxidation or auto-oxidation, and other degradation processes.

Natural flavor compounds are either fat soluble or water soluble. Fat can also serve as a solvent for aroma compounds, or as a carrier for flavor compounds. Fat can however mask certain flavors or volatiles (Hornstein & Wasserman 1987, Mottram 1998). To evaluate meat in a semisensory manner, the meat from older animals is more intense in flavor than meat from younger animals. Cowy/grainy and serumy/bloody flavor aromatics and metallic descriptors have been associated with increased levels of myoglobin in meat from older animals. The traditional mutton flavor associated with meat from mature sheep is a classic example of how some flavor attributes concentrate in meat from older animals. Diet can also contribute to flavor differences in meat from young versus older animals (Miller 1994a, b).

The terms lamb and mutton flavor are used interchangeably and refer to the characteristic flavor of all sheep meat, regardless of age. Terms like "lamb" and "wooly" have been used for cooked lamb. Fat-rich mutton broths were "lamb-like," "sweet," "oily," and "barny" (Young & Braggins 1998).

3.9 LIPID (FAT) CONTRIBUTION TO MEAT FLAVOR

The sensory profile of meat depends largely upon the contribution of fat to the overall flavor complexity.

Lipids can contribute both desirable and undesirable flavors to meats. During cooking, the lipids undergo thermally induced oxidation to give a range of volatiles that contribute to meat aroma. Lipids may also react with the components from the lean tissue to give other flavor compounds. In addition, they may act as a solvent for aroma compounds accumulated during production, processing, and cooking of meat.

Lipids not only impart flavor volatiles by oxidation but their oxidation products can also enter the Maillard reaction, resulting in additional volatiles that are not formed by lean meat precursors. For the most part, the phospholipids of meat rather than triglycerols contribute the fatty acids that interact with Maillard reactions.

According to Miller (1994a, b), most of the volatiles derived from Maillard interactions have a higher boiling point than most of the volatiles identified in the meat flavor. The possibility exists that some of these compounds might have a meaty odor that is species specific.

3.10 MAILLARD REACTION

According to research done by Bailey (1994), the Maillard reaction (nonenzymatic browning) involves the reaction of aldehydes with amines and through numerous reactions; food flavor compounds and dark pigments (melanoidins) are formed. Factors affecting the progress of the

Maillard reaction include temperature, time, moisture content, pH, concentration, and nature of the reactants. The Maillard reaction between reducing sugars and amino compounds does not require the very high temperatures associated with sugar caramelization and protein pyrolysis and readily produces aroma compounds at the temperatures associated with the cooking of food (Mottram 1994).

3.11 SULFUR COMPOUNDS

According to Mottram (1994), meat would have an entirely different flavor in the absence of sulfur compounds. Large quantities of hydrogen sulfide are produced during the heating of meat. Under these conditions, concentration of hydrogen sulfide is very prominent. Sulfur compounds are the most important volatiles formed during meat cookery, and sulfur precursors are used in essentially all meat-flavor mixtures.

Even though sulfur is not considered one of the pleasant flavor compounds, its very existence defines perceived quality and acceptance of many foods (i.e., sharp cheddar cheese) including muscle foods.

3.12 UMAMI FLAVOR OF MEAT

Meat is known to possess a sensory property that many say is one of the innate basic tastes of humans. There has been some argument to this proposed theory, but it is indeed a unique taste sensation, which some can detect while others cannot.

The term "umami" was first proposed by Ikeda (1909). Umami means "deliciousness" in Japanese. Umami can be defined as the taste properties resulting from the natural occurrence or intentional addition of compounds such as monosodium glutamate (MSG) and certain 58-nucleotides such as 58-inosine monophosphate (IMP) and 58-guanosine monophosphate (GMP). These nucleotides have also been referred to, respectively, as inosinic and guanylic acids, 58-inosine and 58-guanylic acids, or disodium 58-guanylate (Maga 1998).

Umami compounds have an ability to act synergistically when used in combination with each other. According to Maga (1983), if MSG is assigned a umami intensity of 1.0, the addition of an equal amount of GMP increases the flavor intensity 30 times. Even addition of as little as 1% GMP to MSG significantly increases flavor intensity.

Intentional addition of MSG and nucleotides significantly modifies meat-flavor perception. For example, the addition of MSG to a variety of processed meats resulted in products that were preferred over the same products to which MSG was not added (Girardot & Peryam 1954).

The addition of either MSG or the nucleotides, thus providing the "umami" taste sensation, greatly enhances the acceptance as well as the preference of select meat items, whether utilized in an emulsion system or directly applied to the muscle surface during cooking.

3.13 TENDERNESS

Tenderness is one of the most important sensory attributes when discussing the acceptance of a muscle food. Tenderness entails many parameters to the consumer. The more tender a muscle food is, the better perception of freshness and quality resides in the mind. However, muscle foods' experts know that this is not always the case.

Tenderness of meat may be simply defined as the ease of teeth to cut meat fibers during mastication. For intact or noncomminuted meat, tenderness to toughness is determined by two groups of meat components: the connective tissues and the muscle fibers (Koohmaraie *et al.* 1988). Tenderness of muscle foods strongly influences and drives consumer's perceptions of acceptability and quality of muscle foods. Muscle tenderness can be segmented into the influence of connective tissue and muscle fiber characteristics. Total amount of connective tissue from within different muscles of the same animal affects tenderness between muscles. For example, the muscles from the round that are used in movement are higher in connective tissue content than muscles from the loin that are mainly used for structural support. Muscles have been generally classified into tender, intermediate, and tough categories. Between animals, the age-associated increase in collagen solubility significantly affects muscle tenderness. As animals increase in age, their meat is tougher. Identification or measurement of animal age, either through ossification of the verbal column of an animal, identification of dentition, or evaluation of animal chronological age, has been used as a benchmark for the effect of connective tissue on meat quality between animals (Cross 1987).

3.14 CONNECTIVE TISSUE

Connective tissue is a fibrous structure composed of primarily collagen fibrils. There are three types of connective tissue in meat: epimysium, perimysium, and endomysium. The epimysium is a thick sheath of connective tissue surrounding the entire muscle; the perimysium is a thin layer enveloping muscle bundles or muscle fibers (myofibers). The perimysium and endomysium connective tissues present a realistic toughness problem to meat and meat products. The specific influence of both connective tissues depends upon their thickness (i.e., the amount of collagen present as well as the density and type of cross-linkages between collagen fibrils). Muscles of movement contain larger quantities of connective tissue than muscles that support the structure (Koohmaraie *et al.* 1988).

Muscles in mature animals have higher contents of collagen cross-links than the same muscles from young animals. The type of collagen cross-links is also different; as the animal's age increases, many of the collagen cross-links are transformed from a soluble type to an insoluble type, resulting in increased toughness of meat (Koohmaraie *et al.* 1988).

3.15 MYOFIBERS

Myofibers regulate meat tenderness in several aspects. Collagen in the connective tissue experiences minimal chemical and biochemical changes and exhibits little structural changes other than slight improvement in solubility (Koohmaraie *et al.* 1988).

3.16 JUICINESS

Juiciness is a sensory term, which refers to mouthfeel of the moisture released from food during mastication. Thus, juiciness is indicative of the moisture content in meat, which is critically affected by the water-holding capacity as well as the hydration ability of the meat. Water in meat is confined via capillarity in the spaces between myofilaments, between the myofibrils, and outside the fibers (cells). Myofibril proteins are believed to be largely responsible for water immobilization in meat. The hydration is attributed to the expansion of the interfilamental spaces within the myofibrils and is facilitated by mechanical actions such as massaging and tumbling (Cross 1987).

3.17 TEXTURAL PROPERTIES

According to Kauffman and Marsh (1987), texture is an important characteristic of muscle foods especially for processed meats that require some degree of comminution. For intact meat, texture of muscle refers to the definition and fineness of muscle fibers and the amount and distribution of fat in the muscle. In intact meat, the texture is determined by the age of the animal, the type of muscle, gender, and the growth condition. In comminuted meats, however, textural properties are characterized by smoothness and homogeneity of the product, which can be visualized, as well as by the rheological properties (hardness, deformability, elasticity, etc.) and cohesiveness of the bind formed between meat particles. These texture-related attributes are closely related to the functionality of muscle proteins particularly their gel-forming and emulsification properties.

3.18 CONCLUSION

In general, the sensory attributes of muscle food products provide great satisfaction in eating and satiety to the end consumer. Acceptance is the predominant factor applied to all of the sensory attributes of muscle foods. The attributes of appearance, aroma, flavor, juiciness, tenderness, and texture, to mention a few, must all be found acceptable when a muscle food is consumed. Muscle foods, unlike other foods, are completely scrutinized by the end consumer and there is very little, if any, tolerance for any deviation of complete perception of quality for all the pertinent sensory attributes. Sensory evaluation of muscle foods can be one of the most difficult analyses to perform due to the inherent differences of the muscle(s). However, it is one of the most important functions the industry utilizes to determine estimates of perception(s) ultimately to be derived by the end consumer.

REFERENCES

Bailey, M.E. (1994) Maillard reactions and meat flavour development. In: *Flavor of Meat and Meat Products* (ed. F. Shahidi), 1st edn., Blackie Academic and Professional, London, U.K.

Castro, C.E. (1971) Theory of hemeprotein reactivity. *Journal of Theoretical Biology*, **33**, 475.

Cross, H.R. (1987) Sensory characteristics of meat. Part I. Sensory factors an evaluation. In: *The Science of Meat and Meat Products*, 3rd edn., Food and Nutrition Press, Westport, Connecticut.

Fox, J.B. (1987) The pigments of meat. In: *The Science of Meat and Meat Products*, 3rd edn., Food and Nutrition Press, Westport, Connecticut.

Fox, J.B., Townsend, W.E., Ackerman, S.A. & Swift, C.E. (1967) Cured color development during frankfurter processing. *Food Technology*, **21**(3a), 68a.

George, P. & Stratmann, C.J. (1952) The oxidation of myoglobin to metmyoglobin by oxygen. *Biochemical Journal*, **51**, 103.

Girardot, N.F. & Peryam, D.R. (1954) MSG's power to perk up foods. *Food Engineering*, **26**(12), 71–74.

Hornstein, I. & Wasserman, A. (1987) Part 2—Chemistry of meat flavor. In: *The Science of Meat and Meat Products*, 3rd edn., Food and Nutrition Press, Westport, Connecticut.

Ikeda, K. (1909) On a new seasoning. *Journal of Tokyo Chemistry*, **34**, 751–755.

Kauffman, R.G. & Marsh, B.B. (1987) Quality characteristics of muscle as food. In: *The Science of Meat and Meat Products*, 3rd edn., Food and Nutrition Press, Westport, Connecticut.

Koohmaraie, M., Babiker, A.S., Shroeder, A.L., Merkel, R.A. & Dutson, T.R. (1988) Factors associated with the tenderness of three bovine muscles. *Journal of Food Science*, **53**, 407.

Maga, J.A. (1983) Flavor potentiators. *CRC Critical Reviews in Food Science and Nutrition*, **18**, 231–232.

Maga, J.A. (1998) Umami flavour of meat. In: *Flavor of Meat, Meat Products, and Seafoods*, 2nd edn., Blackie Academic and Professional, London, U.K.

Miller, R.K. (1994a) Sensory methods to evaluate muscle foods. In: *Muscle Foods, Meat, Poultry and Seafood Technology* (eds D.M. Kinsman, A.W. Kotula & B.C. Breidenstein), Chapman & Hall, New York.

Miller, R.K. (1994b) Quality characteristics. In: *Muscle Foods, Meat, Poultry and Seafood Technology* (eds D.M. Kinsman, A.W. Kotula & B.C. Breidenstein), Chapman & Hall, New York.

Mottram, D.S. (1994) Some aspects of the chemistry of meat flavour. In: *Flavor of Meat and Meat Products* (ed. F. Shahidi), 1st edn., Blackie Academic and Professional, London, UK.

Mottram, D.S. (1998) The chemistry of meat flavor. In: *Flavor of Meat, Meat Products, and Seafoods*, 2nd edn., Blackie Academic and Professional, London, U.K.

Shahidi, F. (1998) Flavor of muscle foods—An overview. In: *Flavor of Meat, Meat Products, and Sea-foods*, 2nd edn., Blackie Academic and Professional, London, U.K.

Tarladgis, B.G. (1962a) An interpretation of the spectra of meat pigments. I. Cooked meats. *Journal of the Science of Food and Agriculture*, **13**, 481.

Tarladgis, B.G. (1962b) Interpretation of the spectra of meat pigments. II. Cured meats. The mechanism of color fading. *Journal of the Science of Food and Agriculture*, **13**, 485.

Young, O.A. & Braggins, T.J. (1998) Sheepmeat odour and flavour. In: *Flavor of Meat, Meat Products, and Seafoods*, 2nd edn., Blackie Academic and Professional, London, U.K.

Zimmerman, G.L. & Snyder, H.E. (1969) Meat pigment changes in intact beef samples. *Journal of Food Science*, **34**, 258.

FURTHER READINGS

Cassens, R.G. (1987) Structure of muscle. In: *The Science of Meat and Meat Products*, 3rd edn., Food and Nutrition Press, Westport, Connecticut.

Chen, J. & Ho, C.T. (1998) The flavour of pork. In: *Flavor of Meat, Meat Products, and Seafoods*, 2nd edn., Blackie Academic and Professional, London, U.K.

Farmer, L.J. & Patterson, R.L.S. (1991) Compounds contributing to meat flavor. *Food Chemistry*, **40**, 201–205.

Lanier, T.C., Carpenter, J.A. & Toledo, R.T. (1977) Effects of cold storage environment on color of lean beef surfaces. *Journal of Food Science*, **42**, 860.

MacLeod, G. (1986) The scientific and technological basis of meat flavours. In: *Developments in Food Flavours* (eds G.G. Birch & M.G. Lindley). Elsevier, London, U.K.

MacLeod, G. (1998) The flavour of beef. In: *Flavor of Meat, Meat Products, and Seafoods*, 2nd edn., Blackie Academic and Professional, London, U.K.

Nicol, A.M. & Jagusch, K.T. (1971) The effect of different types of pasture on the organoleptic qualities of lambs. *Journal of the Science of Food and Agriculture*, **22**, 464–466.

Randall, C.J. & Voisey, P.W. (1977) Effect of meat protein fractions on textural characteristics of meat emulsions. *Canadian Institute of Food Science and Technology Aliment Journal*, **10**(2), 88.

Sink, J.D. (1979) Factors influencing the flavor of muscle foods. *Journal of Food Science*, **44**, 1.

Wasserman, A.E. (1979) Chemical basis for meat flavor. A review. *Journal of Food Science*, **44**, 6.

4 Recent Developments in Flavor Measurements

Jean-Luc Le Quéré

Abstract: Recent developments in techniques used for flavor measurements include sorptive extraction methods, solid-phase microextraction (SPME), stir bar sorptive extraction (SBSE), and headspace sorptive extraction (HSSE). Sensory information can be obtained by combining gas chromatography with olfactometry (GC-O). Recent developments in dynamic instrumental methods that can follow the in vivo sequential release of the aroma molecules are valuable tools. Such methods are electronic noses and mass spectrometry-based systems.

Keywords: Flavor measurement, GC-O, MS, SPME, SBSE, electronic noses, pyrolysis mass spectrometry, aroma

4.1 INTRODUCTION

Measuring aroma means analyzing volatile compounds that are sensed in the nose at the olfactory receptors either via the orthonasal (odor) or retronasal (aroma) routes when foods are eaten. Instrumental analysis of aroma molecules has been the subject of important specialized treatises (for the most recent literature on the subject see Ho & Manley 1993; Marsili 1997; Mussinan & Morello 1998; Stephan *et al.* 2000; van Ruth 2001a; Reineccius 2002; Deibler & Delwiche 2004). The aroma compounds are mainly hydrophobic and instrumental analysis of volatiles must, therefore, consider, as a first step, an extraction method suitable for separating these volatiles from the food matrix. Part of this chapter will focus on recent advances made with sorptive extraction methods: solid-phase microextraction (SPME) (Lord & Pawliszyn 2000) and stir bar sorptive extraction (SBSE) (Baltussen *et al.* 1999).

As no single method yields a "true" picture of a food aroma (Reineccius 2002), isolation and analysis of aroma remain challenging (Teranishi 1998). Moreover, the extraction step may lead to artifacts, and the total volatile content in most cases is very difficult to relate to a sensory profile determined by a panel or experienced by a consumer. Therefore, it appears much more efficient to concentrate efforts on the identification of those compounds that are really relevant to the perceived flavor. Because no universal extraction method exists, it appears essential to choose a method that yields an extract representative of the sensory properties of the food (Abbott *et al.* 1993a; Etiévant *et al.* 1994; Etiévant & Langlois 1998). In order to identify the key volatiles that contribute significantly to food flavor, gas chromatography coupled with olfactometry (GC-O) is very often used (Leland *et al.* 2001; Lee 2003). GC-O and associated extract representativeness will be part of this chapter.

However, it is particularly difficult to predict an aroma perception as it is still not known how the various volatiles combine to produce an overall sensory impression. Moreover,

Handbook of Meat, Poultry and Seafood Quality, Second Edition. Edited by Leo M. L. Nollet.

interactions between taste and aroma (Noble 1996) and interactions of trigeminal sensations with taste and aroma (Green 1996) occur and play an important role in global flavor perception. However, methods that allow direct analysis of flavor molecules released in the mouth during consumption have been developed in the last years (Taylor & Linforth 1996; Roberts & Taylor 2000). Development of instrumental techniques and data obtained recently for volatile flavor compounds will be presented with the aim to explain more closely the relationship between aroma perception and food volatiles composition.

Finally, specific instrumental techniques have been developed for the global analysis of food flavor. The methods currently used in quality control of food flavor are still usually based on sensory evaluation by a panel of experts. These panels are able to monitor the quality of a particular food, to detect defects, and to compare samples for classification purposes. Nevertheless, obtaining fast results at low cost using instruments could be advantageous. The so-called electronic noses based on gas sensor technology, despite some important drawbacks for some of them (Schaller *et al.* 2000a), are theoretically able to perform some classification tasks (Schaller *et al.* 1998), and some applications for the analysis of foodstuffs have been developed (for instance, see Mariaca & Bosset 1997; Schaller *et al.* 1999 for cheese applications). Two other global analysis methods based on mass spectrometry (MS) seem interesting for classification purposes. The first one analyzes total headspace using a mass spectrometer, without any prior GC separation (Vernat & Berdagué 1995). This method is often referred to as a mass-based electronic nose. Alternatively, headspace sampling may be replaced by SPME of food volatiles (Marsili 1999). Both sampling methods, followed directly by mass spectrometry, have found applications for the rapid characterization of foods like cheese, for example (Schaller *et al.* 2000b; Pérès *et al.* 2001; Pérès *et al.* 2002a). The second method is pyrolysis mass spectrometry (Aries & Gutteridge 1987), where a small food sample is pyrolyzed at up to 500°C. The resulting volatile fraction, characteristic of the flavor but also of the matrix composition, is analyzed by a mass spectrometer. For all the rapid instrumental methods used for classification, a pattern or fingerprint is obtained for each sample, and extensive data treatment, either by conventional multivariate statistics or artificial neural networks, is necessary for classification and quality control purposes (Aries & Gutteridge 1987; Pérès *et al.* 2002b).

4.2 SORPTIVE EXTRACTION METHODS

All the extraction procedures used to isolate the volatile fraction from the food matrix should be adapted to the analysis of trace levels of lipophilic molecules generally present in a polyphasic medium, while minimizing losses of highly volatile molecules and preventing modification of compounds or the formation of artifacts.

SPME, first developed for the extraction of volatile organic compounds in water (see Lord & Pawliszyn 2000 for a recent review of the technology), has been applied recently to the isolation of aroma compounds from food (Harmon 1997; Kataoka *et al.* 2000; Pillonel *et al.* 2002; Reineccius 2002). SPME partitions analytes between a liquid or a vapor phase and a thin solid-phase adsorbent, of which there are several choices in terms of polarity and film thickness (Kataoka *et al.* 2000), coated on inert fibers, generally associated with a syringe that serves as a direct injection device (Harmon 1997; Kataoka *et al.* 2000). The method, which is an equilibrium method, can be performed either in the direct extraction mode (immersion of the fiber in sample matrix, generally in an aqueous solution or suspension) or in a headspace configuration. It can be automated very easily, but the extraction of the solutes depends on polarity, volatility, partition coefficients, sample volume, temperature, and the nature of the adsorbent-coating material. Therefore, the technique exhibits a certain degree of selectivity, but with the advantages of sensitivity, ease of use, no solvent, and small sample volume (Harmon 1997; Kataoka *et al.* 2000; Pillonel *et al.* 2002; Reineccius 2002). Nevertheless, each extraction step, that is, extraction mode (direct or headspace), selection of fiber coatings, extraction setup (concentration, time, agitation, temperature), and desorption, gains through a careful optimization for each application (Bicchi *et al.* 2000; Kataoka *et al.* 2000; Ferreira & de Pinho 2003). SPME, used for the first time for the analyses of food volatile compounds in the mid-1990s (Yang & Peppard 1994; Pelusio *et al.* 1995; Chin *et al.* 1996), has since been used in significant applications on food aroma (Kataoka *et al.* 2000; Pillonel *et al.* 2002; Le Quéré & Etiévant 2003; and references cited therein). Analyzing volatiles directly by immersion of the fiber in highly complex matrices could damage the fiber, and SPME is therefore used almost always in the headspace mode. Comparison of direct SPME and headspace SPME of Camem-bert cheese volatiles obtained after cryo-trapping of the aqueous phase under vacuum showed only a slight reduction in sensitivity using headspace SPME compared to direct SPME (Jaillais *et al.* 1999). Compared to other headspace extraction procedures, it is very often concluded that SPME is more appropriate for routine quality control due to its simplicity, repeatability, and low cost (Cavalli *et al.* 2003). That is probably why the method has been widely used in recent works on food aroma (Le Quéré & Etiévant 2003 and references cited therein).

The main limitation of SPME is the relatively low extraction yield due to the relatively small amount of sorbent available on the syringe needle (typically approximately 0.5 mL). A novel extraction technique that uses upto 200 mL of the sorbent PDMS (polydimethylsiloxane) was developed to overcome this drawback (Baltussen *et al.* 1999). The new technique called SBSE consists of a glass-coated magnetic stir bar with a cylindrical coating of PDMS (typically 5 mm film thickness, 10 mm length) that is spun in an aqueous medium for a predefined time (Baltussen *et al.* 1999). After completion of the extraction step, the stir bar is transferred in a thermodesorption system and desorbed at the head of a GC column after cryo-refocusing of the extracted material (Baltussen *et al.* 1999). A complete set of coated stir bars (called Twister(tm)) and a thermodesorption system are commercially available from Gerstel GmbH (Gerstel, Müllheim on the Ruhr, Germany). As expected, the recoveries were higher for SBSE than for SPME (Baltussen *et al.* 1999), and the detection limits were found in the low ng/L range for a wide selection of volatile and semivolatile analytes (Baltussen *et al.* 1999). SBSE has been used for the measurement of volatiles from a wide variety of liquid foods (Pillonel *et al.* 2002, and references cited therein) including recently brewed coffee (Bicchi *et al.* 2002) and wine (Kittel *et al.* 2004).

As an extension to SBSE, headspace sorptive extraction (HSSE) has been developed (Tienpont *et al.* 2000) to overcome the limitation of headspace-SPME in terms of extraction capacity. Limits of detection in the ng/L range have been obtained for the analyses of volatiles of some food samples (Tienpont *et al.* 2000). HSSE bars coated with approximately 55 mL PDMS (commercially available from Gerstel GmbH) are suspended in the headspace of the sample (Bicchi *et al.* 2002) and after sampling completion, the bars are thermally desorbed in a thermal desorption unit connected to a gas chromatograph (Bicchi *et al.* 2002). As expected, when comparing headspace extractions of coffee (Bicchi *et al.* 2002) and olive oil (Cavalli *et al.* 2003) volatiles, HSSE bars showed a higher concentration capacity than SPME fibers due to the higher amount of polymeric coating. However, like SBSE, HSSE needs a thermal desorption unit to be handled and therefore requires a significant investment, compared to SPME if used in manual mode (automation is possible at cost price). Nevertheless, because SBSE and HSSE coated bars are less subject to deterioration than SPME fibers, they can be applied easily to the analyses of both headspace and liquid (Bicchi *et al.* 2002). Direct thermal desorption of food samples is also possible (Pfnür *et al.* 2003), but seems more suited to the extraction of semivolatile compounds (Cavalli *et al.* 2003).

4.3 KEY AROMA COMPOUNDS

4.3.1 Representativeness

As already outlined, because there is no universally applicable method, none of the extraction techniques described here or in the previous chapter yields an aroma isolate that truly represents, either qualitatively or quantitatively, the aroma profile of a food (Reineccius 2002). It is therefore necessary to choose the isolation procedure best suited to address the problem faced: determining the complete aroma profile, identifying key odorants or off flavors, monitoring aroma changes with time in foods, or predicting sensory properties (Reineccius 2002). When the ultimate aim of a particular study is the identification of the compounds that are important for flavor (the key odorants), the most reliable results will be obtained if the odor of the extract resembles closely that of the food itself (Etiévant *et al.* 1994; Etiévant & Langlois 1998). It is for instance advisable to prevent oxidation during the extraction step by addition of a suitable antioxidant, especially when oxidation of sensitive compounds may alter the odor of the extracts (Escudero & Etiévant 1999). Different sensory methods, which necessitate a trained sensory panel, can be used to check the sensory representativeness of the food-extract odors (Etiévant *et al.* 1994). When an estimation of the relative importance of key constituents in a single sample is required, a similarity test is preferred. The panelists are asked to score the similarity of the odor of the extracts obtained by different methods to the odor of the food itself used as reference on an unstructured 10-cm scale. This approach was applied to various foodstuffs including wine (Abbott *et al.* 1993a; Etiévant *et al.* 1994; Bernet *et al.* 1999), cheese (Etiévant *et al.* 1994; Le Quéré *et al.* 1996), ham (Guillard *et al.* 1997), butter (Guyot *et al.* 1998), tomato (Etiévant & Langlois 1998), champagne (Escudero & Etiévant 1999), coffee (Sarrazin *et al.* 2000), vinegar (Charles *et al.* 2000), mussels (Le Guen *et al.* 2000), oyster (Pennarun *et al.* 2002), edible algae (Le Pape *et al.* 2004), black currant (Boccorh *et al.* 2002), and apple (Mehinagic *et al.* 2003). The similarity test can be completed by a descriptive analysis of the extracts (Moio *et al.* 1995; Le Quéré *et al.* 1996; Priser *et al.* 1997; Guyot *et al.* 1998; Sarrazin *et al.* 2000) or even by a quantitative descriptive analysis (QDA) of the extracts compared to a QDA of the food samples (Abbott *et al.* 1993b; Le Guen *et al.* 2000; Pennarun *et al.* 2002; Mehinagic *et al.* 2003; Le Pape *et al.* 2004).

When different food samples have to be compared, triangle tests and overall matching tests are preferred. The different samples are presented as control samples and extracts from the samples, presented in random order, have to be matched with controls. This approach was initially done on beer extracts (Abbott *et al.* 1993b). A key point in these evaluations of representativeness is the choice of a suitable matrix for testing the olfactory character of the extracts. For fat-containing food like cheese or butter, the best results have been obtained when the extracts are added to an emulsion, that is, a matrix similar to food in terms of fat composition (Etiévant *et al.* 1994; Guyot *et al.* 1998). Since, generally, a combination of techniques should be used to obtain a reasonably complete view of an aroma profile (Reineccius 2002), it is noteworthy that substantial efforts have been made recently toward sensory evaluation of headspace or SPME extracts. Thus, oyster volatiles desorbed from the Tenax trap used in dynamic headspace have been collected in flasks containing water (Pennarun *et al.* 2002). Edible red algae volatiles have been desorbed from the dynamic headspace Tenax trap and collected in evacuated brown flasks (Le Pape *et al.* 2004), and solvent-free extracts from apples have been collected by preparative gas chromatography in Teflon bags half filled with nitrogen (Mehinagic *et al.* 2003). The easiest and most promising technique in this field is probably "direct GC-olfactometry" (i.e., without a chromatographic column), where a complete headspace or SPME extract is directly evaluated at the sniffing port of a gas chromatograph (Lecanu *et al.* 2002). This has been recently applied for cheese (Lecanu *et al.* 2002), orange juice (Rega *et al.* 2003a, b), and apricot extracts (Guillot *et al.* 2003).

4.4 GAS CHROMATOGRAPHY-OLFACTOMETRY

The separation technique that uses a human nose as a detector and known as gas chromatography-olfactometry (GC-O, sometimes referred to as "GC-sniffing") has received considerable attention during the past 20 years in aroma research (see, e.g., Blank 1997; Leland *et al.* 2001; Reineccius 2002; Lee 2003). The selectivity of this specific detector is based only on the odorous properties of the individual compounds separated by high-resolution gas chromatography. As the most abundant volatiles may have little, if any, odor of significance in a food (Mistry *et al.* 1997), GC-sniffing has been an invaluable tool for identifying key compounds in aroma extracts.

The first aim of the technique is to discriminate the odorous compounds from the many background volatile components. The so-called aromagram constructed from the chromatogram obtained by simply smelling a GC effluent (Blank 1997; Reineccius 2002) is a potential interface with sensory analysis, as odor descriptors detected at the GC-sniffing port can be compared to the descriptors generated in sensory evaluation of the original food. This method is particularly efficient for identifying off flavors. Selection of key odorants or character-impact compounds in a food is another objective of GC-sniffing. Quantitative approaches (the true GC-olfactometry) based on odor detection thresholds or on odor intensity have been developed and are the subject of specialized treatises (Mistry *et al.* 1997; Leland *et al.* 2001; van Ruth 2001b; Reineccius 2002; Lee 2003).

Three different methods have been developed for GC-olfactometry: (1) dilution analyses based on determination of detection thresholds, (2) detection frequency methods, and (3) intensity measurement methods. Original dilution methods, combined hedonic aroma measurement (CHARM) analysis developed by Acree and coworkers (1984) and aroma extract dilution analysis (AEDA) developed by Grosch and coworkers (Ullrich & Grosch 1987), are essentially screening methodologies since the methods, based only on detection thresholds determination, violate certain sensory rules and psychophysical laws (Grosch 2001; Reineccius 2002, and references cited therein). They can be used to identify those single odorous compounds that are most likely to contribute to the complex odor of a food. Originally developed by McDaniel and coworkers (McDaniel *et al.* 1990), the Osme method is basically a cross-modal technique aimed at measuring the perceived odor intensity of eluting volatiles. In OSME and other cross-modality matching methods (Guichard *et al.* 1995; Etiévant *et al.* 1999), results are not based on odor detection thresholds, and only one concentration of the extract is evaluated by a panel, contrarily to dilution methods where several dilutions of the extract are evaluated. Results can be subjected to statistical analysis and more consistent results are obtained when panelists are trained (Callement *et al.* 2001). The detection frequency methods, originally developed by Roozen and coworkers (Linssen *et al.* 1993), and referred to as nasal impact frequency (NIF) or surface of nasal impact frequency (SNIF) since the work of Chaintreau and coworkers (Pollien *et al.* 1997), also use a group of assessors who simply have to note when they detect an odor in a single GC run (i.e., also at only one concentration). The GC peaks being detected as odorous by the greatest number of assessors are considered to be the most important. Not being based on real odor intensities, the method has important drawbacks, especially when all the odorous compounds are present above their sensory threshold for all the assessors (Reineccius 2002). However, a study on gewürztraminer wines from Alsace showed that the odor intensities measured by OSME using the finger-span cross-modality matching method were well correlated to the detection frequencies (Bernet 2000; Etiévant & Chaintreau 2001). According to these authors, the theoretical saturation limitation when using detection frequencies is practically reachable for only 10% of the detected odors (Etiévant & Chaintreau 2001).

Nevertheless, each of the methods described above has its advantages and weaknesses. Only two studies have compared all the methods for their performance (Le Guen *et al.* 2000; van Ruth & O'Connor 2001). In both cases, the results obtained with the different techniques were found to be very similar and well correlated. Finally, the choice of a GC-O method depends on the objective of the study, on the quality of the panel, and on the time scheduled for the analyses (Le Guen *et al.* 2000). Dilution techniques are clearly time consuming, intensity methods give better results with a trained panel (Le Guen *et al.* 2000; Callement *et al.* 2001), while detection frequency methods are the least demanding, but also the least precise (Le Guen *et al.* 2000). A comparative critical review may be found in Etiévant and Chaintreau (2001).

The aim of any GC-O experiment is to determine the relative odor potency of volatiles present in an aroma extract or fraction and to prioritize compounds for identification then usually performed using GC coupled to mass spectrometry (GC/MS). Mass spectrometry is also used for quantification purposes through the use of a stable isotope dilution assay (Milo & Blank 1998; Blank *et al.* 1999, and references cited therein). Such a precise quantification is required for the determination of odor activity values (OAVs), generally calculated when using AEDA (Grosch 1994, 2001). OAVs, calculated as the ratio of concentrations to odor thresholds, despite their limitations in terms of psychophysical validity (Mistry *et al.* 1997), give a good indication of the respective contributions of key odorants to the aroma of foods. They are the basis of the first attempts of using recombination studies to validate impact odorants sensorially in model foods (Grosch 1994). Aroma-recombination studies are the important last step in sensorially validating the analytical data obtained by GC-O and for quantification of key odorants of food (Mistry *et al.* 1997). Many examples of models used in recombination experiments may be found in a recent review (Grosch 2001). Preparation of aroma models was found simpler for liquid foods than for solid foods, because in that case it is not easy to reproduce the composition and distribution of the nonvolatile fraction of the food matrix (Grosch 2001). However, for cheese models for instance, either bland unripened cheese (Grosch 1994; Preininger *et al.* 1996; Kubickova & Grosch 1998) or specially designed odorless model cheese (Salles *et al.* 1995) have been successfully used to incorporate potential key odorants. Thus, the branched-chain volatile fatty acids, 4-methyloctanoic and 4-ethyloctanoic acids, were confirmed to be essential to the typical goaty note of goat cheese (Le Quéré *et al.* 1996) and their retronasal aroma thresholds were determined using a cheese model (Salles & Le Quéré 1998; Le Quéré & Salles 2001; Salles *et al.* 2002). GC-O allows odor evaluation of individual compounds, but a large loss of sensory properties is encountered when odorants are mixed (Grosch 2001). To understand the perceptual interactions of odorants, recombination studies in model foods or psychophysical experiments are necessary. Masking or enhancing properties of important odorants may be evaluated during GC-O with a new method called OASIS for original aroma simultaneously input to the sniffing port (Hattori *et al.* 2003). It consists in evaluating complex odors at the sniffing port by delivering an original aroma directly at the sniffing port while a GC-O experiment is running. It is then possible to know how the individual components separated in the GC and sensed at the sniffing port influence the original odor notes. The authors have demonstrated that even high odor threshold compounds may affect an original Japanese green tea aroma (Hattori *et al.* 2003).

The GC-O methods that have been developed during the past 20 years, combined with aroma extracts, headspace, or even SPME (Dufour *et al.* 2001), have facilitated the identification of potent odorants in numerous foodstuffs. Hundreds of scientific papers have been published on the subject and it is out of the scope of this chapter to review all of them.

4.5 DYNAMIC METHODS FOR FLAVOR CHARACTERIZATION

Supposing that the "best" extraction and identification methods are used, trying to correlate the overall levels of flavor components in a food to the sensory perception experienced when eating this food is very often unsuccessful. In other words, it is not enough to know the exact composition of food in terms of flavor compounds to understand perfectly the perception of its flavor. Perception of flavor is a dynamic process (Piggott 2000). During the consumption of food, the concentration of aroma compounds at the olfactory epithelium varies with time as they are released progressively from the food matrix during chewing. Release kinetics depends on the nature of the food matrix and on individual mastication patterns. Sensory methods, such as time-intensity, have been used to study the time-related aspects of flavor perception (Piggott 2000).

4.6 RELEASE OF VOLATILES IN VIVO

Methods that measure volatiles directly in the mouth or in the nose have been developed to obtain data that could reflect the pattern of aroma molecules released from food and that are effectively present at the olfactory receptors during consumption, reviewed recently in an edited book (Roberts & Taylor 2000). Among the various approaches aimed at sampling aroma from the nose (nosespace), the collection of expired air samples on Tenax® traps provided the first robust results (Linforth & Taylor 1993; Linforth *et al.* 1994). Applied to cheddar cheese, GC-O of buccal headspace showed a number of volatile compounds that have been suspected to contribute primarily and most likely to the original flavor (Delahunty *et al.* 1996a). It was presumed that the buccal headspace extract was representative of the aroma compounds that a consumer perceives during consumption (O'Riordan & Dela-hunty 2001).

By overlapping the sampling time periods, release curves can be constructed and temporal changes reflecting relative concentrations of volatiles at a particular moment during consumption can be determined (Linforth *et al.* 1996). Correlation of accumulated data with sensory time-intensity data has been demonstrated (see for instance Delahunty *et al.* 1996b).

Real-time in vivo flavor release was demonstrated some time ago using mass spectrometric breath-by-breath analysis with an optimized membrane separator interfaced to a mass spectrometer operated in electron impact mode (Overbosch 1987). Sensory time-intensity data measured in parallel for the perception of 2-pentanone in vegetable oil showed a clear adaptation effect, the stimulus being present in exhaled air long after the perception ended (Overbosch 1987). The method was also used by Soeting and Heridema (1988) and was extensively reviewed by Overbosch (Overbosch *et al.* 1991). However, membrane separator techniques suffer from selectivity problems and from a clear lack of sensitivity (Taylor *et al.* 2000).

More recently, atmospheric pressure chemical ionization mass spectrometry (APCI-MS) has been developed to monitor aroma release during chewing (reviewed by Taylor *et al.* 2000). Air from the nose (nosespace) is sampled directly into the APCI-MS source through an interface making real-time breath-by-breath analysis routinely possible (Linforth *et al.* 1996; Taylor & Linforth 1996; Roberts & Taylor 2000, and references cited therein). Therefore, by combining time-intensity sensory studies with nosespace analysis, it is now possible to relate temporal parameters of aroma release to perception (de Kok & Smorenburg 1999; Linforth *et al.* 2000; Salles *et al.* 2003). Perceptual interactions of aroma with sapid compounds may also be studied by this method through controlled delivery of both aroma and sapid molecules to panelists (see Cook *et al.* 2004 for recent published results). The APCI-MS method has been extensively reviewed in detail in specialized treatises (Roberts & Taylor 2000; Taylor 2002; Taylor & Linforth 2003). APCI sources may also be connected to a tandem mass spectrometer (MS-MS), such as an ion trap, with the selectivity and structural capabilities benefits of MS-MS (Haahr *et al.* 2003; Sémon *et al.* 2003).

Another powerful chemical ionization method is proton-transfer-reaction mass spectrometry (PTR-MS). Originally developed by the Lindinger group (Lindinger *et al.* 1993) for on-line trace gas analysis, it consists of a three-chamber system. In the first chamber, a nearly pure H3O1 ion is generated by electrical discharges in H_2O vapor. A small electric field drives H_3O_1 ions through an orifice into a drift tube, where chemical ionization takes place, while neutral volatiles are introduced into the drift tube. Volatile compounds with proton affinities exceeding that of water (166.5 kcal/mol) ionize by proton transfer from H_3O_1 and are accelerated into the third chamber, the mass spectrometer. Specificity of PTR-MS compared to other chemical ionization approaches is that the generation of the reactant ion and the chemical ionization process are spatially and temporally separated. Individual optimization is therefore possible and quantification is made easier (Yeretzian *et al.* 2000a). Since the early works on headspace flavor volatiles that used PTR-MS (Yeretzian *et al.* 2000a, b), applications developed rapidly and the first specific international conference was organized in 2003 in Innsbruck, Austria (First International Conference on Proton Transfer Reaction Mass Spectrometry and its Applications, January 18–23, 2003). Results may be found in the current literature on headspace (Blank *et al.* 2003; Mayr *et al.* 2003a), nosespace (Mayr *et al.* 2003b; Roberts *et al.* 2003, 2004), and model-mouth (van Ruth *et al.* 2003, 2004) applications.

4.7 IN VITRO MEASUREMENTS: MODEL-MOUTH SYSTEMS

Mechanical devices that aim to mimic the processes that occur in the mouth during eating have been developed (Piggott 2000, and references cited therein). These "model mouths" are often variants of dynamic headspace analysis, but their aim is to obtain time-resolved data similar to those obtained during in vivo studies. The various parameters like temperature, airflow, mastication rate, and addition of artificial saliva can be varied to study their effects on volatile flavor release (van Ruth *et al.* 1995; van Ruth & Roozen 2000; Rabe *et al.* 2002). The main advantages of model mouths are the large quantities of food samples that can be handled, overcoming some sensitivity problems encountered when monitoring volatiles at low concentrations (Taylor 2002), and the suppression of interindividual variations, always encountered in vivo, that can be detrimental to a robust interpretation of the data. Release of volatile flavor compounds from the retronasal aroma simulator (RAS), originally developed by Roberts and Acree (1995), has been compared with flavor release in vivo using APCI-MS detection in both cases (Deibler *et al.* 2001). While delivering higher concentrations of volatiles than from human breath, the RAS gave a good approximation of time-averaged flavor release in the mouth, with volatile compounds present at similar ratios (Deibler *et al.* 2001). The model-mouth device originally developed by Roozen and coworkers (van Ruth *et al.* 1994) has been used to investigate the relationships between the gross, nonvolatile, and volatile compositions and the sensory attributes of Swiss-type cheeses (Lawlor *et al.* 2002). Eight flavor attributes were found to be correlated with subsets of volatiles, amino acids, free fatty acids, and gross compositional constituents, with, for instance, the nutty flavor of Emmental that was positively correlated with the concentrations of pro-panoic acid, ethyl acetate, and 2-pentanone (Lawlor *et al.* 2002). In a recent study, the model-mouth system has been compared to the RAS in terms of the effects of oral physiological characteristics on the release of aromas as a function of the physicochemical properties of model emulsions (Geary *et al.* 2004). Both have been found suitable for the study of oral parameters on aroma release (Geary *et al.* 2004), with confirmed limits for the temporal dimension of the release (Deibler *et al.* 2001; Geary *et al.* 2004). Real-time data, comparable to those obtained in vivo, have been obtained recently with a computerized apparatus that follows the temporal dimension of flavor release from liquid food (Rabe *et al.* 2002; Rabe *et al.* 2003).

Flavor release and flavor perception are dynamic processes and must be studied using dynamic methods (Piggott 2000). Dynamic techniques have been developed to study the parameters of flavor release from foods. Parallel increased applications of dynamic sensory methods provide a better understanding of food flavor. However, further work is needed to improve our knowledge of various interactions arising at different levels in the process of food consumption, such as interactions among food ingredients (Taylor 2002), and interactions at the perceptual levels such as taste–aroma interactions (Noble 1996; Given & Paredes 2002; Taylor 2002), or trigeminal interferences (Green 1996; Given & Paredes 2002), as these play a fundamental role in overall flavor perception.

4.8 GLOBAL AND FAST ASSESSMENT OF FLAVOR

The methods currently used to evaluate and control the quality of food flavor are still essentially based on sensory evaluation by a panel of experts. These trained panels are able to handle such difficult tasks as quality monitoring through descriptive analysis, off flavors detection, and comparison of samples for classification purposes. It could be interesting for such tasks to substitute humans by instruments that could give quicker answers at reduced costs.

4.9 "ELECTRONIC NOSE"

Evaluation of aroma release from food using gas sensors, the so-called electronic noses, is theoretically feasible (Mielle 1996; Hodgins 1997; Schaller *et al.* 1998). Electronic noses are generally composed of arrays of nonspecific gas sensors that are based on different physical principles (Mielle 1996; Hodgins 1997; Schaller *et al.* 1998). The most common sensors are semiconducting metal oxides and conducting organic polymers, and they all give rise to nonspecific responses with typical patterns. Therefore, pattern recognition software, using either standard multivariate data analyses or artificial neural network technology, must be used for data treatment and final presentation of the results (Hodgins 1997; Schaller *et al.* 1998). The electronic nose is particularly attractive for quality control applications where conformity/nonconformity answers are expected (Mielle 1996). Discriminative studies have been conducted on all types of food with some success (Schaller *et al.* 1998; Monge *et al.* 2004, and references cited therein). However, some problems occurred with the repeatability of the system that could be possibly related to the product itself, the sampling technique, or the moisture content of the air used for sampling, precluding its use in routine tests (Schaller *et al.* 1998). Metal oxide semiconductor technology, despite some poisoning problems affecting the sensors, seems more reliable than conducting organic polymer sensors that showed a poor sensitivity to volatile components, the main problem of these sensors lying however with their instability (Schaller *et al.* 2000a). However, recently, the ripening of Danish Blue cheese was successfully monitored by means of an electronic nose that contained 14 conducting polymer (polyaniline) sensors; results were found to be highly correlated to those of sensory analysis and GC-MS analysis of volatile compounds during a 5- to 12-week ripening period (Trihaas *et al.* 2003). The close control of the experimental sampling conditions (quality of dry air with a humidity, 0.5%, and equilibration time at controlled temperature) might explain this success

(Schaller *et al.* 1998). Nevertheless, despite some success in some classification tasks when using perfectly controlled sampling conditions (see for instance Monge *et al.* 2004, for a recent application on flavored pectin gels), electronic noses hardly meet the requirements of the food industry in terms of precision, reproducibility, sensitivity, and stability (Mielle *et al.* 2000). Moreover, the sensors are known to deteriorate or can be poisoned, therefore changing their response. Even with frequent calibration, the inherent weaknesses of the technique make perenniality of the built databases problematic. Giving a global response, these instruments cannot be used to identify single odorants or to differentiate samples with subtle differences in distinctive sensory attributes. Therefore, in off flavor studies where identification of the off flavor compound is a prerequisite and in quality control assessment they may be used successfully only after recognizing their inherent weaknesses (Mielle *et al.* 2000; Reineccius 2002).

4.10 MASS SPECTROMETRY-BASED SYSTEMS

For classification purposes, two other global and fast analytical methods, based on mass spectrometry, have been used for food products and seem more powerful and reliable than electronic noses. The first consists of a global analysis of a headspace sample by an MS operated in electron ionization mode, without any GC separation (Vernat & Berdagué 1995; Pérès *et al.* 2003). The feasibility of the method was originally demonstrated for rapid classification of four rather different French cheeses (Vernat & Berdagué 1995). This method is often referred to as a "MS-based electronic nose" (Schaller *et al.* 2000b). The mass patterns obtained, considered as fingerprints of the food products analyzed, also need data treatment, either by conventional multivariate analyses or by artificial neural networks. SPME may be used as a preconcentration technique instead of headspace sampling (Marsili 1999). Applied to rapid characterization of cheeses, SPME has been demonstrated to be a very efficient preconcentration technique, superior to dynamic headspace analysis in terms of repeatability, simplicity, and compatibility with an autosampler (Schaller *et al.* 2000b). However, contradictory results have been published when SPME (Pérès *et al.* 2001) yielded less satisfactory results than those obtained by dynamic headspace analysis (Pérès *et al.* 2002a). The better performance of the dynamic headspace method in that case was attributed to the absence of signal drift (aging of the SPME fibers causes drift, as demonstrated by Pérès *et al.* 2001) and to the automation of the sample injection into the mass spectrometer, a problem that can be easily overcome by automating the SPME injection. A new review on the subject has recently appeared (Pérès *et al.* 2003).

Developed in the 1980s for food applications, direct pyrolysis-MS is another method that delivers "fingerprints" that can be used for classification/authentication purposes (Aries & Gutteridge 1987). With this method, a tiny food sample is pyrolyzed rapidly at up to 530°C and the resulting volatile fraction, characteristic of the flavor but also of the matrix breakdown, is analyzed immediately by a mass spectrometer operated in low energy electron ionization mode. Here again a complex mass pattern is obtained for each sample and several data-preprocessing steps are often necessary to select a reduced number of mass fragments that allow satisfactory classification. Curie-point pyrolysis-mass spectrometry with associated multivariate data analysis techniques is considered as a powerful classification tool in microbiology for the recognition of microorganisms (Talon *et al.* 2002, and references cited therein) and food science (Aries & Gutteridge 1987; Pérès *et al.* 2002b, and references cited therein). A clear advantage of the method is that it provides a specific fingerprint of the food matrix that can be potentially related to textural parameters (Pérès *et al.* 2002b). In a similar approach, the PTR-MS of the static headspace of Mozzarella cheese was found to display a discrimination power comparable to sensory descriptive analysis (Gasperi *et al.* 2001).

4.11 CONCLUSION

Recent developments in techniques used for flavor measurements include extraction methods based on adsorption of aroma molecules on polymers. These sorptive extraction methods (SPME, SBSE, HSSE) are easy to use and fully automatable. Pertinent complementary sensory information can be obtained by combining gas chromatography of representative flavor extract with olfactometry. Nevertheless, the relationships among aromas present in a foodstuff and sensory perception of that food are not straightforward. It is still not well understood how the various flavor-active components combine to produce a particular sensory perception. Recent developments in dynamic instrumental methods that can follow the in vivo sequential release of the aroma molecules are valuable tools that can account for the time-dependent balance of the flavor compounds released from the food matrix in the mouth. With more complete and accurate information, combined flavor chemistry and sensory evaluation should help us understand the relationship between flavor stimuli and perceived flavor and explain the mechanisms of flavor perception.

Authentication of foodstuffs and routine fast analyses are other challenges. Tools developed recently that combine analytical instrumentation for global assessment of flavor with multivariate data analyses have demonstrated their usefulness for classification purposes.

ACKNOWLEDGMENTS

The research conducted by the author in the field is supported by the Burgundy Regional Council. The author is grateful to M Dominique Langlois for helpful aid in the preparation of this manuscript.

REFERENCES

Abbott, N., Bayonove, C., Etiévant, P. & Issanchou, S. (1993a) Production of wine extracts for chemical and sensory analysis with an odour representative of that of the original wine. In: *Proceedings Connaissance Aromatique des Cépages et Qualité des Vins*, Montpellier, France, Février 9–10, 1993.

Abbott, N., Etiévant, P., Langlois, D., Lesschaeve, I. & Issanchou, S. (1993b) Evaluation of the representativeness of the odor of beer extracts prior to analysis by GC eluate sniffing. *Journal of Agricultural and Food Chemistry*, **41**, 777–780.

Acree, T.E., Barnard, J. & Cunningham, D.G. (1984) A procedure for the sensory analysis of gas chromatographic effluents. *Food Chemistry*, **14**, 273–286.

Aries, R.E. & Gutteridge, C.S. (1987) Applications of pyrolysis mass spectrometry to food science. In: *Applications of Mass Spectrometry in Food Science* (ed. J. Gilbert), pp. 377–431. Elsevier Applied Science, London, UK.

Baltussen, E., Sandra, P., David, F. & Cramers, C. (1999) Stir bar sorptive extraction (SBSE), a novel extraction technique for aqueous samples: Theory and principles. *Journal of Microcolumn Separations*, **11**, 737–747.

Bernet, C. (2000) Contribution à la connaissance des composés d'arôme clés des vins du cépage gewurztraminer cultivé en Alsace. PhD thesis, University of Burgundy, Dijon, France.

Bernet, C., Dirninger, N., Etiévant, P. & Schaeffer, A. (1999) Evaluation de la représentativité aromatique d'extraits de vins de gewurztraminer. *Science des Aliments*, **19**, 701–709.

Bicchi, C., Drigo, S. & Rubiolo, P. (2000) Influence of fibre coating in headspace solid-phase microextraction-gas chromatographic analysis of aromatic and medicinal plants. *Journal of Chromatography A*, **892**, 469–485.

Bicchi, C., Iori, C., Rubiolo, P. & Sandra, P. (2002) Headspace sorptive extraction (HSSE), stir bar sorptive extraction (SBSE), and solid phase microextraction (SPME) applied to the analysis of roasted arabica coffee and coffee brew. *Journal of Agricultural and Food Chemistry*, **50**, 449–459.

Blank, I. (1997) Gas chromatography-olfactometry in food aroma analysis. In: *Techni-Ques for Analyzing Food Aroma* (ed. R. Marsili), pp. 293–329. Marcel Dekker, Inc., New York.

Blank, I., Milo, C., Lin, J. & Fay, L.B. (1999) Quantification of aroma-impact components by isotope dilution assay-recent developments. In: *Flavor Chemistry, Thirty Years of Progress* (eds R. Teranishi, E.L. Wick & I. Hornstein), pp. 63–74. Kluwer Academic/Plenum, New York.

Blank, I., Devaud, S., Matthey-Doret, W., Pollien, P., Robert, F. & Yeretzian, C. (2003) Formation of odour-active compounds in Maillard model systems based on proline. In: *Flavour Research at the Dawn of the Twenty-First Century* (eds. J.L. Le Quéré & P.X. Etiévant), pp. 458–463. Lavoisier, Tec & Doc, Paris, France.

Boccorh, R.K., Paterson, A. & Piggott, J.R. (2002) Extraction of aroma components to quantify overall sensory character in a processed blackcurrant (*Ribes nigrum* L.) concentrate. *Flavour and Fragrance Journal*, **17**, 385–391.

Callement, G., Bouchet, M., Langlois, D., Etiévant, P. & Salles, C. (2001) Odor intensity measurements in gas chromatography-olfactometry using cross modality matching: Evaluation of training effects. In: *Gas Chromatography-Olfactometry The State of the Art* (eds J.V. Leland, P. Schieberle, A. Buettner & T.E. Acree), pp. 172–186. American Chemical Society, Washington, District of Columbia.

Cavalli, J.F., Fernandez, X., Lizzani-Cuvelier, L. & Loiseau, A.M. (2003) Comparison of static headspace, headspace solid phase microextraction, headspace sorptive extraction, and direct thermal desorption techniques on chemical composition of French olive oils. *Journal of Agricultural and Food Chemistry*, **51**, 7709–7716.

Charles, M., Martin, B., Giniès, C., Etiévant, P., Coste, G. & Guichard, E. (2000) Potent aroma compounds of two red wine vinegars. *Journal of Agricultural and Food Chemistry*, **48**, 70–77.

Chin, H.W., Bernhard, R.A. & Rosenberg, M. (1996) Solid phase microextraction for cheese volatile compound analysis. *Journal of Food Science*, **61**, 1118–1122.

Cook, D.J., Davidson, J.M., Linforth, R.S.T. & Taylor, A.J. (2004) Measuring the sensory impact of flavor mixtures using controlled delivery. In: *Handbook of Flavor Characterization* (eds K.D. Deibler & J. Delwiche), pp. 135–149. Marcel Dekker, Inc., New York.

Deibler, K.D. & Delwiche, J. (eds) (2004) *Handbook of Flavor Characterization*, p. 493. Marcel Dekker, Inc., New York.

Deibler, K.D., Lavin, E.H., Linforth, R.S.T., Taylor, A.J. & Acree, T.E. (2001) Verification of a mouth simulator by in vivo measurements. *Journal of Agricultural and Food Chemistry*, **49**, 1388–1393.

Delahunty, C.M., Piggott, J.R., Conner, J.M. & Paterson, A. (1996a) Flavor Evaluation of Cheddar Cheese. In: *Flavor-Food Interactions* (eds R.J. McGorin & J.V. Leland), pp. 202–216. American Chemical Society, Washington, District of Columbia.

Delahunty, C.M., Piggott, J.R., Conner, J.M. & Paterson, A. (1996b) Comparison of dynamic flavour release from hard cheeses and analysis of headspace volatiles from the mouth with flavour perception during consumption. *Journal of the Science of Food and Agricultural*, **71**, 273–281.

Dufour, J.P., Delbecq, P. & Perez Albela, L. (2001) Solid-phase microextraction combined with gas chromatography-olfactometry for analysis of cheese aroma. In: *Gas Chromatography-Olfactometry The State of the Art* (eds J.V. Leland, P. Schieberle, A. Buettner & T.E. Acree), pp. 123–137. American Chemical Society, Washington, District of Columbia.

Escudero, A. & Etiévant, P. (1999) Effect of antioxidants on the flavor characteristics and the gas chromatography/olfactometry profiles of Champagne extracts. *Journal of Agricultural and Food Chemistry*, **47**, 3303–3308.

Etiévant, P. & Chaintreau, A. (2001) Les analyses olfactométriques: Synthèse critique et recommandations. In: *Proceedings 20 èmes Journées Internationales Huiles Essentielles et Extraits*, Dignes les Bains, France, pp. 5–7, September.

Etiévant, P.X. & Langlois, D. (1998) L'amélioration de la représentativité olfactive des extraits aromatiques dans la recherche des composés clefs de l'arôme des aliments. *Proceedings 17 èmes Journées Internationales Huiles Essentielles et extraits*. Digne les Bains, France, September 2–5, 1998.

Etiévant, P.X., Moio, L., Guichard, E., *et al.* (1994) Aroma extract dilution analysis (AEDA) and the representativeness of the odour of food extracts. In: *Trends in Flavour Research, Developments in Food Science 35*. (eds H. Maarse & D.G. Van der Heij), pp. 179–191. Elsevier, Amsterdam.

Etiévant, P.X., Callement, G., Langlois, D., Issanchou, S. & Coquibus, N. (1999) Odor intensity evaluation in gas chromatography-olfactometry by finger span method. *Journal of Agricultural and Food Chemistry*, **47**, 1673–1680.

Ferreira, A.C.S. & de Pinho, P.G. (2003) Analytical method for determination of some aroma compounds on white wines by solid phase microextraction and gas chromatography. *Journal of Food Science*, **68**, 2817–2820.

Gasperi, F., Gallerani, G., Boschetti, A., *et al.* (2001) The mozzarella cheese flavour profile: A comparison between judge panel analysis and proton transfer reaction mass spectrometry. *Journal of the Science of Food and Agriculture*, **81**, 357–363.

Geary, M.D., van Ruth, S.M., Delahunty, C.M., Lavin, E.H. & Acree, T.E. (2004) Effects of oral physiological characteristics on the release of aroma compounds from oil-in-water emulsions using two mouth simulator systems. In: *Handbook of Flavor Characterization* (eds K.D. Deibler & J. Delwiche), pp. 345–361. Marcel Dekker, Inc., New York.

Given, P. & Paredes, D. (eds) (2002) *Chemistry of Taste. Mechanisms, Behaviors, and Mimics*, p. 362. American Chemical Society, Washington District of Columbia.

Green, B.G. (1996) Chemesthesis: Pungency as a component of flavor. *Trends in Food Science and Technology*, **7**, 415–420.

Grosch, W. (1994) Determination of potent odourants in foods by aroma extract dilution analysis (AEDA) and calculation of odour activity values (OAVs). *Flavour and Fragrance Journal*, **9**, 147–158.

Grosch, W. (2001) Evaluation of the key odorants of foods by dilution experiments, aroma models and omission. *Chemical Senses*, **26**, 533–545.

Guichard, H., Guichard, E., Langlois, D., Issanchou, I. & Abbott, N. (1995) GC Sniffing analysis: Olfactive intensity measurement by two methods. *Z Lebensm Unters Forsch*, **201**, 344–350.

Guillard, A.S., Le Quéré, J.L. & Vendeuvre, J.L. (1997) Emerging research approaches benefit to the study of cooked cured ham flavour. *Food Chemistry*, **59**, 567–572.

Guillot, S., Boulanger, R., Crouzet, J., Bureau, S., Lepoutre, J.P. & Galindo, S. (2003) Tracers of apricot aromatic quality. In: *Flavour Research at the Dawn of the Twenty-First Century* (eds J.L. Le Quéré & P. Etiévant), pp. 606–609. Lavoisier, Tec & Doc, Paris, France.

Guyot, C., Bonnafont, C., Lesschaeve, I., Issanchou, S. & Spinnler, H.E. (1998) Mise au point et validation d'une technique d'extraction des arômes du beurre permettant de conserver la typicité de l'odeur des produits. *Lait*, **78**, 351–361.

Haahr, A.M., Madsen, H., Smedsgaard, J., Bredie, W.L.P., Stahnke, L.H. & Refsgaard, H.H.F. (2003) Flavor release measurement by atmospheric pressure chemical ionization ion trap mass spectrometry, construction of interface and mathematical modeling of release profiles. *Analytical Chemistry*, **75**, 655–662.

Harmon, A.D. (1997) Solid-phase microextraction for the analysis of flavors. In: *Techniques for Analyzing Food Aroma* (ed. R. Marsili), pp. 81–112. Marcel Dekker Inc., New York.

Hattori, S., Takagaki, H. & Fujimori, T. (2003) Evaluation of Japanese green tea extract using GC/O with original aroma simultaneously input to the sniffing port method (OASIS). *Food Science and Technology Research*, **9**, 350–352.

Ho, C.T. & Manley, C.H. (1993) *Flavor Measurement*, p. 373. Marcel Dekker Inc., New York.

Hodgins, D. (1997) The electronic nose: sensor array-based instruments that emulate the Human nose. In: *Techniques for Analyzing Food Aroma* (ed. R. Marsili), pp. 331–371. Marcel Dekker, Inc., New York.

Jaillais, B., Bertrand, V. & Auger, J. (1999) Cryo-trapping/SPME/GC analysis of cheese aroma. *Talanta*, **48**, 747–753.

Kataoka, H., Lord, H.L. & Pawliszyn, J. (2000) Applications of solid-phase microextraction in food analysis. *Journal of Chromatography A*, **880**, 35–62.

Kittel, K.M., Lavin, E.H., Deibler, K.D., Acree, T.E. & Maillard, A. (2004) Analysis of off-aromas in wines caused by aging. In: *Handbook of Flavor Characterization* (eds K.D. Deibler & J. Delwiche), pp. 455–462. Marcel Dekker, Inc., New York.

de Kok, P.M.T. & Smorenburg, H.E. (1999) Flavor release in the mouth. In: *Flavor Chemistry, Thirty Years of Progress* (eds R. Teranishi, E.L. Wick & I. Hornstein), pp. 397–407. Kluwer Academic/Plenum, New York.

Kubickova, J. & Grosch, W. (1998) Evaluation of flavour compounds of Camembert cheese. *International Dairy Journal*, **8**, 11–16.

Lawlor, J.B., Delahunty, C.M., Wilkinson, M.G. & Sheehan, J. (2002) Relationships between the gross, non-volatile and volatile compositions and the sensory attributes of eight hard-type cheeses. *International Dairy Journal*, **12**, 493–509.

Le Guen, S., Prost, C. & Demaimay, M. (2000) Critical comparison of three olfactometric methods for the identification of the most potent odorants in cooked mussels (*Mytilus edulis*). *Journal of Agricultural and Food Chemistry*, **48**, 1307–1314.

Le Pape, M.A., Grua-Priol, J., Prost, C. & Demaimay, M. (2004) Optimization of dynamic headspace extraction of the edible red algae *Palmaria palmata* and identification of the volatile components. *Journal of Agricultural and Food Chemistry*, **52**, 550–556.

Le Quéré, J.L. & Etiévant, P.X. (eds) (2003) *Flavour Research at the Dawn of the Twenty-First Century*, p. 780. Lavoisier, Tec & Doc, Paris, France.

Le Quéré, J.L. & Salles, C. (2001) Goat cheese flavour compounds. In: *Recent Advances on Goat Milk Quality, Raw Material for Cheesemaking* (ed. G. Freund), pp. 115–122. ITPLC Editions, Surgères, France.

Le Quéré, J.L., Septier, C., Demaizières, D. & Salles, C. (1996) Identification and sensory evaluation of the character-impact compounds of goat cheese flavour. In: *Flavour Science Recent Developments* (eds A.J Taylor . & D.S. Mottram), pp. 325–330. The Royal Society of Chemistry, Cambridge, UK.

Lecanu, L., Ducruet, V., Jouquand, C., Gratadoux, J.J. & Feigenbaum, A. (2002) Optimization of headspace solid-phase microextraction (SPME) for the odor analysis of surface-ripened cheese. *Journal of Agricultural and Food Chemistry*, **50**, 3810–3817.

Lee, S.-J. (2003) Finding key odorants in foods: Gas chromatography olfactometry (GC/O). *Food Science and Biotechnology*, **12**, 597–602.

Leland, J.V., Schieberle, P., Buettner, A. & Acree, T.E. (eds) (2001) *Gas Chromatography-Olfactometry. The State of the Art*, p. 219. American Chemical Society, Washington, District of Columbia.

Lindinger, W., Hirber, J. & Paretzke, H. (1993) An ion/molecule-reaction mass spectrometer used for on-line trace gas analysis. *International Journal of Mass Spectrometry Ion Processes*, **129**, 79–88.

Linforth, R.S.T. & Taylor, A.J. (1993) Measurement of volatile release in the mouth. *Food Chemistry*, **48**, 115–120.

Linforth, R.S.T., Savary, I., Pattenden, B. & Taylor, A.J. (1994) Volatile compounds found in expired air during eating of fresh tomatoes and in the headspace above tomatoes. *Journal of the Science of Food and Agriculture*, **65** (2), 241–247.

Linforth, R.S.T., Ingham, K.E. & Taylor, A.J. (1996) Time course profiling of volatile release from foods during the eating process. In: *Flavour Science: Recent Developments* (eds A.J. Taylor & D.S. Mottram), pp. 361–368. The Royal Society of Chemistry, Cambridge, UK.

Linforth, R.S.T., Hollowood, T. & Taylor, A.J. (2000) Analysis of correlation between temporal aspects of aroma release and perception. In: *Frontiers of Flavour Science* (eds P. Schieberle & K.H. Engels), pp. 275–278, Deutsche Forschungsanstalt für Leb-ensmittelchemie, Garching.

Linssen, J.P.H., Janssens, J.L.G.M., Roozen, J.P. & Posthumus, M.A. (1993) Combined gas chromatography and sniffing port analysis of volatile compounds of mineral water packed in laminated packages. *Food Chemistry*, **46**, 367–371.

Lord, H. & Pawliszyn, J. (2000) Evolution of solid-phase microextraction technology. *Journal of Chromatography A*, **885**, 153–193.

Mariaca, R. & Bosset, J.O. (1997) Instrumental analysis of volatile (flavour) compounds in milk and dairy products. *Lait*, **77**, 13–40.

Marsili, R. (ed.) (1997) *Techniques for Analyzing Food Aroma*, p. 383. Marcel Dekker Inc., New York.

Marsili, R.T. (1999) SPME-MS-MVA as an electronic nose for the study of off-flavors in milk. *Journal of Agricultural and Food Chemistry*, **47**, 648–654.

Mayr, D., Märk, T., Lindinger, W., Brevard, H. & Yeretzian, C. (2003a) In-vivo analysis of banana aroma by proton transfer reaction-mass spectrometry. In: *Flavour Research at the Dawn of the Twenty-First Century* (eds J.L. Le Quéré & P.X. Etiévant), pp. 256–259. Lavoisier, Tec & Doc, Paris, France.

Mayr, D., Hartungen, E., Märk, T., Margesin, R. & Schinner, F. (2003b) Determination of the spoolage status of meat by aroma detection using Proto-Transfer-Reaction Mass-Spectrometry. In: *Flavour Research at the Dawn of the Twenty-First Century* (eds J.L. Le Quéré & P.X. Etiévant), pp. 757–760, Lavoisier, Tec & Doc, Paris, France.

McDaniel, M.R., Miranda-Lopez, R., Watson, B.T., Michaels, N.J. & Libbey, L.M. (1990) Pinot noir aroma: A sensory/gas chromatographic approach. In: *Flavors and Off-Flavors* (ed. G. Charalambous), pp. 23–36. Elsevier, Amsterdam.

Mehinagic, E., Prost, C. & Demaimay, M. (2003) Rep-resentativeness of apple aroma extract obtained by vacuum hydrodistillation: comparison of two concentration techniques. *Journal of Food Science*, **68**, 2411–2415.

Mielle, P. (1996) "Electronic noses": Towards the objective instrumental characterization of food aroma. *Trends in Food Science and Technology*, **7**, 432–438.

Mielle, P., Marquis, F. & Latrasse, C. (2000) Electronic noses: Specify or disappear. *Sensors and Actuators B*, **69**, 287–294.

Milo, C. & Blank, I. (1998) Quantification of impact odorants in food by isotope dilution assay: Strength and limitations. In: *Flavor Analysis Developments in Isolation and Characterization* (eds C.J. Mussinan, & M.J. Morello), pp. 250–259. American Chemical Society, Washington, District of Columbia.

Mistry, B.S., Reineccius, T. & Olson, L.K. (1997) Gas chromatography-olfactometry for the determination of key odorants in foods. In: *Techniques for Analyzing Food Aroma* (ed. R. Marsili), pp. 265–292, Marcel Dekker, Inc., New York.

Moio, L., Chambellant, E., Lesschaeve, I., Issanchou, S., Schlich, P. & Etiévant, P. (1995) Production of representative wine extracts for chemical and olfactory analysis. *Italian Journal of Food Science*, **7**, 265–278.

Monge, M.E., Bulone, D., Giacomazza, D., Bernik, D.L. & Negri, R.M. (2004) Detection of flavour release from pectin gels using electronic noses. *Sensors and Actuators B*, **101**, 28–38.

Mussinan, C.J. & Morello, M.J. (eds) (1998) *Flavor Analysis. Developments in Isolation and Character-ization*, p. 389. American Chemical Society, Washington, District of Columbia.

Noble, A.C. (1996) Taste-aroma interactions. *Trends in Food Science and Technology*, **7**, 439–444.

O'Riordan, P.J. & Delahunty, C.M. (2001) Comparison of volatile compounds released during the consumption of Cheddar cheese with compounds extracted by vacuum distillation using gas chromatography-olfactometry. *Flavour and Fragrance Journal*, **16**, 425–434.

Overbosch, P. (1987) Flavour release and perception. In: *Flavour Science and Technology* (eds M. Martens, G.A. Dalen & H. Russwurm), pp. 291–300. John Wiley & Sons, Chichester.

Overbosch, P., Afterof, W.G.M. & Haring, P.G.M. (1991) Flavor release in the mouth. *Food Reviews International*, **7**, 137–184.

Pelusio, F., Nilsson, T., Montanarella, L., *et al.* (1995) Head-space solid-phase microextraction analysis of volatile organic sulfur compounds in black and white truffle aroma. *Journal of Agricultural and Food Chemistry*, **43**, 2138–2143.

Pennarun, A.L., Prost, C. & Demaimay, M. (2002) Aroma extracts from oyster *Crassostrea gigas*: Comparison of two extraction methods. *Journal of Agricultural and Food Chemistry*, **50**, 299–304.

Pérès, C., Viallon, C. & Berdagué, J.L. (2001) Solid-phase microextraction-mass spectrometry: A new approach to the rapid characterization of cheeses. *Analytical Chemistry*, **73**, 1030–1036.

Pérès, C., Denoyer, C., Tournayre, P. & Berdagué, J.L. (2002a) Fast characterization of cheeses by dynamic headspace-mass spectrometry. *Analytical Chemistry*, **74**, 1386–1392.

Pérès, C., Viallon, C. & Berdagué, J.L. (2002b) Curie point pyrolysis-mass spectrometry applied to rapid characterisation of cheeses. *Journal of Analytical and Applied Pyrolysis*, **65**, 161–171.

Pérès, C., Begnaud, F., Eveleigh, L. & Berdagué, J.L. (2003) Fast characterization of foodstuff by headspace mass spectrometry (HS-MS). *Trends in Analytical Chemistry*, **22**, 858–866.

Pfnür, P., Hässelbarth, A., Schulze, M. & Ullrich, F. (2003) Cocoa liquor aroma quality by rapid thermo-desorption-GC/MS analysis. In: *Flavour Research at the Dawn of the Twenty-first Century* (eds J.L. Le Quéré & P.X. Etiévant), pp. 698–701. Lavoisier, Tec & Doc, Paris, France.

Piggott, J.R. (2000) Dynamism in flavour science and sensory methodology. *Food Research International*, **33**, 191–197.

Pillonel, L., Bosset, J.O. & Tabacchi, R. (2002) Rapid preconcentration and enrichment techniques for the analysis of food volatile. A review. *Lebensmittel-Wissenschaft und -Technologie*, **35**, 1–14.

Pollien, P., Ott, A., Montigon, F., Baumgartner, M., Munoz Box, R. & Chaintreau, A. (1997) Hyphenated headspace-gas chromatography-sniffing technique: Screening of impact odorants and quantitative aromagram comparisons. *Journal of Agricultural and Food Chemistry*, **45**, 2630–2637.

Preininger, M., Warmke, R. & Grosch, W. (1996) Identification of the character impact flavour compounds of Swiss cheese by sensory studies of models. *Z Lebensm Unters Forsch*, **202**, 30–34.

Priser, C., Etiévant, P.X., Nicklaus, S. & Brun, O. (1997) Representative champagne wine extracts for gas chromatography olfactometry analysis. *Journal of Agricultural and Food Chemistry*, **45**, 3511–3514.

Rabe, S., Krings, U., Banavara, D.S. & Berger, R.G. (2002) Computerized apparatus for measuring dynamic flavor release from liquid food matrices. *Journal of Agricultural and Food Chemistry*, **50**, 6440–6447.

Rabe, S., Banavara, D.S., Krings, U. & Berger, R.G. (2003) Real time measurement of flavour release from water and from neutral oil. In: *Flavour Research at the Dawn of the Twenty-First Century* (eds J.L. Le Quéré & P.X. Etiévant), pp. 148–53. Lavoisier, Tec & Doc, Paris, France.

Rega, B., Fournier, N., Brat, P. & Guichard, E. (2003a) Sensory evaluation of headspace extracts by direct GC-Olfactometry: Application to fresh orange juice. In: *Flavour Research at the Dawn of the Twenty-First Century* (eds J.L. Le Quéré & P.X. Etiévant), pp. 706–709. Lavoisier, Tec & Doc, Paris, France.

Rega, B., Fournier, N. & Guichard, E. (2003b) Solid phase microextraction (SPME) of orange juice flavor: Odor representativeness by direct gas chromatography olfactometry (D-GC-O). *Journal of Agricultural and Food Chemistry*, **51**, 7092–7099.

Reineccius, G. (2002) Instrumental methods of analysis. In: *Food Flavour Technology* (ed. A.J. Taylor), pp. 210–251. Sheffield Academic Press, Sheffield.

Roberts, D.D. & Acree, T.E. (1995) Simulation of retronasal aroma using a modified headspace technique: Investigating the effects of saliva, temperature, shearing, and oil on flavor release. *Journal of Agricultural and Food Chemistry*, **43**, 2179–2186.

Roberts, D.D. & Taylor, A.J. (eds) (2000) *Flavor Release*, p. 484. American Chemical Society, Washington, District of Columbia.

Roberts, D.D., Pollien, P., Antille, N., Lindinger, C. & Yeretzian, C. (2003) Comparison of nosespace, head-space, and sensory intensity ratings for the evaluation of flavor absorption by fat. *Journal of Agricultural and Food Chemistry*, **51**, 3636–3642.

Roberts, D.D., Pollien, P., Yeretzian, C. & Lindinger, C. (2004) Nosespace analysis with proton-transfer-reaction mass spectrometry: Intra- and interpersonal variability. In: *Handbook of Flavor Characterization* (eds K.D. Deibler & J. Delwiche), pp. 151–162. Marcel Dekker, Inc., New York.

van Ruth, S.M. (2001a) Aroma measurement: Recent developments in isolation and characterisation. In: *Physics and Chemistry Basis of Biotechnology* (eds M. De Cuyper & J.W.M. Bulte), pp. 305–328. Kluwer Academic Publishers, Amsterdam, the Netherlands.

van Ruth, S.M. (2001b) Methods for gas chromatography-olfactometry: A review. *Biomolecular Engi-neering*, **17**, 121–128.

van Ruth, S.M. & O'Connor, C.H. (2001) Evaluation of three gas chromatography-olfactometry methods: Comparison of odour intensity-concentration relationships of eight volatile compounds with sensory headspace data. *Food Chemistry*, **74**, 341–347.

van Ruth, S.M. & Roozen, J.P. (2000) Influence of mastication and saliva on aroma release in a model mouth system. *Food Chemistry*, **71**, 339–345.

van Ruth, S.M., Roozen, J.P. & Cozijnsen, J.L. (1994) Comparison of dynamic headspace mouth model systems for flavour release from rehydrated bell pepper cuttings. In: *Trends in Flavour Research* (eds H. Maarse & D.G. Van der Heij), pp. 59–64. Elsevier, Amsterdam, the Netherlands.

van Ruth, S.M., Roozen, J.P. & Cozijnsen, J.L. (1995) Volatile compounds of rehydrated French beans, bell peppers and leeks. Part 1. Flavour release in the mouth and in three mouth model systems. *Food Chemistry*, **53**, 15–22.

van Ruth, S.M., Buhr, K., Geary, M., Dings, L. & Odake, S. (2003) Effect of oral parameters on dynamic aroma release under mouth conditions. In: *Flavour Research at the Dawn of the Twenty-first Century* (eds J.L. Le Quéré & P.X. Etiévant), pp. 176–181. Lavoisier, Tec & Doc, Paris, France.

van Ruth, S., Geary, M.D., Buhr, K. & Delahunty, C.M. (2004) Representative sampling of volatile flavor compounds: The model mouth combined with gas chromatography and direct mass spectrometry. In: *Handbook of Flavor Characterization* (eds K.D. Deibler & J. Delwiche), pp. 335–344. Marcel Dekker, Inc., New York.

Salles, C. & Le Quéré, J.L. (1998) Impact of small hydrophobic and hydrophilic molecules on the flavour of goat cheese. In: *Proceedings 25th International Dairy Congress*, Aarhus, Denmark, September 21–24, 1998.

Salles, C., Dalmas, S., Septier, C., *et al.* (1995) Production of a cheese model for sensory evaluation of flavour compounds. *Lait*, **75**, 309–324.

Salles, C., Sommerer, N., Septier, C., *et al.* (2002) Goat cheese flavor: Sensory evaluation of branched-chain fatty acids and small peptides. *Journal of Food Science*, **67**, 835–841.

Salles, C., Hollowood, T.A., Linforth, R.S.T. & Taylor, A.J. (2003) Relating real time flavour release to sensory perception of soft cheeses. In: *Flavour Research at the Dawn of the Twenty-first Century* (eds J.L. Le Quéré & P.X. Etiévant), pp. 170–175. Lavoisier, Tec & Doc, Paris, France.

Sarrazin, C., Le Quéré, J.L., Gretsch, C. & Liardon, R. (2000) Representativeness of coffee aroma extracts: A comparison of different extraction methods. *Food Chemistry*, **70**, 99–106.

Schaller, E., Bosset, J.O. & Escher, F. (1998) "Electronic noses" and their application to food. *Lebensmittel-Wissenschaft und -Technologie*, **31**, 305–316.

Schaller, E., Bosset, J.O. & Escher, F. (1999) Practical experience with "electronic nose" systems for monitoring the quality of dairy products. *Chimia*, **53**, 98–102.

Schaller, E., Bosset, J.O. & Escher, F. (2000a) Instability of conducting polymer sensors in an electronic nose system. *Analusis*, **28**, 217–227.

Schaller, E., Zenhäusern, S., Zesiger, T., Bosset, J.O. & Escher, F. (2000b) Use of preconcentration techniques applied to a MS-based "Electronic Nose." *Analusis*, **28**, 743–749.

Sémon, E., Gierczynski, I., Langlois, D. & Le Quéré, J.L. (2003) Analysis of aroma compounds by atmospheric pressure chemical ionization—Ion trap mass spectrometry. Construction and validation of an interface for in vivo analysis of human breath volatile content. In: *Proceedings 16th International Mass Spectrometry Conference*, Edinburgh. August 31–September 5, 2003.

Soeting, W. & Heridema, J. (1988) A mass spectrometric method for measuring flavour concentration/time profiles in human breath. *Chemical Senses*, **13**, 607–617.

Stephan, A., Bücking, M. & Steinhart, H. (2000) Novel analytical tools for food flavours. *Food Research International*, **33**, 199–209.

Talon, R., Walter, D., Viallon, C. & Berdagué, J.L. (2002) Prediction of *Streptococcus salivarius* subsp. thermophilus and *Lactobacillus delbrueckii* subsp. bulgaricus populations in yoghurt by Curie point pyrolysis-mass spectrometry. *Journal of Microbiological Methods*, **48**, 271–279.

Taylor, A.J. (2002) Release and transport of flavors in vivo: Physicochemical, physiological, and perceptual considerations. *Comprehensive Reviews in Food Science and Food Safety*, **1**, 45–57.

Taylor, A.J. & Linforth, R.S.T. (1994) Methodology for measuring volatile profiles in the mouth and nose during eating. In: *Trends in Flavour Research* (eds H. Maarse & D.G. Van der Heij), pp. 3–14. Elsevier, Amsterdam, the Netherlands.

Taylor, A.J. & Linforth, R.S.T. (1996) Flavour release in the mouth. *Trends in Food Science and Technology*, **7**, 444–448.

Taylor, A.J. & Linforth, R.S.T. (2003) Direct mass spectrometry of complex volatile and non-volatile flavour mixtures. *International Journal of Mass Spectrometry*, **223–224**, 179–191.

Taylor, A.J., Linforth, R.S.T., Harvey, B.A. & Blake, A. (2000) Atmospheric pressure chemical ionisation mass spectrometry for in vivo analysis of volatile flavour release. *Food Chemistry*, **71**, 327–338.

Teranishi, R. (1998) Challenges in flavor chemistry: An overview. In: *Flavor Analysis Developments in Isolation and Characterization* (eds C.J. Mussinan & M.J. Morello), pp. 1–6. American Chemical Society, Washington, District of Columbia.

Tienpont, B., David, F., Bicchi, C. & Sandra, P. (2000) High capacity headspace sorptive extraction. *Journal of Microcolumn Separations*, **12**, 577–584.

Trihaas, J., van den Tempel, T. & Nielsen, P.V. (2003) Ripening monitoring of Danish blue cheese by means of electronic nose system, trained sensory panel and GC-MS. In: *Flavour Research at the Dawn of the Twenty-first Century* (eds J.L. Le Quéré & P.X. Etiévant), pp. 729–732. Lavoisier, Tec & Doc, Paris, France.

Ullrich, F. & Grosch, W. (1987) Identification of the most intense odour compounds formed during the autoxidation of linoleic acid. *Z Lebensm Unters Forsch*, **184**, 277–282.

Vernat, G. & Berdagué, J.L. (1995) Dynamic head-space-mass spectrometry (DHS-MS): A new approach to real-time characterization of food products. In: *Proceedings Bioflavour 95. Dijon*, France, February 14–17, 1995.

Yang, X. & Peppard, T. (1994) Solid-phase microextraction for flavor analysis. *Journal of Agricultural and Food Chemistry*, **42**, 1925–1930.

Yeretzian, C., Jordan, A., Brevard, H. & Lindinger, W. (2000a) Time-resolved headspace analysis by proton-transfer-reaction mass-spectrometry. In: *Flavor Release*(eds D.D. Roberts & A.J. Taylor), pp. 58–72. American Chemical Society, Washington, District of Columbia.

Yeretzian, C., Jordan, A., Brevard, H. & Lindinger, W. (2000b) On-line monitoring of coffee roasting by proton-transfer-reaction mass-spectrometry. In: *Flavor Release* (eds D.D. Roberts & A.J. Taylor), pp. 112–123. American Chemical Society, Washington, District of Columbia.

Part Two
Flavor

5 Sensory Characterization

Karen L. Bett-Garber

Abstract: Flavor results when chemical compounds are perceived by the human senses. Gas chromatography and mass spectrometry can monitor volatile compounds that contribute to flavor. Gas chromatography-olfactometry uses on the one hand an instrumental detector and on the other hand one human smells the effluent.

Descriptive flavor evaluation is a sensory evaluation method that describes individual flavor characters and gives an intensity rating to each descriptor. Various methods of descriptive flavor analysis and various methods used to determine descriptions of flavors are detailed.

Keywords: Descriptive flavor analysis, flavor profile, quantitative descriptive analysis, free-choice profiling, chemical flavor characteristics, gas chromatography-olfactometry.

Flavor, according to Amerine *et al.* (1965), is an attribute of foods, beverages, and seasonings resulting from stimulation of the sensory receptors that are grouped together at the entrance of the alimentary and respiratory tracts, and includes odor and taste. Caul (1957) restricted flavor to the impressions perceived via the chemical senses from a product in the mouth. This would include aromatic compounds perceived retronasally or volatile substances released from product in the mouth via the posterior nares, gustatory perceptions (tastes) caused by soluble substances in the mouth, and chemical feeling factors, which stimulate nerve ends in the soft membranes of the buccal and nasal cavities (Meilgaard *et al.* 1999). For the purpose of this chapter, flavor will include aroma/odor inhaled through the nose in addition to flavor-by-mouth and taste. The underlying principle is that flavor must be perceived by a human to be flavor. This chapter will discuss methods of measuring flavor characters using human beings, as well as methods of defining flavor characteristics.

Flavor results when chemical compounds are perceived by the human senses. Gas chromatographs (GC) and mass spectrometers (MS) can monitor volatile compounds that may or may not invoke a human response. Therefore, they can be used to measure quantities of compounds that contribute to flavor. But, these instruments cannot confirm if compounds are perceivable by humans. Gas chromatography-olfactometry (GC-O) is a method of configuring a gas chromatograph that splits the flow of compounds into two portions. One portion goes to the instrumental detector such as a flame ionization detector or mass spectrometer and the other portion goes to a port where a human smells the effluent laden with molecules. GC-O methods for characterizing flavor will be discussed later in this chapter.

Early in the study of human olfaction, the idea was that odors could be classified or grouped. The search for the grouping system that universally worked was not fully successful. Humans learn from an early age to classify colors, but aroma classifications are not discussed at a

Handbook of Meat, Poultry and Seafood Quality, Second Edition. Edited by Leo M. L. Nollet.

young age in the fashion that color is (Ishii *et al.* 1997). Early efforts to classify odors served to indicate the complexity of odors (Doty 1991). In 1956, Wenger *et al.* concluded that there was not an adequate classification of odors. Ishii *et al.* (1997) found that judges categorizing odors over several sessions did not reach consistency over eight to ten sessions. Use of the perfume industry's categories did not come naturally, but needed considerable training. But, as the cues were increased, judges began to sort closer and closer to the perfume industry's category system (Ishii *et al.* 1997). Within an industry, such as perfume, wine, beer, and drinking water, classification systems have simplified the description of odors, but classification has not helped with elucidating basic mechanisms of the olfactory system (Doty 1991). There is a classic review on aroma classification (Harper *et al.* 1968), and this subject will not be discussed further in this chapter.

Sensory evaluation is the use of humans to evaluate a product using one or more of the five senses (smell, taste, feel, sight, sound). Descriptive flavor evaluation is a sensory evaluation method that describes individual flavor characters and gives an intensity rating to each descriptor. This chapter will contrast the various methods of descriptive flavor analysis and contrast the methods used to determine descriptions of flavors. Panelists are selected on normal abilities to taste and smell and are trained in the methods before data collection takes place. The selection and training process will not be reviewed in the scope of this chapter, but is available in the references cited for each method.

5.1 DESCRIPTIVE FLAVOR ANALYSIS

The birth of descriptive flavor analysis resulted from the development of a laboratory method of analyzing flavor comparable to the expert taster in industrial quality control, namely in the wine, tea, and coffee industries (Caul 1957). The laboratory panels needed to consist of multiple persons (5–20) and needed to be versatile for work on multiple products that may be dissimilar. Methods for selecting and training new panelists needed to be documented so that panels could continue operation by incorporating new panelists after attrition of original members. The Flavor Profile was developed in the 1940s at Arthur D. Little, Inc (Cambridge, Massachusetts) to meet the growing analytical needs of sensory evaluation in the food industry (and in some cases, nonfood industries) (Cairncross & Sjöström 1950; Caul 1957). This method was distinct from expert tasters in quality control, because it made no direct judgment about consumer acceptance (Stone & Sidel 2004). Subsequently, new methods have emerged from these ideas and their applications have grown substantially (Stone *et al.* 1974, Meilgaard *et al.* 1999). These methods differ in their ideologies about characterizing flavor and will be discussed in this chapter.

5.1.1 The Flavor Profile

The method originated from the concept that flavor consists of identifiable attributes of taste, aroma, flavor, and chemical feeling factors. The attributes were called "character notes" and were labeled with descriptive or associative terms. For example, eggy, rubbery, cabbage-like, or skunky terms were used to describe sulfury notes (Cairncross & Sjöström 1950; Caul 1957). The terms were objective terms like "vanilla" and not subjective like "good" (Keane 1992). When a specific chemical compound was identified to be associated with the character note, it was used as the reference. For example, phenylacetic acid has represented the "horsy" note in beer (Caul 1957). When necessary, a character note could be further characterized, such as chemical-sweet and sugar-sweet (Caul 1957), or localized, such as dry mouth, cheek puckering, tongue coating, or tooth roughing for variations of astringency (Cairncross & Sjöström 1950). During character note development, panelists start with a blank sheet and record all aromas, flavors, tastes, and aftertastes they perceive in the product. Then they discuss and develop an agreed upon list, and duplication of a note under different descriptions is eliminated. It may take two to four sessions to finalize the composite list that completely describes the product and that all panelists agree with (Neilson *et al.* 1988). Reference standards are obtained to represent the notes. It is essential that all panelists understand the scope and connotation of each note in the study. The panel members, who are selected according to their abilities to discriminate odor and flavor differences and communicate their perceptions, develop the character notes (Caul 1957).

Within a sample, the character notes are perceived in a specific order. Adding seasoning to the product can affect the order. For example, monosodium glutamate alters the order of appearance of character notes. In a Flavor Profile, the character notes are listed in the order they appear and any changes in order are noted (Caul 1957). When using standardized evaluation techniques for smelling and tasting, the order of appearance for character notes is influenced by location of taste buds on the tongue, volatility of aromatics, and texture of the product (Keane 1992).

Intensity is noted on a 5-point scale (0=not present, just recognizable, or threshold; 1=slight; 2=moderate; and 3=strong). The panelists evaluate the intensity of the observable character notes individually. Upon completion of all samples, they discuss their scores and arrive at a consensus for the intensity of each note in the sample. The panel session is not complete until all panelists in attendance agree individually and as a unit that the consensus represents the sample (Cairncross & Sjöström 1950; Caul 1957). The Flavor Profile was not designed as a numerical system for statistical analysis of data but as a consensus or composite of profile terms and intensities in the final product (Keane 1992). Some variations have emerged that make the Flavor Profile method statistically friendly (Miller 1978), but these are not as widely accepted as some of the newer descriptive methods.

5.1.1.1 Quantitative descriptive analysis

As the food industry expanded, and moved further away from the expert taster concept, a need arose for data that could be analyzed statistically. Scientists at Stanford Research Institute developed the Quantitative Descriptive Analysis (QDA) method (Stone *et al.* 1974; Stone & Sidel 2004). This method screened panelists with discrimination tests, used descriptors developed by the panel, used repeated evaluations in sensory booths to avoid group judgments, and used analysis of variance and graphing of means to interpret data (Zook & Pearce 1988).

QDA starts with panelists that are selected based on their product usage and familiarity, discrimination ability within the product or product category, and task comprehension. During the training process, panelists develop descriptive language, group the attributes (terms) by modality (appearance, aroma, and so on), list them by occurrence, develop definitions for each attribute, identify helpful references, and familiarize themselves with the scoring procedure. The panelists work individually and as a group to ensure that the attributes are fully understood and that all of the product characteristics are fully accounted for. Within the QDA philosophy, the panels develop their own language, without being influenced by the information given them by a panel leader. The group of panelists must decide to include or exclude an attribute. They are encouraged to use their own words, provided they are common everyday language. The meaning of each word-sensation experience is defined so the entire group will be familiar with the word and the meaning. Numerous attributes/terms are encouraged, because having too few may limit panelists' ability to differentiate product differences. Statistical analysis is used to determine which terms/attributes differentiate the samples. Distinct concepts assigned to each attribute are not considered critical, because a magically correct set of independent descriptors may not fully describe the sensory properties. Panelists also become familiar with the order of occurrence of each sensation. References are not necessary, but useful for training and retraining; and can help panelists relate to a particular sensation that is not easily detected or described. The references should not introduce additional fatigue or significantly increase training time and should only be used during language development and not in the initial sessions (Stone 1992; Stone & Sidel 2004).

Relative intensity is measured with a 6 in. or 15-cm open line scale to provide an infinite number of possible options and to avoid number biases. Word anchors are placed one-half inch or 1.5 cm from each end. Low intensity is on the left end, and high intensity is on the right end. To be more confident about reliable results, multiple replicates are evaluated. For most products, four replicates result in the desired sensitivity. Results are analyzed by analysis of variance. Replication allows for characterizing reliability of response patterns and does not generalize the results to a population (Stone 1992; Stone & Sidel 2004).

5.1.1.2 The spectrum descriptive analysis method

The Spectrum method was developed by G.V. Civille in collaboration with a number of companies that were looking for a method to obtain reproducible and repeatable sensory analysis of products (Meilgaard *et al.* 1999). It is flexible enough to evaluate one to all sensory modalities and to evaluate an array of products, including foods, beverages, personal care, home care, paper, and other products. A Spectrum panel describes accurately and in detail the characteristics/parameters of appearance, flavor, and texture and allows statistical treatment of data (Muñoz & Civille 1992; Meilgaard *et al.* 1999). The Spectrum method utilizes a broad selection of references for describing the sensory attributes and for evaluating intensity (Rutledge & Hudson 1990).

To develop descriptors (terms), trained panelists (after being exposed to underlying principles of appearance, flavor, and texture) start by evaluating a broad array of products (commercial brands, pilot plant runs, etc.) that define the product category (Muñoz & Civille 1992; Meilgaard *et al.* 1999). A few sessions are devoted to this process. Panelists individually develop lists of terms that adequately describe the attributes in the product category. The group discusses their lists and the panel leader, in conjunction with the panelists, records and organizes the terms provided by the individuals. The organization is based on the underlying structure of flavor categories. After a few sessions of this type of activity, the panel is led to establish the list of terms that describe the sensory attributes of the product category that will be evaluated. This process includes using references to help all panelists understand each term in the same way. The list of descriptors should be comprehensive, but not overlapping. Several sessions may be required to complete a ballot (Muñoz & Civille 1992; Meilgaard *et al.* 1999). Panel leaders can glean descriptor terms and definitions from the literature to suggest references to complete the ballot. The panel is frequently reminded that the truth is always in the sample(s). A good source of descriptors is ASTM International's book and diskette titled *Aroma and Flavor Lexicon for Sensory Evaluation: Terms, Definitions, References and Examples* (Civille & Lyon 1996).

Panelists generally use a 0–15-point scale, numbered in tenths, for rating intensity of most flavor descriptors. However, panelists may indicate higher intensities (greater than 15) for very strong stimuli. Panelists may also use a line scale with anchors at each end indicating direction (Muñoz & Civille 1992). Extensive use of intensity reference points and specific reference samples are derived from the collective data from several panels at separate locations over several replicates (Meilgaard *et al.* 1999). Training is never over, since panelists keep adding to their knowledge of the product and evaluating their responses. An internal standard or control that is available from a consistent supply of product can be used as a reference for comparison to experimental samples (Rutledge & Hudson 1990). A variety of uni- and multivariate data analysis techniques are used in the many applications of Spectrum descriptive data.

5.1.1.3 Free-Choice Profiling

Free-Choice Profiling is a method developed by Williams and Arnold (1984) in the United Kingdom that allows untrained panelists to invent and use as many terms as they need to describe the sensory characteristics of the product. The use of the generalized Procrustes analysis method makes it possible to reduce the variation in terminology (Williams & Langron 1984). The centroid of each assessor's data space is rotated, expanded, and shrunk until each assessor is made to match as closely as possible the other assessors, but maintaining each assessor's intersample relationships. A consensus configuration is obtained. These configurations (axes) are then related to external factors (Williams & Langron 1984). All of this depends on the individual assessors' ability to precisely and consistently use the terms they generate. Inexperience and isolation from other panelists results in inadequate vocabulary that may be improved by allowing more time for vocabulary development (Piggott & Watson 1992). Rodriguez *et al.* (2000) found that trained panelists using Free-Choice Profiling were more accurate than untrained panelists. Guerrero *et al.* (1997) demonstrated that trained panelists were more efficient in generating terms than semitrained panelists. Gilbert and Heymann (1995) observed that Free-Choice Profiling allowed for quick generation of descriptive terms without the need for extensive training; however, interpretation of results was not as straightforward as descriptive analysis methods and was not recommended for use with untrained panelists. Heymann (1994) reported that terms generated by sensory-savvy panelists were more likely to discriminate among samples. In the end, the experimenter/sensory analyst decides what the terms for each parameter mean (Meilgaard *et al.* 1999). Therefore, this method has undergone changes that cause it to approximate the other descriptive methods (Stone & Sidel 2003).

All of these methods place emphasis on determining the full range of variation in the product during generation of terms for describing flavor. The Flavor Profile method and profile derivatives (e.g., the Spectrum method) use a language that tends to be more technical in nature (developed by a technically trained group) and provide technical guidance to the product developer. The QDA method takes the perspective that the panel should represent consumers and the language should reflect consumer language yet be descriptive (Muñoz & Bleibaum 2001). All of these methods develop definitions for the generated terms except Free-Choice Profiling. Flavor Profile and Spectrum incorporate extensive use of flavor references for both the terms and intensities. Having definitions for descriptive terms allows others outside the panel to understand what is meant by the terms.

5.1.2 Chemical nature of flavor characteristics

There has been much effort devoted to relating flavor compounds to specific flavors. Basic tastes and chemical feeling factors tend to be less complicated to relate to compounds, but natural aromas can be quite complex. To determine the influence of chemical compounds on flavor, the gas chromatograph has been coupled with an olfactometry (GC-O) device for sniffing the compounds after they are separated from each other. There are three main uses: (1) to determine the odor qualities of single compounds, (2) to detect odor-active compounds in complex flavor extracts, and (3) to quantify the odor contribution of individual compounds in flavor systems (Dattatreya *et al.* 2002). Kim *et al.* (2003) devised a method for determining if the aroma of the combined extracted volatiles resembled the aroma of the original sample. The trapped volatiles were injected (splitless mode) into a 0.5-m column at 200°C and evaluated at the sniffing port. If the effluent aroma resembled the original product's aroma, then the essence of the total aroma is being evaluated. If it does not resemble the product's aroma, then one can conclude that key volatiles were lost during the extraction process (Kim *et al.* 2003). There are extraction issues and sensory issues to keep in mind pertaining to GC-O methods.

5.1.2.1 Isolation and extraction of compounds

Numerous techniques are employed to extract aromatic compounds from the various sample matrices. The extraction technique will influence the volatile profiles. Common methods are solvent extraction, static headspace sampling, solid-phase microextraction, simultaneous steam distillation/extraction, and supercritical fluid extraction. A review by Peppard (1999) describes the methods. Some of these methods work better for GC-O than others. Solvent extraction uses organic solvents such as methylene chloride, various pentane/diethyl ether mixtures, or other solvents to extract the flavor compounds from the food or beverage sample. Water-soluble compounds may be incompletely extracted. Flavor compounds are typically present in small amounts and require solvent removal for detection. Concentration of the extract prior to injection may result in loss of some volatiles. Extraction of volatiles from solid particles can be time consuming. Static headspace involves equilibrating the headspace above a sample in a sealed vial under controlled conditions and sampling with a syringe. Headspace sampling is effective for highly volatile flavor compounds. Compounds with low volatility pose a problem and quantitation can be complicated. Dynamic headspace sampling involves passing an inert gas such as helium through or over a sample, then trapping on a charcoal or a Tenax® trap. Volatiles are eluted from the trap with a solvent or desorbed with heat. This is an excellent qualitative method of extracting flavor compounds having a wide range of volatility from difficult-to-extract matrices, but reliable quantitation is difficult due to incomplete extraction. Solid-phase microextraction (SPME) uses a short length of fused silica fiber coated with a thin layer of absorptive material. Several types of coated fiber are available with differing selectivities. SPME can be used in static headspace or the fiber can be inserted into a liquid substance. This extraction method and the subsequent gas chromatography can detect flavor compounds at lower concentrations than traditional static headspace sampling. It is simple, rapid, and requires no solvents. Selectivity depends on the fiber coating. Reliable quantification requires appropriate and consistent care. Simultaneous steam distillation/extraction (Likens–Nickerson) is a steam distillation method with simultaneous extraction of refluxing vapors by a refluxing solvent, such as methylene chloride. It is used with samples that are high in fat or other solvent extractable material. This method is time consuming and requires specific glassware. Only steam distillable volatiles are extracted. High and low volatile and polar compounds are poorly trapped. Extracts tend to contain artifacts due to oxidation, thermal breakdown, and hydrolysis. Supercritical fluid extraction is a powerful method of extracting volatiles from dry, finely ground samples. Liquid or wet samples can be used if the sample is immobilized. Carbon dioxide (CO_2) is pressurized sufficiently to be in the supercritical state so that it has the extracting capabilities of an organic solvent. Adjusting the temperature and/or pressure can change the selectivity by controlling the density of the CO_2. Using adsorbents and CO_2 modifiers, such as methanol, enhance selectivity and aid in capturing flavor compounds (Peppard 1999).

After preparation, samples should be analyzed as soon as possible. Cold on-column injection can avoid thermal induced decomposition of some compounds. Certain thiols dimerize in diethyl ether upon refrigerated storage. Therefore, storage of concentrate should be in pentane at 230°C under inert gas (Blank 1997).

5.1.2.2 Sensory issues with GC-O

According to Sell (2000), a majority of the published odor descriptions of compounds were done on mixtures rather than on single compounds. These mixtures include trace impurities, unresolved peaks on chromatograms, the presence of isomers, etc. Historical data need to be substantiated with recent technologies in chiral capillary gas chromatography coupled with olfactometry to confirm descriptions of compounds (Sell 2000).

Threshold concentration varies among compounds, sometimes as much as 11 orders of magnitude (Weyerstahl 1994). Within a single compound, subjects' relative sensitivities to enantiomers (e.g., b-ionone) diverge widely. For example, some panelists were more sensitive to (1)2 than to (2)-b-ionone and vice versa. The threshold ratios can vary by four orders of magnitude between the two enantiomers (Weyerstahl 1994). These ranges differ from compound to compound.

There are several factors affecting perception other than concentration. Humidity can affect perception and consequently should be controlled. Adaptation (or fatigue) can affect perception. Cross adaptation can affect perception of odor for succeeding compounds (Sell 2000). Time between breaths that may be longer than the compound elution can result in incomplete experiences (Hanaoka *et al.* 2001). Subjects that breathe more rapidly detected odors more often than subjects that breathe slower (Hanaoka *et al.* 2001). The dilution at which the GC-O is carried out affects the sensory detection of compounds, which could be improved via running multiple concentrations for each sample (Ferreira *et al.* 2001). Enantiomeric form can affect flavor. Carvone is a classic example, the (+)–5 carvone has a typical caraway odor and the (−)–5 form has a spearmint odor. Other compounds, such as camphor, have both enantiomeric forms smelling alike (Ohloff 1986).

5.1.2.3 Intensity rating methods for GC-O

In 1957, Patton and Josephson proposed the idea of odor activity value (OAV). This occurred soon after the introduction of the gas chromatograph. The idea was to relate concentration of a compound to its sensory threshold. This early work demonstrated that the most abundant volatiles may have little if any odor significance in a food. These techniques have been invaluable for determining off-odors in foods. Unfortunately, the original ideology has resulted in limited success in flavor duplication (Mistry *et al.* 1997). From this early work, four methods have emerged: dilution analysis method, that is, Aroma Extract Dilution Analysis (AEDA) (Ullrich & Grosch 1987); Charm Analysis (Acree *et al.* 1984); the detection frequency method (Linssen *et al.* 1993); and Osme (McDaniel *et al.* 1990).

The AEDA techniques utilize stepwise diluted extracts of volatile compounds, which reveal the most intense flavor compounds in the extract. This results in D-values (lowest dilution at which a substance is still smelled) that are proportional to the aroma values (Ullrich & Grosch 1987). In Charm Analysis, the beginning and end of each odor are recorded. It uses computerized data collection and a sensory procedure based on odor-detection thresholds rather than psychological estimations of stimulus intensity. The dimensionless measure of odor intensity is called charm (Acree *et al.* 1984; van Ruth & O'Conner 2001a). The charm response chromatogram is essentially a plot of a function of the dilution factor and number of responses at a given dilution factor versus retention index (Acree *et al.* 1984). The detection frequency method monitors nasal impact frequency. Six to eight assessors sniff the headspace of a single dilution of a representative sample. The time it takes the aroma to come through the port is monitored. Nasal impact frequency of 100% means the odorant was detected by all the assessors. Peak intensities are not related to aroma intensities, but to their detection frequencies (Pollien *et al.* 1997).

Osme is a quantitative bioassay method used to measure the perceived odor intensity of a compound eluting from a GC olfactometer. The subject rates the intensity of a compound's odor by using a time-intensity device, providing an odor peak similar to a chromatogram. Verbal descriptions are given for each peak recorded (Miranda-Lopez *et al.* 1992).

A comparison of three ideologies was done by van Ruth and O'Conner (2001a, b). They found similar results with Osme (intensity) and nasal impact frequency (detection frequency). The charm (dilution) method correlated only slightly with the other two methods. This may have been due to large dilution steps used to cover the range of thresholds for the mixture of compounds. They did find a large variability between assessors' thresholds. Charm also required numerous GC runs. In the intensity method (Osme), large quantities of volatile compounds do not necessarily relate to high odor intensities, as this property is due to a difference in thresholds and differences in intensity–concentration relationships. Detection frequencies are limited to the concentration at which the maximum frequency possible is attained and measured by the number of assessors that perceive the aroma. Dilution methods have a large variability due to differences in thresholds (up to a factor of 200) and response criteria of assessors. van Ruth and O'Conner (2001a) recommend using a number of assessors and not relying on one or two judges to achieve reliable GC-O analysis independent of the method used. Both assessors' qualities (e.g., alertness) and analytical conditions clearly influenced the GC-O data and, therefore, require optimization (van Ruth & O'Conner 2001b).

There are inherent problems that occur when panelists try to evaluate aromas that result from a single compound whether it be in a food system or from a GC-O run. In some work done on catfish samples spiked with geosmin or 2-methylisoborneol, the author and colleagues (Bett & Johnsen 1996) found that trained descriptive panelists could distinguish intensity differences in concentration of 2-methyliso-borneol within a sensory session, but could not determine intensity differences between sessions on different days. Meanwhile, they were able to evaluate flavor intensities made up of multiple compounds, such as the chickeny flavor of catfish. These authors, Bett and Dionigi (1997), have also reported that panelists had greater standard deviations when evaluating the off-flavor from one compound (geosmin or 2-methyliosborneol) than they had when evaluating flavors containing multiple compounds, such as chickeny and nutty descriptors. When the author worked with sensory panelists to rate intensities of compounds coming through the GC-O, the same phenomena was observed (unpublished data). We concluded that measuring intensity of a single compound is difficult. Detection thresholds vary from day to day and among panelists. Adaptation, fatigue, and enhancement or suppression by other factors affects repeatability. We have determined that using GC-O is most suited for two purposes: (1) determining compounds with aroma impact and (2) describing the aromas of individual compounds eluting from the GC-O.

5.1.2.4 Relating the compounds of GC-O to food aroma

One purpose of GC-O is to distinguish odor-active compounds from those volatiles without odor impact. GC-O allows one to determine the odor-significant compounds so the scientist can focus on these compounds and can work on their chemical origin or olfactory differences. Correlating GC-O responses to descriptive sensory responses has its limitations. Threshold concentration does not necessarily correlate with aroma intensity. Thresholds can vary with experimental conditions, and the intensity function changes due to concentration differences among volatile compounds. Ideally, intensity ratings of GC-O compounds would correlate better with descriptive aroma (Blank 1997). Kamath *et al.* (2001) compared descriptive analysis of essential oils with GC-O evaluations and found that many descriptors developed in descriptive analysis did not overlap with GC-O descriptors. Generally, only three or four descriptors (out of six to seven descriptive analysis descriptors) were used by both the descriptor panel and the GC-O panel. Meanwhile, more descriptors were generated by the GC-O panel than the descriptive analysis panel (Kamath *et al.* 2001). van Ruth and Roozen (1994) found 10 similar descriptors between GC-O and sensory analysis in rehydrated bell peppers. Meanwhile, there were still 10 GC-O descriptions that did not relate to sensory analysis descriptors (van Ruth & Roozen 1994). Le Fur *et al.* (2003) developed nine aroma categories for French Chardonnay wines. Descriptive panelists and GC-O panelists described aromas for six wines, and all the terms were placed in the nine categories. Correspondence Analysis was used to plot the two sets of data (six samples and nine descriptors). Generalized Procrustes Analysis was used for comparison of the two spatial configurations (six wines and nine descriptor classes). They identified major aroma compounds but could not account for the distinctive sensory profiles of the wines. The deficiency occurred when the descriptive panel observed a spicy character in each wine that was not characterized in the GC-O analysis. Compounds eluted that had an herbaceous character in the GC-O analysis but no herbaceous character was described by the descriptive analysis panel. Therefore, a relationship between herbaceous smelling compounds and the flavor descriptors could not be determined (Le Fur *et al.* 2003).

GC-O has advanced the understanding of compounds that contribute to flavor, but it has not yet made it possible to determine the chemical make-up of common food flavors. Some off-flavors, consisting of few compounds have had their chemical composition identified, but deciphering flavors such as peanutty flavor has not been successful.

In the last half of the twentieth century, many strides were made in understanding flavor characterization. The developments in descriptive sensory methods and advances in the GC-O methods have advanced the understanding of flavor and aroma. Descriptive sensory analysis has made it possible to characterize the many flavor attributes in a product or product line. Flavor-intensity rating methods have made it possible to compare flavor differences due to various treatments, ingredients, and processes. GC-O has made it possible to determine aroma active compounds eluting from the column. It has helped with identification of compounds that make up some flavors, but has shown more success in the identification of off-flavor compounds than in determining compounds responsible for desirable flavors.

REFERENCES

Acree, T.E., Barnard, J. & Cunningham, D.G. (1984) A procedure for the sensory analysis of gas chromatographic effluents. *Food Chemistry*, **14**, 273–286.

Amerine, M.A., Pangborn, R.M. & Roessler, E.B. (1965) *Principles of Sensory Evaluation*, p. 602. Academic Press, New York.

Bett, K.L. & Dionigi, C.P. (1997) Detecting seafood off-flavors: Limitation of sensory evaluation. *Food Technology*, **51** (8), 70–72, 74, 76, 79.

Bett, K.L. & Johnsen, P.B. (1996) Challenges of evaluating sensory attributes in the presence of off-flavors (geosmin and 2-methylisoborneol). *Journal of Sensory Studies*, **11**, 1–17.

Blank, I. (1997) Gas chromatography-olfactometry in food aroma analysis. In: *Techniques for Analyzing Food Aroma* (ed. R. Marsili), pp. 293–329. Marcel Dekker, Inc., New York.

Cairncross, S.E. & Sjöström, L.B. (1950) Flavor profiles—A new approach to flavor problems. *Food Technology*, **4**, 308–311.

Caul, J.F. (1957) The profile method of flavor analysis. *Advances in Food Research*, **7**, 1–40.

Civille, G.V. & Lyon, B.G. (1996) *Aroma and Flavor Lexicon for Sensory Evaluation: Terms, Definitions, References and Examples*, p. 158. *ASTM Data Series: DS66*. ASTM International, West Conshohocken, Pennsylvania.

Dattatreya, B.S., Kamath, A. & Bhat, K.K. (2002) Developments and challenges in flavor perception and measurement—A review. *Food Reviews International*, **18**, 223–242.

Doty, R.L. (1991) Psychophysical measurement of odor perception in humans. In: *The Human Sense of Smell* (eds D.G. Laing, R.L. Doty & W. Breipohl), pp. 95–134. Springer-Verlag, Berlin, Germany.

Ferreira, V., Aznar, M., López, R. & Cacho, J. (2001) Quantitative gas chromatography-olfactometry carried out at different dilutions of an extract. Key differences in the odor profiles of four high-quality Spanish aged red wines. *Journal of Agricultural and Food Chemistry*, **49**, 4818–4824.

Gilbert, J.M. & Heymann, H. (1995) Comparison of four sensory methodologies as alternatives to descriptive analysis for the evaluation of apple essence aroma. *Food Technology*, **24** (4), 28–32.

Guerrero, L., Gou, P. & Arnau, J. (1997) Descriptive analysis of toasted almonds: A comparison between expert and semi-trained assessors. *Journal of Sensory Studies*, **12**, 39–54.

Hanaoka, K., Vallet, N., Giampaloi, P., Heyd, B. & MacLeod, P. (2001) Possible influence of breathing on detection frequency and intensity rating in gas chromatography-olfactometry. *Food Chemistry*, **72**, 97–103.

Harper, R., Bate-Smith, E.C. & Land, D.G. (1968) *Odour Description and Odour Classification*, p. 191. American Elsevier, New York.

Heymann, H. (1994) A comparison of free choice profiling and multidimensional scaling of vanilla samples. *Journal of Sensory Studies*, **9**, 445–453.

Ishii, R., Kemp, S.E., Gilbert, A.N. & O'Mahony, M. (1997) Variation in sensory conceptual structure: An investigation involving the sorting of odor stimuli. *Journal of Sensory Studies*, **12**, 195–214.

Kamath, A., Asha, M.R., Ravi, R., Narasimhan, S. & Rajalakshmi, D. (2001) Comparative study of odour and GC-olfactometic profiles of selected essential oils. *Flavour and Fragrance Journal*, **16**, 401–407.

Keane, P. (1992) The flavor profile. In: *Manual on Descriptive Analysis Testing for Sensory Evaluation*(ed. R.C. Hootman), pp. 5–14. ASTM Manual Series: MNL 13. ASTM, Philadelphia, Pennsylvania.

Kim, T.H., Lee, S.M., Kim, Y.S., Kim, K.H., Oh, S. & Lee, H.J. (2003) Aroma dilution method using GC injector split ratio for volatile compounds extracted by headspace solid phase microextraction. *Food Chemistry*, **83**, 151–158.

Le Fur, Y., Mercurio, V., Moio, L., Blanquet, J. & Meunier, J.M. (2003) A new approach to examine the relationships between sensory and gas chromatography-olfactometry data using generalized Procrustes analysis applied to six French chardonnay wines. *Journal of Agricultural and Food Chemistry*, **51**, 443–452.

Linssen, J.P.H., Janssens, J.L.G.M., Roozen, J.P. & Posthumus, M.A. (1993) Combined gas chromatography and sniffing port analysis of volatile compounds in mineral water packed in laminated packages. *Food Chemistry*, **46**, 367–371.

McDaniel, M.R., Miranda-Lopez, R., Watson, B.T., Micheals, N.J. & Libbey, L.M. (1990) Pinot Noir aroma: A sensory/gas chromatographic approach. In: *Flavor and Off-Flavors, Proceedings of the 6th International Flavor Conference* (ed. G. Charalambous), pp. 23–36. Elsevier Science Publishers BV, Amsterdam, the Netherlands.

Meilgaard, M., Civille, G.V. & Carr, B.T. (1999) *Sensory Evaluation Techniques*, 3rd edn., p. 387. CRC Press, Boca Raton, Florida.

Miller, I. (1978) Statistical treatment of flavor data. In: *Flavor: Its Chemical, Behavioral and Commercial Aspects, Proceedings of the Arthur D. Little, Inc., Flavor Symposium* (ed. C. Apt), p. 229. Westview Press, Boulder, Colorado.

Miranda-Lopez, R., Libbey, L.M., Watson, B.T. & McDaniel, M.R. (1992) Odor analysis of Pinot Noir wines from grapes of different maturities by a gas chromatography-olfactometry technique (Osme). *Journal of Food Science*, **57**, 985–993, 1019.

Mistry, B.S., Reineccius, T. & Olson, L.K. (1997) Gas Chromatography-Olfactometry for the determination of key odorants in foods. In: *Techniques for Analyzing Food Aroma* (ed. R. Marsili), pp. 265–292. Marcel Dekker, Inc., New York.

Muñoz, A.M. & Bleibaum, R.N. (2001) Fundamental descriptive analysis techniques. Exploration of their origins, differences and controversies. *Workshop Presented at "2001: A Sense Odyssey" 4th Pangborn Sensory Science Symposium*, Dijon, France.

Muñoz, A.M. & Civille, G.V. (1992) The Spectrum descriptive analysis method. In: *Manual on Descriptive Analysis Testing for Sensory Evaluation* (ed. R.C. Hootman), pp. 22–34. MNL 13. ASTM, Philadelphia, Pennsylvania.

Neilson, A.J., Ferguson, V.B. & Kendall, D.A. (1988) Profile method: Flavor profile and profile attribute analysis. In: *Applied Sensory Analysis of Foods* (ed H. Moskowitz), Vol. **1**, pp. 21–41. CRC Press, Inc., Boca Raton, Florida.

Ohloff, G. (1986) Chemistry of odor stimuli. *Experientia*, **42**, 271–279.

Patton, S. & Josephson, D.V. (1957) A method for determining significance of volatile flavor compounds in foods. *Food Research*, **22**, 316–318.

Peppard, T.L. (1999) How chemical analysis supports flavor creation. *Food Technology*, **53** (3), 46–51.

Piggott, J.R. & Watson, M.P. (1992) A comparison of free-choice profiling and the repertory grid method in the flavor profiling of cider. *Journal of Sensory Studies*, **7**, 133–145.

Pollien, P., Ott, A., Montigon, F., Baumgartner, M., Moñoz-Box, R. & Chaintreau, A. (1997) Hyphenated headspace-gas chromatography-sniffing technique screening of impact odorants and quantitative aromagram comparisons. *Journal of Agricultural and Food Chemistry*, **45**, 2630–2637.

Rodriguez, I., Jorge, M.C. & Beltran, C. (2000) Comparison of the free profile with the conventional profile for sensory analysis of chocolate bars. *Alimentaria*, **315**, 67–72.

van Ruth, S.M. & O'Conner, C.H. (2001a) Evaluation of three gas chromatography-olfactometry methods: Comparison of odour intensity-concentration relationships of eight volatile compounds with sensory headspace data. *Food Chemistry*, **74**, 341–347.

van Ruth, S.M. & O'Conner, C.H. (2001b) Influence of assessors' qualities and analytical conditions on gas chromatography-olfactometry analysis. *European Food Research and Technology*, **213**, 77–82.

van Ruth, S.M. & Roozen, J.P. (1994) Gas chromatography/sniffing port analysis and sensory evaluation of commercially dried bell peppers after rehydration. *Food Chemistry*, **51**, 165–170.

Rutledge, K.P. & Hudson, J.M. (1990) Sensory evaluation: Method for establishing and training a descriptive flavor analysis panel. *Food Technology*, **44** (12), 78–84.

Sell, C. (2000) Structure/odor correlations: The mechanism of olfaction and the design of novel fragrance ingredients. *Perfumer Flavorist*, **25**, 67–73.

Stone, H. (1992) Quantitative descriptive analysis (QDA). In: *Manual on Descriptive Analysis Testing for Sensory Evaluation* (ed. R.C. Hootman), pp. 15–21. MNL 13. ASTM, Philadelphia, Pennsylvania.

Stone, H. & Sidel, J.L. (eds) (2003) Descriptive analysis. In: *Encyclopedia of Food Science*, 2nd edn., pp. 5152–5161. Academic Press, London, U.K.

Stone, H. and Sidel, J.L. (2004) *Sensory Evaluation Practices*, 3rd edn., p. 408. Elsevier Science and Technology Books, New York.

Stone, H., Sidel, J., Oliver, S., Woolsey, A. & Singleton, R.C. (1974) Sensory evaluation by quantitative descriptive analysis. *Food Technology*, **28** (1), 24, 26, 28, 29, 32, 34.

Ullrich, F. & Grosch, W. (1987) Identification of the most intense volatile flavor compounds formed during autoxidation of linoleic acid. *Zeitschrift für Lebensmittel- Untersuchung und -Forschung*, **184**, 277–282.

Wenger, G.R., Jones, F.N. & Jones, M.H. (1956) *Physiological Psychology*, p. 472. Holt, New York.

Weyerstahl, P. (1994) Odor and structure. *Journal für Praktische Chemie*, **336**, 95–109.

Williams, A.A. & Arnold, G.M. (1984) A new approach to sensory analysis of foods and beverages. In: *Progress in Flavour Research, Proceedings of 4th Weurman Flavour Research Symposium* (ed. J. Adda), pp. 35–50. Elsevier, Amsterdam, the Netherlands.

Williams, A.A. & Langron, S.P. (1984) The use of free-choice profiling for the evaluation of commercial ports. *Journal of the Science of Food and Agriculture*, **35**, 558–568.

Zook, K.L. & Pearce, J.H. (1988) Quantitative descriptive analysis. In: *Applied Sensory Analysis of Foods* (ed. H. Moskowitz), Vol. **I**, pp. 43–71. CRC Press, Inc., Boca Raton, Florida.

6 Chemical Characterization

Neil C. Da Costa and Sanja Eri

Abstract: Methods for isolation of flavor compounds include extraction with a solvent, distillation, or headspace collection. Several techniques can be employed to determine the chemical composition of flavor isolates: mass spectrometry (MS), infrared spectroscopy (IR), and nuclear magnetic resonance spectroscopy. For flavor in mixtures, MS and IR have been coupled with separatory techniques such as gas chromatography and liquid chromatography.

Keywords: GC, GC-O, chiral separations, multidimensional GC, GC-MS, EI, CI, MS-MS, HPLC, IR, GC-IR, NMR

6.1 INTRODUCTION

Because of the complexity of food matrices and the low concentration of flavor compounds within them, to characterize the flavor of a particular food, one has to first isolate and/or concentrate flavor compounds. Methods for isolation of flavor compounds include extraction with a solvent, distillation, headspace collection, etc. (Da Costa & Eri 2004). When using these methods, it is important to keep in mind that instrumental analysis will provide a true picture about flavor profile of a food only if the flavor of the isolate/concentrate resembles that of the starting material.

After a flavor has been isolated and concentrated, it is ready for chemical characterization. Several techniques can be employed to determine the chemical composition of flavor isolates. They include mass spectrometry (MS), infrared spectroscopy (IR), and nuclear magnetic resonance spectroscopy (NMR). These principles can be used to identify the structure of most pure organic molecules. However, in flavor work, we are invariably dealing with chemical mixtures. For that reason, MS and IR have been coupled with separatory techniques, namely gas chromatography (GC) and liquid chromatography (LC), to form a powerful tool in dissecting and characterizing these mixtures. This chapter will outline the main techniques available to the analyst for characterizing flavor chemicals.

6.2 GAS CHROMATOGRAPHY

6.2.1 Principle

The flavor of a food can contain from several hundred to over 1000 chemicals. To obtain structural information about each flavor chemical in a mixture, they need to be separated. Gas chromatography has been successfully used for separation of volatile flavor compounds for

Handbook of Meat, Poultry and Seafood Quality, Second Edition. Edited by Leo M. L. Nollet.

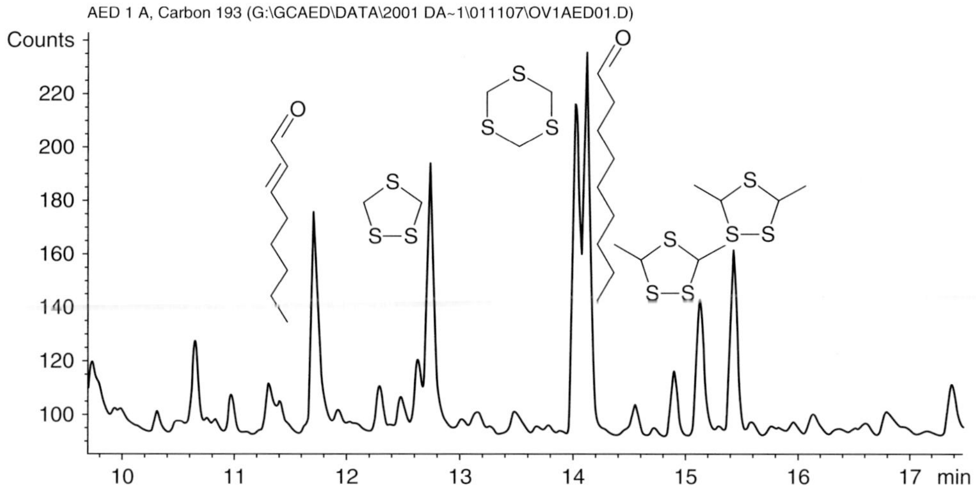

Fig. 6.1 GC-FID chromatogram of fried shrimp steam distillate (from left to right: *trans*-2-octenal, 1,2,4-trithiane, 1,3,5-trithiolane, nonanal, 3,5-dimethyl-1,2,4-trithiane I and II).

decades. In this technique, flavor chemicals are introduced into a gas chromatograph through the injection port and carried by a carrier gas such as hydrogen, helium, or nitrogen through the analytical column to the detector. The GC column contains a polymeric stationary phase that absorbs and subsequently releases flavor compounds. The carrier gas is inert so it is not absorbed by the column, and it does not react with the flavor volatiles. Compounds entering a gas chromatograph get separated due to differences in partitioning behavior between the mobile phase (carrier gas) and stationary phase. Depending on the concentration and thermal stability of the sample, different types of GC introduction systems can be used (e.g., split/splitless injector, on-column injection system, programmed temperature injector). Injection volumes usually range from 0.1 to 5 microliters (mL), depending on the concentration of the flavor sample. Due to different affinities of flavor compounds for the column's stationary phase, some compounds are retained longer than others. When compounds get released from the column, they are carried by the carrier gas toward the detector. Once they reach the detector, a signal is generated and their presence is recorded in a form of a "peak" whose area is proportional to the amount of the compound in the sample. The data are represented as a chromatogram, which is an x, y plot where the x-axis represents retention time (minutes) and the y-axis represents intensity of the signal (Fig. 6.1). The detection of a compound can be based on several different principles; hence, there are several types of detectors.

6.2.2 Detectors

Due to its universality and low limit of detection (minimum detection level 5 picograms [pg] of Carbon/second [C/s]), the most common detector used for flavor analysis is the Flame Ionization Detector (FID). As its name says, this detector responds to molecules that ionize in an air-hydrogen flame. The sensitivity of detection by FID is dependent on the number of carbon atoms—the more carbon atoms in the molecule of the compound, the better its detection. To compensate for the difference in detection, response factors that take into account carbon atoms and molecular weight of the molecule can be used. The disadvantage of FID is that compounds that contain large numbers of sulfur, nitrogen, or oxygen atoms can be undetected. This is especially important in the case of sulfur- and

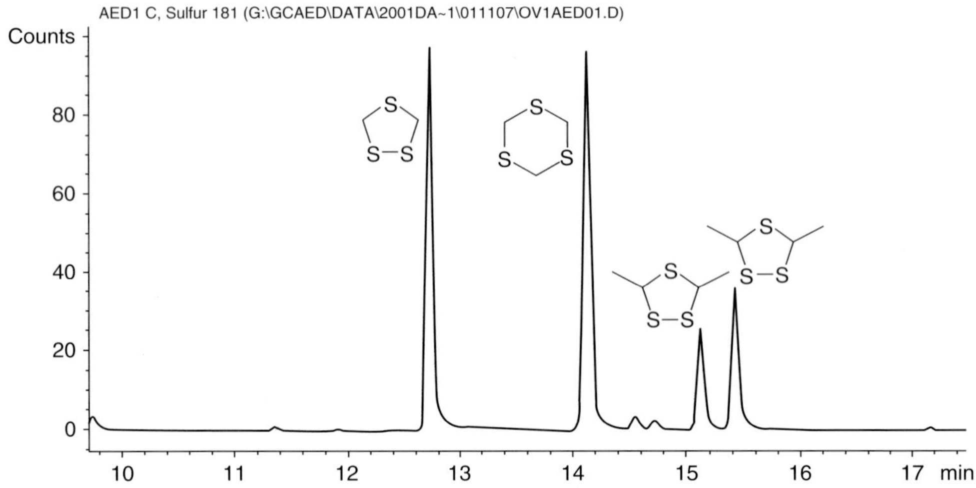

Fig. 6.2 GC-AED chromatogram of fried shrimp steam distillate (from left to right: 1,2,4-trithiane, 1,3,5-trithiolane, 3,5-dimethyl-1,2,4-trithiane I and II).

nitrogen-containing compounds that are among the most powerful flavor chemicals and are often present in trace amounts in foods. Fortunately, there are several selective detectors that can be employed for detection of sulfur and nitrogen compounds. Since these detectors can be tuned so that they respond only to sulfur or nitrogen atoms, they can also provide valuable information toward identification of compounds in a mixture (Westmoreland & Rhodes 1989; Mussinan 1993).

Most commonly used for sulfur compound detection (such as thiophenes, mercaptans, thiols, thioesters) are the Flame Photometric Detector (FPD) (detection limit 20 pg of sulfur[S]/s and 0.9 pg of phosphorus [P]/s), Sulfur Chemiluminescence Detector (SCD), and Atomic Emission Detector (AED) (detection limit 0.1 pg to 1 nanogram [ng]). The most common detectors employed for nitrogen compounds (such as pyrazines, pyridines, pyrroles, amides, amines) are Nitrogen-Phosphorus Detector (NPD) (detection limit 0.4 pg of nitrogen [N]/s, and 0.2 pg of P/s) and AED. Since an AED can be tuned for any element, it can also be used for selectively detecting compounds that contain oxygen (alcohols, aldehydes, etc.) (Mussinan 1993). Figs 6.1 and 6.2 show chromatograms of fried shrimp steam distillate obtained by GC-FID and GC-AED, respectively.

6.2.3 Gas chromatography-olfactometry

Soon after the invention of gas chromatography, flavor chemists started "sniffing" the effluents coming out from a GC column. This technique, called gas chromatography-olfactometry (GC-O), is still a very valuable tool in the identification of flavor molecules. In a standard setup, a gas chromatograph is fitted with an FID and an odor port. The end of the GC column is attached to a "splitter" or Y-connector, which splits the effluent between the FID and odor port. The odor port is also connected to a small water reservoir to humidify the effluent and prevent drying out of nasal passages. In the past, the sample was passed through a thermal conductivity detector (TCD), which was nondestructive and allowed 100% of the effluent to pass through to the odor port. However, this detector was not as sensitive as the destructive FID.

An experienced flavorist sits by the GC and sniffs the carrier gas as it flows from the column. The odor of each component is recorded and makes the basis for creating an "aromagram." At the same time, the FID records a regular chromatogram that can be compared with the "aromagram" to identify areas that contain odor-active compounds. If the results of a GC-O analysis are used in conjunction with GC-MS analysis, the analyst can focus further on identifying the chemicals that are odor-active. The GC-O technique has also been used to identify odor thresholds of flavor compounds by analyzing a series of dilutions (Drawert & Christoph 1984; Leland *et al.* 2001; van Ruth 2001).

6.2.4 Retention times

So far, we have shown that gas chromatography can be used for the separation and detection of volatile compounds. However, gas chromatography can also provide valuable information toward the identification of flavor compounds. Upon introduction to a GC, volatiles spend a finite period of time on the GC column before being detected. If the conditions under which the GC is operated remain constant, the retention time of a compound will remain constant as well. The retention time can therefore be used toward identification of a compound. If a standard of the suspected compound is available, it can be analyzed under the same conditions and should elute at the same time as the unknown. Due to the differences between individual GC instruments and also the length of a GC column (if a GC column gets moved from instrument to instrument, it may be cut a little bit each time), this retention time can be slightly different even if the GC conditions are the same. To overcome that, relative retention times (retention indices) that relate the retention time of a compound to that of a standard compound or a series of standard compounds have been employed. Most commonly used for that purpose is an n-paraffin standard mixture, which is used for calculation of Kovats indices (Kovats 1958), and McReynolds constants (McReynolds 1966; Jennings & Shibamoto 1980). Another example of the standard mixture is an ethyl esters mixture (Mussinan 1993).

As mentioned before, a GC column retains volatiles based on their affinity for the column's polymeric stationary phase. Stationary phases can have different functionalities. Most commonly used for flavor work are methylsilicone-based, nonpolar phases, such as OV-1, and polyethylene glycol-based, polar phases such as Carbowax. The basis for separation on these two phases is quite different. While a nonpolar phase separates volatiles based primarily on their boiling point (the highest boiling compounds elute last), volatiles are separated on polar columns primarily according to their polarity (the most polar compounds elute last). To make use of these two different separation principles, it is standard practice to analyze flavor samples on both columns. That way, co-elutions on a nonpolar phase can be resolved on a polar phase and vice versa. For example, polar compounds such as acids do not elute well on nonpolar columns; they tend to broaden and cover the surrounding peaks. On polar columns, acids form nice, sharp peaks. Moreover, the peaks that were in the vicinity of the acid in question on a nonpolar column most probably elute in a completely different place on a polar column. The number of detected compounds is often much higher when two analyses are combined as compared to the analysis obtained by using only one phase.

With a few exceptions, the standard retention index of a particular compound is rarely the same for polar and nonpolar columns. For identification purposes, both values are used. To unambiguously identify an unknown compound, in addition to mass spectrometric identification, both retention indices have to match that of a standard. The retention indices can be very helpful in cases where two different compounds have very similar mass spectra or in the case of *cis* and *trans* isomers. The retention index then provides crucial evidence for confirmation of the identity of an unknown.

Another example of using GC for identification purposes would be if we already suspect the identity of the compound in a sample. In this case, we can "spike" the sample with the standard of that compound and reanalyze it. If the peak area of the unknown becomes larger on polar and nonpolar columns, there is a pretty good chance that the compound in question is indeed the one we suspected.

6.2.5 Chiral separations

Chiral compounds are compounds whose molecules contain one or more chiral carbon atoms (i.e., a carbon atom connected to four different substituents). For each chiral carbon atom present, chiral compounds have two forms (enantiomers), which differ in the direction in which they rotate plane-polarized light and are nonsuperimposable mirror images of each other.

The importance of chiral compounds in flavor work lies in the fact that enantiomeric pairs can have different organoleptic properties. The best example is carvone, where d-carvone smells like caraway and l-carvone smells like spearmint. It is also possible that enantiomers have the same odor description but different strengths (Pickenhagen 1989).

The ability to resolve enantiomers of a particular chemical can be very helpful in proving authenticity. Most natural materials contain mostly one enantiomer of the chiral compound while synthetic chemicals generally contain a racemic mixture comprising of equal amounts of d- and l-isomers.

Two enantiomers of the same compound elute at the same time on a normal GC capillary column. To resolve them, a special chiral column is needed. It is also possible to make diastereoisomeric derivatives that can be analyzed by achiral columns (Konig 1984; Allenmark 1988). Other nonchromatographic methods that can be used for resolving enantiomers are optical rotation measurements, differential scanning calorimetry, isotope dilution, and NMR spectroscopy (Mussinan 1993).

6.2.6 Multidimensional gas chromatography

If the mixture contains many components that are hard to separate, or many trace components, an analyst may use a technique called multidimensional gas chromatography (MDGC) or GC-GC (Schomburg *et al.* 1984; Cronin & Caplan 1987). In this technique, the GC contains two columns, which usually differ in polarity. Selected fractions from the first column are directed onto the second column (a technique called "heartcutting"), which can provide better resolution of the compounds within that fraction. Also, if multiple injections are performed, trace components can be significantly enriched on a second column and thus the possibility for their identification by MS increased.

A new version of the multidimensional gas chromatography, called comprehensive two-dimensional gas chromatography (GC3GC), has recently been developed. With this technique, instead of transferring only specific fractions of the chromatogram from the first to the second column, the entire effluent from the first column is reanalyzed by the second, short GC column (second "dimension" is fast GC). The interface between the two columns is a modulator that increases the amplitude of the signal that goes from the first column into the second one. The modulator traps and releases signal in portions. Each fraction is analyzed on the second column separately. By this approach, peak capacity becomes significantly increased and a better picture of complex mixtures can be obtained (Dimandja 2003; Dimandja *et al.* 2003).

6.2.7 Preparative gas chromatography

In addition to MS and IR, it is often the case that NMR analysis is needed to confirm the structure of the compound. For NMR analysis, the compound needs to be as pure as possible. Unfortunately, the amount needed for the analysis is usually many times more than the amount present in an extract.

If an unknown component is found at a significant concentration within an extract, and it is well separated from neighboring compounds, it may be possible to use preparatory gas chromatography to isolate enough of a pure component to perform a proton NMR. The preparatory process involves injecting large amounts of a sample into a GC column and collecting the fraction of interest from multiple runs. In a typical setup, the column effluent is split in such a way that one part goes to the FID and the other part can be collected for further analysis. The analyst can monitor the GC display indicating the peak elutions and attach a glass capillary tube (with some sort of cooling involved) to the column outlet at the point at which the unknown elutes. After the unknown stops eluting, the tube is removed from the outlet. Repeating this exercise several times in the same glass tube would usually yield enough sample to rinse with deuterated solvent into an NMR tube and hence determine the structure. This manual process has been made easier by automated preparatory-GC instruments, which can collect the same fraction from many runs with great reproducibility. For other aspects of gas chromatography not described in this section, the reader is encouraged to choose among many books and papers describing GC methods and applications (Grob 1985; Mosandl 1992).

6.3 GAS CHROMATOGRAPHY-MASS SPECTROMETRY

6.3.1 Principle

In gas chromatography-mass spectrometry (GC-MS), compounds are separated on a GC column and subsequently introduced into a mass spectrometer. A mass spectrometer consists of three parts: an ion source, a mass analyzer, and an ion detector. Upon introduction to the mass spectrometer, molecules enter an ion source. There, due to the input of energy, the molecules are fragmented into ions. Each ion has a mass and a charge and therefore its characteristic mass-to-charge ratio (m/z). Based on their mass-to-charge ratios, generated ionic fragments are separated by the mass analyzer and focused toward the ion detector. Several types of mass analyzers available nowadays (i.e., quadrupole, time of flight, ion trap, double-focusing magnetic/electrostatic field analyzers) differ in the way they separate the ions (i.e., by radio frequency, by magnetic field, etc.) and subsequently in their sensitivity and resolution. Once ions reach the detector, a signal that provides a measure of the abundance of the fragments (ions) is produced. At the end of a mass spectrometry run, a Total Ion Chromatogram (TIC), which looks like a gas chromatogram, is generated. Each peak in the TIC, however, contains not only information about the abundance of the compound eluting at a certain time, but its fragmentation fingerprint (mass spectrum) as well. A mass spectrum is a representation of the ions observed by the mass spectrometer. The spectrum is normalized so that the most abundant fragment has 100% intensity. It could be in the form of a written report or a graphic representation (a bar graph where x-axis is the mass-to-charge ratio and the y-axis is the intensity scale). Fig. 6.3 shows a graphic representation of a mass spectrum of the monoterpene, carvone.

Due to the fact that organic compounds have a characteristic fragmentation pattern by which they can be recognized, mass spectrometry can be used for identification purposes. Libraries containing mass spectra of organic compounds are commercially available. Most of the flavor volatiles can be identified by comparing individual spectra with those in a reference spectrum library. If a compound has *cis* and *trans* isomers, a retention index will come in handy. If, however, the mass spectrum of the unknown cannot be found in the library, its structure has to be deduced using knowledge about the rules that govern fragmentation (McLafferty & Turecek 1993; Kitson *et al.* 1996; Lee 1998).

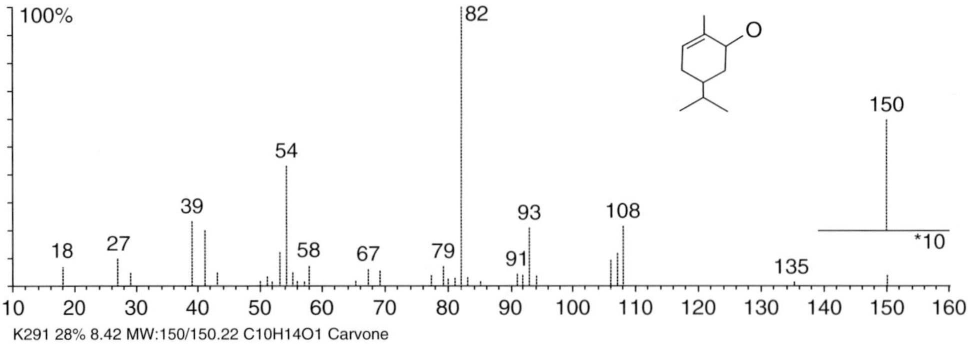

Fig. 6.3 EI mass spectrum of carvone.

6.3.2 Electron ionization (EI)

There are two ionization techniques used in mass spectrometry, namely electron ionization and chemical ionization. The most common ionization technique is electron ionization. In this ionization mode, molecules in an ion source are bombarded by a stream of electrons having energy of 70 electron volts (eV). Due to such high-energy input, this technique produces many fragments and therefore, each compound has a very detailed "fingerprint" that is used for identification purposes (commercially available libraries are EI spectra libraries).

However, with the use of EI, some compounds may undergo extensive fragmentation and may lack their molecular ion in the spectrum. This is often the case with long-chain aliphatic compounds. When an EI-MS spectrum cannot provide information about molecular weight, a different, softer ionization technique called chemical ionization (CI) can be employed.

6.3.3 Chemical ionization

In CI, instead of high-energy electrons, the sample is bombarded with ions of a reagent gas such as ammonia, methane, or isobutene, which are initially ionized by high-energy electrons. During the collision, reagent gas molecules can transfer a proton to the molecule or take a proton from them (depending on the molecule and the reagent gas), hence positive ion and negative ion chemical ionization. In positive-ion CI, the molecular fragment produced is a protonized molecular ion having molecular weight increased by 1 (M11). The fragmentation of the molecules in CI is significantly reduced when compared to that occurring when EI is used. Therefore, CI spectra can provide useful information about the molecular weight of the compound but much less about its structure. Note that methane is a more energetic molecule than isobutane and when used in CI tends to cause more fragmentation than isobutane, but less than in EI.

Figs 6.4, through 6.6 show the mass spectra of lavandulyl acetate as obtained by EI and by CI using ammonia and isobutane as reagent gases, respectively. For more information about the use of mass spectrometry in flavor characterization and specific applications, please refer to the many excellent books and papers published on this subject (Cronin & Caplan 1987; Gilbert 1987; Herderich 1999).

6.3.4 Tandem mass spectrometry

In tandem mass spectrometry (MS-MS), two or more mass analyzers are coupled together. The first mass analyzer separates the ions produced in the ion source of mass spectrometer and provides a mass spectrum of the mixture. After selecting one ion of interest, the first mass

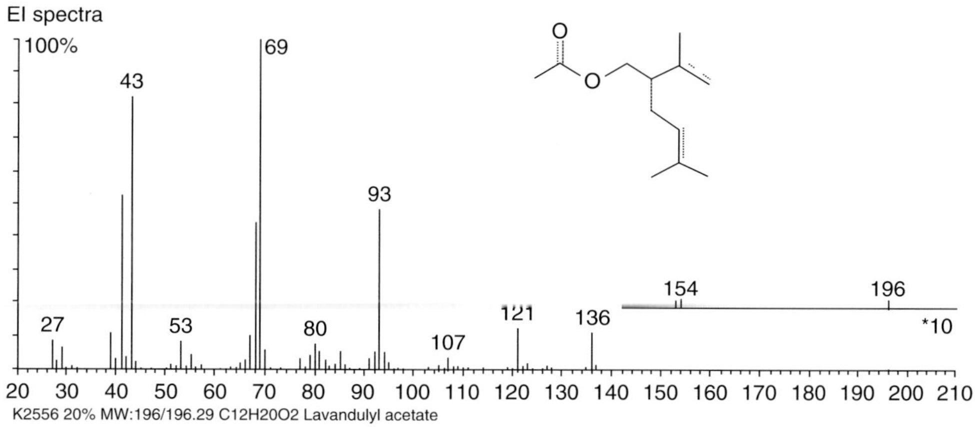

Fig. 6.4 EI mass spectrum of lavandulyl acetate (MW 196).

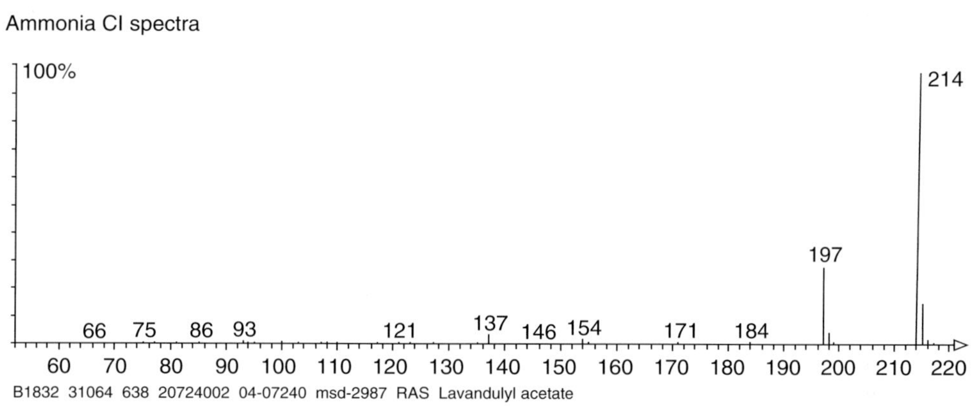

Fig. 6.5 Ammonia CI mass spectrum of lavandulyl acetate (note the ammonia adduct 214 *m/z*).

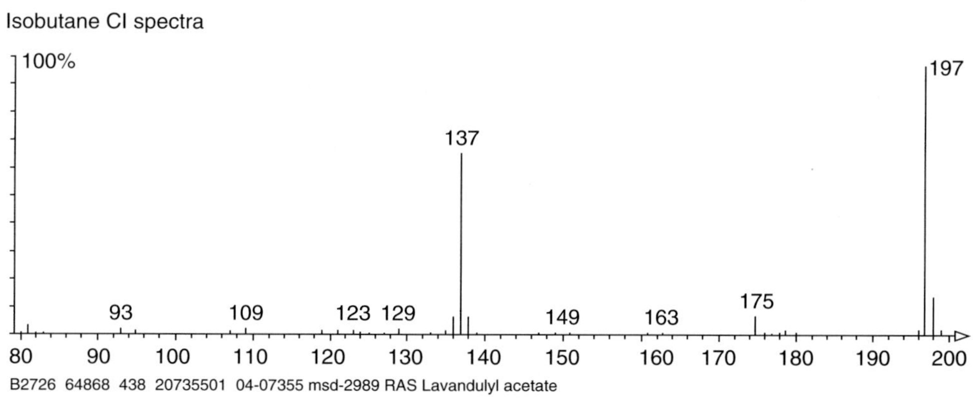

Fig. 6.6 Isobutane CI mass spectrum of lavandulyl acetate (note the proton addition 197 *m/z*).

analyzer is set to pass only that ion. The selected ion (parent ion) travels to the collision chamber where it undergoes collisionally activated dissociation (CAD) and produces "daughter ions." Daughter ions get separated by a second mass analyzer and subsequently recorded. The daughter ions spectrum can be used to identify the compounds by comparison to reference spectra. An advantage of this technique is that it can be used for the analysis of mixtures without prior separation of flavor compounds, since fragment ions derived from a parent ion can be recorded without interference from other ions in the primary spectrum. In addition to structural elucidation, this technique is very useful when looking for a specific compound in the mixture (e.g., in stability studies). Moreover, MS-MS could be helpful for the functional group analysis by looking at the losses of specific neutral molecules from the parent ion (Startin 1987).

6.4 HIGH-PERFORMANCE LIQUID CHROMATOGRAPHY AND LIQUID CHROMATOGRAPHY-MASS SPECTROMETRY

So far, we have described instrumental methods used for characterization of volatile compounds. However, nonvolatile compounds such as sugars, proteins, and fats play an important role in the perception of food flavor. The identification techniques used for characterization of nonvolatile compounds include the ones described for volatiles (MS, IR, NMR). The difference is in the technique used to separate nonvolatiles before their detection. For that purpose, the technique of choice is generally high-performance liquid chromatography (HPLC). As in gas chromatography, a sample is introduced onto a column and carried toward the detector, except in HPLC it is carried by a liquid phase (solvent or mixture of solvents) rather than a gas phase. The detectors used for HPLC are generally nondestructive (such as ultraviolet-visible [UV-VIS] detector) in which case fractions of the sample can be collected after detection and used for sensory evaluation and spectroscopic identification. By using standard dilutions of known compounds, it is possible to determine the concentration of these compounds in food. If, however, we need the identification of an unknown compound, mass spectrometry coupled with HPLC (known as LC-MS) would be the method of choice (Games 1987).

In addition to the determination of nonvolatiles, HPLC and LC-MS are often used for identification and quantitation of very polar or semivolatile compounds (such as vanillin, maltol, furaneol). These compounds do not extract well into a nonpolar solvent during sample preparation for GC analysis. Also, they are usually under-detected by an FID or in the case of vanillin can sublimate in the column reappearing in subsequent runs.

Due to the high temperatures used for volatilization in GC injector port, thermally labile compounds could break down. For characterization of these compounds, in addition to the on-column injection, HPLC and LC-MS may be the methods of choice (Hartman *et al.* 1989).

6.5 INFRARED SPECTROSCOPY AND GAS CHROMATOGRAPHY-INFRARED SPECTROSCOPY

Infrared spectroscopy has been a valuable technique to the analytical chemist for many decades. In this technique, IR radiation is passed through a sample so that when the frequency of the radiation delivered and absorbed corresponds to that of a vibrational mode of a molecule, an IR band in the spectrum is produced. The spectrum usually consists of a plot of wavelength

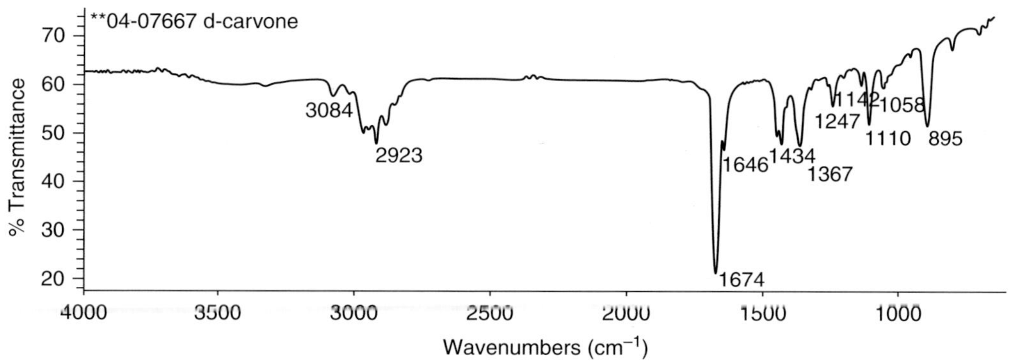

Fig. 6.7 FTIR spectrum of 1-carvone.

versus transmittance or absorbance. Commercial libraries are available for matching against (Sadtler Standard Spectra). The development of Fourier Transform Infrared (FTIR) has led to faster and better quality spectra. FTIR refers to a fairly recent development in the manner in which the data are collected and converted from an interference pattern to a spectrum.

Infrared spectroscopy gives important information about the functional groups attached to any given molecule. It is able to determine the presence of alcohols, amines, carbonyl compounds, etc., as well as differentiate between aldehydes and ketones and *cis* and *trans* isomers, which is not always possible by mass spectrometry. Fig. 6.7 shows an FTIR spectrum of the monoterpene, carvone.

To get a clean interpretable spectrum, a pure substance is usually required. Mixtures are difficult if not impossible to interpret. The sample may be run on a spectrophotometer in several ways. Traditionally, potassium bromide (KBr) plates were used for running liquid samples. This material had few if any interfering bands in the infrared region. A drop applied between two plates produced a perfect spectrum. Solids were similarly handled by forming a thin paste or "mull" using nujol, organic oil. However, this gave interfering bands in the IR spectrum. A more convenient technique for solids is a KBr disc. A small amount of sample is ground into a fine powder with crystalline KBr and poured into a small die within a press. Physical pressure is applied and reduced gas pressure removes any low boiling vapors and moisture. Practice is required to produce a fine translucent disk, which gives a clean spectrum.

More recently, IR has been combined with GC and even MS to give a more powerful technique for analysis of mixtures. In these techniques, all spectra obtained are vapor-phase spectra. They still give important molecular data, but tend to be less detailed than liquid or solid spectra. When compared to mass spectroscopy, a much higher concentration of sample is required to give detailed spectra. However, the technique is nondestructive, thus enabling recovery of the sample for further use (Idstein & Schreier 1985).

6.6 NUCLEAR MAGNETIC RESONANCE SPECTROSCOPY

NMR is a technique that makes use of the fact that nuclei of a material placed in a strong magnetic field absorb radio waves supplied by a transmitter at particular frequencies. The energy of the radio-frequency photons is used to promote the nucleus from a low-energy state, in which the nuclear spin is aligned parallel to the strong magnetic field, to a higher energy state in which the spin is opposed to the field. When the source of the radio waves is turned

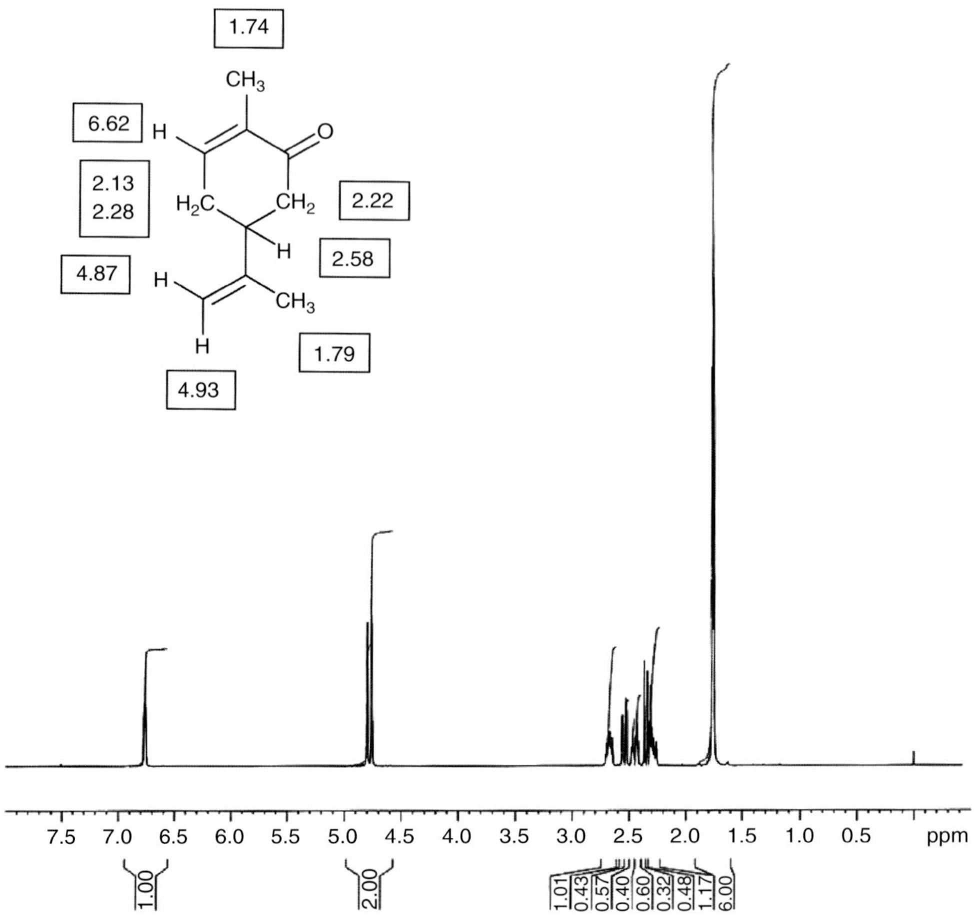

Fig. 6.8 Proton 1 H spectrum of carvone.

off, many nuclei will revert to the lower energy state by emitting photons at characteristic resonance frequencies, thus providing information about the sample structure.

NMR, like x-ray crystallography, can determine the absolute structure of an unknown molecule. In its main mode of operation, it yields information about the position of the hydrogen 1 H and carbon 13 C atoms in a molecule (Kubeczka & Formacek 1984). NMR is generally used to identify pure unknowns since, unlike IR and MS, there are no current commercial instruments available that link NMR with GC. The time required to obtain sensitive and detailed NMR spectra is too long for normal GC techniques. However, NMR has been successfully coupled with liquid chromatography. Fourier transform application to data acquisition has greatly enhanced this technique.

Like IR, NMR is also a nondestructive technique enabling the sample to be used for further tests. Solvent is required for each sample and that can cause interfering bands and peaks. Solvents most often used are deuterated chloroform, methylene chloride, and dimethyl sulfoxide. In practice, about 20 mg of sample is required for proton NMR and 300 mg for carbon NMR. Figs. 6.8 and 6.9 show proton and carbon spectra of carvone in deuterated chloroform (CDCl3), respectively. Fig. 6.10 shows the 135° Distortionless Enhancement by Polarization Transfer (DEPT) spectrum of carvone. DEPT is a Fourier transform data

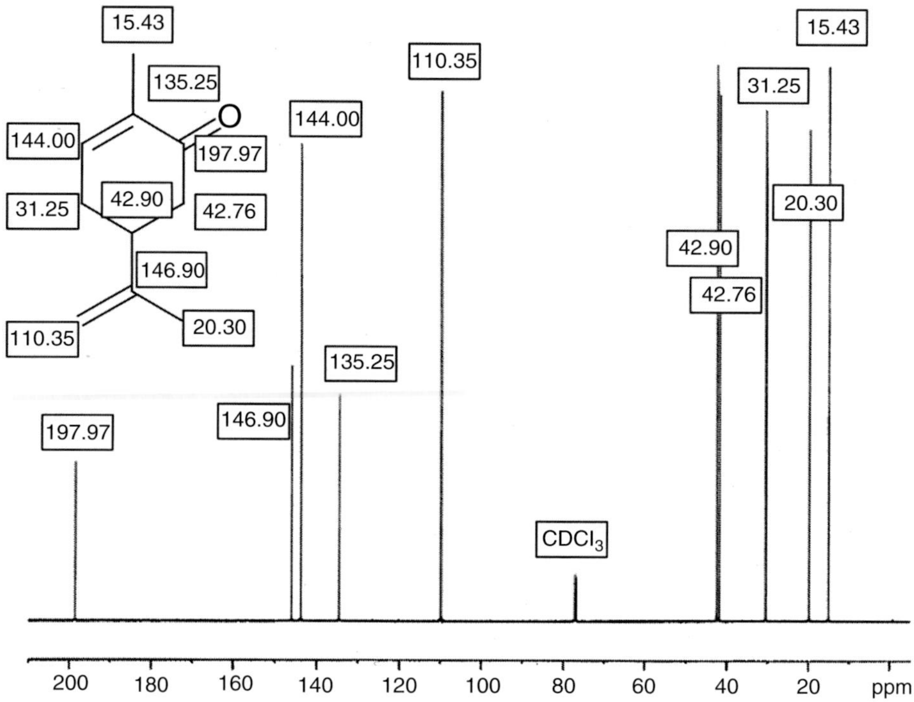

Fig. 6.9 Carbon 13 C spectrum of carvone.

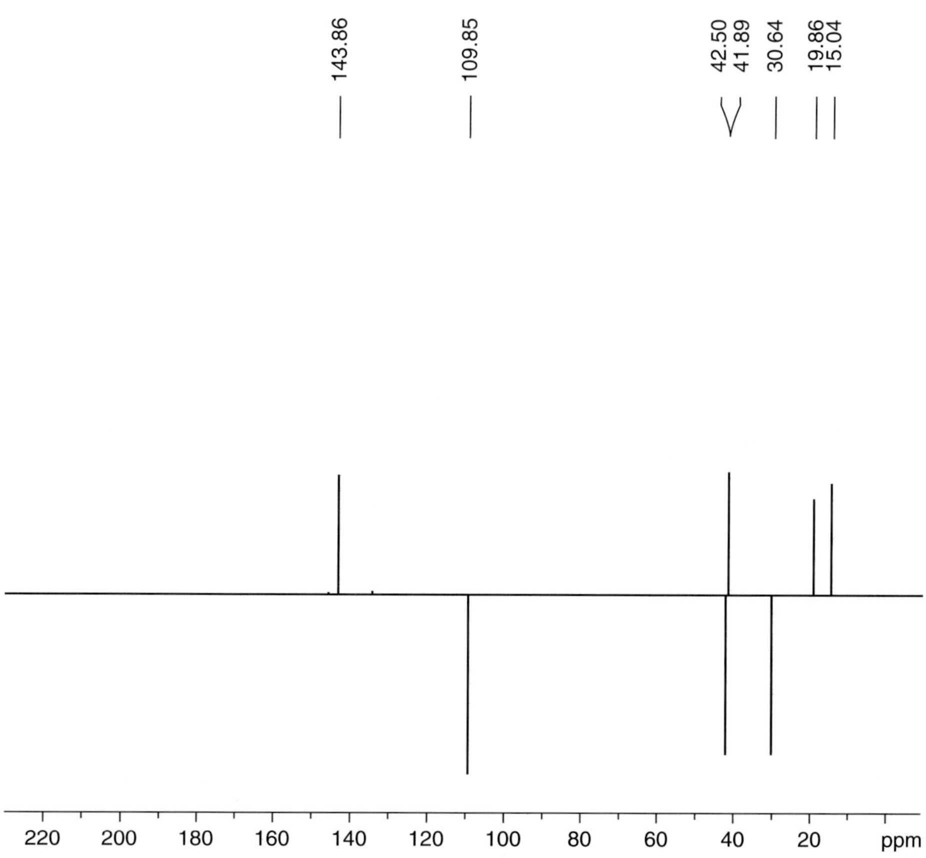

Fig. 6.10 DEPT spectrum of carvone.

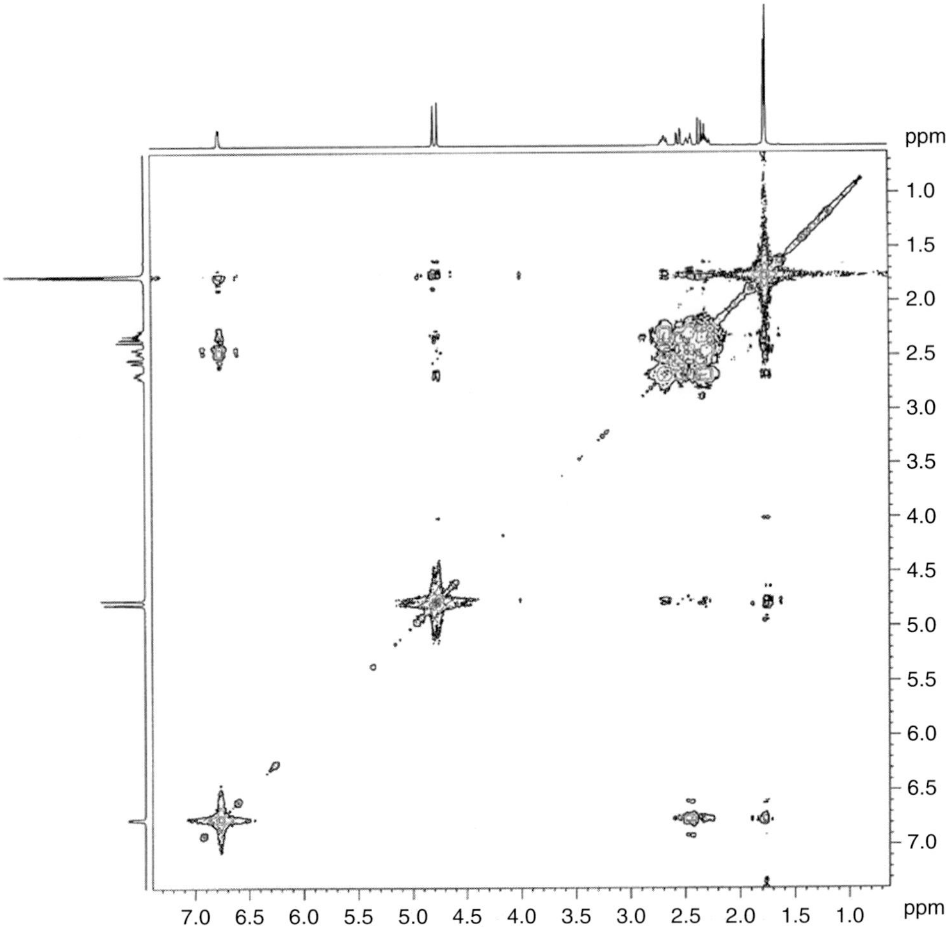

Fig. 6.11 2-Dimensional COSY NMR spectrum of carvone.

processed spectrum technique that shows all methyl (CH3) and methine (CH) protons above the baseline; all methylenes (CH_2) below the baseline and quaternary carbons disappear altogether. Thus, it provides valuable information about molecular structure. Fig. 6.11 shows a 2-D proton COSY (Correlation Spectroscopy) spectrum of carvone. A 2-D proton and carbon correlation spectrum, which shows which protons are directly linked with which carbons in a molecule, is also obtainable.

Carbon-13 and deuterium are the most common nuclei from which NMR spectra are generated. However, by using a multinuclear probe, it is also possible to generate similar spectra for compounds containing 15N, 17O, 19F, and 31P (Brown *et al.* 1988).

A recent development in NMR techniques is Site Specific Natural Isotope Fractionation (SNIF-NMR). This technique has been used to determine the authenticity of foods and natural materials such as wines, fruit juices, and vanilla extracts. It is dependent upon the natural abundance and distribution of deuterium atoms within a molecule. The distribution of deuterium atoms has been found to differ depending upon the source of the molecule. Therefore, an isotopic fingerprint can give information about the origin of the sample. A drawback is the relatively large amount of sample required for the analysis. The technique is described in greater detail by Martin and Martin (1999a, b).

ACKNOWLEDGMENTS

We would like to thank Dr Barry Pope (NMR/IR), Sharon Brown (Mass Spectra), and Donna Lindsey (Graphics) for their valuable help in writing this chapter.

REFERENCES

Allenmark, S.G. (1988) *Chromatographic Enantioseparation: Methods and Applications*, p. 224. Wiley, New York.

Brown, D.W., Floyd, A.J. & Sainsbury, M. (1988) *Organic Spectroscopy*, p. 250. John Wiley & Sons Ltd., Chichester, U.K.

Cronin, D.A. & Caplan, P.J. (1987) Application of GC-MS to identification of flavour compounds in foods. In: *Applications of Mass Spectrometry in Food Science* (ed. J. Gilbert), pp. 1–71. Elsevier Applied Science, New York.

Da Costa, N.C. & Eri, S. (2004) Identification of aroma chemicals. In: *Chemistry and Technology of Flavours and Fragrances* (ed. D. Rowe), pp. 12–34. Black-well Publishing Ltd., Oxford, U.K.

Dimandja, J.M.D. (2003) A new tool for the optimized analysis of complex volatile mixtures: Comprehensive two-dimensional gas chromatography/time-of-flight mass spectrometry. *American Laboratory*, **35**, 42–53.

Dimandja, J.M.D., Clouden, G.C., Colon, I., Focant, J.F., Cabey, W.V. & Parry, R.C. (2003) Standardized test mixture for the characterization of comprehensive two-dimensional gas chromatography columns: The Phillips mix. *Journal of Chromatography A*, **1019**, 261–272.

Drawert, F. & Christoph, N. (1984) Significance of the sniffing technique for the determination of odour thresholds and detection of aroma impacts of terrace volatiles. In: *Analysis of Volatiles: Methods-Applications* (ed. P. Schreirer), pp. 269–293. Walter de Gruyter, New York.

Games, D.E. (1987) Applications of high-performance liquid chromatography/mass spectrometry (LC/MS) in food chemistry. In: *Applications of Mass Spectrometry in Food Science* (ed. J. Gilbert), pp. 193–237. Elsevier Applied Science, New York.

Gilbert, J. (1987) *Applications of Mass Spectrometry in Food Science*, p. 440. Elsevier Applied Science, New York.

Grob, R.L. (1985) *Modern Practice of Gas Chromatography*, p. 330. John Wiley & Sons, New York.

Hartman, T.G., Ho, C.-T., Rosen, J.D. & Rosen, R.T. (1989) Modern techniques in mass spectrometry for the analysis of nonvolatile of thermally labile flavor compounds. In: *Thermal Generation of Aromas* (eds T.H. Parliament, R.J. McGorrin & C.-T. Ho), pp. 73–92. American Chemical Society, Washington, District of Columbia.

Herderich, M. (1999) Mass spectrometry in flavor chemistry: Past and prospect. In: *Flavor Chemistry: Thirty Years of Progress* (eds R. Teranishi, E.L. Wick & I. Hornstein), pp. 9–17. Kluwer Academic/Plenum Publishers, New York.

Idstein, H. & Schreier, P. (1985) High-resolution gas chromatography-Fourier transform IR spectroscopy in flavor analysis: Limits and perspectives. In: *Characterization and Measurement of Flavor Compounds* (eds D.D. Bills & C.J. Mussinan), pp. 109–120. American Chemical Society, Washington, District of Columbia.

Jennings, W. & Shibamoto, T. (1980) *Qualitative Analysis of Flavor and Fragrance Volatiles by Glass Capillary Gas Chromatography*, p. 472. Academic Press, New York.

Kitson, F.G., Larsen, B.S. & McEwen, C.N. 1996. *Gas Chromatography and Mass Spectrometry, a Practical Guide*, p. 381. Academic Press, San Diego, California.

Konig, W.A. (1984) New developments enantiomer separation by capillary gas chromatography. In: *Analysis of Volatiles: Methods-Applications* (ed. P. Schreirer), pp.77–93. Walter de Gruyter, New York.

Kovats, E. 1958. Gas chromatographic characterization of organic compounds. I. Retention indexes of aliphatic halides, alcohols, aldehdyes, and ketones. *Helvetica Chimica Acta*, **41**, 1915–1932.

Kubeczka, K.H. & Formacek, V. (1984) Application of direct carbon-13 NMR spectroscopy in the analysis of volatiles. In: *Analysis of Volatiles: Methods-Applications* (ed. P. Schreirer), pp. 219–230. Walter de Gruyter, New York.

Lee, T.A. (1998) *A Beginner's Guide to Mass Spectral Interpretation*, p. 200. John Wiley & Sons, New York.

Leland, J.V., Schieberle, P., Buettner, A. & Acree, T.E. (2001) *Gas Chromatography-Olfactometry: The State of Art*, p. 219. American Chemical Society, New York.

Martin, G. & Martin, M.L. (1999a) Thirty years of flavor NMR. In: *Flavor Chemistry: Thirty Years of Progress* (eds R. Teranishi, E.L. Wick & I. Hornstein), pp. 18–30. Kluwer Academic/Plenum Publishers, New York.

Martin, M.L. & Martin, G.J. (1999b) Site-specific isotope effects and origin inference. *Fractional Recovery Analysis*, **27**, 209–213.

McLafferty, F.W. & Turecek, F. (1993) *Interpretation of Mass Spectra*, p. 371. University Science Books, Mill Valley, California.

McReynolds, W.O. (1966) *Gas Chromatographic Retention Data*. Preston Technical Abstract Company, Evanston, Illinois.

Mosandl, A. (1992) Capillary gas chromatography in quality assessment of flavours and fragrances. *Journal of Chromatography*, **624**, 267–292.

Mussinan, C.J. (1993) Instrumental analysis in the flavor industry. In: *Flavor Science: Sensible Principles and Techniques* (eds T.E. Acree & Teranishi, R.), pp. 169–223. American Chemical Society, Washington, District of Columbia.

Pickenhagen, W. (1989) Enantioselectivity in odor perception. In: *Flavor Chemistry: Trends and Developments* (eds R. Teranishi, R.G. Buttery & F. Shahidi), pp. 151–158. American Chemical Society, Washington, District of Columbia.

van Ruth, S.M. (2001) Methods for gas chromatography-olfactometry: A review. *Biomolecular Engineering*, **17**, 121–128.

Schomburg, G., Husmann, H., Podmaniczky, L., Weeke, F. & Rap, A. (1984) Coupled gas chromatographic methods for separation, identification, and quantitative analysis of complex mixtures: MDGC, GC-MS, GC-IR, LC-GC. In: *Analysis of Volatiles: Methods-Applications* (ed. P. Schreirer), pp. 121–151. Walter de Gruyter, New York.

Startin, J.R. (1987) Applications of mass spectrometry/mass spectrometry in food analysis. In: *Applications of Mass Spectrometry in Food Science* (ed. J. Gilbert), pp. 289–341. Elsevier Applied Science, New York.

Westmoreland, D.G. & Rhodes, G.R. (1989) Analytical techniques for trace organic compounds II. Detectors for gas chromatography. *Pure and Applied Chemistry*, **61**, 1147–1160.

7 Process Flavors

Hyung Hee Baek

Abstract: Process flavor is a food flavoring manufactured by heating various precursors together under controlled reaction conditions, such as temperature, reaction, pH, and moisture content. The basic theory involved in process flavor is Maillard reaction. In addition to browning, reduction in nutritional value, toxicity, and antioxidant properties, this reaction is of great importance in food processing regarding flavor generation.

Keywords: Sulfur-containing compounds, nitrogen-containing compounds, furanones, lipid-derived compounds, meat aroma-active compounds, flavor precursors, reducing sugars, thiamine, ribonucleotides, phospholipid

7.1 INTRODUCTION

Process flavor is a food flavoring manufactured by heating various precursors together under controlled reaction conditions, such as temperature, reaction time, pH, and moisture content, which mimic conventional cooking process. The term "reaction flavor" (instead of "process flavor") has been widely used since its introduction to the food industry. However, process flavor is a more preferred term in the flavor industry because it is more related to a thermal process than to chemical synthesis or to reaction chemistry (Manley & Ahmedi 1995). Process flavoring, thermal process flavoring, and processed flavor (flavoring) are also interchangeably used (May 1991). The International Organization of the Flavor Industry (IOFI) defines process flavor as a thermal process flavoring prepared for its flavoring properties by heating food ingredients and/or ingredients permitted for use in foodstuffs or in process flavorings. The basic theory involved in process flavor is a Maillard reaction. In addition to browning, reduction in nutritional value (e.g., loss of lysine), toxicity, and antioxidant properties, this reaction is of great significance in food processing regarding flavor generation. Reaction mechanism and application of this reaction to the process flavor have been the main topics in food processing (MacLeod 1986; Nagodawithana 1995). In general, the term "process flavor" is used in describing flavors produced by (1) processing (e.g., cocoa/chocolate), (2) enzymatic modification or fermentation, or (3) Maillard reaction (Heath & Reineccius 1986; Manley 1994).

Numerous reviews on process flavors have already been published (Heath & Reineccius 1986; MacLeod 1986; May 1991; Manley 1994; Manley & Ahmedi 1995; Nagodawithana 1995; Weenen & de Rooij 1998; Manley *et al.* 1999). Process flavor is important in imparting a savory or meaty flavor to ready-to-eat foods. Meat process flavors are manufactured by

Handbook of Meat, Poultry and Seafood Quality, Second Edition. Edited by Leo M. L. Nollet.

Table 7.1 Potent sulfur-containing aroma-active compounds found in cooked meats and meatlike process flavors.

Compounds	Occurrence	References
2-Methyl-3-furanthiol	B, C, P	2–4, 6–9
3-Mercapto-2-pentanone	B, P	2–4, 6, 9
Dimethyl trisulfide	B, C, P	2, 5–8
2-Furfurylthiol	B, C, P	2–5, 7, 9
Methional	B, C, P	1, 3–9
2,5-Dimethyl-3-furanthiol	C	9
2-Acetylthiazole	B, P	1, 2, 8, 9
2-Acetyl-2-thiazoline	B, C, P	1–5, 7,
Bis-(2-methyl-3-furyl) disulfide	B, P	2, 4, 9

B, beef; C, chicken; P, process flavor.
References: 1: Cerny and Grosch (1992); 2: Hofmann and Schieberle (1995a); 3: Wu and Cadwallader (2002); 4: Kerscher and Grosch (1997); 5: Guth and Grosch (1994); 6: Baek *et al.* (2001); 7: Kerler and Grosch (1997); 8: Gasser and Grosch (1988); 9: Gasser and Grosch (1990).

reacting precursors, hydrolyzed vegetable protein (HVP), meat extracts, and yeast extracts under controlled conditions, and then by adding spices and flavor enhancers. Enzymatically hydrolyzed vegetable protein (EVP) is used as an alternative to HVP due to possible concern on monochloropropanediols (MCPD) or dichloropropanols (DCP). Process flavor can be developed by heating EVP (Aaslyng *et al.* 1999a, b; Wu *et al.* 2000). Inexpensive protein sources from seafood by-products can also be used as precursors to develop process flavor (Baek & Cadwallader 1995, 1999).

Enzymatic modifications of butter, coffee, chocolate, cereal products, and roasted nuts are also considered process flavors, but meat flavor is the most important among the process flavors. In this chapter, meat flavors will be discussed extensively because the flavors of cooked meats are typical examples of process flavors.

7.2 AROMA-ACTIVE COMPOUNDS FOUND IN MEATS AND PROCESS FLAVORS

To date, many hundreds of volatile flavor compounds have been identified in beef, pork, chicken, and lamb (Mottram 1991; Shahidi 1998). Early research on volatile flavor compounds of cooked meats was not successful in characterizing the meaty flavor because these compounds do not necessarily represent the character-impact compounds or aroma-active compounds of cooked meats. Most important meaty flavors were in low thresholds, which made them difficult to be identified as potent aroma compounds by instrument analysis. Recent development of gas chromatography-olfactometry (GC-O) techniques revealed important aroma-active compounds from cooked meats (Gasser & Grosch 1988, 1990; Guth & Grosch 1994; Kerscher & Grosch 1997). Flavor chemistry of meat flavors has recently been reviewed intensively (Shahidi 1998). Lists of significant aroma-active components frequently found in cooked meats and meatlike process flavors are given in Tables 7.1 and 7.2. Sulfur- and nitrogen-containing (or heterocyclic) compounds as well as furanones and lipid-derived compounds are important in meats and process flavors. Of these, sulfur-containing compounds contribute most significantly to process flavors. Classes of heterocyclic aroma-active compounds found in meats and process flavors are shown in Fig. 7.1.

Table 7.2 Potent aroma-active compounds frequently found in cooked meats and meatlike process flavors.

Compounds	Occurrence	References
3-Methylbutanal	B, C, P	3, 4, 5, 7
2,3-Butanedione	B, P	1–6
Hexanal	B, C, P	1, 3–5, 7
1-Octen-3-one	B, C, P	1, 3–5, 7
2-Acetyl-1-pyrroline	B, P	3, 4
(E,E)-2,4-decadienal	B, C	1, 4, 5, 7
4-Hydroxy-2,5-dimethyl-3(2 H)-furanone	B, C, P	1–5, 7
3-Hydroxy-4,5-dimethyl-2(5 H)-furanone	B, C, P	1, 2, 4, 5, 7
γ-dodecalactone	C	9

B, beef; C, chicken; P, process flavor.
References: 1: Cerny and Grosch (1992); 2: Hofmann and Schieberle (1995a); 3: Wu and Cadwallader (2002); 4: Kerscher and Grosch (1997); 5: Guth and Grosch (1994); 6: Baek *et al.* (2001); 7: Kerler and Grosch (1997); 8: Gasser and Grosch (1988); 9: Gasser and Grosch (1990).

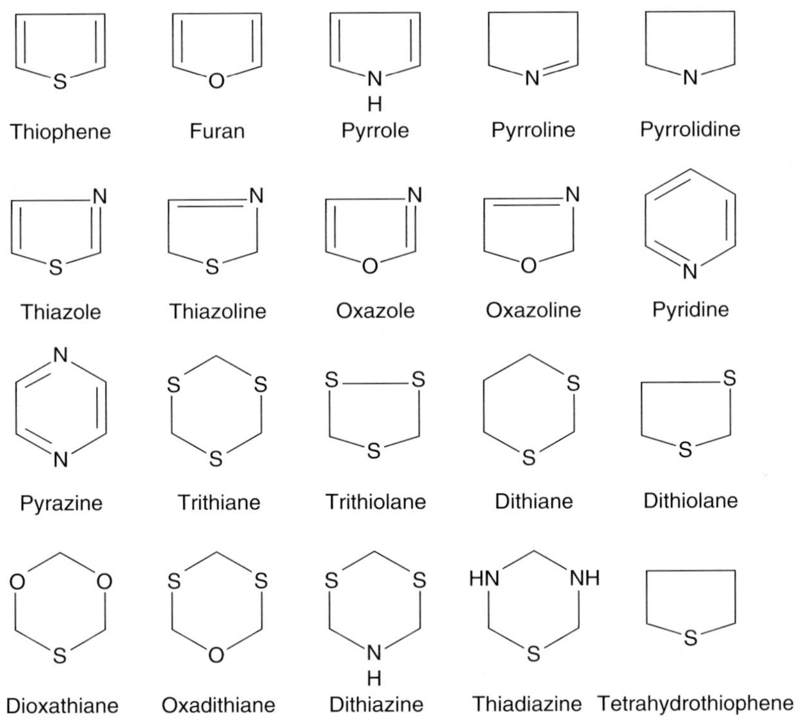

Fig. 7.1 Classes of heterocyclic aroma-active compounds found in meats and process flavors.

7.2.1 Sulfur-containing compounds

Sulfur-containing (or heterocyclic) compounds are most important in cooked meat flavors (Mussinan & Keelan 1994). Sulfur-containing aroma-active compounds found in beef, chicken, and process flavors are listed in Table 7.1. MacLeod (1986) found that most meaty aromas were sulfur-containing compounds. 2-Methyl-3-furanthiol (2-MF), 2-furfurylthiol, and 2,5-dimethyl-3-furanthiol were identified as primary odorants of chicken broth (Gasser & Grosch 1990). 2-MF and its oxidative dimer bis-(2-methyl-3-furyl) disulfide play very important roles in cooked beef (Gasser & Grosch 1988). These compounds are formed by thiamine degradation and reactions involving cysteine (Zhang & Ho 1991b; Hofmann & Schieberle 1995a). These sulfur-containing compounds have been identified from process flavors made from enzyme-hydrolyzed soybean protein (Baek *et al.* 2001; Wu & Cadwallader 2002). Odor thresholds of 2-MF and bis-(2-methyl-3-furyl) disulfide were known to be 0.005–0.01 and 0.00002 ppb, respectively (Gasser & Grosch 1988; Buttery 1999). Furans with a sulfur atom at the 3- or 4-position are known to have a particularly strong meat character (Weenen & de Rooij 1998). A number of sulfur-containing furans have been previously identified in yeast extracts (Ames & MacLeod 1985; Ames 1994).

The aroma difference between beef and chicken might be explained by the fact that bis-(2-methyl-3-furyl) disulfide and methional predominated in beef whereas lipid-derived (*E,E*)-2,4-decadienal and g-dodecalactone were more potent in chicken (Gasser & Grosch 1990).

2-Acetyl-2-thiazoline is considered to be a key contributor to meat process flavors with roasty and popcorn-like aroma note (Hofmann & Schieberle 1995b). This heterocyclic compound has also been identified as a potent aroma-active compound in beef, chicken, and process flavors.

7.2.2 Nitrogen-containing compounds

Pyrazines are one of the major groups of nitrogen-containing volatile compounds generated by Maillard reaction, which are responsible for the roasted aroma of meats. These compounds also contribute to the roasted aroma note of process flavors. Few pyrazines including 2-ethyl-3,5-dimethylpyrazine have been identified in process flavors from enzyme-hydrolyzed soybean protein (Wu *et al.* 2000; Baek *et al.* 2001). 2-Ethyl-3,5-dimethylpyrazine and 2-ethyl-3,6-dimethylpyrazine have odor thresholds of 0.04 and 8.6 ppb, respectively (Buttery & Ling 1997).

2-Acetyl-1-pyrroline (2-AP) is an important aroma-active compound in heated foods. L-Proline is a responsible precursor for this compound (Roberts & Acree 1994). The formation of 2-AP from L-proline was affected by pH, which prefers neutral and slightly alkaline conditions (Blank *et al.* 2003).

7.2.3 Furanones

Furanones, such as 4-hydroxy-2,5-dimethyl-3(2 H)-furanone (HDMF) and 3-hydroxy-4,5-dimethyl-2(5 H)-furanone (sotolon), were identified as key aroma-active compounds of stewed beef (Guth & Grosch 1994). Enolones, such as maltol, cyclotene, and furanones, are important contributors to meat flavors to give rise to sweet and caramel-like aromas (Belitz & Grosch 1999). 4-Hydroxy-5-methyl-3(2 H)-furanone (HMF) and HDMF were first isolated from beef broth (Tonsbeek *et al.* 1968). These two furanones play an important role in meat flavor as precursors for beef flavor (Shu *et al.* 1985; Shu & Ho 1988; Whitfield & Mottram 1999).

7.2.4 Lipid-derived compounds

The Maillard reaction products can be classified into four groups, which include sulfur-, oxygen-, and nitrogen-containing heterocyclic compounds, cyclic enolones, polycarbonyls, and monocarbonyls, depending on their aroma types, chemical structures, molecular shapes, and processing parameters (Heath & Reineccius 1986). Fat degradation products appear to determine the species-specific flavor of meat. (*E,E*)-2,4-Decadienal, which is formed by autoxidation of linoleic acid, is known to be a characteristic aroma of chicken (Gasser & Grosch 1990; Chen & Ho 1998b). However, this lipid-derived compound was of minor importance in beef broth (Gasser & Grosch 1990).

Regardless of meat species, characteristic meaty aroma is associated with all meats. Species-specific aroma-active compounds of meats are believed to be derived from lipid sources. 12-Methyltridecanal has been identified as a species-specific odorant of stewed beef (Guth & Grosch 1993).

7.3 OTHER VOLATILE COMPOUNDS FOUND IN MEATS AND PROCESS FLAVORS

Thiophenes, thiazoles, thiazolines, dithianes, dithiolanes, trithiolanes, and trithianes are other important meat volatiles found in meat flavors or process flavors (Shu *et al.* 1986; Mottram 1991; Zhang & Ho 1991b; Chen & Ho 1998a). These sulfur-containing heterocyclic compounds are considered to contribute to the meat aroma. Of these, 3-thiophenethiol having meaty aroma note plays an important role in meat flavor. Thiophenes with long alkyl chains are believed to arise from lipid sources or phospholipids by the reaction of hydrogen sulfide with unsaturated fatty acids (Whitfield *et al.* 1988; Ho *et al.* 1989). Large numbers of thiazoles have been identified from cooked meat flavors (Mottram 1991). Of these, 2-acetylthiazole was identified as an aroma-active compound of meats.

Oxazoles and oxazolines have been found in boiled beef, pork, and fried chicken (Mottram 1991). Although 2,4,5-trimethyl-3-oxazoline and other alkyl-substituted oxazoles are found in meats, the contribution of these oxygen-containing heterocyclic compounds to meat aroma is probably not as important as sulfur- and nitrogen-containing heterocyclic compounds. Some important heterocyclic compounds of cooked meats and meatlike process flavors are shown in Fig. 7.1.

Heterocyclic compounds with long-chain alkyl substitution are formed from the reaction of lipid or lipid oxidation products, for which the formation mechanism has been proposed (Ho *et al.* 1989). The formation of 2,5-dimethyl-3-pentylpyrazine can be explained by the reaction of 2,5-dimethyldihydropyrazine with pentanal. 2-Pentylpyridine is formed by the interaction of 2,4-decadienal and amino acids.

7.4 PRECURSORS USED IN THE MANUFACTURE OF PROCESS FLAVORS

Water-soluble components from meat extracts contain a complex mixture of precursors including nucleic acids, nucleotides, peptides, amino acids, sugars, lipids, and vitamins. When these precursors react, meat flavor would be theoretically well duplicated. However, because of the cost, availability, or labeling restrictions, other precursors have also been considered.

7.4.1 Amino acids

Amino acid is the most important factor determining aroma property for process flavors. Lane and Nursten (1983) reported that thermally generated aromas are basically the same for each particular amino acid on the basis of the assessment of aroma properties generated from more than 400 model systems.

Cysteine is the most important amino acid as a source of sulfur in the generation of meat flavor (Fig. 7.2). Precursor mixtures used in the industrial production of process flavors

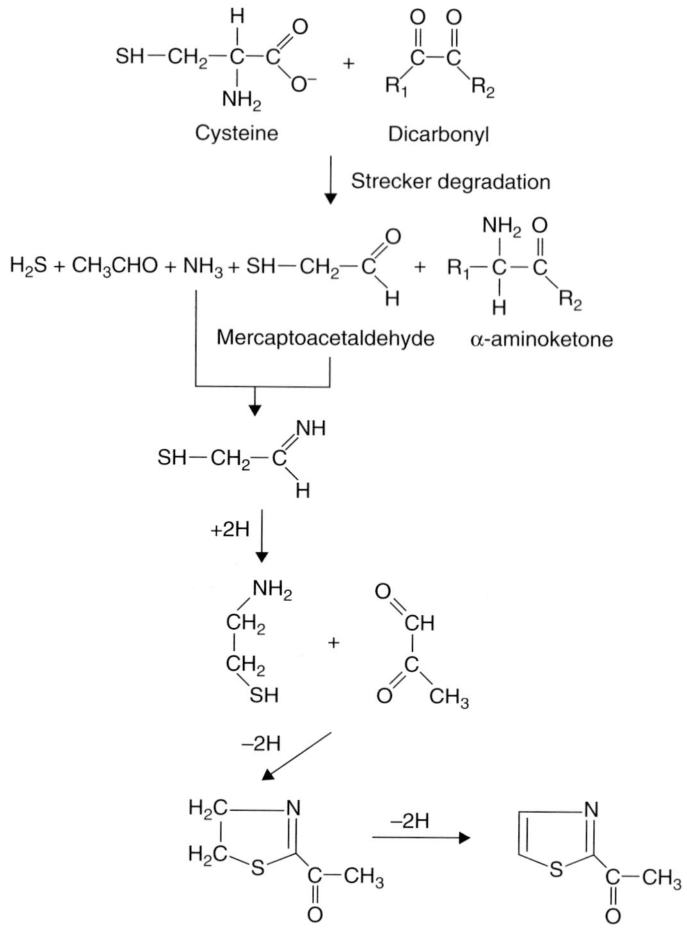

Fig. 7.2 The formation of sulfur-containing heterocyclic compounds via Strecker degradation of cysteine. *Source*: Adapted from Nagodawithana (1995).

usually contain cysteine. Numerous research has proved that this amino acid is essential in the production of process flavors (Hofmann & Schieberle 1995a; Schieberle & Hofmann 1996, 1998; Wu *et al.* 2000). More than 200 volatile compounds have been identified from the reaction of cysteine and reducing sugar (Schieberle & Hofmann 1998). Cysteine reacted with ribose to produce meat flavor comprising mainly furanthiols (Hofmann & Schieberle 1995a).

Other sulfur-containing amino acids, such as methionine and cysteine, and glutathione (Glu-Cys-Gly) are also important precursors. From a practical point of view, cysteine is the most satisfactory sulfur compound in manufacturing process flavors.

Proline, which is an amino acid, reacts with reducing sugars to produce 2-AP (Roberts & Acree 1994). Proline does not produce pyrazines because this amino acid cannot release ammonia in Strecker degradation (Nursten 1986). Maillard reaction products differ depending on 1,2- and 2,3-enolization, which is dependent on pH. At lower pH, 1,2 enolization occurs, and at higher pH, 2,3-enolization occurs. In contrast, hydroxyamino acids, such as serine and threonine, produce pyrazines (Nursten 1986). Glutamic acid and monosodium glutamate (MSG) are also known for their ability to serve as precursors for process flavors as well as flavor enhancers (Nagodawithana 1995).

7.4.2 Reducing sugars

Ribose and ribose-5-phosphate increased during aging from inosine-59-monophosphate (IMP) decomposition. The most significant loss occurred for cysteine and ribose during the heating of meats, which led to the first process flavor patent by Morton *et al.* (1960). A reaction mixture of cysteine and ribose is generally accepted to produce meat flavor.

Previously, it was known that overall odor of Maillard reaction products is not influenced by sugar moiety (Lane & Nursten 1983). However, recently it was reported that the sugar moiety significantly influenced the volatile pattern formed (Hofmann & Schieberle 1997, 1998; Schieberle & Hofmann 1998). When cysteine is reacted with different reducing sugars, specific aroma-active compounds are formed depending on the source of reducing sugar. For example, rhamnose produced 5-methyl-2-furfurylthiol and 3-hydroxy-6-methyl-2(2H) pyranone, whereas 2-(1-mercaptoethyl) furan was exclusively formed from glucose. In addition to monosaccharides, reducing disaccharides, such as maltose and lactose, can participate in Maillard reaction (Hurrell 1982). Pentoses are more reactive than hexoses, with xyloses being the most reactive.

7.4.3 Thiamine

Thiamine (vitamin B1) is a very important nonvolatile flavor precursor, which decomposes thermally to give rise to sulfur-containing compounds important to meat flavor. Reaction mechanism of thermal degradation of thiamine can be found elsewhere (MacLeod 1986; Nagodawithana 1995; Weenen & de Rooij 1998). Thermal degradation of thiamine depends on pH (Fig. 7.3). Thermal degradation at neutral or slightly alkaline pH appears to favor the breakdown of thiazole ring, which leads to the formation of H_2S. Breakdown of the methylene bridge is predominant under acidic conditions (Nagodawithana 1995).

7.4.4 Ribonucleotides

Ribonucleotides are known to serve as a source of ribose as well as flavor enhancers. Nucleotides, such as IMP, which is formed from adenosine-59-triphosphate (ATP) during aging, are the most important aroma precursors in the development of cooked meat flavor.

IMP increased the meaty note of process flavor by thermal reaction (Zhang & Ho 1991b; Farmer *et al.* 1996). The role of IMP is more than just a ribose source. By thermal reaction of cysteine and IMP, a lot of sulfur-containing compounds including 2-MF and 2-furfurylthiol were formed (Zhang & Ho 1991b). IMP is considered an important precursor to generate sulfur-containing compounds responsible for meat flavor. A possible mechanism for the formation of 2-MF from thermal reaction of IMP with cysteine has been proposed (Zhang & Ho 1991b). The major compounds were furfural and 2-furfurylthiol, indicating that liberation of ribose from IMP is the main pathway and these sulfur compounds were subsequently produced by dehydration and oxidation at acidic conditions.

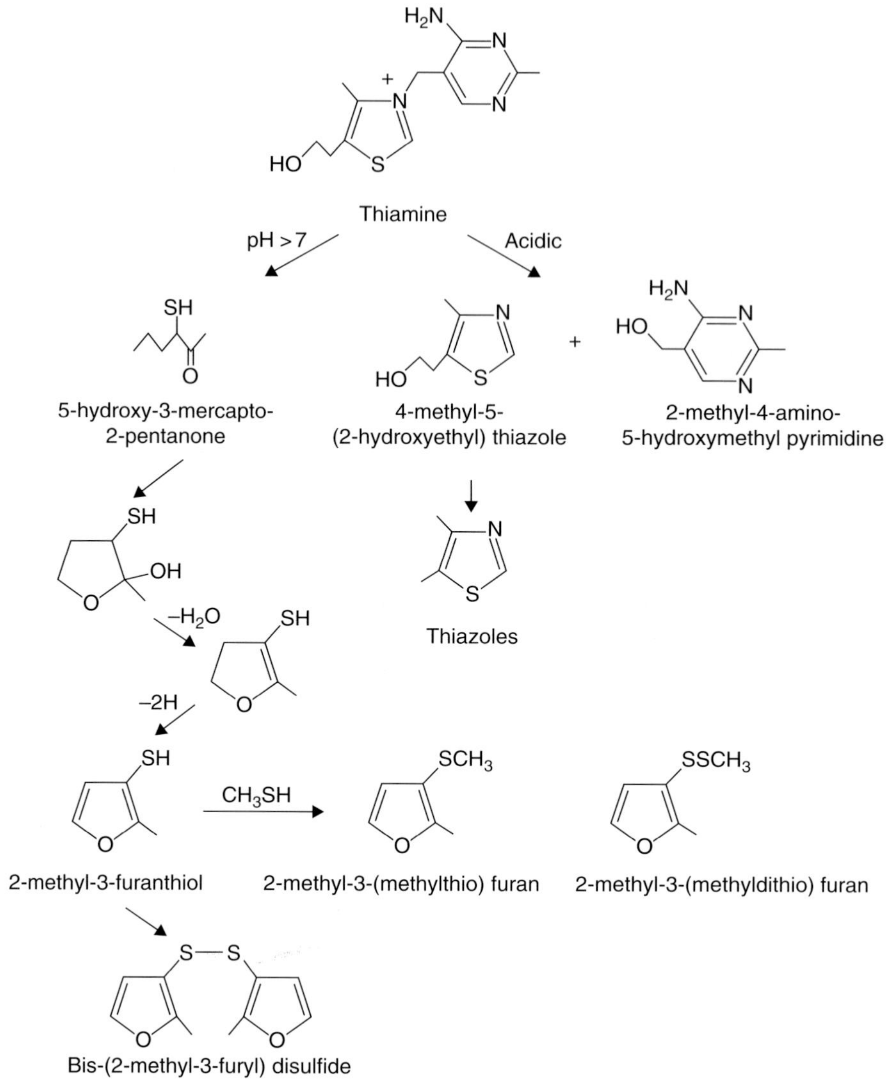

Fig. 7.3 The formation of sulfur-containing heterocyclic compounds from thermal degradation of thiamine at different pH levels.
Source: Adapted from Nagodawithana (1995).

7.4.5 Phospholipid

Mottram and Edwards (1983) reported that the characteristic aroma of cooked meat disappeared by the removal of phospholipid from meat, indicating that phospholipids played a significant role in the generation of cooked meat flavor. Phospholipids contribute to the aroma of cooked meat through thermally induced lipid oxidation and interaction of lipid with Maillard reaction products (Farmer & Mottram 1990; Mottram & Whitfield 1995a, b). Phospholipids inhibited the formation of heterocyclic compounds, such as alkylpyrazines, by participating in Maillard reaction in the aqueous system (Whitfield *et al.* 1988; Farmer *et al.* 1989). Sulfur-containing compounds, such as methylthio-substituted furans and thiophenes, were generated by the addition of phospholipid in nonaqueous systems (Mottram & Whitfield 1995a). Water contents also influenced volatiles produced in Maillard reaction systems with phospholipid (Mottram & Whitfield 1995b).

Aromas generated by thermal degradation of phospholipids are mainly due to lipid-derived volatiles, such as trans-4,5-epoxy-(*E*)-decenal and (*E,E*)-2,4-decadienal (Lin & Blank 2003). Recently, it was hypothesized that molecular organization of phospholipids played an important role in thermal generation of aromas (Vauthey *et al.* 2000; Lin *et al.* 2004).

7.4.6 Others

Meat extracts, which contain precursor amino acids, are usually major ingredients in many process flavors. Protein hydrolysates are good sources of amino acids for process flavors. HVP is an economical source; however, the use of HVP has decreased in light of safety concerns. Flavors without HVP have been regarded as "clean-label" flavors in the USA (Manley *et al.* 1999). EVP can be used as an alternative to HVP. Autolyzed yeast extract is a very useful reactant in creating process flavors. Yeast extract generates sulfur-containing compounds, such as 2-MF and bis-(2-methyl-3-furyl) disulfide, which impart meatlike aroma (Ames & MacLeod 1985). Glutathione is also a good source for process flavor (Zhang & Ho 1991a).

Peptides can react with reducing sugars via complex reactions. Taking into consideration the raw material price, meat by-products are good sources of amino acids and peptides. Protein hydrolysates from plants or animal origins provide a good nitrogen source to process flavor. HVP and autolyzed yeast extracts are widely used as nitrogen sources. Incorporation of fats into the mixture before reaction gives species-specific process flavors, such as mutton, chicken, and pork.

HMF and HDMF are considered to be involved in the meaty flavors through their reactions with either hydrogen sulfide or sulfur-containing amino acids (Shu *et al.* 1986; Shu & Ho 1988; Whitfield & Mottram 1999). 3,5-Dimethyl-1,2,4-trithiolane and other heterocyclic compounds described as roasted and meaty were formed by the thermal reaction of HDMF and cysteine. The formation of HDMF from rhamnose has been proposed.

Maillard reaction intermediates, such as acetaldehyde, pyruvaldehyde (2-oxopropanal), and 2,3-butanedione, have been used as precursors for the development of process flavors (MacLeod 1986).

7.5 FORMATION MECHANISM

Numerous reviews on the formation mechanisms of meat flavor have been published (MacLeod 1986; Nursten 1986; Nagodawithana 1995). Since sulfur-containing compounds play a critical role in the perception of meat flavor, a possible source of sulfur-containing compounds and their formation mechanisms have been extensively reviewed (Mussinan & Keelan 1994). Hydrogen sulfide evolved from Strecker degradation of cysteine plays a central role in the generation of meaty flavor. In addition, mercaptoacetaldehyde, acetaldehyde, and ammonia produced from Strecker degradation of cysteine serve as intermediates in the generation of sulfur-containing heterocyclic compounds, such as 2-acetyl-2-thiazoline and 2-acetylthiazole (Fig. 7.2).

The basic chemistry of process flavors has been reviewed in recent articles (Mottram 1998; Weenen & de Rooij 1998; Manley *et al.* 1999). The most important reactions involved in process flavors are Maillard reaction, Strecker degradation, sugar degradation/fragmentation, thermal degradation of thiamine, and lipid oxidation (Nagodawithana 1995; MacLeod 1986). The formation mechanism relating to process flavor has been reviewed elsewhere (Nagodawithana

1995; MacLeod 1986). Fig. 7.3 shows the thermal degradation of thiamine at different pHs to generate aroma-active compounds important to process flavors, such as 2-MF, 2-methyl-3-(methylthio) furan, 2-methyl-3-(methyldithio) furan, and bis-(2-methyl-3-furyl) disulfide.

Thermal breakdown mechanisms of sugars and amino acids have been described in detail (Scarpellino & Soukup 1993). 5-Hydroxymethyl-2-furfural and 2-furfural are formed under acidic conditions through 1,2-enolization of Amadori compounds derived from thermal breakdown of pentose and hexose, respectively. Under alkaline conditions, the formation of HMF and HDMF are through 2,3-enolization of Amadori compounds derived from the thermal breakdown of pentose and hexose, respectively. Sulfur-substituted furans and thiophenes were formed as main volatile compounds from the reaction of HMF and hydrogen sulfide (van den Ouweland & Peer 1975). Amino acids are thermally degraded through Strecker degradation, which is a reaction of an amino acid with a dicarbonyl to produce Strecker aldehyde that has one less carbon than the original amino acid. Strecker aldehydes play important roles in process flavors as aroma-active compounds as well as precursors for further reactions.

7.6 REACTION PARAMETERS

The aroma property of process flavor depends on (1) the nature and ratio of amino acids/protein and reducing sugars present, (2) the time/temperature conditions of the reaction, and (3) the nature of any thermal degradation reactions. Free amino acids are presumably more important than peptides or proteins. Of these, temperature has the greatest influence on the reaction.

Lipid oxidation products play an important role in the development of some thermally generated aromas as described by Ho *et al.* (1989).

Moisture content is one of the important reaction parameters in the generation of aromas from process flavors. Volatile profiles of process flavors showed significantly different patterns depending on moisture contents of reactants. Dry heating of precursor mixtures generated higher yield of aroma-active compounds than heating under the aqueous system (Schieberle & Hofmann 1998). Some odorants increased in dry heating, whereas some decreased under dry heating. For example, 2-furfurylthiol, 3-mercapto-2-pentanone, 2-MF, and 5-acetyl-2,3-dihydro-1,4-thiazine were the most intense odorants generated in aqueous solution (Hofmann & Schieberle 1995a). 2-Furfurylthiol and 2-acetyl-2-thiazoline were considered the most potent aroma-active compounds in dry-heated mixture, whereas 5-acetyl-2,3-dihydro-1,4-thiazine and 3-mercapto-2-pentanone significantly decreased (Schieberle & Hofmann 1998). Also, a comparison has been made in dry-heated and the aqueous rhamnose/cysteine systems (Hofmann & Schieberle 1997). Caramel- and seasoning-like aroma notes prevailed in the aqueous system, whereas roasty note predominated in the dry-heated system.

The pH effect on the volatile formation in process flavors is important. Cysteine produces meat flavor with the reaction of ribose at acidic pH. In the reaction of cysteine-ribose, pH remarkably influenced the overall aroma (Schieberle & Hofmann 1996). As pH increased from 3.0 to 7.0, 2-MF and 2-furfurylthiol decreased significantly, whereas the HDMF (furaneol) and 3-hydroxy-4,5-dimethyl-2(5 H)-furanone (sotolon) increased, which resulted in significant changes of overall aroma. pH plays a key role in manipulating the volatile compositions in producing process flavors. The production of sulfur-substituted furans with meatlike aromas from the reaction of cysteine is dependent on pH, which is favored at acidic pH (Zhang & Ho 1991a, 1991b; Whitfield & Mottram 1999). pH influenced the volatile formation from the reaction of cysteine and HDMF with the generation of the best meat flavor at pH 2.2 (Shu & Ho 1988). The formation of pyrazines was favored at alkaline pH due to the increased reactivity of amino groups (Hwang *et al.* 1994). In general, browning is more favored at alkaline pH.

7.7 SAFETY CONCERN

Heterocyclic amines formed by heat treatment and MCPD or DCP in HVP are the major concern of safety in process flavors. However, these harmful materials can be reduced by processing conditions and technological achievements. Furthermore, the levels of process flavors used in foods are below levels of safety concern.

Recently, acrylamide has been detected in heated foodstuffs including potato chips and meats. Moderate levels of acrylamide (5–50 mg/kg) were measured in heated protein-rich foods, such as boiled beef, whereas higher contents (150–4000 mg/kg) were measured in carbohydrate-rich foods (Tareke *et al.* 2002). Acrylamide is formed by the reaction of asparagines and glucose (Yaylayan *et al.* 2003). However, process flavor is used as an ingredient in small amounts, normally less than 2%, so it does not seem to be harmful to food. Process flavors are on the list of generally recognized as safe (GRAS) (Manley *et al.* 1999).

The IOFI guideline defines process flavor as a product or a mixture prepared for its flavoring properties and that is produced from ingredients or mixtures of ingredients that are permitted for use in foodstuffs, or are present naturally in foodstuffs, or are permitted for use specifically in process flavorings by a process used for the preparation of foods for human consumption. The ingredients can include (1) a protein nitrogen source, (2) a carbohydrate source, (3) a fat or fatty acid source, and (4) other ingredients including herbs and spices, thiamine, organic acids, sodium chloride, and lecithin. The processing conditions should not exceed 15 min at 180°C or proportionately longer at lower temperatures. The pH should not exceed 8.0.

REFERENCES

Aaslyng, M.D., Larsen, L.M. & Nielson, P.M. (1999a) Development of chemical and sensory characteristics during enzymatic hydrolysis of soy. *Zeitschrift für Lebensmittel- Untersuchung und -Forschung A*, **208**, 50–56.

Aaslyng, M.D., Poll, L., Nielson, P.M. & Flyge, H. (1999b) Sensory, chemical and sensometric studies of hydrolyzed vegetable protein produced by various processes. *European Food Research and Technology*, **209**, 227–236.

Ames, J.M. (1994) Volatile sulfur compounds in yeast extracts. In: *Sulfur Compounds in Foods* (eds C.J. Mussinan & M.E. Keelan), pp. 147–159. American Chemical Society, Washington, District of Columbia.

Ames, J.M. & MacLeod, G. (1985) Volatile components of a yeast extract composition. *Journal of Food Science*, **50**, 125–131.

Baek, H.H. & Cadwallader, K.R. (1995) Enzymatic hydrolysis of crayfish processing by-products. *Journal of Food Science*, **60**, 929–935.

Baek, H.H. & Cadwallader, K.R. (1999) Optimization of the enzymatic hydrolysis of crab processing by-products using Flavourzyme. *Food Science and Biotechnology*, **8**, 43–46.

Baek, H.H., Kim, C.J., Ahn, B.H., Nam, H.S. & Cadwallader, K.R. (2001) Aroma extract dilution analysis of a beeflike process flavor from extruded enzyme-hydrolyzed soybean protein. *Journal of Agricultural and Food Chemistry*, **49**, 790–793.

Belitz, H.D. & Grosch, W. (1999) *Food Chemistry*, 2nd edn., p. 992. Springer-Verlag, Berlin, Germany.

Blank, I., Devaud, S., Matthey-Doret, W. & Robert, F. (2003) Formation of odorants in Maillard model systems based on L-proline as affected by pH. *Journal of Agricultural and Food Chemistry*, **51**, 3643–3650.

Buttery, R.G. (1999) Flavor chemistry and odor thresholds. In: *Flavor Chemistry: Thirty Years of Progress* (eds R. Teranishi, E.L. Wick & I. Hornstein), pp. 353–365. Kluwer Academic/Plenum Publishers, New York.

Buttery, R.G. & Ling, L.C. (1997) 2-Ethyl-3,5-dimethylpyrazine and 2-ethyl-3,6-dimethylpyrazine: Odor thresholds in water solution. *Lebensmittel- Wissenschaft und-Technologie*, **30**, 109–110.

Cerny, C. & Grosch, W. (1992) Evaluation of potent odorants in roasted beef by aroma extract dilution analysis. *Zeitschrift für Lebensmittel- Untersuchung und -Forschung*, **194**, 322–325.

Chen, J. & Ho, C.T. (1998a) The flavor of pork. In: *Flavor of Meat, Meat Products and Seafoods* (ed. F. Shahidi), 2nd edn., pp. 61–83. Blackie Academic and Professional, London, U.K.

Chen, C.W. & Ho, C.T. (1998b) The flavor of poultry meat. In: *Flavor of Meat, Meat Products and Seafoods* (ed. F. Shahidi), 2nd edn., pp. 84–100. Blackie Academic and Professional, London, U.K.

Farmer, L.J. & Mottram, D.S. (1990) Interaction of lipid in the Maillard reaction between cysteine and ribose: The effect of a triglyceride and three phospholipids on the volatile products. *Journal of the Science of Food and Agriculture*, **53**, 505–525.

Farmer, L.J., Mottram, D.S. & Whitfield, F.B. (1989) Volatile compounds produced in Maillard reactions involving cysteine, ribose and phospholipid. *Journal of the Science of Food and Agriculture*, **49**, 347–368.

Farmer, L.J., Hagan, T.D.J. & Paraskevas, O. (1996) A comparison of three sugars and inosine monophosphate as precursors of meat aroma. In: *Flavor Science, Recent Developments* (eds A.J. Taylor & D.S. Mottram), pp. 225–230. The Royal Society of Chemistry, Cambridge, U.K.

Gasser, U. & Grosch, W. (1988) Identification of volatile flavor compounds with high aroma values from cooked beef. *Zeitschrift für Lebensmittel- Untersuchung und -Forschung*, **186**, 489–494.

Gasser, U. & Grosch, W. (1990) Primary odorants of chicken broth. *Zeitschrift für Lebensmittel- Untersuchung und -Forschung*, **190**, 3–8.

Guth, H. & Grosch, W. (1993) 12-Methyltridecanal, a species-specific odorant of stewed beef. *Lebensmittel-Wissenschaft und-Technologie*, **26**, 171–177.

Guth, H. & Grosch, W. (1994) Identification of the character impact odorants of stewed beef juice by instrumental analyses and sensory studies. *Journal of Agricultural and Food Chemistry*, **42**, 2862–2866.

Heath, H.B. & Reineccius, G. (1986) Flavoring materials made by processing. In: *Flavor Chemistry and Technology* (eds H.B. Heath & G. Reineccius), pp. 252–289. An AVI Book, New York.

Ho, C.T., Bruechert, L.J., Zhang, Y. & Chiu, E.M. (1989) Contribution of lipids to the formation of heterocyclic compounds in model systems. In: *Thermal Generation of Aromas* (eds T.H. Parliament, R.J. McGorrin & C.T. Ho), pp. 105–113. American Chemical Society, Washington, District of Columbia.

Hofmann, T. & Schieberle, P. (1995a) Evaluation of the key odorants in a thermally treated solution of ribose and cysteine by aroma extract dilution techniques. *Journal of Agricultural and Food Chemistry*, **43**, 2187–2194.

Hofmann, T. & Schieberle, P. (1995b) Studies on the formation and stability of the roast-flavor compound 2-acetyl-2-thiazoline. *Journal of Agricultural and Food Chemistry*, **43**, 2946–2950.

Hofmann, T. & Schieberle, P. (1997) Identification of potent aroma compounds in thermally treated mixtures of glucose/cysteine and rhamnose/cysteine using aroma extract dilution analysis. *Journal of Agricultural and Food Chemistry*, **45**, 898–906.

Hofmann, T. & Schieberle, P. (1998) Identification of key aroma compounds generated from cysteine and carbohydrates under roasting conditions. *Zeitschrift für Lebensmittel- Untersuchung und -Forschung A*, **207**, 229–236.

Hurrell, R.F. (1982) Maillard reaction in flavor. In: *Food Flavors: Part A. Introduction* (eds I.D. Morton & A.J. MacLeod), pp. 399–437. Elsevier, Amsterdam, the Netherlands.

Hwang, H.I., Hartman, T.G., Rosen, R.T., Lech, J. & Ho, C.T. (1994) Formation of pyrazines from the Maillard reaction of glucose and lysine-a-amine-15N. *Journal of Agricultural and Food Chemistry*, **42**, 1000–1004.

Kerler, J. & Grosch, W. (1997) Character impact odorants of boiled chicken: Changes during refrigerated storage and reheating. *Zeitschrift für Lebensmittel- Untersuchung und -Forschung A*, **205**, 232–238.

Kerscher, R. & Grosch, W. (1997) Comparative evaluation of potent odorants of boiled beef by aroma extract dilution and concentration analysis. *Zeitschrift für Lebensmittel- Untersuchung und -Forschung A*, **204**, 3–6.

Lane, M.J. & Nursten, H.E. (1983) The variety of odors produced in Maillard model systems and how they are influenced by reaction conditions. In: *The Maillard Reaction in Foods and Nutrition* (eds G.R. Waller & M.S. Feather), pp. 141–158. American Chemical Society, Washington, District of Columbia.

Lin, J. & Blank, I. (2003) Odorants generated by thermally induced degradation of phospholipids. *Journal of Agricultural and Food Chemistry*, **51**, 4364–4369.

Lin, J., Leser, M.E., Löliger, J. & Blank, I. (2004) A new insight into the formation of odor active carbonyls by thermally-induced degradation of phospholipids in self-assembly structures. *Journal of Agricultural and Food Chemistry*, **52**, 581–586.

MacLeod, G. (1986) The scientific and technological basis of meat flavors. In: *Developments in Food Flavors* (eds G.G. Birch & M.G. Lindley), pp. 191–223. Elsevier Applied Science, London, U.K.

Manley, C.H. (1994) Process flavors. In: *Source Book of Flavors* (ed. G. Reineccius), 2nd edn., pp. 139–154. Chapman & Hall, New York.

Manley, C.H. & Ahmedi, S. (1995) The development of process flavors. *Trends in Food Science & Technology*, **6**, 46–51.

Manley, C.H., Choudhury, B.H. & Mazeiko, P. (1999) Thermal process flavorings. In: *Food Flavorings* (ed. P.R. Ashurst), 3rd edn., pp. 283–325. Aspen Publishers, Inc., Gaithersburg, Maryland.

May, C.G. (1991) Process flavorings. In: *Food Flavorings* (ed. P.R. Ashurst), 1st edn., pp. 257–286. Blackie and Son Ltd., London, U.K.

Morton, I.D., Akroyd, P. & May, C.G. (1960) *Flavoring substance*. U.K. Patent 836694.

Mottram, D.S. (1991) Meat. In: *Volatile Compounds in Foods and Beverages* (ed. H. Maarse), pp. 107–177. Marcel Dekker, Inc., New York.

Mottram, D.S. (1998) The chemistry of meat flavor. In: *Flavor of Meat, Meat Products and Seafoods* (ed. F. Shahidi), 2nd edn., pp. 5–26. Blackie Academic and Professional, London, U.K.

Mottram, D.S. & Edwards, R.A. (1983) The role of triglycerides and phospholipids in the aroma of cooked beef. *Journal of the Science of Food and Agriculture*, **34**, 517–522.

Mottram, D.S. & Whitfield, F.B. (1995a) Maillard-lipid interactions in nonaqueous systems: Volatiles from the reaction of cysteine and ribose with phosphatidylcholine. *Journal of Agricultural and Food Chemistry*, **43**, 1302–1306.

Mottram, D.S. & Whitfield, F.B. (1995b) Volatile compounds from the reaction of cysteine, ribose, and phospholipid in low-moisture systems. *Journal of Agricultural and Food Chemistry*, **43**, 984–988.

Mussinan, C.J. & Keelan, M.E. (1994) *Sulfur Compounds in Foods*, p. 301. American Chemical Society, Washington, District of Columbia.

Nagodawithana, T.W. (1995) *Savory Flavors*, p. 468. Esteekay Associates, Inc., Milwaukee, Wisconsin.

Nursten, H.E. (1986) Aroma compounds from Maillard reaction. In: *Developments in Food Flavors* (eds G.G. Birch & M.G. Lindley), pp. 173–190. Elsevier Applied Science, London, U.K.

van den Ouweland, G.A.M. & Peer, H.G. (1975) Components contributing to beef flavor. Volatile compounds produced by the reaction of 4-hydroxy-5-methyl-3(2H)-furanone and its thio analog with hydrogen sulfide. *Journal of Agricultural and Food Chemistry*, **23**, 501–505.

Roberts, D.D. & Acree, T.E. (1994) Gas chromatography-olfactometry of glucose-proline Maillard reaction products. In: *Thermally Generated Flavors: Maillard, Microwave, and Extrusion Processes* (eds T.H. Parliament, M.J. Morello & R.J. McGorrin), pp. 71–79. American Chemical Society, Washington, District of Columbia.

Scarpellino, R. & Soukup, R.J. (1993) Key flavors from heat reactions of food ingredients. In: *Flavor Science: Sensible Principles and Techniques* (eds T.E. Acree & R. Teranishi), pp. 309–335. American Chemical Society, Washington, District of Columbia.

Schieberle, P. & Hofmann, T. (1996) Identification of the key odorants in processed ribose-cysteine Maillard mixtures by instrumental analysis and sensory studies. In: *Flavor Science, Recent Developments* (eds A.J. Taylor & D.S. Mottram), pp. 175–181. The Royal Society of Chemistry, Cambridge, U.K.

Schieberle, P. & Hofmann, T. (1998) Characterization of key odorants in dry-heated cysteine-carbohydrate mixtures: Comparison with aqueous reaction systems. In: *Flavor Analysis: Developments in Isolation and Characterization* (eds C.J. Mussinan & M.J. Morello), pp. 320–330. American Chemical Society, Washington, District of Columbia.

Shahidi, F. (1998) *Flavor of Meat, Meat Products and Seafoods*, 2nd edn., p. 429. Blackie Academic and Professional, London, U.K.

Shu, C.K. & Ho, C.T. (1988) Effect of pH on the volatile formation from the reaction between cysteine and 2,5-dimethyl-4-hydroxy-3(2H)-furanone. *Journal of Agricultural and Food Chemistry*, **36**, 801–803.

Shu, C.K., Mookherjee, B.D. & Ho, C.T. (1985) Volatile components of the thermal degradation of 2,5-dimet hyl-4-hydroxy-3(2H)-furanone. *Journal of Agricultural and Food Chemistry*, **33**, 446–448.

Shu, C.K., Hagedorn, M.L. & Ho, C.T. (1986) Two novel thiophenes identified from the reaction between cysteine and 2,5-dimethyl-4-hydroxy-3(2H)-furanone. *Journal of Agricultural and Food Chemistry*, **34**, 344–346.

Tareke, E., Rydberg, P., Karlsson, P., Eriksson, S. & Törnqvist, M. (2002) Analysis of acrylamide, a carcinogen formed in heated foodstuffs. *Journal of Agricultural and Food Chemistry*, **50**, 4998–5006.

Tonsbeek, C.H.T., Plancken, A.J. & van de Weerdhof, T. (1968) Components contributing to beef flavor: Isolation of 4-hydroxy-5-methyl-3(2H)-furanone and its 2,5-dimethyl homolog from beef broth. *Journal of Agricultural and Food Chemistry*, **16**, 1016–1021.

Vauthey, S., Milo, C., Frossard, P., Garti, N., Leser, M.E. & Watzke, H.J. (2000) Structured fluids as microreactors for flavor formation by the Maillard reaction. *Journal of Agricultural and Food Chemistry*, **48**, 4808–4816.

Weenen, H. & de Rooij, J.F.M. (1998) Process flavorings. In: *Flavorings: Production, Composition, Applications, Regulations* (eds E. Ziegler & H. Ziegler), pp. 233–258. Wiley-VCH, Weinheim, Germany.

Whitfield, F.B. & Mottram, D.S. (1999) Investigation of the reaction between 4-hydroxy-5-methyl-3(2H)-furanone and cysteine or hydrogen sulfide at pH 4.5. *Journal of Agricultural and Food Chemistry*, **47**, 1626–1634.

Whitfield, F.B., Mottram, D.S., Brock, S., Puckey D.J. & Salter, L.J. (1988) Effect of phospholipid on the formation of volatile heterocyclic compounds in heated aqueous solutions of amino acids and ribose. *Journal of the Science of Food and Agriculture*, **42**, 261–272.

Wu, Y.F. & Cadwallader, K.R. (2002) Characterization of the aroma of a meatlike process flavoring from soybean-based enzyme hydrolyzed vegetable protein. *Journal of Agricultural and Food Chemistry*, **50**, 2900–2907.

Wu, Y.F., Baek, H.H., Gerard, P.D. & Cadwallader, K.R. (2000) Development of a meat-like process flavoring from soybean-based enzyme-hydrolyzed vegetable protein (E-HVP). *Journal of Food Science*, **65**, 1220–1227.

Yaylayan, V.A., Wnorowski, A. & Locas, C.P. (2003) Why asparagines needs carbohydrates to generate acrylamide. *Journal of Agricultural and Food Chemistry*, **51**, 1753–1757.

Zhang, Y. & Ho, C.T. (1991a) Comparison of the volatile compounds formed from the thermal reaction of glucose with cysteine and glutathione. *Journal of Agricultural and Food Chemistry*, **39**, 760–763.

Zhang, Y. & Ho, C.T. (1991b) Formation of meatlike aroma compounds from thermal reaction of inosine 59-monophosphate with cysteine and glutathione. *Journal of Agricultural and Food Chemistry*, **39**, 1145–1148.

8 Savory Flavors

Christoph Cerny

Abstract: Flavor of a food is not caused by a single compound or a single compound class, but by the sensory impression stimulated by a multitude of differently structured chemicals. These components can be divided into three categories: volatile aroma compounds (smell), nonvolatile compounds (five basic tastes), and synergistic intensifying compounds.

These savory flavors are detailed for meat (beef, chicken, pork, sheep) and for fish and seafood.

Keywords: Savory flavors, meat flavor, fish flavor, seafood flavor, taste compounds

Savory-tasting food like meat and fish form a substantial part of our diet and are important nutrient sources for biologically high-grade proteins, iron, and vitamins. Vegetables add minerals, vitamins, dietary fibers, and antioxidants to our diet. The pleasure of eating savory dishes depends on good quality of the foodstuffs as well as on a well-balanced and stimulating recipe and finally on the perfectly executed preparation and cooking. The resulting flavor is not caused by a single compound or a single compound class, but by the sensory impression stimulated by a multitude of differently structured chemicals. These components can be divided into three categories (Dwivedi & Snell 1975):

- Aroma compounds, which are volatile and perceived by the sense of smell through the olfactory receptors in the nasal mucosa
- Nonvolatile compounds, which contribute the five basic tastes (sweet, sour, salty, bitter, and umami) as well as tactile properties by interacting with receptors in the oral cavity
- Synergistic compounds intensifying the effect of other flavor sensations

More than 7000 volatiles have been identified in food to date (Boelens 2000), whereof only a fraction actually contributes to the aroma. Analytical techniques, which combine instrumental analysis with sensory methods, are suitable to distinguish the aroma-active volatiles from the others. Most of these techniques are based on the aroma-value concept (Rothe & Thomas 1963) and use gas chromatography-olfactometry (GC-O). Aroma extract dilution analysis (AEDA) is a further development of GC-O (Gasser & Grosch 1988). Successively diluted aroma extracts are analyzed by GC-O until the odor is no longer perceived. The flavor dilution (FD) factor indicates the highest dilution at which a compound is still detected at the sniffing port. The aroma value or odor activity value (OAV) is defined as the quotient of concentration and odor threshold of a compound in a given matrix.

Handbook of Meat, Poultry and Seafood Quality, Second Edition. Edited by Leo M. L. Nollet.

Table 8.1 Identified volatiles and aroma compounds in boiled beef.

	Volatiles*	Aroma compounds†
Hydrocarbons	73	0
Alcohols	32	1
Aldehydes	57	20
Ketones	49	11
Acids	9	5
Esters and lactones	11	0
Bases	91	1
Sulfur compounds	124	19
Ethers	5	0
Halogens	6	0
Nitriles and amides	1	0
Phenols	3	3
Furans	22	2
Oxazoles	5	0

*Number of identified volatiles according to Boelens (2000).
†Number of aroma-active compounds identified by GC-O (Gasser & Grosch 1988; Konopka & Grosch 1991; Kerscher & Grosch 1997, 1998; Kerscher 2000).

Table 8.1 compares the identified volatiles in boiled beef (Boelens 2000) with the aroma impact compounds (Gasser & Grosch 1988; Konopka & Grosch 1991; Kerscher & Grosch 1997, 1998; Kerscher 2000). It becomes obvious that the volatiles outnumber the aroma-active ones among them and that not all chemical classes are equally important. The objective of this chapter is to concentrate the discussion on the aroma-active volatiles, and in addition on the nonvolatiles that contribute to the taste of meat, fish, and seafood.

8.1 MEAT FLAVOR

Meat flavor has been the subject of many review articles (Dwivedi & Snell 1975; Ohloff & Flament 1978; MacLeod *et al.* 1981; Baines & Mlotkiewicz 1984; Werkhoff *et al.* 1993; Mottram 1998) and books dedicated to this subject (Shahidi 1994, 1998). The most popular and most commonly produced meats worldwide are pork (86 million tons per year), poultry (58 million tons), beef (54 million tons), and lamb (7 million tons) (Belitz *et al.* 1999).

The precursors responsible for the basic aroma of meat are generally water soluble and found in the lean part of the meat, whereas the species-specific aroma compounds stem mainly from the lipids (Hornstein & Crowe 1960; Hornstein *et al.* 1963). Table 8.2 shows the concentrations of potential water-soluble aroma precursors in beef (top round) meat. Besides the species of the meat, the way of cooking has a great influence on the aroma. While, for example, acetaldehyde, methylpropanal, methional, octanal, nonanal, and 2(*Z*)-nonenal dominate the aroma of roasted chicken meat (Kerscher 2000), the odor of boiled chicken is caused mainly by 2-furfurylthiol, hexanal, 2(*E*)-nonenal, 2,4(*E*,*E*)-nonadienal, 2,4(*E*,*E*)-decadienal, and 2,4(*E*,*Z*)-decadienal (Kerler & Grosch 1997). In chicken skin, 3-mercapto-2-pentanone, acetaldehyde, methylpropanal, 2-methylbutanal, 3-methylbutanal, and methional show the highest OAVs (Kerscher 2000). Methanthiol is a key aroma compound in all three types of chicken aroma. The cooking time equally affects the formation of aroma compounds. The concentration of 4-hydroxy-2, 5-dimethyl-3(2H)-furanone (HDMF) in beef increases continuously up to 1.6 mg/kg during 15 min of roasting, and even reaches 11.6 mg/kg when stewed for 4 h (Grosch *et al.* 1993).

Table 8.2 Free amino acids, thiamine, and sugars in beef.

	Beef (mg/kg)*,†
Alanine	313
Arginine	76
Aspartic acid	7
Cysteine	37
Glutamic acid	103
Glycine	96
Histidine	48
Isoleucine	44
Leucine	80
Lysine	56
Methionine	23
Phenylalanine	52
Proline	39
Serine	48
Threonine	32
Tyrosine	48
Valine	80
Glutamine	706
Thiamine	0.8‡
Glucose-6-phosphate	1834
Glucose	1072
Fructose-6-phosphate	503
Fructose	283

*Based on moist mass of raw meat; beef: top round; chicken: breast.
†Grosch *et al.* (1993).
‡Zeiler-Hilgart (1994).

8.1.1 Beef

Table 8.3 lists the volatiles that have been shown by GC-O methods to be aroma active in roasted, boiled, and stewed beef, respectively. Fig. 8.1 shows some of their structures. The quantitative data were obtained by stable isotope dilution assay (Sweeley *et al.* 1966; Schieberle & Grosch 1987), which is a reliable method to analyze delicate aroma compounds, which are unstable or occur just in minute concentrations.

Sulfur-containing compounds have been thought to contribute to the meaty character of cooked beef (Wilson *et al.* 1973), in particular sulfides like 2-methyl-3-furanthiol (compound 1 in Fig. 8.1), 2-furfurylthiol (2), and methanthiol (3) (Gasser & Grosch 1988; Guth & Grosch 1994). According to van den Ouweland *et al.* (1989), compounds with meaty character possess a five-membered ring, a double bond, and a methyl and a thiol group. While the role of sulfur compounds for the aroma of boiled beef is important, roasted beef is more dominated by roasted notes deriving from certain pyrazines together with aldehydes from lipid degradation (Cerny & Grosch 1992; Kerler & Grosch 1996).

Key reactions for the generation of 1 and 2 in meat are, on the one hand, the Maillard reaction between the amino acid cysteine and the reducing sugar ribose (Farmer *et al.* 1989; Farmer & Mottram 1990), and on the other hand, the degradation of thiamine (Arnold *et al.* 1969; Dwivedi & Arnold 1973; van der Linde *et al.* 1979). Aldehydes are mainly formed via oxidation of unsaturated fatty acids like oleic, linoleic, and linolenic acid present in the lipid fraction. Strecker degradation of valine, isoleucine, leucine, methionine, and

Table 8.3 Aroma-active compounds in beef, chicken, and pork.

No.		Beef		Chicken			Pork	
		Roasted*	Boiled†	Stewed‡	Boiled§	Roasted**	Boiled††	Stewed‡‡
	Aldehydes							
9	Acetaldehyde	+§§	1817	6400	4700	3815	3953	1500
11	Methylpropanal	+	117		39	83	90	
13	2-Methylbutanal	+			+			
12	3-Methylbutanal	16	26	10	5	17	27	21
6	Methional	9–33	36	13	15	53	11	23
15	Phenylacetaldeyhde	+	+	+	+			
	Pentanal				+			
23	Hexanal	269	+	72	1596	283		15
	2(*E*)-Hexenal		+					
	Heptanal	+	+		+			
	2(*E*)-Heptenal		+		+			
25	Octanal	+	382	+	148	190	154	+
	2(*E*)-Octenal		+					
26	Nonanal	+	1262	+	+	534	643	
27	2(*E*)-Nonenal	42	32	+	75	23	15	
	2(*Z*)-Nonenal	+	6	+		6	+	+
	3(*Z*)-Nonenal	+						
28	2,4(*E*,*E*)-Nonadienal	+	+	+	57			
30	2,6(*E*,*Z*)-Nonadienal	+	+			3		
	Decanal			+	+			
	2(*E*)-Decenal				+			
	2(*Z*)-Decenal		+					
	4,5-Epoxy-2(*E*)-decenal	1.5		+	+			+
31	2,4(*E*,*E*)-Decadienal	+	+	12	302	11	7	10
	2,4(*E*,*Z*)-Decadienal		27		31		7	
	2-Undecenal				+			
	2(*E*)-Dodecenal	+						
32	12-Methyltridecanal		962	52				

	Ketones							
14	2,3-Butanedione	+	+	46				38
	2,3-Pentanedione	+						
	2-Heptanone		+					
	2-Octanone		+					
33	1-Octen-3-one	+	9	+	6	7	5	+
34	1,5(*Z*)-Octadien-3-one	+	+		+			
	2-Nonanone		+					
	1-Nonen-3-one				+			
	2-Decanone		+					
	2-Undecanone		+					
	2-Dodecanone		+					
	2-Tridecanone		+					
	b-Ionone		+		+			
	Alcohols							
	1-Octen-3-ol		+					
	Tridecanol				+			
	Acids/phenols							
	Acetic acid	+	+	200 000	22 100			270 000
	Butyric acid	+	7 074	6 100	3 800	8 119	17 200	3 900
	2-/3-Methylbutyric acid		+	74	+			95
	Valeric acid		+					
	Phenylacetic acid		+					
	Guaiacol	2–8	+		+			
	p-Cresol		+		+			
	p-Vinylguaiacol			+				+
	Vanilline		+					
	Sulfur compounds							
	Hydrogen sulfide				+	290		
3	Methanthiol	+	311	300	1456	202	278	500
4	Dimethylsulfide	+	105					
5	Dimethyl trisulfide	+	+	+	0.8			+

(*Continued*)

Table 8.3 (*Continued*)

No.		Beef			Chicken		Pork	
		Roasted*	Boiled†	Stewed‡	Boiled§	Roasted**	Boiled††	Stewed‡
+	2-Methyl-3-furanthiol		7–28		1–5	0.4–4	6–9	
	2,5-Dimethyl-3-furanthiol				+			
	2-Methyl-3-(methylthio)furan		+		+			
	2-Methyl-3(methyltrithio)furan				+			
8	3-Mercapto-2-pentanone		55–73	+	23–100	27–31	66–117	
7	2-Mercapto-3-pentanone		20–44		2–13	0.5–3.4	11–14	
2	Furfurylthiol		13–42	0.5	0.4–5	0.1–1.9	8–10	0.6
	Benzyl mercaptane		+					
	Bis(2-methyl-3-furyl)disulfide		+	+				
	Bis(2-furfuryl)disulfide				+			
	2-Methylthiophene				+			
	2-Formylthiophene				+			
	2-Formyl-5-methylthiophene		+		+			
	2-Acetylthiophene		+		+			
	3-Acetyl-2,5-dimethylthiophene	+						
	2,4,6-Trimethyltetrahydro-1,3,5-thiadiazine				+			
	3,5-Dimethyl-1,2,4-trithiolane				+			
	5,6-Dihydro-2,4,6-trimethyl-4H-1,3,5-dithiazine				+			
	Lactones							
	γ-Octalactone	+						
	γ-Decalactone				+			
	γ-Dodecalactone			+	+			
17	4,5-Dimethyl-3-hydroxy-2(5H)-furanone	9	+	5	5			3
	Other Heterocycles							
	Trimethylpyrazine	+						
20	2-Ethyl-3,5-dimethylpyrazine	2–5		+	+			

21	2,3-Diethyl-5-methylpyrazine	1–27		+			+
	2-Acetylthiazole	+	+		+		
19	2-Acetyl-2-thiazoline	12–28	+	+	14		5
	2,4,5-Trimethylthiazole				+		
	Benzothiazole		+				
18	2-Acetyl-1-pyrroline	+	+	+	+		
	Indol				+		
16	2,5-Dimethyl-4-hydroxy-3-659–1108 9075 (2H)-furanone	8000	1232	50	2170	2700	
	4-Hydroxy-5-methyl-3(2H)-furanone			+			

*Literature data (Cerny & Grosch 1992, 1993; Kerler & Grosch 1996).
†Literature data (Gasser & Grosch 1988; Kerscher & Grosch 1997, 1998; Kerscher 2000).
‡Literature data (Guth & Grosch 1993, 1994, 1995).
§Literature data (Gasser & Grosch 1990; Farkas *et al.* 1997; Kerscher & Grosch 1997, 1998; Farmer *et al.* 2000).
**Literature data (Kerscher & Grosch 1998; Kerscher 2000).
††Literature data (Kerscher & Grosch 1998; Kerscher 2000).
‡‡Literature data (Guth & Grosch 1995).
§§"1": detected by GC-olfactometry; concentration in (mg/kg), as determined by isotope dilution analysis (IDA).

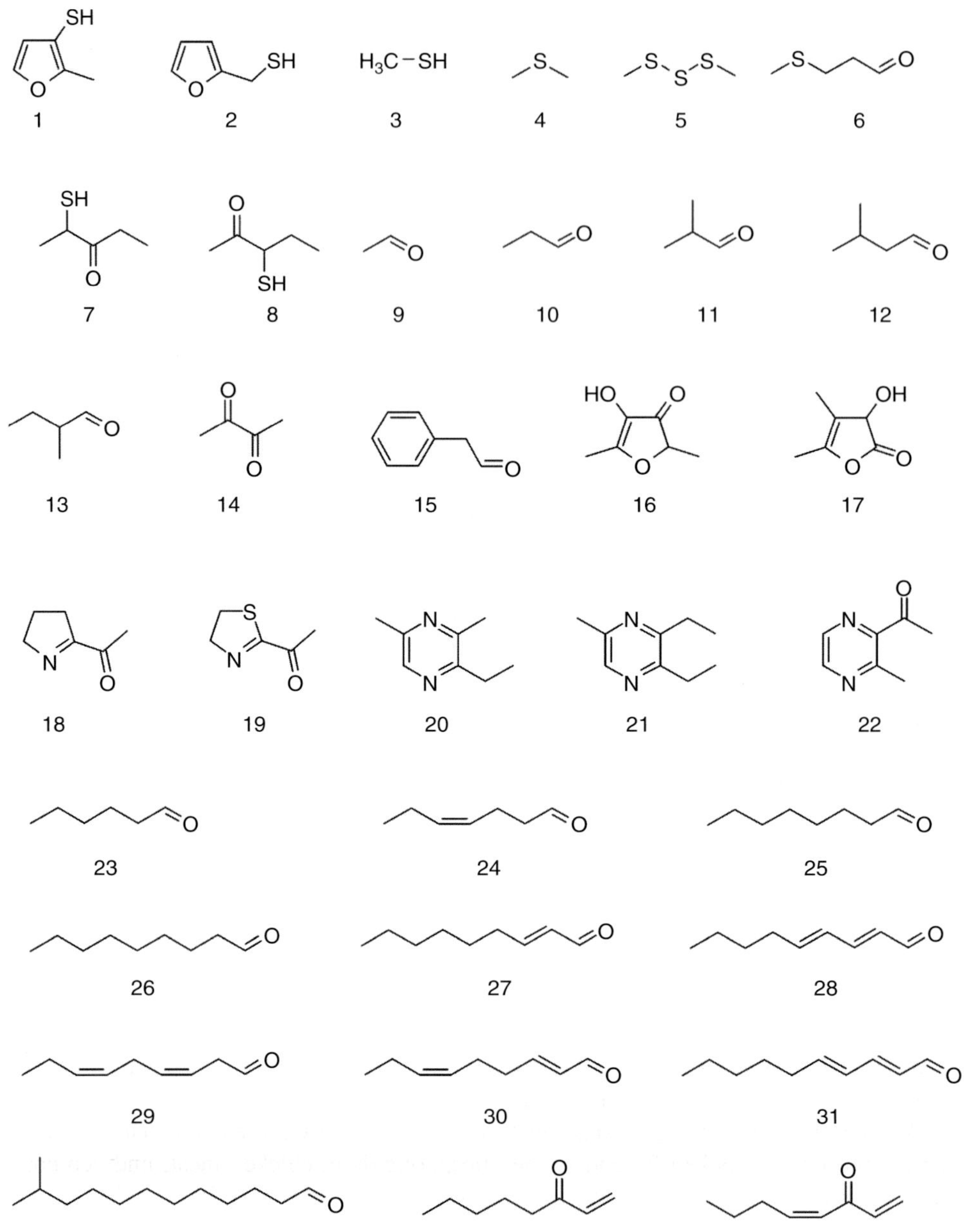

Fig. 8.1 Chemical structure of some key aroma compounds of meat, fish, and seafood.

phenylalanine causes the formation of methylpropanal (11), 2-methylbutanal (13), 3-methylbutanal (12), methional (6), and phenylacetaldehyde (15). The long-chain branched aldehyde 12-methyltridecanal (32) was considered a species-specific aroma compound in boiled and stewed beef (Guth & Grosch 1993). The compound 32 occurs also in higher concentrations in lamb, springbok, and deer, but only at a very low level in pork and chicken, and hence seems to be characteristic for ruminant animals. Possibly it is formed by the rumen

Table 8.4 Recombination model mixture of boiled beef aroma (Kerscher 2000).*

No.†	Odorant	Odor	µg/kg	Odor activity value‡
32	12-Methyltridecanal	Tallowy, fatty	962	9620
1	2-Methyl-3-furanthiol	Meaty	24	3429
2	2-Furfurylthiol	Sulfury, roasted	29	2900
3	Methanthiol	Sulfury, rotten	311	1555
26	Nonanal	Fatty	1262	1262
16	4-Hydroxy-2,5-dimethyl-3(2H)-furanone	Caramel	9075	908
25	Octanal	Citrus	382	546
4	Dimethylsulfide	Cooked asparagus	105	350
	2(*Z*)-Nonenal	Fatty, green	6	310
33	1-Octen-3-one	Mushroom	9	188
9	Acetaldehyde	Fruity, pungent	1817	182
6	Methional	Boiled potato	36	180
11	Methylpropanal	Malty	117	167
31	2,4(*E*,*E*)-Decadienal	Fatty, fried	27	135
27	2(*E*)-Nonenal	Fatty, cardboard	32	128
8	3-Mercapto-2-pentanone	Sulfury, catty	69	99

*Values taken from Kerscher (2000).
†Numbering corresponds to Fig. 8.1.
‡OAVs were calculated by dividing the concentration of the odorants by their nasal odor threshold in water.

micro flora, absorbed, and later incorporated into the plasmalogens of the muscle membranes from where it is released during longer cooking (Guth & Grosch 1993).

Kerscher (2000) has, based on the concentrations of the key aroma compounds found in boiled beef, prepared a synthetic recombination model mixture of the aroma (Table 8.4) and found it highly similar compared to the original beef sample. In subsequent omission tests, the absence of 32 did not change the aroma character significantly, indicating that its contribution to beef flavor is much lower than previously assumed (Kerscher 2000).

8.1.2 Chicken

Chicken meat has in general a higher fat content than beef (Belitz *et al.* 1999). Chicken fat contains more unsaturated fatty acids, especially linoleic acid (19.5 %), than do both beef (3.1 %) and pork fat (9.5 %) (United States Department of Agriculture 2003). The fatty acids, which are present in chicken fat and in the phospholipids in chicken meat, undergo partial oxidation during cooking. Aldehydes and ketones belong to the major volatiles formed. Consequently, numerous review articles have assigned carbonyl compounds an important role in chicken aroma (Wilson & Katz 1972; Schliemann *et al.* 1988; Shi and Ho 1994), together with sulfur-containing compounds, which are believed to be another important chemical class.

Sulfur-containing compounds are more important for the aroma of boiled chicken than for roasted chicken. The concentrations of 1–3 and 3-mercapto-2-pentanone (8) are overall higher in boiled chicken (1–5 mg/kg; 0.4–5 mg/kg; 1456 mg/kg; 23–100 mg/kg) than in roasted chicken meat (0.4–4 mg/kg; 0.1–1.9 mg/kg; 202 mg/kg; 27–31 mg/kg) (Kerscher & Grosch 1998; Kerscher 2000). When compared to boiled beef, the concentrations of 1 and 2 are generally lower in boiled chicken (see Table 8.3).

Compounds 23 and 31 belong to the primary volatile products that are formed by autoxidation of linoleic acid at elevated temperatures (Grosch 1987). While 23 results from the breakdown of the corresponding 13-monohydroperoxide, 31 stems from the degradation of the 9-monohydroperoxide. Due to their low odor thresholds (23: 10 mg/L water; 31: 0.2 mg/L water) (Grosch *et al.* 1993), both compounds contribute to the flavor of boiled and roasted chicken. Pippen and Nonaka (1958) recognized the contribution of decadienal to chicken flavor. Roasted chicken skin, which is much more greasy, contains even higher levels than the meat with a concentration of 893 *versus* 283 mg/kg for 23 and a content of 711 *versus* 11 mg/kg for 31 (Kerscher 2000). Other volatiles with high impact on chicken aroma include the oleic acid degradation compounds heptanal, octanal (25), nonanal (26); the linoleic and arachidonic acid oxidation products 2(*E*)-nonenal (27), 2,4(*E*,*E*)-nonadienal (28), 1-octen-3-one (33), and 2,4(*E*,*Z*)-decadienal; as well as 2,6(*E*,*Z*)-nonadienal (30) and 1,5(*Z*)-octadien-3-one (34), which originate from linolenic acid oxidation.

Besides carbonyls from lipid autoxidation and sulfur compounds from Maillard reaction, aldehydes from the Strecker degradation of valine, leucine, isoleucine, and methionine contribute with high OAVs to the aroma of chicken: 6 and 11–13. Gasser and Grosch (1990) identified 14 primary odor compounds in chicken broth by AEDA: 1, 2, 6, 26–28, 31, 2,4,5-trimethylthiazole, 2-formyl-5-methylthiophene, *p*-cresol, 2-undecenal, b-ionone, γ-decalactone, and γ-dodecalactone. Kerler (Kerler & Grosch 1997) who quantified chicken volatiles and determined their OAVs found additionally 23, 25, 33, 34, acetic acid, butyric acid, 2/3-methylbutyric acid, 2,4(*E*,*Z*)-decadienal, 2-acetyl-2-thiazoline (19), 4-hydroxy-2,5-dimethyl-3(2H)-furanone (16), and 3-hydroxy-4,5-dimethyl-2(5H)-furanone (17), as well as the very volatile odorants hydrogen sulfide, methanthiol, and acetaldehyde to be contributing to chicken flavor. The key odorants in roasted chicken skin were analyzed by Kerscher (2000), who prepared a model solution of 20 aroma compounds with high OAVs in oil–water emulsion, which agreed well with the aroma of the original chicken skin (See Table 8.5). Volatile trithiolane derivatives, which have been identified in chicken flavor by different authors (Tang *et al.* 1983; Hartman *et al.* 1984), did not show up as aroma impact compounds in Kerscher's study. As with beef aroma, the very volatile compounds methanthiol and acetaldehyde had a high impact on the aroma of chicken.

8.1.3 Pork

Pork is the most consumed meat worldwide, particularly in China (Belitz *et al.* 1999), with negligible consumption in countries with a large Muslim or Jewish population. Nonetheless, its aroma has been studied much less than beef. While fat degradation products form the majority of the volatile fraction of ground pork meat (Mottram 1982), pyrazines dominate in grilled pork.

Gasser and Grosch (1991) found compounds 1 and 2 to be key aroma compounds of boiled pork, and the earthy, roasted smelling 2,4-dimethyl-5-ethylthiazole to be a pork-specific odorant. Later, Kerscher (2000) confirmed the importance of 1 and 2, and added 3, 9, 11, 16, 25, 26, 33, and 2,4(*E*,*Z*)-decadienal as key odorants. Pork aroma lacks the beef-specific aldehyde 32, but apart from that, differences between beef and pork aroma are more due to differences in the concentration of Maillard reaction products (Kerscher 2000). The differences in the aroma of braised pork and beef can be ascribed to the absence of 32 in pork and a considerably higher concentration of 16 in stewed beef (Guth & Grosch 1995).

In cured pork, the concentration of carbonyl compounds in the volatile fraction is drastically reduced versus uncured pork, for example, hexanal occurring at 12.66 mg/kg *versus* 0.03 mg/kg (Ramarathnam *et al.* 1991). The typical aroma note of cured pork is not the result of a single odorant, but the interaction of several aroma compounds. In Parma-type ham, 6, 15–18, 27, *p*-cresol, and phenylacetic acid belong to the relevant aroma compounds (Blank

Table 8.5 Recombination model mixture of chicken skin aroma.*

No.†	Odorant	Odor	µg/kg	Odor activity value‡
9	Acetaldehyde	Fruity, pungent	3 287	14 941
3	Methanthiol	Sulfury, rotten	164	2 733
6	Methional	Boiled potato	97	485
13	2-Methylbutanal	Malty	455	207
11	Methylpropanal	Malty	538	158
8	3-Mercapto-2-pentanone	Sulfury, catty	27	142
12	3-Methylbutanal	Malty	668	124
18	2-Acetyl-1-pyrroline	Roasted	2.9	29
16	4-Hydroxy-2,5-dimethyl-3(2H)-furanone	Caramel	395	16
25	Octanal	Citrus	535	9.6
17	3-Hydroxy-4,5-dimethyl-3(2H)-furanone	Seasoning-like	1.9	9.5
	2-Propionyl-1-pyrroline§	Roasted	0.8	8
1	2-Methyl-3-furanthiol	Meaty	4.1	7.3
21	2,3-Diethyl-5-methylpyrazine	Roasted	2.5	5
2	2-Furfurylthiol	Sulfury, roasted	1.9	4.8
28	2,4(*E*,*E*)-Decadienal	Fatty, fried	711	4
19	2-Acetyl-2-thiazoline	Roasted	5.8	3.2
20	2-Ethyl-3,5-dimethylpyrazine	Roasted	4.3	2
26	Nonanal	Fatty	832	≤1
27	2(*E*)-Nonenal	Fatty, cardboard	147	≤1

*Values and odor descriptors taken from Kerscher (2000).
†Numbering corresponds to Fig. 8.1.
‡OAVs were calculated by dividing the concentration of the odorants by their nasal odor threshold in water.
§2-Propionyl-1-pyrroline was substituted by 2-acetyl-1-pyrroline on the basis of OAVs.

et al. 2001). Interestingly, thiols like 1 and 2 as well as the fatty smelling lipid degradation product 31 did not seem to play a role for this flavor. Stahnke (2000) identified earlier the heterocyclic, popcorn-like smelling odorant 2-acetyl-1-pyrroline (18) in several sausages. She could show that the molded surface of sausages contained higher amounts than the center, indicating that the compound was microbially formed.

8.1.4 Sheep

In many countries, sheep meat has a low consumer acceptance due to an objectionable smell. Branched and unsaturated fatty acids with 8–10 carbon atoms have been identified in cooked mutton mince, and, because of their odor properties, were thought to contribute to the undesirable flavor (Wong *et al.* 1975). Metabolic processes in the rumen of the animals are probably involved in their formation. These acids include 4-methyloctanoic and 4-ethyloctanoic acids, both of which possess low odor thresholds and were therefore regarded as important contributors of sheep flavor (Brennand & Lindsay 1992; Brennand *et al.* 1989). Sutherland and Ames (1996) analyzed the level of branched-chain fatty acids in the adipose tissue of lambs and found 3–5 mg/kg 4-methyloctanoic acid. Ha and Lindsay (1990) made the higher level of 4-ethyloctanoic acid in the meat of older animals responsible for the stronger characteristic mutton odor. Other authors suggest aldehydes and ketones, which were identified in the neutral volatile fraction of ovine adipose tissue, as important contributors to mutton flavor (Caporaso *et al.* 1977). Buttery *et al.* (1977) suspect alkyl pyridines, found in roasted lamb fat, to be the reason for its rejection by some consumers.

A more recent study (Schieberle & Rota 2003) analyzed raw and cooked sheep meat using GC-O and isotope dilution analysis to quantify the aroma-relevant compounds. According to their OAVs, compounds 3 and 34 contribute most to the aroma of the cooked meat. Compound 34 was already present before cooking and was, together with nonanal, the most potent odorant of the raw meat. Interestingly, the level of the lipid degradation-derived aroma compounds did not change significantly during cooking. The role of the intensely sheep-like smelling 4-methyloctanoic and 4-ethyloctanoic acids could be confirmed. Both acids, which are located in the membrane lipids, are already present in the raw meat.

8.2 FISH AND SEAFOOD FLAVOR

8.2.1 Fish

Fish has a high nutritional value because it is rich in polyunsaturated fatty acids (PUFA), especially v-3-fatty acids; has high value protein and iodine (50–200 mg/100 g in seafish, also known as saltwater fish); and in general contains less fat than meat.

Raw fresh fish usually has a green, seaweed-like aroma (Josephson *et al.* 1983). In the headspace of fresh whitefish, 2(*E*)-nonenal (27), 2,6(*E*,*Z*)-nonadienal (30), 6-nonen-1-ol, 1-octen-3-ol, 1-octen-3-one (33), 1,5-octadien-3-ol, 1,5-octadien-3-one (34), 2,5-octadien-1-ol, and other compounds were identified by GC-O. The single compounds were described as having cucumber-like, melon-like, mushroom-like, and plant-like aromas (Josephson *et al.*1983). Hirano *et al.* (1992) analyzed the aroma of fresh Ayu fish, a small Japanese river trout. GC-O analysis revealed 27, 30, and 3,6-nonadien-1-ol to be important for the cucumber and melon-like flavor of the fish. Another unsaturated aldehyde, 4(*Z*)-heptenal (24), was found to be important for the aroma of cold-stored cod (McGill *et al.* 1974, 1977). The authors explained its formation by the oxidation of PUFA present in the fish phospholipid fraction. The 12-lipoxygenase, present in fish tissues, essentially in gills, is able to generate carbonyl compounds from arachidonic and eicosapentaenoic acids (Hsieh & Kinsella 1989), yielding 1-octen-3-ol, 2-octenal, 2-nonenal, 2,6-nonadienal, 1,5-octadien-3-ol, and 2,5-octadien-1-ol as the main volatiles. Distinct differences in flavor exist between different fish species and even between fish of the same species but different geographic origin (Prell & Sawyer 1988). Sérot *et al.* (2001) could show that the diet also affected the aroma composition of carnivore fish. Turbot under a high PUFA diet contained more 30, 2(*E*)-pentenal, and 1,3,5(*E*,*E*,*Z*)-octatriene, whereas a vegetable oil diet resulted in a higher hexanal and decanal level in the fish. Trimethylamine has often been related to fish odor, especially to the amine-like note. However, it is less present in fresh fish, but formed by bacterial spoilage from its N-oxide at temperatures above 0 °C. The precursor trimethylamine oxide level in the serum of marine fishes is between 0.2 and 47 millimols per liter (mmol/L), depending on the species (Shibamoto & Horiuchi 1997).

Cooking changes the flavor profile of fish. Milo *et al.* have carried out several studies to identify and quantify the volatile compounds that are relevant for the aroma of boiled salmon, cod, and trout (Milo & Grosch 1993, 1995, 1996, 1997; Milo 1994). Their data are summarized in Table 8.6. The most important aroma compounds in cooked salmon are 34 (OAV 750), 30 (OAV 485), 10 (OAV 240), 9 (OAV 230), 6 (OAV 200), and 31 (OAV 120) (Milo 1994; Milo & Grosch 1996).

The key odorants of cooked salmon, with the exception of 3,6(*Z*,*Z*)-nonadienal, are already present in raw fish. The concentration of 6 increases by a factor of 2–3, the concentration of compounds 23 and 24 doubles. Compound 10 is reduced to half its concentration, but the other

Table 8.6 Aroma-active compounds in cooked fish.

No.		Salmon (boiled)*	Cod (boiled)†	Trout (boiled)‡
	Aldehydes			
9	Acetaldehyde	2300	1300	3600
10	Propionanldehyde	1700		2300
11	Methylpropanal		27	
13	2-Methylbutanal		20	
12	3-Methylbutanal		51	100
6	Methional	1	11	6
23	Hexanal	58	115	37
25	Octanal	+§		+
	3(*Z*)-Hexenal	4	1	2
24	4(*Z*)-Heptenal	1	2	3
	2(*E*)-Octenal			+
27	2(*E*)-Nonenal	2.7		3
	2(*Z*)-Nonenal	+		+
	4,5-Epoxy-2(*E*)-decenal	+	+	+
	2,4(*E*,*E*)-Heptadienal			+
28	2,4(*E*,*E*)-Nonadienal	3	3	5
30	2,6(*E*,*Z*)-Nonadienal	10	4	8
29	3,6(*Z*,*Z*)-Nonadienal	6	1	+
31	2,4(*E*,*E*)-Decadienal	6	4	
	2,4(*E*,*Z*)-Decadienal			+
34	1,5(*Z*)-Octadien-3-one	0.3	0.1	0.4
	3-Methylnonane-2,4-dione			+
	Ketones			
14	2,3-Butanedione	52	200	268
	2,3-Pentanedione	234	86	140
33	1-Octen-3-one	0.4	0.7	0.6
	Acids			
	Butyric acid	1137		
	2-/3-Methylbutyric acid	122		
	Sulfur compounds			
3	Methanthiol		100	
4	Dimethylsulfide		77	
5	Dimethyl trisulfide	+	0.2	
	Dimethyl tetrasulfide		+	
	Heterocycles			
19	2-Acetyl-2-thiazoline			+
18	2-Acetyl-1-pyrroline	+	+	+

*Literature data (Milo 1994; Milo & Grosch 1996).
†Literature data (Milo 1994; Milo & Grosch 1995; Milo & Grosch 1996).
‡Literature data (Milo & Grosch 1993; Milo 1994; Milo & Grosch 1995).
§"1": detected by GC-olfactometry; concentration in (mg/kg), as determined by isotope dilution analysis.

odorants are less affected. Most of the important aroma compounds in trout are also found in salmon. Exceptions are 31, which adds less to the aroma of trout, but is important for the fatty note of salmon, and 12 (OAV 400), which plays a major role in trout flavor (Milo & Grosch 1993, 1995; Milo 1994). The sulfur compounds 3 (OAV 50) and 4 (OAV 39) are involved in the aroma of boiled cod but not of salmon or trout, in addition to 6 (OAV 275), 34 (OAV 250), 30 (OAV 175), 12 (OAV 204), and 9 (OAV 130). Recombination studies have shown that already an aqueous solution containing only two aroma compounds, namely 6 (10 mg/kg) and 34 (0.16 mg/kg), has the ability to evoke a fresh fish-like odor, along with potato- and

Table 8.7 Aroma-active compounds of cooked seafood.

No.		Crab (cooked)*	Crayfish (cooked)†	Lobster (cooked)‡
	Aldehydes			
9	Acetaldehyde		+§	+
11	Methylpropanal		+	+
13	2-Methylbutanal			+
12	3-Methylbutanal		+	+
6	Methional	+	+	+
23	Hexanal			+
	Heptanal			+
24	4(*Z*)-Heptenal	+	+	+
25	Octanal			+
	2(*E*)-Octenal		+	+
27	2(*E*)-Nonenal		+	+
	4(*E*)-Decenal	+		
28	2,4(*E,E*)-Nonadienal		+	
30	2,6(*E,Z*)-Nonadienal		+	+
29	3,6(*Z,Z*)-Nonadienal			+
31	2,4(*E,E*)-Decadienal		+	+
	Ketones			
14	2,3-Butanedione	+	+	+
	2-Heptanone			+
	1-Hexen-3-one			+
33	1-Octen-3-one		+	+
34	1,5(*Z*)-Octadien-3-one		+	+
	b-Damascenone			+
	o-Aminoacetophenone			+
	3-Methyl-1-butanol			+
	2-Methylisoborneol		+	
	Phenols, amines			
	Guaiacol			+
	p-Cresol		+	
	Trimethylamine	+	+	+
	Sulfur compounds			
	Hydrogen sulfide		+	+
3	Methanthiol		+	+
4	Dimethylsulfide			+
5	Dimethyl trisulfide		+	+
	Dimethyl tetrasulfide		+	
1	2-Methyl-3-furanthiol		+	+
	Heterocycles			
	Trimethylpyrazine			+
20	2-Ethyl-3,5-dimethylpyrazine			+
	2-Acetylpyrazine			+
22	2-Acetyl-3-methylpyrazine			+
	2-Acetylpyridine			+
	2,4,6-Trimethylpyridine			+
	2-Acetylthiazole			+
19	2-Acetyl-2-thiazoline		+	
	2,4,5-Trimethylthiazole			+
	Benzothiazole			+
18	2-Acetyl-1-pyrroline	+	+	+
	Pyrrolidine	+		
	3-Methylindol		+	+
	Benzothiophene			+
16	2,5-Dimethyl-4-hydroxy-3-(2H)-furanone			+

*Literature data (Chung & Cadwallader 1994; Chung *et al.* 1995).
†Literature data (Cadwallader & Baek 1998).
‡Literature data (Cadwallader *et al.* 1995; Lee *et al.* 2001).
§"+": detected by GC-olfactometry.

geranium-like notes (Belitz *et al.* 1999). Thiols such as 1 and 2, which represent key odorants of boiled meat, seem to be unimportant for the aroma of fish. Only tuna, which contains 1, is an exception in this respect (Withycombe & Mussinan 1988).

The most potent aroma compounds in ripened anchovy are 6, 9, 11, 12, and 34 (Triqui & Guth 1997). In addition to these, Cha *et al.* (1997) found high odor values in salt-fermented anchovy for 6, 24, 30, and 34, as well as miscellaneous esters and 3-methyl-1-butanol and 2-phenylethanol, which obviously stem from fermentation.

8.2.2 Seafood

Crustaceans and mollusks are rich in protein. Crab, lobster, and crayfish, respectively, contain 15 %–19 % protein; this is more than 80 % of the dry matter content. Table 8.7 lists the relevant aroma compounds in cooked crab, crayfish, and lobster, which have been identified by the research group of Cadwallader (Chung & Cadwallader 1994; Cadwallader *et al.* 1995; Chung *et al.* 1995; Cadwallader & Baek 1998, Lee *et al.* 2001) using GC-O. Similar to most fish, degradation products of PUFA, contribute to the aroma of seafood. On the other hand, 2-acetyl-1-pyrroline (18), a known Maillard degradation product of the amino acid proline with a cracker-like odor, is an important aroma impact compound and plays a major role for the aroma of seafood products, which is not the case with fish.

Together with 6, 14, and 24, 18 was found to be the most important odorant in cooked crabmeat (Chung & Cadwallader 1994; Cadwallader *et al.* 1995). Likewise, in crayfish, 18 and 6 dominate the aroma, together with 19, 28, 30, and 31 (Cadwallader & Baek 1998). In the headspace above the cooked crayfish tail meat, 3, 33, 34, hydrogen sulfide, trimethylamine, and dimethyltrisulfide were detected as aroma-active compounds.

In cooked lobster meat, again 18 was the most intense odorant (Cadwallader *et al.* 1995). Other dominant aroma compounds are 6, 14, 24, 33, 2-acetyl-3-methylpyrazine, and trimethylamine. The authors held the latter three volatiles responsible for a negative impact on lobster aroma. Lee *et al.* (2001) confirmed the role of 6, 14, 18, and 33 for lobster aroma. In addition, 12, 33, 34, and the very volatile sulfur compounds 3 and 4 belonged to the predominant odorants of lobster.

8.3 TASTE COMPOUNDS

Taste compounds, molecules that cause a taste sensation when interacting with receptors on the tongue and in the mouth cavity, are in general nonvolatile, polar, and water soluble. The existence of five basic tastes is broadly agreed: sweet, sour, salty, bitter, and umami. The word "umami" originates from the Japanese language and means "delicious." Umami taste can be described as glutamate-like or bouillon-like. A short time ago, a corresponding receptor was identified in the tongue tissue thereby confirming that the umami is actually a basic taste (Chaudhari *et al.* 2000). As with aroma compounds, the analysis of taste compounds requires a combination of instrumental and sensory methods. A systematic approach uses the quantification of potentially taste-active, water-soluble compounds in the respective food, recombination trials, and sensorial comparison with the original food to verify if all major taste compounds are included, followed by omission tests to narrow down the number of relevant tastants (Hayashi *et al.* 1981; Warendorf *et al.* 1992; Schlichtherle-Cerny & Grosch 1998; Warmke *et al.* 2004).

Amino acids and peptides are regarded as important taste compounds in meat (Mabrouk 1976; van den Ouweland *et al.* 1978). They show a ternary synergism—that is, they have no

Table 8.8 Taste-active compounds in beef juice and boiled crab extract.

No.	Concentration in mg/kg	Stewed beef juice*	Boiled crab extract†
35	Glycine		6230
36	Glycine betaine		3570
37	Alanine	840	1870
38	Arginine		5790
39	Glutamic acid	250	190
40	Aspartic acid	20	
41	Cysteine	60	
42	Carnosine	5 280	
43	L-Lactic acid	14 000	
44	Succinic acid	260	
45	5′-AMP	240	320
46	5′-CMP		60
47	5′-GMP		40
48	5′-IMP	2 800	
49	Creatinine	4 900	
50	Creatine	3 030	
51	Hypoxanthine	500	
	K^{1}	6 020	1971
	Na^{1}	940	993
	Mg^{21}	410	
	$PO4^{3-}$	5 670	2169
	Cl^{-}	670	1946

*Literature data (Schlichtherle-Cerny & Grosch 1998).
†Literature data (Hayashi *et al.* 1981; Konosu & Yamaguchi 1987).

characteristic taste on their own, but enhance the basic taste in combination with other compounds like glutamic acid and 5′-nucleotides. Alongside amino acids, other organic acids play an essential role (van den Ouweland *et al.* 1978). Warendorf *et al.* (1992) analyzed taste-active compounds in a beef bouillon and found that only low-molecular compounds (<500 dalton [Da]) were taste active, the most important ones being glutamic acid (39), 5′-adenosine monophosphate (45, AMP), 5′-inosine monophosphate (48, IMP), aspartic acid (40), the dipeptides carnosine (42) and anserine, lactic acid (43), carnitine and sodium, potassium, magnesium, calcium, chloride, and phosphate ions. In beef juice, only 47 low-molecular weight compounds were taste active (Schlichtherle-Cerny & Grosch 1998). Their number could be narrowed down by screening trials to 17 compounds, which are listed in Table 8.8. Fig. 8.2 shows the corresponding chemical structures.

The two 5′-nucleotides 45 (240 mg/kg) and 48 (2800 mg/kg), 39 (250 mg/kg), and 40 (20 mg/kg) were responsible for the bigger part of the brothy, umami taste of beef juice. Compound 39 had been already identified in 1908 by Ikeda (1909) as the taste-active principle in Konbu seaweed (*Laminaria japonica*). It is present in relatively high concentrations in Parmesan cheese (12 000 mg/kg), peas (2 000 mg/kg), tomatoes (1 400 mg/kg), and cured ham (3 370 mg/kg) (Loeliger 2000; Ninomiya 2002). Compound 48 was for the first time isolated from dried bonito (Katsuobushi) by Kodama (1913). It shows synergism with 39 (Kuninaka 1960), small amounts of 48 (1–10 %) relative to 39 already provoking strong taste enhancement (Yamaguchi 1967). The 5′-guanosin monophosphate (47, GMP), which has a similar effect as 48, is even 2.3 times more potent (Yamaguchi *et al.* 1971). Examples of foods rich in 5′-nucleotides are tuna (2860 mg/kg IMP), pork (2000 mg/kg IMP), chicken (2010 mg/kg IMP), and dried shiitake mushrooms (1500 mg/kg GMP) (Ninomiya 1998).

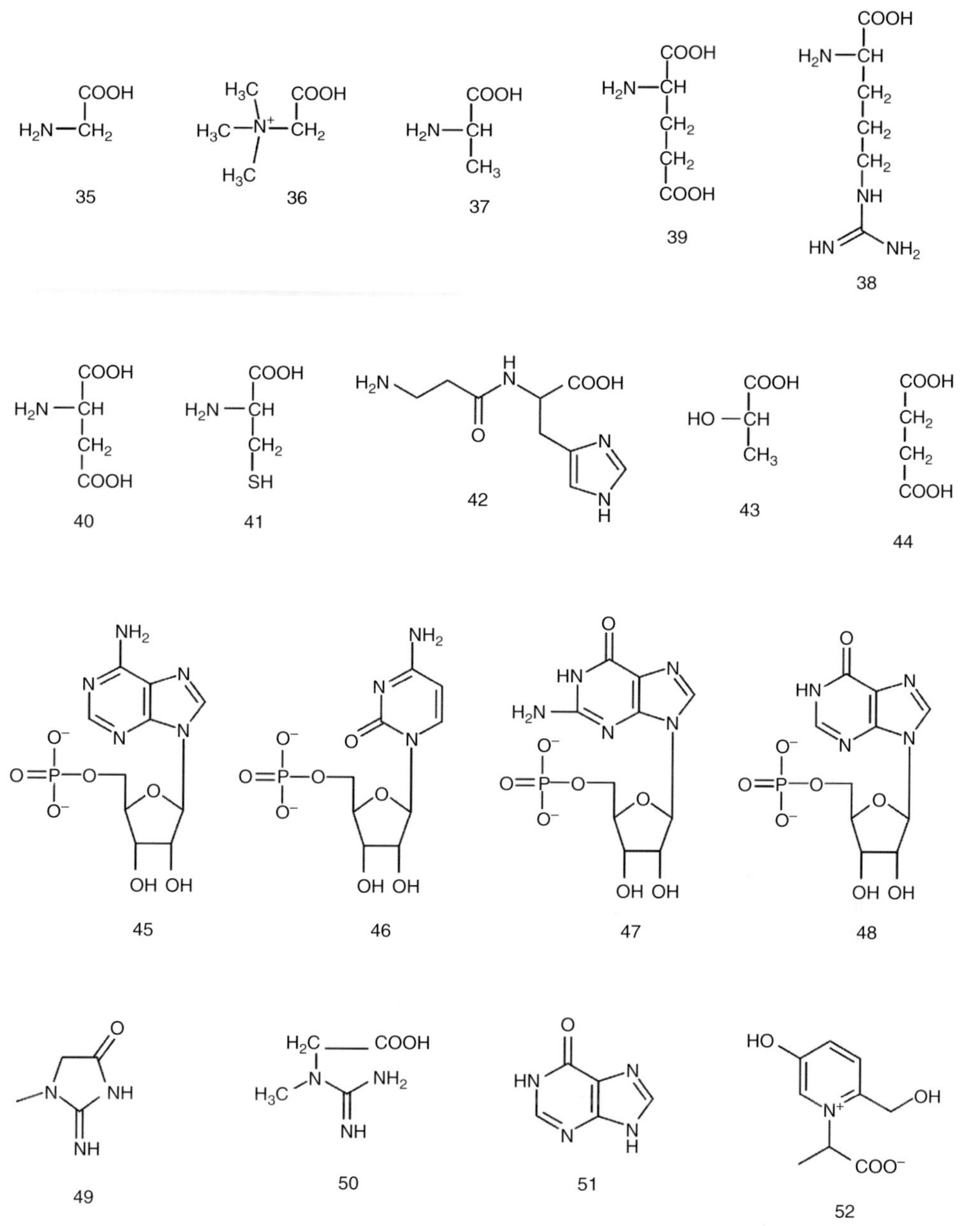

Fig. 8.2 Chemical structure of taste compounds in meat and seafood.

The group of 43 and 44, potassium, sodium, magnesium, chloride, and phosphate salts contributed to the umami, salty, and sour taste of beef juice (Schlichtherle-Cerny & Grosch 1998). Sodium succinate on its own tastes glutamic acid-like, but also bitter and salty at the same time (Velisek *et al.* 1978). The dipeptide 42 adds sourness and mouthfeel to beef juice. At a pH value of 5.7, which corresponds to the pH in meat, carnosine has a distinct sour note, whereas at higher values of 6.8–7.6, its taste becomes sweet and shows stronger mouthfeel (Suyama & Shimizu 1982). Creatine (50), creatinine (49), hypoxanthine (51), alanine (37), and cysteine (41) have less impact on the taste of beef juice than the other tastants mentioned above. Compounds 49–51 have a bitter taste, 37 a sweet savor, and 41 a meaty note (Schlichtherle-Cerny & Grosch 1998). Recently, the new compound alapyridaine (52) was identified in beef broth (Ottinger & Hofmann 2003). It is claimed to enhance both sweet and umami character in beef broth.

The taste-relevant compounds in snow crab extract have been investigated by Hayashi, Konosu, and Yamaguchi (Hayashi *et al.* 1981; Konosu & Yamaguchi 1987). They quantified 44 nonvolatiles and, using sensory studies including omission tests, found only 12 of these compounds to contribute to the characteristic taste (see Table 8.8): 35–39, 45–47, potassium, sodium, chloride, and phosphate ions. A synthetic extract of the 12 compounds simulated fairly well the taste profile of the original extract. Omission of 35 led to reduced sweet and umami taste. Umami taste was largely lost without 39, as well as sweetness. When the nucleotides were left out, the umami taste also decreased distinctively, the taste became flat, and the boiled crab taste disappeared completely. Likewise, without 38, the recombination solution tasted flatter and less crab-like. Omission of the minerals caused a decrease in the overall taste, sweetness, and umami and resulted in an inharmonic taste. Interestingly, already with a recombination of the taste compounds alone, the typical crabmeat character could be perceived, which underlines the importance of taste compounds for the flavor of a food. Most probably, due to better and more readily available analytical methods, such as high-performance liquid chromatography coupled with tandem mass spectrometry (LC-MS/MS), more taste-relevant compounds will be identified in our food.

REFERENCES

Arnold, R.G. Libbey, L.M. & Lindsay, R.C. (1969) Volatile flavour compounds produced by heat degradation of thiamine (vitamin B1). *Journal of Agricultural and Food Chemistry*, **17**, 390–392.

Baines, D.A. & Mlotkiewicz, J.A. (1984) The chemistry of meat flavour. In: *Recent Advances in the Chemistry of Meat* (ed. *A.J. Bailey*), pp.119–164. Royal Society of Chemistry, London, U.K.

Belitz, H.-D. Grosch, W. & Schieberle, P. (1999) *Food Chemistry*. Springer, Berlin, Germany.

Blank, I. Devaud, S. Fay, L.B. Cerny, C. Steiner, M. & Zurbriggen, B. (2001) Odor-active compounds of dry-cured meat: Italian-type salami and Parma ham. In: *Aroma Active Compounds in Foods* (ACS Symposium Series), Chapter 2, pp. 9–20. American Chemical Society, Washington, District of Columbia.

Boelens, M.H. (2000) VCF 2000 *Database of Volatile Compounds in Food*. Boelens Aroma Chemical Information Service, Huizen, the Netherlands.

Brennand, C.P. Ha, J.K. & Lindsay, R.C. (1989) Aroma properties and thresholds of some branched-chain and other minor volatile fatty acids occurring in milk fat and meat lipids. *Journal of Sensory Studies*, **4**, 105–120.

Brennand, C.P. & Lindsay, R.C. (1992) Influence of cooking on concentrations of species-related flavour compounds in mutton. *Lebensmittel Wissenschaft Technologie*, **25**, 357–364.

Buttery, R.G. Ling, L.C. Teranishi, R. & Mon, T.R. (1977) Roasted lamb fat: Basic volatile components. *Journal of Agricultural and Food Chemistry*, **25**, 1227–1229.

Cadwallader, K.R. & Baek, H.H. (1998) Aroma-impact compounds in cooked tail meat of freshwater crayfish (*Procambarus clarkii*). *Food flavors: Formation, Analysis and Packaging influences*, **40**, 271–278.

Cadwallader, K.R. Tan, Q. Chen, F. & Meyers, S.P. (1995) Evaluation of the aroma of cooked spiny lobster tail meat by aroma extract dilution analysis. *Journal of Agricultural and Food Chemistry*, **43**, 2432–2437.

Caporaso, F. Sink, J.D. Dimick, P.S. Mussinan, C.J. & Sanderson, A. (1977) Volatile flavor constituents of ovine adipose tissue. *Journal of Agricultural and Food Chemistry*, **25**, 1230–1234.

Cerny, C. & Grosch, W. (1992) Evaluation of potent odorants in roasted beef by aroma-extract dilution analysis. *Zeitschrift für Lebensmittel-Untersuchung und-Forschung*, **194**, 322–325.

Cerny, C. & Grosch, W. (1993) Quantification of character-impact odor compounds of roasted beef. *Zeitschrift für Lebensmittel-Untersuchung und-Forschung*, **196**, 417–422.

Cha, Y.J. Lee, G.H. & Cadwallader, K.R. (1997) Aroma-active compounds in salt-fermented anchovy. In: *Flavor and Lipid Chemistry of Seafoods* (eds F. Shahidi & K.R. Cadwallader), pp. 131–147. American Chemical Society, Washington, District of Columbia.

Chaudhari, N. Landin, A.M. & Roper, S.D. (2000) A metabotropic glutamate receptor variant functions as a taste receptor. *Nature Neuroscience*, **3**, 113–119.

Chung, H.Y. & Cadwallader, K.R. (1994) Aroma extract dilution analysis of blue crab claw meat volatiles. *Journal of Agricultural and Food Chemistry*, **42**, 2867–2870.

Chung, H.Y. Chen, F. & Cadwallader, K.R. (1995) Cooked blue crab claw meat aroma compared with lump meat. *Journal of Food Science*, **60**, 289–291.

Dwivedi, B.K. & Arnold, R.G. (1973) Chemistry of thiamine degradation in food products and model systems: A review. *Journal of Agricultural and Food Chemistry*, **21**, 54–60.

Dwivedi, B.K. & Snell, F.D. (1975) Meat flavour. *CRC Critical Reviews in Food Technology*, **5**, 487–535.

Farkas, P. Sadecka, J. Kovac, M. Siegmund, B. Leitner, E. & Pfannhauser. W. (1997) Key odourants of pressure-cooked hen meat. *Food Chemistry*, **60**, 617–621.

Farmer, L.J. & Mottram, D.S. (1990) Interaction of lipid in the Maillard reaction between cysteine and ribose: The effect of a triglyceride and three phospholipids on the volatile products. *Journal of the Science of Food and Agriculture*, **53**, 505–525.

Farmer, L.J. Mottram, D.S. & Whitfield, F.B. (1989) Volatile compounds produced in Maillard reactions involving cysteine, ribose and phospholipid. *Journal of the Science of Food and Agriculture*, **49**, 347–368.

Farmer, L.J. Nolan, M. & McAlinden, E. (2000) Studies on the origin of chicken flavour. In: *Frontiers of Flavour Science* (eds P. Schieberle & K.H. Engel), pp. 21–28. Deutsche Forschungsanstalt für Lebensmittelchemie, Garching, Germany.

Gasser, U. & Grosch, W. (1988) Identification of volatile flavor compounds with high aroma values from cooked beef. *Zeitschrift für Lebensmittel-Untersuchung und-Forschung*, **186**, 489–494.

Gasser, U. & Grosch, W. (1990) Primary odorants of chicken broth. A comparative study with meat broths from cow and ox. *Zeitschrift für Lebensmittel-Untersuchung und-Forschung*, **190**, 3–8.

Gasser, U. & Grosch, W. (1991) Aroma of cooked pork. *Lebensmittelchemie*, **45**, 15.

Grosch, W. (1987) Reactions of hydroperoxides—Products of low molecular weight. In: *Autoxidation of Unsaturated Lipids* (ed. H.W. Chan), pp. 95–137. Academic Press, London, U.K.

Grosch, W. Zeiler-Hilgart, G. Cerny, C. & Guth, H. (1993) Studies on the formation of odorants contributing to meat flavors. In: *Progress in Flavour Precursors Studies* (eds P. Schreier & P. Winterhalter), pp. 329–342. Allured Publishing, Carol Stream, Illinois.

Guth, H. & Grosch, W. (1993) 12-Methyltridecanal, a species-specific odorant of stewed beef. *Lebensmittel Wissenschaft Technologie*, **26**, 171–177.

Guth, H. & Grosch, W. (1994) Identification of the character impact odorants of stewed beef juice by instrumental analyses and sensory studies. *Journal of Agricultural and Food Chemistry*, **42**, 2862–2866.

Guth, H. & Grosch, W. (1995) Comparison of the juices of stewed beef and stewed pork by instrumental analyses of the odorants and by sensory studies. *Bioflavour 95*. INRA, Paris, France, pp. 201–205.

Ha, J.K. & Lindsay, R.C. (1990) Distribution of volatile branched-chain fatty acids in perinephric fat of various red meat species. *Lebensmittel Wissenschaft Technologie*, **23**, 433–440.

Hartman, G.J. Carlin, J.T. Hwang, S.S. Bao, Y. Yang, J. & Ho, C.T. (1984) Identification of 3,5-diisobutyl-1,2,4-trithiolane and 2-isobutyl-3,5-diisopropylpyridine in fried chicken flavor. *Journal of Food Science*, **49**, 1398–1400.

Hayashi, T. Yamaguchi, K. & Konosu, S. (1981) Sensory analysis of taste-active components in the extract of boiled snow crab meat. *Journal of Food Science*, **46**, 479–483, 493.

Hirano, T. Zhang, C.T. Morishita, A. Suzuki, T. & Shirai, T. (1992) Studies on the odor of fishes. I. Identification of volatile compounds in ayu fish and its feeds. *Nippon Suisan Gakkaishi*, **58**, 547–557.

Hornstein, I. & Crowe, P.F. (1960) Flavor studies on beef and pork. *Journal of Agricultural and Food Chemistry*, **8**, 494–498.

Hornstein, I. Crowe, P.F. & Sulzbacher, W.L. (1963) Flavour of beef and whale meat. *Nature*, **199**, 1252–1254.

Hsieh, J.H. & Kinsella, J.E. (1989) Lipoxygenase generation of specific volatile flavor carbonyl compounds in fish tissues. *Journal of Agricultural and Food Chemistry*, **37**, 279–286.

Ikeda, K. (1909) New seasonings. *Journal of the Tokyo Chemical Society*, **30**, 820–826.

Josephson, J.B. Lindsay, R.C. & Stuiber, D.A. (1983) Identification of compounds characterizing the aroma of fresh whitefish (*Coregonus clupeaformis*). *Journal of Agricultural and Food Chemistry*, **31**, 326–330.

Kerscher, R. (2000) *Objectivization of species-specific differences in aroma of heated meat.* Dissertation Technical University Munich. Available from Dr. Hut, Munich, Germany.

Kerler, J. & Grosch, W. (1996) Odorants contributing to warmed-over flavor (WOF) of refrigerated cooked beef. *Journal of Food Science*, **61**, 1271–1274, 1285.

Kerler, J. & Grosch, W. (1997) Character impact odorants of boiled chicken: Changes during refrigerated storage and reheating. *Zeitschrift für Lebensmittel-Untersuchung und-Forschung*, **205**, 232–238.

Kerscher, R. & Grosch, W. (1997) Comparative evaluation of potent odorants of boiled beef by aroma extract dilution and concentration analysis. *Zeitschrift für Lebensmittel-Untersuchung und-Forschung*, **204**, 3–6.

Kerscher, R. & Grosch, W. (1998) Quantification of 2-methyl-3-furanthiol, 2-furfurylthiol, 3-mercapto-2-pentanone, and 2-mercapto-3-pentanone in heated meat. *Journal of Agricultural and Food Chemistry*, **46**, 1954–1958.

Kodama, S. (1913) On a procedure for separating inosinic acid. *Journal of the Tokyo Chemical Society*, **34**, 751–755.

Konopka, U.C. & Grosch, W. (1991) Potent odorants causing the warmed-over flavour in boiled beef. *Zeitschrift für Lebensmittel-Untersuchung und-Forschung*, **193**, 123–125.

Konosu, S. & Yamaguchi, K. (1987) Role of extractive components of boiled crab in producing the characteristic flavor. In: *Umami: A Basic Taste* (eds Y. Kawamura & M.R. Kare), pp. 236–253. Marcel Dekker, New York.

Kuninaka, A. (1960) Studies on taste of ribonucleic acid derivatives. *Nippon Nogeikagaku Kaishi*, **34**, 489–492 (in Japanese).

Lee, G.H. Suriyaphan, O. & Cadwallader, K.R. (2001) Aroma components of cooked tail meat of American lobster (*Homarus americanus*). *Journal of Agricultural Food Chemistry*, **49**, 4324–4332.

van der Linde, L.M. van Dort, J.M. de Valois, P. Boelens, H. & de Rijke, D. (1979) Volatile components from thermally degraded thiamine. In: *Progress in Flavour Research* (eds D.G. Land & H.E. Nursten), pp.219–224. Applied Sciences, London, U.K.

Loeliger, J. (2000) Function and importance of glutamate for savory foods. *The Journal of Nutrition*, **130**(4S), 915S–920S.

Mabrouk, A.F. (1976) Nonvolatile nitrogen and sulfur compounds in red meats and their relation to flavor and taste. In: *Phenolic, Sulfur, and Nitrogen Compounds in Food Flavors* (eds G. Charalambous & I. Katz), pp. 146–183. American Chemical Society, Washington, District of Columbia.

MacLeod, G. Seyyedain-Ardebili, M. & Chang, S.S. (1981) Natural and simulated meat flavors (with particular reference to beef). *CRC Critical Reviews in Food Science and Nutrition*, **14**, 399–425.

McGill, A.S. Hardy, R. & Burt, J.R. (1974) Hept-cis-4-enal and its contribution to the off-flavour in cold stored cod. *Journal of the Science of Food and Agriculture*, **25**, 1477–1489.

McGill, A.S. Hardy, R. & Gunstone, F.D. (1977) Further analysis of the volatile components of frozen cold stored cod and the influence of these on flavour. *Journal of the Science of Food and Agriculture*, **28**, 200–205.

Milo, C. (1994) *Aroma compounds of cooked trout, cooked cod and salmon before and after deep frozen storage of the raw material.* Dissertation Technical University Munich, Munich, Germany.

Milo, C. & Grosch, W. (1993) Changes in the odorants of boiled trout (*Salmo fario*) as affected by the storage of the raw material. *Journal of Agricultural and Food Chemistry*, **41**, 2076–2081.

Milo, C. & Grosch, W. (1995) Detection of odor defects in boiled cod and trout by gas chromatography-olfactometry of headspace samples. *Journal of Agricultural and Food Chemistry*, **43**, 459–462.

Milo, C. & Grosch, W. (1996) Changes in the odorants of boiled salmon and cod as affected by the storage of the raw material. *Journal of Agricultural and Food Chemistry*, **44**, 2366–2371.

Milo, C. & Grosch, W. (1997) Potent odorants in boiled cod as affected by the storage of raw material. In: *Flavor and Lipid Chemistry of Seafoods* (eds F. Shahidi & K.R. Cadwallader), pp. 110–119. American Chemical Society, Washington, District of Columbia.

Mottram, D.S. (1982) A comparison of the flavour volatiles from cooked beef and pork meat systems. *Journal of the Science of Food and Agriculture*, **33**, 934.

Mottram, D.S. (1998) Flavour formation in meat and meat products: A review. *Food Chemistry*, **62**, 415–424.

Ninomiya, K. (1998) Natural occurrence. *Food Reviews International*, **14**, 177–211.

Ninomiya, K. (2002) Umami: A universal taste. *Food Reviews International*, **18**, 23–28.

Ohloff, G. & Flament, I. (1978) Heterocyclic constituents of meat aroma. *Heterocycles*, **11**, 663–695.

Ottinger, H. & Hofmann, T. (2003) Identification of the taste enhancer alapyridaine in beef broth and evaluation of its sensory impact by taste reconstitution experiments. *Journal of Agricultural and Food Chemistry*, **51**, 6791–6796.

van den Ouweland, G.A.M. Olsman, H. & Peer, H.G. (1978) Challenges in meat flavor research. In: *Agricultural and Food Chemistry: Past, Present, Future* (ed. R. Teranishi), pp. 292–314. AVI Publishing, Westport, Connecticut.

van den Ouweland, G.A.M. Demole, E.P. & Enggist, P. (1989) Process meat flavor development and the Maillard reaction. In: *Thermal generation of aromas* (eds T.H. Parliment, R.J. McGorrin & C.T. Ho), pp. 433–441. American Chemical Society, Washington, District of Columbia.

Pippen, E.L. & Nonaka, M. (1958) A convenient method for synthesizing normal aliphatic 2,4-dienals. *Journal of Organic Chemistry*, **23**, 1580–1582.

Prell, P.A. & Sawyer, F.M. (1988) Flavor profiles of 17 species of North Atlantic fish. *Journal of Food Science*, **53**, 1036–1042.

Ramarathnam, N. Rubin, L.J. & Diosady, L.L. (1991) Studies on meat flavor. II. A quantitative investigation on the volatile carbonyls and hydrocarbons in uncured and cured beef and chicken. *Journal of Agricultural and Food Chemistry*, **39**, 1839–1847.

Rothe, M. & Thomas, B. (1963) Aroma compounds of bread. *Zeitschrift für Lebensmittel-Untersuchung und-Forschung*, **119**, 302–311.

Schieberle, P. & Grosch, W. (1987) Quantitative analysis of aroma compounds in wheat and rye bread crusts using a stable isotope dilution assay. *Journal of Agricultural and Food Chemistry*, **35**, 252–257.

Schieberle, P. & Rota, V. (2003) On the role of lipid derived aroma compounds in raw and cooked sheep meat. In: *226th ACS National Meeting New York (NY)—Conference Proceedings*, 7–11 September, AGFD 168.

Schlichtherle-Cerny, H. & Grosch, W. (1998) Evaluation of taste compounds of stewed beef juice. *Zeitschrift für Lebensmittel-Untersuchung und-Forschung*, **207**, 369–376.

Schliemann, J. Wölm, G. Schröder, R. & Ruttloff, H. (1988) Chicken flavour—Formation, composition and production. Part 2. Flavour substances. *Nahrung*, **32**, 595–607.

Sérot, T. Regost, C. Prost, C. Robin, J. & Arzel, J. (2001) Effect of dietary lipid sources on odour-active compounds in muscle of turbot (*Psetta maxima*). *Journal of the Science of Food and Agriculture*, **81**, 1339–1346.

Shahidi, F. (1994) *Flavor of Meat and Meat Products*. Blackie Academic & Professional, London, U.K.

Shahidi, F. (1998) *Flavor of Meat, Meat Products and Seafoods*. Blackie Academic & Professional, London, U.K.

Shi, H. & Ho, C.T. (1994) The flavour of poultry meat. In: *Flavor of Meat and Meat Products* (ed. F. Shahidi) Blackie Academic & Professional, London, U.K.

Shibamoto, T. & Horiuchi, M. (1997) Role of aldehydes in cooked fish flavors. In: *Flavor and Lipid Chemistry of Seafoods* (eds F. Shahidi & K.R. Cadwallader), pp. 20–30. American Chemical Society, Washington, District of Columbia.

Stahnke, L.H. (2000) 2-Acetyl-1-pyrroline—Key aroma compound in mediterranean dried sausages. In: *Frontiers of Flavour Science* (eds P. Schieberle & K.H. Engel), pp. 361–365. Deutsche Forschungsanstalt für Lebensmittelchemie, Garching, Germany.

Sutherland, M.M. & Ames, J.M. (1996) Free fatty acid composition of the adipose tissue of intact and castrated lambs slaughtered at 12 and 30 weeks of age. *Journal of Agricultural and Food Chemistry*, **44**, 3113–3116.

Suyama, M. & Shimizu, T. (1982) Buffering capacity and taste of carnosine and its methylated compounds. *Nippon Suisan Gakkaishi*, **48**, 89–95.

Sweeley, C.C. Elliott, W.H. Fries, I. & Ryhage, R. (1966) Mass spectrometric determination of unresolved components in gas chromatographic effluents. *Analytical Chemistry*, **38**, 1549–1553.

Tang, J. Jin, Q.Z. Shen, G.H. Ho, C.T. & Chang, S.S. (1983) Isolation and identification of volatile compounds from fried chicken. *Journal of Agricultural and Food Chemistry*, **31**, 1287–1295.

Triqui, R. & Guth, H. (1997) Determination of potent odorants in ripened anchovy (*Engraulis encrasicholus L.*) by aroma extract dilution analysis and by gas chromatography-olfactometry of headspace samples. In: *Flavor and Lipid Chemistry of Seafoods* (eds F. Shahidi & K.R. Cadwallader), pp. 31–38. American Chemical Society, Washington, District of Columbia.

United States Department of Agriculture. (2003) *USDA National Nutrient Data Base for Standard Reference*. http://www.nal.usda.gov/fuic/foodcomp/Data/index.html

Velisek, J. Davidek, J. Kubelka, V. Tran-Thi, B.T. & Hajslova, J. (1978) Succinic acid in yeast autolysates and its sensory properties. *Nahrung*, **22**, 735–743.

Warendorf, T. Belitz, H.D. Gasser, U. & Grosch, W. (1992) The flavor of bouillon. Part 2. Sensory analysis of non volatiles and imitation of a bouillon. *Zeitschrift für Lebensmittel-Untersuchung und-Forschung*, **195**, 215–223.

Warmke, R. Belitz, H.-D. & Grosch, W. (2004) Evaluation of taste compounds of Swiss cheese (Emmentaler). *Zeitschrift für Lebensmittel-Untersuchung und-Forschung*, **203**, 230–235.

Werkhoff, P. Bruening, J. Emberger, R. Guentert, M. & Hopp, R. (1993) Flavor chemistry of meat volatiles: new results on flavor components from beef, pork and chicken. In: *Recent developments in flavor and fragrance chemistry* (eds R. Hopp & K. Mori), pp. 183–213. VCH, Weinheim, Germany.

Wilson, R.A. & Katz, I. (1972) Review of literature on chicken flavor and report of isolation of several new chicken flavor components from aqueous cooked chicken broth. *Journal of Agricultural and Food Chemistry*, **20**, 741–747.

Wilson, R.A. Mussinan, C.J. Katz, I. & Sanderson, A. (1973) Isolation and identification of some sulfur chemicals present in pressure-cooked beef. *Journal of Agricultural and Food Chemistry*, **21**, 873–876.

Withycombe, D.A. & Mussinan, C.J. (1988) Identification of 2-methyl-3-furanthiol in the steam distillate from canned tuna fish. *Journal of Food Science*, **53**, 658–660.

Wong, E. Nixon, L.N. & Johnson, C.B. (1975) Volatile medium chain fatty acids and mutton flavor. *Journal of Agricultural and Food Chemistry*, **23**, pp. 495–498.

Yamaguchi, S. (1967) The synergistic taste effect of monosodium glutamate and disodium 5′-inosinate. *Journal of Food Science*, **32**, 473–478.

Yamaguchi, S. Yoshikawa, T. Ikeda, S. & Ninomiya, T. (1971) Measurement of the relative taste intensity of some L-alpha-amino acids and 5′-nucleotides. *Journal of Food Science*, **36**, 846–849.

Zeiler-Hilgart, G. (1994) *Model experiments for the formation of 2-methyl-3-furanthiol and 2-furfurylthiol in boiled meat*. Dissertation Technical University Munich, Munich, Germany.

9 Off Flavors and Rancidity in Foods

Ronald B. Pegg and Fereidoon Shahidi

Abstract: Off flavor is the term used to describe unpleasant odors or tastes resulting from the natural deterioration of a food, with extensive deterioration giving rancidity. Taint is regarded as unpleasant odors or tastes arising from contamination of a food by some foreign chemicals. Detecting any of these involves organoleptic appraisal of the off flavor quality of a food. Warmed-over flavor is a common phenomenon referred to the rapid development of lipid-derived oxidized flavor in refrigerated cooked-meat products.

Keywords: Rancidity, oxidative rancidity, enzymatic rancidity, photooxidation, warmed-over flavor (WOF), antioxidants, packaging, flavor masking

9.1 INTRODUCTION

Off flavor is the term used to describe unpleasant odors or tastes resulting from the natural deterioration of a food, with extensive deterioration giving rancidity. A taint, on the other hand, is generally regarded as unpleasant odors or tastes arising from contamination of a food by some foreign chemical with which it has accidentally come into contact (Baigrie 2003). Detecting any of these "conditions" involves organoleptic appraisal of the off flavor quality of a food. This is subjective in nature, as the ability to perceive an off flavor varies from person to person. To illustrate this point, Baigrie (2003) noted that there are approximately 1000 olfactory receptor genes influencing an individual's perception of odors, making it possible for the human nose to detect very low concentrations of some of these compounds (i.e., below 0.01 ppb). Consequently, identifying an off flavor or taint, both qualitatively and quantitatively, can be a challenging process.

Although microbial spoilage of food can generate an unpleasant odor in a product, this chapter will focus solely on the chemical and enzymatic processes resulting in off flavor and rancidity development. For example, when considering off flavors in lipid-rich foods, there are at least four types of mechanisms that need to be taken into consideration, namely hydrolytic, enzymatic, oxidative, and photooxidative rancidity (Palm 2002). Of these mechanisms, the oxidation of unsaturated lipids has been extensively studied because it relates to the deterioration of lipid-containing foods, production of both desirable and undesirable scission products, and numerous reactions with other food constituents (Wong 1989). The impact of lipid oxidation on muscle foods and edible oils has been the subject of numerous investigations. In fact, it has been over 45 years since the term "warmed-over flavor" was first coined in reference to a marked deterioration in the quality of cooked meat products that have been chill-stored

Handbook of Meat, Poultry and Seafood Quality, Second Edition. Edited by Leo M. L. Nollet.

(Tims & Watts 1958). Warmed-over flavor or "WOF" is a common phenomenon that originally referred to the rapid development of lipid-derived oxidized flavor in refrigerated cooked-meat products. In meats, this distinctive off flavor can become readily apparent within a few hours of thermal processing, and is most noticeable when refrigerated cooked-meat products are reheated. Products develop a flavor (i.e., odor and taste) that consumers do not like, and at the same time desirable meaty flavor notes are lost. The stale off flavors so formed have often been described as "cardboard-like," "painty," and "rancid" (Love 1988, Vega & Brewer 1994).

Although WOF development was first recognized as being a dynamic process of flavor change in cooked meat, principally because of a cascade of oxidation events, the original definition has been expanded to include fresh meat that has been stored in a freezer. While freezing can delay the onset of WOF development, it does not prevent it. Stale or off flavor notes such as "ice box," "rancid," and "freezer-burn" have been used to describe the phenomenon. Although the aroma profile related to stored fresh meat is somewhat different from the characteristic WOF of reheated meats, the flavor compounds involved are qualitatively the same but present at different concentrations. When the specter of WOF is raised, it often is in relation to comminuted products such as meat loaves, chicken nuggets, and precooked burgers. Thus, it is generally accepted that any process involving disruption of the integrity of muscle tissue, such as cooking, grinding, mechanical deboning, massaging, restructuring, or freezing, will enhance the development of WOF. WOF is not solely a result of lipid oxidation. Scientists in the 1980s suggested that protein degradation reactions contributed toward WOF development (St. Angelo & Bailey 1987). In an effort to better describe the complex series of chemical reactions that contribute to an overall increase in off flavor notes and a loss in desirable meaty ones, the term "meat flavor deterioration" was proposed.

9.2 ABSORPTION/CONTAMINATION

Foods can develop taints simply by coming in contact with an off flavor constituent during processing, packaging, or storage. Even the development of a slightly unpleasant aroma or taste may be sufficient for a consumer to reject a food. The lipid component of a food or an edible oil acts as a good reservoir or sink into which volatile off flavor compounds can dissolve. The amount of volatile material absorbed by a food product is likely to be minute; nevertheless, this can have a significant impact on the product by causing a taint in the flavor. Whether a chemical can cause an unpleasant taste or odor is a function of the food product in question: if the taint imparts a taste or odor in keeping with the food, a deleterious effect may not be perceived (Lord 2003). As an example, an acidic taint in a fermented meat product will not impact its flavor, whereas the same taint in a cooked chicken product will have a deleterious impact on its flavor.

The contribution of a single organic volatile compound to an overall smell is based on its odor quality and odor intensity. The odor intensity is dependent on both the concentration of the component in question and the odor threshold in the food matrix. Odor quality and odor threshold data are, therefore, useful tools to gain insight into the influence of a volatile compound on a given food flavor (Rychlik *et al.* 1998). In various resources, flavor threshold data of chemicals in a specified medium are available, but in most cases the threshold values are quoted simply against water. As a result, detection thresholds of off flavors or taints in foods may be lower; therefore, they have been defined in sensory studies as the minimum concentration which a pure compound can be perceived by 50% of the members of a trained panel. Some general flavor thresholds for volatile classes of compounds found in foods are presented in Table 9.1. A more specific listing of odor thresholds of key food odorants has been compiled by Rychlik *et al.* (1998).

Table 9.1 Threshold values of different volatile compound classes.*

Compound class	Threshold values in water (mg/kg)
Hydrocarbons	90–2150
Substituted furans	2–27
Vinyl alcohols	0.5–3
1-Alkenes	0.02–9
2-Alkenals	0.04–2.5
Alkanals	0.04–1.0
2-*t*,4-*t*-Alkadienals	0.04–0.3
Isolated alkadienals	0.002–0.3
Isolated *c*-alkenals	0.0003–0.01
2-*t*,4-*c*-Alkadienals	0.002–0.006
Vinyl ketones	0.00002–0.007

**t* 5 *trans* and *c* 5 *ci*.
Source: Adapted from Palm (2002).

9.3 HYDROLYTIC AND ENZYMATIC RANCIDITY

Most off flavors in food develop as a result of hydrolytic, enzymatic, and oxidative rancidity. Hydrolytic rancidity is associated with the presence of water. At typical food production temperatures, water can dissolve in oil from 0.1% to 0.5%. If a suitable catalyst such as a lipase from bacterial contamination is present and the temperature is sufficient, hydrolytic rancidity can take place. The process involves the hydrolysis of triacyglycerol molecules to first diacylglycerols, then monoacylglycerols, and finally free fatty acids. The release of short-chain fatty acids has a direct effect on food aroma. These fatty acids have their own flavor threshold values and can impart specific flavors to food. In some instances, release of free fatty acids can combine with sodium ions in food to form soaps, thereby giving a soapy flavor to some food products. Lipoxygenases and cyclooxygenases are enzymes found in many plants. They catalyze the oxidation of unsaturated fatty acids containing a 1-*cis*,4-*cis*-pentadiene system to their corresponding monohydroperoxides and endoperoxides, respectively. Cyclooxygenase incorporates two molecules of oxygen into the fatty acid and can be considered as a special case of lipoxygenase that incorporates only one molecule of oxygen. The preferred substrates are linoleic and linolenic acids for the enzyme in plants and arachidonic acid in animals (Belitz & Grosch 1987). In fruits and vegetables, the hydroperoxides formed from lipoxygenase breakdown into volatile secondary products with a variety of flavors. These volatiles are responsible for the taste of bananas, grapes, apples, cucumbers, and mushrooms (Palm 2002). In general, lipoxygenase produces similar flavor volatiles to those formed during autoxidation, although the relative proportions of the products may vary widely depending on the specificity of the enzyme conditions (Hamilton 2003). Apart from these wanted flavors, however, oxidation continues and many of the resultant volatiles serve as signal substances warning us from eating rancid food.

9.4 OXIDATIVE RANCIDITY

The major reaction in the formation of off flavors is a free-radical chain reaction referred to as autoxidation. Autoxidation of lipid molecules (LH) can be catalyzed by the presence of heat, light, ionizing radiation, metal ions, and metalloporphyrins (Wong 1989). There are many factors related to the propensity of lipids to oxidize, but an important difference between autoxidation in muscle foods and in refined vegetable oils is the presence of several catalytic systems for oxygen activation in meat. Although formation of the first lipid free radical may occur by several different means, one possibility is via metal ion (e.g., M^{3+}) catalysis. Autoxidation then follows initiation, propagation, and termination steps, as outlined below.

$$LH \rightarrow L\cdot + H\cdot$$
$$\left(\text{e.g.}, M^{3+} + LH \rightarrow L\cdot + H^{+} + M^{2+}\right)$$

$$L\cdot + {}^{3}O_2 \rightarrow LOO\cdot$$
$$LOO\cdot + LH \rightarrow LOOH + L\cdot$$

$$L\cdot + L\cdot \rightarrow L-L$$
$$L\cdot + LOO\cdot \rightarrow LOOL$$
$$LOO\cdot + LOO\cdot \rightarrow LOOL + O_2$$

Hydroperoxides are the primary products of lipid autoxidation. Even though they are odorless and tasteless, their degradation to alkoxy radicals and then additional scission products leads to the formation of complex mixtures of low molecular-weight compounds with distinctive aromas. These secondary products of lipid oxidation principally include alkanes, alkenes, aldehydes, ketones, alcohols, esters, furans, lactones, epoxy compounds, polymers, and acids (Shahidi *et al.* 1986, Shahidi 1989). It should be stressed, however, that not all flavors derived from oxidation give rise to unpleasant off odors. Short-chain aldehydes with unsaturation at the 2-position are described as sweet and pungent while longer chain analogues have been described as sweet, fatty, and green. Some characteristic flavors and associated compounds isolated and identified in oxidized fats of various foodstuffs are presented in Table 9.2.

In the case of cooked meat, aldehydic breakdown products play a significant role in the flavor, as they possess low odor threshold values and are responsible for the development of WOF and rancidity. Meat lipids are made up of adipose and intramuscular tissues, and they contain both saturated and unsaturated fatty acids. However, it is the membrane lipids (i.e., the phospholipids) that tend to possess the lion's share of the polyunsaturated fatty acids (PUFAs), and these are most prone to oxidation. When meat is ground, chopped, or cooked, membranes are disrupted releasing cell contents; PUFAs are then exposed to oxidative stress. Consequently, the process of WOF development can begin within several hours of cooking the meat product compared to regular lipid oxidation, which can take several days to develop. Research has shown that subcutaneous fat from meat can produce about 50 volatile compounds during WOF development, whereas intramuscular lipids can generate more than 200. There seems to be no question that oxidation of phospholipids is the primary source of off flavor notes generated during WOF development.

Oxidation of the unsaturated C_{18} fatty acids of meat, namely oleic ($18{:}1n-9$), linoleic ($18{:}2n-6$), and -linolenic ($18{:}3n-3$) acids, has been reported to produce low molecular-weight aldehydes (C_3–C_{12}) such as pentanal, hexanal, and *trans,trans*-2,4-decadienal as well as oxocompounds like the very potent *trans*-4,5-epoxy(*E*)-2-decenal (Konopka & Grosch 1991); these are believed to be partially responsible for WOF and rancidity development of

Table 9.2 Characteristic flavors and associated compounds isolated and identified in oxidized fats from various foodstuffs*.

Flavor note	Compounds identified
Cardboard, tallow	Octanal; alkanals (C_9–C_{11}); alk-2-enal (C_8, C_9); 2,4-dienals (C_7, C_{10}); nona2-*t*, 6-*t*-dienal
Fatty/oily	Alkanals (C_5–C_7); hex-2-enal; 2,4-dienals (C_5–C_{10}); 2-*t*-pentenylfuran
Painty	Alkanals (C_5–C_{10}); alk-2-enals (C_5–C_{10}); hepta-2-*t*,4-*t*-dienal
Oxidized	2-heptanone; pent-2-*t*-enal; but-2-*t*-enal
Fishy	oct-1-ene-3-one; octanal; hept-2-enal; 2,4-heptadienal; alkanals (C_2–C_9)
Grassy	Alkanals (C_5–C_{10}); alk-2-enals (C_5–C_{10}); hepta-2-*t*,4-*t*-dienal; 2-alkanones (C_3–C_{11}); oct-1-en-3-one; deca-2-*t*,4-*c*,7-*t*-trienal; pent-1-en-3-one
Mild, pine-like	2-*t*-hexenal; nona-2,6-dienal; 2-*c*-pentenylfuran; hexanal; heptanal
Green-beany	3-*t*-hexenal
Beany	3-*c*-hexenal
Deep-fried fat	Alkanals, non-2-enal
Sweet aldehydic	2-*t*,4-*t*-decadienal
Mushroom, moldy	2-*t*,4-*c*-decadienal
Metallic	oct-1-en-3-one; oct-1-en-3-ol
Rancid	pent-1-en-3-one; oct-1-en-3-one; 2-*t*-pentenylfuran; 1-*c*,5-octadien-3-one
Nutty	2-*t*-nonenal; volatile fatty acids (C_4–C_{10}) 2-*t*,4-*t*-octadienal; 2-*t*,2-*c*-octenal

**t* 5 *trans* and *c* 5 *cis*.
Source: Adapted from Kochhar (1996).

Table 9.3 Dominant volatile aldehydes derived from the autoxidation of methyl linolate.

		Odor threshold value (μg/kg)	
Aldehyde	**Quantity* (μg/g)**	**In water**	**In oil**
Pentanal	55	10	100
Hexanal	5100	4.5	150
Heptanal	50	30	45
Trans-2-heptenal	450	50	14 000
Octanal	45	40	50
Cis-2-octenal	990	—	—
Trans-2-octenal	420	4	7 000
Cis-3-nonenal	30	—	—
Trans-3-nonenal	30	—	—
Cis-2-decenal	20	—	—
Trans-2, *trans*-4-nonadienal	30	90	460
Trans-2, *cis*-4-decadienal	250	—	20
Trans-2, *trans*-4-decadienal	150	0.1	200

*One gram of linoleate was autoxidatized at 20°C by an uptake of 0.5 mol of oxygen per mole linoleate.
Source: Adapted from Belitz and Grosch 1987.

cooked meats during storage. The impact of dominant aldehydic scission products from the autoxidation of methyl linoleate and their odor threshold values are shown in Table 9.3. It can be reasonably hypothesized that meats containing higher levels of PUFAs are more susceptible to oxidation. This is indeed so, because certain muscle tissue exhibits more of a propensity toward the development of WOF. Tichivangana and Morrissey (1985) have shown that the autoxidation of muscle foods occurs in the following order:

$$\text{fish} \rightarrow \text{poultry}\left(\text{i.e., chicken and turkey}\right) \rightarrow \text{pork} \rightarrow \text{beef} \rightarrow \text{lamb}$$

9.5 PHOTOOXIDATION

An alternate mechanism resulting in oxidation can occur in the presence of light; thus, the term photooxidation has been given to it. Photooxidation involves the interaction of singlet oxygen with lipid constituents. Singlet oxygen can react directly with unsaturated lipid moieties in contrast to autoxidation. The presence of sensitizers such as myoglobin, riboflavin, and chlorophyll is necessary, as this allows triplet oxygen to interact with light to produce singlet oxygen. In turn, the singlet oxygen reacts with unsaturated lipid constituents and forms hydroperoxides. These build up quickly as soon as oxidation begins because there is little or no inductive period. The photosensitized oxidation process is illustrated as follows, where Sens, $^1Sens^*$, and $^3Sens^*$ represent the sensitizer, excited singlet state, and excited triplet state, respectively (Kochhar 1996).

$$
\begin{aligned}
&\text{light, hv} \\
&\text{Sens} \rightarrow {}^1\text{Sens}^* \rightarrow {}^3\text{Sens}^* \\
&{}^3\text{Sens}^* + {}^3\text{O}_2 \rightarrow {}^1\text{Sens} + {}^1\text{O}_2^* \\
&{}^1\text{O}_2^* + \text{LH} \rightarrow \text{LOOH}
\end{aligned}
$$

There are two types of photooxidation: types I and II. In type I, the typical hydroperoxides are generated as in autoxidation, but for type II, the intermediates formed are different. In type II photooxidation, an ene addition reaction occurs whereby a molecule of singlet oxygen adds across the lipid double bond. This has the effect of generating hydroperoxides that are different from those formed by autoxidation (Hamilton 2003). This in turn leads to different volatile breakdown products and off flavors than those generated by autoxidation. Photooxidation cannot be curbed by synthetic antioxidants such as BHA, which is a great concern for the edible oil industry. Carotenoids, such as b-carotene, are naturally present in many crude oils and some foods, and can quench singlet oxygen and protect lipids against light-induced oxidation, as the carotenoid reacts more quickly with singlet oxygen than triacylglycerol molecules. Refining and bleaching of edible oils remove, at least in part, the endogenous photosensitizers. This along with using suitable packaging that is absorbent to the light energy necessary for photosensitization or storage in the dark will protect oils against deterioration by singlet oxygen.

9.6 EXAMINATION OF WOF: AN OFF FLAVOR IN MEAT PRODUCTS

The contribution of sensory analysis toward the development of descriptors, definitions, and references to describe the phenomenon of WOF in cooked meat products has come a long way since the time of Tims and Watts. Today, sensory descriptive vocabularies for WOF in meats of different species have been prepared. An example lexicon of flavor descriptors employed for pork is presented in Table 9.4. Such a vocabulary can be used by panelists to describe perceived sensory characteristics in a sample set; the resultant profile is a perceptual map of the variations in a sample type that can be employed alone or in combination with chemical/instrumental data to help explain and elucidate underlying sensory and chemical relationships. It is important, however, that any analytical technique employed to assess causes and to find solutions for the problem of WOF must be conducted in combination with sensory analysis. The descriptors need to be well defined to allow trained panelists to accurately track the development of WOF with time. Data from the mid-1980s indicate that the sensory perception of WOF was similar across meat patties of beef, pork, chicken, and turkey, but the intensity of samples varied. Some specific data on beef indicated that when it is just cooked, the intensity of fresh cooked beef notes is strong. Over time, however, cardboard notes develop and then disappear; this generally happens before oxidized notes become apparent. At about the same time, there is a marked reduction in fresh beefy notes. For samples that had been stored for 3–7 days, oxidized/rancid/painty notes were dominant (Johnson & Civille 1986). From recent studies on WOF development in pork, sensory profile data clearly demonstrated the specific association and distinct relationship of at least two groups of odor and flavor terms (see Table 9.4) with increasing days of chill-storage and WOF development (i.e., rancid-like flavor, linseed oil-like odor and fish-like flavor, rubber-like odor, and cardboard-like odor) (Byrne *et al.* 2001).

Various chemical methods to semiquantify WOF development in meat and meat products include measuring changes in conjugated dienes and carbonyl values as well as the more recent employment of electronic noses. Malonaldehyde is a relatively minor production of autoxidation of PUFAs in muscle tissue; yet, its presence and concentration in meat products is monitored as an indicator of lipid oxidation by the classical 2-thiobarbituric acid (TBA) test. This assay involves the reaction of malonaldehyde in oxidized foods with the TBA reagent under acidic conditions; a pink adduct forms with a distinctive absorption maximum at 532 nm (Tarladgis *et al.* 1960, Siu & Draper 1978). The TBA test was once believed to be specific for

Table 9.4 Sensory descriptive terms with definitions developed for the evaluation of WOF in cooked, chill-stored, and reheated pork meat.

Term	Definition with reference material*
Odor	*Odor associated with*
Cardboard	Shredded wet cardboard
Linseed oil	Warmed linseed oil/linseed oil-based paint
Rubber/sulfur	Warmed rubber or the white of a boiled egg
Nut	Crushed fresh hazelnuts
Green	Fresh French green beans
Fatty	Pig back fat (fresh, nonoxidized)
Taste	*Taste associated with*
Sweet	Sucrose, 1 g/L aqueous solution
Salt	Sodium chloride, 0.5 g/L aqueous solution
Sour	Citric acid monohydrate, 0.3 g/L aqueous solution
Bitter	Quinine chloride, 0.05 g/L aqueous solution
Umami/MSG	Monosodium glutamate, 0.5 g/L aqueous solution
Flavor	*Aromatic taste sensation associated with*
Metallic/bloody	Ferrous sulfate, 0.1 g/L aqueous solution
Fresh cooked pork	Oven cooked pork without browning
Rancid	Oxidized vegetable oil
Lactic acid/fresh sour	Natural yoghurt
Vegetable oil	Fresh vegetable oil
Piggy/animal	Skatole, 0.06 mg/mL refined vegetable oil
Fishy	Fish stock in boiling water
Tinny	Stainless steel strip
Livery	Cooked beef liver
Aftertaste	*Feeling factor on skin surfaces of the oral cavity associated with*
Astringent	Aluminum sulfate, 0.02 g/L aqueous solution

*Definitions of sensory terms, as derived during vocabulary development by O'Sullivan *et al.* (2002).

malonaldehyde, but this is not so. In fact, the TBA method has been criticized as lacking specificity and adequate sensitivity toward the dialdehyde. Due to the ambiguity concerning the identity of compounds that can react with the TBA reagent, the term "thiobarbituric acid-reactive substances" (TBARS) is now employed in lieu of TBA number or value (Ke *et al.* 1984, Gray & Pearson 1987). Nevertheless, determining the content of TBARS (i.e., often reported as mg malonaldehyde equivalents/kg meat) appears to be a useful indicator of meat quality deterioration. Recent studies have shown that TBARS are highly correlated with many of the sensory terms related to WOF from pork (Byrne *et al.* 2001). On the other hand, the importance of hexanal, a dominant volatile oxidation product of linoleic acid, as an indicator of WOF development by sensory perception has come into question. Hexanal has a characteristic "tallowy" or "green leafy" aroma, but this odor term was not strongly perceived as being associated with WOF odor terms (e.g., linseed oil-like and cardboard-like) during vocabulary development for pork and chicken meat (Byrne *et al.* 1999a, b). Results from gas chromatography/mass spectrometry (GC/MS) analyses have indicated that oxidation compounds such as pentanal, 2-pentylfuran, octanal, nonanal, 1-octen-3-ol, and hexanal covaried with the sensory descriptor green.

9.7 STRATEGIES TO AVOID WOF DEVELOPMENT

Important questions facing the food scientist and technologist include how one controls or limits WOF development in meat and meat products and what arsenal of countermeasures is available to address WOF. By examining the causes of off flavor development, strategies can be designed to improve the situation.

9.7.1 Meat quality

The choice of meat is an important factor with regard to the problems of oxidation. Fish and other muscle tissues containing high levels of PUFAs exhibit more of a propensity toward off flavor development. Chicken has less tendency to develop oxidized flavors than turkey, because the higher level of vitamin E in chicken fat retards oxidation; however, the problem will be worse in chicken thigh meat than in white meat, as the darker meat contains more lipid and heme iron. Fresh meat used in product formulations shows less of a tendency to develop WOF than older meat. If the meat is fresh then little enzymatic oxidation has taken place, which can generate autocatalytic compounds that propagate oxidation even after enzymes have been inactivated by thermal processing. Antioxidant enzymes, including catalase, glutathione peroxidase, and superoxide dismutase, continue to function postmortem at curbing lipid oxidation in uncooked muscle foods; however, their efficacy diminishes with increasing age of the meat. Ensuring that high quality meat is used in product formulations is critical. Incorporating antioxidants into the product and reducing the time from cooking to plate are other means by which food service operators can minimize WOF.

The actual cooking method employed for precooked products can also influence the extent of WOF. One might assume that grilling of meat would exaggerate the potential for WOF on account of the high temperatures to which lipid and protein constituents are subjected, but this is not so. In fact, thermal processes that employ very high temperatures like grilling seem to inhibit WOF development through the formation of Maillard-browning intermediates (Bailey *et al.* 1987, van Ruth *et al.* 1998). Similarly, conditions that favor browning, such as addition of glucose or smoke intermediates, may help to retard or inhibit WOF development.

9.7.2 Primary antioxidants: synthetic and natural

Food technologists attempt to reduce the problem of WOF in meat products by adding food-grade antioxidants, when permitted, or ingredients that impart antioxidant properties. Antioxidants extend the induction period and delay the onset of fatty acid oxidation by acting as free-radical acceptors or hydrogen atom donors. Such antioxidants (AH) trap free radicals directly and delay the free-radical chain reaction in a concentration-dependent manner. The antioxidant radical (A·) formed from the reactions shown below is much more stable than the alkyl (L·), alkoxy (LO·), or hydroperoxy (LOO·) free radicals and does not partake in the propagation reactions.

$$L\cdot + AH \rightarrow LH + A\cdot$$
$$LO\cdot + AH \rightarrow LOH + A\cdot$$
$$LOO\cdot + AH \rightarrow LOOH + A\cdot$$

Synthetic compounds such as tert-butylhydroquinone (TBHQ), butylated hydroxyanisole (BHA), butylated hydroxytoluene (BHT), and propyl gallate (PG) are commonly used antioxidants by the food industry. Their usage levels are strictly regulated, and in the past few years, concerns have been raised about the safety of these compounds (Valentão *et al.* 2002). As "chemical soup" is not customer-friendly, a wide variety of natural compounds can offer protection against oxidation in the form of food ingredients. The most common natural antioxidants utilized to fight WOF are vitamin E, rosemary and sage extracts, and carotenoids such as b-carotene and lycopene. In addition, many other spices, fruits, and vegetables contain constituents with antioxidant properties, and these might provide benefits to a meat formulation. Reducing exposure of these antioxidants to oxygen and light will enhance their effectiveness in minimizing WOF development.

Addition of spice and herb extracts to meat products is a popular means of incorporating natural antioxidants. Rosemary, oregano, and sage extracts as well as their combinations with tocopherols and/or ascorbate have been utilized. Rosemary in particular appears to be the most effective (Barbut *et al.* 1985, Lai *et al.* 1991); it contains a number of antioxidant compounds called diterpenes, with carnosic acid being the most prevalent and also possessing the greatest antioxidant potency. Because carnosic acid and rosmanol, another antioxidant constituent of rosemary, are odorless, manufacturers can develop spice-based ingredients with reduced flavor impact and increased protection against oxidation, and the subsequent WOF development. Nevertheless, spice companies have been working to develop odorless extracts possessing antioxidant activity for addition to meat systems.

9.7.3 Dietary antioxidants

Typically, antioxidants are added to a meat product during its formulation, but scientists are experimenting with means to increase the antioxidant capacity of muscle tissue before the animal is harvested. One such approach is by supplementing the feed of domesticated species with dietary antioxidants. Vitamin E's supplementation in the form of a-tocopherol acetate to the feed of animals prior to their slaughter can minimize the potential for WOF development (Buckley *et al.* 1989, Asghar *et al.* 1991). This effect has been seen in a number of studies where the basal diets of hogs have been supplemented with increasing levels of vitamin E (O'Sullivan *et al.* 1997). A progressive increase in the concentration of a-tocopherol has been found in the muscle tissue, mitochondria, and microsomes of hogs. a-Tocopherol migrates into muscle cell membranes where it lies adjacent to highly oxidizable phospholipids; this localization makes a-tocopherol a particularly effective antioxidant. Sensory studies of meat products have shown that vitamin E supplementation can prolong flavor freshness, inhibit WOF development, and positively influence tenderness and juiciness.

9.7.4 Secondary antioxidants

Another method of reducing oxidation is by the incorporation of chelators into meat products. A chelator or sequestrant is a food additive, which reacts with trace metal ions in foods and forms tightly bound complexes, thereby preventing the metal ion's interaction and subsequent catalytic action on lipid constituents. Typical chelators added to processed meat products are alkaline phosphates. Not only do alkaline phosphates improve the functionality of the meat product in question (e.g., via water-binding capacity, chewiness, and other textural attributes), but they also have the ability to complex with or "chelate" free iron ions in the meat matrix. Furthermore, sodium tripolyphosphate, tetrasodium pyrophosphate, and sodium hexametaphosphate have the ability to complex with iron ions that are released from the heme moiety of myoglobin during thermal processing and the subsequent denaturation of meat proteins. Besides these phosphates, citric, ascorbic, and ethylenediaminetetraacetic acids are common food-grade additives, which help to stabilize metal ions by reducing their activity and ability to act as oxidants. Ascorbic acid, its sodium salt, and isomer (erythorbate) also function synergistically with other antioxidants and added polyphosphates to give protection to meats against oxidative degradation.

9.7.5 Nitrites and nitrates

Nitrites and nitrates are unique additives to meat products, in that they perform a multifunctional role in the meat matrix. One role is to retard lipid oxidation and WOF development (Pegg and Shahidi 2000). The mechanism by which nitrite prevents or retards the oxidation of meat lipids, and consequently WOF development, is still a matter of discussion. From the literature, however, four different mechanisms have been proposed for the antioxidative effect of nitrite in meats: (1) formation of a stable complex between heme pigments and nitrite, thereby preventing the release of iron ions from the porphyrin molecule; (2) stabilization of unsaturated lipids within tissue membranes against oxidation; (3) interaction of nitrite as a metal ion chelator so that it ties up trace metals in meat as well as any liberated nonheme iron from denatured heme pigments; and (4) formation of nitroso and nitrosyl compounds in meat which possesses antioxidative properties by acting as radical scavengers.

9.7.6 Natural and liquid smoke

Smoking is another popular meat-preservation technique. Like nitrites, constituents of smoke impart antioxidant properties to meat and meat products. Additionally, incomplete combustion of gases during natural smoke generation from the pyrolysis of hardwood can result in the formation of various nitrogen oxides; some of these function as nitrite does in the curing process. Research has shown that certain smoke flavorings can reduce the occurrence of WOF and extend shelf life when added to fresh, precooked, and processed meats. Formulators can add liquid smoke to their products directly by atomizing, dipping, drenching, spraying, or injecting. Smoke ingredients with strong flavors can help mask WOF, whereas flavorless smoke fractions can be employed at low levels in marination systems to reduce the extent of WOF. In some cases, the incorporation of smoke flavors has reduced lipid oxidation by 20–30%. It is the phenols present in smoke that act as antioxidants. The flavors also contain certain carbonyls that react with amino groups of meat proteins to inhibit WOF production.

9.7.7 Packaging

Packaging is a physical means to reduce off flavor development in meat and meat products. Because light and the presence of oxygen can accelerate oxidation, eliminating their exposure through packaging technologies will help. Vacuum packaging controls oxygen interactions at the meat surface, and thereby minimizes oxidation. This technique, coupled with nitrogen flushing or modified atmosphere packaging (MAP; e.g., 70% N_2 and 30% CO_2) techniques, can give substantial shelf life to finished products. Other packaging technologies, such as oxygen scavengers, can be added to the package either as stand-alone or be incorporated into the packaging film. An approach used by the packaging industry has been the development of films that act as oxygen barriers; this can help decrease WOF development. In fact, packages of the future may become part of the product itself. Edible films containing natural antioxidants have been examined as a means to control WOF development in cooked meat products. The films were composed of modified cornstarch, soy protein, wheat gluten, corn zein, and natural antioxidants. Some problems encountered, however, had to do with solubility issues; that being, the natural antioxidant was fat soluble (i.e., hydrophobic), whereas the film itself was hydrophilic in nature. Therefore, the release and delivery of the active components need to be clarified. In some cases, it is more effective to mix the antioxidant directly into the meat formulation, but this does not work for whole-muscle and some restructured meat products.

9.7.8 Flavor masking

Flavor masking or other methods using flavors that modify the perception of off notes can be effective tools in the fight against oxidative off flavors in meat products. An example of which has already been eluded to is that of smoking. Addition of complementary (i.e., savory) flavors from mild herbs and spices or their extracts can potentiate meat flavors. These inhibit lipid oxidation and help stabilize desirable meaty flavors, which fade during processing and refrigerated storage. Beer flavoring is another ingredient that helps eliminate flavor problems that occur during the reheating of meats. At low levels, it functions as an enhancer, but at higher concentrations it imparts strong beer flavor notes. Its anti-WOF effect may be a function of the typical yeast notes found in the brew, because yeast-based flavors can improve and enhance flavors, mask bitterness, increase aroma, and provide some protection against oxidation. Still, masking agents that have been developed to cover off flavors, such as metallic notes in high-intensity sweeteners or beany notes in soybean products, have also been proven effective in masking WOF. A good masking agent will not have much flavor on its own when employed at low levels. However, coprocessing with other flavors, especially desirable meaty notes, can tailor the mask to specific applications, such as masking WOF. In vacuum-packaged refrigerated products where one is looking for a 60-day shelf life, it holds up well and is stable in retort products. Because the flavor can also mask certain desirable meaty notes if used at too high of a concentration, product testing is necessary to develop the best application level. Merely disguising the oxidative notes is another option. That is, highly flavored systems such as those found in spicy Mexican or East Indian seasonings can distract the consumer from any off notes. The sensations of heat and tanginess dominate consumers' palates. Heavy spicing of meat products is common practice in countries where proper refrigerated storage is an issue.

In developing a flavor system designed to combat WOF, it is important to consider the overall flavor profile of the finished product, along with the prevention of off flavor development. Both temperature and hold time during the cooking process will influence the overall taste. Therefore, to achieve an optimal flavor, spices and flavoring agents should be incorporated at a slightly higher level than would be needed for immediate consumption. Once an appropriate flavor system is developed, it is incorporated into the product by means of a mix-in seasoning, rub, or marinade, depending on the type of product formulated. In a whole-muscle product, marination is recommended to enhance protection and minimize the possibility of off notes developing within the internal area of the product. Because the product's surface is more prone to oxidation than the interior of the meat, topical application would certainly help to minimize the formation of WOF notes.

9.8 SUMMARY

Off flavors have always been a problem in lipid-rich foods. In particular, WOF has often been a known adversary in prepared food service entrées. As far as oxidation is concerned, it is not possible to prevent the decomposition of hydroperoxides once they are formed; therefore, the goal is to prevent or retard their formation as long as possible. This may be achieved to some degree by the incorporation of synthetic or natural antioxidants and chelators to foods (i.e., extended shelf life and reduced off flavor development). From the point of view of taints, using proper packaging to prevent the absorption of chemical and microbial contaminants from the environment will be advantageous.

REFERENCES

Asghar, A., Gray, J.I., Booren, A.M., *et al.* (1991) Influence of supranutritional dietary vitamin E levels on subcellular deposition of alpha-tocopherol in the muscle and on pork quality. *Journal of the Science of Food and Agriculture*, **57**, 31–41.

Baigrie, B. (2003) Introduction. In: *Taints and Off-Flavours in Food* (ed. B. Baigrie), pp. 1–4. CRC Press LLC, Boca Raton, Florida.

Bailey, M.E., Shin-Lee, S.Y., Dupuy, H.P., St. Angelo, A.J., and Vercellotti, J.R. (1987) Inhibition of warmed-over flavour by Maillard reaction products. In: *Warmed-Over Flavour of Meat* (eds. A.J. St. Angelo & M.E. Bailey), pp. 237–266. Academic Press, London, U.K.

Barbut, S., Josephson, D.B. & Naurer, A.J. (1985) Antioxidant properties of Rosemary Oleoresin in Turkey sausage. *Journal of Food Science*, **50**, 1356–1359.

Belitz, H.-D. & Grosch, W. (1987) Lipids. In: *Food Chemistry*, pp. 128–200. Springer-Verlag, Berlin, Germany.

Buckley, D.J., Gray, J.I., Ashghar, A., *et al.* (1989) Effects of dietary antioxidants and oxidized oil on membranal lipid stability and pork product quality. *Journal of Food Science*, **54**, 1193–1197.

Byrne, D.V., Bak, L.S., Bredie, W.L.P., Bertelsen, G. & Martens, M. (1999a) Development of a sensory vocabulary for warmed-over flavour: Part I. In porcine meat. *Journal of Sensory Studies*, **14**, 47–65.

Byrne, D.V., Bredie, W.L.P. & Martens, M. (1999b) Development of a sensory vocabulary for warmed-over flavour: Part II. In chicken meat. *Journal of Sensory Studies*, **14**, 67–78.

Byrne, D.V., Bredie, W.L.P., Bak, L.S., Bertelsen, G., Martens, H. & Martens, M. (2001) Sensory and chemical analysis of cooked porcine meat patties in relation to warmed-over flavor and pre-slaughter stress. *Meat Science*, **59**, 229–249.

Gray, J.I. & Pearson, A.M. (1987) Rancidity and warmed-over flavor. In: *Advances in Meat Research. Vol. 3: Restructured Meat and Poultry Products* (eds A.M. Pearson & T.R. Dutson), pp. 221–269, Van Nostrand Reinhold Co, New York.

Hamilton, R.J. (2003) Oxidative rancidity as a source of off-flavours. In: *Taints and Off-Flavours in Food* (ed. B. Baigrie), pp. 140–161. CRC Press LLC, Boca Raton, Florida.

Johnson, P.B. & Civille, G.V. (1986) A standardized lexicon of meat WOF descriptors. *Journal of Sensory Studies*, **1**, 99–104.

Ke, P.J., Cervantes, E. & Robles-Martinez, C. (1984) Determination of thiobarbituric acid reactive substances (TBARS) in fish tissue by an improved distillation-spectrophotometric method. *Journal of the Science of Food and Agriculture*, **35**, 1248–1254.

Kochhar, S.P. (1996). Oxidative pathways to the formation of off-flavours. In: *Food Taints and Off-Flavours* (ed. M.J. Saxby), 2nd edn., pp. 165–225. Blackie Academic & Professional, Glasgow, U.K.

Konopka, U.C. & Grosch, W. (1991) Potent odorants causing the warmed-over flavour in boiled beef. *Zeitschrift Für Lebensmitteluntersuchung Und -Forschung*, **193**, 123–125.

Lai, S.M., Gray, J.I., Smith, D.M., Booren, A.M., Crackel, R.L. & Buckley, D.J. (1991) Effects of rosemary oleoresin, tertiary butylhydroquinone, and sodium tripolyphosphate on the development of oxidative rancidity in restructured chicken nuggets. *Journal of Food Science*, **56**, 616–626, 635.

Lord, T. (2003) Packaging materials as a source of taints. In: *Taints and Off-Flavours in Food* (ed. B. Baigrie), pp. 64–111. CRC Press LLC, Boca Raton, Florida.

Love, J.D. (1988) Sensory analysis of warmed-over flavor in meat. *Food Technology*, **42**(6), 140–143.

O'Sullivan, M.G., Kerry, J.P., Buckley, D.J., Lynch, P.B. & Morrissey, P.A. (1997) The distribution of dietary vitamin E in the muscle of the porcine carcass. *Meat Science*, **45**, 297–305.

O'Sullivan, M.G., Byrne, D.V., Martens, H. & Martens, M. (2002) Data analytical methodologies in the development of a vocabulary for evaluation of meat quality. *Journal of Sensory Studies*, **17**, 539–558.

Palm, H. (2002) Detection & characterisation of unwanted off-flavors caused by rancidity through analysis of volatiles. *Malaysian Oil Science and Technology*, **11**, 14–18.

Pegg, R.B. & Shahidi, F. (2000) Nitrite curing of meat. *The N-Nitrosamine Problem and Nitrite Alter-natives*. Food & Nutrition Press, Inc, Trumbull, Connecticut.

van Ruth, S.M., Cheraghi, T. & Roozen, J.P. (1998) Lipid-derived off-flavours in meat by-products: Effect of antioxidants and Maillard reactants. In: *Flavor of Meat, Meat Products and Seafoods.* (ed. F. Shahidi), 2nd edn., pp. 257–266. Blackie Academic & Professional, London, U.K.

Rychlik, M., Schieberle, P. & Grosch, W. (1998) *Compilation of Odor Thresholds, Odor Qualities and Retention Indices of Key Food Odorants*. Deutsche Forschungsanstalt für Lebensmittelchemie and Institut für Lebensmittelchemie der Technischen Universität München, Garching, Germany.

Shahidi, F. (1989). Flavor of cooked meats. In: *Flavor Chemistry: Trends and Developments*, (eds R. Teranishi, R.G. Buttery & F. Shahidi), ACS Symposium Series 388, pp. 188–201. American Chemical Society, Washington, District of Columbia.

Shahidi, F., Rubin, L.J. & D'Souza, L.A. (1986) A review of the composition, techniques, analyses and sensory evaluation. *CRC Critical Reviews in Food Science & Nutrition*, **24**, 141–243.

Siu, G.M. & Draper, H.H. (1978) A survey of the malonaldehyde content of retail meats and fish. *Journal of Food Science*, **43**, 1147–1149.

St. Angelo, A.J. & Bailey, M.E. (eds) (1987) *Warmed-Over Flavor of Meat*. Academic Press, Inc, Orlando, Florida.

Tarladgis, B.G., Watts, B.M., Younathan, M.T. & Dugan, L.R., Jr. (1960) A distillation method for the quantitative determination of malonaldehyde in rancid foods. *Journal of the American Oil Chemists' Society*, **37**, 44–48.

Tichivangana, J.Z. & Morrissey, P.A. (1985) Metmyoglobin and inorganic metals as pro-oxidants in raw and cooked muscle systems. *Meat Science*, **15**, 107–116.

Tims, M.J. & Watts, B.M. (1958) Protection of cooked meats with phosphates. *Food Technology*, **12**, 240–243.

Valentão, P., Fernandes, E., Carvalho, F., Andrade, P.B., Seabra, R.M. & Bastos, M.L. (2002) Antioxidative properties of cardoon (*Cynara cardunculus L.*). Infusion against superoxide radical, hydroxyl radical, and hypochlorous acid. *Journal of Agricultural and Food Chemistry*, **50**, 4989–4993.

Vega, J.D. & Brewer, M.S. (1994) Detectable odor thresholds of selected lipid oxidation compounds at various temperatures in a gelatin model system. *Journal of Food Lipids*, **1**, 229–245.

Wong, D.W.S. (1989) *Mechanism and Theory in Food Chemistry*, pp. 1–47. Van Nostrand Reinhold, New York.

10 Land Animal Products

Terri Boylston

Abstract: Over 1000 flavor compounds in meats have been identified. However, a single compound or group of compounds responsible for "meaty flavor" has not been identified due to the overall complexity of meat flavor. Meat flavor is dependent on the pool of flavor precursors in the meat tissue, the chemical reactions during processing, and subsequent storage. The precise flavor precursors vary between and within species, beef, pork, lamb, and poultry.

Keywords: Volatile flavor compounds, Maillard reaction, lipid oxidation, taste compounds, sensory evaluation, curing, irradiation

Extensive research over the past years has identified over 1000 flavor compounds in meats. However, a single compound or group of compounds responsible for "meaty flavor" has not and perhaps never will be identified due to the overall complexity of meat flavor. Meat flavor is dependent on the pool of flavor precursors in the meat tissue and the chemical reactions that occur during processing. Processing and subsequent storage contribute to the development of the characteristic flavors of meats. Because the precise flavor precursors vary between and within species, beef, pork, lamb, and poultry each have distinctive flavor characteristics. In this chapter, the major contributors to meat flavor, from an instrumental and sensory point of view, and factors that affect the final flavor quality of the meat when it is consumed will be discussed.

10.1 CONTRIBUTORS TO FLAVOR

Flavor is composed of aroma or volatile flavor compounds, which are detected by specialized cells of the olfactory epithelium of the nasal cavity and nonvolatile taste compounds, which are detected by the taste buds on the tongue and back of the oral cavity. The integration of aroma and taste is critical to the overall perception of meat flavor.

10.1.1 Volatile flavor compounds

Numerous complex reactions occur during the cooking of meat to contribute to flavor development. The meaty flavor of muscle foods is attributed to the lean portion of the muscle, which is similar between the different species (Wasserman 1979). The unique flavor characteristics that distinguish the different species of land animals from each other is attributed to differences in the fatty acid composition of these muscles (Hornstein & Crowe 1960; Sink

Handbook of Meat, Poultry and Seafood Quality, Second Edition. Edited by Leo M. L. Nollet.

1979; Wasserman 1979) and the lipid oxidation reactions that occur during cooking (Shahidi *et al.* 1986; Mottram 1998). The volatile flavor compounds that contribute to the desirable meaty and browned flavors are present at parts per billion levels, while grassy, cardboardy, and painty flavors, which contribute to undesirable flavors, are present at parts per million levels (Vercelloti *et al.* 1987). The differences in the concentrations, thresholds, and aroma qualities of these compounds have a significant impact on the flavor characteristics and desirability of meat flavor.

10.1.1.1 Maillard reaction

The Maillard reaction is a nonenzymatic browning reaction involving carbonyl and amine compounds. Through a series of complex reactions, aromas with roasted, toasted, and meaty flavor characteristics and polymerized brown pigments are formed to contribute to the characteristic aroma of cooked meat. Chapter 20 discusses the chemistry of the Maillard reaction and its contribution to flavor in foods in more detail.

Heterocyclic and sulfur-containing compounds formed through the Maillard reaction are important contributors to desirable meaty flavors (Fig. 10.1). The flavor descriptions of the volatile flavor compounds that are formed through the Maillard reaction are presented in Table 10.1. Although these compounds are present at parts per billion levels, they are critical to the desirable flavors of meat because of their low aroma thresholds (Min *et al.* 1979; MacLeod & Seyyedain-Ardebili 1981; Hsu *et al.* 1982; Hartman *et al.* 1983; Galt & MacLeod 1984; MacLeod & Ames 1986; St. Angelo *et al.* 1987; Shahidi 1989; Ramarathnam *et al.* 1993; Mottram 1998). The importance of sulfur-containing volatile flavor compounds to the roasted meaty flavor characteristics has been demonstrated in model systems consisting of cysteine, cystine, thiamine, reducing sugars, and other meat flavor precursors (Güntert *et al.* 1990; Werkhoff *et al.* 1990; Mottram & Nobrega 2002). The flavor quality of the sulfur-containing compounds is dependent on the concentration, with pleasant meaty aromas exhibited at low concentrations and objectionable aromas exhibited at high concentrations (Shahidi 1989).

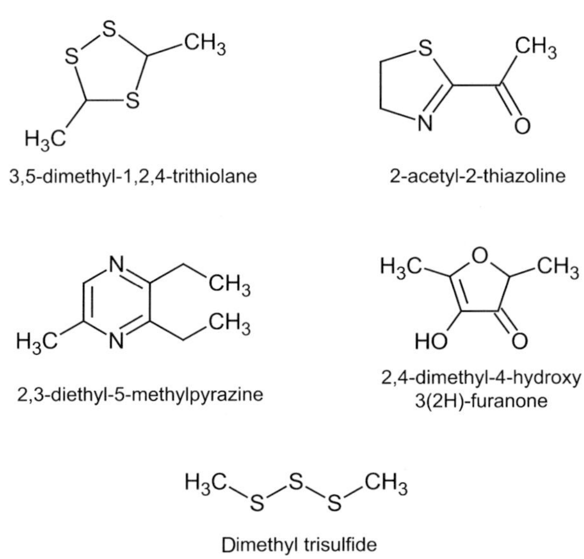

Fig. 10.1 Examples of heterocyclic and sulfur-containing compounds found in meats.

Table 10.1 Flavor characteristics of volatile flavor compounds formed through Maillard reactions.

Volatile compound	Flavor description	Reference*
Aliphatic sulfur compounds		
Hydrogen sulfide	Boiled or rotten eggs	E
Methanethiol	Rotten eggs, meat or fish, cheesy, cooked cabbage	A E
Dimethyldisulfide	Strong onion, garlic-like, pungent	B
Dimethyl trisulfide	Roasty, rotting green vegetables, cabbage	A B
Methional	Cooked potatoes, meat broth	B C
3-(Methylthio) propanal	Cooked potato	D
3,5-Dimethyl-1,2,4-trithiolane	Putrid	D
Heterocyclic sulfur compounds		
4,5-Dimethylthiazole	Smoky, roasty, nutty	B
2-Acetyl-2-thiazoline	Roasty, floral	A C
2,4,6-Trimethyltetrahydro-1,3, 5-thiadiazine	Popcorn, burnt	D
5,6-Dihydro-2,4,6-trimethyl-4 H, 1,3,5-dithiazine	Roasty, musty	D
2-Methyl-3-(methylthio)furan	Meaty (not roasted)	A
2-Methyl-3-furanthiol	Meaty, sweet	D
2-Furfuryl thiol	Roasty, coffee	D
Heterocyclic nitrogen compounds		
Methylpyrazine	Green, fragrant	A
2,3-Dimethylpyrazine	Meaty	A
2,5 (and/or 2,6)-Dimethylpyraine	Meaty roast, green	A
Trimethylpyrazine	Roasty, earthy, caramel	A C
Tetramethylpyrazine	Walnuts, green	A
2-Ethyl-3-methylpyrazine	Green, floral, fresh nuts	A
2-Ethyl-5-methylpyrazine	Fruity, sweet, toffee, pungent	A B
2-Ethyl-3,5-dimethylpyrazine	Roasty, earthy, nutty	C D
2-Ethyl-3,6-dimethylpyrazine	Burnt, pungent, roasty, nutty	A B
2,3-Diethyl-5-methylpyrazine	Meaty, roasty, fragrant, sweet, earthy, crushed leaves	A B C
2,5-Diethyl-3-methylpyrazine	Sweet, caramel	A
6,7-Dihydro-(5 H)cyclopentapyrazine	Green, fatty	A
2-acetyl-1-pyrroline	Roasty, overheated meat-like, cured ham-like	C E
Heterocyclic oxygen compounds		
2-Furanmethanol	Meaty roast	A
4-Hydroxy-5-methyl-3(2 H)-furanone	Caramel, fried chicken	D
2,5-Dimethyl-4-hydroxy-3(2 H)-furanone	Roasted almonds, sweet, caramel	B C D
3-Hydroxy-4,5-dimethyl-2(5 H)-furanone	Seasoning-like	C
4,5-Dihydro-5-propyl-2(3 H)-furanone	Fruity, fatty, sweet, roasty	B
4,5-Dihydro-5-butyl-2(3 H)-furanone	Caramel-like, pungent, roasty, fruity	B

*References for aroma characterization: A: MacLeod and Ames (1986); B: Specht and Baltes (1994); C: Kerler and Grosch (1996); D: Farkaš *et al.* (1997); E: Carrapiso *et al.* (2002).

The addition of Maillard reaction products, formed through reactions between sugars and amino acids, has been shown to function as a meat flavor enhancer and as an antioxidant when added to meat products (Lingnert & Eriksson 1981; Bailey & Um 1992). The basic amino acids have been shown to form the most effective antioxidants and the choice of amino acid rather than sugar is most critical in the formation of Maillard reaction products with antioxidant activity (Lingnert & Eriksson 1981).

10.1.1.2 Lipid oxidation

Lipid oxidation is the major cause of loss of desirable meat flavor. Tims and Watts (1958) first described the development of off flavors in cooked meats following refrigerated storage and coined the term "warmed-over flavor" (WOF). Research since then has shown that WOF can develop whenever the muscle tissue is disrupted by processes such as grinding, restructuring, or cooking of the muscle (Greene 1969; Sato & Hegarty 1971). A detailed discussion of the mechanism of WOF formation and processing techniques, including use of antioxidants to minimize WOF formation, are discussed in Chapter 17.

The unsaturated fatty acids associated with the phospholipids in the muscle tissue are the major precursors of the flavor compounds associated with WOF. The oxidation of these lipids, which is catalyzed by iron, results in the formation of a complex mixture of aldehyes, ketones, and alcohols (Younathan & Watts 1960; Sato & Hegarty 1971; Igene *et al.* 1979). These lipid oxidation compounds are present at parts per million concentrations and are the predominant peaks in the gas chromatogram of meat samples (St. Angelo *et al.* 1987). Pentanal, hexanal, heptanal, 2,3-octanedione, 1-octen-3-ol, nonanal, and 2-pentylfuran are among the major lipid oxidation products identified in meats as contributors to WOF (Fig. 10.2). The carbonyl compounds have the greatest impact on flavor due to their high rate of formation during lipid oxidation and low flavor threshold in comparison to the other oxidation products (St. Angelo *et al.* 1987; Drumm & Spanier 1991; Decker & Hultin 1992). Table 10.2 summarizes the aroma characteristics of volatile flavor compounds formed through lipid oxidation reactions.

Thiobarbituric acid reactive substances (TBARS) and hexanal contents are frequently used as indicators of lipid oxidation in meats. Thiobarbituric acid reacts with carbonyl compounds formed through lipid oxidation to form a colored product. Although this assay lacks specificity, it is widely used to monitor lipid oxidation reactions, especially in meats. Hexanal, which is the major product of linoleic acid oxidation, is present in the highest intensity and is formed at significantly higher rates than the other oxidation products (St. Angelo *et al.* 1987, 1992; Drumm & Spanier 1991; Decker & Hultin 1992; Kerler & Grosch 1996). These chemical indicators have been shown to be positively correlated with the intensity of sensory descriptors of undesirable flavor notes, such as cardboard, painty, warmed-over, oxidized, sour, and bitter and negatively correlated with desirable flavor notes, such as cooked beef/brothy, meaty, browned/caramel, serum, and sweet (Hwang *et al.* 1990; St. Angelo *et al.* 1987, 1992). However, because the rates of formation for the individual oxidation products are different and independent of the initial concentration, the use of a single compound, such as hexanal, does not adequately characterize the overall flavor (Ullrich & Grosch 1987; Drumm & Spanier 1991).

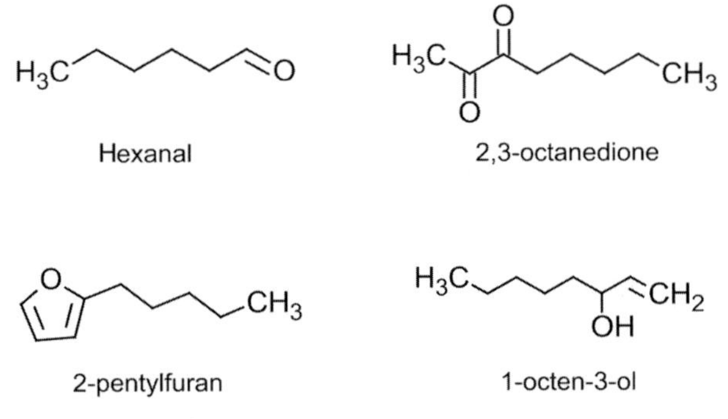

Fig. 10.2 Examples of lipid oxidation products found in meats.

Table 10.2 Flavor characteristics of volatile flavor compounds formed through lipid oxidation reactions.

Volatile compound	Flavor description	Reference*
Aldehydes		
2-Methylpropanal	Toasted, fruity, pungent	E
3-Methylbutanal	Green, burnt, caramel, malty, pungent, fruity, toasted	A B C E
2-Methylbutanal	Sweet, floral, fruity, roasty, malty	B C
Hexanal	Green, grassy, fatty	A B C
Heptanal	Fruity, fatty, sweet, caramel, buttery	A B
Octanal	Soapy, sweet, fruity, nutty	A C
Nonanal	Fragrant, sweet, fatty, green, tallowy	B C
Trans-2-nonenal	Pungent, fragrant, fruity, roasty, tallowy, fatty	B C
Cis-3-nonenal	Fatty	C
Cis-2-nonenal	Soapy	C
Trans-2, cis-6-nonadienal	Cucumber-like	C
Trans, trans-2,4-Nonadienal	Fatty, soapy	C
Decanal	Sweet, fruity, roasty	B
Trans-2-decenal	Pungent, green, fragrant, fruity	B
Trans-2, trans-4-decadienal	Sweet, fatty, pungent, fried potato, chickeny	B C D
Trans-2-undecenal	Sweet, fruity, fatty	B
Trans-2-dodecenal	Sweet, fruity, roasty, pungent	B
Benzaldehyde	Nutty, almonds	A
Phenylacetaldehyde	Sweet, fruity, flowery, honey-like	B C
Ketones		
2,3-Butanedione	Sweet, buttery, caramel	A B
2,3-Pentanedione	Buttery, lemon-like, sweet fruity, caramel	A B
1-Penten-3-one	Rotten, fruity	E
Octan-2-one	Caramel	A
1-Octen-3-one	Mushroom-like	B C D
1-Octadien-3-one	Green, geranium-like	C
Nonan-2-one	Boiled milk, slightly burnt	A
3-Methylcyclopentanone	Meaty, fatty, fruity, caramel	A
Alcohols		
Hexanol	Fragrant	A
Heptanol	Sweet, fruity, butterscotch	A
1-Octen-3-ol	Green	A
Acids		
Butanoic acid	Cheese-like, rancid	A

*References for aroma characterization: A: MacLeod and Ames (1986); B: Specht and Baltes (1994); C: Kerler and Grosch (1996), D: Farkaš *et al.* (1997), E: Carrapiso *et al.* (2002).

10.1.1.3 Flavors formed through the interaction between lipid oxidation and Maillard reactions

The interaction of carbonyl compounds generated through lipid oxidation reactions and Maillard reaction products also contributes to the formation of heterocyclic compounds with meaty flavor characteristics. Alkylthiazoles, alkylthiophenes, alkylpyridines, alkylpyrazines, and trithiolanes have been proposed to form through reactions between aliphatic aldehydes and hydroxyketones formed from lipid oxidation products and ammonia and hydrogen sulfide from the decomposition of amino acids (Farmer *et al.* 1989; Farmer & Mottram 1990; Mottram 1998; Elmore *et al.* 1999). Phospholipids are critical to the formation of a pronounced "cooked meat, beefy" aroma, as shown in comparing model systems consisting of cysteine 1 ribose with and without phospholipids (Farmer *et al.* 1989; Farmer & Mottram 1990). In beef, increases in polyunsaturated fatty acid content of the muscle were correlated with increases in content of alkylthiophenes and ethylthiapyran (Elmore *et al.* 1999). The greater contribution of phospholipids, rather than triglycerides, in the formation of meaty aromas was attributed to the higher degree of unsaturation and higher water solubility of the phospholipids (Mottram & Edwards 1983; Farmer & Mottram 1990).

10.1.2 Taste compounds

Meat constituents that contribute to the basic tastes include inorganic salts (salty), hypoxanthine (bitter), sugars (sweet), and organic acids (sour). In addition, the amino acids contribute sweet, sour, bitter, and salty tastes to meat. Hydrophilic amino acids, such as hydroxyproline, glycine, serine, and threonine, contribute sweet tastes. Hydrophobic amino acids, such as histidine, arginine, methionine, valine, isoleucine, leucine, phenylalanine, and tryptophan, contribute bitter tastes. Aspartic acid, glutamic acid, and asparagine are noted for their sour taste characteristics, and the sodium salts of glutamic and aspartic acids have salty taste characteristics (MacLeod & Seyyedain-Ardebili 1981; Spanier *et al.* 1988, 1990). Sour taste characteristics in meat have also been attributed to lactic acid formed from glycogen through anaerobic glycolysis (Larick & Turner 1990). Proteolytic activity during postmortem aging produces peptides that can increase the overall taste intensity and palatability of the meat (MacLeod & Seyyedain-Ardebili 1981). Degradation products of ribonucleotides, including inosine 5′-monophosphate and guanosine 5′-monophosphate, also function as flavor potentiators to enhance meaty flavor and contribute a umami taste characteristic (Nagodawithana 1995).

10.2 SENSORY EVALUATION OF MEAT FLAVOR

Sensory evaluation provides the human perception aspect of meat flavors. During the past 20 years, there has been a substantial effort to develop lexicons to describe beef (Johnson & Civille 1986; Lynch *et al.* 1986), lamb (St. Angelo *et al.* 1991), pork (Jeremiah *et al.* 1990; Byrne *et al.* 1999a), cured ham (Flores *et al.* 1997), and poultry (Lyon 1987; Chambers *et al.* 1992; Byrne *et al.* 1999b) flavor. The basic tastes, sweet, sour, salty, and bitter, as well as universal flavor descriptors, such as meaty, serum, browned, painty, and cardboard are common across all species. In addition to these universal flavor descriptors, descriptors and definitions that are unique to each species, such as beefy, gamey/muttony, porky, and chickeny/poultry, provide species-specific descriptors (St. Angelo *et al.* 1991; Bett 1993; Imafidon & Spanier 1994).

Principal factor analysis of sensory and instrumental measures of flavor in beef patties has shown the sensory attributes sweet, cooked beef/brothy, and browned/caramel to be associated with the desirable freshly cooked patties. On the other hand, the sensory attributes, painty, sour, bitter, and cardboard, were associated with cooked and stored patties that did not receive treatments to reduce lipid oxidation and had undesirable flavor characteristics. High contents of the volatile flavor compounds—hexanal, nonanal, 2,3-octanedione, propanol, pentanal, decanol, and butanol—and high TBARS values were also associated with those patties with undesirable, oxidized flavor characteristics (Spanier *et al.* 1992a, b).

Gas chromatography-olfactometry (GC-O) is a useful tool to establish the relationship between flavor compounds detected by instrumental techniques and sensory evaluation. A panelist sniffs the volatile compounds using a sniffer apparatus as these compounds are eluted from the gas chromatograph and provides descriptive terms and intensities for the individual flavor compounds. Due to differences in the thresholds of volatile flavor compounds, the use of GC-O allows food scientists to determine the impact of individual compounds on the flavor of the food. Several researchers have used GC-O and other techniques to describe the flavor characteristics of the individual compounds that contribute to meat flavor, as shown in Tables 10.1 and 10.2 (MacLeod & Ames 1986; Specht & Baltes 1994; Kerler & Grosch 1996; Farkaš *et al.* 1997; Carrapiso *et al.* 2002).

10.3 FACTORS INFLUENCING FLAVOR CHARACTERISTICS OF MEATS

Species, breed, sex, and diet have a significant impact on the fatty acid composition of the meat. Following slaughter, processing and storage conditions contribute to the development of desirable and undesirable flavors through reactions involving the lipids and other constituents of the muscle tissue.

10.3.1 Preslaughter factors

10.3.1.1 Species

The basic precursors for meat flavor, that is, the proteins, sugars, and lipids, are similar across all land animal species. Carbonyl compounds and hydrocarbons are most often recognized for flavor differences between species (Ramarathnam *et al.* 1993). The degree of unsaturation of the lipids in muscle tissues varies greatly from species to species. Because the susceptibility of fatty acids to oxidation increases with the degree of unsaturation, the rate of lipid oxidation reactions and the flavors that are formed through these reactions are species dependent.

Shahidi *et al.* (1986) reviewed meat flavor and identified the species from which specific volatile flavor compounds had been isolated. Many of these compounds were identified in several different species. However, failure to identify a specific compound may be attributed to differences in processing treatments and methods for isolation of the volatile flavor compounds and may not be necessarily due to species differences.

10.3.1.2 Breed

A majority of the research focused on the effects of breed on flavor characteristics has been conducted in beef animals. Lean muscle from bison has a higher polyunsaturated fatty acid content than lean muscle from either Hereford or Brahman cattle. The more intense off flavors and aftertastes, described as ammonia, bitter, gamey, liverish, and sour, detected in bison was attributed to differences in the fatty acid composition between these three beef species (Larick *et al.* 1989). A comparison of flavor characteristics of Pirenaica (beef type) and Friesian (dairy type) breeds of beef from the Mediterranean area resulted in the beef from the Friesian breed as having a higher intensity of fatty flavor and aftertaste, and a higher content of aliphatic hydrocarbons, but lower content of butanal and dimethylsulfide. These breeds are characterized as having low-fat carcasses, although the Friesian breed has a higher content of saturated intramuscular fat (Gorraiz *et al.* 2002). Higher contents of hexanal and other lipid oxidation products in Japanese Wagyu and American Wagyu beef cooked and stored for 3 days, in comparison to beef from Angus or Longhorn breeds, were attributed to significantly higher total lipid and unsaturated fatty acid contents in the Wagyu breeds (Boylston *et al.* 1996). Genetic control of animal lipid metabolism affects the fatty acid composition of the muscle tissue, and thus contributes to the observed differences in flavor characteristics between breeds (Larick *et al.* 1989).

10.3.1.3 Sex

Beef from bulls is characterized as having a stronger livery and bloody flavor, while beef from heifers has a more common meaty flavor. Contents of 2-propanone and dimethylsulfide were significantly higher in the beef from the bulls (Gorraiz *et al.* 2002). Heifers typically have higher triglyceride and lower phospholipids percentages in the intramuscular fat than in the fat of bulls, which may be attributed to the role of sex hormones and the genetic control of animal development on lipid composition (Sink 1979).

10.3.1.4 Diet

Nutritional awareness of the importance of decreases in saturated fatty acids and increases in polyunsaturated fatty acids in the diet has led to an interest in manipulating the fatty acid composition of meat tissue through the animals' diet. In addition to changing the nutritional value, changes in the diet of land animals have an impact on the flavor of the cooked meat. The fatty acid content of the diet has a more direct impact on the fatty acid content of the muscle tissue in pork and poultry than in beef and lamb due to the biohydrogenation reactions in the rumen (Wood *et al.* 2004).

The supplementation of pig diets with polyunsaturated fatty acids, such as safflower oil (Larick *et al.* 1992), canola or flaxseed (Shackelford *et al.* 1990), and rapeseed oil and fish oil (Leskanich *et al.* 1997), resulted in increases in the content of hexanal and other lipid oxidation products. The formation of off-flavors is further increased if the meat undergoes oxidative stress during processing (Shackelford *et al.* 1990). Similarly, in broilers, the fatty acid composition of dietary oils is reflected in the fatty acid composition of the neutral lipid and phospholipid fractions (Lin *et al.* 1989). The higher content of polyunsaturated fatty acids in the tissues of pork and poultry contributes to greater susceptibility of oxidation reactions and off flavor formation during refrigerated storage following cooking.

Beef from grain-fed cattle is described as having a higher intensity of roasted beef, beefy, blood-like, tallowy, cooked beef fat, and sour flavors and a lower intensity of sweet, grassy, and gamey flavors than beef from forage-fed cattle (Larick & Turner 1990; McMillin *et al.* 1991; Maruri & Larick 1992). Grain-fed beef have higher percentages of linoleic acid and lower percentages of oleic and linolenic acids than pasture-fed beef (Yang *et al.* 2002). The intensity of the grassy and gamey flavors is correlated with the content of 2,3-octanedione and several diterpenoid compounds in beef steaks (Larick *et al.* 1987; Maruri & Larick 1992). Lactones have been shown to be present in higher concentrations in grain-fed cattle and are positively correlated with roasted and blood-like flavor characteristics (Maruri & Larick 1992). The higher glycogen content of corn-fed cattle could contribute to higher lactic acid contents and the increased sour taste (Larick & Turner 1990). Steaks with increased polyunsaturated fatty acid contents, due to the feeding of different supplemental fats, result in higher contents of aldehydes and alcohols formed through lipid oxidation reactions (Elmore *et al.* 1999).

The supplementation of diets with a-tocopherol results in an increased content of the antioxidant in the membranes and has been shown to be effective in protecting unsaturated phospholipids from oxidation. The presence of tocopherols in the tissues stabilizes the membrane lipids and minimizes lipid oxidation and off flavor development that occurs during refrigerated and frozen storage of beef, pork, and poultry (Lin *et al.* 1989; Asghar *et al.* 1990, 1991; Yang *et al.* 2002).

10.3.2 Postslaughter processing

10.3.2.1 Aging

Aging of beef contributes to the development of desirable flavor characteristics through proteolytic, lipolytic, and other enzyme reactions, resulting in increases in the sugars, peptides, and free amino acids, and nucleotides (Spanier *et al.* 1990). Proteolytic activity has a significant effect on subsequent meat quality. Top round beef aged for up to 4 days was judged to have the optimal flavor by a sensory evaluation panel (Spanier *et al.* 1997). The concentration of peptides in the 3–10 kDa (kilodalton) molecular weight range increased with aging time and was shown to be correlated with taste intensity in beef and pork (Claeys *et al.* 2004). Amino acids generated through the proteolytic reactions can directly function as flavor compounds or react with sugars and other carbonyl compounds in the meat through Maillard reactions to produce roasted and meaty flavors. Lipid oxidation, however, has only a minor impact on flavor during the aging process (Spanier *et al.* 1997).

10.3.2.2 Cooking

Raw meat has very little flavor, but contains a number of nonvolatile flavor precursors, including amino acids, peptides, reducing sugars, and vitamins. It is only through the cooking process that the cooked, roasted, and meaty flavors associated with meats are developed through the Maillard reaction and other heat-dependent reactions. Cooking methods, such as boiling, roasting, frying, and pressure-cooking, vary in their temperature of heating and moisture and thus affect the interactions between carbonyl and amine compounds to form Maillard reaction products and influence overall meat flavor.

Sensory evaluation has shown an increased intensity of browned and species-specific flavor notes and a decreased intensity of bloody-serumy, metallic, and sour flavor notes as the final internal temperature of the meat increased (Bowers *et al.* 1987; Heymann *et al.* 1990). Higher cooking temperatures, longer cooking times, or cooking treatments that increase surface browning increase the formation of Maillard reaction products. These products contribute to browned and meaty flavors. In addition, the Maillard reaction products (MRP) have been shown to have antioxidative properties, and thus provide additional benefits in minimizing off flavor formation during subsequent storage (Huang & Greene 1978; Gros *et al.* 1986; Specht & Baltes 1994). However, lipid autoxidation reactions are also enhanced at elevated temperatures due to enhanced myoglobin degradation and release of free iron (Spanier *et al.* 1990, 1992b). Above approximately 70°C, the Maillard reaction predominates, with MRP antioxidants contributing to a reduction in lipid oxidation and off-flavor formation (Spanier *et al.* 1990).

10.3.2.3 Treatments to minimize lipid oxidation reactions

Processing of meats through mechanical deboning, grinding, restructuring, or cooking disrupts the tissue membranes and allows the catalysts of lipid oxidation to react with the unsaturated fatty acids to initiate lipid oxidation reactions. The effects of these processing treatments on lipid oxidation reactions have been shown as increases in the contents of TBARS, hexanal, and other lipid oxidation products, and increases in the intensity of cardboard and painty flavors assessed by sensory evaluation panels (Gray *et al.* 1996). The addition of synthetic and natural antioxidants during the processing of meat products has been shown to be effective in reducing the lipid oxidation reactions and maintaining desirable meat quality. Antioxidants may function as free radical terminators or free radical preventers. Phenolic antioxidants, such as butylated hydroxytoluene (BHT), butylated hydroxyanisole (BHA), *tert*-butyl hydroquinone (TBHQ), and tocopherol, are widely used in meat products to inhibit oxidation by donating hydrogens to the free radicals formed in lipid oxidations. Other antioxidants, such as ethylenediamine tetraacetic acid (EDTA), citric acid, and phosphates, bind metal cations to inhibit the initiation of the lipid oxidation reaction (Shahidi *et al.* 1987; Ladikos & Lougovois 1990).

Consumer interest in increasing the use of natural rather than synthetic additives in foods has led to interest in the identification of natural antioxidants. The antioxidant activity of numerous plant extracts, including rosemary, sage, tea catechins, and rice hull extracts (Nissen *et al.* 2000; McCarthy *et al.* 2001a, b; Kim *et al.* 2003), has been studied in meat products. The antioxidant activity of these plant extracts is attributed to the ability of the polyphenolic compounds present in these plant extracts to quench free radicals (Foti *et al.* 1996). Carnosine, a naturally occurring skeletal muscle dipeptide, functions as a chelator, free radical scavenger, and hydrogen donor in the aqueous phase of the muscle tissue to inhibit lipid oxidation (Decker & Faraji 1990; Decker & Crum 1991, 1993). Honey has also been shown to be an effective antioxidant in processed meats. The antioxidant activity of honey is attributed to the presence of reducing sugars, which can participate in Maillard reactions, phenolic compounds, and other natural antioxidants (McKibben & Engeseth 2002).

Because oxygen is critical to the lipid oxidation reactions and the development of WOF, packaging treatments that exclude oxygen from the environment have a significant impact on the flavor quality of meat during storage. Vacuum packaging has been shown to effectively reduce WOF formation and maintain desirable meaty flavor characteristics in raw ground beef (Lynch *et al.* 1986), cooked beef (McDaniel *et al.* 1984; Hwang *et al.* 1990; Spanier *et al.* 1992a, b), and cooked turkey and pork (Nolan *et al.* 1989) products. Hot vacuum packaging following cooking effectively excludes oxygen from the packaging system so that prooxidants, fat content, and fatty acid composition have no effect on lipid oxidation during storage (Ahn *et al.* 1992).

10.3.2.4 Curing

Curing of meats with sodium nitrite was originally developed as a means of preservation. However, the nitrite also has a significant impact on the meat flavor. The addition of nitrite (50–500 mg/kg) to meat contributed to an increase in ham aroma and flavor and a decrease in off odors and off flavors, as detected by a sensory evaluation panel (MacDonald *et al.* 1980a). However, in a comparison of the volatile flavor profiles of cured and uncured meats, the curing process contributed to a significant decrease in the concentration of hexanal and other carbonyl compounds, but did not result in the formation of new flavor compounds that would contribute to the cured meat flavor. Nitrite stabilizes the microsomal lipids and heme pigments to reduce lipid oxidation reactions and the resulting off-flavor formation (MacDonald *et al.* 1980b; Shahidi *et al.* 1987; Ramarathnam *et al.* 1991). Shahidi (1989) suggested that the flavor of nitrite-cured meats reflects the natural flavor of the meat without the flavor compounds contributed by the lipid oxidation reactions.

10.3.2.5 Irradiation

Irradiation is an effective processing method to improve the safety of meats through decreasing the load of spoilage and pathogenic microorganisms (Lefebvre *et al.* 1992; Fu *et al.* 1995). Irradiation doses of 1–10kGy (kilogray) have been approved by the United States Department of Agriculture for raw meat and poultry products (USDA 1999). However, the irradiation process produces reactive hydroxyl radicals that react with proteins and lipids in the meat to produce off-flavors. The formation of dimethyl disulfide, dimethyl sulfide, and other volatile sulfur compounds from the radiolysis of sulfur-containing amino acids in irradiated meats contributes to an undesirable irradiation off-odor, described as rotten egg, sweet, bloody, cooked meat, or barbecued corn, burnt, sulfur, metallic, alcohol, or acidic (Ahn *et al.* 2000; Lee & Ahn 2003; Zhu *et al.* 2004). The presence of oxygen during irradiation accelerates lipid oxidation reactions through the formation of free radicals and hydroperoxides, breakdown of hydroperoxides, and destruction of antioxidants (Nawar 1996). These degradative oxidation reactions and the subsequent off-flavor formation are minimized when meats are vacuum packaged prior to irradiation (Luchsinger *et al.* 1996; Ahn *et al.* 1998). The addition of antioxidants to the meats has also been shown to be effective in reducing the formation of sulfur compounds and lipid oxidation products through the ability of the antioxidants to scavenge and quench free radicals (Nam *et al.* 2002; Nam & Ahn 2003; Lee & Ahn 2003).

10.3.2.6 Storage

Storage has an adverse effect on the flavor quality of cooked meat products. In cooked beef patties, following 4 days of refrigerated storage, the intensity of beefy and brothy sensory attributes decreased, while the intensity of painty, sour, and bitter sensory attributes increased. The increased intensity of the painty attribute is attributed to lipid oxidation reactions and correlates with increases in hexanal and TBARS contents. During storage of cooked meat, the content of the hydrophilic peptide residues decreases and the content of the hydrophobic peptide residues remains unchanged. These changes in the distribution of peptides in the cooked and stored meat resulted in an increase in sour and bitter sensory attributes and a decrease in cooked beef and brothy flavor attributes (Spanier *et al.* 1990). Antioxidant and vacuum treatments designed to reduce lipid oxidation and off flavor formation also minimize losses in the desirable meaty attributes (Spanier *et al.* 1992b). During the storage of beef patties, contents of heterocyclic compounds, which contribute to desirable meaty characteristics, were stable, while contents of lipid oxidation products and aliphatic and cyclic sulfur compounds increased. The increases in the lipid oxidation and sulfur-containing compounds were attributed to free-radical degradation reactions (Drumm & Spanier 1991). Because the flavor characteristics of the sulfur-compounds are dependent on concentration (Shahidi 1989), these sulfur compounds may contribute undesirable flavor characteristics following extended refrigerated storage. The decrease in intensity of desirable flavor notes during refrigerated storage may be attributed to masking of the desirable flavor notes by the increased content of undesirable flavor compounds, rather than a degradation of desirable flavor compounds (Drumm & Spanier 1991).

10.4 CHALLENGES IN MEAT FLAVOR RESEARCH

Flavor has an important impact on the consumer acceptability of all foods and meat from land animals is no exception. Food scientists will continue their research to identify and characterize those compounds that impact meat flavor and further develop production and processing methods to enhance the desirable flavor characteristics of meats.

ACKNOWLEDGMENT

This chapter of the Iowa Agriculture and Home Economics Experiment Station, Ames, Iowa, Project No. 3574, was supported by the Hatch Act and State of Iowa funds.

REFERENCES

Ahn, D.U., Wolfe, F.H., Sim, J.S. & Kim, D.H. (1992) Packaging cooked turkey meat patties while hot reduces lipid oxidation. *Journal of Food Science*, **57**, 1075–1077, 1115.

Ahn, D.U., Olson, D.G., Lee, J.I., Jo, C., Wu, C. & Chen, C. (1998) Packaging and irradiation effects on lipid oxidation and volatiles in pork patties. *Journal of Food Science*, **63**, 15–19.

Ahn, D.U., Jo, C., Du, M., Olson, D.G. & Nam, K.C. (2000) Quality characteristics of pork patties irradiated and stored in different packaging and storage conditions. *Meat Science*, **56**, 203–209.

Asghar, A., Lin, C.F., Gray, J.I., Buckley, D.J., Booren, A.M. & Flegal, C.J. (1990) Effects of dietary oils and a-tocopherol supplementation on membranal lipid oxidation in broiler meat. *Journal of Food Science* **55**, 46–50, 118.

Asghar, A., Gray, J.I., Booren, A.M., *et al.* (1991) Effects of supranutritional dietary vitamin E levels on subcellular disposition of a-tocopherol in the muscle and on pork quality. *Journal of the Science of Food and Agriculture*, **57**, 31–41.

Bailey, M.E. & Um, K.W. (1992) Maillard reactions products and lipid oxidation. In: Lipid Oxidation in Food (ed. A.J. St. Angelo), pp. 123–139. American Chemical Society, Washington, District of Columbia.

Bett, K.L. (1993) Measuring sensory properties of meat in the laboratory. *Food Technology*, **47** (11), 121–126, 134.

Bowers, J.A., Craig, J.A., Kropf, D.H. & Tucker, T.J. (1987) Flavor, color, and other characteristics of beef longissimus muscle heated to seven internal temperatures between 55° and 85°C. *Journal of Food Science*, **52**, 533–536.

Boylston, T.D., Morgan, S.A., Johnson, K.A., Wright R.W. Jr., Busboom, J.R. & Reeves, J.J. (1996) Volatile lipid oxidation products of Wagyu and domestic breeds of beef. *Journal of Agricultural and Food Chemistry*, **44**, 1091–1095.

Byrne, D.V., Bak, L.S., Bredie, W.L.P., Bertelsen, G. & Martens, M. (1999a) Development of a sensory vocabulary for warmed-over flavor. I. In porcine meat. *Journal of Sensory Studies*, **14**, 47–65.

Byrne, D.V., Bredie, W.L.P. & Martens, M. (1999b) Development of a sensory vocabulary for warmed-over flavor. *II. In chicken meat. Journal of Sensory Studies*, **14**, 67–78.

Carrapiso, A.I., Jurado, Á., Timón, M. & García, C. (2002) Odor-active compounds of Iberian hams with different aroma characteristics. *Journal of Agricultural and Food Chemistry*, **50**, 6453–6458.

Chambers, E. IV., Bowers, J.R. & Smith, E.A. (1992) Flavor of cooked, ground turkey patties with added sodium tripolyphosphate as perceived by sensory panels with differing phosphate sensitivity. *Journal of Food Science*, **57**, 521–523.

Claeys, E., De Smet, S., Balcaen, A., Raes, K. & Demeyer, D. (2004) Quantification of fresh meat peptides by SDS-PAGE in relation to ageing time and taste intensity. *Meat Science*, **67**, 281–288.

Decker, E.A. & Crum, A.D. (1991) Inhibition of oxidative rancidity in salted ground pork by carnosine. *Journal of Food Science*, **56**, 1179–1181.

Decker, E.A. & Crum, A.D. (1993) Antioxidant activity of carnosine in cooked ground pork. *Meat Science*, **34**, 245–253.

Decker, E.A. & Faraji, H. (1990) Inhibition of lipid oxidation by carnosine. *Journal of the American Oil Chemist's Society*, **67**, 650–652.

Decker, E.A. & Hultin, H.O. (1992) Lipid oxidation in muscle foods via redox iron. In: *Lipid Oxidation in Food* (ed. A.J. St. Angelo), pp. 33–54. American Chemical Society, Washington, District of Columbia.

Drumm, T.D. & Spanier, A.M. (1991) Changes in the content of lipid autoxidation and sulfur-containing compounds in cooked beef during storage. *Journal of Agricultural and Food Chemistry*, **39**, 336–343.

Elmore, J.S., Mottram, D.S., Enser, M. & Wood, J.D. (1999) Effect of the polyunsaturated fatty acid composition of beef muscle on the profile of aroma volatiles. *Journal of Agricultural and Food Chemistry*, **47**, 1619–1625.

Farkaš, P., Sádecká, J., Kovác, M., Siegmund, B., Leitner, E. & Pfannhauser, W. (1997) Key odourants of pressure-cooked hen meat. *Food Chemistry*, **60**, 617–621.

Farmer, L.J. & Mottram, D.S. (1990) Interaction of lipid in the Maillard reaction between cysteine and ribose: The effect of a triglyceride and three phospholipids on the volatile products. *Journal of the Science of Food and Agriculture*, **53**, 505–525.

Farmer, L.J., Mottram, D.S. & Whitfield, F.B. (1989) Volatile compounds produced in Maillard reactions involving cysteine, ribose and phospholipids. *Journal of the Science of Food and Agriculture*, **49**, 347–368.

Flores, M., Ingram, D.A., Bett, K.L., Toldra, F. & Spanier, A.M. (1997) Sensory characteristics of Spanish "Serrano" dry-cured ham. *Journal of Sensory Studies*, **12**, 169–179.

Foti, M., Piattelli, M., Baratta, M.T. & Ruberto, G. (1996) Flavonoids, coumarins, and cinnamic acids as antioxidants in a micellar system. Structure–activity relationship. *Journal of Agricultural and Food Chemistry*, **44**, 497–501.

Fu, A.-H., Sebranek, J.G. & Murano, E.A. (1995) Survival of *Listeria monocytogenes* and *Salmonella typhimurium* and quality attributes of cooked pork chops and cured ham after irradiation. *Journal of Food Science*, **60**, 1001–1005.

Galt, A.M. & MacLeod, G. (1984) Headspace sampling of cooked beef aroma using Tenax GC. *Journal of Agricultural and Food Chemistry*, **32**, 59–64.

Gorraiz, C., Beriain, M.J., Chasco, J. & Insausti, K. (2002) Effect of aging time on volatile compounds, odor, and flavor of cooked beef from Pirenaica and Friesian bulls and heifers. *Journal of Food Science*, **67**, 916–922.

Gray, J.I., Gomaa, E.A. & Buckley, D.J. (1996) Oxidative quality and shelf life of meats. *Meat Science*, **43**, S111–S123.

Greene, B.E. (1969) Lipid oxidation and pigment changes in raw beef. *Journal of Food Science*, **34**, 110–113.

Gros, J.N., Howat, P.M., Younathan, M.T., Saxton, A.M. & McMillin, K.W. (1986) Warmed-over flavor development in beef patties prepared by three dry heat methods. *Journal of Food Science*, **51**, 1152–1155.

Güntert, M., Brüning, J., Emberger, R., *et al.* (1990) Identification and formation of some selected sulfur-containing flavor compounds in various meat model systems. *Journal of Agricultural and Food Chemistry*, **38**, 2027–2041.

Hartman, G.J., Jin, Q.Z., Collins, G.J., Lee, K.N., Ho, C.-T. & Chang, S.S. (1983) Nitrogen-containing heterocyclic compounds identified in the volatile flavor constituents of roasted beef. *Journal of Agricultural and Food Chemistry*, **31**, 1030–1033.

Heymann, H., Hedrick, H.B., Karrasch, M.A., Eggeman, M.K. & Ellersieck, M.R. (1990) Sensory and chemical characteristics of fresh pork roasts cooked to different endpoint temperatures. *Journal of Food Science*, **55**, 613–617.

Hornstein, I. & Crowe, P.F. (1960) Flavor studies on beef and pork. *Journal of Agricultural and Food Chemistry*, **8**, 494–498.

Hsu, C.-M., Peterson, R.J., Jin, Q.Z., Ho, C.-T. & Chang, S.S. (1982) Characterization of new volatile compounds in the neutral fraction of roasted beef flavor. *Journal of Food Science*, **47**, 2068–2069, 2071.

Huang, W.H. & Greene, B.E. (1978) Effect of cooking method on TBA numbers of stored beef. *Journal of Food Science*, **43**, 1201–1203, 1209.

Hwang, S.-Y., Bowers, J.A. & Kropf, D.H. (1990) Flavor, texture, color, and hexanal and TBA values of frozen cooked beef packaged in modified atmosphere. *Journal of Food Science*, **55**, 26–29.

Igene, J.O., King, J.A., Pearson, A.M. & Gray, J.I. (1979) Influence of heme pigments, nitrite, and non-heme iron on development of warmed-over flavor (WOF) in cooked meat. *Journal of Agricultural and Food Chemistry*, **27**, 838–842.

Imafidon, G.I. & Spanier, A.M. (1994) Unraveling the secret of meat flavor. *Trends Food Science Technology*, **5**, 315–321.

Jeremiah, L.E., Murray, A.C. & Gibson, L.L. (1990) The effects of differences in inherent muscle quality and frozen storage on the flavor and texture profiles of pork loin roasts. *Meat Science*, **27**, 305–327.

Johnson, P.B. & Civille, G.V. (1986) A standardized lexicon of meat WOF descriptors. *Journal of Sensory Studies*, **1**, 99–104.

Kerler, J. & Grosch, W. (1996) Odorants contributing to warmed-over flavor (WOF) of refrigerated cooked beef. *Journal of Food Science*, **61**, 1271–1274, 1284.

Kim, J.-S., Suh, M.-H., Yang, C.-B. & Lee, H.G. (2003) Effect of γ-oryzanol on the flavor and oxidative stability of refrigerated cooked beef. *Journal of Food Science*, **68**, 2423–2429.

Ladikos, D. & Lougovois, V. (1990) Lipid oxidation in muscle foods: A review. *Food Chemistry*, **35**, 295–314.

Larick, D.K. & Turner, B.E. (1990) Flavor characteristics of forage- and grain-fed beef as influenced by phospholipids and fatty acid composition differences. *Journal of Food Science*, **55**, 312–317, 368.

Larick, D.K., Hedrick, H.B., Bailey, M.E., *et al.* (1987) Flavor constituents of beef as influenced by forage- and grain-feeding. *Journal of Food Science*, **52**, 245–251.

Larick, D.K., Turner, B.E., Koch, R.M. & Crouse, J.D. (1989) Influence of phospholipid content and fatty acid composition of individual phospholipids in muscle from Bison, Hereford and Brahman steers on flavor. *Journal of Food Science*, **54**, 521–526.

Larick, D.K., Turner, B.E., Schoenherr, W.D., Coffey, M.T. & Pilkington, D.H. (1992) Volatile compound content and fatty acid composition of pork as influenced by linoleic acid content of the diet. *Journal of Animal Science*, **70**, 1397–1403.

Lee, E.J. & Ahn, D.U. (2003) Effect of antioxidants on the production of off-odor volatiles and lipid oxidation in irradiated turkey breast meat and meat homogenates. *Journal of Food Science*, **68**, 1631–1638.

Lefebvre, N., Thibault, C. & Charbonneau, R. (1992) Improvement of shelf-life and wholesomeness of ground beef by irradiation. 1. Microbial aspects. *Meat Science*, **32**, 203–213.

Leskanich, C.O., Matthews, K.R., Warkup, C.C., Noble, R.C. & Hazzledine, M. (1997) The effect of dietary oil containing n-3 fatty acids on the fatty acid, physiochemical and organoleptic characteristics of pig meat and fat. *Journal of Animal Science*, **75**, 673–683.

Lin, C.F., Gray, J.I., Asghar, A., Buckley, D.J., Booren, A.M. & Flegal, C.J. (1989) Effects of dietary oils and a-tocopherol supplementation on lipid composition and stability of broiler meat. *Journal of Food Science*, **54**, 1457–1460, 1484.

Lingnert, H. & Eriksson, C.E. (1981) Antioxidative effects of Maillard reaction products. *Progress in Food and Nutrition Science*, **5**, 453–466.

Luchsinger, S.E., Kropf, D.H., García Zepeda, C.M., *et al.* (1996) Color and oxidative rancidity of gamma and electron beam-irradiated boneless pork chops. *Journal of Food Science*, **61**, 1000–1005, 1093.

Lynch, N.M., Kastner, C.L., Caul, J.F. & Kropf, D.H. (1986) Flavor profiles of vacuum packaged and polyvinyl chloride packaged ground beef: A comparison of cooked flavor changes during product display. *Journal of Food Science*, **51**, 258–262, 267.

Lyon, B.G. (1987) Development of chicken flavor descriptive attribute terms aided by multivariate statistical procedures. *Journal of Sensory Studies*, **2**, 55–67.

MacDonald, B., Gray, J.I., Stanley, D.W. & Usborne, W.R. (1980a) Role of nitrite in cured meat flavor: Sensory analysis. *Journal of Food Science*, **45**, 885–888, 904.

MacDonald, B., Gray, J.I., Kakuda, Y. & Lee, M.L. (1980b) Role of nitrite in cured meat flavor: Chemical analysis. *Journal of Food Science*, **45**, 889–892.

MacLeod, G. & Ames, J.M. (1986) Capillary gas chromatography-mass spectrometric analysis of cooked ground beef aroma. *Journal of Food Science*, **51**, 1427–1434.

MacLeod, G. & Seyyedain-Ardebili, M. (1981) Natural and simulated meat flavors (with particular reference to beef). *Critical Reviews in Food Science and Nutrition*, **14**, 309–437.

Maruri, J.L. & Larick, D.L. (1992) Volatile concentration and flavor of beef as influenced by diet. *Journal of Food Science*, **57**, 1275–1281.

McCarthy, T.L., Kerry, J.P., Kerry, J.F., Lynch, P.B. & Buckley, D.J. (2001a) Assessment of the antioxidant potential of natural food and plant extracts in fresh and previously frozen pork patties. *Meat Science*, **57**, 177–184.

McCarthy, T.L., Kerry, J.P., Kerry, J.F., Lynch, P.B. & Buckley, D.J. (2001b) Evaluation of the antioxidant potential of natural food/plant extracts as compared with synthetic antioxidants and vitamin E in raw and cooked pork patties. *Meat Science*, **58**, 45–52.

McDaniel, M.C., Marchello, J.A. & Tinsley, A.M. (1984) Effect of different packaging treatments on microbiological and sensory evaluation of precooked beef roasts. *Journal of Food Protection*, **47**, 23–26.

McKibben, J. & Engeseth, N.J. (2002) Honey as a protective agent against lipid oxidation in ground turkey. *Journal of Agricultural and Food Chemistry*, **50**, 592–595.

McMillin, K.W., Bidner, T.D., Felchle, S.E., Dugas, S.M. & Koh, K.C. (1991) Flavor and oxidative stability of ground beef patties as affected by source and storage. *Journal of Food Science*, **56**, 899–902.

Min, D.B.S., Ina, K., Peterson, R.J. & Chang, S.S. (1979) Preliminary identification of volatile flavor compounds in the neutral fraction of roast beef. *Journal of Food Science*, **44**, 639–642.

Mottram, D.S. (1998) Flavor formation in meat and meat products: A review. *Food Chemistry*, **62**, 415–424.

Mottram, D.S. & Edwards, R.A. (1983) The role of triglycerides and phospholipids in the aroma of cooked beef. *Journal of the Science of Food and Agriculture*, **34**, 517–522.

Mottram, D.S. & Nobrega, I.C.C. (2002) Formation of sulfur aroma compounds in reaction mixtures containing cysteine and three different forms of ribose. *Journal of Agricultural and Food Chemistry*, **50**, 4080–4086.

Nagodawithana, T.W. (1995) Development of meat-like flavors through reaction flavor technology. In: *Savory Flavors*, pp. 164–224. Esteekay Associates, Milwaukee, Wisconsin.

Nam, K.C. & Ahn, D.U. (2003) Use of antioxidants to reduce lipid oxidation and off-odor volatiles of irradiated pork homogenates and patties. *Meat Science*, **63**, 1–8.

Nam, K.C., Prusa, K.J. & Ahn, D.U. (2002) Addition of antioxidant to improve quality and sensory characteristics of irradiated pork patties. *Journal of Food Science*, **67**, 2625–2630.

Nawar, W.W. (1996) Lipids. *Food Chemistry* (ed. O.R. Fennema), Chapter 5, 3rd edn., pp. 226–319. Marcel Dekker, New York.

Nissen, L.R., Månsson, L., Bertelsen, G., Huynh-Ba, T. & Skibsted, L.H. (2000) Protection of dehydrated chicken meat by natural antioxidants as evaluated by electron spin resonance. *Journal of Agricultural and Food Chemistry*, **48**, 5548–5556.

Nolan, N.L., Bowers, J.A. & Kropf, D.H. (1989) Lipid oxidation and sensory analysis of cooked pork and turkey stored under modified atmospheres. *Journal of Food Science*, **54**, 846–849.

Ramarathnam, N., Rubin, L.J. & Diosady, L.L. (1991) Studies on meat flavor. 1. Qualitative and quantitative differences in uncured and cured pork. *Journal of Agricultural and Food Chemistry*, **39**, 344–350.

Ramarathnam, N., Rubin, L.J. & Diosady, L.L. (1993) Studies on meat flavor. 4. Fractionation, characterization, and quantitation of volatiles from uncured and cured beef and chicken. *Journal of Agricultural and Food Chemistry*, **41**, 939–945.

Sato, K. & Hegarty, G.R. (1971) Warmed-over flavor in cooked meats. *Journal of Food Science*, **36**, 1098–1102.

Shackelford, S.D., Reagan, J.O., Haydon, K.D. & Miller, M.F. (1990) Effects of feeding elevated levels of monounsaturated fats to growing-finishing swine on acceptability of boneless hams. *Journal of Food Science*, **55**, 1485–1487, 1517.

Shahidi, F. (1989) Flavor of cooked meats. In: *Flavor Chemistry: Trends and Developments* (eds R. Teranishi, R.G. Buttery & F. Shahidi), pp. 188–201. American Chemical Society, Washington, District of Columbia.

Shahidi, F., Rubin, L.J. & D'Souza, L.A. (1986) Meat flavor volatiles: A review of the composition, techniques of analysis, and sensory evaluation. *CRC Critical Reviews in Food Science and Nutrition*, **24**, 141–243.

Shahidi, F., Rubin, L.J. & Wood, D.F. (1987) Control of lipid oxidation in cooked ground pork with antioxidants and dinitrosyl ferrohemochrome. *Journal of Food Science*, **52**, 564–567.

Sink, J.D. (1979) Factors influencing the flavor of muscle foods. *Journal of Food Science*, **44**, 1–5.

Spanier, A.M., Edwards, J.V. & Dupuy, H.P. (1988) The warmed-over flavor process in beef: A study of meat proteins and peptides. *Food Technology*, **42** (6), 110–118.

Spanier, A.M., McMillin, K.W. & Miller, J.A. (1990) Enzyme activity levels in beef: Effect of postmortem aging and end-point cooking temperature. *Journal of Food Science*, **55**, 318–322, 326.

Spanier, A.M., Vercellotti, J.R. & James, C. Jr. (1992) Correlation of sensory, instrumental, and chemical attributes of beef as influenced by meat structure and oxygen exclusion. *Journal of Food Science*, **57**, 10–15.

Spanier, A.M., St. Angelo, A.J. & Shafer, G.P. (1992) Response of beef flavor to oxygen depletion and an antioxidant/chelator mixture. *Journal of Agricultural and Food Chemistry*, **40**, 1656–1662.

Spanier, A.M., Flores, M., McMillin, K.W. & Bidner, T.D. (1997) The effect of post-mortem aging on meat flavor quality in Brangus beef. Correlation of treatments, sensory, instrumental, and chemical descriptors. *Food Chemistry*, **59**, 531–538.

Specht, K. & Baltes, W. (1994) Identification of volatile flavor compounds with high aroma values from shallow-fried beef. *Journal of Agricultural and Food Chemistry*, **42**, 2246–2253.

St. Angelo, A.J., Vercellotti, J.R., Legendre, M.G., *et al.* (1987) Chemical and instrumental analysis of warmed-over flavor in beef. *Journal of Food Science*, **52**, 1163–1168.

St. Angelo, A.J., Koohmaraie, M., Crippen, K.L. & Crouse, J. (1991) Acceleration of tenderization/inhibition of warmed-over flavor by calcium chloride-antioxidant infusion into lamb carcasses. *Journal of Food Science*, **56**, 359–362.

St. Angelo, A.J., Spanier, A.M. & Bett, K.L. (1992) Chemical and sensory evaluation of flavor in untreated and antioxidant-treated meat. In: *Lipid Oxidation in Food* (ed. A.J. St. Angelo), pp. 140–160. American Chemical Society, Washington, District of Columbia.

Tims, M.J. & Watts, B.M. (1958) Protection of cooked meats with phosphates. *Food Technology*, **12**, 240–243.

U.S. Department of Agriculture (USDA). (1999) Irradiation of meat food products. Final rule. 9CFR381 & 424 Docket No. 97–076C, Washington, District of Columbia.

Ullrich, F. & Grosch, W. (1987) Identification of the most intense volatile flavour compounds formed during autoxidation of linoleic acid. *Zeitschrift für Lebensmittel-Untersuchung und-Forschung*, **184**, 277–282.

Vercelloti, J.R., Kuan, J.W., Liu, R.H., Legendre, M.H., St. Angelo, A.J. & Dupuy, H.P. (1987) Analysis of volatile heteroatomic meat flavor principles by purge-and-trap/gas chromatography-mass spectrometry. *Journal of Agricultural and Food Chemistry*, **35**, 1030–1035.

Wasserman, A.E. (1979) Chemical basis for meat flavor: A review. *Journal of Food Science*, **44**, 6–11.
Werkhoff, P., Brüning, J., Emberger, R., *et al.* (1990) Isolation and characterization of volatile sulfur-containing meat flavor components in model systems. *Journal of Agricultural and Food Chemistry*, **38**, 777–791.
Wood, J.D., Richardson, R.I., Nute, G.R., *et al.* (2004) Effects of fatty acids on meat quality: A review. *Meat Science*, **66**, 21–32.
Yang, A., Brewster, M.J., Beilken, S.L., Lanari, M.C., Taylor, D.G. & Tume, R.K. (2002) Warmed-over flavor and lipid stability of beef: Effects of prior nutrition. *Journal of Food Science*, **67**, 3309–3313.
Younathan, M.T. & Watts, B.M. (1960) Oxidation of tissue lipids in cooked pork. *Food Research*, **25**, 538–543.
Zhu, M., Lee, E.J., Mendonca, A. & Ahn, D.U. (2004) Effect of irradiation on the quality of turkey ham during storage. *Meat Science*, **66**, 63–68.

11 Marine Animal and Plant Products

Narendra Narain and Maria Lúcia Nunes

Abstract: The marine environment incorporates an immense diversity of plants and animals constituting an enormous amount of the food chain. Marine animals are generally classified as invertebrates and vertebrates. Among all these animals, there may be hundreds of species varying in flavor and nutritional characteristics.

The most common marine plants are algae. They form the basis of the marine food chain serving as potential sources for foods and food ingredients.

Keywords: Aquaculture, seaweed, sea grass, crustacea, fish off flavors, surimi, fish oil

11.1 INTRODUCTION

The marine environment is an extremely large geographical area, comprising about 71% of the surface of this planet. This gigantic proportion incorporates an immense diversity of plants and animals, which thrive in this environment and constitute an enormous amount of the food chain, catering to the human population with a great variety of seafood, nutrients, pharmaceuticals, and other important chemical compounds. The entire system of both marine plants and animals is quite intricate in its nature and is subjected to a very complex survival under a habitat full of variations and limitations when compared with those from the land resources. However, marine plants are absolutely essential for the survival and existence of animals living in this environment. Both produce organic compounds that regulate and control the activity of potential predators, pathogens, and other stresses caused by the lack of solar radiation, oxygen, and other inherent deficiencies of physicochemical parameters.

Marine animals are generally classified as invertebrates and vertebrates. Invertebrates possess a shell or a hard exoskeleton and lack a backbone, which is a basic characteristic of vertebrates. Some of the invertebrates are sponges, jellyfish, corals, hydroids, clams, snails, comb jellies, crabs, shrimps, lobsters, copepods, squids, octopi, starfish, brittle stars, and marine worms. Vertebrates possess a backbone and represent thousands of fish species such as bony fish, snappers, hagfish, angelfish, sharks, dolphins, sea turtles, alligators, whales, etc. Moreover, among each of these animals, there may be hundreds of species varying in their flavor and nutritional characteristics. It has been estimated that the total number of fish species lies between 15 000 and 40 000, most of which are captured at least to some extent (de Groot 1995). Some fish are considered dangerous and should not to be touched (jellyfish, corals, and hydroids) as these are known to contain toxic poisonous substances.

Marine plants have evolved unique biochemical processes and structures in adapting to their chemical, physical, and biological environments. Plants usually grow only in the depths

Handbook of Meat, Poultry and Seafood Quality, Second Edition. Edited by Leo M. L. Nollet.

to which light penetrates the water. Some coral produces certain amino acids, which absorb solar radiation and serve as a sunscreen that consequently protects algae and their hosts from sunburn. The most common marine plants are algae, which form the basis of the marine food chain serving as potential sources of pharmaceuticals, agricultural chemicals, foods and food ingredients, industrial chemical feedstock, and other useful products.

Several reviews on flavor generation in seafood have been published (Heath & Reineccius 1986; Josephson & Lindsay 1986; Lindsay 1990; Josephson 1991; Arganosa & Flick Jr. 1992; Hsieh 1994; Lindsay 1994; Kawai 1996). Initially, Josephson and Lindsay (1986) presented a mechanism for enzymatic degradation of eicopentaenoic acid (EPA) leading to the biosynthesis of volatile compounds such as 1-penten-3-ol, (*E*)2-hexenal, (*Z*)3-hexenal, various non-adienals, (*Z*)1,5-octadien-3-one, and alcohols such as (*Z*)1,5-octadien-3-ol and (*Z*)3-hexen-1-ol. Later, an excellent review was published by Josephson (1991) who described precursor-based detailed mechanisms for the formation of volatile compounds such as lipid-derived aroma compounds, which included enzyme-catalyzed conversion and autoxidative deterioration of polyunsaturated fatty acids, carotenoids, and other chemical constituents (sulfur- and nitrogen-containing compounds) serving as precursors for various chemical reactions or microbial degradation leading to the formation of volatiles (dimethyl sulfide and trimethylamine), which contributed to aroma or off flavor in seafood. Several volatile carbonyl compounds, enzymatically derived from the lipid moiety, contributed to fresh, plant-like aromas in freshly harvested fish. Furthermore, these reviews also detailed the change in volatile profiles during processing such as salting, pickling, drying, frying, boiling, baking, and smoking of seafood. The more abundant volatiles generated during lipid autoxidation were hexanal and various isomers of 2,4-heptadienals, 3,5-octadien-2-ones, 2,4-decadienals, and 2,4,7-decatrienals. Some character-impact compounds formed during thermal processing of seafood were identified such as isomers (*Z,Z,Z* and *E,Z,Z*) of 5,8,11-tetradecatrien-2-one and dimethyl-2-phenylethylamine in boiled shrimps, dimethyl sulfide in stewed clams and oysters, and 2-methyl-3-furanthiol in canned tuna fish. It was also reported that thermally processed flavors could be generated from nonvolatile constituents such as amino acids, nucleotides, sugars, oxygenated lipids, and partly due to the Maillard reactions between sugars, chiefly glucose and ribose, and free amino acids such as serine, alanine, and lysine, commonly found in the flesh of many fish species.

The reviews or texts cited above were published before 1997, and since then, in spite of the vast amount of literature published over the last decade, there has been very little data assembled in the form of reviews dealing specifically on marine flavors. Moreover, the topic of flavor resources from marine plants and animals is very broad in nature and hence it is beyond the scope of this chapter to delve exhaustively into details. It is therefore proposed to present initially a characterization of marine environment and aquaculture systems from the flavor standpoint; the potential it represents from its plants and animals; the state of knowledge generated on the flavor aspects of these organisms emphasizing on new data generated over the past decade, concentrating mainly on the identification of volatile aroma compounds; and, finally, present the development of products based on potential marine resources.

11.2 MARINE ENVIRONMENT VERSUS AQUACULTURE SYSTEMS

The marine environment constitutes a basic medium of saltwater and is composed of, on average, 96.5% water and 3.5% dissolved substances, mainly salts, which include ions of chloride (55.04%), sodium (30.61%), sulfate (7.68%), magnesium (3.69%), calcium (1.16%), and potassium (1.10%). Some minor ions such as bicarbonate (0.41%), bromide (0.19%), boric acid (0.07%), and strontium (0.04%) are also present (Nybakken 1993). Several metal ions such as sodium, potassium, magnesium, and calcium are essential to sustain marine biological life. The taste threshold for the calcium ion varies from 100 to 300 milligrams per consumers. The taste threshold of the chloride anion in water is also dependent on the associated cation, and concentrations above 250 mg/L lead to detectable taste in water. Taste thresholds for sodium chloride and calcium chloride in water vary between 200 and 300 mg/L (Zoeteman 1980). The guideline value of 0.5 mg/L for magnesium is considered adequate while sodium may affect the taste of drinking water at levels above 200 mg/L (WHO 1979).

Besides the taste attributes imparted in marine living organisms by the water medium, salt also serves as a flavor enhancer and is used as a preservative for the improvement of texture and color of food products. Typically, salt increases the headspace concentration of nonpolar volatile compounds, which consequently affects the intensity of aroma attributes perceived by sniffing due to the change in concentration of these volatiles. In general, from the flavor point of view, sodium salts also characterize as suppressing agents for bitter-tasting compounds (Breslin 1996).

The habitat of marine species is determined both by the physical environment and by the organisms themselves. Traditionally, tidal migration and food selection by crustacean and fish species have been attributed to the innate ability of animals to sense, through taste and smell, specific flavors, nutrients, and toxins in plants. The survival of some of these species is being threatened due to unfavorable conditions in their habitat or because of excessive fishing. The overexploitation of the fishery resources with high commercial values also induces a leveling off of the total capture volume, resulting in a consecutive quantitative decline of certain species and size of fish landed. On the other hand, in the last two decades, there has been a great upsurge in the development of aquaculture systems all over the world. These systems constitute a cultivation of shrimp (Keys 2003) and some other high-priced, commodity-specific species such as crawfish, salmon, tilapia, trout, and catfish. However, it is now well known that the ocean fish possesses its peculiar taste due to a diet rich in bromophenols, the chemical compounds that occur throughout the marine environment and are present in algae, sea mosses, sandworms, and sea salt. A distinct difference is still observed in iodine-like or marine-like flavors appreciated very much in fish and shrimp obtained from marine resources as opposed to those which are cultivated in aquaculture and in which there is lack of its characteristic flavor, although both types of products look alike in appearance.

11.3 MARINE PLANTS

The most common marine plants belong to algae and sea grasses. There is a very wide difference between them. While algae are simple plants, sea grasses are known to be much more complex in nature. Algae are mainly green (chlorophyta), brown (phaeophyta), red (rhodophyta), blue-green (cyanophyta), and diatoms (chrysophyta). Seaweed are large marine algae that grow in coastal water. People of Asian cultures eat many kinds of seaweed including kelp, which are appreciated for their flavor characteristics and nutrients such as vitamins and iodine.

11.3.1 Seaweed

Generally, seaweed does not possess any aroma at the time of harvest. However, some volatile compounds such as ammonia, methylamine, and trimethylamine are generated by enzymatic or chemical reactions or by microorganisms. Trimethylamine oxide, the precursor

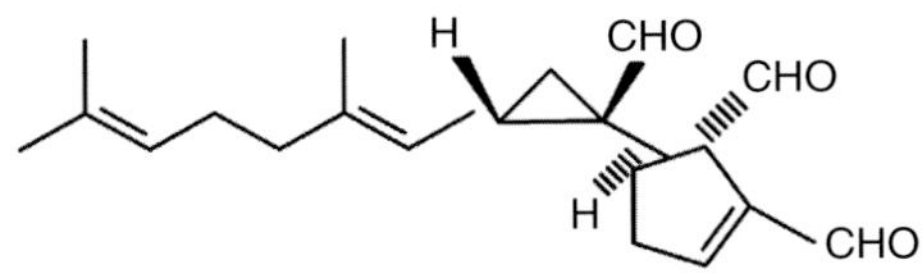

Halimedatrial [1]

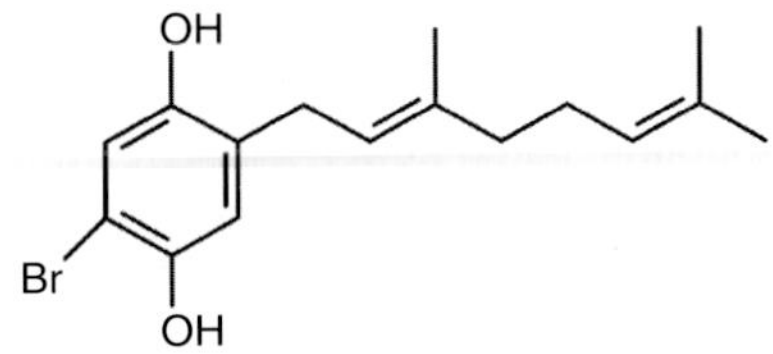

Cymopole [2]

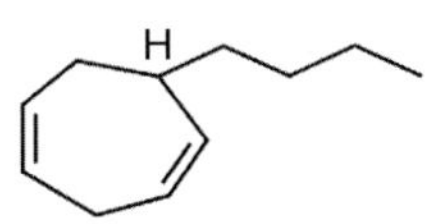

(–) -dictyopterene C′ [3]

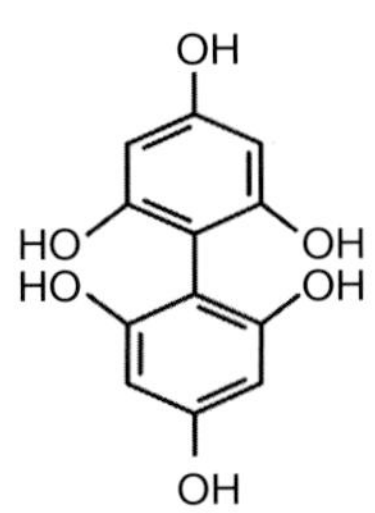

Difucol [4]

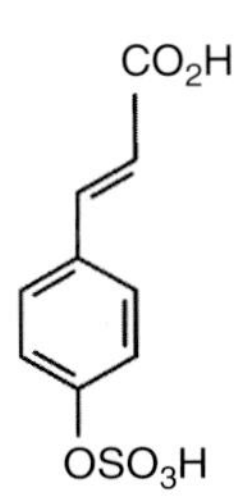

Zosteric acid [5]

Fig. 11.1 Some seaweed compounds.

for the formation of trimethylamine, is also found in seaweed. On roasting dry green and purple seaweed, a stimulating aroma, much appreciated by Japanese, is generated by the formation of dimethyl sulfide from dimethyl-b-propiothetin (Ito & Hori 1989).

Some seaweed defend themselves against herbivorous fishes and invertebrates by producing a diterpene known as halimedatrial (see [1] on Fig. 11.1), a secondary metabolite that is very distasteful to these fish. Ito and Hori (1989) in their review on seaweed described the presence of several volatile compounds in different types of seaweed. Some important ones are discussed here. Green seaweed produces cymopole (see [2] on Fig. 11.1), a condensation product of bromophenols and monoterpenes. Brown seaweed produces C11-hydrocarbons, sulfur-containing compounds, and a series of diterpenes, the most authentic being (-)-dictyopterene C′ [3] (see Fig. 11.1), which serves as a sex pheromone and difucol [4] (see Figure 11.1).

Red seaweed is known to synthesize halogenated substances containing bromine and iodine. Several species including *Asparagopsis taxiformis* (commonly known as "Limu Kohu"), *Bonnemaisonia hamifera*, and *B. asparagoides* contain high concentrations of bromine and generate a series of polyhalogenated ketones (1,1,3,3-tetrabromo-2-heptanone and 1-iodo-3,3-dibromo-2-heptanone), halogenated sesquiterpenes, and several bromophenols. Some of these halogenated compounds containing bromine impart a sweet flavor, especially in red seaweed when wet. The red seaweed *Palmaria palmata* is known to possess flavor characteristics peculiar to marine, crustacean, and green notes (Michel *et al.* 1997).

The volatile compounds (iodoethane and iodopentane) from the headspace of red algae were considered to be character-impact compounds (Pape *et al.* 2004); however, isoprene (2-methyl-1,3-butadiene) production has been reported in several species of red (*Chondrus crispus* and *Asparagopsis armata*), green (*Ulva intestinalis*), and brown (*Laminaria digitata, Ascophyllum nodosum, Pelvetia canaliculata, Fucus vesiculosus, Fucus serratus, Halidrys siliquosa*, and *Laminaria saccharina*) algae (Broadgate *et al.* 2004).

Some volatile halogenated organic compounds (VHOC) such as dibromomethane, bromoform, dibromochloromethane, bromodichloromethane, diiodomethane, and chloroiodomethane were identified in 28 different species of Antarctic macroalgae (Laturnus *et al.* 1996). Besides the contribution to the flavor characteristics of the algae, these compounds play an important role in chemical defense of Antarctic macroalgae against herbivores. The algae species *Desmarestia anceps* and *D. menziesii*, which release high quantities of VHOCs, are an important food for a variety of marine herbivores. In seaweed, an increased production of iodinated, brominated, or chlorinated low-molecular-weight VHOCs are associated with oxidative reactions (Mtolera *et al.* 1996). In general, the compounds responsible for retention of the marine flavor of algae are better retained when stored in artificial seawater at 4°C for 15 days as compared to frozen algae, which develops a marked green flavor (Pape *et al.* 2002).

11.3.2 Sea grass

Very little data are available on flavor aspects of sea grasses. However, it is noteworthy that many marine organisms remain quite clean in spite of the abundance of algal spores and other larvae that could settle and grow on them. It is now known that one way some seaweed and invertebrates remain clean is by producing a potent antifouling compound, zosteric acid [5] (see Fig. 11.1), that was isolated from young shoots of sea grass and that functions to deter or kill settling larvae and spores.

11.4 MARINE ANIMALS

In general, fresh saltwater fish are almost odorless because they contain a small quantity of volatiles while freshwater fish give off pyrrolidine and other earthy-odor compounds, which are responsible for their maturity and surrounding water pollution (Kawai 1996).

The compounds responsible for the development of flavor during seafood cooking can be classified in two groups. One, which represents the pleasant cucumber/green, almond/nutty, and potato aroma notes, consists of highly volatile, low-molecular-weight compounds belonging to various chemical classes such as aldehydes, ketones, alcohols, esters, nitrogen, phenols, and sulfur-containing compounds (Kubota *et al.* 1996; Sekiwa *et al.* 1997; Le Guen *et al.* 2001; Lee *et al.* 2001). The second is due to water soluble, low-molecular-weight free amino acids (taurine, glutamic acid, glycine); nucleotides (purine derivatives); organic acids (lactic acid); and inorganic salts (Na, K, Cl) (Konosu & Yamaguchi 1982).

Biogenic amines are nitrogen-containing compounds, which are present at very low levels in fresh fish. However, during storage and deterioration, biogenic amines can be produced by amino acid decarboxylation from bacterial enzymes. Among biogenic amines formed, putrescine and cadaverine have a putrid flavor while histamine and phenylethylamine have a pungent and fishy flavor, respectively. Biogenic amines are thermally stable and, therefore, have been used as indices to determine fish freshness (Gloria *et al.* 1999; Glória 2003). Volatile amines such as trimethylamine (TMA) or dimethylamine (DMA) are formed from trimethylamine oxide (TMAO), and these compounds also serve as a quality index for marine fish (Duflos *et al.* 1999; Baixas-Nogueras *et al.* 2003).

11.4.1 Fish

Several studies have been undertaken to identify volatile compounds in different fish species, among them are channel catfish (Grimm *et al.* 2004), cod (Milo & Grosch 1995, 1996), hake (Baixas-Nogueras *et al.* 2001), herring (Aro *et al.* 2003), salmon (Milo & Grosch 1996; Whitfield *et al.* 1996; Yoshiwa *et al.* 1997), sardine (Yoshiwa *et al.* 1997), squid (Koizumi *et al.* 1990; Kubota *et al.* 1996; Morita *et al.* 2002), trout (Milo & Grosch 1995), and some Mediterranean fish (Simeonidou *et al.* 1998). Some important features of these fishes are presented.

11.4.1.1 Catfish

Farm-raised channel catfish are known for their mild flavor and lack of the fishy odor that is peculiar to marine and wild freshwater fish. Flavor quality is the most important criterion for the acceptability of fish by the channel catfish processing plants. Catfish are primarily rejected for having woody, grassy, rotten, diesel, pesticide, pine, and earthy and muddy aromas. Approximately 80% of the rejected fish have the earthy, muddy aroma. Ploeg (1991) stated that the most common off flavor descriptors in catfish were earthy and musty/muddy, which are represented by the compounds geosmin (GSM) and 2-methylisoborneol (MIB), respectively. The threshold concentration of these compounds is 0.02 micrograms per kilogram (mg/kg). These compounds are produced by certain species of blue-green algae and Actinomycetes bacteria.

Grimm *et al.* (2004) reported that the odor threshold of channel catfish varies from 0.1 to 0.2 mg/kg for 2-methylisoborneol and from 0.25 to 0.50 mg/kg for GSM. The odor threshold values for both MIB and GSM in pure water are around 15 nanograms per kilogram (ng/kg), while for catfish the values for both these compounds range from 35 to 40 ng/kg.

11.4.1.2 Cod

The most prominent volatile compounds responsible for the flavor of boiled cod were characterized on their odor activity value (OAV). According to Milo and Grosch (1996), the OAVs of different flavor compounds in decreasing order are methional (OAV 275, concentration 11 mg/kg); (*Z*)-1,5-octadien-3-one (OAV 250, 0.1 mg/kg); 3-methylbutanal (OAV 204, 51 mg/kg); (*E*,*Z*)-2,6-nonadienal (OAV 175, 3.5 mg/kg); acetaldehyde (OAV 130,

1300 mg/kg); and (*E,E*)-2,4-decadienal (OAV 117, 3.5 mg/kg). Methional, 1-octen-3-ol, and 2,6-nonadienal were identified as the most important compounds contributing to ripened cod odor. Spoilage flavors were mainly 3-methyl-1-butanol and 3-methylbutanal, which can be determined by an electronic nose and are suggested as quality indicators for objective assessing of cod ripening.

11.4.1.3 Hake

The main amines present in fresh hake (*Merluccius merluccius*) are the natural polyamines, spermine, and spermidine (Baixas-Nogueras *et al.* 2001). During storage, the levels of volatile and nonvolatile amines change as affected by temperature. In samples stored at 220°C, changes only occurred in dimethylamine and agmatine, while at higher temperatures, major changes were observed in cadaverine, histamine, putrescine, and tyramine.

11.4.1.4 Salmon

Salmon contains carotenoids, astaxanthin (3,3′-dihydroxy-b2b-carotene-4,4′-dione), and canthaxanthin (b,b-carotene-4,4′-dione), which are responsible for its attractive red color. Volatile compounds generated from carotenoid-related oxidation reactions played an important role in cooked salmon flavor (Josephson 1991). Differences in the overall eating quality occurred between river-caught and sea-caught salmon. River-caught salmon tended to score high on earthy flavor and aftertaste, and scored low for salmon-like odor and flavor (Farmer *et al.* 2000).

11.4.1.5 Sardine

Prost *et al.* (2004) monitored qualitative and quantitative changes in the volatile aroma compounds during storage of raw sardine (*Sardina pilchardus*). A sensory panel generated a list of 20 odorant descriptors for raw sardine. Thirty-four volatile components were identified as potent odorants using an olfactometric method. (*E,E*)-2,4-octadienal, (*E*)-2-penten-1-ol, and 2,3-butanedione were the most potent odorants of raw sardine. The odor-active compounds responsible for oxidized and rancid flavors increased during storage while sulfur-containing compounds associated with marine odors decreased. The decrease in marine odor could be attributed to the decrease in dimethyl sulfide during storage, while the increase in rancid odor was related to the increase in heptanal and (*Z*)-3-hexen-1-ol in the initial period of storage and in 2,3-pentanedione, hexanal, (*Z*)-4-heptenal, (*E*)-2-heptenal, methional, and 2-nonanol on longer storage.

11.4.1.6 Squid

More than 100 compounds have been identified in boiled squid (Koizumi *et al.* 1990; Kubota *et al.* 1996). Koizumi *et al.* (1990) reported that the compounds in neutral and basic fractions contributed to cooked squid flavor. Furthermore, Kubota *et al.* (1996) found that ten compounds including pyrazines, thiazoles, thiazoline, and a hydroxy furanone were potent odorants. Based on ten sensory attributes, Morita *et al.* (2002) evaluated the effect of pH and different parts of squid on aroma characteristics of boiled squid. Aroma notes such as sweet, roasted shrimp, cooked fish, and cooked squid scored higher in both muscle and muscle with skin at pH 6.5.

11.4.2 Shrimp and crustacean

11.4.2.1 Shrimp

Potent flavor-active compounds in cooked tail meat of freshwater prawn (*Macrobrachium rosenbergii*) were identified by flavor extract dilution analysis (Cadwallader *et al.* 1996). The most intense compounds identified were 2-acetyl-1-pyrroline, 3-(methylthio)-propanal, and 2,3-butanedione representing popcorn-like, baked potato-like, and buttery flavors, respectively.

The aroma components of processed by products (heads, shells, and tails) of Northern pink shrimp (*Pandalus borealis*) and spotted shrimp (*Trachypenaeus curvirostris*) caught near Tongyeong, South Korea, were reported by Heu *et al.* (2003). Volatile basic nitrogen of the by-products was lower than that of the edible parts. The total content of free amino acids of the processed by-products (2 g/100 g) was 15% higher than that of the edible parts (1.7 g/100 g). Major free amino acids were taurine, threonine, leucine, tryrosine, and phenylalanine. Based on taste threshold values in the by-products, the authors suggested that aspartic and glutamic acids most influenced the taste.

Bromophenols such as 2- and 4-bromophenol, 2,4- and 2,6-dibromophenol, and 2,4,6-tribromophenol are key flavor compounds in seafood (Whitfield *et al.* 1988, 1997, 1998). In water, the strongest of these flavored compounds, 2,6-dibromophenol, 2-bromophenol, and 2,4,6-tribromophenol, have flavor threshold concentrations of 5×10^{-4}, 3×10^{-2}, and 0.6 ng/g, respectively (Whitfield *et al.* 1996). At these concentrations, the flavors of 2,6-dibromophenol and 2,4,6-tribromophenol are iodoform-like and that of 2-bromophenol as phenolic/iodine-like. However, at levels below the flavor threshold concentrations, these bromophenols contribute recognizable marine- or ocean-like flavors to seafood and enhance the intensities of existing seafood flavors (Whitfield *et al.* 1998). Whole wild ocean prawns had a total bromophenol content that ranged from 9.5 to 1114 ng/g, depending on species, location, and harvest time. The concentration was higher in heads than in tail meats. Cultivated or farmed prawns contained a much lower tribromophenol content that ranged from 0.31 to 1.3 ng/g (Whitfield *et al.* 1997). As a consequence, the flavor of cultivated prawns is frequently described as bland and lacking the natural flavors of wild crustaceans.

Volatile components of boiled prawns were analyzed and correlated with the sensory evaluation of its aroma by Morita *et al.* (2001). Seafood-like odors were related to some unsaturated methyl ketones among more than 100 compounds identified, which included 40 sulfur- and/or nitrogen-containing compounds in cooked shrimp volatiles. Odor descriptions for 2-acetyl-1-pyrroline and 2-actylthiazole were nutty/popcorn-like, and these compounds were found to be aroma active in cooked shrimp (Baek & Cadwallader 1997). Ketones, alcohols, and pyrazines were selected for roasted shrimp; dimethyl trisulfide for green vegetable note; benzaldehyde for creamy and nutty (Tanchotikul & Hsieh 1989), and almond and fruity (Girard & Durance 2000); and 2,4-dimethylthiazole for meaty and cocoa-like. Every one of these compounds was considered important for boiled prawn flavor.

Live kuruma prawn (*Penaeus japonicus*), which is rich in adenosine 5′-triphosphate (ATP), is considered to be delicious but not sweet. However, on boiling the prawn, a strong sweet taste develops due to the thermal degradation of ATP leading to the formation of adenosine 59 monophosphate (AMP) (200 mg/100 g of muscle) and inosine 5′-monophosphate (IMP) (4 mg/100 g of muscle). The two substances together impart the characteristic umami taste to the boiled prawns (Fuke & Ueda 1996).

11.4.2.2 Crustacea and others

The flavor profile of several crustacean species have been reported; among them are clams (Tanchotikul & Hsieh 1991; Chung & Cadwallader 1994; Baek & Cadwallader 1996; Sekiwa *et al.* 1997), lobsters (Lee *et al.* 2001), mussels (Le Guen *et al.* 2000a, 2000b), and oysters (Kim *et al.* 2000). Chung and Cadwallader (1994) analyzed several volatile extracts from cooked blue crab (*Callinectes sapidus*) claw meat obtained by different extraction techniques and characterized four compounds with high flavor dilution factors: 2,3-butanedione (sour, creamy); (*Z*)4-heptenal (potato-like); 2-acetyl-l-pyrroline (nutty, popcorn-like); and 3-(methylthio)propanal (salty, soy sauce-like). Baek and Cadwallader (1997) reported 23 aroma active compounds in cooked crustaceans by using the aroma extract dilution analysis, while Whitfield *et al.* (1988) characterized an off flavor in Australian crustacean due to the presence of 2,6-dibromophenol.

Volatile generation in oysters (*Crassostrea gigas*) was related to compounds formed from fatty acid oxidation (86%), particularly to v-3 polyunsaturated fatty acid oxidation (66%), while only one compound was related to amino acid degradation. The main flavors characterized were green/sulfur/crustacean, represented by dimethyl sulfide, 1-penten-3-one, and hexanal; mushroom/citrus-like represented by (*E,E*)-2,4-heptadienal, 1-octen-3-one, 1-octen-3-ol, 6-methyl-5-hepten-3-one, and octanal; and marine/cucumber notes attributed to (*E,Z*)-2,6-nonadienal and (*E*)-2-octenal, and decanal.

Three olfactometric methods (olfactometry global analysis, Osme, and aroma extract dilution analysis) were used to evaluate the main impact flavor compounds of cooked mussels (Le Guen *et al.* 2000b). Six compounds seemed to contribute actively to the aroma of mussels: 2,3-butanedione (buttery, caramel-like odor); (*Z*)-4-heptenal (boiled potato-like odor); (*E*)-2-penten-1-ol (mushroom-like odor); 2-ethylpyrazine (nutty odor); methional (boiled potato-like odor); and (*E,E*)-2,4-octadienal (cucumber-like odor).

Wash water from clam-processing plants is usually used to obtain a marketable natural clam flavoring agent. Kim *et al.* (2000) concentrated oyster cooker effluent, which, on enzymatic hydrolyzes, led to an increase in free amino acids, with taurine comprising about 20% of the total. Inosine monophosphate was predominant (456 mg/100 g) among nucleotides and related compounds. Trimethylamine, trimethylamine oxide, and total creatinine levels were not affected by enzyme treatment. The predominant aroma-active compounds of enzyme-hydrolyzed oyster effluent were 2-acetyl-1-pyrroline and 3-(methylthio) propanal.

11.4.3 Off flavors in fish and products

Off flavors in seafood may originate from environmental pollutants, the growth of microorganisms, oxidation of lipids, or endogenous enzymatic decomposition in the seafood, or from the water assimilated by the fish. The off flavors most commonly detected are petroleum and blackberry-like flavors in salmon and cod; garlic-like flavors in prawns and sand lobsters; iodoform-like flavors in prawns, shrimp, and fish; and muddy and earthy flavors in brackish-water fish and shrimp (Wilkes *et al.* 2000). GSM and MIB are bicyclic tertiary alcohols produced by microorganisms, namely actinomycetes, cyanobacteria, and fungi (Nilsson *et al.* 1999). Both compounds cause undesirable flavors (Young *et al.* 1996). Even fresh, healthy, pond-raised catfish may absorb off flavors from microbial metabolites in the water and sediments. Different catfish in the same pond may temporarily assimilate different amounts of GSM and MIB and then release them at different rates. Purge-and-trap GC has been used to determine 2-methylisoborneol in water, mud, and cooked channel catfish samples.

The bioconcentration factor (concentration in water/concentration in fish) reported for 2-methylisoborneol was 28.1 ± 14.0 (Jackson *et al.* 1997).

Other off flavors, especially in fish, result from the production of volatile microbial metabolites including volatile bases. Secondary products, such as propanal, can serve as a reliable indicator of flavor deterioration for fish products. Dimethyl sulfide and 2-pentanone can be used to monitor the flavor quality of cultured catfish and for their likelihood to occur in fish samples from contaminated environments, for distinctive odor characteristics that could be distinguished from inherent marine fresh fish flavor (Wilkes *et al.* 2000).

In squid, spoilage was attributed to autolytic enzymes rather than microbial degradation (Gill 1990). Autolytic decomposition also occurs in shrimp that produces indole, a relatively nonvolatile amine, along with ammonia when stored at room temperature and only ammonia when stored at refrigerated temperature (Wilkes *et al.* 2000).

Off flavors in ice-packed fish and seafood are produced during spoilage by several microorganisms, including Pseudomonas spp., Acinetobacter spp., Moraxella spp., and *Shewanella putrefaciens*, evolving several different unpleasant flavors (Jackson *et al.* 1997). A strain of Lactobacillus sake used as an antibacterial agent in cold-smoked salmon was judged unsuitable because it produced sulfurous off flavors (Nilsson *et al.* 1997).

11.4.4 Processed products

Several products prepared from marine animal and plant resources have long been appreciated by people from Japan and other southeastern Asian countries for their flavor quality. Lately, there has been a demand for the development of novel products derived from marine ingredients in the Western Hemisphere. A recent publication (Pszczola 2003) describes the current status on organoleptic, nutritional, therapeutic, and other health potential of several products elaborated from the use of marine or aquatic raw materials. Some of the important products related to the flavor aspects are presented here.

11.4.4.1 Surimi

In the preparation of high quality surimi from dark-fleshed fish mince, discoloration poses a problem. It is considered to be a critical operation that is controlled by optimizing washing conditions in order to remove undesirable materials such as lipids and trimethylamine oxide. Chen *et al.* (1997) reported the beneficial effects of washing horse mackerel mince with ozonized water for 10–20 min. However, it also had limitations such as a marked decrease in pH and undesirable gel strength of the mince, accompanied by oxidation of the fish oil, which occurred during ozone treatment. A market test for surimi from Atlantic blue whiting (*Micromesistius poutassou*) was performed by evaluation of the main price-determining parameters such as gel-strength, whiteness, and taste. The results indicated that the respondents valued the samples at almost the same level as *Alaska pollock* surimi, at the same level as southern blue whiting surimi, and better than Pacific hake surimi (Trondsen 1998).

11.4.4.2 Fish sauce

Fish sauce is commonly used as a flavoring agent and is basically produced from a mixture of fish and salt (3:1) that has been allowed to ferment for a period of more than 6 months at 30–35°C. It is now widely used in many prepared foods and sauces in Japanese and Western markets because of its favorable taste, resulting from an amino acid balance and high quantities of peptides. Based on flavor dilution studies on fish sauce, Fukami *et al.* (2002) identified the presence of seven volatiles, which included 2-methylpropanal, 2-methylbutanal, 2-pentanone, 2-ethylpyridine, dimethyl trisulfide, 3-(methylthio)propanal, and 3-methylbutanoic acid. The first four compounds contributed distinctive sweet rancid flavor, and the last two compounds were characterized by a fishy fecal note in the fish sauce. The concentrations of 2-methylpropanal, 2-methylbutanal, 2-ethylpyridine, and dimethyl trisulfide in sauce were estimated to be 370.7, 38.5, 1.4, and 7.5 ng/mL, respectively. Besides these four compounds, the role of dimethyl disulfide and butanoic acid in the development of fish-sauce flavor was emphasized. Improvement in fish-sauce flavor was enhanced by treatment with bacteria (*Staphylococcus xylosus*) isolated from the fish-sauce mush (Moromi) made from frigate mackerel (Fukami *et al.* 2004).

Deodorization of fish sauce is essential to improve its acceptability in areas outside Southeast Asia. Application of supercritical carbon dioxide for the extraction of lipophilic organic compounds and separation of highly unsaturated fatty acids from fish oil has been tried (Shimoda *et al.* 2000). It was found that bringing microbubbles of supercritical carbon dioxide into contact with the aqueous solution enabled volatiles extraction with high efficiency. After the treatment at a CO_2 flow ratio of 0.29 and 10 mega Pascal (MPa) pressure, the remaining percentage (the concentration in treated sample/that in untreated one) was 5.2% trimethylamine, 8.0% S-methyl ethanethioate, 30% dimethyl disulfide, 55–61% aliphatic aldehydes, and 25–42% of carboxylic acids. Shimoda *et al.* (1997) reported a column treatment using aromatic porous polymer beads to remove odorants in fish sauce.

11.4.4.3 Fish oil

Fish is an important dietary source of polyunsaturated fatty acids (PUFA) such as eicosapentaenoic acid (EPA, 20:5-ω-3) and docosahexaenoic acid (docosahexaenoic acid [DHA], 22:6-ω-3) (Passi *et al.* 2002). Their beneficial effects have been attributed to an increased ratio of -ω-3- to -ω-6-polyunsaturated acids in human blood lipids and cell membrane lipids, and fish having higher ratios of these acids are considered to result in greater health benefits (Yazawa & Kageyama 1991; Kitessa *et al.* 2001). However, DHA and EPA in fish oil are more easily oxidized than the unsaturated fatty acids in vegetable oil. Lipid oxidation generally results in negative attributes of flavor quality and contributes greatly to the off flavors generated on improper use of stored foods (Durrington *et al.* 2001).

The oxidative deterioration of fish oil involves the formation of hydroperoxides from PUFA in the triglycerides, which on further autoxidation lead to the formation of a complex mixture of secondary oxidation products. Although lipid hydroperoxides are known to be flavorless and odorless, the secondary oxidation products are responsible for the changes in the aroma and flavor properties of foods (Hsieh & Kinsella 1999; Venkateshwarlu *et al.* 2004a, 2004b). The oxidative stability of DHA, eicosapentaenoic acid (EPA), and oxidized volatile compounds in two types of enriched fish oil—triacylglycerol and ethyl ester of DHA—was verified during storage at 80°C with aeration (Lee *et al.* 2003). The rate of DHA autoxidation was higher in ethyl ester form, which was more susceptible to autoxidation than the triacylglycerol form. (*E*)-2-pentenal, 2-(1-pentenyl) furan, and (*E,E*)-2,4-heptadienal were commonly detected as oxidized volatile compounds.

Karahadian and Lindsay (1989) used the dynamic headspace purge-and-trap method to extract the volatile compounds generated during oxidation of fish oil. (*E,Z*)-2,6-nonadienal, (*E*)-2-hexenal, 1,5-octadien-3-one, (*Z*)-4-heptenal, (*E,Z,*sZ), (*E,E,Z*)-2,4,7-decatrienal, hexanal, 2,4-heptadienal, and 2,4-decatrienal were the major volatile compounds associated with fishy flavor. In another study, using the dynamic headspace concentration procedure, (*Z*)-2-pentenal, (*E*)-2-hexenal, 2,4-heptadienal, 2,4-decadienal, and (*E,Z,Z*)- and

(*E,E,Z*)-2,4,7-decatrienal were identified as the volatile compounds in menhaden fish oil produced from the oxidation of linolenic acid and other PUFAs (Hsieh *et al.* 1989).

Several attempts have been made to develop functional foods by incorporating fish oil (Kolanowski *et al.* 1999); among them are bread, mayonnaise (Hartvigsen *et al.* 2000), and milk emulsion (Venkateshwarlu *et al.* 2004a, b). The development of objectionable fishy off flavors was an obstacle in the development of fish oil-enriched foods. The volatile (*Z,Z*)-3,6-nonadienal, reported as a potent odorant in boiled cod by Milo and Grosch (1996), was detected neither in the milk emulsion enriched with cod liver oil nor in a commercial cod liver oil (Karahadian & Lindsay 1989).

Venkateshwarlu *et al.* (2004a, 2004b) performed detailed studies on volatiles profiles of fish oil-enriched milk during cold storage (2°C) for 14 days. The potent odorants identified by gas chromatography-olfactometry (GC-O) were 1-penten-3-one, (*Z*)-4-heptenal, 1-octen-3-one, (*Z*)-1,5-octadien-3-one, (*E,E*)-2,4-heptadienal, and (*E,Z*)-2,6-nonadienal. Despite their potency, none of the separated volatiles imparted a fishy or metallic odor. Two isomers (*E,Z,Z*) and (*E,E,Z*) of 2,4,7-decatrienal, were identified in fish oil-enriched milk emulsions (Venkateshwarlu *et al.* 2004a). Though decatrienals were not detected by GC-O, they had sensory significance at higher concentration, which may have impact in combination with other volatiles. The study showed further that the incorporation of fish oil with a very low initial peroxide value (0.1 milliequivalent/kilogram [meq/kg]) did not result in fishy off flavor development upon storage, while the development of distinct fishy off flavor was recorded when cod liver oil with initial peroxide value of 1.5 meq/kg was incorporated into milk. Based on the addition of the volatiles (*E,Z*)-2,6-nonadienal, 1-penten-3-one, (*Z*)-4-heptenal, and (*E,E*)-2,4-heptadienal to milk containing 1.5% fat, Venkateshwarlu *et al.* (2004b) evaluated the individual and combinatory effects of these four potent volatiles on sensory properties. They suggested that (*E,Z*)-2,6-nonadienal and 1-penten-3-one could serve as useful markers for fishy and metallic off flavors in fish oil and fish oil-enriched foods. In fish oil–enriched milk, Aidos *et al.* (2002) had earlier reported that 1-penten-3-one contributes to the off flavor.

Encapsulation of fish oil by an enzymatic gelation process using transglutaminase cross-linked proteins was investigated by Cho *et al.* (2003). They found that fish oil containing DHA could be encapsulated using double emulsification and subsequently an enzymatic gelation method, using microbial transglutaminase cross-linked proteins. Isolated soy protein was selected as a wall material because it showed better emulsion stability and higher reactivity with transglutaminase than other proteins. Microcapsules prepared by this method showed higher stability against oxygen and a low water solubility, which subsequently resulted in a sustained release of fish oil.

11.4.4.4 Other processed products

The physicochemical differences between pork and fish gelatin on the sensory characteristics of a gelatin–water gel were evaluated (Choi & Regenstein 2000). The flavored fish gelatin dessert gel product had less undesirable off flavor and off odor and a more desirable release of flavor and aroma than the same product made with an equal bloom but higher melting point, like pork gelatin.

The concentration of marine flavor is obtained using processes such as spray-drying, freeze-drying, drum-drying, and nanofiltration (Lin & Chiang 1993). Vandanjon *et al.* (2002) used a membrane processing system consisting of various physical operations such as ultrafiltration followed by nanofiltration or reverse osmosis in order to reduce first the pollution load and then to concentrate flavor compounds of cooking juices from shrimp and tuna. Nanofiltration was not suitable for flavor recovery while reverse osmosis functioned very well for shrimp cooking juices. They showed that a good retention of marine flavors requires the use of reverse osmosis membranes. But these membranes presented a strong mass transfer limitation during concentration of cooking waters of shrimp. To maintain acceptable flows, it is necessary to combine a pretreatment by ultrafiltration followed by reverse osmosis.

Jayarajah and Lee (1999) evaluated a membrane concentration system consisting of tubular polysulphone ultrafiltration and polyamide reverse osmosis for concentrating key water-soluble flavor compounds from lobster extracts. Major flavor-giving compounds in the extract were glutamic acid, glycine, arginine, uridine 5-monophosphate, succinic acid, and glucose. The reverse osmosis system retained all dissolved flavor components at its optimized conditions (40°C, 2.8 MPa log mean transmembrane pressure and a flow rate of 15 L/min). A combination of ultrafiltration and reverse osmosis functioned well for the recovery of lobster flavor compounds. Ultrafiltration could be used to separate proteins and large molecular weight nonflavor compounds from the extract while the resulting permeate enriched with low-molecular-weight flavor components could be subsequently concentrated by reverse osmosis.

One of the procedures commonly employed to prevent the incidence of black spots in shrimps' and lobsters' tails is an immersion treatment in sodium metabisulfite solution prior to freezing. However, excessive use of this compound causes formation of high levels of sulfur dioxide (SO_2) in the muscle, which in turn contributes to an accelerated decomposition of trimethylamine oxide, which forms dimethylamine, trimethylamine, and formaldehyde. All of these compounds are highly undesirable in frozen shrimps or lobster tails.

Gennadios *et al.* (1997) reported the advantages of using edible coatings with biopolymers such as gelatin, blood protein, feather keratin, fish myofibrillar protein, and the polysaccharide chitosan produced from crustacean shells for storage of frozen fish products. These coatings reduce moisture loss and lipid oxidation while maintaining better retention of color and flavor attributes. Silver salmon fillets coated by dipping in Myvacet 5-07 acetylated monoglyceride (60°C) lost less moisture and had lower peroxide values than uncoated control samples during frozen storage at −10°C. Carrageenan, a polysaccharide gum obtained from Irish moss (*Chondrus crispus*), and some other species of red seaweed could extend storage life of frozen fatty fish like mackerel.

REFERENCES

Aidos, I., Jacobsen, C., Jensen, B., Luten, J.B., Padt, A. & Boom, R.M. (2002) Volatile oxidation products formed in crude herring oil under accelerated oxidative conditions. *European Journal of Lipid Science and Technology*, **104**, 808–818.

Arganosa, G.C. & Flick, G.J., Jr. (1992) Off-flavors in fish and shellfish. In: *Off-flavors in Foods and Beverages* (ed. G. Charalambous), pp. 103–126. Elsevier, New York.

Aro, T., Tahvonen, R., Koskinen, L. & Kallio, H. (2003) Volatile compounds of baltic herring analyzed by dynamic headspace sampling-gas chromatography-mass spectrometry. *European Food Research and Technology*, **216**, 483–488.

Baek, H.H. & Cadwallader, K.R. (1996) Volatile compounds in flavor concentrates produced from crayfish-processing by-products with and without protease treatment. *Journal of Agricultural and Food Chemistry*, **44**, 3262–3267.

Baek, H.H. & Cadwallader, K.R. (1997) Character-impact aroma compounds of crustaceans. In: *Flavor and Lipid Chemistry of Seafoods* (eds F. Shahidi & K. Caldwallader), pp. 85–94. *American Chemical Society Symposium Series* 674, Washington, District of Columbia.

Baixas-Nogueras, S., Bover-Cid, S., Vidal-Carou, M.C. & Veciana-Nogués, M.T. (2001) Volatile and nonvolatile amines in Mediterranean hake as a function of their storage temperature. *Journal of Food Science*, **66** (1), 83–88.

Baixas-Nogueras, S., Bover-Cid, S., Veciana-Nogués, M.C. & Vidal-Carou, M.T. (2003) Volatile and nonvolatile amines in Mediterranean hake as a function of their storage temperature. *Journal of Food Science*, **68** (5), 1607–1810.

Breslin, P.A.S. (1996) Interactions among salty, sour and bitter compounds. *Trends in Food Science & Technology*, **7**, 390–399.

Broadgate, W.J., Malin, G., Kupper, F.C., Thompson, A. & Liss, P.S. (2004) Isoprene and other non-methane hydrocarbons from seaweeds: A source of reactive hydrocarbons to the atmosphere. *Marine Chemistry*, **88**, 61–73.

Cadwallader, K.R., Baek, H.H. & Chang-Lee, G.H. (1996) Aroma extract dilution analysis of cooked freshwater prawn. *Institute of Food Technologists Annual Meeting*, New Orleans, Louisiana, p. 197.

Chen, H.H., Chiu, E.M. & Huang, J.R. (1997) Color and gel-forming properties of horse mackerel (*Trachurus japonicus*) as related to washing conditions. *Journal of Food Science*, **62** (5), 985–991.

Cho, Y.H., Shim, H.K. & Park, J. (2003) Encapsulation of fish oil by an enzymatic gelation process using transglutaminase cross-linked proteins. *Journal of Food Science*, **68** (9), 2717–2723.

Choi, S.-S. & Regenstein, J.M. (2000) Physicochemical and sensory characteristics of fish gelatin. *Journal of Food Science*, **65** (2), 194–199.

Chung, H.Y. & Cadwallader, K.R. (1994) Aroma extract dilution analysis of blue crab claw meat volatiles. *Journal of Agricultural and Food Chemistry*, **42**, 2867–2870.

Duflos, G., Dervin, C., Malle, P. & Bouquelet, S. (1999) Use of biogenic amines to evaluate spoilage in plaice (*Pleuronectes platessa*) and whiting (*Merlangus merlangus*). *Journal of the Association of Official Agricultural Chemists International*, **82** (6), 1357–1363.

Durrington, P.N., Batnagar, D., Mackness, M.I., *et al.* (2001) An omega-3 polyunsaturated fatty acid concentrate administered for one year decreased triglycerides in simvastin treated patients with coronary heart disease and persisting hypertriglyceridemia. *Heart*, **85**, 544–548.

Farmer, L.J., McConnell, J.M. & Kilpatrick, D.J. (2000) Sensory characteristics of farmed and wild Atlantic salmon. *Aquaculture*, **187**, 105–125.

Fukami, K., Ishiyama, S., Yaguramaki, H., *et al.* (2002) Identification of distinctive volatile compounds in fish sauce. *Journal of Agricultural and Food Chemistry*, **50**, 5412–5416.

Fukami, K., Funatsu, Y., Kawasaki, K. & Watabe, S. (2004) Improvement of fish-sauce odor by treatment with bacteria isolated from the fish-sauce mush (Moromi) made from frigate mackerel. *Journal of Food Science*, **69** (2), 45–49.

Fuke, S. & Ueda, Y. (1996) Interactions between umami and other flavor characteristics. *Trends in Food Science & Technology*, **7**, 407–411.

Gennadios, A., Hanna, M.A. & Kurth, L.B. (1997) Application of edible coatings on meats, poultry and seafoods: A review. *Lebensmittel-Wissenschaft und- Technologie*, **30**, 337–350.

Gill, T.A. (1990) Objective analysis of seafood quality. *Food Reviews International*, **6**, 681–714.

Girard, B. & Durance, T. (2000) Headspace volatiles of sockeye and pink salmon as affected by retort process. *Journal of Food Science*, **65**, 34–39.

Gloria, M.B.A. (2003) Amines. In: *Encyclopedia of Food Science and Nutrition* (eds B. Caballero, L. Trugo & P. Finglas), pp. 173–181. Academic Press, London, U.K.

Gloria, M.B.A., Daeschel, M.A., Craven, C. & Hilderbrand Jr., K.S. (1999) Histamine and other biogenic amines in albacore tuna. *Journal of Aquatic Food Product Technology*, **8**, 55–69.

Grimm, C.C., Steven, W., Lloyd, S.W., Paul, V. & Zimba, P.V. (2004) Instrumental versus sensory detection of off-flavors in farm-raised channel catfish. *Aquaculture*, **236**, 309–319.

de Groot, S.I. (1995) Edible species. In: *Fish and Fishery Products—Composition, Nutritive Properties and Stability* (ed. A. Ruiter), pp. 31–76. CAB International, Wallingford, U.K.

Hartvigsen, K., Lund, P., Hansen, L.F. & Hølmer, G. (2000) Dynamic headspace gas chromatography/mass spectrometry characterization of volatiles produced in fish oil enriched mayonnaise during storage. *Journal of Agricultural and Food Chemistry*, **48**, 4858–4867.

Heath, H.B. & Reineccius, G. (1986) Off-flavors in food. 4. In: *Flavor Chemistry and Technology* (ed. N. Macmillan), p. 121. Academic, New York.

Heu, M.-S., Kim, J.-S. & Shahidi, F. (2003) Components and nutritional quality of shrimp processing by-products. *Food Chemistry*, **82**, 235–242.

Hsieh, R.J. (1994) Contribution of lipoxygenase pathway to food flavours. In: *Lipids in Food Flavours* (eds C.T. Ho & T.G. Hartman), pp. 30–48. American Chemical Society, Washington, District of Columbia.

Hsieh, R. & Kinsella, J. (1999) Oxidation of polyunsaturated fatty acids: Mechanisms, products and inhibition with emphasis on fish. *Advances in Food & Nutrition Research*, **33**, 233–341.

Hsieh, T.C.Y., Williams, S.S., Vejaphan, W. & Meyers, S.P. (1989) Characterization of volatile components of menhaden fish (*Brevoortia tyrannus*) oil. *Journal of the American Oil Chemists' Society*, **66**, 114–117.

Ito, K. & Hori, K. (1989) Seaweed: Chemical composition and potential food uses. *Food Reviews International*, **5** (1), 101–144.

Jackson, T.C., Acuff, G.R. & Dickson, J.S. (1997) Microbial spoilage of foods. In: *Food Microbiology: Fundamentals and Frontiers* (eds M.P. Doyle, L.R. Beuchat & T.J. Montville), p. 83. Blackwell, Oxford, U.K.

Jayarajah, C.N. & Lee, C.M. (1999) Ultrafiltration/reverse osmosis concentration of lobster extract. *Journal of Food Science*, **64** (1), 93–98.

Josephson, D.B. (1991) Seafood. In: *Volatile Compounds in Food and Beverages* (ed. H. Maarse), pp. 179–204. Marcel Dekker, New York.

Josephson, D.B. & Lindsay, R.C. (1986) Enzymic generation of volatile aroma compounds from fresh fish. In: *Biogeneration of Aroma* (eds T.H. Parliament & R. Croteau), pp. 201–219. American Chemical Society, Washington, District of Columbia.

Karahadian, C. & Lindsay, R.C. (1989) Evaluation of compounds contributing characterizing fishy flavors in fish oils. *Journal of the American Oil Chemists' Society*, **66**, 953–960.

Kawai, T. (1996) Fish flavor. *Critical Reviews in Food Science and Nutrition*, **36** (3), 257–298.

Keys, S.J. (2003) Aspects of the biology and ecology of the brown tiger prawn, *Penaeus esculentus*, relevant to aquaculture. *Aquaculture*, **217**, 325–334.

Kim, D.S., Baek, H.H., Ahn, C.B., *et al.* (2000) Development and characterization of a flavoring agent from oyster cooker effluent. *Journal of Agricultural and Food Chemistry*, **48**, 4839–4843.

Kitessa, S.M., Gulati, S.K., Ashes, J.R., Fleck, E., Scott, T.W. & Nichols, P.D. (2001) Utilisation of fish oil in ruminants I. Fish oil metabolism in sheep. *Animal Feed Science and Technology*, **89**, 189–199.

Koizumi, C., Ohshima, T. & Lee, E.-H. (1990) Volatile constituents of processed squid product. *Journal of the Korean Society of Food Science and Nutrition*, **19**, 547–554.

Kolanowski, W., Swiderski, F. & Berger, S. (1999) Possibilities of fish oil application for food products enrichment with omega-3 PUFA. *International Journal of Food Sciences and Nutrition*, **50**, 39–49.

Konosu, S. & Yamaguchi, K. (1982) The flavor components in fish and shellfish. In: *Chemistry and Biochemistry of Marine Food Products* (ed. R.E. Roy), pp. 367–404. AVI, Westport, Connecticut.

Kubota, K., Matsukage, Y., Sekiwa, Y. & Kobayashi, A. (1996) Identification of the characteristic volatile flavor compounds formed by cooking squid (*Todarodes pacificus* Steenstrup). *Food Science and Technology International*, **2**, 163–166.

Laturnus, F., Wiencke, C. & Kliiser, H. (1996) Antarctic macroalgae—Sources of volatile halogenated organic compounds. *Marine Environment*, **41** (2), 169–181.

Le Guen, S., Prost, C. & Demaimay, M. (2000a) Characterization of odorant compounds of mussels (*Mytilus edulis*) according to their origin using gas chromatography-olfactometry and gas chromatography-mass spectrometry. *Journal of Chromatography A*, **896**, 361–371.

Le Guen, S., Prost, C. & Demaimay, M. (2000b) Critical comparison of three olfactometric methods for the identification of the most potent odorants in cooked mussels (Mytilus edulis). *Journal of Agricultural and Food Chemistry*, **48** (4), 1307–1314.

Le Guen, S., Prost, C. & Demaimay, M. (2001) Evaluation of the representativeness of the odor of cooked mussel extracts and the relationship between sensory descriptors and potent odorants. *Journal of Agricultural and Food Chemistry*, **49** (3), 1321–1327.

Lee, G.H., Suriyaphan, O. & Cadwallader, K.R. (2001) Aroma components of cooked tail meat of American lobster (*Homarus americanus*). *Journal of Agricultural and Food Chemistry*, **49** (9), 4324–4332.

Lee, H., Kizito, S.A., Weese, S.J., *et al.* (2003) Analysis of headspace volatile and oxidized volatile compounds in DHA-enriched fish oil on accelerated oxidative storage. *Journal of Food Science*, **68** (7), 2169–2177.

Lin, C.Y. & Chiang, B.H. (1993) Desalting and recovery of flavour compounds from salted shrimp processing waste water by membrane processes. *Journal of Food Science and Technology*, **28**, 453–460.

Lindsay, R.C. (1990) Fish flavours. *Food Reviews International*, **6**, 437–455.

Lindsay, R.C. (1994) Flavour of fish. In: *Sea Foods: Chemistry. Processing Technology and Quality* (eds F. Shahidi & J.R. Botta), pp. 30–48. Blackie, London, U.K.

Milo, C. & Grosch, W. (1995) Detection of odor defects in boiled cod and trout by gas chromatography-olfactometry of headspace samples. *Journal of Agricultural and Food Chemistry*, **43**, 459–462.

Milo, C. & Grosch, W. (1996) Changes in the odorants of boiled salmon and cod as affected by the storage of the raw material. *Journal of Agricultural and Food Chemistry*, **44**, 2366–2371.

Michel, F., Priol, J., Galaup, P. & Demaimay, M. (1997) Effet de deux techniques de séchage sur les composés volatils de deux algues alimentaires Ulva sp et Palmaria palmata. *Sciences des Aliments*, **17**, 601–617.

Morita, K., Kubota, K. & Aishima, T. (2001) Sensory characteristics and volatile components in aromas of boiled prawns prepared according to experimental designs. *Food Research International*, **34**, 473–481.

Morita, K., Kubota, K. & Aishima, T. (2002) Investigating Influence of pH and parts on sensory characteristics and volatile components in boiled squid using experimental designs. *Journal of Food Science*, **67** (2), 848.

Mtolera, M.S.P., Collen, J., Pedersen, M., Ekdahl, A., Abrahamson, K. & Semesi, A.K. (1996) Stress-induced production of volatile halogenated organic compounds in *Eucheuma denticulatum* (Rhodophyta) caused by elevated pH and high light intensities. *European Journal of Phycology*, **31**, 89–95.

Nilsson, L., Huss, H.H. & Gram, L. (1997) Inhibition of *Listeria monocytogenes* on cold-smoked salmon by nisin and carbon dioxide atmosphere. *International Journal of Food Microbiology*, **38**, 217–227.

Nilsson, L., Gram, L. & Huss, H.H. (1999) Growth control of *Listeria monocytogenes* on cold-smoked salmon using a competitive lactic acid bacteria flora. *Journal of Food Protection*, **62** (4), 336–342.

Nybakken, J.W. (1993) *Marine Biology—An Ecological Approach*, pp. 4–22. Harper Collins, New York.

Pape, M.A.L., Priol, J. & Demaimay, M. (2002) Effect of two storage conditions on the odor of an edible seaweed, palmaria palmata, and optimization of an extraction procedure preserving its odor characteristics. *Journal of Food Science*, **67** (8), 3135–3139.

Pape, M.A.L., Priol, J., Prost, C. & Demaimay, M. (2004) Optimization of dynamic headspace extraction of the edible red algae Palmaria palmata and identification of the volatile components. *Journal of Agricultural and Food Chemistry*, **52** (3), 550–556.

Passi, S., Cataudella, S., Marco, P.D., Simone, F.D. & Rastrell, L. (2002) Fatty acid composition and antioxidant levels in muscle tissue of different Mediterranean marine species of fish and shellfish. *Journal of Agricultural and Food Chemistry*, **50** (25), 7314–7322.

Ploeg, M. (1991) *Testing Flavor Quality of Preharvest Channel Catfish*. Southern Regional Aquaculture Center Publication Number 431, Stoneville, Mississippi.

Prost, C., Hallier, A., Cardinal, M., Serot, T. & Courcoux, P. (2004) Effect of storage time on raw sardine (Sardina pilchardus) flavor and aroma quality. *Journal of Food Science*, **69** (5), 199–204.

Pszczola, D.E. (2003) Aquatic ingredients provide a new wave of opportunity. *Food Technology*, **57** (5), 71–79.

Sekiwa, Y., Kubota, K. & Kobayashi, A. (1997) Characteristic flavor components in the brew of cooked clam (*Meretrix lusoria*) and the effect of storage on flavor formation. *Journal of Agricultural and Food Chemistry*, **45**, 826–830.

Shimoda, M., Peralta, R., Yoshino, T. & Osajima, Y. (1997) Selection of resin for column treatment to remove odorants in fish sauce. *Food Science and Technology International*, **3**, 196–199.

Shimoda, M., Yamamoto, Y., Cocunubo-Castellanos, J., *et al.* (2000) Deodorization of fish sauce by continuous-flow extraction with microbubbles of supercritical carbon dioxide. *Journal of Food Science*, **65** (8), 1349.

Simeonidou, S., Govaris, A. & Vareltzis, K. (1998) Quality assessment of seven Mediterranean fish species during storage on ice. *Food Research International*, **30** (7), 479–484.

Tanchotikul, U. & Hsieh, T.C.Y. (1989) Volatile flavor components in crayfish waste. *Journal of Food Science*, **54**, 1515–1520.

Tanchotikul, U. & Hsieh, T.C.Y. (1991) Analysis of volatile flavor components in steamed rangia clam by dynamic headspace sampling and simultaneous distillation and extraction. *Journal of Food Science*, **56**, 327–331.

Trondsen, T. (1998) Blue whiting surimi: New perspectives on the market value. *Fisheries Research*, **34**, 1–15.

Vandanjon, L., Cros, S., Jaouen, P., Quéméneur, F. & Bourseau, P. (2002) Recovery by nanofiltration and reverse osmosis of marine flavours from seafood cooking waters. *Desalination*, **144**, 379–385.

Venkateshwarlu, G., Let, M.B., Meyer, A.S. & Jacobsen, C. (2004a) Modeling the sensory impact of defined combinations of volatile lipid oxidation products on fishy and metallic off-flavors. *Journal of Agricultural and Food Chemistry*, **52**, 1635–1641.

Venkateshwarlu, G., Let, M.B., Meyer, A.S. & Jacobsen, C. (2004b) Chemical and olfactometric characterization of volatile flavor compounds in a fish oil enriched milk emulsion. *Journal of Agricultural and Food Chemistry*, **52**, 311–317.

WHO. 1979. *Sodium, chlorides and conductivity in drinking water*. EURO Reports and Studies 2, Regional Office for Europe, Copenhagen, Denmark.

Whitfield, F.B., Last, J.H., Shaw, K.J. & Tindale, C.R. (1988) 2,6-Dibromophenol: The cause of an iodoform-like off-flavor in some Australian crustacea. *Journal of the Science of Food and Agriculture*, **46**, 29–42.

Whitfield, F.B., Helidoniotis, F. & Drew, M. (1996) The role of diet and environment in the natural flavours of seafoods. In: *Flavour Science—Recent Developments* (eds A.J. Taylor & D.S. Mottram), pp. 3–12. The Royal Society of Chemistry, Cambridge, U.K.

Whitfield, F.B., Helidoniotis, F., Shaw, K.J. & Svoronos, D. (1997) Distribution of bromophenols in Australian wild-harvested and cultivated prawns (shrimp). *Journal of Agricultural and Food Chemistry*, **45**, 4398–4405.

Whitfield, F.B., Helidoniotis, F., Shaw, K.J. & Svoronos, D. (1998) Distribution of bromophenols in species of ocean fish from eastern Australia. *Journal of Agricultural and Food Chemistry*, **46**, 3750–3757.

Wilkes, J.G., Conte, E.D., Kim, Y., Holcomb, M., Sutherland, J.B. & Miller, D.W. (2000) Sample preparation for the analysis of flavors and off-flavors in foods. *Journal of Chromatography A*, **880**, 3–33.

Yazawa, K. & Kageyama, H. (1991) Physicochemical activity of docosahexaenoic acid. *Journal of the American Oil Chemists' Society*, **40**, 202–206.

Yoshiwa, T., Morimoto, K., Sakamoto, K., Ishikawa, Y., Tokita, M. & Morita, M. (1997) Volatile compounds of fishy odor in sardine by simultaneous distillation and extraction under reduced pressure. *Nippon Suisan Gakkaishi*, **63**, 222–230.

Young, W.F., Horth, H., Crane, R., Ogden, T. & Arnott, M. (1996) Taste and odour threshold concentrations of potential potable water contaminants. *Water Research*, **30**, 331–340.

Zoeteman, B.C.J. (1980) *Sensory Assessment of Water Quality*, p. 258. Pergamon Press, New York.

Part Three
Beef Quality

12 Sensory Evaluation of Beef Flavor

Rhonda K. Miller

Abstract: Beef flavor is an important sensory attribute when conducting sensory evaluation of beef. While there are multiple sensory tools available for assessing the level of beef flavor, the fact that beef flavor is not a single attribute, but multiple attributes, makes the sensory evaluation of its components challenging. While discriminative tests can be useful sensory tools for determining differences in beef flavor between beef products, they often do not provide quantitative differences in specific attributes. Descriptive sensory tools, mainly Meat Descriptive Attribute, Spectrum® Descriptive Attribute, and Quantitative Descriptive Attribute testing methods, are the most common tools used to assess beef flavor differences. Some use the Free-Choice Profiling Method, but it is not common. Consumer sensory evaluation of beef can provide important information on consumer acceptance of beef products.

Keywords: Beef flavor, sensory evaluation, testing environment, sample preparation, discriminative tests, descriptive tests, meat descriptive analysis, Spectrum® analysis, free-choice profiling method, consumer evaluation

12.1 INTRODUCTION

Beef flavor is an integral component of the overall acceptability of beef, and it is a very complex and dynamic sensory component. Initial beef flavor in a product changes with fresh and frozen storage, is influenced by microbial growth, is altered due to chemical components and their reactions that are inherent in muscle fibers or adipose cells, can be affected by the environment, is influenced by cooking and cooking method, and can be altered by suppressing or masking some components and enhancing other attributers. Beef flavor can be segmented into general categories of positive- and negative-flavor attributes. However, what truly constitutes a positive and/or negative attribute is very subjective and differs by individual. The major component of beef flavor—cooked beefy/brothy—can be generally agreed upon as the most important positive flavor attribute. It is the beefy/brothy flavor that is associated with beef and is the major flavor attribute in canned beef broth or in broth made from cooking beef in water. However, beef flavor is composed of multiple attributes and is a blend of aromatics, basic tastes, mouthfeel, and aftertaste. When consumers eat beef, they tend to identify beef flavor as one sensory component: beef flavor. However, when an off component of beef flavor is presented in beef, consumers can identify that "something is wrong" and some may be able to identify or express the attribute. So this is the challenge. In reality, beef flavor is a very complex sensory element and has multiple components. It is identified using three of the five senses. Chemically, quantifying the chemical constituents of beef flavor is very challenging as

Handbook of Meat, Poultry and Seafood Quality, Second Edition. Edited by Leo M. L. Nollet.

gas chromatograph and mass spectroscopy (GC-MS) can help identify thousands of individual chemical compounds that make up beef flavor. But to the consumer, beef flavor is one attribute.

In the sensory detection of beef flavor, there are multiple tools that can be used. Each tool has strengths and weaknesses, and the interpretation of the sensory results provides useful information. The key for the sensory professional is to understand the objective of their test and then to select the sensory tool that will adequately measure beef flavor without external biases. In this chapter, factors affecting beef flavor and the sensory evaluation of beef flavor will be discussed. First, components of beef flavor will be defined, and then specific aspects of environmental and sample preparation will be discussed to assure that the sensory evaluation of beef flavor is not influenced by these factors. Then, the use of discriminative, descriptive, and consumer sensory tests to evaluate beef flavor will be presented.

12.2 COMPONENTS OF BEEF FLAVOR

Beef flavor is identified through the olfactory senses as odor and flavor aromatics, through the taste receptors on the tongue as basic tastes, and through the trigeminal nerve as mouthfeel. After consumption, aftertaste and mouthfeel can also contribute to beef flavor. In addition, there are multiple factors that contribute that can potentially impact beef flavor. Table 12.1 presents a general beef flavor lexicon. This lexicon was first presented by Johnsen and Civille (1986) for detection of warmed-over flavor in beef or lipid oxidation components. The base lexicon has been adjusted to include attributes that have been used or identified in beef and is not all inclusive. The lexicon can be easily adjusted to a product-specific basis using Civille and Lyons (1996) to include attributes that are product specific. More recently, Bhumiratana *et al.* (2011) developed a beef lexicon containing major and minor beef flavor attributes (Table 12.2). Each attribute has a definition and references that can be used to anchor trained panelists to each beef flavor attribute. This lexicon can be used in combination with Johnsen and Civille (1986) if warmed-over flavors are also a concern within the product.

Chemically, beef flavor is comprised of the beefy/brothy/lean component that is derived from the lean or muscle fiber of the beef cut. Beefy/brothy flavors are detected as odor and/or as flavor aromatics by the olfactory receptors in the nasal cavity. There are other flavors associated with the lean or muscle fiber component of beef. These odor and/or flavor aromatics also are volatile compounds and have been described as cowy, grainy, serumy/bloody, livery, browned, and burnt. The basic tastes of salt and sweet are generally recognized as derived from the lean component of beef. Salts, such as sodium and potassium, are components of the muscle fiber and contribute to the salty basic taste of beef. Sweet basic tastes are derived from glycogen and glucose compounds in beef. For example, meat from species that are higher in glycogen is usually described as sweeter (i.e., horse meat is sweeter than beef).

Within the lean component of meat are heme-containing compounds, mainly myoglobin and some free iron, that are water-soluble compounds of the muscle fiber. These components have been associated with serumy and livery aromatics and metallic mouthfeel in beef. An example of this flavor in chicken is the higher serumy flavor of dark chicken meat versus white chicken meat. This same flavor difference can be identified between beef muscles.

The lipid component of beef also contributes significantly to beef flavor. The odor and/or flavor of cooked beef fat contribute to overall beef flavor. This lipid component is what is associated with the species-specific flavor or the flavor that identifies a meat piece as being beef and is referred to as cooked beef fat aromatic. This species-specific flavor is mainly due to the unique fatty acid composition of beef. Beef fat is not only associated with cooked beef

Table 12.1 Lexicon terms used to evaluate beef flavor originally developed by Johnsen and Civille (1986) and expanded with attributes from Civille and Lyons (1996).

Attribute	Description or definition of attributes
Odors	Defined as the volatile compounds detected by the olfactory senses in the nasal cavity during smelling
Cooked beef lean	The aromatic associated with cooked beef muscle meat
Cooked beef fat	The aromatic associated with cooked beef fat
Browned	The aromatic associated with the outside of grilled or broiled beef
Serum/bloody	The aromatic associated with raw beef lean
Grainy/cowy	The aromatic associated with cow meat and/or beef with grain/feed character is detectable
Cardboardy	The aromatic associated with slightly stale beef, refrigerated for a few days only, and associated with wet cardboard and stale oils and fats
Painty	The aromatic associated with rancid oil and fat
Fishy	The aromatic associated with some rancid fats and oils
Livery/organy	The aromatic associated with beef liver and/or kidney
Flavor aromatics	Defined as the volatile compounds detected by the olfactory senses in the nasal cavity while chewing the sample in the mouth
Cooked beef lean	The aromatic associated with cooked beef muscle meat
Cooked beef fat	The aromatic associated with cooked beef fat
Browned	The aromatic associated with the outside of grilled or broiled beef
Serum/bloody	The aromatic associated with raw beef lean
Grainy/cowy	The aromatic associated with cow meat and/or beef with grain/feed character is detectable
Cardboardy	The aromatic associated with slightly stale beef, refrigerated for a few days only, and associated with wet cardboard and stale oils and fats
Painty	The aromatic associated with rancid oil and fat
Fishy	The aromatic associated with some rancid fats and oils
Livery/organy	The aromatic associated with beef liver and/or kidney
Basic tastes	Defined as the compounds detected by the taste receptors on the tongue
Sweet	Taste on the tongue associated with sugars
Sour	Taste on the tongue associated with acids
Salty	Taste on the tongue associated with sodium ions
Bitter	Taste on the tongue associated with bitter agents such as caffeine, quinine, etc.
Mouthfeel	Defined as the senses detected by the trigeminal nerves in the mouth and throat that are feeling factors
Astringent	The chemical feeling factor on the tongue, described as puckering/dry and associated with tannins and alum
Metallic	The chemical feeling factor on the tongue associated with iron and copper
Aftertaste	Defined as any of the above sensory attributes or new attributes that leave residual sensations and can be flavor, basic tastes, mouthfeel, or other attributes after swallowing or expectorating a sample

fat odor and/or aromatic, but the fat component can also contribute other flavors, as volatile compounds can be stored in the water-soluble component of lipid cells that are derived from the animal diet. These compounds may impart odor and/or flavor aromatics to beef flavor. These compounds are usually animal diet-derived and some descriptors are grassy, musty, boiled corn, and nutty. These aromatics are very diet specific, and the presence of these types of compounds is easily influenced by seasonality, forage quality, soil conditions, and availability of alternate forages or feedstuffs.

The lipid component of beef also can be associated with off flavor that is a result of lipid oxidation. The aromatics associated with this chemical reaction are cardboardy, painty, fishy, and livery aromatics. Sometimes these flavor changes are consolidated into the term rancid or

Table 12.2 Lexicon for major flavor and aroma attributes for beef developed by Bhumiratana *et al.* (2011).

Attribute	Definition	Reference
Beefy flavor	Amount of beef flavor identity	Beef brisket = 4.0 (tasted) Swanson's beef broth = 5.0 (tasted) 80% lean ground chuck = 7.0 (tasted)
Brown/roasted	A round, full aromatic generally associated with beef suet that has been broiled	Beef suet (broiled) = 8.5 (tasted) 80% lean ground chuck = 10.0 (tasted)
Bloody/serumy	An aromatic associated with blood on cooked meat products. Closely related to a metallic aromatic	Select strip steak = 5.5 (tasted) Beef brisket = 6.0 (tasted)
Fat-like	Aromatics associated with cooked animal fat	Beef suet (broiled) = 12.0 (tasted) Hillshire Farms Lit'l Beef Smokies = 7.0 (tasted)
Metallic	The impression of slightly oxidized metal, such as iron, copper, and silver spoons	0.10% potassium chloride solution = 1.5 (tasted) Select strip steak = 4.0 (tasted) Dole canned pineapple juice = 6.0 (tasted)
Liver-like	Aromatics associated with cooked organ meat/liver	Beef liver (broiled) = 7.5 (tasted) Brauschweiger liver sausage = 10 (tasted and swallowed)
Green, hay-like	Brown/green dusty aromatics associated with dry grasses, hay, dry parsley, and tea leaves	Dry parsley in medium snifter = 5.0 (smelled) Dry parsley in 1 oz cups = 6.0 (tasted)
Umami	Flat, salty, somewhat brothy. The taste of glutamate, salts of amino acids, and other molecules called nucleotides	0.035% Accent Flavor Enhancer solution = 7.5 (tasted)
Overall sweet	The combination of sweet tastes and sweet aromatics	Post Shredded Wheat Spoon Size = 1.5 (tasted) Hillshire Farms Lit'l Beef Smokies = 3.0 (tasted)
Sweet	Aromatics associated with the impression of sweet or the fundamental taste factor associated with a sucrose solution	SAFC ethyl maltol 99% = 4.5 (smelled) 2.0% sucrose solution = 2.0 (tasted)
Sour	Aromatics and taste associated with sour substances	Dillon's buttermilk (covered) = 5.0 (smelled) 0.015% citric acid solution = 1.5 (tasted) 0.050% citric acid solution = 3.5 (tasted)
Salty	A fundamental taste factor of which sodium chloride is typical	0.15% NaCl solution = 1.5 0.25% NaCl solution = 3.5
Bitter	The fundamental taste factor associated with a caffeine solution	0.01% caffeine solution = 2.0 0.02% caffeine solution = 3.5

rancidity. However, as lipid oxidation proceeds in beef, cardboardy flavor aromatics usually slowly increase, and with later stages of oxidation, painty and fishy aromatics become apparent. Livery aromatics increase as lipid oxidation proceeds as well. Therefore, in beef products where changes in flavor are due to lipid oxidation, measuring individual attributes of flavor changes provides a more detailed description of the rate and extent of lipid oxidation.

Beef flavor also can be impacted by microbial growth. In aerobic packaging systems, spoilage microorganisms can impart putrid, rotten-egg, or sulfur-type aromatics to beef. In vacuum-packaged beef, growth of spoilage microorganisms has been shown to increase soured aromatics. Other off flavors may be identified during fresh storage that affect beef flavor such as nutty, musty, acidic, buttery, and yeasty.

Environmental factors can affect beef flavor. A classic example of this is the effect of plastic flavors associated with packaging materials on beef. During storage in plastic films, beef can absorb plastic aromatics from the packaging materials.

Cooking and cooking methods can influence beef flavor. During cooking, serumy/bloody aromatics are slowly replaced by cooked beefy/brothy aromatics. Also, with increased degree of doneness, livery aromatics increase.

It is apparent that beef flavor is a highly complex entity and it can be influenced by a wide variety of factors. The sensory evaluation of beef flavor should include all components of beef flavor, provide an opportunity to describe flavor attributes that may emerge with time, and provide an environment that will not interfere in sensory detection of beef flavor attributes.

12.3 TESTING ENVIRONMENT AND SAMPLE PREPARATION

The testing environment and sample preparation can easily influence beef flavor. Controls for the sensory testing environment and sample preparation have been discussed in Chapter 2. These components must be considered for all sensory testing, and a protocol must be defined and considered prior to any sensory evaluation, regardless of the sensory tool being used, as beef flavor can be easily influenced by storage and preparation. As the goal of any sensory evaluation, Tables 12.3 and 12.4 provide a checklist of factors to consider, standardize, or control in test samples for sample storage and sample preparation, respectively. Items presented in these tables are not all inclusive, but they are presented to provide guidelines on

Table 12.3 List of important parameters and examples of how these parameters may influence the beef flavor of samples during storage.

Item	Examples of how sensory attributes could be affected
1. Raw material meat source	Raw materials can vary due to animal of origin and harvesting influences
a. Age of animal	Meat from older animals is tougher, darker in color, and more intense in flavor for some attributes; meat from very young animals is more tender, lighter in color, and has lower flavor attributes; should consider what is normal for your product
b. Sex of animal	Meat from intact males may differ in flavor attributes and be more variable in tenderness and/or tougher than meat from castrated males or females; meat from females differs in fat content and/or connective tissue depending on the relationship of growth and the onset of puberty
c. Nutritional background of animal	Meat from animals fed grass can be more variable in flavor and tenderness than meat from animals fed grain. Flavor attributes may be influenced by the type of forage and the duration the specific forage was fed
d. Quality grade of carcass	Meat from carcasses with higher fat content differs in flavor and can differ in tenderness or the variability of tenderness than meat from animals with lower fat content
e. Yield grade of carcass	Meat from carcasses with a higher percentage of fat may differ in tenderness and flavor than meat from carcasses with a lower percentage of fat
f. Elimination of sources with obvious defects (bruises, abscesses, excessive trim, quality defects such as pale meat, dark meat, blood splash, etc.)	Quality defects can influence multiple meat palatability attributes; for example, meat that is pale due to animal stress prior to harvesting may be drier, less flavorful, and tougher than meat from animals not stressed

(Continued)

Table 12.3 *(Continued)*

Item	Examples of how sensory attributes could be affected
2. Definition of handling and/or trimming specifications	Top rounds denuded on one day and used in a study with top rounds not trimmed of external fat from another day may result in flavor differences as microbial growth occurs on the surface of meat during storage. The sour flavor in meat induced by growth of spoilage microorganisms during vacuum packaging may be higher in the denuded top round as the external surface may not be sufficiently trimmed prior to obtaining samples for sensory evaluation
3. Sampling technique of raw material	Meat removed from a carcass by professional meat cutters on the fabrication line of processing facility will usually be more consistent than meat removed by inexperienced personnel in a meat cooler
4. Age or length of storage of raw material	Meat stored for 5 days versus meat stored for 28 days will most likely differ in tenderness, flavor, and juiciness due to differences in microbial growth, lipid oxidation, and postmortem proteolysis in the meat during storage
5. Storage container of raw material	Packaging materials can impart flavors. Beef stored in different plastic liners may differ in flavor attributes
6. Gaseous atmosphere during storage	Gaseous atmosphere affects microbial growth and specific microorganisms can impart specific flavors. Meat stored in vacuum packages can develop sour flavors that are the result of acid producing spoil microorganisms, whereas meat stored aerobically can have sulfur, putrid flavors that are imparted by spoilage organisms
7. Particle size during storage (i.e., whole muscle vs. comminuted)	Once meat tissue has been lysed, it is more susceptible to lipid oxidation and microbial growth, and these factors can affect beef flavor
8. Temperature during storage of raw material	Temperature affects rates of lipid oxidation and microbial growth that subsequently can impact beef flavor
9. Potential temperature fluctuations during storage	Temperature affects rates of lipid oxidation and microbial growth that subsequently can impact meat flavor even with small fluctuations during frozen or refrigerated storage
10. Processing day	Raw material source, temperature, and worker expertise can vary across processing days, and these factors can affect any of the aforementioned items
11. Batch within processing day	Batches differ due to different raw material and potential handling, temperature, equipment, and worker differences, and these factors can affect any of the aforementioned items
12. Transportation parameters	Temperature fluctuations during transportation and loss of packaging integrity can affect rates of lipid oxidation and microbial growth as previously discussed
a. Containers	Containers can impart specific aromatics, either protect or not protect packaging integrity and/or affect temperature stability
b. Temperature	Standardized temperature during transportation is critical. Continuous temperature recorders can verify proper temperature controls
c. Length of time	Standardized length of time during transportation or standardized conditions are critical
d. Special considerations	Any factor that may either impart flavor, alter packaging integrity, affect temperature, or result in an increase in rate of lipid oxidation and microbial growth needs to be preexamined and controlled

factors that may influence sensory verdicts for beef flavor. For individual evaluations and specific products, additional items may need to be considered, added, or deleted. The major point is that consideration and discussion of sample storage and preparation issues occur and a full understanding, elimination, or standardization of their effects on factors of beef flavor must be included in the design of the sensory study.

After standardization and attention to consistent sample storage and preparation, standardization of the sample presentation to sensory panelists is critical to assure that sensory

Table 12.4 Examples of factors that can affect sample preparation and examples of how these factors may influence beef flavor.

Item	Examples of how sensory attributes may be affected
Temperature of meat prior to cooking	Frozen or meat at the freeze/thaw temperature of 28 °C will require longer cooking times that can decrease juiciness and tenderness and subsequently impact beef flavor attributes—higher livery, burnt, and brown aromatics
Sample thickness and size	Thicker meat portions require longer cook times to reach a standard internal degree of doneness. Larger or smaller meat portion sizes may also require more or less cooking time, respectively, that will alter the juiciness, tenderness, and flavor of the meat
Sample fat trim	Fat is an insulator and can impact temperature transfer rates, and, therefore, cook time and cook yields may impact meat palatability
Cooking method	Dry and wet cooking methods, type of grill (electric vs. gas, serrated vs. flat, two-sided or one-sided heat source), air flow (conventional vs. convection vs. impingement), distance from heat source, environment (enclosed versus open), and humidity impact meat palatability attributes
Cooking temperature	The temperature of the environment or the service temperature of the oven or grill impacts the rate of cooking and will subsequently impact beef flavor. For example, higher temperatures may result in higher burnt/browned and livery aromatics while lower temperatures may increase cardboardy aromatics
Cooking endpoint temperature	Meat flavor, tenderness, and juiciness attributes change as meat is cooked from low or rare to high or well done degrees of doneness. Note that endpoint temperature and visual degree of doneness may not always coincide
Monitoring of endpoint cooking temperature	Type of thermocouples and placement of thermocouples may impact the accuracy of monitoring cooking endpoint temperature
Cook time	Longer cook times can result in drier, tougher meat due to dehydration from longer exposure to a heat source. Beef flavor may be lower in serumy/bloody, cowy/grainy and higher in cooked beefy/brothy, livery, and browned/burnt aromatics

attributes detected by the panelists are the true attributes of the sample and that are not present due to sample presentation. The size and number of samples presented to each panelist needs to be standardized. When evaluating beef flavor, multiple attributes are being assessed. Therefore, the number of samples evaluated in a sensory day should be tightly controlled and the number of samples should minimize taste panel sensory fatigue. Additionally, the serving size should be sufficient for panelists to evaluate all attributes. Two serving portions are most commonly used. The serving size for whole-muscle meat products is usually two or three 1.27-cm cubes. For ground beef patties, one or two wedges that may represent 1/6, 1/4, or 1/3 of the patty may be served including the edges, or 1.27- or 2.54-cm squares where the edges are excluded may be served. In some cases, the entire ground beef patty may be served and the panelist cuts the size of the sample that they are accustomed to eating. Ground beef patties also may be served on a bun and panelists eat the sample as they would when eating a sandwich. For processed meat samples, such as sausages or frankfurters, a whole sausage or frankfurter may be served especially when texture attributes are being evaluated, but 2.54-cm sections may be served when only flavor attributes are being determined. The interpretation of the results of the sensory evaluation is dependent on the sample size and form served to panelists. The sample size should be considered in relationship to the purpose of the sensory evaluation and the sample size needed for adequate evaluation by the panelists.

Sample temperature and potential temperature fluctuations of samples within and between panelists are critical issues. As aromatic compounds volatilize at higher temperatures, if sample temperature fluctuates, sensory panelists may detect differences in the amount of beef odor and/or flavor attribute between samples, but the difference is actually due to sample temperature differences. It is important to standardize sample serving temperatures within and between sensory days to remove this potential effect. Samples should be identified for sensory panelists using random, three-digit codes to eliminate subconscious selection of a sample due to identification. The serving container should be neutral in color, nonabsorbent, and not impart flavor or odors to the sample. For example, glass or white, glazed china containers will not have container-specific aromatics and will be visually neutral. Plastic containers can be used, but they should be opaque or neutral in color and tested to assure that odors or flavors associated with the container are not present. For beef samples being evaluated for odor and flavor, a glass custard dish with a concave glass watch glass top provides an opportunity for odor evaluation by gently shaking the sample, then moving the glass back to smell the sample. The container can be preheated to a standard temperature to reduce temperature fluctuations during evaluation and to not burn nasal tissues during evaluation.

Samples should be served to panelists in random order as there are strong first-order bias in central location consumers and trained sensory evaluation tests (either higher or lower ratings in comparison to samples evaluated later) and last-order biases for long-term, in-home use consumer tests especially for preference data. Order effects can either be randomized across samples or balanced (i.e., every sample can be evaluated in each order and order combination sufficiently to account for the effect in the statistical model as a block). Visual attributes of a sample can affect sensory verdicts due to expectation errors. For example, panelists may expect meat cooked to a rare degree of doneness to have higher serumy flavors than meat cooked to a well-done degree of doneness. Therefore, the panelists' evaluation is not independent, but is influenced by the visual attributes of the sample. In this case, the use of red lights during evaluation will assist in controlling this effect.

Guidelines for cookery and sensory analysis of meat (ASTM 1978; AMSA 1995) should be used as standardized techniques for the sensory evaluation of beef flavor.

12.4 DISCRIMINATIVE TESTS

Discriminative tests are used to discriminate or to tell overall differences between one or more products. There are multiple tests to select from and each test has strengths and weaknesses. In the meat science literature, discriminative tests are not commonly used. They are more commonly used in industry to detect differences between one or more samples.

Discriminative tests are useful tools and may not be directed at determining differences in one attribute, like beef flavor, but differences between two or more products. The common discriminative tests used to determine differences in beef flavor will be discussed and information on the type of data and decisions that can be made from each test will be presented. Trained sensory panelists or consumer sensory panelists can be used for discriminative tests. Trained sensory panelists will generally be able to determine smaller differences between samples than consumers, whereas consumer responses more closely mirror differences from the general population.

There are two categories of difference tests: overall difference tests and attribute difference tests. Overall difference tests determine if sensory differences exist between two samples, and attribute difference tests are more specific and ask if samples differ in specific sensory attributes. Triangle tests, two-out-of-five test, duo-trio tests, simple difference test, "A"–"not A" tests, difference-from-control tests, sequential tests, and similarity tests are examples of difference tests. Directional difference tests, pairwise ranking tests, simple ranking test, rating approach, multisample difference tests, and balanced incomplete block (BIB) design-ranking tests are examples of attribute difference tests. Although all of these tests could be used to determine differences in beef flavor, generally, triangle tests, duo-trio tests, and simple difference tests are used. These three tests will be discussed for use in determined differences in beef flavor.

Triangle tests use three samples coded with random three-digit codes, and the panelist is asked to select the sample that is different. Two samples are alike and a third sample is different. Multiple panelists, either trained or consumer, are used and the order of the three samples is randomized, and the sample that differs is alternated between the two samples. Using a predetermined significance level, the correct number of samples versus the total number of tests evaluated is used to determine if differences existed between the two samples. Tables to determine significance are presented in Meilgaard *et al.* (1999). Panelists have a 33% chance of guessing the correct answer in a triangle test, and the greater the number of tests conducted, the more the guessing factor influence is decreased. For beef flavor determination, the two samples being evaluated would be presented in the test following the aforementioned controls. If panelists detected differences based on the predetermined significance level and number of samples served, then it can be concluded that the two samples differed in beef flavor. The challenge with this test is determining what aspect of beef flavor the panelists are evaluating. In the question, the sensory professional may ask the panelists to differentiate samples based on "beef flavor," therefore, leaving the interpretation of what beef flavor is up to the panelists. This provides a general evaluation of beef flavor and may be sufficient for the study. Another approach is to ask a specific question concerning beef flavor. For example, in evaluating irradiated versus nonirradiated ground beef patties, a triangle test could be conducted with consumer panelists to determine if irradiation affects beef flavor or if "rancid-type flavors" differ. The sensory professional should make sure that the panelists understand what "rancid-type flavors" are. These tests can provide valuable information regarding whether samples differ in beef flavor, but they may not provide information on what aspect of beef flavor differs or is affected or how much difference exists in beef flavor between samples.

For duo-trio tests, panelists are given a reference or control sample. They are then given two samples and are asked to identify the sample that is the same as the reference or control sample. Panelists have a 50% chance of selecting the correct sample. Again, the number of correct responses are calculated and compared to the total number of evaluations completed. The probability that the samples differ is determined using tables from Meilgaard *et al.* (1999).

In simple difference tests, panelists are provided two samples and asked if the samples differ or if they are similar. Panelists may be provided one or up to four possible combinations of presentations for the two samples (1 then 2, 2 then 1, 1 then 1, and 2 then 2). The total number of correct responses for each of the four combinations is determined. The placebo effect is compared to the treatment effect using x^2-analysis. The x^2-statistic is calculated by:

$$x^2 = \sum \frac{(\mathrm{O}-\mathrm{E})^2}{\mathrm{E}},$$

where O is the observed value and E is the expected number. The determined x^2-value is compared to an x^2 table at a probability of 0.05. If the calculated value is greater than the table value, a significant difference exists between the two samples.

Duo-trio and simple difference tests, as previously discussed for triangle tests, can provide viable tools for determining differences in beef flavor between two samples. However, the challenge is determining what attributes of flavor differ. It may not be important to know what aspects of beef flavor differ between two samples, but if differences in specific flavor attributes need to be quantified, descriptive sensory tests can be used.

12.5 DESCRIPTIVE TESTS

To describe and quantify the intensity of sensory attributes in meat samples, descriptive sensory tests are an effective tool and they are particularly effective tests for quantifying differences in specific beef flavor attributes. Descriptive tests use trained sensory panels to rate the intensity of attributes that are product specific. The type of test, the development of the attributes to be evaluated, and the rating scale may differ across tests, but the overall objective of descriptive tests are to describe and quantify sensory attributes.

Three types of descriptive tests can be used to determine beef flavor: Meat Descriptive Analysis, Spectrum® Descriptive Analysis, and Quantitative Descriptive Analysis. Free-choice profiling has been used on a limited basis to determine differences in beef flavor. This method can be used as an effective tool in assessing beef flavor differences in meat products.

Descriptive tests use a dictionary or family of attributes called a lexicon. The attributes in a lexicon are defined so that each attribute can be quantified. A lexicon can be product specific or a general product lexicon can be used as a base and other attributes added as defined by trained panelists. A general beef flavor lexicon is presented in Table 12.1. Attributes can be identified by using existing published lexicons, by attributes identified from Civille and Lyons (1996), by free-choice profiling where individual panelists develop their own descriptors, or by identification through ballot development sessions with a trained descriptive attributes sensory panel.

Ballot development sessions can be a very useful tool in descriptive sensory evaluation for identifying the beef flavor attributes. In ballot development sessions, panelists that have shown the ability to describe and provide input into individual sensory attributes are used. The panel leader presents products that are similar to the products that will be used in the study. Sensory panelists are asked to evaluate the product and provide ideas and input into the sensory attributes that they detect. The panel leader can provide standard lexicons for similar products or use resources like Civille and Lyons (1996) that provide a large list of sensory attributes, definitions, and references. Care should be taken to not depend too much on standard lexicons because product-specific attributes may be overlooked. Products should be presented to the panelists that represent the range and average products for the study. The panel leader should help direct the session, but they should not stifle creativity and discussion. A guideline for the sequence of events that could be followed during a ballot development session is presented in Table 12.5. Some descriptive analysis tests do not require extensive ballot development sessions (quantitative descriptive analysis), whereas other methods (meat descriptive analysis and spectrum descriptive analysis) use extensive sessions and a standardized ballot.

12.5.1 Meat descriptive analysis

This descriptive method was developed by meat scientists to evaluate the palatability of whole-muscle red meat products. Palatability attributes for red meat are defined as juiciness, muscle fiber tenderness, connective tissue amount, overall tenderness, and flavor intensity. These attributes are measured using 8-point scales. The American Meat Science Association has published guidelines for use of this method (AMSA 1978, 1995). This is the predominant sensory method used in the scientific literature to determine differences in red meat palatability; however, note that in the original ballots presented in AMSA (1978) and then again in AMSA (1995), the beef flavor attributes are described as beef flavor intensity and are rated using 1—extremely bland, and 8—extremely intense. This method is a very effective tool for

Table 12.5 An example of the sequence of events for ballot development sessions for beef flavor in a whole-muscle beef steak.

Session	Session goal	Exercises used in the session
1	Familiarize panelists with beef flavor product Initiate and stimulate identification of flavor attributes	1. Presentation of a baseline beef steak—use cooking methods as in study; present beef flavor lexicon for discussion 2. Introduce one or two references for beef lexicon to begin anchoring on attributes and to familiarize panelists to the concept of lexicon and attribute identification
2	Continue initiation and stimulation of identification of beef flavor attributes	1. Presentation of the baseline beef steak. Ask panelists to describe the flavor. Begin to narrow down the descriptors into categories that are similar between panelists and to get consensus on some attributes
	Introduce one to two products into categories that are variations from the baseline product	2. Begin using attributes that are being defined to evaluate one or two steaks that are variations from the baseline but that are representative of steaks that will be used in the study. Continue to encourage and provide a free discussion of the flavor attributes
3	Continue with identification of the flavor attributes in the product as in Session 2	1. Present the baseline product. Ask panelists to describe the flavor. Utilize the descriptors identified in Session 2. Continue to fine-tune these descriptors. Bring in references of attributes to clarify what the specific attribute is and how to identify it
	Introduce the baseline product and the new ingredient in its pure form and begin identifying the flavor attributes of the ingredient	2. Examine one or two additional products that are variations from the baseline. Continue to discuss flavor descriptors, identifying new descriptors, or altering existing descriptors 3. Taste the new ingredient in its pure form. Begin identifying the flavor descriptors for the ingredient
4	Repeat Session 3 until panelists are familiar and comfortable with the attributes as defined	1. Same as Session 3 2. As panelists gain experience, confidence, and repeatability on each attribute, add new attribute references and continue to expand the ballot
5	Identify references for the flavor	1. Present the baseline product. Review the current list of descriptors being defined 2. Introduce and examine baseline examples of each descriptor and begin anchoring the descriptors for each panelist
6	Introduce scaling for each descriptor	1. Present the baseline product. Review the current list of descriptors 2. Utilize references of each descriptor to continue anchoring panelists on each attribute 3. Introduce the product with varying concentrations of the ingredient and begin scaling with the descriptors as defined 4. Continue identification of alternate descriptors
7	Develop a final ballot	1. Progressively continue sessions until a final ballot is defined and panelists are familiar with the descriptors and the scale

determining juiciness, muscle fiber tenderness, connective tissue amount, and overall tenderness between samples when the sensory panelists have been adequately trained, but the beef flavor intensity attribute is very general. This method is not highly sensitive to flavor intensity differences, and if the objective of a study is to understand flavor differences, the Spectrum Method for Descriptive Flavor Attributes, the Tragon Quantitative Descriptive Analysis Method, and the Free-Choice Profiling Method are better sensory tools. In my laboratory, the Meat Descriptive Ballot has been altered to provide more meaningful beef flavor sensory information. This ballot is presented in Table 12.6. Note that the beef flavor attributes of cooked beef brothy and cooked beef fat have been added. In addition, attributes for off flavor attributes are listed. Panelists identify off flavor attributes by listing the number and then the intensity rating of the attribute. The problem with rating only beef flavor intensity

Table 12.6 Meat descriptive attribute ballot for whole muscle meat as defined by AMSA (1995).

Name ____________ Session ____________ Date ____________

Sample	Juiciness	Muscle fiber tenderness	Overall tenderness	Connective tissue amount	Overall flavor intensity	Cooked beefy/brothy	Cooked beef fat	Off-flavor characteristics
Warm-up								

Juiciness	Muscle fiber and overall tenderness	Connective tissue amount	Flavor intensity	Off-flavor characteristics			
				X	Other (describe)	M	Metallic
8 Extremely juicy	8 Extremely tender	8 None	8 Extremely intense	A	Acid	MU	Musty
7 Very juicy	7 Very tender	7 Practically none	7 Very intense	B	Bitter	N	Nutty
6 Moderately juicy	6 Moderately tender	6 Traces	6 Moderately intense	BR	Browned	P	Putrid
5 Slightly juicy	5 Slightly tender	5 Slight	5 Slightly intense	C	Cardboard	SA	Salty
4 Slightly dry	4 Slightly tough	4 Moderate	4 Slightly bland	CH	Chemical	SB	Serumy/bloody
3 Moderately dry	3 Moderately tough	3 Slightly abundant	3 Moderately bland	CW	Cowy	SD	Soured
2 Very dry	2 Very tough	2 Moderately abundant	2 Very bland	F	Fish-like	SO	Sour
1 Extremely dry	1 Extremely tough	1 Abundant	1 Extremely bland	L	Liver	SW	Sweet

is that beef flavor is a multicomponent attribute and evaluating only one generalized attribute is not consistent and repeatable and is difficult to define.

For further processed meat products and ground meat, the Meat Descriptive Attribute method can be used; however, attributes need to be changed to reflect the product being tested. Initial juiciness, sustained or final juiciness, initial beef flavor, and sustained or final beef flavor are examples of sensory attributes used for ground beef. For further processed products, attributes for the texture and flavor of the product should be used. The Spectrum Method for Descriptive Flavor Attributes, the Tragon Quantitative Descriptive Analysis Method, and the Free-Choice Profiling Method may be more appropriate methods for assessing sensory attributes for further processed meat products.

After completion of the sensory evaluation, data can be analyzed using analysis of variance to determine treatment differences for each descriptive attribute. Principal component analysis and other multivariate analyses may be used to understand the combined effect of all attributes on differences in the meat products tested.

12.5.2 The Spectrum® Method

This method was developed by Gail Vance Civille to use across a wide variety of products and it is easily adapted for use in assessing beef flavor. This method uses a universal intensity scale from 0 (none) to 15 (extremely intense) to quantify beef flavor differences. The universal scale standardizes the level of intensity for any attribute and is used to standardize panelists when using the scale. The universal scale defines a 2 as the soda flavor in saltless saltine crackers, and a 5 as the apple flavor in Mott's applesauce. The orange flavor in Minute Maid frozen orange juice reconstituted as defined on the package is a 7, and the grape flavor in Welch's grape juice is a 10 for intensity. The cinnamon flavor in Big Red Gum is a 12 on this intensity scale. The scale is then used on specific attributes where the attributes are defined using ballot development sessions or a standard lexicon as presented in Table 12.1 for beef flavor. Note that an attribute defined as other is always included on the ballot. If panelists identify an attribute that is not on the ballot, they provide a descriptor and then rate the attribute for intensity. This provides an avenue for the identification of attributes not defined during ballot development that become apparent during testing. Tables 12.7 and 12.8 are examples of Spectrum® Method ballots used to evaluate beef flavor in whole-muscle beefsteaks and ground beef.

This sensory method is easily adapted across meat products and is useful for whole-muscle and processed meat products. These data are analyzed as defined for the meat descriptive analysis, and differences in specific attributes can be determined or multivariate analyses can be used to examine changes of multiple attributes in meat products.

12.5.3 Quantitative descriptive analysis method

This sensory method is commonly used to determine beef flavor differences between samples and is an especially valuable tool used in industry for value-added or processed beef products. It is a useful tool for determining beef flavor differences. The Tragon Corporation developed the Quantitative Descriptive Analysis Method or QDA Method for descriptive analysis. In this method, a large pool of trained sensory panelists is used. For a particular meat product, panelists who can discriminate beef flavor attributes are selected. Ballot development and training sessions are similar as previously described for meat descriptive and spectrum methods except the panel leader is more of a facilitator and does not take as active a role in directing the discussion. References for attributes are provided, and products

Table 12.7 Descriptive attribute ballot for beef flavor determination in whole-muscle beef steaks using the Spectrum® Method of descriptive analysis.

Name ______________ **Date** ____________ **Session** ____________

Sample Number	______	______	______
Aromatics			
Cooked beef/brothy		______	______
Cooked beef fat		______	______
Serumy/bloody		______	______
Grainy/cowy		______	______
Cardboard		______	______
Painty		______	______
Fishy		______	______
Liver		______	______
Soured		______	______
Browned/burnt		______	______
Dirt		______	______
Medicine		______	______
Old/putrid		______	______
Plastic		______	______
Other (describe)		______	______
Feeling factors			
Metallic		______	______
Astringent		______	______
Basic tastes			
Salt		______	______
Sour		______	______
Bitter		______	______
Sweet		______	______

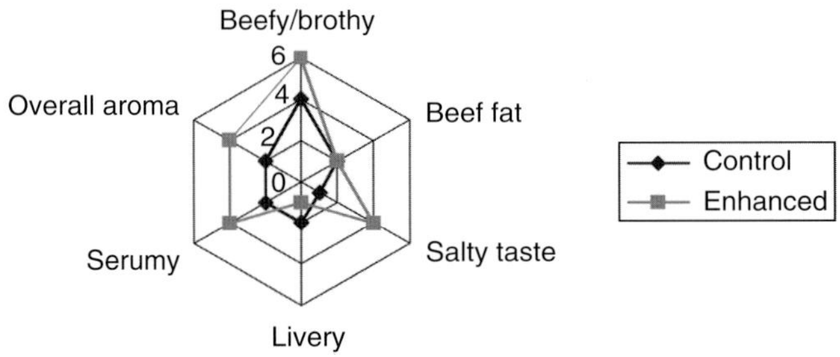

Fig. 12.1 A "spider web" for quantitative descriptive analysis for enhanced and nonenhanced beef steaks.

that emulate products used in the study are presented to the panelists, but panelists are free to develop their own descriptive terms. Descriptive terms are rated using a 15-cm line scale. Panelists will independently rate each product for descriptive attributes and then are provided feedback from the panel leader as to their consistency and performance. Beef flavor attributes can easily be evaluated using this method. The multiple components of beef flavor can be used as defined in Table 12.1 or attributes as defined by the panelists will be assessed. As panelists independently evaluate each product and do not discuss data, the

Table 12.8 Descriptive attribute ballot for beef flavor determination in ground beef patties using the Spectrum® Method of descriptive analysis.

Name ______________ **Date** ______________ **Session** ______________

Sample ID Number	**Warm-Up**	______	______	______	______
Aromatics					
Cooked beef/brothy	______	______	______	______	______
Cooked beef fat	______	______	______	______	______
Serumy/bloody	______	______	______	______	______
Grainy/cowy	______	______	______	______	______
Cardboard	______	______	______	______	______
Painty	______	______	______	______	______
Fishy	______	______	______	______	______
Liver	______	______	______	______	______
Soured	______	______	______	______	______
Browned/burnt	______	______	______	______	______
Other (describe)	______	______	______	______	______
Feeling factors					
Metallic	______	______	______	______	______
Astringent	______	______	______	______	______
Basic tastes					
Salt	______	______	______	______	______
Sour	______	______	______	______	______
Bitter	______	______	______	______	______
Sweet	______	______	______	______	______
Aftertaste					
Astringent	______	______	______	______	______
Fat mouthfeel	______	______	______	______	______
Bitter	______	______	______	______	______
Browned/burnt	______	______	______	______	______
Sour	______	______	______	______	______
Sweet	______	______	______	______	______
Other (describe)	______	______	______	______	______
After feeling factors					
Lip burn	______	______	______	______	______
Metallic	______	______	______	______	______

terminology, or the samples after each taste session, the terminology may differ from project to project. The panel leader provides the feedback to the panelists independently. Data are analyzed statistically using the QDA analysis package, and data are reported in a "spider web." The statistical package organizes and categorizes similar beef flavor attributes into attributes that are presented in the spider web. By looking at differences in intensity of attributes across different products in the test, beef flavor differences can be determined. In Fig. 12.1, a spider web for beef flavor of enhanced and nonenhanced beef steaks is presented. The interpretation of data presented in Fig. 12.1 is that enhanced beef steaks are higher in beefy/brothy, salty taste, serumy flavors, and overall aroma than control steaks. However, enhancement did not increase the beef fat flavor and slightly decreased livery flavor in beef steaks when compared to the control steaks. This type of data presentation is easy to interpret and provides a good snapshot of the attributes that are changing or that do not differ across treatments.

12.5.4 Free-Choice Profiling Method

The Free-Choice Profiling Method gives panelists the freedom to develop their own descriptors and scales. Panelists do not need to be uniformly anchored and efforts are not made to assure that panelists use the scale similarly. Panelist training sessions include only a short session where the sensory procedures are explained. Panelists have not been biased by training and are considered "consumers." Sensory samples are served to the panelists, and they are asked to evaluate the products for beef flavor. They then describe the attributes of beef flavor and rate the intensity as they perceive it. The data are analyzed using the generalized Procrustes analysis where it collates sensory terms that are similar. The sensory professional then interprets differences in beef flavor attributes. This method does not require the extensive time to train and standardize panelists. Panelists' performance is not evaluated and ballot development sessions are not needed. Retaining trained sensory panelists is not needed. Although this method is not commonly used to determine beef flavor differences in beef products, it is an increasingly used sensory tool for food and an acceptable sensory tool for ascertaining beef flavor differences in beef products.

12.6 CONSUMER EVALUATION OF BEEF FLAVOR

Consumer protocol, ballot design, and consumer selection have been discussed in other chapters in this book. Beef flavor is easily included in consumer evaluation of beef acceptability. As consumer studies are a method of assessing consumer preference or acceptance, beef flavor is obviously a component of consumer acceptance. Beef flavor like/dislike should be included on the ballot as a question and structured similarly to the questions for overall like/dislike. Questions regarding the intensity of beef flavor also are viable questions to ask consumers. By asking a flavor like/dislike question followed by a question asking consumers to rate the intensity of beef flavor, relationships between how much beef flavor in the product drives consumer acceptability can be ascertained. Care should be taken in asking consumers questions about specific beef flavor attributes as defined by trained sensory panelists. Consumers may not understand the attribute and while they may rate the attribute, the data will be meaningless. Beef flavor questions are appropriate for both in-home and central location tests and also are very useful when evaluating further processed or value-added beef products.

12.7 CONCLUSION

In conclusion, beef flavor is an important sensory attribute when conducting sensory evaluation of beef.

While there are multiple sensory tools available for assessing the level of beef flavor, the fact that beef flavor is not a single attribute, but multiple attributes, makes the sensory evaluation of its components challenging. While discriminative tests can be useful sensory tools for determining differences in beef flavor between beef products, they, many times, do not provide quantitative differences in specific attributes. Descriptive sensory tools, mainly Meat Descriptive Attribute, Spectrum® Descriptive Attribute, and Quantitative Descriptive Attribute testing methods, are the most common tools used to assess beef flavor differences. Some use the Free-Choice Profiling Method, but it is not common. Consumer sensory evaluation of beef can provide important information on consumer acceptance of beef products. Ascertaining acceptability of beef flavor as a component of a consumer test is a viable sensory tool and should be used when conducting consumer evaluation of beef and beef products.

REFERENCES

AMSA (1978) *Guidelines of Cookery and Sensory Evaluation of Meat*. American Meat Science Association, National Live Stock and Meat Board, Chicago, Illinois.

AMSA (1995) *Research Guidelines for Cookery, Sensory Evaluation and Instrumental Tenderness Measurements of Fresh Meat*. American Meat Science Association, National Live Stock and Meat Board (formerly Chicago, Illinois), Savoy, Illinois.

ASTM (1978) *Compilation of Odor and Taste Threshold Values Data*. American Society for Testing and Materials, Philadelphia.

Bhumiratana, N., Vazquez-Araujo, L., Adhikari, K., Chambers IV, E., Philip, C. & Miller, R.K. (2011) Differentiation of beef flavor across muscles and quality grades. *Journal of Sensory Studies* (submitted).

Civille, G.V. & Lyons, B.G. (1996) *Aroma and Flavor Lexicon for Sensory Evaluation: Terms, Definitions, References, and Examples*. ASTM Data Series Publication DS 66, ASTM West Conshohocken, Pennsylvania.

Johnsen, P.B. & Civille, G.V. (1986) A standardized lexicon of meat WOF descriptors. *Journal of Sensory Studies*, **1**, 99.

Meilgaard, M., Civille, G.V. & Carr, B.T. (1999) *Sensory Evaluation Techniques*, 3rd edn., CRC Press Inc, Boca Raton, Florida.

13 Beef Quality and Tainting

Leo M.L. Nollet

Abstract: The processes associated with the slaughter and fabrication of beef carcasses can contaminate the products with a variety of microorganisms. There are several interventions that can be implemented into the production chain that will effectively reduce the rate of product deterioration. Methods such as chilling, freezing, packaging, organic acid sprays, and steam treatments are commonly used by the beef industry today. Through these intervention steps, tainting or further spoilage can be reduced.

Keywords: Beef quality, beef tainting, spoilage, refrigeration, freezing, microbial tainting

13.1 INTRODUCTION

The quality characteristics of beef, as with any other meat, are very important to the ultimate eating satisfaction of the consumer. Various factors such as tenderness, juiciness, color, and flavor may be influenced by changes occurring during the conversion of muscle to meat. Similarly, the postslaughter fabrication and handling steps by the muscle foods processor can affect the ultimate eating quality of fresh and previously frozen beef products. Beef, as with most fresh meats, provides an excellent medium for the support of microbial growth. Microbial growth if permitted to continue will render meats unacceptable to consumers. The ultimate goal of the beef industry is to produce high quality products that consumers will enjoy and purchase again. The subject of tainting or spoilage is therefore of critical concern to the muscle foods industry. Muscle deterioration begins soon after the process of slaughter as the result of microbial, chemical, and physical processes. These processes must be controlled or possibly eliminated, or they will render beef unfit for consumption. As this chapter progresses, we will examine the causes and consequences of beef tainting in fresh and frozen product. Furthermore, the discussion will include various protective measures that can be used to produce top quality beef products.

When beef is evaluated for its sensory characteristics, all of the previous factors and other possible factors should be considered. The material that follows is not a review of the scientific literature, rather a general compilation of available information. The author acknowledges the vast contributions made to this topic by a variety of researchers and textbook authors. There is a list of several references provided from which the information contained in this chapter was gleaned so that the reader can investigate this topic further if they so desire.

Handbook of Meat, Poultry and Seafood Quality, Second Edition. Edited by Leo M. L. Nollet.

13.2 TAINTING/SPOILAGE INTERVENTIONS

The extension of shelf life is essential for storage of most fresh and processed meat products. By any standard, fresh meat must be considered to be one of the more perishable foods. Preventative measures to stop tainting and spoilage of muscle foods must be applied quickly after slaughter. The means of meat deterioration are diverse and include several processes.

Meat preservation involves application of measures to delay or prevent certain changes, which make meat unusable as food or which downgrade quality characteristics. Most of the edible tissues of healthy animals up to the time of slaughter are either sterile or contain very low concentrations of microorganisms. Postmortem muscle experiences a sudden cessation of the body's defenses against invasion and growth of foreign microorganisms. Apparently, there are very few if any residual bacteriostatic or bacteriocidal properties in the muscles of freshly slaughtered animals. The process of hide removal, evisceration, and carcass fabrication operations can result in microbial contamination of the meat. Perhaps the most universal preventative measure against tainting in beef or any fresh meat is postmortem carcass cooling or refrigeration, and subsequent freezing.

13.3 REFRIGERATION AND FREEZING

Once the slaughter process is completed, the internal temperatures of animal carcasses generally range between 30°C and 39°C. This postmortem body heat must be removed during initial chilling. As a general rule, the internal temperature of the thickest portion of the carcass should be reduced to 5°C or less. Beef carcasses are initially chilled in blast (forced air) coolers or chill tunnels at temperatures generally ranging from −10°C to 0°C. If the temperature is reduced below −2°C, the carcasses will begin to freeze, which changes the physical state of the muscle tissue as well as endogenous enzymatic and chemical changes. There are some considerations with regard to carcass chilling. These factors include specific heat of the beef carcass, carcass size, amount of external fat (which acts as insulation), and the temperature of the chilling environment. Due to the fat cover of most beef carcasses, the chilling process may take up to 48 h or longer to reach an internal chilled temperature of 5°C or lower.

The amount of shrinkage or moisture loss incurred during initial chilling must be held to a minimum for both economic reasons and quality retention. Given adequate cooling of hot beef carcasses and cuts, the deterioration of fresh chilled beef is generally due to surface changes. The natural surface consists of fat and connective tissue, and during cooling the consistency of the latter changes so that further loss of water by evaporation is restricted. Conversely, muscle surfaces continue to lose water at a fairly rapid rate, and this desiccation leads to an increased concentration of salts at the surface, which causes oxidation of the muscle pigment to a brown color of various intensities.

Once beef is removed from the cooler, moisture tends to condense on the cool surfaces, especially when the relative humidity of the atmosphere is high, a phenomenon known as "sweating." This effect can promote microbial growth, if the product is temperature abused (−4.44°C) for periods of about 2 h or longer. It is important to consider that effective refrigeration along with various protective coverings allow longer storage or shipping of beef. Most beef carcasses are fabricated into subprimal cuts that are vacuum packaged and boxed for storage and shipping. Proper packaging materials for fresh retail cuts are important to maintain desirable color. The eye-appealing bright cherry-red color most retail fresh beef displays is dependent on several variables including proper packaging and display temperatures. Additionally, proper packaging can help prevent contamination during retail display.

The duration for which fresh meat can be stored in refrigeration depends on certain characteristics of the muscle. For example, pork, poultry, and fish possess more highly unsaturated fats than beef or lamb and tend to be more susceptible to the development of oxidative rancidity or "warmed-over" flavor. The length of time meat may be kept safely under refrigerated storage in the home is determined by previous handling conditions and refrigerator temperature. However, even under ideal home refrigeration temperatures, fresh meat should be consumed within about 4 days of purchase to not compromise full sensory quality. To prevent further quality deterioration and subsequent spoilage, fresh meat that will not be consumed within this time frame should be frozen. Some deterioration will occur during slow freezing in home refrigerators, but such losses in quality are preferable to bacterial spoilage and discoloration, which would render the beef product useless.

The effectiveness of freezing in reducing the rate of muscle spoilage has long been understood. The advantages of temperatures below the freezing point of fresh meat, about 28°C, are recognized as an excellent method for slowing the rate of spoilage. The process of freezing at a rapid rate has little detrimental effect on flavor, color, odor, or juiciness of meat after cooking. However, over time in frozen storage, a gradual decrease in the sensory qualities of odor and flavor can develop. Several questions should be answered when assessing the ultimate quality of any frozen fresh meat. The manner of freezing, the reason for freezing, the kind of packaging, and the time in frozen storage all affect quality characteristics during frozen storage. Generally, good quality can be maintained for several months (4–12) of frozen storage if some precautions are observed. Tightfitting and moisture-proof packaging and consistent storage temperatures (0°C or lower) are important factors.

Frozen storage that is prolonged in improperly packaged beef products can result in "freezer burn" or surface dehydration. This phenomenon is the result of water loss from the affected surfaces. The resulting appearance will manifest itself as gray patches on the products surface. Variations in frozen storage temperatures can accelerate the onset of freezer burn. This condition can cause accelerated lipid oxidation or the warmed-over flavor most consumers would find objectionable. Surfaces that are desiccated provide sufficient surface area for interaction with oxygen that leads to volatile flavor compounds that produce objectionable odors and flavors.

The development of oxidative rancidity can be of concern during frozen storage. This is generally recognized as a sensory problem with regard to odor and flavor. Frozen storage does not prevent but rather slows the onset of oxidative rancidity. Generally, beef stored in lower frozen temperatures will manifest slower rates of rancidity onset. Oxidative rancidity occurs most rapidly from −2°C to −4°C and almost ceases below −30°C. Fresh pork and poultry meat are at a greater risk for rancidity development in frozen storage primarily due to the characteristics of the fat. Meats with greater degrees of unsaturation tend to develop this negative quality attribute faster than more saturated meats such as beef.

13.4 MICROBIAL TAINTING

Beef, as are all fresh meat products, is a very perishable commodity. It is generally agreed that the internal tissues of healthy slaughter animals are free of bacteria at the time of slaughter. However, when the cut of fresh meat arrives at the retail level, especially if it is in ground form (i.e., hamburger), there can be varying numbers and types of microorganisms found. There can be many avenues by which the fresh beef cut becomes contaminated with microorganisms.

After being stunned and hoisted, beef animals are exsanguinated by severing the jugular vein with what is generally referred to in industry as the "stick knife." A dirty stick knife can introduce organisms into the bloodstream, before cessation of heart activity, where they can be deposited in some areas of the carcass. Organisms from the beef hide may also be deposited onto the carcass or onto freshly cut surfaces. Bovine manure and other organic material tend to remain on the hides of even the most well-kept animals. Transfer of these indigenous organisms can cause contamination. Additionally, some hide microorganisms can become airborne and contaminate adjacent carcasses or cut surfaces. Punctures in the gastrointestinal tract may deposit large inoculums of microorganisms onto the surface of beef carcasses. Meat plant workers can also be a source of bacterial contamination during the slaughtering and fabrication procedures. However, it must be stressed that as the carcass leaves the slaughter floor, there are several steps the beef industry currently has in place to minimize the chance of fecal or other contamination to the beef carcass. These procedures include steps such as steam vacuuming, passing carcasses through steam cabinets, real-time identification of organic residue by biophotonic technique, spraying the carcasses with bactericidal organic solutions, and washing with hot water. Another concern is the placing of fresh beef cuts into nonsterile containers, which inoculate the cut with whatever organism may be present in the container. This can be an effective source of contamination of ground or minced meats.

After slaughter, and subsequent slaughter floor decontamination procedures, beef carcasses are thoroughly chilled and later moved into the food distribution channels. Carcass fabrication procedures to render primal and retail cuts allow increasingly higher numbers of microorganisms to contaminate the product surfaces. The fate of these microorganisms depends upon the ultimate use of the meat. Fresh meat has high water content corresponding to a water activity of approximately 0.99. This environment readily supports and even promotes the growth of contaminating microorganisms, which may be present on the meat surface. Prompt effective carcass chilling creates a selective environment and permits growth of only those microorganisms that are capable of growing at temperatures approaching the freezing point of meat. Vacuum packaging of meats in oxygen-impermeable films results in a further constraint that reduces the microorganism's ability to grow and proliferate.

Microbiological tainting of fresh meats can involve many different organisms. Viruses, molds, yeasts, and bacteria all factor in the spoilage of fresh meat products. Viruses are very small organisms that do not usually contribute to meat spoilage but may be harmful for meat plant workers or consumers. Molds are multicellular organisms that display a variety of colors and are typically recognized by their "fuzzy," cotton-like appearance. Molds can develop numerous very small spores that are spread by air currents and other means, which causes mold growth to appear on other meat surfaces. Yeasts, as mold spores, can be spread through the air or by other means and will contaminate meat and equipment surfaces wherever they settle. Bacteria are unicellular and vary in appearance from elongated and short rods to spherical or spiral forms.

Fresh meat begins to undergo change from the moment of slaughter. The tissues no longer respire due to oxygen deficiency, and reserve glycogen will be metabolized beneath the tissue surface via fermentative pathways, which results in the accumulation of lactic acid and concomitant lowering of the pH. The extent of this pH drop will mediate the ability of the tainting organisms to develop, grow, and proliferate on the carcass surface. Therefore, the spoilage organisms present on dark-cutting beef (high pH) should be different from those present on pale soft and exudative pork (low pH).

Meat products for the most part require refrigerated temperatures to maintain sensory integrity. Therefore, organisms that can grow at refrigerated temperatures cause most types of meat spoilage. These spoilage organisms are generally referred to as psychrotrophs. The problematic attribute of these bacteria is their ability to grow at refrigerated temperatures (−5°C). Different organisms have different abilities to render meat unusable for consumption. Usually the first manifestation to the consumer is some type of visible defect. These defects may include milky exudates, excessive free liquid in the vacuum or over-wrapped package, slime, and generalized off odors. Additionally, there can be discoloration that can be green, yellowish, or brown. Moreover, there are other organisms that can produce quantities of organic compounds such as amines, esters, and acetoin that are further regarded as possessing high spoilage potential.

Microbial spoilage is generally considered either aerobic or anaerobic, depending on the conditions under which it occurs. This also depends on what type of organism causes the spoilage. Slime formation, unappealing flavors and odors, and color changes usually occur as the result of contamination with aerobic bacteria and yeasts. Environmental conditions such as temperature and water activity dictate which species of bacteria or yeasts cause slime formation. Many of these bacteria can produce oxidizing compounds, which then convert oxymyoglobin or the bright cherry-red color of fresh beef to unattractive gray, green, or brown colors.

Spoilage will develop with certain variations, depending on the species of bacteria that are predominant. The following discussion will concentrate on specific groups of organisms that are grossly responsible for tainting and spoilage of beef.

After the slaughter processes are completed, most carcass surfaces tend to have high microbial populations of fecal and soil origin. The majority of these organisms are nonpathogenic. These organisms are referred to as spoilage bacteria, because their presence subsequently leads to diminished sensory qualities such as appearance and odor. Spoilage organisms that thrive in refrigerated temperatures are classified as psychrophiles. These organisms grow best at temperatures lower than 20°C. There are several genera of this type of bacteria that can be present and grow on carcass and cut surfaces at refrigerated temperatures. These include *Pseudomonas, Moraxella*, *Psychrobacter*, and *Acinetobacter. Aeromonas, Altermonas, Shewanella, Micrococcus, Lactobacillus, Streptococcus, Leuconostoc, Pediococcus, Flavobacterium*, and *Proteus* may accompany these bacteria, in smaller numbers.

Fresh beef is susceptible to fungal spoilage by the following genera of molds: *Thamnidium, Mucor*, and *Rhizopus*. These molds are prone to develop "whiskers" on beef; *Penicillium* produces green patches; *Sporotrichum* and *Chrysosporium* produce "white spots"; and *Cladosporium* is a common cause of "black spots" that can be seen on aged beef carcasses. Storage temperatures below −5°C apparently inhibit the growth of molds. Yeasts that have been recovered from refrigerator-spoiled beef include the genera *Candida* and *Rhodotorula*. In contrast to spoilage that can occur on beef carcasses, only bacteria spoil ground beef. The following generas are responsible for the most spoilage problems: *Pseudomonas, Alcaligenes, Acinetobactor, Moraxella*, and *Aeromonas*.

Pathogenic bacteria are those that cause illness. There are two types of illness associated with food poisoning—food intoxication and food infection. Food intoxication results from the consumption of bacterial toxins produced in the food, while food infection occurs as the result of consumption of bacteria, which then produce toxins in the alimentary system. Food that appears perfectly appetizing can contain bacterial toxins that are relatively odorless and tasteless, and therefore readily consumed.

Currently, there is considerable effort being made to implement a system to track beef products from the farm, processing plant, retailer, and finally to the table. Currently, most beef products, if recalled due to foodborne illness, can be traced back to the processor, but not to the producer or individual animal. Most foodborne illness linked to beef consumption are associated with ground beef. However, fresh beef tainting can also encompass all other beef cuts.

Most pathogenic organisms associated with fresh beef tainting live in the intestinal tract of healthy animals. The more prevalent pathogenic organisms are:

- *E. coli* 0157:H7: This pathogen occurs in the gastrointestinal system of beef animals. The early 1980s saw this organism emerge as a source of serious illness in humans who had eaten undercooked ground beef. *E. coli* 0157:H7 causes acute bloody diarrheas, abdominal cramps, and hemolytic uremic syndrome, which may develop into chronic kidney failure.
- The typical mode of product contamination occurs during the skinning and eviscerating part of the slaughter process. Current technology to detect the disease in humans has improved, thus increasing the number of reported cases. It should be noted that increased frequency of testing at the processing plant and strict sanitation and hygiene programs provide the consumer with the safest beef products possible.
- *Salmonella*: *Salmonella* has been recognized as a primary cause of gastroenteritis in humans for over 100 years. This organism is widely dispersed with humans and animals being the primary reservoirs. A recent report of a study discovered 45% of the rumen contents of healthy cattle were found to contain *salmonella*. The Food Safety and Inspection Service (FSIS) of the USDA has started a *salmonella* testing program for slaughter plants and ground beef processors.
- *Campylobacter*: This organism may be the greatest cause of acute bacterial diarrheas in humans. Meat animals such as beef carry the organism in their intestinal tracts. These bacteria can contaminate many types of raw meats, and is no longer considered an emerging pathogen. Advances in technology have made it more effective to test for *Campylobacter*; thus, control of this pathogen can be enhanced.

In 2009, two interesting works were published about analysis techniques of a great number of parameters in muscle foods (Nollet & Toldrá 2009a) and processed meats and poultry (Nollet & Toldrá 2009b).

Fung (2009) enumerated methods for evaluating microbial flora in muscle foods. Different sample preparation methods, total viable cell counts, identification methods, immunological methods, biomass measurements, genetic testing, and biosensors for microbes are fully discussed.

Duffy *et al.* (2009) give an overview of methods to predict spoilage of muscle foods. An analogue overview of methods for processed meats is given by Guerrero-Legarreta (2009).

A review of microbial foodborne pathogens on raw meat is presented by Mataragas *et al.* (Mataragas & Drosinos 2009; Mataragas *et al.*, 2009).

13.5 SUMMARY

In summary, tainting or microbial spoilage can result in slight or catastrophic reduction in beef quality characteristics, especially odor and flavor. The processes associated with slaughter and fabrication of beef carcasses can contaminate the products with a variety of microorganisms. There are several interventions that can be implemented into the production chain that will effectively reduce the rate of product deterioration. Methods such as chilling, freezing, packaging, organic acid sprays, and steam treatments are commonly used by the beef industry today. Through these intervention steps, tainting or further spoilage can be reduced so the consumer can enjoy the attributes of a most nutritious food.

REFERENCES

Duffy, G., Dolan, A. & Burgess, C.M. (2009) Methods to predict spoilage of muscle foods. In: *Handbook of Muscle Foods Analysis* (eds L.M.L. Nollet & F. Toldrá). CRC Press, Boca Raton, Florida.

Fung, D.Y.C. (2009) Methods for evaluating microbial flora. In: *Handbook of Muscle Foods Analysis* (eds L.M.L. Nollet & F. Toldrá). CRC Press, Boca Raton, Florida.

Guerrero-Legarreta, I. (2009) Spoilage detection. In: *Handbook of Processed Meats and Poultry Analysis* (eds L.M.L. Nollet & F. Toldrá). CRC Press, Boca Raton, Florida.

Mataragas, M. & Drosinos, E.H. (2009) Microbial foodborne pathogens. In: *Handbook of Processed Meats and Poultry Analysis* (eds L.M.L. Nollet & F. Toldrá). CRC Press, Boca Raton, Florida.

Mataragas, M., Kagli, D. & Nychas, G.-J.E. (2009) Microbial foodborne pathogens. In: *Handbook of Muscle Foods Analysis* (eds L.M.L. Nollet & F. Toldrá). CRC Press, Boca Raton, Florida.

Nollet, L.M.L. & Toldrá, F. (2009a) *Handbook of Muscle Foods Analysis*. CRC Press, Boca Raton, Florida.

Nollet, L.M.L. & Toldrá, F. (2009b) *Handbook of Processed Meats and Poultry Analysis*. CRC Press, Boca Raton, Florida.

FURTHER READINGS

Hui, Y.H., Nip, W.K., Rogers, R.W. & Young, O.A. (2001) *Meat Science and Applications*. Marcel Dekker, New York.

Jay, J.M. (2000) *Modern Food Microbiology*, 6th edn. Aspen Publishers, Inc., Gaithersburg, Maryland.

Judge, M., Aberle, E., Forrest, J., Hedrick, H. & Merkel, R. (1989) *Principles of Meat Science*, 2nd edn. Kendall/Hunt Publishing Company, Dubuque, Iowa.

Lawrie, R.A. (1998) *Meat Science*, 6th edn. Woodhead Publishing Limited, Cambridge, U.K.

Pearson, A.M. & Dutson, T.R. (1986) *Advances in Meat Research*, Vol. 2, Meat and Poultry Micro-biology. AVI Publishing Company, Inc., Westport, Connecticut.

Price, J.F. & Schweigert, B.S. (1987) *The Science of Meat and Meat Products*, 3rd edn. Food and Nutrition Press, Inc., Westport, Connecticut.

Romans, J.R., Costello, W.J., Carlson, C.W., Greaser, M.L. & Jones, K.W. (2001) *The Meat We Eat*, 14th edn. Interstate Publishers, Inc., Danville, Illinois.

14 Microbiological and Sensory Properties of Beef

Jack Thomas

Abstract: Different sources of bacteria found in beef are reported. The bacteria may come in contact with beef tissues via immobilization, exsanguination, hide removal, evisceration, handling, and further processing into wholesale and retail cuts.

One of the most effective methods to prevent the growth of pathogenic bacteria is to store meat at cold temperatures. Vacuum packaging is also an effective method to control bacterial growth in beef. Other aids to reduce contamination are washing, decontamination with organic acids, and ionizing radiation. Finally vacuum packaging and modified atmospheric packaging are techniques playing a role in preservation and safety of meat products.

Keywords: Meat preservation, bacterial growth, vacuum packaging, modified atmospheric packaging

14.1 INTRODUCTION

While beef has always been an important and popular mainstay of our diet, its popularity has reached new levels in recent years. Demand for beef steadily increased from the beginning of the twentieth century until about 1976 (USDA Agricultural Fact Book 2002). Then high production costs and competition from other animal proteins, most notably chicken, lessened the demand for beef. Beef needed to reposition itself in the market. Through the joint efforts of producers, meat companies, and researchers, intense activities focused on returning beef to the center of the dinner plate. New consumer-friendly products were, and continue to be, developed, but most importantly, the beef industry renewed its focus on quality. Marketing efforts that promoted both new products and high quality traditional beef cuts brought consumers back to beef. They rediscovered how flavorful beef can be. As a result, demand for beef has increased dramatically in the past few years. Much of this growth has been in the food service industry, especially in the white tablecloth restaurant trade. Also, the recent popularity of high protein, low carbohydrate diets has helped maintain the high demand for beef.

The road to today's record high demand has not always been smooth. The 1993 outbreak of food-borne illness in the northwestern U.S. from *Escherichia coli* 0157:H7 (Centers for Disease Control and Prevention 1993) forced the industry to focus on the microbiological safety of beef. Since then, the beef industry has spent over $3 billion developing and implementing methods aimed at reducing the incidents of food poisoning arising from the consumption of beef. Through joint efforts of the federal government and industry, meat inspection was completely revamped with the passage of the Pathogen Reduction,

Handbook of Meat, Poultry and Seafood Quality, Second Edition. Edited by Leo M. L. Nollet.

Hazard Analysis and Critical Control Point Systems Act (United States Department of Agriculture, Food Safety and Inspection Service 1996). Hazard Analysis and Critical Control Point (HACCP) programs were developed for all meat slaughter and processing facilities. Because of this renewed focus on food safety and the success of HACCP, there has been a decrease in pathogenic bacteria in meat products. An added bonus has been a reduction in the general microbiological load on beef and a concomitant increase in shelf life of beef retail cuts. Additional efforts aimed at educating consumers on safe handling of food at home have also been important in reducing food-poisoning outbreaks.

14.2 SOURCES OF BACTERIA IN BEEF

The microbiology of beef has been studied for more than 100 years. Ayers (1955) wrote an excellent review on the microbiology of meat. He cited articles published as far back as 1886 where researchers were concerned about bacteria in beef. Obviously, the concern over the ability of bacteria in meat to possibly cause disease in consumers and waste through spoilage is not new. Ayers (1955) and other researchers (Sofos *et al.* 1999) have reported on the sources of bacteria found in beef. Prior to the slaughter process, the meat from a healthy animal should be sterile. It is through the processes of immobilization, exsanguination, hide removal, evisceration, handling, and further processing into wholesale and retail cuts that bacteria may come in contact with beef tissues.

The hides of cattle are loaded with bacteria (Ayers 1955, Sofos *et al.* 1999). While the total bacterial load on the hides is influenced by the amount of mud and manure present and time of year, suffice it to say that the external surface of beef hides carries a high bacterial load and also a wide variety of bacterial species. Even with extremely careful hide removal, some of these bacteria will be transferred to the exposed surfaces of the carcass. The transfer can come from direct contact of the exterior surface of the hide with the carcass or indirectly through workers and equipment. Also, since the kill-floor environment is generally warm and humid, bacteria that become airborne may survive and later contact exposed carcass surfaces (Sofos *et al.* 1999).

The intestinal tract of cattle is also loaded with bacteria. Bacteria from the intestinal tract may come in contact with the carcass surface if the intestinal tract is ruptured during evisceration. Also, feces voided during slaughter and rumen contents spilled through the esophagus may be additional sources of contamination. After the carcass is chilled, further contamination may occur during the processing into wholesale and retail cuts. This contamination will likely be from the transfer of bacteria on the carcass surface to newly exposed lean surfaces. Also, the possibility exists of freshly cut surfaces being contaminated by equipment and workers.

Ayers (1955), Savell *et al.* (1981, 1986), and Dixon *et al.* (1991) all reported on the microflora of beef carcasses. Generally speaking, numerous bacterial species may initially be present and often in high numbers. Fortunately, most bacterial species found in beef are relatively benign, and do not survive postmortem preservation practices. When it comes to pathogenic and spoilage bacteria of importance, relatively few species are responsible for most damage.

The pathogens of most concern are *Campylobacter jejuni*, Salmonella spp., *Escherichia coli* 0157:H7, and *Listeria monocytogenes*. Fortunately, these bacteria are only present in very low counts on a limited number of carcasses. Unfortunately, very few of these bacteria are necessary to cause sickness, and possibly death, when people consume contaminated products. Therefore, methods to control bacterial growth on beef must consider the growth requirements of these major pathogens. Since pathogen control is of utmost importance, preservation often involves altering the environmental conditions under which beef is stored. The controlled storage conditions inhibit pathogenic bacterial growth but may allow growth of some spoilage bacteria. The spoilage bacteria that usually end up dominating the microflora are Pseudomonas spp. or Lactobacillus spp.

14.3 METHODS OF PRESERVATION

In addition to requiring an adequate nutrient supply, bacteria's ability to grow depends on their environment. Variations in temperature, oxygen, pH, and water availability, and other environmental factors will determine not only how well bacteria can grow but also which species are capable of growing under those conditions. Obviously, beef meets the nutrient requirements of most bacteria. Although somewhat limited, it does supply carbohydrates, which many bacteria utilize for energy. More importantly, it has an abundant supply of protein and bacteria that can metabolize amino acids (putrefactive bacteria) find beef to be an especially favorable growth medium. Beef is also high in water and has a pH (pH approximately 5.7), which is well within acceptable limits for most bacteria.

While it is desirable to limit the growth of all bacteria in fresh beef, controlling the growth of pathogenic bacteria is of primary importance. Many of the pathogenic bacteria are mesophilic (body temperature) and require temperatures between 10°C and 40°C for optimal growth. Obviously, one of the most effective methods to prevent the growth of many pathogenic bacteria is to store meat at cold temperatures. It is common in developed countries to store meat at temperatures of 22–2°C as soon after slaughter as possible. While pathogenic bacteria may still be present, their ability to grow is greatly inhibited by the cold temperature and the competition for nutrients from bacteria that can grow at refrigeration temperatures. Mechanical refrigeration has been in widespread use in industry for more than 100 years. However, it was only after World War II that mechanical refrigeration became readily available at the household level. Unfortunately, refrigeration is still not available in most of the developing countries of the world. Anyone who has visited the "wet" markets in Indonesia, China, Africa, or other developing areas notices meat products being stored at ambient temperatures. The only recourse is to consume this meat before the pathogenic bacteria have time to grow to dangerous levels and to very thoroughly cook the meat prior to consuming.

During refrigerated storage, Pseudomonas spp. becomes dominant because they are the most capable of growing at refrigeration temperatures under aerobic conditions (Newton & Rigg 1979, Savell *et al.* 1986). Shelf life depends on the initial psychrotrophic (cold-loving bacteria) load and storage temperature (Greer & Jeremiah 1981). These workers reported a significant inverse relationship between retail display case blower temperature and bacterial generation time (temperatures close to 0°C greatly decreased bacterial growth rate). Chandran *et al.* (1986) also reported lower levels of bacteria on steaks stored in the dark at 0–1°C versus steaks stored in lighted display at 2–5°C.

Refrigeration was the primary method utilized to preserve fresh meat until the 1960s. This is when vacuum packaging of wholesale and primal cuts of beef began. While the development and acceptance of vacuum packaged beef cuts took many years, and the process is somewhat complicated, the concept is simple—limit the oxygen available at the meat surface during refrigerated storage and the ability of Pseudomonas spp. (or other aerobic psychrotrophs) to grow is greatly limited. Lactobacillus spp. then becomes the dominant microflora (Newton & Rigg 1979, Savell *et al.* 1986, Dixon *et al.* 1991). The advantage is that Lactobacillus spp. grow very slowly at refrigeration temperatures and this greatly extends the storage life of beef. Newton and Rigg (1979) reported an inverse relationship between oxygen permeability of meat packaging film and growth of Lactobacillus spp. Gill and Penney (1988) reported that putrid spoilage caused by aerobic bacteria was reduced by increasing CO_2 concentrations. Even though lactic acid bacteria will eventually dominate within the vacuum package, their metabolic by-products are not as offensive as the by-products produced by putrefying aerobic psychrotrophs.

As effective as refrigeration and vacuum packaging have been at controlling bacterial growth in beef, they are not perfect. With the food poisoning outbreaks from *E. coli* 0157:H7 and continued problems with pathogenic and spoilage bacteria, the beef industry began searching for other intervention methods to further reduce the bacterial numbers in beef.

One obvious step to reduce carcass contamination is to reduce the bacterial load on the exterior of cattle prior to them entering the processing room. Processing plants routinely wash the live animals to eliminate the visible hide contamination (manure and mud). Although washing does not remove all bacteria on the hides, it does reduce the number. Also, eliminating manure and mud reduces the possibility of contaminating equipment and the carcass surface during hide removal. Bowling and Clayton (1992) patented a process to chemically remove hair from cattle hides. The process is applied after exsanguination and prior to the start of hide removal from the carcass. Schnell *et al.* (1995) reported that chemical dehairing reduced visible contamination (hair and feces) on carcasses. Castillo *et al.* (1998) found that chemical dehairing was effective at reducing levels of inoculated Salmonella spp., *Escherichia coli* 0157:H7, and Listeria monocytogenes.

Reagan *et al.* (1996) deliberately contaminated beef carcasses with bovine fecal matter at a level of 104 cfu/cm2 to evaluate the efficacy of washing, trimming, hot water washing (74°C–87.8°C), and rinsing with 0.3–2.3 ppm ozone or 5% hydrogen peroxide solution at reducing bacterial loads. All intervention steps reduced the numbers of bacteria. However, trimming and washing or hot water rinsing were the most effective. In a similar study, Gorman *et al.* (1995) found hot water rinsing was as effective at reducing bacteria numbers as chemical treatments consisting of 5% hydrogen peroxide, 0.5% ozone, 12% trisodium phosphate, or 2% acetic acid. Hardin *et al.* (1995) evaluated washing carcasses with 35°C water followed by rinsing with either a 2% lactic acid or 2% acetic acid solution. Again, all treatments were effective but lactic acid spray was superior. However, it should be noted that the water temperature used was considerably lower than that used in the previous studies (Gorman *et al.* 1995, Reagan *et al.* 1996). Phebus *et al.* (1997) reported on the use of steam pasteurization, steam vacuuming, trimming, washing, and lactic acid to reduce bacterial loads on beef carcasses. They reported all treatments effectively lowered bacterial counts, but were most effective when used in combination. Strict sanitary procedures and a 55°C, 1% lactic acid solution was used by Dixon *et al.* (1991) to effectively reduce the bacteria numbers on beef and extend the shelf life of steaks manufactured from these carcasses. Other researchers (Quartey-Papofio *et al.* 1980, Woolthuis & Smulders 1985, Hamby *et al.* 1987, Prasai *et al.* 1991) have shown reductions in bacterial counts on beef carcasses by treating with lactic and/or acetic acid. Acuff *et al.* (1987) concluded that decontamination with organic acids is most effective when done immediately after slaughter. Other chemical applications that have shown effectiveness at reducing bacterial numbers on beef carcasses include acidified sodium chlorite and activated lactoferrin. Castillo *et al.* (1999) reported sodium chlorite activated with citric acid reduced counts of inoculated *Escherichia coli* 0157:H7 and *Salmonella typhimurium* 4.5 log cycles on beef cuts removed on the slaughter floor. This treatment was more effective than phosphoric acid-activated sodium chlorite or washing with water. Acetic acid solutions are corrosive to equipment and can pose a safety hazard. Acidified sodium chlorite is more user friendly. Throughout the beef industry, most microbial interventions will consist of an acid treatment (lactic, acetic, or acidified sodium chlorite) in combination with steam vacuuming and/or steam pasteurization.

Lücke (2000) reported on the practice of inoculating meat with lactic acid-producing bacteria, commonly referred to as probiotics, to control the growth of spoilage and pathogenic bacteria in meat. Lactic acid bacteria do not compete very well with gram-negative pseudomonas and therefore are not likely to be effective in preserving aerobically packaged fresh meat. However, probiotics may be beneficial in vacuum-packaged meat. Strains of lactic acid bacteria that are only slightly putrefactive may be used to suppress the growth of naturally present, highly putrefactive lactic acid bacteria.

Ionizing radiation also is effective at reducing microbial numbers in meat. However, it is not currently feasible to apply this to a beef carcass. Currently, its greatest promise is in reducing bacteria numbers in frozen ground beef patties.

The beef slaughter, processing, and retail industries operate on very tight profit margins. Each industry segment is continually searching for ways to increase their production efficiencies. An obvious way to increase efficiency at the retail level is to process vacuum-packaged primal cuts into packaged retail cuts at a central location and then distribute these retail-ready cuts to stores for consumer purchase. This system allows concentration of skilled labor into one or perhaps a few locations. Skilled labor is not needed at the retail store, since the only preparation necessary is to remove the packaged retail ready cuts from their shipping container and place them in the display case. In addition to improved labor efficiency, centralized processing offers the potential for increased quality control, including improved sanitation. With all of these advantages, it would seem that all beef offered for retail sale would be centrally processed. However, the shelf life of conventionally packaged (foam tray over wrapped in oxygen-permeable film) retail beef is too short for the process to be successful. After processing and packaging, the cuts still have to be shipped and possibly stored at the retail market before being placed in the display case for sale. Once in the display case, the cuts need to maintain their consumer acceptability for 3–5 days. According to Tewari *et al.* (1999), centrally processed and packaged retail cuts must be able to maintain their quality for 20–30 days of storage prior to being placed in the retail display case.

Even though vacuum packaging offers increased storage time, it is usually applied only to wholesale or primal cuts. It is not an acceptable method of packaging for individual retail cuts. Because of the reduced oxygen levels during vacuum storage, the meat pigment myoglobin remains in the unoxygenated state. The dark color, usually described as purple, associated with unoxygenated myoglobin is unacceptable to consumers. Also, there is formation of oxidized myoglobin, referred to as metmyoglobin. Metmyoglobin is brown in color and is not acceptable to consumers. Oxidation of myoglobin to metmyoglobin naturally happens during storage and retail display of beef. It will usually be the first quality defect noted during retail display and is not associated with microbial activity. If small amounts of metmyoglobin are formed during vacuum storage, the metmyoglobin-reducing activity of the muscle tissue may convert it to myoglobin upon exposure to oxygen. When the meat surface is exposed to oxygen, oxymyoglobin is formed. Oxymyoglobin is the pigment responsible for the bright red color that consumers associate with high quality beef. Altering atmospheric conditions under which the meat is stored so that high quality appearance is maintained while bacterial growth is limited is necessary for the concept of centralized processing to be realized. The process utilized is generally referred to as modified atmospheric packaging (MAP).

With MAP, the atmosphere can be modified within each package or a number of conventionally packaged cuts can be placed in a container, which then has its atmosphere modified. Regardless of the exact method used, the modified atmosphere will usually contain a high level of CO_2 to aid in inhibiting bacterial growth. In systems where MAP is applied to a container holding several cuts, the cuts will be wrapped in an oxygen-permeable film so that when they are removed from the container oxygen can interact with myoglobin. One disadvantage to this system is the necessity for a large container capable of maintaining a high CO_2 concentration. More importantly, depending on the amount of metmyoglobin formed at the meat surface during storage, it may take up to 2 days for oxymyoglobin to be formed after exposure to air (O'Keefe & Hood 1982). That makes this system difficult to employ at retail.

Systems where MAP is applied to each package usually have an atmosphere containing both a high level of CO_2 and O_2. This allows for inhibition of bacterial growth while maintaining acceptable beef color. It should be noted that this combination of CO_2 and O_2 is a compromise and the shelf life of cuts packaged under these conditions are not as great as steaks stored under high CO_2 and/or low O_2. Gill and Jones (1994) reported steaks packaged in CO_2 plus O_2 atmospheres began to deteriorate in appearance after 12 days. Color degraded prior to achieving sufficient numbers of bacteria to have any flavor effects. They noted that the useful storage life of steaks stored in high CO_2 packages is limited by the onset of bacterial spoilage and this may take several weeks. Beef stored in high O_2 environments is also more susceptible to fat oxidation and the resultant undesirable flavors.

Since it is not currently possible to slaughter cattle aseptically, carcasses and the resultant wholesale and retail cuts will carry a bacterial load. These bacteria will eventually affect the odor, color, flavor, and overall acceptability of the meat. Additionally, intervention steps aimed at reducing the bacterial load must also be assessed not only for their ability to limit bacteria but also for their effect on sensory characteristics of beef.

Pseudomonas spp., which eventually dominate on aerobically packaged meat, are putrefactive. While these bacteria will utilize any available glucose for energy, they ultimately metabolize amino acids. This will lead to production of putrid odors, slime, and undesirable flavors. Greer and Jeremiah (1981) and Greer *et al.* (1983) showed that time for putrid odor production was dependent on initial psychrotrophic load and storage temperature. Nassos *et al.* (1988) reported that odor and appearance decreased in acceptability with increasing storage time. Newton and Rigg (1979) reported undesirable putrefactive odors were detected when Pseudomonas spp. reached levels of 106 cfu/cm^2. They found that putrid odors were detectable after 2 weeks of storage at 0°C in highly oxygen-permeable packages. However, meat from these packages was still acceptable in flavor after 11 weeks of storage. Savell *et al.* (1986) found steaks cut from beef knuckles stored in highly oxygen-permeable film scored low in color desirability even after 1 day of retail display. These low scores were associated with aerobic plate counts of 106 cfu/cm^2 and a high percentage of Pseudomonas spp. making up the microflora. Egan and Shay (1982) inoculated sterile, vacuum-packaged meat samples with a homofermentative strain of lactobacillus. They found that objectionable odors and flavor developed when lactobacillus levels reached 108 cfu/cm^2. Many other workers (Savell *et al.* 1986, Acuff *et al.* 1987, Dixon *et al.* 1991) have reported that vacuum packaging prevented growth of most bacteria and greatly increased storage time of beef primals. Dixon *et al.* (1991) showed retail steaks removed from loins and ribs stored for up to 80 days in vacuum packaging were acceptable in color and overall appearance for 2 days of retail display. Egan and Shay (1982) also reported that vacuum packaging greatly extended storage time; however, even vacuum-packaged sterile meat samples had a limited storage time due to development of undesirable flavors.

Quartey-Papofio *et al.* (1980) conducted a study designed to evaluate the effectiveness of potassium sorbate and acetic, formic, and propionic acids as meat sanitizers. They evaluated acid concentration, pH, and exposure time on 11 different microorganisms. While acetic acid at pH 1.0 was more effective than acetic acid at pH 2.0, they concluded the high amount of hydrogen chloride (HCl) required to lower the pH was impractical. All further tests were conducted at normal pH and at pH adjusted to 2.0. Potassium sorbate, either alone or in combination with the other acids, offered little benefit in reducing bacterial numbers. They found 2% formic acid or a combination of 1% formic acid and 1% acetic acid were the most effective at eliminating bacteria. However, meat treated with formic acid (up to 2% solution) turned brown shortly after treatment. The degree of discoloration was believed to be dependent on concentration. They attributed this to oxidation of the iron in myoglobin and the resultant formation of metmyoglobin. To overcome this discoloration, they added 1% or 5% ascorbic acid to the 2% formic acid solution. The mixture of 1% ascorbic acid did not improve color. But, the addition of 5% ascorbic acid resulted in minimal discoloration and an improvement in antimicrobial activity. Bell *et al.* (1986) tested inoculated strips of beef by dipping them in various concentrations of acetic acid (0.6%, 1.2%, 1.8%, or 2.4%) or acetic acid with formic acid (0.6% acetic with 0.046%, 0.092%, 0.184%, or 0.23% formic acid). They found an increase in discoloration with an acetic acid concentration greater than 1.8% and with all acetic acid treatments exposed for 10 min versus 1 min. When 0.6% acetic acid was combined with formic acid concentrations of 0.046% or greater, color was less desirable than controls. Taste panelists were able to detect differences in flavor between control samples (dipped in distilled water) and samples treated with 1.2% acetic acid. However, panelists were not able to distinguish flavor differences between control samples and samples treated with 0.6% acetic acid combined with 0.046% formic acid.

Woolthuis and Smulders (1985) treated calf carcasses with up to 4.0% (vol/vol) lactic acid. Solutions up to 1.25% did not produce unacceptable color and solutions up to 2% had no effect on flavor of roasted Longissimus cuts. However, solutions of 4% resulted in a slightly sour taste. They also reported high concentrations of lactic acid created rusty brown spots on carcasses. They concluded that 1.25% lactic acid solutions could effectively be used to reduce bacterial numbers and have little or no effect on lean color and flavor. Mikel *et al.* (1996) treated beef strip loins with an equal mixture of 2% lactic acid and 2% acetic acid. They reported that muscle ultrastructure was degraded with the acid treatment. However, shear force value was not affected. They also found no differences in Hunter color values, fat oxidation, or cooking loss. Prasai *et al.* (1991) sprayed 1% lactic acid on carcasses immediately after hide removal, after evisceration, or at both times. Samples for bacterial enumeration were removed from the carcasses just prior to entering the chill cooler 0 h and after 72 h of chilled storage. In all 0 h samples and in all but one 72 h samples, lactic acid treatment reduced bacterial numbers. Generally, the greatest reductions in bacterial counts were on carcasses treated after evisceration or after a combination of both hide removal and evisceration. Treating after evisceration alone was more effective than treating only after hide removal. Loins removed from treated and control carcasses were stored for 14 days in vacuum package and evaluated for color, odor, and overall acceptability. No differences were detected for these sensory attributes. Hamby *et al.* (1987) treated carcasses with intermittent sprays of water, lactic acid, or acetic acid (alone and in combination) for up to 12 h following slaughter. They reported no differences in odor, color, or purge from treated and control samples. Acuff *et al.* (1987) removed strip loins from chilled beef carcasses that had not been treated with any decontamination procedure postslaughter. The loins were treated with either a 1% lactic acid solution; 1% acetic acid solution; or a solution containing 1% lactic acid, 2% acetic acid, 0.25% citric acid, and 0.1% ascorbic acid. None of the treatments had a consistent advantage over others for lean color, fat color, surface discoloration, or odor.

Microorganisms play an important role in storage, preservation, consumer acceptability, and safety of beef products. Undesirable changes in beef's appearance often occur prior to bacterial numbers reaching levels sufficient to cause spoilage. Nonmicrobial color changes are used by consumers to determine product freshness. Probably unknown to most consumers, this offers them a margin of safety between undesirable color and unacceptable bacterial numbers. However, pathogenic bacteria do not need to be present in large numbers to be dangerous and pose a human health hazard. Therefore, efforts to reduce and eliminate pathogens are necessary. Industry programs aimed at reducing pathogens have the added benefit of lowering numbers of all bacteria. This has not only led to beef being safer but also to increasing its shelf life. These changes have come without lowering of beef's appearance or eating qualities.

REFERENCES

Acuff, G.R., Vanderzant, C., Savell, J.W., Jones, D.K., Griffin, D.B. & Ehlers, J.G. (1987) Effect of acid decontamination of beef subprimal cuts on the microbiological and sensory characteristics of steaks. *Journal of Meat Science*, **19**, 217–226.

Ayers, J.C. (1955) Microbiological implications in the handling, slaughtering, and dressing of meat animals. *Advances in Food Research*, **6**, 109–161.

Bell, M.F., Marshall, R.T. & Anderson, M.E. (1986) Microbiological and sensory tests of beef treated with acetic and formic acids. *Journal of Food Protection*, **49**(3), 207–210.

Bowling, R. & Clayton, R.P. (1992) Method for dehairing animals. US Patent No. 5149295.

Castillo, A., Dickson, J.S., Clayton, R.P., Lucia, L.M. & Acuff, G.R. (1998) Chemical dehairing of bovine skin to reduce pathogenic bacteria and bacteria of fecal origin. *Journal of Food Protection*, **61**, 623–625.

Castillo, A., Lucia, L.M., Kemp, G.K. & Acuff, G.R. (1999) Reduction of *Escherichia coli* O157:H7 and *Salmonella typhimurium* on beef carcass surfaces using acidified sodium chlorite. *Journal of Food Protection*, **62**, 580–584.

Centers for Disease Control and Prevention. (1993) *Morbidity and Mortality Weekly Report*, **42**(14), 258.

Chandran, S.K., Savell, J.W., Griffin, D.B. & Vanderzant, C. (1986) Effect of slaughter-dressing, fabrication and storage conditions on the microbiological and sensory characteristics of vacuum-packaged beef steaks. *Journal of Food Science*, **51**(1), 37–39.

Dixon, Z.R., Acuff, G.R., Lucia, L.M., *et al.* (1991) Effect of degree of sanitation from slaughter through fabrication on the microbiological and sensory characteristics of beef. *Journal of Food Protection*, **54**, 200–207.

Egan, A.F. & Shay, B.J. (1982) Significance of lactobacilli and film permeability in the spoilage of vacuum-packaged beef. *Journal of Food Science*, **47**, 1119–1126.

Gill, C.O. & Jones, T. (1994) The display life of retail-packaged beef steaks after their storage in master packs under various atmospheres. *Meat Science*, **38**, 385–396.

Gill, C.O. & Penney, N. (1988) The effect of initial gas volume to meat weight ratio on the storage life of chilled beef packaged under CO_2. *Meat Science*, **22**, 53–63.

Gorman, B.M., Sofos, J.N., Morgan, J.B., Schmidt, G.R. & Smith, G.C. (1995) Evaluation of hand-trimming, various sanitizing agents, and hot water spray-washing as decontamination interventions for beef brisket adipose tissue. *Journal of Food Protection*, **58**, 899–907.

Greer, G.G. & Jeremiah, L.E. (1981) Proper control of retail case temperature improves beef shelf life. *Journal of Food Protection*, **44**, 297–299.

Greer, G.G., Jeremiah, L.E. & Weiss, G.M. (1983) Effects of wholesale and retail contamination on the case life of beef. *Journal of Food Protection*, **46**, 842–845.

Hamby, P.L., Savell, J.W., Acuff, G.R., Vanderzant, C. & Cross, H.R. (1987) Spray-chilling and carcass decontamination systems using lactic and acetic acid. *Journal of Meat Science*, **21**, 1–14.

Hardin, M.D., Acuff, G.R., Lucia, L.M., Oman, J.S. & Savell, J.W. (1995) Comparison of methods for decontamination from beef carcass surfaces. *Journal of Food Protection*, **58**(4), 368–374.

Lücke, F. (2000) Utilization of microbes to process and preserve meat. *Journal of Meat Science*, **56**, 105–115.

Mikel, W.B., Goddard, B.L. & Bradford, D.D. (1996) Muscle microstructure and sensory attributes of organic acid-treated beef strip loins. *Journal of Food Science*, **61**(5), 1058–1093.

Nassos, P.S., King, A.D. & Stafford, A. (1988) Lactic acid concentration as an indicator of acceptability in refrigerated or freeze-thawed ground beef. *Applied and Environmental Microbiology*, **54**(3), 822–823.

Newton, K.G. & Rigg, W.J. (1979) The effect of film permeability on the storage life and microbiology of vacuum-packed meat. *Journal of Applied Bacteriology*, **47**, 433–441.

O'Keefe, M. & Hood, D.E. (1982) Biochemical factors influencing metmyoglobin formation on beef from muscles of differing colour stability. *Meat Science*, **7**, 209–228.

Phebus, R.K., Nutsch, A.L., Schafer, D.E., *et al.* (1997) Comparison of steam pasteurization and other methods for reduction of pathogens on surfaces of freshly slaughtered beef. *Journal of Food Protection*, **60**(5), 476–484.

Prasai, R.K., Acuff, G.R., Lucia, L.M., Hale, D.S., Savell, J.W. & Morgan, B. (1991) Microbiological effects of acid decontamination of beef carcasses at various locations in processing. *Journal of Food Protection*, **54**(11), 868–872.

Quartey-Papofio, E.A., Marshall, R.T. & Anderson, M.E. (1980) Short-chain fatty acids as sanitizers for beef. *Journal of Food Protection*, **43**(3), 168–171.

Reagan, J.O., Acuff, G.R., Buege, D.R., *et al.* (1996) Trimming and washing of beef carcasses as a method of improving the microbiological quality of meat. *Journal of Food Protection*, **59**(7), 751–756.

Savell, J.W., Hanna, M.O., Vanderzant, C. & Smith, G.C. (1981) An incident of predominance of Leuconostoc sp. in vacuum-packaged beef strip loins—Sensory and microbial profile of steaks stored in O_2-CO_2-N_2 atmospheres. *Journal of Food Protection*, **44**(10), 742–745.

Savell, J.W., Griffin, D.B., Dill, C.W., Acuff, G.R. & Vanderzant, C. (1986) Effect of film oxygen transmission rate on lean color and microbiological characteristics of vacuum-packaged beef knuckles. *Journal of Food Protection*, **49**(11), 917–919.

Schnell, T.D., Sofos, J.N., Littlefield, V.G., *et al.* (1995) Effects of postexsanguination dehairing on the microbial load and visual cleanliness of beef carcasses. *Journal of Food Protection*, **58**, 1297–1302.

Sofos, J.N., Kochevar, S.L., Bellinger, G.R., *et al.* (1999) Sources and extent of microbiological contamination of beef carcasses in seven United States slaughtering plants. *Journal of Food Protection*, **62**(2), 140–145.

Tewari, G., Jayas, D.S. & Holley, R.A. (1999) Centralized packaging of retail meat cuts: A review. *Journal of Food Protection*, **62**, 418–425.

United States Department of Agriculture, Food Safety and Inspection Service. (1996) Procedures for the safe and sanitary processing and importing offish and fishery products: final rule. *Federal Register*, **61**(144), 38805–38989, docket no. 93N-0195.

United States Department of Agriculture. Office of Communications. (2002) *Fact Book*, For Sale by the U.S. Government Printing Office. Superintendent of Documents, Mail Stop: SOOP, Washington, District of Columbia 20402–9328.

Woolthuis, C.H. & Smulders, F.J.M. (1985) Microbial decontamination of calf carcasses by lactic acid sprays. *Journal of Food Protection*, **48**(10), 832–837.

15 Quality Measurements in Beef

Susan Brewer

Abstract: Different sensory attributes may be evaluated by different methods. Panelists should be trained in the proper techniques to conduct the evaluation. Data generated by sensory methods are statistically processed. pH, water-holding capacity, drip loss and cooking loss, and color are elements in the physical–chemical analysis. Near-infrared spectroscopy is used for the assessment of fat, moisture, and protein.

Texture analysis comprises Warner–Bratzler, tensile methods, penetrometer, texture profile analysis, and computer vision.

The analysis of fatty acids and the oxidative quality are elements of flavor analysis of beef.

Keywords: Beef sensory attributes, beef sensory evaluation, panelists, beef pH, water-holding capacity, drip loss, beef color, near-infrared spectroscopy, texture analysis, Warner–Bratzler, penetrometer, texture profile analysis, computer vision, beef fatty acids, oxidative quality

15.1 INTRODUCTION

Beef quality is ultimately defined by consumer acceptability. Visual characteristics (color, amount of fat, amount of visible water, textural appearance), tenderness, and flavor have significant impacts on consumer expectations of satisfaction with the product (Brewer *et al.* 1998, 2001a, b; NCBA 2001; Robbins *et al.* 2003b). A number of quality aspects of beef including tenderness, juiciness, intramuscular fat content, water-holding capacity (WHC), color, and flavor can be measured instrumentally and/or by sensory evaluation (Fischer *et al.* 1976; Honikel *et al.* 1981; Chambaz *et al.* 2003). The meat industry needs reliable meat quality information during the entire production process in order to guarantee high-quality meat products for consumers; however, they would like to use objective measures as soon postslaughter as possible in order to segregate carcasses, cuts, etc., to qualitatively and economically optimize the use of both.

In the research arena, objective measures are used to evaluate treatment differences, or differences that are known and can be controlled. There are several general characteristics that should be standardized for comparison of quality data (Honikel 1998). These include species, sex, age, feeding regime, transport and preslaughter handling, slaughter conditions, chilling, aging, pH, and pH decline rate (Asghar & Pearson 1980).

Handbook of Meat, Poultry and Seafood Quality, Second Edition. Edited by Leo M. L. Nollet.

15.2 TEXTURE

In general, consumers rate tenderness as the major factor that determines the eating quality of meat (Brewer & Novakofski 2008). A primary goal in meat texture research has been to develop or refine objective measures that accurately and consistently measure textural characteristics important in consumer quality judgments (Moskowitz 1987; Wheeler *et al.* 1997).

"Texture" is a group of physical properties derived from the structure of the food. Texture is ultimately a sensory property of food that embodies all of the mouth-feel characteristics perceived kinesthetically, both prior to mastication (particle size, oiliness), those perceived during mastication (tenderness, juiciness), and after mastication (fibrous residue, mouth coating) (Civille & Liska 1975; Chauvin *et al.* 2009). In the case of meat texture, mastication involves primarily biting (shearing) and grinding (compression).

Muscle foods have an inherent set of textural characteristics due to the nature of the raw material (Chrystall 1994). These include fibers, fluid/fat exudation, and connective tissue and can be measured physically (fluid/fat), chemically (collagen), physically (fiber length, drip loss), and using sensory evaluation. Textural characteristics can be objectively measured as a function of mass and distance; most use some type of deforming force.

Because sensory observations are a combination of all of the deforming forces plus all of the additional mouth-feel characteristics, measuring it is complex. Any one objective measure will likely fall short of a high relationship with a specific sensory characteristic *regardless of the circumstances*. In addition, the various parameters may be difficult for sensory panelists to separate. However, there are some aspects of texture measurement that can be defined. In addition to toughness/tenderness, force to compress at first bite, chewiness, and juiciness are often evaluated. Some of these sensory characteristics can be modeled using objective measures (shear force, compression; Bourne 1992; Chrystall 1994; Lorenzen *et al.* 2003). For the measures to be useful, the attributes must be well-defined in sensory terms (what is being measured, when it is being measured) and samples must be handled so that extraneous factors are standardized (fiber direction, serving temperature).

Tenderness, as a primary factor in eating quality can be measured in a variety of ways. Tenderness is also an integrated texture property composed of mechanical (hardness, cohesiveness, and elasticity), particulate (grittiness and fibrousness), and chemical (juiciness and oiliness) components (Bourne 1992). Sarcomere length, muscle or connective tissue proteins, and proteolytic degradation are also important factors in meat tenderness and can be measured physically or chemically (Koohmaraie *et al.* 2002).

Fatness and marbling have been used as indicators of meat quality. Because of the effects of fat on meat flavor and tenderness, highly marbled meat has been considered to be the ideal. The United States Department of Agriculture (USDA) quality grade, based on marbling and maturity, has been used to predict beef quality and tenderness. Since consumer assessment is the gold standard, it is important to understand the relationship between USDA grades and/or instrumental evaluations and sensory assessment of quality (Brewer & Novakofski 2008). USDA grades do not consistently predict consumer acceptability. This should come as no surprise because tenderness is a complex attribute, as a sensory characteristic and as a combination of physical properties (Neely *et al.* 1990; Lorenzen *et al.* 2003). Muscle water content has been shown to be a better predictor of tenderness and juiciness than marbling score (Novakofski & Brewer 2006; Brewer & Novakofski 2008). This underscores the disconnection between the measures of adiposity used for quality grading and the sensory attributes important in meat quality. Measuring water content may be a more useful measure than fat content.

15.2.1 Sensory evaluation of texture

Because the consumer is the ultimate judge of textural acceptability, it is of primary importance to choose the textural modalities that best relate to product acceptability (Civille & Liska 1975). Some sensory terms commonly used for sensory evaluation of texture include the following:

Hardness: force required to compress food between the molar teeth
Adhesiveness: force required to remove food compressed against the palate using the tongue
Viscosity: rate of flow of a liquid under force (gravity)
Cohesiveness: force required to deform food between the molar teeth
Springiness: amount of time required for sample to recover from compression between the molar teeth or between the tongue and palate
Chewiness: masticating at a rate of 1 chew/second required to reduce the sample to a steady state for swallowing
Gumminess: amount of manipulation required using the tongue to cause disintegration
Fracturability: brittleness

15.2.1.1 Sensory texture profile analysis

Munoz (1986) and Meilgaard *et al.* (1991) have outlined sensory methods for using texture profile analysis (TPA) to evaluate textural characteristics of a variety of food products. The various characteristics evaluated are based on the anchors for that characteristic. Anchors for the various scale points for a variety of characteristics are shown in Table 15.1.

15.2.2 Instrumental measures of texture and tenderness

Objective measures of texture attempt to duplicate the forces required to reduce a sample to a condition acceptable for swallowing (Bourne 1992) and to predict meat quality characteristics such as tenderness and juiciness. The mechanical forces most commonly measured are compressive forces, tensile forces, and shearing forces. These forces are a function of stress (change in force per unit area) and strain (change in sample dimensions such as volume or length in response to applied force). Forces acting on materials produce stress and strain. Stress measures the change in force/unit area. Strain measures the change in dimensions of the object (volume, length) in response to applied force. A "perfect" solid obeys Hooke's Law: stress is directly proportional to strain ([change in force/unit area] = [change in dimensions]).

Textural quality can be instrumentally assessed by mechanical (Warner–Bratzler (WB) shear force), optical (fluorescence), electrical probing using ultrasonic measurements, electromagnetic waves, nuclear magnetic resonance (NMR), or near-infrared (NIR) reflectance

Table 15.1 Texture profile scale.

Texture attribute	Low	Medium	High
Hardness	Cream cheese	Frankfurter	Hard candy
Springiness	Cream cheese	Marshmallow	Gelatin dessert
Cohesiveness	Corn muffin	Dried fruit	Chewing gum
Denseness	Whipped topping	Malted milk balls	Fruit jellies
Adhesiveness	Hydrogenated Vegetable oil	Marshmallow topping	Peanut butter

Source: Adapted from Munoz (1986) and Meilgaard *et al.* (1991).

spectroscopy (Damez & Clerjon 2008). Mechanical shearing devices (WB shear, Kramer shear-press), biting devices (General Foods texturometer), penetrating instruments (slice tenderness evaluator, US Army meat penetrometer, CSIRO penetrometer, Armour tenderometer), instruments that measure tensile properties, and compression (tenderness press) can be used (Herring 1976).

15.2.2.1 Shearing methods

Eating meat involves initial biting with the front teeth followed by grinding with the molars. Biting is primarily the force exerted during the first phase. Many systems attempt to mimic these actions; however, the deformation rates may not correspond to human mastication, which tends to be variable. "Shearing" involves cutting action and is probably the most common objective method of measuring meat texture.

The Kramer shear-press, the Kramer shear, and the Instron Universal Testing Machine measure the resistance of the product to compression, shear-pressure, division, and traction (Maltini 1972). These methods can detect tenderness differences and correlate to the sensory tenderness of meat (Bianchi *et al.* 2007). There is an exponential dependence between maximum compression force and degree of compression (Werlein & Watkinson 1998). For texture measurement via a shearing method, meat should be cooked then cooled overnight at 2–5°C prior to removing samples. Strips or cores ($1\,cm^2 \times 3\,cm$) are removed parallel to the muscle fibers, then sheared perpendicular to the long axis of the muscle fibers.

Sample dimension, muscle fiber orientation, and sample temperature all affect these measurements. Other conditions of importance during mechanical measurement include whether or not the configuration of the sample is confined or allowed to change with the deforming force (compression may be conducted such that lateral spread is restricted); whether the meat is raw or cooked, and, if cooked, what cooking method is used (moist vs dry); temperature; amount of fat and connective tissue; and homogeneity of the sample(s) (Szczeniak 2002).

The Kramer shear-press is composed of a slotted container that holds the sample and a container with up to ten blades that are spaced to fit into the slots. When the blades move downward, the sample is first compressed then sheared such that sample slices exit through the bottom of the device.

The WB shear is composed of a single rectangular blade with a triangular notch in the center. The blade is connected to a scale. The sample is placed in the triangular notch, and then two bars are lowered onto the sample pressing it into the notch. The peak force (in kg) required to shear through the sample is recorded. Results can also be expressed as force to shear/shear area.

WB shear values are usually determined with an Instron Universal Testing Machine (Instron, Canton, MA) or the Texture Analyzer (Texture Technologies, Scarsdale, NY) fitted with a WB shear attachment. Sample cores ($100^2\,mm \times 30\,mm$) are removed parallel to the muscle fibers usually from cooked, cooled samples (Honikel 1998). Samples are then sheared at right angles to the muscle fibers using a specified load in the load cell (10 kg) moving at a specified rate (100–250 mm/min). Data are reported as peak force (kg) required to shear the sample. For the purposes of conversion to tenderness, "categories" can then be set up (tender = 2–3.5 kg; intermediate tenderness = 3.5–5 kg; tough > 5 kg; Boleman *et al.* 1997). However, reported correlations between WB shear and tenderness vary greatly among laboratories (Lorenzen *et al.* 2003; Aalhus *et al.* 2004; Dikeman *et al.* 2005). The Instron Universal Testing Machine and the Texture Analyzer have blade designs of both the WB and Kramer shear systems. Because differences in crosshead speed can affect shear results, the recommended crosshead speed is 200–250 mm/min. The load cells and full scale load range should be selected such that they are between 20% and 80% of the range (Petracci & Baeza 2009).

Tenderness, as a specific textural characteristic, can be measured instrumentally with compression TPA, penetration TPA, and WB shear values using either triangular or square shear devices (Pollak *et al.* 2005). The square-cutting device is preferable as it gives results in N/cm^2. Sensory and instrumental analyses of tenderness are predictive indexes of each other; however, they do not measure the same physical properties of the meat (Hansen *et al.* 2004). Sensory tenderness is a result of the type and rate of deformation and the heterogeneity of the sample assessed, while instrumental measures are a result of resistance to shearing.

The WB shear force value (SFV) measures only the shear force value maximum (SFVmax). The SFVmax of almost all intact meat occurs at the final shearing point of the stroke while that of emulsified products (sausage, frankfurters) occurs at the center of the stroke. Muscle fiber toughness maximum (MFTmax) of intact meat, sausage and ham, could be considered as SFV at the center of the stroke (Izumimoto & Doi 2000).

Slice shear force (SSF) using a blade with a half-round bevel on the shearing edge is more strongly correlated with tenderness assessed by a sensory panel than is WB shear force determined with a V-shaped blade (Shackelford *et al.* 1999a). Beef carcasses can be classified into tenderness groups based on SSF (tender<23 kg, intermediate tenderness=23–40 kg, tough>40 kg) (Shackelford *et al.* 1999b). SSF requires 10–15 min, the period that carcasses are usually held to allow the rib eye to bloom for quality grading. This makes this method a feasible system for use in commercial situations.

Shear mapping is a method where 1 cm^2 cross-sectional samples are cut parallel to the muscle fiber orientation across the complete cross-sectional area (e.g., steak; Zuckerman *et al.* 2002). Each sample is identified by a coordinate reference grid code that corresponds to its location in the steak. Shear force measurements can be determined and compared before and after a given treatment and then compared based on their relative position in a steak or other cut.

15.2.2.2 Compression systems

These systems compress a sample, usually to some percentage of its original thickness, measuring the force required to do so. Sufficient compression will cause a brittle sample to fracture. The sample is usually compressed in one direction (uniaxial). It may be allowed to spread out laterally or it may be confined as it is subjected to the compressive force. Compression systems are often used for evaluating the texture of processed products (frankfurters) and gels. When the WB shear force is used to predict tenderness, a decrease in shearing velocity, from 200 to 100 mm/min, and a replacement of the cutting blade with an aperture of 60° by one with an aperture of 30° improves the correlation between the two (Van Oeckel *et al.* 1999a, b). Of the WB systems, the square-cutting device is preferable as it gives results in N/cm^2.

15.2.2.3 Penetration

A penetrometer attempts to duplicate the mastication process. The amount of pressure required to penetrate (puncture) the tissue can be determined with a sharp single needle or multineedle device. Puncture data correlate with raw muscle firmness but not with cooked meat texture. A flat-ended plunger (1.3 cm in diameter) is plunged 80% of the way through a meat sample (1 cm thick) at a rate of 100 mm/min (Honikel 1998). The sample is punctured twice and a force–deformation curve is plotted (Honikel 1997). The work–deformation and force–deformation curves are obtained and can be used to calculate hardness, cohesiveness, and gumminess:

$$Hardness = \text{Maximum force}(\text{N})\text{ for the first deformation}$$
$$Cohesiveness = \text{Work during the second penetration/Work during the first penetration}$$
$$Gumminess = \text{Hardness} \times \text{cohesiveness}$$

Johnston (1991) patented a device (WO9119976-A1) consisting of a load cell and probe with four tip pins. Force supplied to the probes is registered by the load cell. The probe is surrounded by a movable piston with prongs to allow locating and securing the devise in the correct position. Depth of penetration of a ten-needle probe into fresh beef is highly correlated with tenderness scores of cooked beef (Krzywicki 1977). WB shear, penetration, and compression tests do not measure the same properties in the same proportions.

15.2.2.4 Tensile strength assessments

Devices that measure tensile strength measure the ability of the sample to adhere to itself or to surfaces so the force exerted on the sample is one of pulling it apart. Honikel (1998) describes a method whereby samples (3.5 mm thick and 28 mm long) are removed perpendicular to the muscle fibers. Samples are then subjected to extension of 56 mm/min at a strain rate (strain rate = extension rate/sample length) of 2 min parallel to the long axis of the muscle fibers. The sample should fracture parallel to the fibers. A load–deformation curve is obtained and breaking stress is calculated:

$$\text{Breaking stress} = \text{Peak force/Measured width} \times \text{Thickness}$$

In meat, pressure exerted at a right angle to the muscle fibers gives a measure of fiber adhesion (which may relate to connective tissue). The WB shear can measure tensile strength. Samples are compressed by 20% and 80% of the sample height. The values obtained are considered to be the mechanical resistance of the muscle fiber and of the muscle fibers plus collagen (Culioli *et al.* 1990). The information is collected as a time–deformation curve.

15.2.2.5 Grinding assessments

Eating involves grinding of tissue with variable muscle fiber orientations and amounts of connective tissue and fat. Results depend on the condition of the grinding surface (or auger and blades if an actual grinder is used) and the amount and rate at which material is fed into the system. Meat texture can be evaluated using an electric grinder equipped with a torque/speed sensor to evaluate maximum grinding torque (Tmax) and total energy (TE) necessary to grind 150 g of uncooked meat (Kamdematk 1995). Grinding parameters (Tmax and TE) are highly correlated with TPA, relaxation, and cutting parameters.

15.2.2.6 Structural methods

Meat is fibrous; therefore, description of the fibers may be useful to assess texture.

Changes in the myofibrillar protein structure of muscle can have significant effects on tenderness (Muir *et al.* 2000). Muscle fiber diameter, degree of fiber shortening, degree of fiber breakdown, myofibrillar fractionation index (MFI), and sarcomere length can all be evaluated to assess tenderness. Breaks in the myofibrils themselves, particularly within the I-band, lead to increased myofibrillar fragility and fragmentation. Fragmentation measures the ease with which the fibers of the meat are fragmented. The increase in myofibrillar fragmentation is indicative of the amount of tenderization that has taken place in meat (Sañudo *et al.* 2004).

Sarcomere length is the distance between adjacent Z-discs on a myofibril. Sarcomere length can be measured using phase contrast microscopy (Silva *et al.* 1999) on cryosectioned tissue samples that are fixed on slides and stained for evaluation. Glutaraldehyde is replaced with sucrose and fixed a second time, and then measured by diffraction using a helium–neon laser (Cross *et al.* 1980; Koolmees *et al.* 1986; Swatland & Irie 1992). In this method, light from a xenon arc is focused to a point (about 2 mm diameter) on the upper surface of samples cut to a set thickness (about 25 mm) (Swatland & Irie 1992). Reflectance is measured in the optical axis with a perpendicular path length of 25 mm through the sample. The spatial measurement of scattering is then calculated from the slope of the line of data points collected over the 400–700 region (Swatland & Irie 1992). Transmitted light is collected by an optical fiber moved by a scanner across the underside of the sample. It then passes through a monochromator to a photomultiplier.

Multiangle light scattering can be used to determine myofibrillar fragmentation (Lametsch *et al.* 2007). This method is based on measuring the myofibril particle size distribution using an optical unit containing several individual detectors. This method is well correlated with the WB shear force, is fast, and sample preparation is simple.

Sarcomere length can also be measured using standard electron microscopy.

15.2.2.7 Instrumental texture profile analysis

Texture profile analysis (TPA) is an instrumental measurement method that imitates the action of the jaw on a bite-sized piece of meat (Bourne 1978). Viscoelastic characteristics can be determined using a Texture Analyzer. Usually, two bites are imitated using two compression and decompression cycles of a Texture Analyzer (Instron, Canton, MA). The mechanical parameters are then quantified from the force–deformation curves (Fig. 15.1) (Szczeniak 1992). A plastic probe is allowed to descend into a meat sample with a known force for a known period of time (AOAC 1990). Force is measured in grams and distance traveled by the probe is determined based on seconds (Fig. 15.1). Texture characteristics are calculated as described by Bourne (1978):

Hardness: the force necessary to attain a 50% deformation or rupture. This is the peak force of the TPA curve.

Springiness: the rate at which a deformed sample returns to its undeformed condition after the deforming force is removed. Springiness is calculated as the time difference between the beginning and the end of the second peak divided by the time difference between the beginning and the end of the first peak (Fig. 15.1; Time 3–4/Time 1–2).

Cohesiveness: the simulated force that reflects the internal bond strength of the body of the sample. Cohesiveness is calculated as the area of the second peak divided by the area of the first peak (Fig. 15.1; A2/A1).

Gumminess: the simulated energy required to disintegrate a semisolid to a steady state for swallowing. Gumminess is calculated as hardness/time (Fig. 15.1; F1 × [A2/A1]).

Chewiness: the simulated energy required to masticate a semisolid sample to a steady state for swallowing. Chewiness is calculated as hardness × cohesiveness × springiness:

$$([\mathrm{F1} \times \mathrm{A2/A1}] \times [\mathrm{Time}\ 3-4/\mathrm{Time}\ 1-2])$$

Meat fibers have anisotropic characteristics—properties that differ according to the direction of measurement. Samples should be compressed by at least 70% to assure that the force causes an irreversible destruction of the structure in order to measure hardness, gumminess, and chewiness, which correlate with sensory texture parameters (Schubring 1999). Adhesiveness, hardness, gumminess, chewiness, and springiness measurement vary due to the speed of the

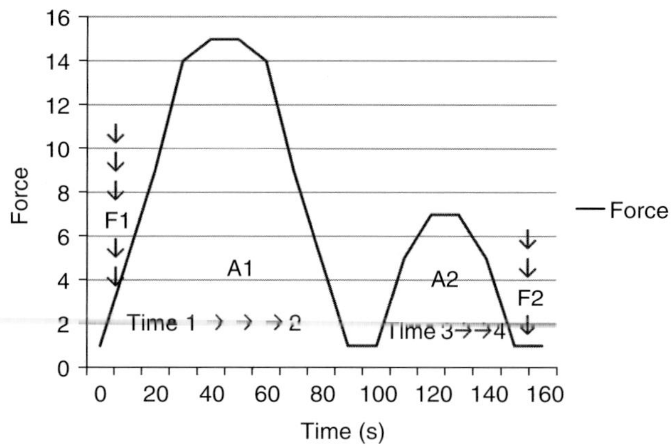

Fig. 15.1 Texture profile analysis curve. Hardness = F1, cohesiveness = A2/A1, springiness = Time 3–4/Time 1–2, chewiness = (F1 × A2/A1) × Time 3–4/Time 1–2, gumminess = F1 × (A2/A1).
source: Adapted from Bourne (1978).

compression instrument in the range from 0.4 to 2.4 mm/s (Schubring 2000). Springiness and hardness are influenced by the diameter/length ratio of the cylindrical sample being evaluated. A ratio of 1.5 is important to relate these characteristics to sensory texture evaluations. Recommendations for TPA include samples with a diameter/length ratio of 1.5, a compression ratio of 75%, and compression rate of 1–2 cm/min (Mittal *et al.* 1992). The TPA parameters, hardness, fracturability, and adhesiveness, have been used successfully to classify grades of meat (Tao *et al.* 2010).

15.2.2.8 Ultrasound

Texture can be assessed using ultrasound. Elastography is the quantitative imaging of strain and elastic modulus distributions in soft tissues using ultrasound (Solomon *et al.* 2008). Ultrasonic strain images are obtained perpendicular to the muscle fibers of a 8-mm^3 meat sample (Berg *et al.* 1999). Radio-frequency data are obtained from a 40 × 30 mm^2 field-of-view and digitized for each compression step. Tissue displacements are then calculated for each compression step. Tissue strains are computed from displacement data located in the field-of-view that represents areas of harder and softer muscle tissue. Softer (more elastic) tissue appears as light-colored regions (Morse *et al.* 2008). Elastographic measurements are well correlated with shear values (Berg *et al.* 1999).

Lull *et al.* (2002) used ultrasonic velocity measurements consisting of a pair of narrow-band ultrasonic transducers, a pulse-receiver, and a digital storage oscilloscope linked to a computer. They reported that the ultrasonic technique accurately measured initial texture as well as changes in textural properties over time or due to treatments.

15.2.2.9 Spectroscopy

Reflectance spectra between 750 and 1098 nm can be used as a measure of quality uniformity in beef (Byrne *et al.* 1998). Reflectance spectral data were more highly correlated with sensory quality scores than WB shear values. Jensen and Haagensen (1990) patented a process (US4908703) to measure meat texture whereby meat (fish) is illuminated in ≥2 directions perpendicular to the base of the meat. A loose texture (deep cracks) result in electronic image information can be used to control equipment.

There is a high correlation between fluorescence spectroscopy and texturometer results (for texture; Frencia *et al.* 2003). Fluorescence spectroscopy is a reliable, rapid, and nondestructive method for the determination of meat tenderness. Autofluorescence spectra can also be used to generate information relevant to WB peak values (Egelandsdal *et al.* 2002). Wavelengths around 375 nm, excitation 332 nm, appear to be related to a component in the perimysial tissue (most likely collagen I or III). Using excitation at 380 nm, the wavelength 460–480 nm range appears to be important and is presumed to be due to collagens in the perimysium (Egelandsdal *et al.* 2002).

Recently, near-infrared (NIR) reflectance measurements have been used to predict meat quality. The NIR region is the 700–2500-nm region between the visible and mid-infrared (IR) regions. Reflectance in this region indicates quality changes due to changes in the state of water and H-bonds (Torgersen *et al.* 2002). NIR measures have been correlated to reference measures of pre- and postrigor pH, prerigor glycogen content, postrigor shear force, and postrigor WHC (Rosenvold *et al.* 2009). NIR appears to be measuring specific biochemical and physical changes, most likely changes in glycolytic metabolites, muscle shortening, and/or proteolysis.

15.2.2.10 Electrical impedance

Electronic carcass grading probes which measure total tissue thickness over the 12th rib to define fatness and total tissue depth between the ribs compare well with all measures for carcass water (indicating muscle) and fat content (Kirton *et al.* 1995). Swatland (2007) used a pair of hypodermic needles curved to touch at their tips to evaluate meat texture and exudate. Separation of needles creates tension along the lengths of myofibers to cause stress perpendicularly across the myofibers. Electrical impedance is reported to be better than fiber-optic spectrophotometry for detecting the meat exudate released as the needles were moved apart.

15.2.3 Relationship between objective and sensory methods

Attempts to relate subjective to objective measures are often driven by the desire to establish a specific point (objective number) that serves as the cutoff point for consumer acceptability. Most attempts to relate subjective and objective assessments rely on linear correlation analysis assuming that consumer and mechanical devices give the same linear responses as the characteristic changes (e.g., tenderness and WB shear). Sensory responses have a low detection limit and an upper measurable limit resulting in a curvilinear response. Most instruments do not respond in a curvilinear way. Testing samples within a narrow range often gives high correlations between subjective and objective measures; however, at the extremes of the scale, the relationship often breaks down.

15.3 CHEMICAL CHARACTERISTICS

Chemical characteristics such as connective tissue content, myofibrillar proteins, glucose content, glycolytic potential (GP), and pH can affect meat texture.

15.3.1 Measuring pH, glucose content, and glycolytic potential

After harvest, the loss of circulatory competency requires that the muscle tissue shifts to anaerobic metabolism. This results in the accumulation of metabolic by-products, especially lactic acid, resulting in a pH decline from about 6.8 to 5.7. As the pH approaches 5.4, endogenous proteinases (cathepsins B and L) become active (Spanier & Miller 1993; Spanier

& Miller 1993). These enzymes tenderize meat proteins. Monitoring the decline of antemortem pH can provide an effective measure of endpoint quality. Toughness is negatively correlated with initial pH and rate of pH decline (Silva *et al.* 1999).

The pH is usually determined in fresh, rather than cooked, meat samples. To determine pH using a standard pH meter (glass-body combination pH electrode), the device should be standardized at pH 4 and 7 prior to use (Pigott *et al.* 2000). The sample (10 g) is homogenized in distilled water (100 mL), and then the pH is determined by placing the tip of the electrode into the homogenate and allowing it to equilibrate. Alternatively, the pH can be determined directly by piercing the meat to a depth of at least 1 cm with a calibrated probe (Young *et al.* 2004). The meat can also be pierced with a knife, and then the electrode can be inserted directly into it (O'Neill *et al.* 2004). As an online method to predict tenderness, pH is significantly better than most other methods (Van Oeckel *et al.* 1999a, b).

The glucose content and GP of a tissue are an indicator of potential ultimate pH because, under anaerobic conditions, glucose is broken down only as far as lactic acid. Glucose content can be determined with a standard blood glucose meter according to the manufacturer's instructions on drip collected for drip loss determination.

Postmortem GP can be determined by adding sample to amyloglucosidase, ATP/NADP/glucose-6-phosphate, and hexokinase, and then adding the combined mixture to a lactate solution. Absorbance is determined at 340 nm and compared with a blank containing no lactate. GP is calculated based on the equations of Monin and Sellier (1985) and Miller *et al.* (2000):

$$\text{Glycolytic potential} = 2[\text{Glycogen} + \text{glucose} + \text{glucose-6-phosphate}] + [\text{Lactate}]$$

An increase in lactic acid concentration increases the GP value, as well as lowers the pH, resulting in a negative relationship between pH and GP.

15.3.2 Proximate analyses

Fat and moisture content are usually determined on lean tissue samples (Novakofski *et al.* 1989). Moisture content is determined gravimetrically by the difference between the wet weight and the dry weight of a sample. A sample is oven dried to a constant weight (125°C, 2–4 h) (AOAC Method 950.46). Fat content is commonly determined gravimetrically by extracting the dried sample with a solvent such as ether (AOAC Method 960.30) or methylene chloride (AOAC Method 985.15). Chloroform: methanol (4:1) will extract phospholipids as well (Novakofski *et al.* 1989).

Near-infrared (NIR) spectroscopy can be used to assess moisture, fat, and protein. Fresh samples are cut to a thickness of 3 mm, and then reflectance is measured at 2–4-nm intervals over the 400–2500-nm range (Leroy *et al.* 2003; Liu *et al.* 2003).

Protein content can be determined in several ways. It can be determined spectrophotometrically as described by Segel (1976) using the difference in absorbance at 280 and 260 nm:

$$\text{Protein} = 1.55(A_{280}) - 0.76(A_{260})$$

There are several methods for determining protein content, for example, by determining nitrogen and then converting that amount back to total protein. Protein content can be determined using the Kjeldahl method (AOAC Method 976.05) wherein the sample is heated in concentrated sulfuric acid, treated with alkali, and then distilled. The NH_3 concentration in the distillated is determined by titration and converted to total protein using the factor 6.25 (because protein is 16% nitrogen).

The sample can be combusted at 850°C and then the released nitrogen can be quantified (AOAC Method 992.15).

15.3.3 Measuring collagen content

Collagen makes up 20–25% of the protein in mammals. It is located primarily in connective tissues and in the epimysial tissue that binds the muscle fibers together. Increasing intermolecular cross-linking among collagen molecules decreases their extensibility and their solubility (Forrest *et al.* 1975), which ultimately contributes to toughness. Muscles that contain higher amounts of collagen are tougher. According to Hocquette *et al.* (2007), one of the major differences among cattle of different breeds or among different genotypes of the same breed with tenderness differences is their connective tissue characteristics (collagen cross-linking and solubility); therefore, measuring collagen content and/or solubility can be an indicator of beef quality.

Hydroxyproline is usually used as an indicator of collagen content in meat tissue. Hydroxyproline content can be determined by acid hydrolysis of the sample, defatting with acetone/chloroform (which also removes nonprotein N compounds and carbohydrate), oxidation of hydroxyproline in the defatted hydrolysate with chloramine-T, and then reaction of the oxidation product with 4-dimethylaminobenzaldehyde. The result is a red derivative, with an extinction value that can be determined spectrophotometrically at 558 nm (Germany, Bundesamt fuer Verbraucherschutz und Lebensmittelsicherheit 2010).

An alternative method is to convert the hydroxyproline to a fluorescent derivative by reacting with 7-chloro-4-nitrobenzofurazan, followed by reverse phase high-performance liquid chromatography and the measurement of fluorescence (excitation = 465 nm, fluorescence = 535 nm; Vazquez-Ortiz *et al.* 2004).

Hydroxyproline content in meat can also be determined using gas chromatography-mass spectrometry (Johnson 1988). The hydroxyproline is converted to the N(o)-trifluoroacetyl *n*-propyl ester, then separated gas chromatographically by vitreous silica capillary columns, followed by mass selective detection in the selective ion monitoring mode.

15.4 WATER-HOLDING CAPACITY

WHC is the ability of meat to hold on to its own or added water when force (heat, pressure) is applied (Hamm 1960; Lawrie & Ledward 2006). WHC has been reported to be one of the most important factors affecting the juiciness of meat during mastication. Most of the water in muscle is held by capillary forces between the thick and thin filaments (Offer & Trinick 1983; Offer & Knight 1988). Water is lost due to myofibrillar shrinkage as the pH declines.

WHC can be measured by the filter paper method, filter paper press method, by drip loss, and by centrifugation. Online measures include light scattering, conductivity, and double density light transmission (Van Oeckel *et al.* 1999a, b). WHC determined by the filter paper press method and drip loss is correlated with juiciness while the online methods are less useful.

15.4.1 Measuring WHC by fluid loss

WHC is generally reflected in the amount of fluid lost from the meat (in the package, during cooking, etc.). To determine fluid losses, initial weights and surface area of samples being evaluated should be approximately the same. Gravimetric measures to measure WHC include drip loss, purge loss, and cooking loss. A relatively simple method of measuring WHC is a gravimetric method where a weighed sample (80–100 g) is suspended inside a netting in a plastic bag or other container, sealed, held at 4°C for 24 h, and then blotted dry and reweighed (Honikel 1998). The weight lost is considered to be the drip or water that the tissue was unable to retain. WHC is the amount of water that the tissue did not lose:

$$\text{WHC} = \left(\frac{\text{Fresh meat weight} - \text{Drip loss weight}}{\text{Fresh meat weight}} \right) \times 100$$

Drip loss is the weight difference between fresh raw samples and samples that have been allowed to release fluids (drip) for a defined period of time expressed as a percent of the raw meat weight:

$$\%\text{Drip loss} = \left(\frac{\text{Fresh meat weight} + \text{Weight of drip}}{\text{Fresh meat weight}} \right) \times 100$$

The amount of liquid in the package is a major determiner for the consumer's decision to purchase meat products (Jeremiah, 1982, 2001; Robbins *et al.* 2003a). For this reason, determining purge and drip losses can be useful. Package purge is the weight difference between packaged meat samples and fresh meat samples removed from the package expressed as a percent of the raw meat weight. To determine purge loss, the meat in the package must be weighed, and then the meat must be removed from the package and weighed separately:

$$\%\text{Purge loss} = \left(\frac{\text{Weight of pkg} + \text{meat} - \text{Weight of meat}}{\text{Weight of pkg} + \text{meat}} \right) \times 100$$

Cook loss occurs as proteins denature and shrink squeezing fluid out of the tissue. Cook loss is the weight difference of meat before and after cooking expressed as a percent of the raw meat weight. A freshly cut meat sample of standard weight (no more than 50 mm thick) is placed in a thin-walled plastic bag, thermocoupled to monitor temperature, submerged in boiling water, and cooked to an internal temperature of 75°C. After reaching the final internal temperature, the bag is removed and cooled in a water bath (1–5°C). The sample is removed from the bag, blotted, and reweighed (Honikel 1998). For comparison purposes, samples must all be of the same weight and dimensional size prior to cooking, and the cooking rate and endpoint temperature must be consistent and monitored. This method is subject to variation due to temperature, which can affect fat loss (high temperatures), cooking method, which may help retain water (braising, boilable bags), and evaporative losses (convection cooking):

$$\%\text{Cook loss} = \left(\frac{\text{Cooked weight} - \text{Raw weight}}{\text{Raw weight}} \right) \times 100$$

15.4.2 Measuring WHC using press methods

Press methods determine the amount of water that can be squeezed from a sample (Kauffman *et al.* 1986). The press methods for determining WHC measure three types of deformations—plastic, elastic, and permanent. The sample is compressed between parallel plates and the released moisture is taken up by preweighed filter paper (Zhang *et al.* 1993). Recommended test conditions are as follows: sample size = 1 g sample, force = 20 kN for 2 min (Wang *et al.* 1993). The compensating polar planimeter is used to measure the area of fluid on the filter paper that has been released from the meat sample. Using a caliper to determine the area is also an acceptable method of determining the press fluid area (Zamorano & Gambaruto 1997). However, a range of forces, times, and temperatures have been used. This contributes to a lack of consistency in reported results because WHC decreases with additional applied force and test duration.

Video image analysis (VIA) combined with meat/liquid ratios can be used for measuring WHC by the press method (Irie *et al.* 1996). For VIA measurement of WHC, a meat sample (400–600 mg) is placed on filter paper, weighed, and sandwiched between translucent plastic plates. The sample is pressed (35 kg/cm^2 for 1 min), and then the meat and liquid areas are measured using VIA and expressed in cm^2.

Brendl and Klein (1971) reported a modified press method wherein a homogenized meat sample is sandwiched between filter paper and glass and then subjected to a pressure of 1 kg for 5 min. The water-binding capacity is calculated from the weight changes in the sample and by planimetry.

There is a nonplanimetric filter paper press method based on measuring the ratio (M/T) of the pressed meat film area (M) to the total moist area (T) on the filter paper, rather than the conventional measurement of the liquid ring zone (Sakata *et al.* 1993).

15.4.3 Measuring WHC using centrifugation

Another physical method of measuring WHC is centrifugation. A weighed meat sample is placed in a centrifuge tube then centrifuged to separate the water from the solids (proteins). The supernatant is poured off and weighed. The weight lost is considered to be the drip or water that the tissue was unable to retain. WHC is the amount of water that the tissue did not lose, calculated as above. Centrifugation at 27 000–31 000 × g for 15–20 min at 2–4°C is a common range used for measuring WHC (Regenstein *et al.* 1980; Hernandez *et al.* 2004). However, centrifuging using much less centrifugal force (860–25 000 × g) has been reported to give acceptable results (Zhang *et al.* 1995; Ramirez *et al.* 2004).

15.4.4 Measuring WHC using other methods

WHC can be measured using low field NMR transverse relaxation time (T2; Bertram *et al.* 2002, 2004). Transverse relaxation reflects pH-induced structural changes occurring in muscles postmortem (Bertram *et al.* 2001; Bertram & Andersen 2007).

X-ray diffraction at 680 nm can be used to measure interfilament separation and is well correlated with WHC (Swatland 1986). Differential interference contrast microscopy at 400 nm for interfiber area is also well correlated with WHC (Swatland *et al.* 1989).

Diode array near-infrared (NIR) spectroscopy can be used to measure WHC over time (Rosenvold *et al.* 2009). It appears to measure changes in specific biochemical and physical characteristics (i.e., changes in glycolytic metabolites, muscle shortening, and proteins).

15.5 MEASURING COLOR

"Color" is the visual integration and perception of light in the 390–760-nm region of the spectrum (Brewer *et al.* 1998). Color is important in the meat industry because consumers associate a bright pink/red color with meat freshness (Zhu & Brewer 1999). Meat color is contributed primarily by the heme proteins, hemoglobin, and myoglobin. Often, the pigment forms are separated for measurement. Myoglobin and hemoglobin in beef can be almost completely separated using an anion exchange column and a column based on hydrophobic interaction (Oellingrath *et al.* 1990). Electron paramagnetic resonance (EPR) at −150°C without pigment extraction is highly correlated with known metmyoglobin concentrations determined by visible spectrophotometry at lower pigment concentrations (Viriyarattanasak *et al.* 2008).

The interior of a whole cut of meat has low oxygen tension. The iron in the myoglobin is in the reduced state (Fe^{2+}) but oxygen is not bound to it. Reduced myoglobin is red-purple. The bright red color is the result of bloom that occurs when myoglobin at the surface of the meat is in the reduced state (Fe^{2+}) and then reversibly binds oxygen forming oxymyoglobin.

The extent of bloom in meat depends on a variety of factors, including the myoglobin content of the muscle, oxidation and denaturation of myoglobin, which determines the ability of myoglobin to reversibly oxygenate, availability of oxygen, availability of reducing equivalents, and tissue pH (Dean & Ball, 1960; Bendall & Swatland 1988). Low pH alters oxygen solubility increasing the depth of oxygen diffusion that increases the degree of bloom (Ledward 1992).

In the living tissue, reducing equivalents (NADH) maintain the iron in its reduced state. When the reducing equivalents are consumed and ATP is no longer available, the Fe^{2+} converts to Fe^{3+}. If the myoglobin reversibly binds to O_2, it is converted to oxymyoglobin, which is red; if it irreversibly binds to oxygen (oxidizes), it is converted to metmyoglobin, which is brown, a generally unacceptable color for fresh meat.

The optical properties of meat can be used to sort or grade meat. Visible light can be used to measure dark-cutting beef, yellowing of fat by dietary carotene, and high myoglobin content in older animals. It can also be used to assess the state of myoglobin, which may be associated with microbiological quality.

15.5.1 Reflectance

Pigment concentration is most often measured by absorbance. Actual meat color is commonly measured by reflectance using one of two broad approaches—reflectance spectroscopy and colorimetry. It is reflected light that we see. Light in the 400-nm range appears blue, in the 500-nm range appears green, in the 570–590-nm range appears yellow, and in the 600-nm range appears red. The terms in which the color values are determined may be expressed in one of several color evaluation systems. These include the Munsell color solid, the Commission Internationale de l'Éclairage (CIE) color solid, reflectance spectroscopy, tristimulus colorimetry, and the Hunter color solid. The Munsell system defines color in terms of hue (red, yellow, green, blue, purple) and chroma (intensity).

Reflectance spectroscopy determines reflected light at wavelengths in the visible range (400–700 nm) and then converts them into CIE X, Y, and Z values. The CIE used the spectral aspect of illuminating with three different colors (light passed through filters) to derive tristimulus values. Reflectance spectroscopy defines color in terms of the additive mixture of the primary colors, X (red), Y (green), and Z (blue), required to match a color as viewed by a standard observer. Colorimeters use filters placed between a light source and a reflecting sample so that a known amount of energy passes through the filter and is subsequently reflected from the sample back to a detector in the instrument. The amount of energy reflected (at a group of wavelengths) is then used in an algorithm to calculate the color values. Hunter L*, a*, and b* values are based on this type of tristimulus colorimetry (Hunter & Harold 1987).

Typically, the instrument is calibrated (with standard colored tiles) and the spectral curve is determined over the 400–700-nm range using a specified illuminant (A, F, D65) and a specified standard observer (2°, 10°). The illuminant corresponds to the light source (incandescent, fluorescent, noon daylight) and should be as close to the light under which the sample is evaluated (visually or instrumentally). The light must contain the color of the sample if it is to be reflected to the viewer (or detector). The standard observer corresponds to the angle from which the sample is being viewed. The older method of calculation that results in L, a, and b values was described by Hunter in 1948 (Hunter 1948) while the newer method that results in L* (light/dark), a* (red/green), and b* (yellow/blue) values was adopted by CIE in 1978 (CIE 1978) in an attempt to make the color "spaces" more uniform such that objects of various colors could be described in equally accurate and precise terms.

The difference between these two methods is in the calculation of the values from the reflected light data.

If meat color is to be measured through packaging material, the instrument should be calibrated by determining L*, a*, and b* of the standardization tile through the packaging material. Total color difference between the sample in the package and out of the package can be calculated as follows:

$$\text{Delta E} = [\text{Delta L*}]^2 + [\text{Delta a*}]^2 + [\textit{Delta b*}]^2$$

Determination of these color values depends on the specific combination of light source, resolution of the instrument, and its detection system (Hunter & Harold 1987; Starrenberg 1989). These calculated values (L*, a*, b* values and L, a, b values) may or may not be the same for a particular object and may or may not reflect color changes to the same degree. Because different color measurement devices use different methods to collect and calculate color parameters, their measurements of meat color often are not the same and cannot be interconverted (Brewer *et al.* 2001a). Hue angle ($\tan^{-1}$[b*/a*]) is a calculated value that measures the departure, in degrees, from the true red axis (0) of the CIE color space (CIE 1978). Chroma ($\sqrt{[a^2+b^2]}$) measures the degree of saturation (intensity of the hue).

Color assessment performed by instrumental analyses should be consistent and must reflect, as nearly as possible, the conditions of human visual appraisal, which ultimately determines consumer acceptability (Solomon *et al.* 2008a). A sufficient amount of time after cutting and before color measurement should be allowed for bloom to occur (30–60 min at <4°C; Brewer & Harbers 1991; Honikel 1998). Because bloom is a major factor in meat color, it is important to understand the ability of instruments to assess changes in the purple-red region. The light source, photometric conditions, viewing geometry, and background for measuring color are listed in ASTM Method D1729. The CIE L*, a*, b* scale using illuminant D65 and a 10° observer is recommended (HunterLab 2000). Despite high sensitivity, instrumental methods may not completely model human visual evaluation as indicated by low correlations among the various visual and instrumental measures (Brewer *et al.* 1998).

Swatland (1986) determined meat color from its internal reflectance measured over the 400–600-nm range with a portable fiber-optic system. Using a xenon flash to illuminate samples and a prism coupled to a synchronized photodiode array spectrophotometer, this method was very sensitive to wavelengths near the absorbance peak of myoglobin (555 nm). Sensitivity was sufficient to accept/reject carcasses based on paleness.

15.5.2 Calculation of myoglobin forms from reflectance data

When light is directed at the surface of meat, it is reflected from the uppermost layer but also penetrates to a depth of 3–4 mm. It is absorbed by meat pigments and by colorless structural elements of the muscle. The isosbestic point, 525 nm, is the point at which the absorption spectra for deoxy-, oxy-, and metmyoglobin intersect (and are equal; Stewart *et al.* 1965). The isosbestic point for myoglobin and oxymyoglobin is 572 nm (Stewart *et al.* 1965). Percent metmyoglobin can be calculated from reflectance data derived from the surface of raw meat. The reflectivity data, when calculated as the corresponding ratios of the absorption coefficient, K, to the scattering coefficient, S, are linearly related to total pigment extract from the meat. The K/S ratio uses the Kubelka–Munk equation, $[(1-R)^2+2R]$, where R is the light reflectance data (at a particular wavelength). The percent of the surface color composed of metmyoglobin can be determined from the ratio and the percent deoxymyoglobin can be calculated from K/S ratios. Oxymyoglobin can then be calculated by difference:

Metmyoglobin = (K/S 572 nm) + (K/S 525 nm)
Deoxymyoglobin = (K/S 574 nm) + (K/S 525 nm)
Oxymyoglobin = 100% – (Metmyoglobin + Deoxymyoglobin)

More recently, Mancini and Hunt (2003) reported that direct ([K/S 610]/[K/S 525]) oxymyoglobin determination on the surface of ground beef is more accurate and repeatable than indirect determination (100% – % deoxymyoglobin + % metmyoglobin).

Krzywicki (1979) reported that relative content of myoglobin, metmyoglobin, and oxymyoglobin at the surface of beef can be calculated based on measurements of reflex attenuance of incident light at the isosbestic points 572, 525, 473, and 730 nm. Reflex attenuance of incident light is determined (measured at 90° to the meat surface, diffuse reflected light at 45° to the surface; Snyder 1965). The reflex attenuance at 730 nm corresponds to the achromatic attenuance of light at the meat surface, and may be used as a measure of lightness of the meat. Subsequently, Krzywicki's (1979) equations have been modified using wavelength maxima at 503, 557, and 582 nm for metmyoglobin, deoxymyoglobin, and oxymyoglobin, respectively, to attain more accurate determinations of total myoglobin concentration (Tang *et al.* 2004).

Using a spectrophotometer with an integration sphere, the light reflected at 580 and 630 nm (%R 630/%R 580) can be used as an indicator of oxymyoglobin (Van Den Oords & Wesdorp 1971; Perez-Alvarez *et al.* 2000). Higher values indicated more redness (AMSA 2001). The maximum reflectance of oxymyoglobin occurs at 630 nm and the minimum reflectance occurs at 580 nm (Strange *et al.* 1974). The difference between reflectance and 630 and 580 nm is also an indicator of red color (AMSA 1991).

Instrumental measures of color can be highly variable within a group of samples if there are substantial differences in intramuscular fat (marbling), collagen (connective tissue), or spotty discoloration. Because different color measurement devices use different methods to collect and calculate color parameters, their measurements of meat color often are not the same and cannot be interconverted (Brewer *et al.* 2001a).

15.6 FLAVOR

15.6.1 Meat flavor

Raw meat has little aroma and only a blood-like taste. However, it is a reservoir of compounds that serve as aroma and flavor precursors. Flavor and aroma ultimately depend on the amounts and proportions of precursor compounds present (Shahidi 1994). Meat flavor is thermally derived. Cooking results in the development of Maillard reaction products from amino acids and reducing sugars. These components contribute water-soluble flavor components. Heating unsaturated fatty acids induces oxidation producing intermediate hydroperoxides that decompose via free radical mechanisms producing aldehydes, unsaturated alcohols, ketones, and lactones that have low detection thresholds (Mottram 1998; Shahidi 1994). At least 17 compounds contribute to the aroma of cooked beef specifically. Seven of these compounds (bis[2-methyl-3-furyl]disulfide, methional, 4-ethyl-1-methylhexane, 1,1,3-trimethylcyclohexane, alpha-pinene, 4-ethyl-1,2-dimethylbenzene, and 3,6-dimethylundecane) are unique to beef (Ramarathnam *et al.* 1993; Imafidon & Spanier 1994; Insausti *et al.* 2005). Sulfur-containing compounds derived from sulfur-containing amino acids contribute to meaty flavor, often have very low detection thresholds, and are objectionable at high levels (Mottram *et al.* 1982; Vega & Brewer 1995; Mottram 1998; Hogan 2002; Elmore *et al.* 2004). The relationship between some of the more common volatiles in beef and their respective flavors is shown in Table 15.2.

Table 15.2 Flavors and aromas associated with volatile compounds in beef.

Compound	Flavors and aromas
Hexanal	Green, grassy, fatty
Heptanal	Green, fatty, oily
Nonanal	Soapy
12-Methyltridecanal	Beefy
Nona-2(E)-enal	Tallowy, fatty
Deca-2(E), 4(E)-dienal	Fatty, fried potato
Butanoic acid	Rancid
Hexanoic acid	Sweaty
2,3-Octanedione	Warmed-over flavor, lipid oxidation
2-Pentyl furan	Metallic, green, earthy, beany
2-Methyl-3-[methylthio]furan	Meaty, sweet, sulfurous
4-Hydroxy-5-methyl-3(2 H)-furanone (HMF)	Meaty
Pyrazines	Nutty, cracker-like, roasted
Bis(2-methyl-3-furyl) disulfide	Roasted meat
2-Methyl-3-furanthiol	Roasted meat

Source: Spanier and Miller (1993), Mottram (1998), Shahidi (1998), Rowe (2002), Insausti *et al.* (2005), Hodgen *et al.* (2006).

15.6.2 Off-flavors

Off-flavors can occur because of the original composition of the meat (iron content) or because of changes that occur during processing, cooking, or storage. Lipid oxidation, which can be initiated by light, heat, metals (iron and copper), myoglobin, and some ingredients, produces a variety of aldehydes (hexanal, heptenal, pentanal, 2, 4-decadienal) from phospholipids and polyunsaturated fatty acids. Cooking beef prior to storage accelerates lipid oxidation often precipitating warmed-over flavor (WOF), a stale, cardboard-like, painty, or rancid odor, due to a loss of desirable odorants (furanones) and an increase in lipid oxidation products (Kerler & Grosch 1996).

Thiobarbituric reactive substances (TBARS), as a measure of lipid oxidation, can be determined by extraction (Witte *et al.* 1970; Dupuy *et al.* 1987; Miller 1998; Wang *et al.* 2010). TBARS methods measure malondialdehyde. The meat sample is ground with butylated hydroxy toluene (BHT) and trichloroacetic acid. The mixture is filtered, aliquots (5 mL) are transferred to test tubes, and 2-thiobarbituric acid is added. Tubes are capped, held in the dark (22°C, 20 h), and the absorbance is determined at 530 nm. Sample blank absorbances are subtracted from test sample readings. The concentration of malondialdehyde is calculated from a standard curve using solutions of tetraethoxypropane. TBARS can also be determined by distillation as described by Tarladgis *et al.* (1960).

Sulfur-containing compounds (thiols, sulfides, thiazoles, sulfur-substituted furans) interact with carbonyl compounds to produce livery flavor (Werkhoff *et al.* 1993, 1996; Campo *et al.* 1999). Livery flavor also correlates with 16–20-carbon fatty acids (Hodgen *et al.* 2006; Yancey *et al.* 2006) and phospholipids (phosphatidylcholine, phosphatidylethanolamine; Larick *et al.* 1989; Elmore *et al.* 1999). Lactones are positively correlated with roasted beef flavor and negatively correlated with gamy/stale off-flavor; diterpenoids are positively correlated with gamy/stale off-flavor (Maruri & Larick 1992). Determining these compounds indicates flavor quality.

Livery flavor increases and beef flavor decreases as iron content increases (Calkins, 2006; Calkins & Cuppett 2006). Therefore, determination of iron content can be an indicator for the propensity of beef to develop this off-flavor. The standard methods of determining iron content are flame ionization atomic absorption and wet ashing (AOAC method 14.013). A more rapid method is to extract iron in HCl, trichloroacetic acid, and hydroxylamine hydrochloride, and then heat (100°C, 15 min), cool, and filter (Kosse *et al.* 2001). The filtrate is then mixed with bathophenanthroline disulfonic acid, and the iron concentration is determined at 535 nm.

Volatile constituents can be analyzed using solid-phase microextraction-gas chromatography-mass spectrometry (SPME-GC-MS). Samples (1.5 cm^3) are frozen by immersion in liquid nitrogen, ground to a fine powder, and transferred to headspace vials. The vial is equilibrated (60°C, 20 min) with agitation and then a SPME fiber is exposed to the headspace. The fiber is desorbed (260°C, 1 min) and compounds are separated using gas/liquid chromatography. Compounds are identified by comparison of their mass spectra and retention indices (Van den Dool & Kratz 1963) to those of standards, and then quantitated (Zhou *et al.* 2002).

REFERENCES

Aalhus, J., Jeremiah, L., Dugan, M., Larsen, I.L. & Gibson, L.L. (2004) Establishment of consumer thresholds for beef quality attributes. *Canadian Journal of Animal Science*, **84**, 631–638.

AMSA (American Meat Science Association Color Standards. 16.55) (1991) National livestock and meat board. In: *Proceedings of the International Meat Science and Technology,* Chicago, Illinois.

AMSA (American Meat Science Association) (2001) Guidelines for meat color evaluation. In: *Proceedings of the 44th Annual Reciprocal Meat Conference,* Kansas State University, Manhattan, Kansas, June 9–12, 2001.

AOAC 14.013 (1984) *Official Method of Analysis*, 14th edn. Association of Official Analytical Chemists, Washington, District of Columbia.

AOAC (1990) *Official Method of Analysis*, 15th edn. Association of Official Analytical Chemists, Washington, District of Columbia.

Asghar, A. & Pearson, A. (1980) Influence of ante- and post-mortem treatments on muscle composition and meat quality. *Advances in Food Research*, **26**, 53–61.

Bendall, J.R. & Swatland, H.J. (1988). A review of the relationships of pH with physical aspects of pork quality. *Meat Science*, **24** (2), 85–126.

Berg, E., Kallel, F., Hussain, F., Miller, R., Ophir, J. & Kehtarnavaz, N. (1999) The use of elastography to measure quality characteristics of pork semimembranosus muscle. *Meat Science*, **53** (1), 31–35.

Bertram, H.C. & Andersen, H. (2007) NMR and the water-holding issue of pork. *Journal of Animal Breeding and Genetics*, **124** (1), 35–42.

Bertram, H., Andersen, H. & Karlsson, A. (2001) Comparative study of low-field NMR relaxation measurements and two traditional methods in the determination of water holding capacity of pork. *Meat Science*, **57** (2), 125–132.

Bertram, H., Donstrup, S., Karlsson, A. & Andersen, H. (2002) Continuous distribution analysis of T2 relaxation in meat—An approach in the determination of water-holding capacity. *Meat Science*, **60** (3), 279–285.

Bertram, H., Hu, J., Rommereim, D., Wind, R.A. & Andersen, H. (2004) Dynamic high-resolution 1 H and 31P NMR spectroscopy and 1 HT2 measurements in postmortem rabbit muscles using slow magic angle spinning. *Journal of Agricultural and Food Chemistry*, **52** (9), 2681–2688.

Bianchi, A., Petracci, M., Pascual, M. & Cavani, C. (2007) Comparison between Allo–Kramer and Warner–Bratzler devices to assess rabbit meat tenderness. *Italian Journal of Animal Science*, **6**, 749–751.

Boleman, S.J., Boleman, S.L., Miller, R., *et al.* (1997) Consumer evaluation of beef of known categories of tenderness. *Journal of Animal Science*, **75**, 1521–1524.

Bourne, M. (1978) Texture profile analysis. *Food Technology*, **32**, 62–66.

Bourne, M. (1992) Calibration of rheological techniques used for foods. *Journal of Food Engineering*, **16**, 151–163.

Brendl, J. & Klein, S. (1971) Determination of water-binding capacity of meat and statistical evaluation of the results. *Deutsche Lebensmittel-Rundschau*, **67** (10), 353–356.

Brewer, M.S. & Harbers, C.A.Z. (1991) Effect of packaging material on color and physical characteristics of ground pork in long term frozen storage. *Journal of Food Science*, **56** (2), 363–370.

Brewer, M.S. & Novakofski, J.E. (2008). Consumer quality evaluation of aging of beef. *Journal of Food Science*, **73** (1), S78–S82.

Brewer, S., Lan, H.Y. & McKeith, F. (1998) Consumer evaluation of pork appearance with differing physiological and packaging conditions. *Journal of Muscle Foods*, **9** (2), 173–183.

Brewer, S., Zhu, L., Bidner, B., Meisinger, D.J. & McKeith, F.K. (2001a) Measuring pork color: Effects of bloom time, muscle, pH and relationship to instrumental parameter. *Meat Science*, **57** (2), 169–176.

Brewer, S., Zhu, L. & McKeith, F. (2001b) Marbling effects on quality characteristics of pork loin chops: Consumer purchase intent, visual and sensory characteristics. *Meat Science*, **59** (2), 153–163.

Brewer, M.S., Jensen, J., Prestat, C., Zhu, L.G. & McKeith, F.K. (2002) Visual acceptability and consumer purchase intent of pumped pork loin roasts. *Journal of Muscle Foods*, **13** (1), 53–68.

Brewer, S., Ryan, K., Jensen, J., *et al.* (2004) Enhancement effects on quality characteristics of pork derived from pigs of various commercial genetic backgrounds. *Journal of Food Science*, **69** (1), 5–10.

Byrne, C., Downey, G., Declan, T. & Buckley, D. (1998) Non-destructive prediction of selected quality attributes of beef by near-infrared reflectance spectroscopy between 750 and 1098 nm. *Meat Science*, **49** (4), 399–409.

Calkins, C. (2006) *Mitigation of Flavor in Fed and Non-fed Cow Beef.* Final Report to the National Cattlemens Beef Association, Centennial, Colorado.

Calkins, C. R. & Cuppett, S. (2006). *Volatile compounds in beef and their contribution to off-flavors*. Final report to the National Cattlemens Beef Association, Centennial, Colorado.

Campo, M.M., Sanudo, C., Panea, B., Alberti, P. & Santolaria, P. (1999) Breed type and ageing time effects on sensory characteristics of beef strip loin steaks. *Meat Science*, **38**, 443–451.

Chambaz, A., Scheeder, M., Kreuzer, M. & Dufey, P. (2003) Meat quality of Angus, Simmental, Charolais and Limousin steers compared at the same intramuscular fat content. *Meat Science*, **63**, 491–500.

Chauvin, M., Parks, C., Ross, C. & Swanson, B. (2009) Multidimensional representation of the standard scales of food texture. *Journal of Sensory Studies*, **24** (1), 93–110.

Chrystall, B. (1994) Meat texture measurement. *Advances in Meat Research*, **9**, 316–336.

CIE (Commission Internationale de L'eclairage) (1978) Recommendations on uniform color spaces—Color difference equations, psychometric color terms. Supplement No. 2 to CIE Publication No. 15 (E-1.3.1) 1971/(TC-1-3), Paris, France.

Civille, G. & Liska, I. (1975) Modifications and applications to foods of the General Foods sensory texture profile technique. *Journal of Texture Studies*, **6** (1), 19–31.

Civille, G.V. & Liska, I.H. (1976) Modifications and applications of foods to the General Foods texture profile technique. *Journal of Texture Studies*, **6**, 19.

Cross, R., West, R.L. & Dutson, T.R. (1980) Comparison of methods for measuring sarcomere length in beef semitendinosus muscle. *Meat Science*, **5**, 261–266.

Culioli, J., Touraille, C., Bordes, P. & Girard, J.-P. (1990) Caracteristiques des carcasses et de la viande du poulet label fermier. *Archives Fur Geflugelkunde*, **53**, 237–245.

Damez, J.-L. & Clerjon, S. (2008) Meat quality assessment using biophysical methods related to meat structure. *Meat Science*, **80** (1), 132–149.

Dean, R. & Ball, C.O. (1960) Analysis of the myoglobin fractions on the surfaces of beef cuts. *Food Technology*, **14**, 271–280.

Dikeman, M.E., Pollak, J., Zhang, Z., Moser, D.W., Gill, C.A. & Dressler, E.A. (2005) Phenotypic ranges and relationships among carcass and meat palatability traits for fourteen cattle breeds, and heritabilities and expected progeny differences for Warner-Bratzler shear force in three beef cattle breeds. *Journal of Animal Science*, **83** (10), 2461–2467.

Dupuy, H.P., Bailey, M.E., St. Angelo, A.J., Vercellotti, J.R. & Legendre, M.G. (1987) Instrumental analysis of volatiles related to warmed-over flavor of cooked meats. In: *Warmed-over Flavor of Meat* (eds A. St. Angelo & M.E. Bailey), pp. 165–192. Academic Press, New York.

Egelandsdal, B., Wold, J., Sponnich, A., Neegard, S. & Hildrum, K. (2002) On attempts to measure the tenderness of Longissimus Dorsi muscles using fluorescence emission spectra. *Meat Science*, **60** (2), 187–202.

Elmore, S., Mottram, D., Enser, M. & Wood, J.D. (1999) Effect of the polyunsaturated fatty acid composition of beef muscle on the profile of aroma volatiles. *Journal of Agricultural and Food Chemistry*, **47** (4), 1619–1625.

Elmore, S., Warren, H.E., Mottram, D.S., *et al.* (2004) Comparison of the aroma volatiles and fatty acid compositions of grilled beef muscle from Aberdeen Angus and Holstein Friesian steers fed diets based on silage or concentrates. *Meat Science*, **68**, 27–33.

Fischer, C., Hofmann, K. & Hamm, R. (1976) Hofmann's capillary volumeter method for determining the water binding capacity of meat. *Fleischwirtschaft*, **56** (1), 91–95.

Forrest, J.C., Aberle, E.D., Hedrick, H.B., Judge, M.D. & Merkel, R. (1975) Part II. Muscle and associated tissues. In: *Principles of Meat Science*, pp. 25–144. W.H. Freeman & Company, San Francisco, California.

Frencia, J.P., Thomas, E. & Dufour, E. (2003) Measure of meat tenderness using front-face fluorescence spectroscopy. *Sciences des Aliments*, **23** (1), 142–145.

Germany, Bundesamt fuer Verbraucherschutz und Lebensmittelsicherheit. (2010) Analysis of foods. Determination of hydroxyproline in meat and meat products. Photometric method after acid digestion. Amtliche Sammlung von Untersuchungsverfahren nach. Paragraph 64 LFGB Paragraph 35 Vorlaeufiges Tabakgesetz Paragraph 28b GenTG, L 06.00-8.

Hamm, G. (1960) Biochemistry of meat hydration. *Advances in Food Research*, **10**, 355–359.

Hansen, S., Hansen, M.D., Aaslyng, M. & Byrne, D. (2004) Sensory and instrumental analysis of longitudinal and transverse textural variation in pork Longissimus Dorsi. *Meat Science*, **68** (4), 611–629.

Hernández, P., Aliaga, S., Pla, M. & Blasco, A. (2004) The effect of selection for growth rate and slaughter age on carcass composition and meat quality traits in rabbits. *Journal of Animal Science*, **82**, 3138–3143.

Herring, H. (1976) Evaluation of beef texture. In: *Food Evaluation Symposium*, pp. 7–27. National Research Council, Committee on Food Stability, Montreal, Quebec, Canada.

Hocquette, J.F., Lehnert, S., Berendse, I., Cassar-Malek, I. & Picard, B. (2007) Recent advances in cattle functional genomics and their application to beef quality. *Animal*, **1**, 159–173.

Hodgen, J., Calkins, C. & Cuppett, S. (2006) Identification of off-flavored compounds in beef. In: *Proceedings of the Reciprocal Meat Conference*, American Meat Science Association, Chicago, Illinois, p. 11.

Hogan, B. (2002) Putting punch in meat flavor profiles. *Food Product Design*, July.

Honikel, K. (1997) Reference methods supported by OECD and their use in Mediterranean meat products. *Food Chemistry*, **59**, 573–582.

Honikel, K. (1998) Reference methods for the assessment of physical characteristics of meat. *Meat Science*, **49** (4), 447–457.

Honikel, K., Fischer, C. & Hamm, R. (1981) Influence of postmortem changes in bovine muscle on the water-holding capacity of beef. Postmortem storage of muscle at 20°C. *Journal of Food Science*, **46**, 1–6.

Hunter, R. (1948). Accuracy, precision and stability of new photoelectric color difference meter. *Journal of the Optical Society of America*, **38**, 1094A.

Hunter, R. & Harold, R. (1987) *The Measurement of Appearance*, 2nd edn. John Wiley & Sons, New York.

HunterLab. (2000) What is color and how is measured? *Insight on Color*, **12** (5). http://www.hunterlab.com/appnotes/an05_00.pdf (accessed June 28, 2005).

Imafidon, G.I. & Spanier, A. (1994) Unraveling the secret of meat flavor. *Trends in Food Science and Technology*, **15**, 315–321.

Insausti, K., Goni, V., Petri, E., Gorraiz, C. & Beriain, M.J. (2005) Effect of weight at slaughter on the volatile compounds of cooked beef from Spanish cattle breeds. *Meat Science*, **70**, 83–90.

Irie, M., Izumo, A. & Mohri, S. (1996) Video image analysis for determining water holding capacity. *Meat Science*, **42** (1), 95–102.

Izumimoto, M. & Doi, M. (2000) Analysis and comparison of shear force profile for meat and meat products. *Animal Science Journal*, **71** (10), J498–J504.

Jensen, S. & Haagensen, P. (1990) Method of measuring meat texture. Patent Assignee: Lumetech A.S. Patent Number: US4908703. United States Patent Office.

Jeremiah, L. (1982) A review of factors influencing consumer selection and acceptability of meat purchases. *Journal of Consumer Studies and Home Economics*, **6**, 137–154.

Jeremiah, L. (2001) Packaging alternatives to deliver fresh meats using short- or long-term distribution. *Food Research International*, **34**, 749–772.

Johnson, S.K. (1988) The determination of hydroxyproline in meat using gas chromatography-mass spectrometry. *Meat Science*, **2** (3), 221–227.

Johnston, G. (1991) Measuring meat tenderness. Patent Number: WO9119976-A1. PCT International Patent Application.

Kamdematk, H.J. (1995) Grinding as a method of meat texture evaluation. *Meat Science*, **39** (2), 225–236.

Kauffman, R., Eikelenboom, G., van der Wal, P.G. & Zaar, M. (1986) A comparison of methods to estimate water-holding capacity in post-rigor porcine muscle. *Meat Science*, **18**, 307–322.

Kerler, J. & Grosch, W. (1996) Odorants contributing to warmed over flavor WOF of refrigerated cooked beef. *Journal of Food Science*, **61**, 1271–1274, 1284.

Kirton, A.H., Mercer, G.J.K., Duganzich, D. & Uljee, A.E. (1995) Use of electronic probes for classifying lamb carcasses. *Meat Science*, **39** (2), 167–176.

Koohmaraie, M., Kent, M.P., Shackelford, S.D., Veiseth, E. & Wheeler, T. (2002) Meat tenderness and muscle growth: Is there any relationship. *Meat Science*, **62**, 345–352.

Koolmees, P., Korteknie, F. & Smulders, F. (1986) Accuracy and utility of sarcomere length assessment by laser diffraction. *Food Microscopy*, **5**, 71–76.

Kosse, J., Yeung, A.D., Gil, A.I. & Miller, D. (2001) A rapid method for iron determination in fortified foods. *Food Chemistry*, **75** (3), 371–376.

Krzywicki, K. (1977) Assessment of meat tenderness on the basis of penetration measured in raw tissue. *Acta Alimentaria*, **6** (2), 119–122.

Krzywicki, K. (1979) Assessment of relative content of myoglobin, oxymyoglobin and metmyoglobin at the surface of beef. *Meat Science*, **3** (1), 1–10.

Lametsch, R., Knudsen, J., Ertbjerg, P., Oksbjerg, N. & Therkildsen, M. (2007) Novel method for determination of myofibril fragmentation post-mortem. *Meat Science*, **75** (4), 719–724.

Larick, D., Turner, B., Koch, R. & Crouse, J. (1989) Influence of phospholipid content and fatty acid composition of individual phospholipids in muscle from bison, Herford and Brahman steers on flavor. *Journal of Food Science*, **54**, 521–526.

Lawrie, R.A. (1991) The eating quality of meat. In: *Meat Science*, 5th edn., pp. 190–199. Pergamon Press, New York.

Lawrie, R.A. & Ledward, D.A. (2006) Chemical and biochemical constituents of meat. In: *Meat Science* (eds R.A. Lawrie & D.A. Ledward), 7th edn, Chapter 4. Woodhead Publishing, Cambridge, England.

Ledward, D.A. (1992) Color of raw and cooked meat. In: *The Chemistry of Muscle-based Foods* (eds D.E. Johnston, M.K. Knight & D.A. Ledward), pp. 128–144. Royal Society of Chemistry, Thomas Graham House, Cambridge, England.

Leroy, B., Lambotte, S., Dotreppe, O., Lecocq, H., Itassec, K. & Clinquart, A. (2003) Prediction of technological and organoleptic properties of beef *Longissimus thoracis* from near infrared reflectance and transmission spectra. *Meat Science*, **66**, 45–54.

Liu, Y., Lyon, B., Windham, W., Realini, C., Pringle, D. & Ducket, C. (2003) Prediction of color, texture, and sensory characteristics of beef steaks by visible near infrared reflectance spectroscopy . A feasibility study. *Meat Science*, **65**, 1107–1115.

Lorenzen, C., Miller, R., Taylor, J.F., *et al.* (2003) Beef customer satisfaction: Trained sensory panel ratings and Warner–Bratzler shear force values. *Journal of Animal Science*, **81**, 143–149.

Lull, P., Simal, S., Femenia, A., Benedito, J. & Rossello, C. (2002) The use of ultrasound velocity measurement to evaluate the textural properties of sobrassada from Mallorca. *Journal of Food Engineering*, **52** (4), 323–330.

Maltini, E. (1972) Use of dynamometric techniques in studying the textural properties of food products. *Annali dell'Istituto Sperimentale per la Valorizzazione Tecnologica dei Prodotti Agricoli*, **3**, 233–248.

Mancini, R. & Hunt, M. (2003) Reflectance at 610 nm estimates oxymyoglobin content on the surface of ground beef. *Meat Science*, **6**, 157–162.

Maruri, J.L. & Larick, D.K. (1992). Volatile concentration and flavor of beef as influenced by diet. *Journal of Food Science*, **57** (6), 1275–1281.

Meilgaard, M., Civille, G. & Carr, T. (1991) *Sensory Evaluation Techniques*, 2nd edn., pp. 166–167. CRC Press, Boca Raton, Florida.

Miller, D. (1998) *Food Chemistry. A Laboratory Manual*, pp. 65–67. John Wiley & Sons, New York.

Miller, D., Ellis, M., Sutton, D., McKeith, F. & Wilson, E. (2000) Influence of live animal sampling and storage technique on glycolytic potential of longissimus samples taken by spring loaded biopsy equipment. *Journal of Muscle Foods*, **11**, 61–67.

Mittal, G., Nadulski, R., Barbut, S. & Negi, S. (1992) Textural profile analysis profile analysis test conditions for meat-products. *Food Research International*, **25** (6), 411–417.

Monin, G. & Sellier, P. (1985) Pork of low technological quality with a normal rate of muscle pH fall in the immediate post-mortem period: The case of the Hampshire breed. *Meat Science*, **13**, 49.

Moskowitz, H. (1987) *Food Texture: Instrumental and Sensory Measurement*, pp. 239–245. CRC Press, Boca Raton, Florida.

Mottram, D. (1998) Flavor formation in meat and meat products: A review. *Food Chemistry*, **62**, 415–424.

Mottram, D., Edwards, R.A. & MacFie, H.J.H. (1982) A comparison of the flavor volatiles from cooked beef and pork meat systems. *Journal of the Science of Food and Agriculture*, **33**, 934–944.

Muir, P., Wallace, G., Dobbie, P.M. & Bown, M. (2000) A comparison of animal performance and carcass and meat quality characteristics in Hereford, Hereford x Friesian, and Friesian steers grazed together at pasture. *New Zealand Journal of Agricultural Research*, **43**, 193–205.

Munoz, A.M. (1986) Development and application of texture reference scales. *Journal of Sensory Studies*, **1** (1), 55–83.

NCBA (National Cattlemen's Beef Association) (2001) *Beef Customer Satisfaction II: The Marketplace and Beyond What Factors Drive Customer Satisfaction of Beef?* National Cattlemen's Beef Association, Englewood, Colorado, pp. 1–11.

Neely, T.R., Lorenzen, C., Miller, R., *et al.* (1990) Beef customer satisfaction: Role of cut, USDA quality grade, and city on in-home consumer ratings. *Journal of Animal Science*, **76**, 1027–1033.

Novakofski, J. & Brewer, M.S. (2006) The paradox of toughening during the aging of tender steaks. *Journal of Food Science*, **71**, S473–S479.

Novakofski, J., Park, S., Bechtel, P. & McKeith, F. (1989) Composition of cooked pork chops. The effect of removing subcutaneous fat before cooking. *Journal of Food Science*, **54**, 15–17.

Oellingrath, I., Iversen, A. & Skrede, G. (1990) Quantitative determination of myoglobin and haemoglobin in beef by high-performance liquid chromatography. *Meat Science*, **28** (4), 313–320.

Offer, G. & Knight, P. (1988) The structural basis of water holding in meat. Part 2: Drip losses. In: *Developments in Meat Science* (ed. R. Lawrie), 4th edn. Elsevier, Oxford.

Offer, G. & Trinick, J. (1983) On the mechanism of water-holding in meat: The swelling and shrinkage of myofibrils. *Meat Science*, **8**, 245–281.

O'Neill, D., Troy, D. & Mullen, A. (2004) Determination of potential inherent variability when measuring beef quality. *Meat Science*, **66**, 765–770.

Perez-Alvarez, J., Sayas-Barbera, E. & Aranda-Catala, V. (2000) Characterization of the different states of myoglobin in pork using color parameters and reflectance rations. *Journal of Muscle Foods*, **11** (3), 157–167.

Petracci, M. & Baeza, E. (2009) Harmonization of methodology of assessment of poultry meat quality features. *Working Paper of WPSA Working Group 5 Poultry Meat*. World's Poultry Science Association, U.K., pp. 1–18.

Pigott, R.S., Kenney, P.B., Slider, S. & Head, M.K. (2000) Formulation protocol and dicationic salt affect protein functionality of model system beef batters. *Journal of Food Science*, **65**, 1151–1154.

Pollak, J., Zhang, Z., Moser, D., Gill, C. & Dressler, E. (2005) Phenotypic ranges and relationships among carcass and meat palatability traits for fourteen cattle breeds, and heritabilities and expected progeny differences for Warner–Bratzler shear force in three beef cattle breeds. *Journal of Animal Science*, **8**, 2461–2467.

Ramarathnam, N., Rubin, L.J. & Diosady, L.L. 1993.Studies on meat flavor. 3. A novel method for trapping volatile components from uncured and cured pork. *Journal of Agriculture and Food Chemistry.* **41** (6), 933–938.

Ramírez, J.A., Oliver, M.A., Pla, M., *et al.* (2004) Effect of selection for growth rate on biochemical, quality and texture characteristics of meat from rabbits. *Meat Science*, **67** (4), 617–624.

Regenstein, J., Gorimar, T.S. & Sherbon, J. (1980) Measuring the water-holding capacity of natural actomyosin from chicken breast muscle in the presence of pyrophosphate and divalent cations. *Journal of Food Biochemistry*, **3** (4), 205–211.

Robbins, K., Jensen, J., Ryan, K., Homco-Ryan, C., McKeith, F. & Brewer, S. (2003a) Consumer attitudes towards beef and acceptability of enhanced beef. *Meat Science*, **65** (2), 721–729.

Robbins, K., Jensen, J., Ryan, K., Homco-Ryan, C., McKeith, F. & Brewer, S. (2003b) Dietary vitamin E supplementation effects on the color and sensory characteristics of enhanced beef steaks. *Meat Science*, **64** (3), 279–285.

Rosenvold, K., Micklander, E., Hansen, P.W., *et al.* (2009) Temporal, biochemical and structural factors that influence beef quality measurement using near infrared spectroscopy. *Meat Science*, **82** (3), 379–388.

Rowe, D. (2002) High impact aroma chemicals. Part 2: The good, the bad, and the ugly. *Perfumer Flavorist*, **27**, 24–29.

Sakata, R., Deguchi, T. & Nagata, Y. (1993) Effectiveness of the filter paper press method for determining the water holding capacity of meat. *Fleischwirtschaft*, **73** (12), 1425–1426.

Sañudo, C., Macie, E.S., Villarroel, J.L., Olleta, M., Panea, B. & Alberti, P. (2004) The effects of slaughter weight, breed type and ageing time on beef meat quality using two different texture devices. *Meat Science*, **66**, 925–932.

Schubring, R. (1999) Investigation of factors influencing the instrumental texture profile analysis (TPA) of fishery products. 1. Influence of compression. *Deutsche Lebensmittel-Rundschau*, **95** (9), 373–386.

Schubring, R. (2000) Investigation of factors influencing the instrumental texture profile analysis (TPA) of fishery products. 2. Influence of test speed and compression. *Deutsche Lebensmittel-Rundschau*, **96** (2), 45–50.

Segel, I. (1976) *Biochemical Calculations*, 2nd edn. John Wiley & Sons, New York.

Shackelford, S., Wheeler, T. & Koohmaraie, M. (1999a) Evaluation of slice shear force as an objective method of assessing beef longissimus tenderness. *Journal of Animal Science*, **77** (10), 2693–2699.

Shackelford, S., Wheeler, T. & Koohmaraie, M. (1999b) Tenderness classification of beef: II. Design and analysis of a system to measure beef longissimus shear force under commercial processing conditions. *Journal of Animal Science*, **77** (6), 1474–1481.

Shahidi, F. (1998) Flavor of muscle foods: An overview. In: *Flavor of Meat, Meat Products, and Seafoods* (ed. F. Shahidi), pp. 1–4. Blackie Academic and Professional, London, U.K.

Shahidi, F. & Pegg, R.B. (1994) Hexanal as an indicator of the flavor deterioration of meat and meat products. *ACS Symposium Series*, **558**, 256–279.

Silva, A., Patarata, L. & Martins, C. (1999) Influence of ultimate pH on bovine meat tenderness during ageing. *Meat Science*, **52**, 453–450.

Snyder, H.E. (1965) Analysis of pigments at the surface of fresh beef with reflectance spectrophotometry. *Journal of Food Science*, **30** (3), 457–463.

Solomon, M., Eastridge, J.S., Paroczay, E.W. & Bowker, B. (2008a) Measuring meat texture. In: *Handbook of Muscle Foods Analysis* (eds M.L. Nollet & F. Toldrá), pp. 4.3–4.5, Chapter 27. CRC Press, Boca Raton, Florida. ddr.nal.usda.gov/bitstream/10113/25625/1/IND44161372.pdf

Solomon, M.B., Eastridge, J.S., Paroczay, E.W. & Bowker, B.C. (2008b). Measuring meat texture. In: *Handbook of Muscle Foods Analysis* (ed. F. Toldra), pp. 479–480. Bookshop Krisostomus Tartu, Estonia.

Spanier, A. & Miller, J.A. (1993) Role of proteins and peptides in meat flavor. In: *Food Flavor and Safety* (eds A.M. Spanier, H. Okai & M. Tamura), pp. 78–97. ACS Symposium Series No. 528. American Chemical Society, Washington, District of Columbia.

Starrenberg, P. (1989) Influence of sample illumination and viewing on colour measurement of translucent materials like meat. In: *Proceedings of the International Congress of Meat Science and Technology*, Copenhagen, Denmark, pp. 610–613.

Stewart, M.R., Zipser, M. & Watts, B. (1965) The use of reflectance spectrophotometry for the assay of raw meat pigments. *Journal of Food Science*, **30**, 464.

Strange, E., Benedict, R., Gugger, R.E., Metzger, V.G. & Swift, C. (1974) Simplified methodology for measuring meat color. *Journal of Food Science*, **39**, 988–993.

Swatland, H. (1986) Color measurements on pork and veal carcasses by fiber optic spectrophotometry. *Canadian Institute of Food Science and Technology Journal*, **19** (4), 170–173.

Swatland, H. (2007) Feasibility of measuring meat texture and exudate using paired hypodermic needles for rheology, spectrophotometry and electrical impedance. *Journal of Texture Studies*, **30** (2), 217–230.

Swatland, H. & Irie, M. (1992) Effect of wavelength on spatial measurements of light scattering for the measurement of pork quality. *Journal of Animal Science*, **70**, 2138–2143.

Swatland, H., Irving, T.C. & Millman, B.M. (1989) Fluid distribution in pork, measured by x-ray diffraction, interference microscopy and centrifugation compared to paleness measured by fiber optics. *Journal of Animal Science*, **67** (6), 1465–1470.

Szczeniak, A. (2002) Texture is a sensory property. *Food Quality and Preference*, **13**, 215–225.

Tang, J., Faustman, C. & Hoagland, T.A. (2004) Krzywicki revisited: Equations for spectrophotometric determination of myoglobin redox forms in aqueous meat extracts. *Journal of Food Science*, **69**, C717–C720.

Tao, H.H., GaiMing, Z., YunLin, Y.X., MiaoYun, L., GuangHui, Z. & Kun, F. (2010) Discriminant analysis of the grades of ham sausages based on hardness, fracturability and adhesiveness properties. *Scientia Agricultura Sinica*, **43** (10), 2182–2188.

Tarladgis, B., Watts, B., Younathan, M. & Dugan, L. (1960) A distillation method for quantitative determination of malonaldehyde in rancid foods. *Journal of the American Oil Chemists' Society*, **37**, 44–48.

Torgersen, G., Rodbotten, R. & Hildrum, K. (2002) Recent advances in prediction of meat quality by NIR spectroscopy. In: *Research Advances in the Quality of Meat and Meat Products* (ed. F. Toldra), pp. 1–17. Research Signpost, Trivandrum, India.

USDA Quality Grades and Yield (2012) *Standardization division USDA*. AMS, Washington, District of Columbia. www.ams.usda.gov/AMSv1.0/getfile?dDocName=STELDEV3067087.

Van Den Dool, H. & Kratz, P.D. (1963) A generalization of the retention index system including linear temperature programmed gas liquid partition chromatography. *Journal of Chromatography*, **11**, 463–471.

Van Den Oords, A.H. & Wesdorp, J.J. (1971) Analysis of pigments in intact beef samples. *Journal of Food Technology*, **6**, 1–8.

Van Oeckel, M., Warnants, N. & Boucque, C.V. (1999a) Pork tenderness estimation by taste panel, Warner–Bratzler shear force and on-line methods. *Meat Science*, **53** (4), 259–267.

Van Oeckel, M., Warnants, N. & Boucque, C.V. (1999b) Comparison of different methods for measuring water holding capacity and juiciness of pork versus on-line screening methods. *Meat Science*, **51** (4), 313–320.

Vázquez-Ortiz, F.A., Moron-Fuenmayor, O.E. & Gonzalez-Mendez, N.F. (2004) Hydroxyproline measurement by HPLC: Improved method of total collagen determination in meat samples. *Journal of Liquid Chromatography and Related Technologies*, **27** (17), 2771–2780.

Vega, J. & Brewer, S. (1995) Detectable odor thresholds of selected lipid oxidation compounds in a meat model system. *Journal of Food Science*, **60** (3), 592–595.

Viriyarattanasak, C., Matsukawa, S., Hamada-Sato, N., Watanabe, M. & Suzuki, T. (2008) Quantitative measurement of metmyoglobin in tuna flesh via electron paramagnetic resonance. *Food Chemistry*, **111** (4), 1050–1056.

Wang, C., Zhu, L. & Brewer, M.S. (1993) Comparison of TBARS determination methods in various types of fresh, frozen meat. In: *IFT Annual Meeting,* June 23–26, New Orleans, Louisiana.

Wang, W., Peng, Y. & Wu, J. (2010) Prediction of pork water-holding capacity using hyperspectral scattering technique. American Society of Agricultural and Biological Engineers, St. Joseph, Michigan. www.asabe.org. Pittsburgh, Pennsylvania, June 20–June 23, 2010, 1008571.

Werkhoff, P., Bruening, J., Emberger, R., Guentert, M. & Hopp, R. (1993) Flavor chemistry of meat volatiles: New results on flavor components from beef, pork, and chicken. In: *Recent Developments in Flavor and Fragrance Chemistry. Weinheim Conference Proceedings*, Weinheim, Germany, pp. 183–213.

Werkhoff, P., Bruening, J., Guentert, M., Kaulen, J., Krammer, G. & Sommer, H. (1996) Potent mercapto/methylthio substituted aldehydes and ketones in cooked beef liver. *Advances in Food Science*, **18**, 19–27.

Werlein, H.-D. & Watkinson, B. (1998) Standardized measuring of textures. *Fleischwirtschaft*, **78** (10), 1085–1088.

Wheeler, T., Shackelford, S., Johnson, L., Miller, M., Miller, R. & Koohmaraie, M. (1997) A comparison of Warner–Bratzler shear force assessment within and among institutions. *Journal of Animal Science*, **75**, 2423–2432.

Witte, V., Krause, G. & Bailey, M. (1970) A new extraction method for determining 2-thiobarbituric acid values of pork and beef during storage. *Journal of Food Science*, **35**, 582–585.

Yancey, E., Grobbel, J.P., Dikeman, M., *et al.* (2006) Effects of total iron, myoglobin, hemoglobin, and lipid oxidation of uncooked muscles on livery flavor development and volatiles of cooked beef steaks. *Meat Science*, **73**, 680–686.

Young, O., West, J., Hart, A.L. & van Otterdijk, F.F.H. (2004) A method for early determination of meat ultimate pH. *Meat Science*, **66**, 493–498.

Zamorano, J. & Gambaruto, M. (1997) Contribution to improving the meat water holding capacity test by the filter paper press method. A comparison of three methods for measuring areas. *Meat Science*, **46** (1), 129–137.

Zhang, M., Mittal, G. & Barbut, S. (1993) Optimum conditions for determining water holding capacity of beef by the press method. *Journal of Muscle Foods*, **4** (4), 255–268.

Zhang, M., Mittal, G. & Barbut, S. (1995) Effects of test conditions on the water holding capacity of meat by a centrifugal method. *Food Science and Technology*, **28** (1), 50–55.

Zhou, Q., Wintersteen, C. & Cadwallader, K. (2002) Identification and quantification of aroma-active components that contribute to the distinct malty flavor of buckwheat honey. *Journal of Agricultural and Food Chemistry*, **50** (7), 2016–2021.

Zhu, L. & Brewer, S. (1999) Relationship between instrumental and sensory color evaluation in a raw, fresh beef and chicken model system. *Journal of Muscle Foods*, **10**, 131–146.

Zuckerman, H., Berry, B., Eastridge, J. & Solomon, M. (2002) Shear force mapping: A tool for tenderness measurement. *Journal of Muscle Foods*, **13** (1), 1–12.

16 Shelf Life of Meats

Leo M.L. Nollet

Abstract: To extend shelf life, several postharvest techniques are used to decontaminate carcasses and fresh meat. In addition to microbial deterioration, lipid oxidation is one of the primary mechanisms of quality deterioration in meat products. Color fading in displayed meat is another issue that affects the shelf life of meat. Different detection and enumeration methods exist for spoilage and pathogenic microorganisms. Preservation technologies to extend the shelf life of meats are chilling, freezing, vacuum packaging, modified atmosphere packaging, irradiation, ultrahigh pressure, and the use of antimicrobials.

Keywords: Meat shelf life, lipid oxidation color fading, spoilage microorganisms, pathogenic microorganisms, vacuum packaging, modified atmosphere packaging, irradiation, ultrahigh pressure, antimicrobials

16.1 SHELF LIFE

Food security measurements in the meat industry include veterinary inspection prior to slaughter, laboratory screening for unusual levels of compounds in animal tissues, use of metal detectors, carcass treatment with hot water or steam, postpackaging treatment, and microbiological testing (Hueston 2002). Skeletal muscle from healthy animals has been considered sterile prior to slaughter (Huffman 2002). However, there are intrinsic factors such as pH, acidity, buffering power, water activity, redox potential, presence of antimicrobials, identity, and distribution of natural microflora that affect the shelf life of meat. Additionally, there are extrinsic factors such as temperature, relative humidity, light intensity and wavelength, atmospheric gas composition and ratio, packaging characteristics, processing characteristics, storage, distribution, and display considerations that will also affect the shelf life of meat (McDonald & Sun 1999). Extrinsic factors are by far the greatest contributors to carcass and meat contamination (Huffman 2002). Intrinsic factors like pH and the concentration of L-lactate and fat affect the shelf life (rate of spoilage) of vacuum-packed pork and beef (Blixt & Borch 2002).

To extend shelf life, several postharvest techniques are used to decontaminate carcasses and fresh meat. These techniques include chemical dehairing, hot water rinse, steam pasteurization, steam vacuum, chemical rinses, and use of lactoferrin. Chemical dehairing involves applying sodium sulfide, using hydrogen peroxide, and rinsing with lactic acid. The use of hot water (>74°C) produces a sanitizing effect in the reduction of aerobic plate counts, total coliforms, *Salmonella typhimurium*, and *Escherichia coli* O157:H7. Steam pasteurization

Handbook of Meat, Poultry and Seafood Quality, Second Edition. Edited by Leo M. L. Nollet.

is an extension of the use of hot water. This technique reduces total aerobic plate counts and *E. coli*. However, steam pasteurization can lead to meat surface discoloration. Steam vacuum involves steam pasteurization or hot water followed by vacuuming. This technique reduces aerobic plate counts and total coliform counts. Also, organic acids are used for chemical rinses. Acetic acid and citric acid in concentrations of 1.5–2.5% are recommended. Commercial compounds such as Safe2OTM are available for chemical rinsing. Other compounds classified as microbial blocking agents are used to decontaminate carcasses. Among these compounds is lactoferrin, an iron-binding protein that is effective against *E. coli* O157:H7, *Listeria monocytogenes*, *Salmonella* spp., and *Campylobacter* (Huffman 2002).

In addition to microbial deterioration, lipid oxidation is one of the primary mechanisms of quality deterioration in meat products. Lipid oxidation in meat changes quality in color, flavor, texture, and nutritive value. Lipid oxidation affects shelf life, which can be extended through dietary vitamin E supplementation (Gray *et al.* 1996). Ruminant meats are relatively a good source of n-3 polyunsaturated fatty acids (PUFA) due to the presence of 18:3 in grass. These compounds affect flavor producing a "grass-fed" taste and oxidation due to points of unsaturation (Wood *et al.* 2003).

Color fading in displayed meat is another quality issue that affects the shelf life of meat. Color fading of beef steaks displayed in refrigerated cabinets is caused by the combination of oxygen tension, surface microbial growth, temperature, and type of lighting. Light plays a role in pigment photooxidation. Studies suggest that lighting without ultraviolet (UV) radiation leads to a delay of meat spoilage as assessed by surface color, lipid oxidation, bacterial counts, and sensory evaluation (discoloration and color). The shelf life of meat can be extended when low UV lamps or fluorescent lamps with UV filter are used (Djenane *et al.* 2001). When beef is pasteurized using hot water (85°C) for 60 s, vacuum packaged and stored at 2°C, the color of pasteurized meat is rated paler as compared to nonpasteurized meat. Odors are perceived similarly (Gill & Badoni 2002).

16.2 SPOILAGE MICROORGANISMS

Meat is described as spoiled if organoleptic changes make it unacceptable to the consumer. The most common changes are appearance, off odors, and slime formation (Ellis & Goodacre 2001). Growth of spoilage bacteria results in the reduced shelf life of meat causing major economic losses (Venkitanarayanan *et al.* 1997). The colonization and growth of microorganisms on meat surfaces occurs gradually. First, the bacterial cells need to attach to the meat surface, which may be a reversible stage. The next step is an irreversible attachment in which the bacterium produces an adhesive extracellular polysaccharide (Ellis & Goodacre 2001). Recontamination with pathogens during cutting or packaging may result in higher growth on decontaminated cuts than on untreated meat due to the lack of competing nonpathogenic microorganisms (Nissen *et al.* 2001). On the other hand, there is a phenomenon called sticky aging, which is an abacterial enzymatic process that occurs in beef when maintained at body temperature during the first 14–18 h postmortem. During this process, butyric acid is produced from the meat (Stephan *et al.* 1997).

Off odors and off flavors are a common result of microbial and chemical spoilage. Indole is attributed to growth of enterobacteriaceae in aerobic and vacuum-packed stored bovine tripe. Gluconic and 2-oxogluconic acids are produced as glucose is oxidized by *Pseudomonas* during aerobic storage (Dainty 1996). Biogenic amine is considered a freshness marker. Amines are promoted by enzymatic reactions, microbial activity, or endogenous tissue activities. The most common amines produced in meat are tryptamine, putrescine, cadaverine, serotonin, tyramine, spermidine, and spermine. In red meat, the level of biogenic amines are low during the first 9 days of storage (≤30 mg/kg), but after 36 days the levels of biogenic amines, especially cadaverine and tyramine, become high (≥120 mg/kg) (Vinci & Antonelli 2002). Concentration of biogenic amines such as spermine increases after 76 days of storage in vacuum-packaged meat at 0°C (Lee & Yoon 2001).

During deboning, meat undergoes extensive handling and is exposed to the environment, which makes it susceptible to contamination. After deboning, the cuts are trimmed, vacuum packaged, and boxed. Microbial contamination may be overcome with vacuum packaging; however, spoilage bacteria such as *Pseudomonas* spp. (Nel *et al.* 2004), *Acinetobacter*, *Aeromonas*, and *Lactobacillus* (Huffman 2002) can contaminate the meat surface before packing. Other possible spoilage bacteria include *Moraxella* and *Psychrobacter* (Ellis & Goodacre 2001), enterobacteriaceae, *Psychrobacter*, *Shewanella*, *Carnobacterium, Lactobacillus, Leuconostoc, Brochothrix*, and *Kurthia* spp. (Pin & Baranyi 1998). *Leuconostoc mesenteroides, Lactobacillus sake*, and *L. curvatus* are common components of the microflora of meat subjected to vacuum packaging (Dykes *et al.* 1995). At levels of 107 colony-forming unit (cfu)/square centimeter (cm^2), off odors become evident and once the meat surface reaches 108 cfu/cm^2, recognizable off odors are produced. The off odor produced is recognized as dairy, buttery, fatty, and cheesy to a sickly sweet, fruity aroma, and finally putrid odor at 109 cfu/cm^2 (Ellis & Goodacre 2001). The type of bacteria is more important than the numbers when it comes to lipid oxidation in meat. Fresh raw beef has a weak lactic acid or bloodlike odor. This odor changes to an old, stale, oxidized, putrid, and rancid odor. Bacteria such as *Pseudomonas, Acinetobacter, Moraxella, and Lactobacillus* are most common on chilled meat (Chung *et al.* 2002). Lactic acid bacteria can interfere with spoilage and pathogenic bacteria by producing bacteriocins that extend shelf life or inhibit food-borne pathogens (Hugas 1998). For example, cooked ham is a highly perishable meat product. *When Lactobacillus alimentarius* or *Staphylococcus xylosus* are used as protective cultures, ham produced with the *L. alimentarius* is acceptable for up to 28 days compared with ham without the lactic culture with a shelf life of 21 days. These protective cultures reduce total aerobic bacteria, micrococci, and staphylococci (Kotzekidou & Bloukas 1996). Enterococci form part of the lactic acid bacteria group. They are present in unprocessed red meats and can spoil processed meats (Evans *et al.* 2003; Franz *et al.* 2003).

In fresh ground beef stored aerobically, the dominant spoilage organism is *Pseudomonas* spp. This genus undergoes a phenomenon called quorum sensing, where microorganisms produce certain small molecules that pass outside of producing cells and accumulate in the environment. When a critical level (a quorum) is reached, the inducer substances induce pigment production and slime formation. In Gram-negative bacteria, the inducer substances are acylhomoserine lactones (Jay *et al.* 2003).

16.3 PATHOGENIC MICROORGANISMS

The USDA Food Safety Inspection Service (FSIS) has proposed that all slaughter establishments should apply at least one antimicrobial treatment or other approved intervention procedures to livestock carcasses (Shimoni & Labuza 2000) in order to avoid potential pathogens such as *Bacillus cereus, Staphylococcus aureus, Listeria monocytogenes, Escherichia coli*, and *Salmonella* spp. (Lee and Yoon 2001; Huffman 2002). Other pathogenic bacteria of concern are *Aeromonas* spp., *Arcobacter* spp., and *Campylobacter* spp. (Borch & Arinder 2002). USDA requires mandatory testing by the manufacturer for generic *E. coli* and random testing for specific pathogens (Shimoni & Labuza 2000). *Yersinia enterocolitica* grows slower under anaerobic conditions (Huffman 2002). *C. perfringens* is also found in meat. The spores of this organism are heat resistant, surviving normal cooking or thermal processing conditions (Huang 2003).

L. monocytogenes is found in soil, water, and the digestive system of warm-blooded animals. It is commonly found in unprocessed foods of animal origin like raw milk, meat, poultry, and fish (Doyle *et al.* 2003; Sofos *et al.* 2003). FSIS and the Food and Drug Administration (FDA) have set "zero tolerance" for *L. monocytogenes* (Doyle *et al.* 2003). *L. monocytogenes* is a Gram-positive facultative anaerobe bacterium (Huffman 2002; Novak & Juneja 2003). The versatility of *Listeria* to grow in a broad temperature range (2–45°C) limits the effectiveness of restrictive measures to control this pathogen. Refrigerated and frozen storage of heat-adapted *L. monocytogenes* in ground beef does not decrease the potential of this pathogen to survive additional low temperature cooking of food (Novak & Juneja 2003).

E. coli O157:H7 is an uncommon contaminant in beef and other foods. Cattle may carry this pathogen in their intestines but exhibit no signs of illness (Doyle & Sofos 2002). *E. coli* O157:H7 was first identified as a human pathogen in 1982. It is found in the feces of healthy cattle, and it is transmitted to humans through contaminated food, water, and direct contact with infected people or animals. This pathogen causes bloody diarrhea or hemolytic uremic syndrome (Mead & Griffin 1998). *E. coli* can survive in acidic conditions (Mustapha *et al.* 2002). Mathematical models for exposure assessment have been developed specifically for *E. coli* O157:H7. These mathematical models consider five factors: (1) processing and grinding (mass of package, mass of trimmings, cross contamination, reduction due to spray washing, and growth during processing); (2) storage conditions (time in storage, maximum temperature experience during storage, maximum population density); (3) cooking (cooking preference, final internal temperature); (4) consumption (mass consumed/age dependent); and (5) dose-response (host susceptibility, probability of infection) (Cassin *et al.* 1998).

16.4 DETECTION AND ENUMERATION METHODS

To estimate spoilage bacterial load on meat, a fluorescein diacetate hydrolysis test has been proposed. This test involves the swabbing of a 25 cm^2 meat surface with cotton swabs and then washing in a tube with 9 mL of peptonated water. Resazurin reagent (1 mL) is added to 8 mL of the test solution where the swab was washed. The mixture is incubated at 25°C, and the time to change color from violet to pink is monitored. Results indicate that resazurin reduction time and fluorescein diacetate hydrolysis can be used as rapid methods to estimate spoilage bacterial load in aerobically stored meat (Venkitanarayanan *et al.* 1997).

For enumeration of aerobic plate counts, plate count agar plates are incubated at 25°C for 72 h or 30°C for 48 h (Hinton *et al.* 1998). To enumerate enterobacteriaceae (coliforms), the overlaid technique is used with Violet Red Bile agar plates incubated at 32°C for 24–48 h (Lee & Yoon 2001) or at 37°C for 24 h (Hinton *et al.* 1998). To quantify *Bacillus cereus*, selective agar plates are used. Plates are incubated at 30°C for 18–24 h. Typical colonies are blue, with egg yolk precipitation (Lee & Yoon 2001). To cultivate and enumerate *Staphylococcus aureus*, Baird–Parker agar plates are incubated at 35°C for 24–26 h. Typical colonies are black with white margins surrounded by clear zones (Cassin *et al.* 1998; Lee & Yoon 2001). Cetrimide agar plates are used to cultivate and enumerate *Pseudomonas* spp. Plates are incubated at 25°C for 24 h. Colonies are fluorescent under UV light (Lee & Yoon 2001).

To detect *Listeria monocytogenes*, a preenrichment step is necessary. After preenrichment, Oxford agar is inoculated and plates are incubated at 37°C for 48 h (Lee & Yoon 2001). A rapid method to detect *Listeria* and next-day confirmation of *L. monocytogenes* has been developed. This methodology consists of 6 h preenrichment, followed by overnight incubation in selective broth at 35°C. Changes in light transmittance in the selective broth are registered by an optical sensor. *Listeria* hydrolyzes esculin, resulting in black coloration and a drop in light transmittance (Peng & Shelef 2000).

Violet Red Bile plates with 4-methylumbelliferone glucuronide (MUG) are used to enumerate *Escherichia coli* at 37°C for 18–24 h. Typical colonies are fluorescent blue under UV light (Lee & Yoon 2001). Also, immunomagnetic separation can be used to enumerate *E. coli* O157:H7 (35). To isolate *E. coli* O157:H7, Sorbitol MacConkey agar is used. Plates are incubated at 37°C for 24 h (Dykes & Moorhead 2001). A simple method for detecting presumptive *E. coli* on fresh retail beef has been developed. The method is based on the use of double-strength lauroyl sulfate tryptone broth plus 5-bromo-4 chloro-3-indolyl β-D-glucuronide and a Durham tube (Bolling *et al.* 2002). To estimate *Salmonella*, xylose lysine desoxycholate agar is used. Plates are incubated at 37°C for 24 h (Dykes & Moorhead 2001).

Current rapid enumeration methods for microorganisms are based on microscopy, adenosine-5′-triphosphate (ATP) bioluminescence, or the measurement of electrical phenomena. Microscopy techniques are based on the use of fluorescent dyes. ATP bioluminescence acts by measuring ATP levels in bacterial cells in culture. Current detection methods are based on immunological or nucleic acid–base procedures. The most common form of these methods is the enzyme-linked immunosorbent assays (ELISA). Nucleic acid–based procedures use probes that are small segments of single-stranded complementary nucleic acid that are used to detect either DNA or RNA sequences to identify a specific microorganism. The most common nucleic acid detection method used is the polymerase chain reaction (PCR) (Ellis & Goodacre 2001).

Rapid and accurate pathogen testing is vital in the prevention of outbreaks. A combination of enzyme immunoassays (EIA) and culture methods are used in testing for *Salmonella*; this technique gives reliable results within 48 h. However, a presumptive positive EIA result must be confirmed requiring an additional 72 h. On the other hand, PCR tests for *Salmonella* are used, but some tests can take up to several days. An alternative for this is the use of real-time PCR with fluorescent hybridization probes. This methodology is ideal for rapid detection of *Salmonella* in raw and ready-to-eat meat products and gives results in less than 12 h (Ellis & Goodacre 2001). Real-time PCR is also used for quantitative meat species testing (Ellingson *et al.* 2004) and includes incubation at 35°C–37°C for 6 h of the sample homogenate in peptone water with novobiocin. An aliquot of the sample is transferred to a tube to centrifuge at 2500 × gravity (*g*) for 10 min. The pellets are extracted with the QIAGEN QIAmp DNA mini kit. Oligonucleotides used to detect *Salmonella* amplify the 251 base pair products spanning from base 2305 to base 2555. Extracted DNA is used as template for PCR amplification. The conditions to amplify DNA are 95°C for 2 min and 45 heating/cooling cycles. DNA is denatured using hybridization probes (Ellis & Goodacre 2001).

Recent concerns about prion diseases has led scientists to develop rapid methods to detect infectious agents responsible for transmitting bovine spongiform encephalopathy (BSE) and variant Creutzfeldt–Jakob Disease (vCJD). A sensitive and semiquantitative TaqManTM real-time PCR system has been developed to detect prions. This methodology relies on three basic steps: DNA extraction, use of oligonucleotide primers and probes, and the PCR conditions to amplify DNA (Sawyer *et al.* 2003).

Two comprehensive works were published in 2009 about analysis techniques of a great number of parameters in muscle foods (Nollet & Toldrá 2009a) and processed meat and poultry (Nollet & Toldrá 2009b).

Fung (2009) enumerated methods for evaluating microbial flora in muscle foods. Different sample preparation methods, total viable cell counts, identification methods, immunological methods, biomass measurements, genetic testing, and biosensors for microbes are fully discussed.

Duffy *et al.* (2009) give an overview of methods to predict spoilage of muscle foods. An analogue overview of methods for processed meats is given by Guerrero-Legarreta (2009).

A review of microbial food-borne pathogens on raw meat is presented by Mataragas *et al.* (2009a,b).

Wolffs and Radström (2006) give a description of real-time PCR as a methodology for detection of microorganisms in meat.

16.5 PRESERVATION TECHNOLOGIES

Preservative packaging for raw meats delays appearance of deterioration and retards the onset of bacterial spoilage. Browning is slowed in atmospheres poor in oxygen, and persistent browning is prevented entirely when meat is packed under oxygen-depleted atmospheres. To control bacterial spoilage, temperature is also a major factor that requires to be monitored (Brodman & Moor 2003). Modified atmosphere (MA) is used to extend the shelf life of meat products. The two major concerns for MA-packed products is microbiological and appearance shelf life. Gas mixtures of 60%–70% carbon dioxide (CO_2), 30%–40% nitrogen (N_2), and 0.3%–0.5% carbon monoxide (CO) have been used. However, temperature plays a major role. High concentrations of CO_2 and 4°C prolong the shelf life of ground beef. *Yersinia enterocolitica* is almost inhibited at 4°C and 10°C using high CO_2 concentrations. The same conditions cause *L. monocytogenes* to grow slower but do not affect *Salmonella* in ground beef at 10°C (Gill 1996).

Multilayered antimicrobial polyethylene (PE) films have been used to package ground beef using a grapefruit seed extract as an antimicrobial agent. This antimicrobial film has shown activity against *E. coli, S. aureus*, and *Bacillus subtilis*. In addition, these films contribute to a reduction of the growth rates of aerobic and coliform bacteria when compared to plain PE films (Nissen *et al.* 2000). Antimicrobial films are designed to slowly release bacterial inhibitors. These are used in combination with vacuum packaging under refrigerated conditions. Acetic and propionic acids do not have any effect on lactic acid bacteria but delay the growth of enterobacteriaceae and *Serratia liquefaciens* (Ha *et al.* 2001). Nisin, an antimicrobial peptide, is used as an inhibitor when incorporated into PE-based plastic film used to wrap meat cuts. They have proven to control indicator bacteria such as *Lactobacillus helveticus* and *Brochothrix thermosphacta* as compared with meat cuts vacuum packaged without nisin and held under the same storage conditions (Ouattara *et al.* 2000).

Microbial contamination of carcasses results from processing to conditioning. Processing influences not only the quantity of microorganisms but also the type. Meat stored under aerobic conditions is rapidly spoiled by bacteria that cause discoloration and off odors. The most common bacterium responsible for these changes is *Pseudomonas*. Additionally, other factors that influence bacterial growth and therefore shelf life are oxygen and water activity. Under refrigerated conditions, *Brochothrix thermosphacta* is a common spoilage microorganism. Vacuum or MA also affects microorganisms. In MA, several lactic acid bacteria are present, with the most common species being *Lactobacillus sakei* and *L. curvatus* (Siragusa *et al.* 1999).

16.5.1 Chilling

To extend the shelf life of meat, chilled storage of fresh beef under vacuum or carbon dioxide is used commercially. Meat packed under vacuum or carbon dioxide kept at low temperatures and inoculated with *E. coli* and *Salmonella* does not support the growth of these two pathogens (Dykes & Moorhead 2001). The objectives of postslaughter carcass and product handling are to minimize contamination and to restrict subsequent proliferation. Temperature is one of the most important factors influencing the growth of microorganisms. When hot-boned meats cool in vacuum packs or in bulk-packed cartons, they cool slowly, therefore, increasing the chances of microbial proliferation. However, compared to cooling/boning regimes, where the temperature of carcasses is reduced to 7°C or less, studies indicated that both processes are comparable and that microbiologically they are equivalent (Labadie 1999).

The rate of heat removal and the resulting rate of temperature reduction of the carcass have a substantial influence on storage life and eating quality of the meat. All meat temperatures within the carcass must be below 7°C before the carcass is further processed. After chilling, carcass meat is stored for a period of hours and up to 2 weeks. Low temperatures, minimal air movement, and high relative humidity should be maintained to maximize storage life (Bell *et al.* 1998). Once the initial microbial population increases in refrigerated beef packed with high gaseous permeability film (PE) or low-gaseous permeability films (ethyl vinyl acetate, SARAN, polyvinylidene chloride copolymer), the adaptation period of bacterial growth decreases regardless of the packaging film (James 1996).

16.5.2 Freezing

The main reason batches of frozen foods have different microbial counts and profiles are variations in the initial load and the type of bacteria. In addition, the preparatory processes, freezing conditions, and storage temperature affect the fluctuations of microbial counts during freezing (Giannuzzi *et al.* 1998). Freezing initiates several physical and physicochemical changes that lead to deterioration of meat. Manufacturing beef from exporting countries is still packed in cartons and frozen to −18°C. Studies conducted on the rate of freezing show that this rate alone does not affect protein solubility or meat color. Slowly frozen semimembranosus have more drip than fast frozen muscles. In addition, storage at a temperature below −18°C affects sarcoplasmic protein solubility and myofibrillar protein solubility. Current practice of blast freezing and storage at −18°C to −20°C is sufficient to maintain quality of manufacturing beef (Nussinovitch & Peleg 2000).

16.5.3 Vacuum packaging

Discoloration or browning results from oxidation of deoxymyoglobin or oxymyoglobin to metmyoglobin. The principal function of preservative packaging is to delay microbial spoilage by restricting the growth of spoilage organisms. Growth of aerobic organisms is inhibited when the atmosphere inside a package is modified. Meat surfaces exposed to oxygen-containing atmospheres sustained accelerated bacterial growth and discoloration due to oxidation of myoglobin. Films with minimum oxygen permeability (<100 mL/[m^2-24 h] atmosphere at 23°C and 0% relative humidity [rh]) can generally be used for vacuum or MA. Storage life can be increased by 10–15% by using a film with a permeability of <2 m/(m^2-24 h) atmosphere at 23°C and 0% rh. Most multilayer meat packaging films are of multilayer construction since gas transmission rate is a function of polymer type, film thickness, temperature, and humidity (Farouk *et al.* 2003).

The following microflora have been identified in vacuum-packaged beef under chilling storage: *Brochothrix thermosphacta, Carnobacterium divergens, Carnobacterium piscicola, Lactobacillus algidus, Lactobacillus* spp., *Lactococcus piscium, Leuconostoc gelidum, Acinetobacter, Aeromonas, Bacillus, Corynebacterium, Enterobacteriaceae, Pseudomonas*, and *Psychrobacter* (Jeremiah 2001). Slicing of meat and repacking influence shelf life (Sakala *et al.* 2002). Cuts of meat that are vacuum packaged and stored for 52 days at 0°C exceed total aerobic plate counts (107 cfu/cm^2) but enterobacteriaceae and *Pseudomonas* show retardation (Lee & Yoon 2001).

Evacuation of air from the package is clearly beneficial. Meat can be vacuum packaged in a heat-shrinkable plastic bag, with low permeability, and sealed. The effectiveness of vacuum packaging depends on close contact between a packaging film of low gas permeability and all product surfaces. Vacuum packaging achieves its preservative effect by maintaining an oxygen-depleted atmosphere. It also reduces off-odor development and moisture losses. Beef processed and vacuum packaged under appropriate conditions has a shelf life of 70 days, and vacuum packaged beef stored at 1°C has a storage life of 11 weeks. However, off odors and discoloration can be perceived after 11 weeks (Farouk *et al.* 2003).

16.5.4 Modified atmosphere packaging

Modified atmosphere packaging (MAP) is a food preservation method important in the meat industry. This technique involves modifying an atmosphere by permitting air to be enclosed or by injecting a desired initial gas mixture (Holley & McKellar 1996). An indicator of freshness for meat is the bright red color. Concentration of CO affects the color of meat cuts. For instance, ground beef, and loin eye, inside round and tenderloin steaks, stored for up to 35 days at 2°C results in typical bloomed color (Skandamis & Nychas 2002). Even when using MAP, oxygen is not always completely removed or may permeate through the packaging material. The presence of small amounts of oxygen in cooked MAP products accelerates oxidation. If MAP products contain high levels of oxygen (>1.2%), this results in quality deterioration and reduced shelf life (Hunt *et al.* 2004).

Temperature, CO_2, and water activity are factors that influence the growth of spoilage microorganisms in cooked meat subjected to MAP (Smiddy *et al.* 2002). All atmospheres containing CO_2 reduced total aerobic population numbers; however, lactic acid bacteria were not affected. Concentrations of 0.5–0.75% of CO_2 are able to extend shelf life by 5–10 days at 1 ± 1°C (Devlieghere *et al.* 1999). A 100% CO_2 concentration and low storage temperature inhibit bacterial growth, especially *Pseudomonas* and *Brochothrix thermosphacta*. Lactic acid bacteria (105–106) also inhibit growth of *Pseudomonas* and *B. thermosphacta* (Luño *et al.* 1998). Lactic acid bacteria increase to a maximum of 5.5 log cfu/cm^2 after 8 weeks in beef rib eyes kept under 100% CO_2 at 2°C. After 6 weeks, the flavor of samples drops to an inappropriate level. Flavor descriptors include "off barny" aromatic and "off barny" aftertaste, suggesting that growth rates of these organisms are related to the organoleptic changes in meat under MA and refrigerated storage conditions (Nissen *et al.* 1996). Meat cuts undergoing hot deboning and that are chilled at 7°C, electrically stimulated, and MA-packed produce meat of acceptable tenderness within 8 days of slaughter. Steaks after 24 h show dull dark lean tissue and gray to green discoloration of the fat. The aroma also deteriorates as an indication of spoilage (Nattress & Jeremiah 2000).

Utilization of spices, spice mixtures, or spice extracts such as marjoram, caraway, sage, basil, thyme, and ginger in semiprepared products intended to be frozen for up to 6 months is recommended to prolong the shelf life of the food and to inhibit lipid peroxidation during heat treatment and chilling storage (Bell *et al.* 1996). The use of antioxidant mixtures like rosemary and vitamin C together with the absence of UV radiation reduced the rate of metmyoglobin formation, lipid oxidation, and microbial growth under MA (70% O_2, 20% CO_2, and 10% N_2). These conditions extended the shelf life from about 10 to 20 days (El-Alim *et al.* 1999).

Sodium lactate is a compound that can be used to extend the shelf life of meat products packed under MA and low refrigeration temperatures. There is a synergistic effect between sodium lactate and CO_2 that extends the meat products' shelf life; this can be explained by the reduction of pH in the medium. In addition, CO_2 changes the permeability and rigidity of the microbial cell membrane and decreases the activity of membrane-bound proteins (Djenane *et al.* 2003). Meat pH values under MA remain between 5.3 and 5.6. For beef cuts under MA, L* color value increases with storage time while a* color value decreases. Shelf life (measured as color and odor degradation) can be as short as 5–15 days depending on the breed and mostly due to lipid oxidation when samples are kept at 2 ± 1°C with 90–95% relative humidity (Devlieghere *et al.* 2000).

Finally, the attention of the reader is directed to two articles published in 2006: one on MAP by Sebranek and Houser (2006) and another on perspectives for the active packaging of meat products by Coma (2006).

16.5.5 Irradiation

Food irradiation can extend the shelf life of perishable foods like beef and eliminate food-borne pathogens. High dose sterilization (10–50 kilogray [kGy]) is used to sterilize foods and eliminate food-borne viruses. Medium dose pasteurization (1–10 kGy) reduces spoilage microorganisms and can also be used to treat raw or frozen meat. Both doses can be applied to meat (Insausti *et al.* 2001). Doses of 5.0 kGy reduce bacteria counts by two to three log cycles, and when samples of meat are heated in a microwave oven, bacterial counts are reduced by one log cycle in 2 s and by two log cycles in 30 s exposure (Crawford & Ruff 1996). After treatment with 5 kGy to ground beef, lipids suffer oxidative damage, especially ground beef with high content of fat (30%) (Al-Bachir & Mehio 2001; Aziz *et al.* 2002). Irradiation reduces the counts of microorganisms and increases the shelf life of luncheon meat from 10 to 14 weeks (Aziz *et al.* 2002). A dose of 2 kGy causes ground beef patties to develop more off odors and off colors, and lower International Commission on Illumination (CIE) *L**, *a**, and *b** after 4 days of storage at 0 ± 1°C (Poon *et al.* 2003). Incorporation of antioxidants to minced beef irradiated with 4 kGy results in better retention of color. Irradiation increases Hunter "a" values up to 4 days when tocopheryl acetate is used and up to 6 days when rosemary extracts are added. Thiobarbituric reacting substance (TBARS) values increase with increasing irradiation dose; however, supplemented samples with antioxidants have lower TBARS (Montgomery *et al.* 2003).

To produce ready-to-use shelf life stable meat products, it is common to use a combination of technologies like irradiation, reduced water activity, and vacuum packaging. These hurdles are used to prevent the growth of *Clostridium sporogenes*, *Staphylococcus aureus*, and *Bacillus cereus*. Radiation treatments of 2.5 kGy result in complete elimination of inoculated *S. aureus* and *B. cereus*. Water activity of 0.85 combined with vacuum packaging prevents the growth of these three microorganisms. In general, these three technologies used in combination result in microbiologically safe and shelf-stable meat products (Formanek *et al.* 2003).

Ahn *et al.* (2006) discussed in depth meat decontamination by irradiation.

16.5.6 Ultrahigh pressure

High pressure can modify the enzymatic system, texture ultrastructure, and myofibrillar proteins of meat. High pressure has an effect on color. The increase in *L** values begins at 200 MPa. These color changes can be explained as a result of pigment denaturation and an increase in drip losses or damage of the porphyric ring and protein coagulation. Pressure intensity is more significant than holding time for redness, total color difference, and metmyoglobin content. Pressures higher than 300 MPa induce modification of meat color. Pressures of 520 MPa with contact times of 260 s decrease the total flora (Chawla & Chander 2003).

The application of high hydrostatic pressure in processed meat is treated by Suzuki *et al.* (2006) and the application of hydrodynamic pressure by Solomon *et al.* (2006).

16.5.7 Antimicrobials

Multiple intervention technology involves the use of different barriers or hurdles, such as pH changes, oxidizing environments, or other environmental changes, to cause disruption of microbial cells or cellular metabolism to either destroy or retard cell growth. Treatments using combinations of acetic acid and cetylpyridinium chloride, chlorine dioxide and cetylpyridinium chloride, and cetylpyridinium chloride and trisodium phosphate reduce *E. coli, Salmonella*, coliforms, and aerobic plate counts; however, cetylpyridinium chloride and trisodium phosphate have been shown to preserve the color of meat surfaces better than the other treatments (Jung *et al.* 2003). Active packaging is one that changes to extend shelf life or improve safety or sensory properties. Antimicrobial packaging is an active packaging concept. Antimicrobial compounds have been tested in active packaging; some of these compounds are organic acids and their salts, enzymes, bacteriocins, and compounds such as triclosan, silver zeolites, and fungicides (Pohlman *et al.* 2002).

One of the problems in the meat industry is *E. coli* O157:H7. To reduce the incidence of food-borne pathogens, two types of interventions are commonly used: one dealing with the whole carcass, the other dealing with the use of antimicrobial agents. However, there is no single intervention that will result in 100% pathogen-free meat. Bacterial attachment occurs the first minute of bacterial contact with tissue surfaces. Therefore, physical removal of contamination is the primary means for controlling *E. coli*. The most common methodologies to accomplish this include trimming and washing before slaughter (Quintavalla & Vicini 2002) and low-intensity ultrasound (Brown 2003). The use of antimicrobial agents include lactoferrin, peroxy acids, acidified sodium chloride and ozone (Quintavalla & Vicini 2002), acetic acid, lactic acid, trisodium phosphate, citric acid, formic acid, and gluconic acid (Brown 2003).

Chemical treatments commonly used in the meat industry are peroxyacetic acid (0.02%), acidified sodium chloride (0.16%), and lactic acid (2% and 4%). There are widely used alternatives to decontaminate carcasses. Peroxyacetic acid at 0.02% and acidified sodium chlorite solutions have little effect on aerobes and coliforms. Lactic acid (4%) results in bacterial reductions higher than 1.5 log cycles. Treatment with 2% lactic acid is similar to lactic acid at 4%. The major concern with these acid treatments is injured bacteria that may recover during storage and that become resistant to these types of acid treatments (Gill & Landers 2003; Pohlman & McElyea 2003; Gill & Badoni 2004).

Aymeric *et al.* (2006) discuss the use of bacteriocins against meat-borne pathogens.

REFERENCES

Ahn, D.U., Lee, E.J. & Mendonca, A. (2006) Meat decontamination by irradiation. In: *Advanced Technologies for Meat Processing* (eds L.M.L. Nollet & F. Toldrá). CRC Press, Boca Raton, Florida.

Al-Bachir, M. & Mehio, A. (2001) Irradiated luncheon meat: Microbiological, chemical and sensory characteristics during storage. *Food Chemistry*, **75**, 169–175.

Aymeric, T., Garriga, M., Jofré, A., Martín, B. & Monfort, J.M. (2006) The use of bacteriocins against meat-borne patogens. In: *Advanced Technologies for Meat Processing* (eds L.M.L. Nollet & F. Toldrá). CRC Press, Boca Raton, Florida.

Aziz, N.H., Mahrous, S.R. & Youssef, B.M. (2002) Effect of gamma-ray and microwave treatment on the shelflife of beef products stored at 5°C. *Food Control*, **13**, 437–444.

Bell, R., Penney, N., Gilbert, K., Moorhead, S. & Scott, S. (1996) The chilled storage life and retail display performance of vacuum and carbon dioxide packed hot deboned beef striploins. *Meat Science*, **42**, 371–386.

Bell, R., Harrison, C., Moorhead, S. & Jones, R. (1998) Microbiological quality of cold and hot processed chilled and frozen beef. *Food Research International*, **31**, 167–174.

Blixt, Y. & Borch, E. (2002) Comparison of shelf life of vacuum-packed pork and beef. *Meat Science*, **60**, 371–378.

Bolling, B., Schmidt, D. & Ingham, S. (2002) Development of a simple method for detecting presumptive *Escherichia coli* on fresh retail beef. *Journal of Food Science*, **67**, 258–261.

Borch, E. & Arinder, P. (2002) Bacteriological safety issues in read meat and ready-to-eat meat products, as well as control measures. *Meat Science*, **62**, 381–390.

Brodmann, P. & Moor, D. (2003) Sensitive and semiquantitative TaqManTM real-time polymerase chain reaction systems for the detection of beef (*Bos taurus*) and the detection of the family *Mammalia* in food and feed. *Meat Science*, **65**, 599–607.

Brown, H.G. (2003) Microbial problems, causes, and solutions in meat and poultry processing operations. In: *Proceedings of the 56th American Meat Science Association Reciprocal Meat Conference*, Columbia, Missouri, pp. 89–93.

Cassin, M., Lammerding, A., Todd, E., Ross, W. & McColl, R. (1998) Quantitative risk assessment for *Escherichia coli* O157:H7 in ground beef hamburgers. *International Journal of Food Microbiology*, **41**, 21–44.

Chawla, S. & Chander, R. (2003) Microbiological safety of shelf-stable meat products prepared by employing hurdle technology. *Food Control*, **15**, 559–563.

Chung, M., Lee, J. & Min, D. (2002) Effects of *Pseudomonas putrafaciens* and *Acinetobacte*r spp on the flavor quality of raw ground beef. *Journal of Food Science*, **67**, 77–83.

Coma, V. (2006) Perspectives for the active packaging in meat products. In: *Advanced Technologies for Meat Processing* (eds L.M.L. Nollet & F. Toldrá). CRC Press, Boca Raton, Florida.

Crawford, L.M. & Ruff, E.H. (1996) A review of the safety of cold pasteurization through irradiation. *Food Control*, **7**, 87–97.

Dainty, R.H. (1996) Chemical/biochemical detection of spoilage. *International Journal of Food Microbiology*, **33**, 19–33.

Devlieghere, F., van Belle, B. & Debevere, J. (1999) Shelf life of modified atmosphere packed cooked meat products: A predictive model. *International Journal of Food Microbiology*, **46**, 57–70.

Devlieghere, F., Geeraerd, A., Versyck, K., Bernaert, H., van Impe, J. & Debevere, J. (2000) Shelf life of modified atmosphere packed cooked meat products: Addition of Na-lactate as a fourth shelf life determinative factor in a model and production validation. *International Journal of Food Microbiology*, **58**, 93–106.

Djenane, D., Sanchez, A., Beltran, J. & Roncales, P. (2001) Extension of the retail display life of fresh beef packaged in modified atmosphere by varying lightning conditions. *Journal of Food Science*, **66**, 181–186.

Djenane, D., Sanchez, A., Beltran, J. & Roncales, P. (2003) Extension of the shelf life of beef steaks packaged in a modified atmosphere by treatment with rosemary and displayed under UV-free lightning. *Meat Science*, **64**, 417–426.

Doyle, M. & Sofos, J. (2002) Pathogen control in ground beef. Fact Sheet. American Meat Institute. www.meatami.org (retrieved on March 15, 2004).

Doyle, M., Donelly, C. & Wiedmann, M. (2003) *Listeria monocytogenes* fact sheet. American Meat Institute. www.meatami.com (retrieved on March 15, 2004).

Duffy, G., Dolan, A. & Burgess, C.M. (2009) Methods to predict spoilage of muscle foods. In: *Handbook of Muscle Foods Analysis* (eds L.M.L. Nollet & F. Toldrá). CRC Press, Boca Raton, Florida.

Dykes, G. & Moorhead, S. (2001) Survival of *Campylobacter jejuni* on vacuum or carbon dioxide packaged primal beef cuts stored at 1.5°C. *Food Control*, **12**, 553–557.

Dykes, G., Cloete, T. & von Holy, A. (1995) Taxonomy of lactic acid bacteria associated with vacuum-packaged processed meat spoilage by multivariate analysis of cellular fatty acids. *International Journal of Food Microbiology*, **28**, 89–100.

El-Alim, S., Lugasi, A., Hovari, J. & Dworschak, E. (1999) Culinary herbs inhibit lipid oxidation in raw and cooked minced meat patties during storage. *Journal of the Science of Food and Agriculture*, **79**, 277–285.

Ellingson, J.L., Anderson, J.L., Carlson, S.A. & Sharma, V.K. (2004) Twelve hour real-time PCR technique for the sensitive and specific detection of *Salmonella* in war and ready-to-eat-meat product. *Molecular and Cellular Probe*, **18**, 51–57.

Ellis, D. & Goodacre, R. (2001) Rapid and quantitative detection of the microbial spoilage of muscle foods: Current status and future trends. *Trends in Food Science & Technology*, **12**, 414–424.

Evans, J., Russell, S., James, C. & Corry, J. (2003) Microbial contamination of food refrigeration equipment. *Journal of Food Engineering*, **62**, 225–232.

Farouk, M., Wielickxo, K. & Merts, I. (2003) Ultra-fast freezing and low storage temperatures are not necessary to maintain the functional properties of manufacturing beef. *Meat Science*, **66**, 171–179.

Formanek, Z., Lynch, A., Galvin, K., Farkas, J. & Kery, J.P. (2003) Combined effects of irradiation and the use of natural antioxidants on the shelf-life stability of overwrapped minced beef. *Meat Science*, **63**, 433–440.

Franz, C., Stiles, M., Schleifer, K. & Holzapfel, W. (2003) Enterococci in foods—A conundrum for food safety. *International Journal of Food Microbiology*, **88**, 105–122.

Fung, D.Y.C. (2009) Methods for evaluating microbial flora. In: *Handbook of Muscle Foods Analysis* (eds L.M.L. Nollet & F. Toldrá). CRC Press, Boca Raton, Florida.

Giannuzzi, L., Pinotti, A. & Zaritzky, N. (1998) Mathematical modeling of microbial growth in packaged refrigerated beef stored at different temperatures. *International Journal of Food Microbiology*, **39**, 101–110.

Gill, C. & Landers, C. (2003) Microbiological effects of carcass decontaminating treatments at four beef packaging plants. *Meat Science*, **65**, 1005–1011.

Gill, C. & Badoni, M. (2002) Microbiological and organoleptic qualities of vacuum-packaged ground beef prepared from pasteurized manufacturing beef. *International Journal of Food Microbiology*, **74**, 111–118.

Gill, C. & Badoni, M. (2004) Effects of peroxyacetic acid, acidified sodium chlorite or lactic acid solutions on the microflora of chilled beef carcasses. *International Journal of Food Microbiology*, **91**, 43–50.

Gill, C. (1996) Extending the storage life of raw chilled meats. *Meat Science*, **43**, S99–S109.

Gray, J., Gomaa, E. & Buckley, D. (1996) Oxidative quality and shelf life of meats. *Meat Science*, **43**, S111–S123.

Guerrero-Legarreta, I. (2009) Spoilage detection. In: *Handbook of Processed Meats and Poultry Analysis* (eds L.M.L. Nollet and F. Toldrá). CRC Press, Boca Raton, Florida.

Ha, J., Kim, Y. & Lee, D. (2001) Multilayered antimicrobial polyethylene films applied to the packaging of ground beef. *Packaging Technology and Science*, **15**, 55–62.

Hinton, M., Holder, J., Hudson, W., Coombs, E., Allen, V. & Corry, J. (1998) The bacteriological quality of British feed 3. Frozen primal joints. Meat Science, **50**, 403–409.

Holley, R. & McKellar, R. (1996) Influence of unsliced delicatessen meat freshness upon bacterial growth in subsequently prepared vacuum packed slices. *International Journal of Food Microbiology*, **29**, 297–309.

Huang, L. (2003) Estimation of growth of *Clostridium perfringens* in cooked beef under fluctuating temperature conditions. *Food Microbiology*, **20**, 549–559.

Hueston, W. (2002) Food security in the meat and poultry industry. Fact Sheet. American Meat Institute. www.meatami.org (retrieved on March 15, 2004).

Huffman, R.D. (2002) Current and future technologies for the contamination of carcasses and fresh meat. *Meat Science*, **62**, 285–294.

Hugas, M. (1998) Bacteriocinogenic lactic acid bacteria for the biopreservation of meat and meat products. *Meat Science*, **49**, S139–S150.

Hunt, M., Mancini, R., Hachmeister, K., Kropf, D., Merriman, M., Delduca, G. & Milliken, G. (2004) Carbon monoxide in modified atmosphere packaging affects color, shelf life, and microorganisms of beef steaks and ground beef. *Journal of Food Science*, **69**, 45–52.

Insausti, K., Beriain, M., Purroy, A., Alberti, P., Gorraiz, C. & Alzueta, M. (2001) Shelf life of beef from local Spanish cattle breeds stored under modified atmosphere. *Meat Science*, **57**, 273–281.

James, S. (1996) The chill chain "from carcass to consumer." *Meat Science*, **43**, S203–S216.

Jay, J., Vilai, J. & Hughes, M. (2003) Profile and activity of the bacterial biota of ground beef held from freshness to spoilage at 5–7°C. *International Journal of Food Microbiology*, **81**, 105–111.

Jeremiah, L. (2001) Packaging alternatives to deliver fresh meats using short- or long-term distribution. *Food Research International*, **34**, 749–772.

Jung, S., Ghoul, M. & de Lamballerie, M. (2003) Influence of high pressure on the color and microbial quality of beef meat. *Lebensmittel-Wissenschaft & Technologie*, **36**, 625–631.

Kotzekidou, P. & Bloukas, J. (1996) Effect of protective cultures and packaging film permeability on shelf-life of sliced vacuum-packed cooked ham. *Meat Science*, **42**, 333–345.

Labadie, J. (1999) Consequences of packaging on bacterial growth. Meat is an ecological niche. *Meat Science*, **52**, 299–305.

Lee, K. & Yoon, C. (2001) Quality changes and shelf life of imported vacuum-packaged beef chuck during storage at 0°C. *Meat Science*, **59**, 71–77.

Luño, M., Beltran, J. & Roncales, P. (1998) Shelf-life extension and color stabilization of beef packaged in a low O_2 atmosphere containing CO: Loin steaks and ground meat. *Meat Science*, **48**, 75–84.

Mataragas, M. & Drosinos, E.H. (2009a) Microbial foodborne pathogens. In: *Handbook of Processed Meats and Poultry Analysis* (eds L.M.L. Nollet & F. Toldrá). CRC Press, Boca Raton, Florida.

Mataragas, M., Kagli, D. & Nychas, G.-J.E. (2009b) Microbial foodborne pathogens. In: *Handbook of Muscle Foods Analysis* (eds L.M.L. Nollet & F. Toldrá). CRC Press, Boca Raton, Florida.

McDonald, K. & Sun, D. (1999) Predictive food microbiology by the meat industry: A review. *International Journal of Food Microbiology*, **52**, 1–27.

Mead, P. & Griffin, P. (1998) *Escherichia coli* O157:H7. *The Lancet*, **352**, 1207–1212.

Montgomery, J., Parrish, F., Olson, D., Dickson, J. & Niebuhr, S. (2003) Storage and packaging effects on sensory and color characteristics of ground beef. *Meat Science*, **64**, 357–363.

Mustapha, A., Ariyapitipun, T. & Clarke, A. (2002) Survival of *Escherichia coli* O157:H7 on vacuum packaged raw beef treated with polyactic acid, lactic acid, and nisin. *Journal of Food Science*, **67**, 262–267.

Nattress, F. & Jeremiah, L. (2000) Bacterial mediated off-flavors in retail-ready beef after storage in controlled atmospheres. *Food Research International*, **33**, 743–748.

Nel, S., Lues, J.F., Buys, E.M. & Venter, P. (2004) Bacterial populations associated with meat from the deboning room of a high throughput red meat abattoir. *Meat Science*, **66**, 667–674.

Nissen, H., Sorheim, O. & Dainty, R. (1996) Effects of vacuum, modified atmospheres and storage temperature on the microbial flora of packaged beef. *Food Microbiology*, **13**, 183–191.

Nissen, H., Alvseike, O., Bredholt, S., Holck, A. & Nesbakken, T. (2000) Comparison between the growth of *Yersinia enterocolitica, Listeria monocytogenes, Escherichia coli* O157:H7 and *Salmonella* spp. in ground beef packed by three commercially used packaging techniques. *International Journal of Food Microbiology*, **59**, 211–220.

Nissen, H., Mugusten, T. & Lea, P. (2001) Survival and growth of *Escherichia coli* O157:H7, *Yersinia entercololitica* and *Salmonella enteritidis* on decontaminated and untreated meat. *Meat Science*, **57**, 291–298.

Nollet, L.M.L. & Toldrá, F. (2009a) *Handbook of Muscle Foods Analysis*. CRC Press, Boca Raton, Florida.

Nollet, L.M.L. & Toldrá, F. (2009b) *Handbook of Processed Meats and Poultry Analysis*. CRC Press, Boca Raton, Florida.

Novak, J.S. & Juneja, V.K. (2003) Effects of refrigeration or freezing on survival of *Listeria monocytogenes* Scott A in under-cooked ground beef. *Food Control*, **14**, 25–30.

Nussinovitch, A. & Peleg, P. (2000) Analysis of the fluctuating patterns of microbial counts in frozen industrial food products. *Food Research International*, **33**, 53–62.

Ouattara, B., Simard, R., Piette, G., Begin, A. & Holley, R. (2000) Inhibition of surface spoilage bacteria in processed meats by application of antimicrobial films prepared with chitosan. *International Journal of Food Microbiology*, **62**, 139–148.

Peng, H. & Shelef, L.A. (2000) Rapid detection of low levels of *Listeria* in food and next-day confirmation of *L. monocytogenes*. *Journal of Microbiological Methods*, **41**, 113–120.

Pin, C. & Baranyi, J. (1998) Predictive models as means to quantify the interactions of spoilage organisms. *International Journal of Food Microbiology*, **41**, 59–72.

Pohlman, F.W. & McElyea, K. (2003) Postharvest pathogen intervention for meat and poultry. In: *Proceedings of the 56th American Meat Science Association Reciprocal Meat Conference*, Columbia, Missouri, pp. 39–47.

Pohlman, F.W., Stivarius, M.R., McElyea, K.S., Johnson, Z.B. & Johnson, M.G. (2002) Reduction of microorganisms in ground beef using multiple intervention technology. *Meat Science*, **61**, 315–322.

Poon, P., Wong, P., Dubeski, P., Durance, T. & Kitts, D. (2003) Application of electron-beam irradiation pasteurization of ground beef, from steers fed vitamin E fortified diets: Microbial and chemical effects. *Journal of the Science of Food and Agriculture*, **83**, 542–549.

Quintavalla, S. & Vicini, L. (2002) Antimicrobial food packaging in meat industry. *Meat Science*, **62**, 373–380.

Sakala, R., Hayashidami, H., Kato, Y., *et al.* (2002) Change in the composition of the microflora on vacuum-packaged beef under chiller storage. *International Journal of Food Microbiology*, **74**, 87–99.

Sawyer, J., Wood, C., Shahanan, D., Gout, S. & McDowell, D. (2003) Real-time PCR for quantitative meat species testing. *Food Control*, **14**, 579–583.

Sebranek, J.G. & Houser, T.A. (2006) Modified atmosphere packaging. In: *Advanced Technologies for Meat Processing* (eds L.M.L. Nollet & F. Toldrá). CRC Press, Boca Raton, Florida.

Shimoni, E. & Labuza, T. (2000) Modeling pathogen growth in meat products: Future challenges. *Trends in Food Science and Technology*, **11**, 394–402.

Siragusa, G., Cutter, C. & Willet, J. (1999) Incorporation of bacteriocin in plastic retains activity and inhibits surface growth of bacteria in meat. *Food Microbiology*, **16**, 229–235.

Skandamis, P. & Nychas, G. (2002) Preservation of fresh meat with active modified atmosphere packaging conditions. *International Journal of Food Microbiology*, **79**, 35–45.

Smiddy, M., Papkovsky, D. & Kerry, J. (2002) Evaluation of oxygen content in commercial modified atmosphere packs (MAP) of processed cooked meats. *Food Research International*, **35**, 571–575.

Sofos, J., Skandamis, P., Stopforth, J. & Bacon, T. (2003) Current issues related to meatborne pathogenic bacteria. In: *Proceedings of the 56th American Meat Science Association Reciprocal Meat Conference*, Columbia, Missouri, pp. 33–37.

Solomon, M.B., Liu, M.N., Patel, J.R., Bowker, B.C. & Sharma, M. (2006) Hydrodynamic pressure processing to improve meat quality and safety. In: *Advanced Technologies for Meat Processing* (eds L.M.L. Nollet & F. Toldrá). CRC Press, Boca Raton, Florida.

Stephan, R., Stierli, F. & Untermann, F. (1997) Chemical attributes characterizing sticky post-mortem ageing in beef. *Meat Science*, **47**, 331–335.

Suzuki, A., Kim, K., Tanji, H., Nishiumi, T. & Ikeuchi, Y. (2006) Application of high hydrostatic pressure to meat and meat processing. In: *Advanced Technologies for Meat Processing* (eds L.M.L. Nollet & F. Toldrá). CRC Press, Boca Raton, Florida.

Venkitanarayanan, K., Fasutman, C., Hoagland, T. & Berry, B. (1997) Estimation of spoilage bacterial load on meat by fluorescein diacetate hydrolysis or resazurin reduction. *Journal of Food Science*, **62**, 601–604.

Vinci, G. & Antonelli, M. (2002) Biogenic amines: Quality index of freshness in red and white meat. *Food Control*, **13**, 519–524.

Wolffs, P. & Radström, P. (2006) Real-time PCR for the detection of pathogens in meat. In: *Advanced Technologies for Meat Processing* (eds L.M.L. Nollet & F. Toldrá). CRC Press, Boca Raton, Florida.

Wood, J., Richardson, R., Nute, G., *et al.* (2003) Effects of fat on meat quality: A review. *Meat Science*, **66**, 21–32.

17 Packaging and Freezing of Beef as Related to Sensory Properties

Leo M.L. Nollet

Abstract: The packaging material used, the storage conditions, and the package graphics are the three most important factors that affect the sensory properties and data analysis as related to packaging of beef.

Freezing is an acceptable method to extend the shelf life of beef provided it has been properly packaged, quickly frozen, stored at the proper temperature, and the packaged material has not been damaged. However, for optimum quality, the length of the storage should not exceed 9–12 months.

Keywords: Beef packaging, beef freezing, modified atmosphere packaging, controlled atmosphere packaging

17.1 INTRODUCTION

When evaluating the sensory properties of beef, various preslaughter factors such as the animal's diet, vitamin E supplementation, age, breed, sex, health status, etc., should be considered. Likewise, there are various postslaughter factors that may play a role in the sensory characteristics of beef. These factors may include (a) electrical stimulation of the carcass; (b) the length of aging of the carcass prior to cutting the meat and evaluating the sensory properties; (c) whether the meat was removed from the bone in a prerigor state or after the completion of rigor; (d) chilling rate of the carcass; (e) amount of external fat covering on the carcass; (f) the amount of internal fat (marbling) in the meat; (g) the degree of muscling of the carcass; (h) the particular cut or muscle being evaluated; (i) the packaging material used to package the product; (j) the length of time of storage of the packaged product; (k) the state of the product while in storage (fresh, cured, refrigerated, frozen); (l) the method of cooking (broiling, roasting, frying, microwave cooking, etc.); (m) distance of the meat from the heat source; (n) the presence or absence of forced air movement during cooking; (o) cooking temperature; (p) the length of time from cooking to evaluation; (q) the matter of serving the product immediately upon cooking or waiting and reheating the product before serving; (r) the presence or absence of salt and/or seasoning on or in the product; (s) the presence or absence of visual fat and connective tissue in the sample; (t) whether or not the meat has been subjected to irradiation and if so, how much; (u) the microbiological condition of the fresh meat before and after cooking, etc.

When evaluating beef for its sensory properties, all of the above factors and other possible conditions should be considered. The emphasis of this chapter is to discuss the major relevant factors that affect the sensory properties of beef as they relate to the packaging and freezing

Handbook of Meat, Poultry and Seafood Quality, Second Edition. Edited by Leo M. L. Nollet.

prior to preparing and presenting it for sensory evaluation by a trained or expert panel, a consumer panel, or the public at large. In order to make valid comparisons and sound conclusions about the sensory properties of beef, there are several factors that should be considered. Some of these factors include controlling the raw materials, the overall packaging, and handling of the food during preparation and presentation, to name a few. These factors must be standardized as much as possible in order to eliminate unexplained variations so the main effects of the study can be properly evaluated.

The material presented in this chapter is primarily what is considered to be well-accepted information and is not a review of the scientific literature on the subjects discussed. However, a list of several good references is provided so that those interested in more detail of the many scientific principles involved will be able to access that information with relative ease.

17.2 PACKAGING

The main purpose of the primary package of beef, as for any food, is for the protection of the food during handling, storage, shipping, and sales of the product. In addition to the primary package, there may also be secondary and tertiary packaging materials and systems involved. Consumer packaging of meat began with the development of the supermarket industry. However, since this material covers matters only relating to the sensory properties of beef, the functions of the other levels of packaging, above the primary package, will not be discussed in any detail. The primary package is the one that actually comes in direct contact with the food product and is generally considered to be the single most important component of the packing process. However, for the purpose of the sensory evaluation of beef products, the proper identification of the meat is also essential, so one might say the graphics identifying the product properly are also extremely important.

The packaging material chosen should be appropriate for the type of storage to which the product is to be exposed. If the product is to be used as a fresh (not frozen) item, then the appearance of the package in the retail counter is of extreme importance. The type of packaging material used will determine the visual appearance, a sensory trait of the beef. Consumers expect beef to be a "bright cherry red" color in order to indicate freshness and to present the appearance to which they have become accustomed. Therefore, a packaging film that allows the passage of some oxygen is essential if the color is to be as expected. The packaging film companies that supply packaging materials for fresh beef offer different types and thickness of film, but they all possess the ability to allow for the passage of enough oxygen, at atmospheric pressure, to keep the "bloom" or "bright cherry red" color for a short period of time, usually up to 48–72 h, or maybe slightly longer. After prolonged exposure to oxygen, the pigment "oxymyoglobin" will become oxidized to "metmyoglobin," which produces an undesirable brown color that is formed when the iron atom in the myoglobin pigment becomes oxidized, loses an electron, and changes from the Fe^{2+} (ferrous) to the Fe^{3+} (ferric) state.

The time that the product will remain in the desirable state can vary considerably, depending on the microbiological condition of the product, the temperature, and the light intensity under which the product is stored and if the animals had been subjected to vitamin E supplementation in their diets prior to harvesting, as this can significantly improve the color of beef over a greater period of time. Beef products offered for typica l retail sale that have been kept very clean, in low temperature storage (0°C–3°C) and in the dark or under very low light intensity, will have a much longer shelf life, being defined here as the time that the visual color is acceptable, than will identical products where the microbiological numbers are high and the temperature and light intensity are high. Of course, it is understood that the proper packaging film has been used to display these products as fresh items. The packaging materials commonly available for the packaging of all meats are often composed of different layers of materials that perform different functions such as retard water movement; regulate air movement; add strength; provide heat sealing capability, thermoformability, sparkle, etc.; and are designed to perform the functions desired in the packaging process.

It should be noted, however, that the packaging materials used for fresh meat display should not be used to package meat that is to be frozen, as the material will allow the passage of oxygen, which will cause excessive oxidation of the product and produce undesirable sensory characteristics of the products when cooked and served.

The packaging materials for beef that is to be frozen should have essentially the same physical characteristics as the material for fresh beef, except they should have a very good oxygen barrier, a very good sealing capacity, and be "tougher" so as not to crack or break under freezer conditions. These products are what is commonly called "barrier bags" or "barrier film" and are commonly used for vacuum packaging (VP). In order to preserve the native sensory properties of the frozen meat, the oxygen must be removed and kept out during storage. If this is accomplished, the shelf life of beef can easily be from 9 to 12 months without any significant loss of desirable organoleptic traits. However, if the seal breaks or the package is punctured, the shelf life is minimal, as the oxygen will cause oxidation and the presence of undesirable sensory properties as well as the occurrence of dehydration of the surface, known as freezer burn. Likewise, leaving loose packaging material around the meat allows for "frosting" inside the package, which produces a very unattractive package and presents a situation for possible loss of desirable sensory properties of the meat.

Beef may also be packaged in a modified atmosphere package or controlled atmosphere package normally called MAP or CAP packaging. The principle involved here is that the oxygen is removed and replaced with other gases, normally carbon dioxide and nitrogen. These packages may have a fluffed or pillow appearance and leave a lot of headspace where frosting can occur; thus, this type of packaging is generally only used for storage and display of nonfrozen products.

If beef is to be frozen for the purpose of accumulating samples in research projects, for example, the proper identification of the material is essential. Likewise, if commercially packaged beef products are to be evaluated for their sensory properties, having the graphics to identify the cut of beef and any additives present is essential in order to properly evaluate the sensory properties as they might well be influenced by the cut of meat, the presence of additives, etc.

In summary, the packaging material used, the storage conditions used, and the package graphics are the three most important factors that will affect the sensory properties and data analysis as related to the packaging of beef.

In the next paragraphs, some selected recent papers on packaging of beef and meat related to sensory properties are discussed.

Jayasingh *et al.* (2001) experimented to obtain >21 days red color stability for carbon monoxide (CO)-treated beef steaks in VP. In preliminary tests, pretreatment for 24 h in a 5% CO-MAP was needed to maintain redness after repackaging in VP. Pressure pretreatment with 5% CO for 2 h developed redness, but was impractical for large-scale application. Color stability was then compared after treatment of steaks in 5% CO-MAP for 24 h and next in VP; 100% CO-MAP for 1 h and next in VP; steaks and ground beef in 0.5% CO-MAP; and steaks and ground beef in polyvinyl chloride (PVC) wrap. Steaks remained red for 5, 6, 8, and <1-week storage at 2°C, respectively.

The conclusion was that extended color stability in VP was achieved by pretreatment with 5% CO for 24 h or 100% CO for 1 h.

Quality aspects of sliced, dry-cured beef "Cecina de León" (CL) preserved in vacuum and gas mixtures (20%/80% CO_2/N_2 and 80%/20% CO_2/N_2) were studied by Rubio *et al.* (2006). The evolution of microbiological, physicochemical, and sensory parameters were analyzed during storage (210 days) at 6°C. No changes were observed ($p>0.05$) in lightness (L^*), redness (a^*), and yellowness (b^*) in vacuum- and gas-packaged samples during storage. However, sensorially evaluated color showed lower values in gas-packaged samples during 30 days storage. This difference was decisive in establishing the shelf life of CL slices preserved in gas mixtures (20%/80% CO_2/N_2 and 80%/20% CO_2/N_2). Here, the conclusion was that VP allows longer storage than gas packaging as it maintains a good visual appearance of CL, the main parameter in consumers' perception of meat quality.

The effects of packaging method (aerobic packaging [AP], VP, or MAP); the form of pastirma; air-dried cured meat (sliced or non-sliced); and storage time (0, 15, 30, 60, 90, or 120 days) on the chemical, microbiological, and sensory properties of a Turkish pastirma were investigated (Gök *et al.* 2008). Overall, MAP preserved chemical, microbiological, and sensory properties of Turkish pastirma better than AP or VP. Very high correlation coefficients (almost all >0.90) were observed between subjective quality parameters (sensory properties) and objective quality parameters (thiobarbituric acid reactive substances [TBARS], hexanal content, L^*, a^*, and b^*), which suggests that the sensory panel was able to determine the quality changes over storage time precisely. Based on the results of this study, MAP should be the preferred choice of packaging in order to preserve overall quality of Turkish pastirma.

The aim of the study of Rubio *et al.* (2007) was to investigate the shelf life of commercial CL packaged in vacuum (VP) and in CO_2/N_2 atmospheres (20/80% CO_2/N_2 and 80/20% CO_2/N_2). Packaged product was stored at 6°C, measuring microbiological, physicochemical, and sensory parameters during 210 days. A pronounced initial fading was observed in redness (a^*) within the first 15 days in the VP samples and within the second month in the gas-packaged samples, attributed to the presence of white film on the surface of CL portions. No differences were observed ($p>0.05$) in texture parameters evaluated between packaging methods during storage, and the values found are within the range for CL not packaged. The sensory properties of CL stored in 20/80% CO_2/N_2 were slightly less acceptable than the samples packed under vacuum and under 80/20% CO_2/N_2 at 210 days of storage. The conclusion of this study was that no clear advantage of the gas packaging was observed compared to the VP of CL.

Ground beef was either packaged in an atmosphere of 0.4% CO, 30% CO_2, and 69.6% N_2 (CO-MAP) or vacuum (Jeong & Claus 2011). After storage (48 h, 2–3°C), packages of CO-MAP and vacuum were opened and overwrapped with PVC. Other CO-MAP and vacuum packages were left intact. Packages were initially displayed for 7 days (2–3°C). Intact packages were further displayed up to 35 days before being opened and displayed (1 or 3 days). Intact CO-MAP ground beef was always more red than intact VP ground beef. Color was relatively stable for both types of intact packages over 35 days of display. Upon opening CO-MAP ground beef, the red color decreased slower than in ground beef from VP.

Finally, the reader is directed to a webpage with interesting information on packaging of fresh and processed meat of the Food and Agriculture Organization of the United Nations (FAO).

17.3 FREEZING

The freezing of beef for storage, or as some call it, preservation, has been in existence for many years. Factors relating to the quality of frozen beef were mentioned even in the earliest textbooks used to teach about processing and storage of beef at colleges and universities in the early 1900s. Many refer to freezing as a method of preservation of meat, but others only consider freezing to be a processing method. The fact that freezing under commercial or household conditions requires a continual input of energy does not meet the test of keeping the product in an acceptable fashion under whatever conditions to which it might be exposed. When considering freezing as a preservation method, one should compare freezing to dehydration or canning where there is no demand for continued energy input in order to keep the products in an acceptable fashion, microbiologically or physically. The reader will have to decide whether or not to consider freezing a preservation or a processing method. In either case, freezing and subsequent freezer storage is a good method to extend the sensory properties of beef if the process is properly performed.

Freezing is a chilling procedure that results in the formation of ice crystals in meat. Due to the presence of solutes in meat, it generally is considered to be frozen only when it reaches a temperature of −2°C or lower. The exact amount of water remaining in solution, or as free water, depends on the exact temperature of the product. At −5°C, approximately 85% of the water is in the form of ice, whereas at −30°C, almost 100% of the water is frozen.

The purpose of freezing is to keep the original characteristics of the product, over an extended period of time, as close as possible to those of the fresh product. The freezing process is not expected to improve any of the characteristics of beef, so the ultimate quality of the frozen product depends heavily upon the state or condition of the product prior to freezing. Frozen meat used to carry a poor image by many because it was the routine practice to freeze meats only at the end of their natural shelf life as fresh meats. This practice led to producing frozen products that had been exposed to conditions that allowed for microbiological growth, oxidation, drying, etc., and were of minimal quality at best. However, if done properly, frozen beef products can be routinely produced that will keep the basic characteristics of fresh products for extended periods of time (i.e., 9–12 months). It should be kept in mind that in addition to the original meat quality characteristics, the packaging, the actual freezing process, and the storage conditions can all have significant effects on the sensory and nutritional properties of frozen beef. For example, ground beef will have a shorter shelf life than intact cuts due to the greater surface area that is exposed which increases the potential for oxidation (rancidity). Likewise, beef from animals fed on high forage diets tend to have a shorter shelf life, due to a greater amount of unsaturated fats being deposited as compared to meat from animals fed high concentrate (grain) diets. It is the areas of the double bonds in the fatty acids that allow for the oxidation of the fat, thus causing rancidity. It should be noted here that the freezing and storage of precooked beef will produce products with relatively short shelf lives as these products will normally display "warmed-over flavor" upon reheating, due to oxidation of the double bonds during the cooking process. It should also be noted that meat that contains salt, for example, will become rancid much sooner due to the oxidizing ability of the salt. The oxidation of frozen meats appears to be related to low water activity. Fresh meat has a high water activity level (0.97–0.99), whereas frozen meat has water activity levels of around 0.6–0.8. At these low levels of water activity, the heme pigments appear to initiate autoxidation.

The freezing of beef, or any meat, is to primarily stop the growth or replication of microorganisms. Freezing simply "ties up" the water so that it is unavailable for the microorganisms to use for their life processes. Some organisms may be killed during this process, but freezing is not a reliable method to destroy microorganisms, except for parasites. Freezing slows chemical reactions but does not stop them, so it should be kept in mind that there is also some loss of vitamins during long-term storage of beef, as well as for all foods.

The speed at which beef is frozen is an important factor to consider when evaluating the quality of the product upon thawing and cooking. This is especially true at and near the point of freezing (−1°C to −4°C). Water allowed to stay in this range significantly affects the size of ice crystals formed. If frozen slowly, the ice crystals formed are large crystals (140–150 m in diameter) and damage the muscle cells, thus allowing a greater amount of "drip" or "purge" upon thawing of the product. However, when the critical temperature of −1°C to −4°C is passed within a short period of time (1.5–2.0 h), the water in the cells freeze before substantial diffusion of the water to the outside of the cell can occur. Today, it is common in industry to freeze beef by cryogenic methods using either liquid nitrogen or liquid carbon dioxide (CO_2). These materials produce very low temperatures (i.e., −73°C to −129°C in liquid nitrogen tunnels and about −62°C in CO_2 tunnels) and allow for ideal conditions for freezing beef. However, if there are significant temperature fluctuations during storage, small ice crystals can thaw and recrystallize, which can also lead to extensive damage to the cell membranes and cause excessive drip or purge as well as increase rancidity. The storage of frozen beef for extended periods of time in so-called "frost-free" refrigerator units, which defrost themselves often, can magnify the problem related to temperature fluctuations and the shelf life of frozen beef.

The time that has elapsed between the harvesting of the animal and placing the beef in the freezer is also a very important factor. For optimum quality, beef should be frozen only after the completion of rigor mortis, which is the stiffening of the body after death. If beef is frozen prior to the completion of rigor, the muscles will contract an abnormal amount and upon thawing demonstrate "thaw rigor," and this will in turn produce meat with a higher level of contraction and greater toughness. Likewise, the rapid chilling or freezing of a carcass after harvesting causes what is known as "cold shortening," which also produces increased toughness of the meat. This toughness is due to the method of handling alone and does not reflect other true differences in the tenderness of the meat had the imposed toughness due to freezing or rapid chilling not been present. In research projects, it is common to store beef from animals harvested over a particular period of time. The length of time from harvesting to freezing must be kept constant, as different times of aging under refrigeration can be a significant factor in the ultimate level of tenderness of the meat. Most meat scientists generally accept that a fixed time of 7–14 days from harvesting to freezing should elapse. During this time, it is customary to leave the muscles intact in the carcass or wholesale cut, whereupon the cuts are obtained, packaged, identified, frozen rapidly, and stored in the freezer at about −17°C until evaluated.

Freezing of beef also causes the pigment, myoglobin, to use greater amounts of oxygen, thus showing the product to be dark in color rather than bright red. This dark color has caused many problems in trying to merchandise frozen cuts of beef, as consumers generally do not accept this dark color as an indicator of fresh or high quality beef. However, the color of the meat returns to normal upon cooking as the myoglobin turns to denatured metmyoglobin, the pigment responsible for the formation of the brown color of cooked beef.

When using frozen beef, the product should always be thawed under refrigerated conditions (1°C–4°C) in order to maintain the optimum quality and safety of the product. Likewise, it is generally accepted that once meat has been thawed it should not be refrozen. Each time the product is frozen, the ice crystals can cause additional damage to the cells and allow for a greater amount of water loss (purge), and thus reduce the juiciness of the product. However, if the product has been thawed under refrigeration, there is generally no concern for the safety of the meat, just a loss of quality.

In summary, the freezing of beef is an acceptable method to extend the shelf life of the product, provided it has been properly packaged, quickly frozen, stored at the proper temperature, and the packaging material has not been damaged so as to allow the entrance of oxygen or the loss of moisture. However, for optimum quality, the length of the storage should not exceed 9–12 months. Otherwise, significant loss of the desirable sensory properties related to odor and taste can occur. The length of storage time should not normally affect the degree of tenderness or juiciness of beef.

As examples, two studies on the effects of freezing of beef on its sensorial properties are discussed. In the study of Lagerstedt *et al.* (2008), the authors determined how sensory quality, shear force, and water loss differ between beef stored either chilled or frozen before cooking. Meat tenderness was analyzed instrumentally and sensorially using both a consumer panel and a semi-trained panel. Both *M. longissimus dorsi* (LD) from eight young Holstein bulls were cut into eight samples, weighed, vacuum packed, and aged at 4°C for 2, 7, or 14 days. After aging, the frozen samples were kept at −20°C prior to heat treatment. The sensory panel experienced the chilled meat to be more tender, juicier, and having a more intense meat taste than the frozen meat, whereas the consumers could not find any significant difference in degree of liking.

The results indicate that conclusions from studies concerning sensory quality of beef will depend on whether the meat has been kept chilled or frozen before testing.

The effect of frozen storage conditions on meat from 36 Morucha × Charolais crossbred yearlings was studied (Vieira *et al.* 2009). Slices of M. *Longissimus thoracis* were randomly assigned to groups arising from the combination of experimental factors. These factors were: aging extent (3 and 10 days); length of frozen storage (0, 30, 75, and 90 days); and temperature (−20°C and −80°C). As frozen storage period increased, colorimetric parameters L^*, a^*, and C^* decreased, but H^* increased. No effect of freezing temperature was observed in any of the parameters studied.

REFERENCES

FAO (Food and Agriculture Organization of the United Nations), http://www.fao.org/docrep/010/ai407e/AI407E21.htm

Gök, V., Obuz, E. & Akkaya, L. (2008) Effects of packaging method and storage time on the chemical, microbiological, and sensory properties of Turkish pastirma—A dry cured beef product. *Meat Science*, **80** (2), 335–344.

Jayasingh, P., Cornforth, D.P., Carpenter, C.E. & Whittier, D. (2001) Evaluation of carbon monoxide treatment in modified atmosphere packaging or vacuum packaging to increase color stability of fresh beef. *Meat Science*, **59** (3), 317–324.

Jeong, J.Y. & Claus, J.R. (2011) Color stability of ground beef packaged in a low carbon monoxide atmosphere or vacuum. *Meat Science*, **87** (1), 1–6.

Lagerstedt, A., Enfält, L., Johansson, L. & Lundström, K. (2008) Effect of freezing on sensory quality, shear force and water loss in beef *M. longissimus dorsi*. *Meat Science*, **80** (2), 457–461.

Rubio, B., Martínez, B., González-Fernández, C., García-Cachán, D., Rovira, J. & Jaime, I. (2006) Influence of storage period and packaging method on sliced dry cured beef "Cecina de Leon": Effects on microbiological, physicochemical and sensory quality. *Meat Science*, **74** (4), 710–717.

Rubio, B., Martínez, B., González-Fernández, C., García-Cachán, D., Rovira, J. & Jaime, I. (2007) Effect of modified atmosphere packaging on the microbiological and sensory quality on a dry cured beef product: "Cecina de León". *Meat Science*, **75** (3), 515–522.

Vieira, C., Diaz, M.T., Martínez, B. & García-Cachán, MD. (2009) Effect of frozen storage conditions (temperature and length of storage) on microbiological and sensory quality of rustic crossbred beef at different states of ageing. *Meat Science*, **83** (3), 398–404.

FURTHER READINGS

Hui, Y.H., Nip, W.K., Rogers, R.W. & Young, O.A. (2001) *Meat Science and Applications*. Marcel Dekker, New York.

Judge, M., Aberle, E., Forrest, J., Hedrick, H. & Merkel, R. (1989) *Principles of Meat Science*, 2nd edn. Kendall/Hunt Publishing Company, Dubuque, Iowa.

Kinsman, D.M., Kotula, A.W. & Breidenstein, B.C. (1994) *Muscle Foods: Meat*, Poultry and Seafood Technology. Chapman-Hall, New York.

Lawrie, R.A. (1974) *Meat Science*, 2nd edn. Pergamon Press, Oxford, U.K.

Price, J.F. & Schweigert, B.S. (1987) *The Science of Meat and Meat Products*, 3rd edn. Food and Nutrition Press, Inc., Westport, Connecticut.

Robertson, G.L. (1993) *Food Packaging*, Principles and Practice. Marcel Dekker, Inc., New York.

Romans, J.R., Costello, W.J., Carlson, C.W., Greaser, M.L. & Jones, K.W. (2001) *The Meat We Eat*, 14th edn. Interstate Publishers, Inc., Danville, Illinois.

Part Four
Pork Quality

18 Sensory Evaluation of Pork Flavor

Veronika Válková

Abstract: Flavor perception is a highly complex system. Flavorful foodstuffs can include hundreds of volatile compounds. In addition, these volatiles are not uniformly released from the food, but may interact with macro ingredients such as fat, carbohydrates, and proteins of the food base in many different patterns.

Flavor perception can be divided into three stages: odor assessment, flavor in the mouth assessment, and after-taste assessment.

In sensory analysis, a panel of judges/assessors is used. Generally, it is necessary to train the panel. The degree of training depends on the type of test that should be applied. If the opinion of the consumers is being investigated, then an untrained consumer panel is used. Methods to evaluate pork flavor are discriminative methods, descriptive sensory methods, consumer's evaluation, or dynamic sensory methods. All data must be statistically processed. Different factors influence pork flavor.

Keywords: Sensory analysis, discriminative methods, descriptive sensory methods, flavor profile method, quantitative descriptive analysis, the quantitative flavor profiling technique, sensory spectrum, free-choice profiling, preference mapping, dynamic sensory methods, generalized Procrustes analysis, ANOVA, MANOVA, boar taint

18.1 INTRODUCTION

Flavor is one of the most important attributes of pork meat quality. It is still very popular with consumers, although its consummation differs within many factors, for example, geographical origin of the consumer as well as age and gender. Several definitions of flavor exist:

> Flavor is a complex combination of the olfactory, gustatory and trigeminal sensations perceived during tasting. The flavor may be influenced by tactile, thermal, painful and/or kinesthetic effects (ISO 5492:2008 2009).

> Flavor is a complex group of sensations compromising olfactory, taste, and other chemical sensations such as irritation or chemical heat (Lawless & Heymann 1998).

A number of receptors can be involved in flavor perception, these being the gustatory (taste buds), olfactory, touch (tactile), thermal (heating and cooling), and pain receptors. Flavor has a major role in determining the acceptability of foods, and with texture it is a dynamic process, as changes in intensity occur over time. Flavor perception is the result of a number of steps, starting with ingestion and continuing even after the food has been swallowed (Carpenter *et al.* 2000).

Handbook of Meat, Poultry and Seafood Quality, Second Edition. Edited by Leo M. L. Nollet.

Flavor perception is a highly complex system. Flavorful foodstuffs can include hundreds of volatile compounds. In addition, these volatiles are not uniformly released from the food, but may interact with macro ingredients such as fat, carbohydrates, and proteins of the food base in many different patterns (Chung *et al.* 2003). Flavor can be perceived retronasally only when the odor-active volatiles are released from the food to the buccal headspace during mastication (Laing & Jinks 1996).

Flavor perception can be divided into three stages (Carpenter *et al.* 2000):

1. Odor assessment—sniffing food before it enters the mouth
2. Flavor in the mouth assessment—when food is in the mouth
3. Aftertaste assessment—the sensations perceived after a sample has been swallowed

In sensory analysis, a panel of judges/assessors is used. There are some criteria for the selection of assessors (health status, allergies, availability personality, verbal creativity, concentration, motivation, team player, smoker, dietary habits, education, sensitivity, specific anosmias, previous experience, dentures, medication, user of products, supplements) (Murray *et al.* 2001). Generally, it is necessary to train the panel. The degree of training depends on the type of test that should be applied. If the opinion of the consumers is being investigated, then an untrained consumer panel is used.

18.2 SENSORY ANALYSIS—GENERAL CONDITION

Sensory analysis as an analytical method has to be conducted exactly in the defined and reproducible conditions. Hence, a laboratory sensory method should be carried out in a sensory laboratory. The sensory laboratory should be in a silent part of the building with climate control. The preparation room has to be separated from other parts of the sensory laboratory. During pork flavor evaluation, red light is used to minimize the influence of the appearance of the meat sample (red or brown color of the pork meat could evoke in panelists the degree of doneness or intensity of flavor, which could influence panelists' evaluation). Troublesome effects like noise, odors, and expressive decoration should be minimized so that assessors do not feel stressed. Each assessor should have their own space so as to minimize communication among them, except during discussions. This can be achieved by creating two different areas—a booth area and a discussion area. Right before the sensory analysis, the assessor must not smoke (although using smokers as assessors is controversial, except with consumer testing), drink alcoholic beverages or coffee, or eat any spicy food. When pork flavor is evaluated, assessors must clean their palate between samples with nonsalted crackers and water. Dishes are generally recommended to be white and odor resistant.

The sample should be designed with an anonymous three-digit code. Assessors should get an appropriate amount of sample corresponding to the method used. If it is necessary to taste a sample multiple times, then the quantity of the sample should be adequately higher. For odor evaluation, it is advisable to serve the sample in closed packaging (either in the same packaging in which the sample has been prepared, or that in which it has been covered immediately post preparation). Assessors open the packaging just before evaluation.

A typical scale for the evaluation of pork flavor is an unstructured line scale anchored with minimum and maximum terms. The length of this scale is usually 10 or 15 cm.

In sensory analysis, three types of tests are usually used—discriminative, descriptive, and hedonic.

18.3 DISCRIMINATIVE SENSORY METHODS

Discriminative methods are used for determining the existence of differences between samples. These differences can be investigated between whole samples or within one property of the samples (e.g., in the saltiness of two samples). The most used tests are the paired test, duo-trio test, triangle test, two-out-of-five test, "A"—"not A" test, simple ranking test, and difference-from-control test. Discriminative methods can be a useful tool in the product innovation process, especially when it is necessary to assess if small changes in product composition have an impact on the overall quality, for example, meat products made from pork meat fortified with vitamins or trace elements versus nonfortified, the use of a cheaper ingredient, and an increase or decrease in the quantity of an ingredient (low-fat meat products). The aim of the study would be: Do panelists detect differences? For discriminative testing, it is possible to use both a trained as well as a consumer panel. A trained panel will be able to detect a greater number of minor differences between samples than a consumer panel; for the industry, however, useful information is more often acquired from the real consumer.

A paired preference test is used when difference in the preference of consumers is obtained. The consumer gets two anonymous samples simultaneously and has to determine which one is preferred. Although there can be a modification with a no-preference answer, it usually complicates the statistical interpretation.

Discriminative methods can detect differences in intensity or the acceptance/preference of the sample depending on the examined question; however, these differences cannot be quantified.

18.4 DESCRIPTIVE SENSORY METHODS

Descriptive methods quantify differences between samples. They are useful tools in determining the properties of the sample, the importance of the properties, and the intensity of these properties.

Descriptive analysis is a class of sensory tests in which a trained panel rates the specified attributes of a product on scales of perceived intensity (Lawless & Heymann 1998). Descriptive sensory analyses are distinguished from other sensory-testing methods in that they seek to profile a product on all of its perceived sensory characteristics (Murray *et al.* 2001). They give us information about the sensory character of the product, the main properties for the preference and/or acceptance of the product, and the important properties in the sensory profile of the product including their intensities.

First, the qualitative aspects of a product are determined, for example, all aroma, appearance, flavor, texture, aftertaste, and sound properties that distinguish it from others. Sensory judges then quantify these product aspects in order to facilitate the description of the perceived product attributes (Murray *et al.* 2001). Sensory descriptive analysis usually has two main phases (Carpenter *et al.* 2000):

1. Qualitative, in which attributes are identified
2. Quantitative, in which ratings/scores are assigned

In praxis, it means that first the assessors develop descriptors for the sensory characteristics of a product and then use these descriptors to quantify differences between products.

A major strength of descriptive analysis is its ability to allow relationships between the descriptive sensory and the instrumental or consumer preference measurements to be determined. Knowledge of a "desired composition" allows for product optimization, and validated models between the descriptive sensory and the relevant instrumental and/or preference measures are highly desirable and are increasingly being utilized in the food industry (Murray *et al.* 2001).

Descriptive sensory analyses are also used for quality control, for the comparison of product prototypes to understand consumer responses in relation to products' sensory attributes, and for sensory mapping and product matching (Gacula 1997). It may also be used to track product changes over time with respect to understanding shelf life and packaging effects, to investigate the effects of ingredients or processing variables on the final sensory quality of a product, and to investigate consumer perceptions of products (e.g., free-choice profiling [FCP]) (Murray *et al.* 2001).

Descriptive analysis requires panelists who are trained (degree of training depends on the method used, except FCP where untrained consumer panel is used) and have adequate sensory skills. Panelists are recruited based on the control of sensory skills, because not all panelists have sufficient sensitive sensory skills for descriptive tests.

18.4.1 Flavor profile method

The flavor profile method (FPM) was the first reported descriptive method developed in the late 1940s and early 1950s at Arthur D. Little and Co by Lören Sjöstrom, Stanley Cairncross, and Jean Caul to complement the existing formal and informal sensory techniques for the expanding food industry (Lawless & Heymann 1998; Piggott *et al.* 1998; Murray *et al.* 2001).

The vocabulary development and rating sessions are carried out during group discussions, with panel members considering aspects of the overall flavor and the detectable flavor components of food (Murray *et al.* 2001). The vocabulary used to describe the product and the product evaluation itself is achieved by reaching agreement among the panel members. The profile describes the overall flavor, and the flavor notes and estimates the intensity and amplitude (overall impression) of these descriptors. The technique provides a tabulation of the perceived flavors, their intensities, their order of perception, their aftertastes, and their overall impression (amplitude) (Lawless & Heymann 1998).

A panel of four to six judges is trained to precisely define the flavors of the product category in a 2–3-week period. The panel is exposed to a wide range of samples in the product category, and during training panelists review and refine the flavor vocabulary. Reference standards and definitions for each descriptor are also created during the training phase. The temporal order of attributes is recorded (Lawless & Heymann 1998; Murray *et al.* 2001).

The original flavor profile results either used numbers and symbols or were represented graphically by the "sunburst" diagram. The data set is not analyzable by conventional statistical procedures. Therefore, the FPM is classified as a qualitative descriptive technique.

However, with the introduction of numerical scales, FPM became the profile attributes analysis (PAA) (Murray *et al.* 2001). The use of numerical scales allows researchers employing this method to use statistical techniques to facilitate data interpretation (Lawless & Heymann 1998).

An advantage of the FPM/PAA is that the panel is highly trained and therefore sensitive even to small product differences. The major disadvantage of FPM is that it depends on a small number of highly trained experts. Hence, the departure of even one panel member can have a severe impact on the sensory program. The technical language used by assessors may also be difficult to interpret by marketing personnel, in terms of relating the data to consumer preference (Murray *et al.* 2001).

Discussion leads to data generation and the panel's view of the product. The panel leader participates actively. Samples are served to the panelists in the same form that they would be served to the consumers.

18.4.2 Quantitative descriptive analysis

Quantitative descriptive analysis (QDA®) was developed with the intention of including strategies to account for behavioral aspects of perception. The use of an interval scale and emphasis on the statistical evaluation of results allowed for some behavioral effects to be accommodated. Data are averaged across the panel as a whole, as opposed to the group discussing the result to reach a consensus (Piggott *et al.* 1998).

In contrast to FPM, in QDA® the panel leader is not an active participant in order to prevent bias. Unstructured line scales are used for the evaluation intensity of attributes.

Assessors generate a list of terms that describe differences among the products and discuss them. The result of the discussion is a standardized vocabulary describing the sensory differences among the samples. The reference standards or verbal definition of descriptive terms as well as the order of attributes during evaluation are established. The result is often presented as a cobweb or spider diagram.

Unlike many other methods, QDA® assumes that judges will use different parts of the scale to evaluate product attributes; therefore, it is the relative differences among products and not the absolute differences that provide information. The design for descriptive analysis is based on repeated measures and the statistical analysis is generally conducted using the analysis of variance (ANOVA). A limitation of QDA® is that it is difficult to compare results between panels, between laboratories, and from one time to another with this technique (Murray *et al.* 2001).

18.4.3 Quantitative flavor profiling technique

Quantitative flavor profiling (QFP) technique is a modified version of QDA®. In contrast to QDA® in QFP, only the flavor attributes of products are profiled.

The descriptive language used in QFP is a common standardized flavor language, developed by a panel of six to eight people who are typically flavorists and not directly involved in the project. The language used is technical and a proposed advantage of this method is that no erroneous terms will be included in the vocabulary as the flavorists have a wide technical knowledge. QFP depends to a large degree on the use of reference standards to demonstrate concepts and estimated intensity (Murray *et al.* 2001).

Stampanoni (1994) used this method for the evaluation of flavor profiles of dairy products.

18.4.4 Sensory spectrum®

This method was developed on the principles of the texture profile method and QDA®; it is concentrated on the complete spectrum of product characters. The panelists use a standardized lexicon of terms.

Panel training takes much more time than previous methods, approximately 100–130 h. The scales are standardized and anchoraged with multiple reference point. The scales are based on the extensive use of reference points along their range which correspond to food reference samples. The use of these points purportedly greatly reduces panel variability allowing better correlations with other data, for example, instrumental data (Murray *et al.* 2001). The panelists are trained to use the scales identically, and proponents of this method state that the resultant data values are absolute. This means that it should be possible to design experiments that include only one sample and to compare data from that sample with those derived in a different study (Lawless & Heymann 1998).

The panelists should have basic knowledge of the physiology and psychology of sensory perception and training, which is conducted with regard to the ability to describe each modality (flavor, appearance, taste, etc.) with respect to technical principles.

The intensity scales are said to be absolute, that is, they are created to have equi-intensity across scales; therefore, 5 on a sweetness scale is of equal strength as 5 on a salty scale, and so on. In addition, it is thought that absolute calibration is feasible for most attributes (Lawless & Heymann 1998; Murray *et al.* 2001).

18.4.5 Free-choice profiling

A variation of descriptive analysis is free-choice profiling (FCP). Unlike QDA® and the Spectrum method®, the data cannot be averaged across a panel, as each descriptor has a unique meaning to a particular assessor (Piggott *et al.* 1998).

FCP was developed to assist the demands of the marketing and product development teams who required information on target consumers' perceptions of products rather than the more technical descriptions of the products typically produced by trained sensory panels (Murray *et al.* 2001).

FCP was described by Williams and Langron (1984). The main advantage of FCP over conventional descriptive methods is that an agreed descriptive vocabulary need not be defined and, therefore, untrained consumers may be used as descriptive assessors (Delahunty *et al.* 1997). Each panelist creates their own idiosyncratic list of descriptive terms (Lawless & Heymann 1998). The method allows panelists (consumers) to use any number of their own attributes to describe and quantify product attributes, and it is based on the assumption that panelists differ not in perceptions but merely in the way in which they describe them (Murray *et al.* 2001).

The number of used descriptors by individual panelists differs. Some of them use less descriptors while some use more. That is why classical statistical methods are not suitable for this method.

Individual descriptive information may be rationalized by generalized Procrustes analysis (GPA). However, although products are separated according to differences in perceived sensory character, results are often difficult to interpret due to the large diversity of individual vocabularies. But some descriptive terms, although undefined, may have similar meaning (e.g., sweet flavor) for some of the assessors, and it is likely that these most commonly used descriptors would be included in a developed and agreed vocabulary. It is also likely that two or more descriptors may be measures of a common stimulus and such ambiguity should be taken into account (Delahunty *et al.* 1997).

The GPA allows the analyst to determine the terms used by individual panelists that appear to be measuring the same sensory attributes as other judges. With this technique, each judge's data are transformed into individual spatial configurations. These configurations of individual judges are then matched to a consensus configuration (Lawless & Heymann 1998).

The advantage of the FCP is the small demand on panelists. Participants need only be able to use a scale and be consumers of the product under evaluation. It also provides a more "consumer-oriented" opinion on the product (Murray & Baxter 2003).

Because no training method is relatively fast, Delahunty *et al.* (1997) used FCP for assessing the sensory characteristics of commercial hams and the relation of the product's character to the consumers' preference. Untrained ham consumers easily discriminated between hams in terms of appearance, texture, flavor, and overall preference using FCP. The consumers used terms such as flavorful, salty, sweet, meaty, sour, intense, processed, smoked, and hammy for the evaluation of flavor. For better consensus of the difference between products, the soft independent modeling of class analogy (SIMCA) was used. This approach helped show how similar descriptors were used by different assessors and helped identify descriptors that could be used to discriminate between hams.

FCP has various phases. First, the untrained panel of real consumers is used. Each assessor gets several samples of evaluated foodstuff and in turn creates their own list of terms for the characterization of the foodstuff. Then, each assessor creates their own scale that is used for evaluating the intensity of perceived property. Second, the obtained data are analyzed by GPA to determine parameters from individual assessors that are similar (different expressions from different assessors are uniformed). A two- or three-dimensional space is created.

18.5 CONSUMERS' EVALUATION

Meat quality parameters can provide useful information when observations made by consumers are interpreted. The link between meat quality measurements and consumers' criteria for good eating quality is especially valuable for the meat industry in order to meet consumer preferences and demands (Meinert *et al.* 2008).

Consumers are naive assessors with limited or no training. Their evaluation, though simple with little ability to describe the product, gives irreplaceable information about consumers' acceptance, preference, and perception. A trained descriptive panel is able to provide extensive information about sensory properties of the product; such information is detailed and accurate. Nevertheless, a highly trained panel cannot give information about consumer opinion. Very often, the consumer as well as the trained/expert panel are used to get the complete description of the product.

There exist two types of consumer studies—qualitative and quantitative. In the qualitative test, it is possible to investigate consumers' opinion and behavior. What motivates consumers to choose and buy the specific product can be ascertained. In some studies, it is necessary to investigate the quantity; therefore, the quantitative consumer test is used. To get complete information, both types of tests should be used.

The most important characteristics of quantitative consumer studies of meats are as follows (Muñoz 1998):

- The naive user or potential user of the meat product participates.
- A relatively large group of consumers participates (depending on the statistical power required).
- Careful consideration to the meat product amount, consistency, and preparation for the large number of participating consumers must be given.
- A written questionnaire is used.
- The questionnaire presents attributes and scales to collect the perceived levels of sensory attributes by consumers.
- Sensory questions may include overall liking, liking and perceived intensity of attributes (e.g., juiciness, tenderness), and preference.

- The selection of attributes in the questionnaire is critical.
- Data are statistically analyzed.

A big consumer research program was conducted by Moeller *et al.* (2010a)—2280 consumers in three US cities evaluated pork chops. Consumer recruitment was conducted via telephonic interview. Participants were selected according to criteria like primary household, grocery shoppers, age, household income, and presence of children, and of course all respondents had to be regular pork consumers. 8-point scales were used for evaluating seven parameters:

Overall like/dislike (1 = dislike extremely and 8 = like extremely);
Juiciness like/dislike (1 = dislike extremely and 8 = like extremely);
Level of juiciness (1 = extremely dry and 8 = extremely juicy);
Tenderness like/dislike (1 = dislike extremely and 8 = like extremely);
Level of tenderness (1 = extremely tough and 8 = extremely tender);
Flavor like/dislike (1 = dislike extremely and 8 = like extremely);
Level of flavor (1 = extremely bland or no flavor and 8 = extremely flavorful).

The final ballot question asked: How likely would it be for you to purchase this sample if it were available at a reasonable price in your area? Likelihood of purchase response options were: definitely would not buy, probably would not buy, may or may not buy, probably would buy, and definitely would buy.

18.5.1 Preference mapping

The external preference mapping (PREFMAP) consists of a regression, where consumers' acceptability (hedonic ratings) is the response (dependent) variable, and the main dimensions of a PCA (principal component analysis) carried out on the sensory data, as stated by trained assessors, are the unique explanatory variables. Hence, this technique can be used to map consumers' data onto the assessors' space and obtain the sensory properties that influence consumer's acceptability. The main PREFMAP limitation is that the number of consumers significantly fitted in the map might be low, usually under 50%, even when high significance levels (20% or 25%) are employed (Resano *et al.* 2010).

18.5.2 Focus grouping

Focus grouping is a qualitative sensory method that is based on the moderated discussion of 6–12 real consumers. A moderator is nondirective. The principle of focus grouping is that consumers get the sample and discuss its properties. After discussion, consumers get the sample again to summarize their opinion.

18.6 DYNAMIC SENSORY METHODS

Time–intensity (T–I) measurement is a special case of descriptive analysis, where a single characteristic is tracked as it changes over a period of time (Piggott *et al.* 1998). T–I procedures have been used to characterize persistent sensations and to monitor perceived intensity as flavor is released during ingestion and mastication (Noble & Lesschaeve 2006). The technique was primarily developed as a means of studying the persistence of tastes such as sweetness,

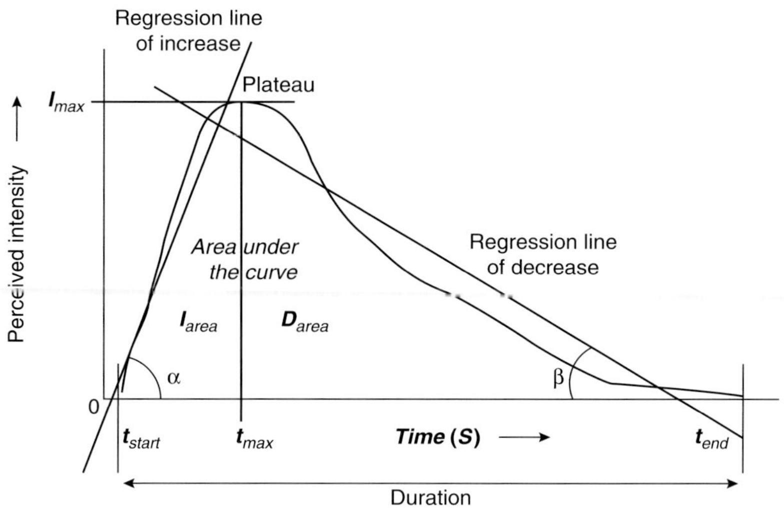

Fig. 18.1 Typical time–intensity curve.
Source: Peyvieux and Dijksterhuis (2001).

bitterness, and astringency (Piggott *et al.* 1998; Piggott 2000; Dijksterhuis & Piggott 2001). The result of T–I measurement is typically a curve showing how the reported intensity of the sensation rises and falls when tasting a sample (Dijksterhuis & Piggott 2001).

An average curve can be constructed from three parameters (maximum intensity (I_{max}), the point at which I_{max} is reached (t_{max}), and the first point of no more perception) that are the averages of the corresponding individual perception. The ascending and descending parts of the curve are treated separately and the curves are normalized both in the intensity and time directions (Dijksterhuis & Piggott 2001).

Software enables sensory subjects to use the mouse to record their sensory response by moving a cursor along a horizontal or vertical line (depending on the type of program being used) displayed on a computer screen. The computer is programmed to record intensity data at fixed time intervals; this results in T–I curves from which several parameters can be extracted (Koeferli *et al.* 1998). A typical T–I curve is shown in Fig. 18.1.

Parameters that can be extracted from T–I curves are as follows (Koeferli *et al.* 1998; Peyvieux & Dijksterhuis 2001):

Area under curve (*AUC*)	Total area under the T–I curve, total sensory intensity
Maximum intensity (I_{max})	Maximum intensity of sensation
Time to I_{max} (T_{max})	Time to maximum intensity (s)
Reaction time (t_{start})	Time necessary for onset of sensory perception (s)
Time at end (t_{end})	Time at which sensory perception is extinguished (s)
Intensity at end (I_{end})	Intensity at the end of T–I measurement
Increasing angle (α)	Angle of increase to I_{max}, i.e., rate of onset
Decreasing angle (β)	Angle of decrease from I_{max}, i.e., rate of decrease
Increasing area (I_{area})	Area under the increasing part of the curve
Decreasing area (D_{area})	Area under the decreasing part of the curve

Even with extensive training, individual panelists have characteristic curves with idiosyncratic, yet reproducible, patterns (Noble & Lesschaeve 2006). The training of the sensory panel for T–I was discussed by Peyvieux and Dijksterhuis (2001). T–I assessments result in data that have a common shape, a large amount of data points, and large interindividual differences but intraindividual consistencies. The panel was trained in three steps:

1. Introduction of the method to the panelists.
2. Familiarization of the panelists with the computer system and the T–I task using basic taste solutions.
3. Training of the panelists using the real product:
   a. Sensory profiling
   b. Pilot T–I study

Two main problems were found in the training stage with basic taste stimuli:

1. Initially, not all assessments are made according to the theoretical and generally expected shape of the T–I curve.
2. Panelists seem to have problems replicating the intensity (mainly with the height of the curve).

GPA was used to analyze the line scale profiling data in order to check the consistency of the use of attributes by the panel. During the training with the real product, most panelists showed that the quality of their curves increased. The PCA was used for analyzing the data.

Reinbach *et al.* (2007) used T–I evaluation to investigate into the effect of textures as well as two different chili products on the intensity of oral burn and meat flavor of pork patties. The panelists tasted two pork patties, one with chili and one without, and agreed upon the descriptors "chili burn" and "meat flavor." "Chili burn" was defined as the total intensity of oral burn in the mouth (oral cavity, palate, and tongue) and in the throat. The flavor perceived from pork patties heated in the oven was defined as "meat flavor." The pork patties spiced with chili powder were perceived significantly hotter and had less pronounced meat flavor than the minced chili patties.

18.6.1 Temporal dominance of sensations technique

When T–I is used for several attributes, it becomes very time consuming as each attribute requires one run (Pineau *et al.* 2009). In the temporal dominance of sensations (TDS) technique, the monitor displays all attributes and panelists continuously record the intensity of the dominant sensation. The TDS method has permitted better discrimination among products than standard T–I and has potential application in the formulation of flavors. It permits modifying the flavor essence so that the temporal profile of the low-fat product resembles the original product more closely (Noble & Lesschaeve 2006).

TDS and T–I methodologies globally provide compliant and coherent results but also have some specificity. TDS is a descriptive multi-attribute methodology that deals with interactions among attributes, whereas T–I focuses on the evolution of the intensity of one attribute (Pineau *et al.* 2009).

18.7 STATISTICAL ANALYSIS OF SENSORY DATA

Although a large number of flavor volatiles are often involved in stimulating human senses, the number of sensory flavor descriptors generated during food consumption is much smaller, which makes the direct correlation of volatiles with flavor descriptors difficult. However, multivariate techniques can be effectively applied to drawing relationships between volatile profiles and sensory descriptive data (Chung *et al.* 2003).

The ANOVA indicates which of the main interaction effects are significant. However, if the sample main effect is significant, the ANOVA does not indicate which samples differ from another. MANOVA (multivariate analysis of variance) tests for differences between two or more treatments. While ANOVA evaluates one dependent variable at a time, MANOVA involves all of the dependent variables simultaneously (Lawless & Heymann 1998). MANOVA can provide significance tests of the interaction between the within-subject effect and significance tests of the overall difference between the levels of the within-subject factor (Hummer 1998).

Multivariate statistics are powerful and useful tools for evaluating sensory data. Multivariate analysis can be used for exploratory investigations (to reveal general patterns in data), correlative purposes (to establish links between parameters or to predict specific values of parameters), and classification (to reveal or predict specific groupings among samples) (Ruiz *et al.* 2002). Multivariate statistical analyses utilize the relationships among response variables included in the analysis to better describe differences occurring among the experimental treatments (Spanier *et al.* 1998).

The most popular multivariate statistical methods in sensory analysis are PCA, PLSR (partial least squares regression), HCA (hierarchical cluster analysis), and GPA. These methods can relate sensory data to sensory, instrumental, chemical, or physical data.

PCA is based on the correlation among variables. It maps samples through scores and variables by the loadings in a new space defined by principal components (PCs) (Ferreira *et al.* 2000). Mathematical procedure transforms a large amount of (eventually) correlated variables into a small amount of noncorrelated variables (PCs) (Meloun *et al.* 2005). The PCs are a simple linear combination of original variables. Score plots allow sample identification, checking if they are similar or dissimilar, typical or outlier. The most important variables are identified from loadings. First, a few PCs are required to describe most of the information contained in the original data. An important point here is that the maximum variability in the data set can be explained in a reduced variable set (Ferreira *et al.* 2000).

Interpretation of the loading vectors (PCs) obtained through PCA often becomes easier when the loading vectors are rotated to better match with the original variable's direction. The final effect of this rotation is to decrease the effect of those variables with intermediate loadings and increase the effect of those variables with large (positive and negative) loadings in each factor (Ferreira *et al.* 2000; Westad *et al.* 2003).

PLSR is a technique of multivariate regression analysis that can be used to compare two blocks of variables (Toscas *et al.* 1999). This technique can be used to understand the relationship between two data sets by predicting one data set (Y) from the other (X). It not only provides solutions for both X and Y variables, but also attempts to find the best solution of X to explain the variation of Y (Cheng & Sun 2005).

HCA's primary goal is to display the data in such a way so as to emphasize their natural clusters and patterns in a two-dimensional space. The objects in each cluster should be "similar" in some sense, making the clusters more homogeneous than the full set of data. The results, qualitative in nature, are usually presented graphically in tree structures called dendrograms, allowing the visualization of clusters and correlations among samples or variables (Ferreira *et al.* 2000; Dahl & Næs 2004). A hierarchical clustering process is based on distances or dissimilarities between the objects. In addition to these basic distances, during the clustering process one will need a measure of the distance between clusters G_i and G_j. This can be constructed in a number of ways, depending on the nature of the objects and the purpose of the clustering (Dahl & Næs 2004).

GPA extracts a common structure from multiple data sets composed of nonidentical variables among the data sets (Chung *et al.* 2003). A consensus plot is generated in GPA that matches the configuration among the multiple data sets by centering, rotating, and adjusting the scales of each data set. GPA does not necessarily require equal numbers of response variables among different data sets and thus is favorably used in analyzing the data obtained by the FCP method (Piggott *et al.* 1998; Morzel *et al.*,1999). This method is also suitable for understanding the relationship between sensory responses and instrumental measurements and has been used in this manner (Chung *et al.* 2003; Meyners 2003; Morand & Pagès 2006).

18.8 FACTORS INFLUENCING PORK FLAVOR

Intrinsic pork quality characteristics, such as leanness, taste, odor, tenderness, and juiciness, can be affected at different stages throughout the pork chain. Factors of influence include: (1) selection, genotype choice, and stress susceptibility at the breeding level; (2) production conditions such as castration, diet composition, feeding and housing system, and end weight at the farm level; and (3) transport, and different pig and carcass handling techniques at the slaughterhouse level. In this way, a number of tools exist to change the intrinsic pork quality parameters and eating quality in certain directions. However, modifications of one parameter can have negative effects on other parameters (Verbeke *et al.* 1999).

Many studies were conducted to determine factors that influence pork flavor. Meaty flavor is created by compounds that are derived from either lean or fat tissues and can be divided into two categories—the characteristic meat flavor common to all species of animals and the specific flavor of beef, pork, lamb, or other species. Meaty flavor, shared by all species, is traditionally associated with the lean tissue. Species-specific flavors, however, have been traditionally associated with the fat tissue. They are influenced by compounds that form during lipid oxidation as well as fat-soluble volatile aromatic compounds and phospholipids. However, some results indicate that the lean tissue in meat products may be the principal contributor to species-specific flavors (Myers *et al.* 2009).

18.8.1 Rearing system

Dransfield *et al.* (2005) considered if there is any difference in the eating quality of pork meat from outdoor production and indoor production. The consumers found no differences in appreciation and trained panelists found little difference in tenderness, juiciness, odor, and flavor. Using a trained panel, Jonsäll *et al.* (2001) found that the rearing system has a significant effect on the acidulous taste, which is higher in pork meat from pigs reared indoor, in contrast with consumer preference when the rearing system did not influence the consumer preference for pork loin (Jonsäll *et al.* 2002).

Carrapiso *et al.* (2003) investigated the influence of the rearing system on the sensory quality of Iberian ham. Iberian ham is an expensive Spanish dry-cured product produced from Iberian pigs and is characterized by a prolonged dry-curing process and excellent consumer acceptance. They found that scores for aroma characteristics were larger in hams from pigs reared outdoors than indoors, especially differences in cured and rancid aroma.

The other situation is in the evaluation of consumers. The preference of consumers depends on information that consumers get on pork meat. Dransfield *et al.* (2005) conducted a study in which consumers were asked about their preference of buying pork meat. When information was provided in the form of labels, the vast majority of consumers preferred pork labeled as originating from their own country as opposed to imported, and pork labeled as pigs raised outdoors as opposed to indoors. There was no difference in the taste of grilled pork from indoor and outdoor production systems, but pork labeled “home produced” or “outdoor” were more appreciated.

18.8.2 Effect of genotype

Genotype has a significant effect on meat taste intensity and odor intensity. Ham from the RN– carriers showed a higher meat taste intensity and odor intensity than ham from noncarriers. There was also a tendency toward a more acidulous taste in ham from RN carriers than in ham from noncarriers (Jonsäll *et al.* 2001). In study conducted by Jonsäll *et al.* (2002), consumers preferred RN^- loins to loins from noncarriers.

18.8.3 Slaughter weight

Meinert *et al.* (2009) were interested in the effect of slaughter weight on pork chop flavor. The chops were fried at a pan temperature of 155°C and a core temperature of 70°C. Pork chops from pigs that weighed around 110 kg had a clear tendency toward a more intense fried meat flavor and a less piggy (melted fat) flavor than pork chops from pigs that weighed around 85 kg, which indicated a higher eating quality (Meinert *et al.* 2009).

18.8.4 Breed

Between breeds, it is possible to find differences in flavor. Meinert *et al.* (2008) used young consumers between the age of 20 and 30 years, who were regular pork eaters, for focus grouping five different breeds (DLY, Duroc × Landrace × Yorkshire; BS, Black spotted; H, Hampshire; WP, Wild pigs; D, Duroc). Among other attributes, flavor was the attribute that the focus groups discussed the most and found relatively hard to describe. The focus groups had the highest preference for D, because the chops were tender and juicy, and the lowest preference for H. DLY, BS, and WP were intermediate, though not similar (Meinert *et al.* 2008). The evaluation for flavor is shown in Table 18.1.

18.8.5 Boar taint

Sensory boar taint is widely reported to have a distinctive and unpleasant character that evokes repulsion and rejection when perceived through a combination of odor, flavor, and taste in pork and pork products during cooking and eating. It has been described as consisting of key sensory characteristics reminiscent of a “male pig,” “animal,” “urine,” “feces,” and/or “sweat,” which are largely causal in terms of the negative reaction with respect to human sensory perception (Dijksterhuis *et al.* 2000; Byrne *et al.* 2008).

Two chemical substances are largely responsible for boar taint: skatole and androstenone. Androstenone is a testicular steroid with a definite urinary odor. Skatole is a breakdown product of tryptophan, which exhibits an intense fecal odor (Babol & Squires 1995)

Consumers prefer pork meat from castrated male pigs to meat from entire male pigs. Castration eliminates boar taint, and the aroma and taste of freshly cooked meat is better (Bañón *et al.* 2004).

Byrne *et al.* (2008) compiled a comprehensive list of terms for descriptive evaluation of cooked pork meat with impact on boar taint and on influence of pigs feeding. Among others, the following flavor terms were used: piggy/animal (cooked pork meat from entire male pigs),

Table 18.1 Flavor evaluation of different breeds by focus grouping young consumers.

Breed	Commentary
DLY	Not very intensive, but neutral in flavor
	Like pork roast without a dominating aftertaste. Intensive piggy flavor. A little boring in flavor
	Good flavor but a little boring. A little acidic. Very ordinary
	Not too intensive meat flavor. No off-flavor
BS	Different flavor from DLY. Not so much aftertaste. A different taste of meat—more genuine
	As fried fat and water. More taste than DLY. Less taste than DLY
	Intensive taste. Not sour. Fatty flavor. No piggy flavor as in DLY
	More intensive odor, flavor, and aftertaste. Different flavor
H	Boring taste, not very intensive
	Mild aftertaste. Taste more of meat than BS
	Hardly any flavor. Taste a little more piggy
	Not really meat flavor. Almost only salt and pepper taste—no flavor
WP	A really different flavor. Good in its own manner
	A very good meat flavor as real meat fried on grill
	Bloody, metallic flavor, sour, and piggy
	A good flavor. More spicy than the other
D	Good flavor. A little like DLY. The flavor is something between the other samples (breeds)
	Intensive taste. Not sour and not too much of a piggy flavor
	Fine and good flavor. Not too intensive

DLY, Duroc × Landrace × Yorkshire; BS, Black spotted; H, Hampshire; WP, Wild pigs; D, Duroc.
Source: Meinert *et al.* (2008).

gamey (freshly cooked game meat as exemplified by deer, pheasant, or wild boar), parsnip (cooked parsnip/earthy/sweet), and manure (reference to male pig excrement/feces presented in a sealed vessel with a perforated cover for assessment with the aim to allow it to evoke "flavor").

There are differences between gilts and castrated males as well. Pork loin from gilts has less off-flavor than loin from castrated males (Jonsäll *et al.* 2002).

18.8.6 Fat

It is important to have some fat cover for cooking, moisture retention, and taste, but not too much, because of health-related concerns mentioned by consumers (Ngapo *et al.* 2003).

There are mixed results in the literature regarding increased fat content in pork and pork flavor. However, most of the research investigated the correlation of intramuscular fat (IMF) levels to pork flavor intensity (Myers *et al.* 2009).

Low positive relationship between IMF and flavor was found in pork meat (Channon *et al.* 2004; Fortin *et al.* 2005). Brewer *et al.* (2001) observed that pork loin chops containing 3.46% IMF were more flavorful than chops containing 1.05% IMF. In pork patties, there was an increase in pork flavor intensity as additional pork fat concentration increased (Myers *et al.* 2009).

Fat flavor is the fat-specific flavor in pork. Moeller *et al.* (2010b) used a 10-point categorical intensity scale for fat flavor evaluation in ground pork patties. Panelists were provided cooked pork fat as a reference point for a 10 on the scale, and ground pork patties containing less than 5% fat (3 on the scale) to 30% fat (6–7 on the scale) were used to scale panelists on cooked pork fat flavor.

It is generally accepted that a higher level of marbling or IMF has a positive influence on the sensory experience associated with eating pork. However, this has to be balanced with health concerns associated with high fat in the diet. This is a challenge for the pork industry (Fortin *et al.* 2005).

18.8.7 Conjugated linoleic acid

Pigs are monogastric animals and many dietary components are consequently readily transferred from the feed to the muscle and fat tissues, which subsequently affect pork quality (Rosenvold & Andersen 2003).

A high content of polyunsaturated fatty acids may lead to meat and meat products that can be characterized as "soft" and therefore of inferior quality (Rosenvold & Andersen 2003).

Gatlin *et al.* (2006) evaluated the combined effects of dietary conjugated linoleic acid (CLA) and supplemental fat (SF) sources on the organoleptic characteristics of bacon and pork loin samples in lean-genotype gilts. The trained sensory panel found small differences between bacon and pork loins from pigs fed and not fed CLA, but also indicated that consumer acceptance of bacon and pork products from pigs fed CLA would not likely differ from commodity pork products.

18.8.8 Salting

Consumers perceived solution-enhanced chops (6% or 12% with a solution of sodium tripolyphosphate) to have a more desirable flavor than untreated chops that had a more intense, bitter, and oxidized or cardboard flavor (Baublits *et al.* 2006). Such flavor characteristics negatively influence consumer liking (Bryhni *et al.* 2003).

18.8.9 Cooking condition

Cooking is one of the common ways of preparing meat in the kitchen. The temperature in the meat core while cooking is approximately to 60–85°C.

A core temperature of 65°C is preferred if the focus is on flavor components, whereas 75°C is recommended if a discrimination of the samples in overall sensory properties is desired (Bejerholm & Aaslyng 2003).

Pork flavor is influenced by the cooking temperature. However, results of studies that investigate the influence of cooking temperature on pork chops flavor differ. Baublits *et al.* (2006) investigated that consumers reported more desirable overall flavor when chops were cooked at a higher temperature as compared to when chops were cooked at a lower temperature. The increased flavor desirability with a higher cooked temperature could be due to decreased blood serum intensity at that temperature. It could also be due to preferences of doneness. That is, 68% of consumers stated they preferred pork cooked to a "medium well" or "well done" doneness. Consumers in the USA prefer pork chops cooked at a lower temperature as well (Moeller *et al.* 2010a). Conversely, Bryhni *et al.* (2003) reported that Scandinavian consumers preferred pork cooked at a lower temperature rather than pork cooked at a higher temperature.

Variation in frying temperature causes changes in flavor. Frying at 250°C compared with 150°C generally increased the intensity of both flavor and odor of fried meat and burnt caramel. At the same time, the intensity of boiled meat, piggy and sour flavors and odor decreased (Meinert *et al.* 2007). A more intense roasted flavor was found at 250°C compared with 150°C, and this is most likely due to the formation of Maillard-derived flavon compounds. The aroma analysis showed that the aroma profile at 150°C correlated to lipid oxidation products, while the aroma profile at 250°C correlated to Maillard-reaction products (Meinert *et al.* 2007).

Table 18.2 Sensory definitions of odors, flavors, and tastes for the evaluation of cooked pork.

Odor	*Odor associated with*
Meat	Meat complex/lean/brothy
Pig	Piggy/animal
Metallic	Bloody/iron/metallic
Sweet	Brothy/sweet
Off	Cheesy/liver/old
Acidic/sour	Sour/soured/acidic
Warmed over	Oxidized/butter/cod liver fat/wet cardboard
Flavor	*Aromatic taste sensation associated with*
Meat	Meat complex/lean/brothy/organy/umami
Pig	Piggy/animal
Metallic	Bloody/iron
Warmed over	Oxidized/butter/cod liver fat/wet cardboard
Taste	*Taste associated with*
Sweet	Sweet taste (sucrose solution-like)
Bitter	Bitter taste (quinine chloride solution-like)/bile
Acidic/sour	Sour taste (citric acid solution-like)

Source: Bryhni *et al.* (2003).

Which cooking method is suitable for sensory analysis? This question was tackled by Bejerholm and Aaslyng (2003). In the experiment, five different cooking techniques were used: mincing and preparing as patties on a pan, steaks prepared on a pan, whole roasts prepared in a pot, whole roasts prepared in an oven at 140°C, and whole roasts prepared in an oven at 90°C. The results of the sensory analyses indicated that cooking as steaks on a pan gave a good all-round sensory analysis with focus on appearance, flavor, and texture, while cooking in an oven at a low temperature gave a more detailed description of variations in texture attributes.

Aaslyng *et al.* (2007) evaluated the impact of the sensory quality of pork on consumer preference. Two different temperatures (65°C and 75°C) and two cooking methods (chops fried on a pan and whole roasts cooked in an oven) were used. The attributes describing flavor nuances were as follows: "boiled meat flavor," "roasted meat flavor," "piggy flavor" (as the flavor of a live pig/melted pork fat), "sour flavor" (fresh sour as buttermilk), "metallic flavor" (like a silver spoon), "mushroom flavor" (1-octene-3-ol), "sweet flavor," "fatty flavor," "flavor not common in pork" (other flavor).

Examples of sensory definitions of odors, flavors, and tastes used for the evaluation of cooked pork are presented in Table 18.2.

18.8.10 Warmed-over flavor

Increased consumer demand for precooked and "ready-to-eat" products has raised the issue of a warmed-over flavor (WOF) that results in loss of meatiness and the development of rancid off-flavors with cooked meat chill storage (Byrne *et al.* 2001a). Consumers prefer pork meat without WOF, metallic, or pig off-flavor. Consumers' awareness of WOF was detected in the questionnaires where the consumers pointed out that "The taste is not as good when pre-heated" and "The flavor becomes different" (Bryhni *et al.* 2002).

WOF in meat is mainly believed to be a result of the oxidation of membrane phospholipids, a process triggered by hemoproteins and other iron species during cooking (Tikk *et al.* 2008). However, degradation of proteins and some minor components, for example, Maillard-derived aroma volatiles associated with the aroma of freshly cooked meat, may also contribute to the

Table 18.3 List of sensory descriptive terms with definitions developed for the evaluation of warmed-over flavor in pork meat by trained panelists.

Sensory term	**Definition with reference material**
Odor	*Odor associated with*
Cardboard-like*	Shredded wet cardboard
Linseed oil-like*	Warmed linseed oil
Rubber-like/sulfur-like*	Warmed rubber or the white of a warm boiled and peeled egg
Nut[†]	Crushed fresh hazelnuts
Green[†]	Fresh green French beans
Fatty[†]	Pig back fat (fresh, nonoxidized)
Flavor	*Aromatic taste sensation associated with*
Cooked pork meat-like*	Cooked pork meat
Fresh cooked pork[†]	Oven-cooked pork without browning
Rancid*	Oxidized vegetable oil
Bread-like*	The surface of "French-style" bread
Vegetable oil-like*	Fresh vegetable oil
Fish-like*/fish[†]	Cod liver oil*/fish stock in boiling water[†]
Nut-like*	Whole hazelnuts
Metallic/bloody[†]	Ferrous sulfate 0.1 g/L aqueous solution
Lactic acid/fresh sour[†]	Natural yoghurt
Piggy/animal[†]	Skatole 0.06 mg/mL refined vegetable oil
Tinny[†]	Stainless steel strip
Livery[†]	Cooked beef liver[†]
Taste	*Taste associated with*
Monosodium glutamate/umami*	Monosodium glutamate, 0.05% solution in water
Metallic*	Ferrous sulfate, 0.01% solution in water
Bitter*	Quinine chloride, 0.005% solution in water
Sweet*	Sucrose, 0.1% solution in water
Salty*	Sodium chloride, 0.05% solution in water
Sour*	Citric acid, 0.03% solution in water
	Chemical feeling factor on skin surfaces of the oral cavity described as dry associated with
Aftertaste	Aluminum sulfate 0.002% solution in water
Astringent*	

*Byrne *et al.* (1999).
[†]O'Sullivan *et al.* (2003a).

formation of WOF, as this results in the disappearance of desirable meat flavor notes present in freshly cooked meat (Byrne *et al.* 2001a). Nonetheless, subsequent storage of the cooked meat and eventually reheating of pork containing high levels of unsaturated lipids is known to enhance the formation of oxidized off-flavors, for example, WOF and a rancid flavor. The development of WOF is also associated with the development of thiobarbituric acid–reactive substances, hexanal, pentanal, pentanol, and nonanal in cold-stored, reheated pork (Tikk *et al.* 2008). Byrne *et al.* (2001a) found increasing preslaughter stress, resulting in elevated ultimate pH and reduced development of WOF in cooked pork.

Byrne *et al.* (1999) investigated WOF in pork meat. A list of terms was developed to describe the aroma/flavor/taste character notes of the meat samples. The main properties of the terms were (1) to have relevance to the product, (2) to discriminate clearly between samples, (3) to be nonredundant, and (4) to have cognitive clarity to assessors. Criteria fulfillments by terms was determined via representative sample assessment, reference assessment, panel discussions, and a multivariate statistical method, PCA. Terms used for sensory profiling are presented in Table 18.3; the table is coupled with terms applied by O'Sullivan *et al.* (2003a).

How to develop a list of terms for evaluation was described by Byrne *et al.* (2001b). First, an initial list of terms is needed. The common way is to use literature sources. A sample set is developed in this preliminary stage by a product expert. Panel members are chosen and their discriminatory ability of experimental sensory dimension is verified, for example, triangle test. In the first session, panel members are familiarized with the process of vocabulary development. During the second and following sessions, the initial list of terms is reduced via a verbal term explanation/line scale scoring. A discussion is conducted before and after each session. In the next stage, reference materials for each term are presented to panel members. The reduction of terms is being continued. A flavor wheel could be presented. The wheel contains terms from previous sessions as a series of concentric rings, for example, moving from the first session terms as the outer ring and other sessions as inner rings. Associated terms within each ring should be grouped (basic tastes, WOF-related terms, etc.), and between the rings these groups are aligned. Descriptive sensory terms are removed or merged after each session based on selection criteria, representative sample and reference assessment, panel discussion, and panel term grouping suggestion. Selection criteria in each session are that terms should have relevance to the product, discriminate clearly between samples, be nonredundant, and have cognitive clarity to the assessors.

There are differences between muscles in the development of WOF. In a study performed by O'Sullivan *et al.* (2003b), *Musculus psoas major* was more susceptible to WOF development than *Musculus longissimus dorsi*, as determined by sensory profiling. The WOF development can be affected by feeding, for example, supplementation feeding with vitamin E or iron (O'Sullivan *et al.* 2003b).

It is possible to monitor the level of WOF by an electronic nose as a potential future quality control tool in the meat industry. Tikk *et al.* (2008) found a significant, positive correlation between the gas sensor signals and the WOF-associated sensory attributes and the levels of secondary lipid oxidation products.

Brøndum *et al.* (2000) used the spectroscopic method to predict sensory parameters and WOF development. VIS reflectance was found to be superior to other techniques in predicting the sensory terms, but LF-NMR (low field ^{1}H nuclear magnetic resonance) and fluorescence were also successful. In general, VIS reflectance and LF-NMR proved to be the best techniques for probing WOF development in meat.

Because of its predictive ability of sensory quality, a spectroscope can be used in rapid online industrial situations to predict the sensory quality of cooked meats (Brøndum *et al.* 2000).

18.8.11 Irradiation

Irradiating fresh meat, even at low doses, can result in off-odors and flavors that have been described as rotten egg, bloody, fishy, barbecued corn, burnt, sulfuric, metallic, alcoholic, or acetic acid-like. The odors vary with the type of meat, temperature during irradiation, oxygen exposure during and/or after the irradiation process, packaging, and presence of antioxidative substances. Irradiation can induce the formation of isooctane-soluble carbonyl compounds in the lipid fraction and low molecular weight, acid-soluble carbonyls in the protein fraction of meat. Increasing the irradiation dose increases these compounds. However, cooking reduces them. Methods to decrease the detrimental effects of irradiation include oxygen exclusion (vacuum packaging), replacement with inert gases (nitrogen), addition of protective agents (antioxidants), and postirradiation storage to allow flavor to return to near-normal levels (repackaging or double packaging in oxygen-permeable film) (Brewer 2009).

The trained panelists easily distinguish between odors of irradiated and nonirradiated pork meat. The acceptance of the meat odor is adverse to the irradiation odor intensity. As the irradiation odor intensity increases, the preference of meat odor decreases (Nam *et al.* 2002).

18.8.12 Meat products

A very good example of descriptors for the flavor evaluation of meat products was given by Pham *et al.* (2008) for American dry-cured ham. They used both the trained and the consumers' panel. Flavor and aroma descriptors for the trained panel and their definitions are shown in Table 18.4. Consumers were asked to evaluate each ham sample based on the overall acceptability and the acceptability of flavor using a 9-point hedonic scale (Meilgaard *et al.* 1999):

1. Dislike extremely
2. Dislike very much
3. Dislike moderately
4. Dislike slightly
5. Neither like nor dislike
6. Like slightly
7. Like moderately
8. Like very much
9. Like extremely

Table 18.4 Sensory descriptors and definitions for American dry-cured ham evaluated by trained panel.

Sensory descriptor	Definition
Aroma	
Rancid*	Aroma associated with extremely oxidized fat or oil
Molasses‡	Aroma associated with molasses; has a sharp, slight sulfur and/or caramelized character
Fermented‡	Aroma associated with lactic or spoilage bacteria as in soured meat
Caramelized‡	Sweet aromatic, characteristic of brown sugar and some other carbohydrates
Pork complex‡	Aromatic associated with a live pig or its habitat, or wet pig hair
Smoky‡	Perception of any type of smoke aroma
Earthy‡	Aroma characteristic of damp soil, wet foliage, or slightly undercooked boiled potato
Savory‡	Aroma associated with savoriness or meatiness
Flavor	
Cured†	Intensity of the typical flavor from cured meat products (very low to very high)
Rancid†	Intensity of the rancid flavor (very low to very high)
Salty*	The taste on the tongue associated with sodium ions
Aftertaste†	Intensity and time extension of the flavor after the sample is swallowed (very low to very high)
Pork complex‡	Flavor associated with a live pig or its habitat, or wet pig hair
Sweet*	Level of sweetness (not sweet to very sweet)
Savory‡	Flavor associated with savoriness or meatiness
Bitter*	Taste on the tongue associated with caffeine
Astringent*	Chemical feeling factor on the tongue described as puckering/dry and associated with strong tea
Mouthdrying‡	Chemical feeling factor on the tongue or other skin surfaces of the oral cavity described as puckering/dry and associated with tannins or alum
Saltburn‡	Burning sensation on the tongue

Source: Pham *et al.* (2008) compiled sensory descriptors and definitions from three sources:
*Armero *et al.* (1999), †Ruiz *et al.* (1998), and ‡Civille and Lyon (1996).

The dry-cured ham products that received higher consumer acceptability scores had more intense "sweet," "savory," "smoky," "caramelized," and "molasses" flavor and aroma attributes. The dry-cured hams that received lower consumer acceptability scores had more intense "rancid," "cured," "fermented," "pork complex," "earthy," "bitter," "salty," "saltburn," and "aftertaste" flavor and aroma attributes (Pham *et al.* 2008).

18.8.13 Social effects

The region of origin of food products affects consumer evaluation in two different ways. First, the origin can act as a quality cue hinting to other good characteristics. Second, the origin can directly affect the value of food due to its symbolic or affective role (Stefani *et al.* 2006).

In four European countries, research was conducted into the main indicators of good quality of pork. Four factors were mentioned in all four countries: fat cover, price, country of origin, and place of purchase. With the exception of country of origin, these characteristics were usually evident at the time of purchase. Other often-mentioned factors were, for example, color of meat, quality of label, overall appearance, taste, tenderness, freshness, and juiciness or dryness (Ngapo *et al.* 2003).

In Scandinavia, the consumers' liking of the tested pork samples is related to consumption frequency. Consumers with a higher score for liking ate pork more often (three times a week or more), while those who had lower liking score seldom ate pork (Bryhni *et al.* 2003). The important sensory quality characteristics of pork are flavor, juiciness, and the absence of off-flavors such as rancidity and WOF. The single-most important attribute is flavor. Suitability for a variety of dishes and flavor are the main reasons for buying pork by Scandinavian consumers (Bryhni *et al.* 2002).

The age of consumers plays a major role in the evaluation of meat quality. There are differences in opinion about the "healthiness" of the meat. Middle-aged and young consumers perceive white meats and fish as "healthy" in contrast to older consumers who describe red meats as "healthy" (Russell & Cox 2004).

Young consumers are more aware of fat content ("healthy image") and show a lower pork consumption than older consumers. Overall, older people like pork more than younger people do (Bryhni *et al.* 2002). Moreover, differences are also seen between genders. Men eat pork more often than women do (Bryhni *et al.* 2002, 2003).

REFERENCES

Aaslyng, M.D., Oksama, M., Olsen, E.V. *et al.* (2007) The impact of sensory quality of pork on consumer preference. *Meat Science*, **76** (1), 61–73.

Armero, E., Flores, M., Toldrá, F. *et al.* (1999) Effects of pig sire type and sex on carcass traits, meat quality and sensory quality of dry-cured ham. *Journal of the Science of Food and Agriculture*, **79** (9), 1147–1154.

Babol, J. & Squires, E.J. (1995) Quality of meat from entire male pigs. *Food Research International*, **28** (3), 201–212.

Bañón, S., Andreu, C., Laencina, J. & Garrido, M.-D. (2004) Fresh and eating pork quality from entire versus castrate heavy males. *Food Quality and Preference*, **15** (3), 293–300.

Baublits, R.T., Meullenet, J.-F., Sawyer, J.T., Mehaffey, J.M. & Saha, A. (2006) Pump rate and cooked temperature effects on pork loin instrumental, sensory descriptive and consumer-rated characteristics. *Meat Science*, **72** (4), 741–750.

Bejerholm, C. & Aaslyng, M.D. (2003) The influence of cooking technique and core temperature on results of a sensory analysis of pork—Depending on the raw meat quality. *Food Quality and Preference*, **15** (1), 19–30.

Brewer, M.S. (2009) Irradiation effects on meat flavor: A review. *Meat Science*, **81** (1), 1–14.

Brewer, M.S., Zhu, L.G. & McKeith, F.K. (2001) Marbling effects on quality characteristics of pork loin chops: Consumer purchase intent visual and sensory characteristics. *Meat Science*, **59** (2), 153–163.

Brøndum, J., Byrne, D.V., Bak, L.S., Bertelsen, G. & Engelsen, S.B. (2000) Warmed-over flavour in porcine meat—A combined spectroscopic, sensory and chemometric study. *Meat Science*, **54** (1), 83–95.

Bryhni, E.A, Byrne, D.V., Rødbotten, M. *et al.* (2002) Consumer perceptions of pork in Denmark, Norway and Sweden. *Food Quality and Preference*, **13** (5), 257–266.

Bryhni, E.A., Byrne, D.V., Rødbotten, M. *et al.* (2003) Consumer and sensory investigations in relation to physical/chemical aspects of cooked pork in Scandinavia. *Meat Science*, **65** (2), 737–748.

Byrne, D.V., Bak, L.S., Bredie, W.L.P., Bertelsen, G. & Martens, M. (1999) Development of a sensory vocabulary for warmed-over flavor: Part I. In porcine meat. *Journal of Sensory Studies*, **14**, 47–65.

Byrne, D.V., Bredie, W.L.P., Bak, L.S., Bertelsen, G., Martens, H. & Martens, M. (2001a) Sensory and chemical analysis of cooked porcine meat patties in relation to warmed-over flavour and preslaughter stress. *Meat Science*, **59** (3), 229–249.

Byrne, D.V., O'Sullivan, M.G., Dijksterhuis, G.B., Bredie, W.L.P. & Martens, M. (2001b) Sensory panel consistency during development of a vocabulary for warmed-over flavour. *Food Quality and Preference*, **12** (3), 171–187.

Byrne, D.V., Thamsborg, S.M. & Hansen, L.L. (2008) A sensory description of boar taint and the effects of crude and dried chicory roots (*Cichorium intybus* L.) and inulin feeding in male and female pork. *Meat Science*, **79** (2), 252–269.

Carpenter, R.P., Lyon, D.H. & Hasdell, T.A. (2000) *Guidelines for Sensory Analysis in Food Product Development and Quality Control*. Aspen Publishers, Inc., Gaithersburg, Maryland.

Carrapiso, A.I., Bonilla, F. & García, C. (2003) Effect of crossbreeding and rearing system on sensory characteristics of Iberian ham. *Meat Science*, **65** (1), 623–629.

Civille, G.V. & Lyon, B.G. (1996) *Aroma and Flavor Lexicon for Sensory Evaluation: Terms, Definitions, References, and Examples*. ASTM Data Series Publication DS 66, ASTM, Philadelphia, Pennsylvania.

Channon, H.A., Kerr, M.G. & Walker, P.J. (2004) Effect of Duroc content, sex and ageing period on meat and eating quality attributes of pork loin. *Meat Science*, **66** (4), 881–888.

Cheng, Q. & Sun, D.-W. (2005) Application of PLSR in correlating physical and chemical properties of pork ham with different cooling methods. *Meat Science*, **70** (4), 691–698.

Chung, S.-J., Heymann, H. & Grün, I.U. (2003) Application of GPA and PLSR in correlating sensory and chemical data sets. *Food Quality and Preference*, **14** (5–6), 485–495.

Dahl, T. & Næs, T. (2004) Outlier and group detection in sensory panels using hierarchical cluster analysis with the Procrustes distance. *Food Quality and Preference*, **15** (3), 195–208.

Delahunty, C.M., McCord, A., O'Neill, E.E. & Morrissey, P.A. (1997) Sensory characterisation of cooked hams by untrained consumers using free-choice profiling. *Food Quality and Preference*, **8** (5–6), 381–388.

Dijksterhuis, G.B. & Piggott, J.R. (2001) Dynamic methods of sensory analysis. *Trends in Food Science & Technology*, **11** (8), 284–290.

Dijksterhuis, G.B., Engel, B., Walstra, P. *et al.* (2000) An international study on the importance of androstenone and skatole for boartaint: II. Sensory evaluation by trained panels in seven European countries. *Meat Science*, **54** (3), 261–269.

Dransfield, E., Ngapo, T.M., Nielsen, N.A. *et al.* (2005) Consumer choice and suggested price for pork as influenced by its appearance, taste and information concerning country of origin and organic pig production. *Meat Science*, **69** (1), 61–70.

Ferreira, M.M.C., Morgano, M.A., de Queiroz, S.C.N. & Mantovani, D.M.B. (2000) Relationships of the minerals and fatty acid contents in processed turkey meat products. *Food Chemistry*, **69** (3), 259–265.

Fortin, A., Robertson, W.M. & Tong, A.K.W. (2005) The eating quality of Canadian pork and its relationship with intramuscular fat. *Meat Science*, **69** (2), 297–305.

Gacula, M.C. (1997) *Descriptive Sensory Analysis in Practice*. Food and Nutrition Press, Trumbull, Connecticut.

Gatlin, L.A., See, M.T., Larick, D.K. & Odle, J. (2006) Descriptive flavor analysis of bacon and pork loin from lean-genotype gilts fed conjugated linoleic acid and supplemental fat. *Journal of Animal Sciences*, **84**, 3381–3386.

Hummer, S. (1998) Application of multivariate analysis of variance. *Food Quality and Preference*, **9** (3), 83–85.

ISO 5492:2008 (2009) Sensory analysis—Vocabulary. Czech version of the European Standard EN ISO 5492:2009.

Jonsäll, A., Johansson, L. & Lundström, K. (2001) Sensory quality and cooking loss of ham muscle (*M. biceps femoris*) from pigs reared indoors and outdoors. *Meat Science*, **57** (3), 245–250.

Jonsäll, A., Johansson, L., Lundström, K., Andersson, K.H., Nilsen, A.N. & Risvik, E. (2002) Effects of genotype and rearing system on sensory characteristics and preference for pork (M. Longissimus dorsi). *Food Quality and Preference*, **13** (2), 73–80.

Koeferli, C.S., Schwegler, P.P. & Hong-Chen, D. (1998) Application of classical and novel sensory techniques in product optimization. *Lebensmittel Wissenschaft und Technologie*, **38**, 407–417.

Laing, D.G. & Jinks, A. (1996) Flavor perception mechanisms. *Trends in Food Science & Technology*, **7** (12), 387–389.

Lawless, H.T. & Heymann, H. (1998) *Sensory Evaluation of Food: Principles and Practices*. Chapman & Hall, New York.

Meilgaard, M., Civille, G.V. & Carr, B.T. (1999) *Sensory Evaluation Techniques, 3rd edn*. CRC Press, Boca Raton, Florida.

Meinert, L., Andersen, L.T., Bredie, W.L.P., Bjergegaard, C. & Aaslyng, M.D. (2007) Chemical and sensory characterisation of pan-fried pork flavour: Interactions between raw meat quality, ageing and frying temperature. *Meat Science*, **75** (2), 229–242.

Meinert, L., Christiansen, S.C., Kristensen, L., Bjergegaard, C. & Aaslyng, M.D. (2008) Eating quality of pork from pure breeds and DLY studied by focus group research and meat quality analyses. *Meat Science*, **80** (2), 304–314.

Meinert, L., Tikk, K., Tikk, M. *et al.* (2009) Flavour development in pork. Influence of flavour precursor concentrations *in longissimus dorsi* from pigs with different raw meat qualities. *Meat Science*, **81** (1), 255–262.

Meloun, M., Militky, J. & Hill, J. (2005) *Počítačová analýza vícerozměrných dat.* Academia, Praha.

Meyners, M. (2003) Methods to analyse sensory profiling data—A comparison. *Food Quality and Preference*, **14** (5–6), 507–514.

Moeller, S.J., Miller, R.K., Aldredge, T.L. *et al.* (2010a) Trained sensory perception of pork eating quality as affected by fresh and cooked pork quality attributes and end-point cooked temperature. *Meat Science*, **85** (1), 96–103.

Moeller, S.J., Miller, R.K., Edwards, K.K. et al. (2010b) Consumer perceptions of pork eating quality as affected by pork quality attributes and end-point cooked temperature. *Meat Science*, **84** (1), 14–22.

Morand, E. & Pagès, J. (2006) Procrustes multiple factor analysis to analyse the overall perception of food products. *Food Quality and Preference*, **17** (1–2), 36–42.

Morzel, M., Sheehan, E.M., Delahunty, C.M. & Arendt, E.K. (1999) Sensory evaluation of lightly preserved salmon using freechoice profiling. *International Journal of Food Science and Technology*, **34** (2), 115–123.

Muñoz, A.M. (1998) Consumer perceptions of meat. Understanding these results through descriptive analysis. *Meat Science*, **49** (S1), S287–S295.

Murray, J.M. & Baxter, I.A. (2003) *Sensory Evaluation/Food Acceptability and Sensory Evaluation. Encyclopedia of Food Sciences and Nutrition*, 2nd edn. Academic Press, Amsterdam, the Netherlands.

Murray, J.M., Delahunty, C.M. & Baxter, I.A. (2001) Descriptive sensory analysis: Past, present and future. *Food Research International*, **34** (6), 461–471.

Myers, A.J., Scramlin, S.M., Dilger, A.C., Souza, C.M., McKeith, F.K. & Killefer, J. (2009) Contribution of lean, fat, muscle color and degree of doneness to pork and beef species flavor. *Meat Science*, **82** (1), 59–63.

Nam, K.C., Du, M., Jo, C. & Ahn, D.U. (2002) Effect of ionizing radiation on quality characteristics of vacuum-packaged normal, pale-soft-exudative, and dark-firm-dry pork. *Innovative Food Science and Emerging Technologies*, **3** (1), 73–79.

Ngapo, T.M., Dransfield, E., Martin, J.F., Magnusson, M., Bredahl, L. & Nute, G.R. (2003) Consumer perceptions: Pork and pig production. Insights from France, England, Sweden and Denmark. *Meat Science*, **66** (1), 125–134.

Noble, A.C. & Lesschaeve, I. (2006) Sensory analysis of food flavor. In: *Flavor in Food* (eds A. Voilley & P. Etiévant). Woodhead Publishing, Cambridge. Online version available at: http://knovel.com/web/portal/browse/display?_EXT_KNOVEL_DISPLAY_bookid = 1477&VerticalID = 0

O'Sullivan, M.G., Byrne, D.V., Jensen, M.T., Andersen, H.J. & Vestergaard, J. (2003a) A comparison of warmed-over flavour in pork by sensory analysis, GC/MS and the electronic nose. *Meat Science*, **65** (3), 1125–1138.

O'Sullivan, M.G., Byrne, D.V., Nielsen, J.H., Andersen, H.J. & Martens, M. (2003b) Sensory and chemical assessment of pork supplemented with iron and vitamin E. *Meat Science*, **64** (2), 175–189.

Peyvieux, C. & Dijksterhuis, G. (2001) Training a sensory panel for TI: Case study. *Food Quality and Preference*, **12** (1), 19–28.

Pham, A.J., Schilling, M.W., Mikel, W.B., Williams, J.B., Martin, J.M. & Coggins, P.C. (2008) Relationships between sensory descriptors, consumer acceptability and volatile flavor compounds of American dry-cured ham. *Meat Science*, **80** (3), 728–737.

Piggott, J.R. (2000) Dynamism in flavour science and sensory methodology. *Food Research International*, **33** (3–4), 191–197.

Piggott, J.R., Simpson, S.J. & Williams, S.A.R. (1998) Sensory analysis. *International Journal of Food Science and Technology*, **33** (1), 7–18.

Pineau, N., Schlich, P., Cordelle, S. *et al.* (2009) Temporal dominance of sensations: Construction of the TDS curves and comparison with time–intensity. *Food Quality and Preference*, **20** (6), 450–455.

Reinbach, H.C., Meinert, L., Ballabio, D. *et al.* (2007) Interactions between oral burn, meat flavor and texture in chili spiced pork patties evaluated by time-intensity. *Food Quality and Preference*, **18** (6), 909–919.

Resano, H., Sanjuán, A.I., Cilla, I., Roncalés, P. & Albisu, L.M. (2010) Sensory attributes that drive consumer acceptability of dry-cured ham and convergence with trained sensory data. *Meat Science*, **84** (3), 344–351.

Rosenvold, K. & Andersen, H.J. (2003) Factors of significance for pork quality—A review. *Meat Science*, **64** (3), 219–237.

Ruiz, J., Ventanas, J., Cava, R., Timón, M.L. & García, C. (1998) Sensory characteristics of Iberian ham: Influence of processing time and slice location. *Food Research International*, **31** (1), 53–58.

Ruiz, J., García, C., Muriel, E., Andrés, A.I. & Ventanas, J. (2002) Influence of sensory characteristics on the acceptability of dry-cured ham. *Meat Science*, **61** (4), 347–354.

Russell, C.G. & Cox, D.N. (2004) Understanding middle-aged consumers' perceptions of meat using repertory grid methodology. *Food Quality and Preference*, **15** (4), 317–329.

Spanier, A.M., Vinyard, B.T., Bett, K.L. & Angelo, A.J.S. (1998) Sensory and statistical analyses in meat flavor research. In *Flavor of Meat, Meat Products and Seafoods* (ed. F. Shahidi), 2nd edn., pp. 395–419. Springer-Verlag, New York.

Stampanoni, C.R. (1994) The use of standardised flavour languages and quantitative flavour profiling technique for flavoured dairy products. *Journal of Sensory Studies*, **9** (4), 383–400.

Stefani, G., Romano, D. & Cavicchi, A. (2006) Consumer expectations, liking and willingness to pay for specialty foods: Do sensory characteristics tell the whole story? *Food Quality and Preference*, **17** (1–2), 53–62.

Tikk, K., Haugen, J.-E., Andersen, H.J. & Aaslyng, M.D. (2008) Monitoring of warmed-over flavour in pork using the electronic nose—Correlation to sensory attributes and secondary lipid oxidation products. *Meat Science*, **80** (4), 1254–1263.

Toscas, P.J., Shaw, F.D. & Beilken, S.L. (1999) Partial least squares (PLS) regression for the analysis of instrument measurements and sensory meat quality data. *Meat Science*, **52** (2), 173–178.

Verbeke, W., Van Oeckel, M.J., Warnants, N., Viaene, J. & Boucqué, Ch.V. (1999) Consumer perception, facts and possibilities to improve acceptability of health and sensory characteristics of pork. *Meat Science*, **53** (2), 77–99.

Westad, F., Hersleth, M., Per, L. & Martens, H. (2003) Variable selection in PCA in sensory descriptive and consumer data. *Food Quality and Preference*, **14** (5–6), 463–472.

Williams, A.A. & Langron, S.P. (1984) The use of free-choice profiling for the evaluation of commercial ports. *Journal of the Science of Food and Agriculture*, **35**, 558–568.

19 Pork Taint

William Benjy Mikel

Abstract: Although pork taint can be the result of improper dry curing or usage of meat from intact males, the cause in most countries outside the USA is due to the swine production methods that exclude the use of castration for various reasons. The meat from intact males is detectable by a large portion of the population regardless of cookery or processing method and even country of origin. Although both androstenone and skatole contribute to off odors and off flavors, it appears that more people tend to object to lower levels of skatole present in the product.

Keywords: Pork taint, bone taint, boar taint, androstenone, skatole, off odor, off flavor

19.1 INTRODUCTION

Tainting in pork is normally thought of in two senses: the off flavor present when meat from intact males is consumed versus males that have been castrated and the off flavor that normally develops as the larger portions or cuts, such as the hams, are dry-cured improperly. Although both taints are undesirable and vary by production method and geographic location, there are methods by which each can be overcome. This section will attempt to discuss both causes of tainting; however, the major emphasis will be placed on tainting that is found predominantly in intact males from European sources.

19.2 BONE TAINT

Perhaps the most important cause of spoilage in dry-cured hams, amounting to losses of almost 5% annually, is bone taint (Blanco *et al.* 1997). This phenomenon, which occurs in larger muscle cuts with greater depth thereby slowing the internal chilling process, is the result of anaerobic bacterial growth. These bacteria (*C. perfringens*, Enterobacteriaceae) predominately create an off odor and off flavor deep within the muscle next to the bone that is often referred to as "bone taint" or "bone sour" (Giolitti *et al.* 1971). The first indication of such a problem is normally noticed as a result of the foul-smelling odor emitted from the muscle mass adjacent to the bone (Garcia *et al.* 2000). Often, the unpleasant odor found in spoiled hams is related to a higher dimethyldisulphide level (Ames & Macleod 1985). These odors possibly originate from the microbial metabolism to higher levels of ketones, alcohols, and esters (Garcia *et al.* 2000).

Handbook of Meat, Poultry and Seafood Quality, Second Edition. Edited by Leo M. L. Nollet.

Although this problem is normally noticed in fully aged or ripened hams, the water activity of the tissue at that time (0.85) suggests that the actual spoilage occurs much earlier in the curing stage of the process (Cordoba *et al.* 1994). As a safeguard against this issue, many country ham processors in the USA probe hams with small metal rods around the center of the deepest part of the ham to try to detect any off odor development. In fact, most official competition judges of intact dry-cured hams will commonly probe the hams because off odor constitutes a large portion of the final product score. Even though this problem is rather small in relation to the entire industry at only 5% occurrence in Europe and at similar or lower levels in the USA (Mikel, unpublished industry data), bone tainting continues to create substantial economic loss to the dry-cured industry.

19.3 BOAR TAINT

"Boar taint" is an issue very seldom noted in the USA due to the mainstream production practices of male castration of swine at an early age. In fact, usually the off-odor occurrence is minimal in the USA except in the cases where old boars are salvaged once they are past a useful breeding age. However, typically, the meat from these animals is included in some type of highly seasoned, further processed item for the purpose of overshadowing any detectable off odor and flavor. In addition, the use of such meat is in such small quantities that the offending odor and pursuant flavor is diluted.

However, the problem with sex odor from boar carcasses has been well known for many decades according to Lerche (1936), as cited in Weiler *et al.* (2000). European countries, on the other hand, often produce their swine without castration methods being employed for many reasons—efficiency, animal welfare, consumer perception, etc. Therefore, quite an effort has been made to determine factors involved in boar tainting and how best to minimize the occurrence and subsequent effects.

Androstenone and skatole were first reported by Paterson (1968) as the two compounds responsible for boar taint in intact males. This taint is most strongly associated with odor from the fat but also has an effect, although to a lesser extent, on the flavor of the product as well (Malmfors & Lundstrom 1983). In addition, Paterson (1968) also reported that the androstenone, a testicular steroid, exhibited a "urine-like" odor. Skatole, on the other hand, exhibits more of a "fecal" or "manure-like" odor according to research by Vold (1970), as cited in Bonneau *et al.* (2000a, b). Dijksterhuis *et al.* (2000) also reported the distinct difference in odors with androstenone having the urine odor and skatole the manure odor with a somewhat lesser naphthalene odor. In their research, they utilized trained panels from seven European countries to evaluate intact male pork in the form of loin chops. Fat samples were analyzed for variations in the content of the two compounds present and assigned to panelists accordingly. Any panelist insensitive to 0.5 ppm of androstenone (presented in a lard sample) was removed from the study. They used the following sensory attributes shown in Table 19.1 to assess taint both in odor and flavor. Use of these qualifiers dramatically enhanced sensory results. They found that all the attributes listed were beneficial in determining either androstenone or skatole content except for "sweat" and "sweet." They also found differences in sensitivity among countries; however, this is to be expected as different cultures have varying acceptance with respect to eating and liking fresh and processed pork from pigs produced under various production schemes.

Walstra *et al.* (1997) reported that although there are no seasonal trends in androstenone and skatole levels, the contents did increase with increased carcass weight and decreased with increasing percentage of lean meat indicating a concentration of the compounds in the fat.

Table 19.1 Sensory attributes used to describe boar taint from intact male swine.

	Odor	Flavor
1	Pig	Pig
2	Urine	Urine
3	Manure/stable	Manure/stable
4	Naftaline/mothballs	Naftaline/mothballs
5	Rancid	Rancid
6	Sweet	Sweet
7	Sweat	Sweat
8	Abnormal odor	Abnormal flavor

Source: From Dijksterhuis *et al.* (2000).

Andersson *et al.* (1997) also reported that the level of androstenone increased as sexual maturation advanced in the intact male with age. Bonneau *et al.* (2000a, b) reported that in the same European study, sensory panelists could distinguish three groups: boars with high skatole contents were easily separated from gilts and boars with low skatole contents, and an intermediate group of boars with high androstenone levels. They also determined that skatole was easier to perceive than androstenone and is the major contributor to off odors in boar meat.

In a consumer study spanning seven European countries, Mathews *et al.* (2000) found a significant difference in acceptability between consumers from different countries. Overall, 22% of the intact male samples were disliked for their flavor and 34% for their odor. In addition, as both skatole and androstenone levels increased in concentration, dissatisfaction levels increased for both odor and flavor. However, skatole concentrations had a much greater impact than androstenone levels on odor with similar effects on flavor scores.

Weiler *et al.* (2000) also compared consumer acceptance as affected by country of origin. Consumers compared the acceptability of intact male loin samples segregated into nine categories based on androstenone (A) and skatole (S) concentrations (low A/low S, low A/medium S, low A/high S, medium A/low S, medium A/medium S, medium A/high S, high A/low S, high A/medium S, high A/high S). Eighteen percent of German and 31% of Spanish consumers were highly sensitive to androstenone with a larger portion being women, 19% versus 16% in Germany and 37% versus 24% in Spain (Table 19.2). For highly sensitive participants, androstenone was the major discriminant, while skatole had a greater effect in insensitive and mildly sensitive consumers (Table 19.3). This agreed with Griffith and Patterson (1970) who stated that some consumers were not capable of perceiving androstenone even at high concentrations, although their ability to smell other substances is normally developed.

In a study by de Kock *et al.* (2001), significant differences in sensitivity were noted among consumers from different ethnic groups in South Africa (black, white, and colored). White females were more negative toward boar odor than white males and blacks, with a higher percentage of coloreds responding negatively to boar odors. In addition, black males were more critical than black females toward samples with detectable skatole. However, they stated that it was possible that the ethnic differences in this study could be compounded by age, socioeconomic status, and other culture-linked differences.

Cookery also has been shown to have an impact on the acceptability of products from intact males (Agerhem & Tornberg 1995). Skatole contribution to boar odor is greatest when

Table 19.2 Distribution of consumers (%) in the different androstenone sensitivity classes (all consumers/country = 100%).

	Highly sensitive (score 5–7)	Mildly sensitive (score 4)	Insensitive (score 1–3)
Germany			
Total (n = 472)	17.6	14.7	67.7
Men (n = 217)	15.6	14.7	69.7
Women (n = 255)	19.3	14.7	65.9
Spain			
Total (n = 480)	30.9	15.5	53.6
Men (n = 230)	23.7	16.6	59.7
Women (n = 250)	37.3	14.6	48.1

Odor and flavor scores* in cooked pork loin of German and Spanish consumers by androstenone sensitivity and skatole and androstenone level†

	Androstenone sensitivity		Skatole level			Androstenone level			Interaction‡		
	High	Mild or insensitive	Low	Medium	High	Low	Medium	High	RSD	Skat And	And Sensitivity
Germany											
N	405	1899	796	818	690	1266	415	623			
Odor	4.47 a	3.99 b	3.94 c	4.19 b	4.55 a	4.18	4.15	4.35	1.49	NS	NS
Flavor	3.58	3.32	3.36 b	3.37 b	3.63 a	3.29	3.48	3.59	1.32	NS	*
Spain											
N	726	1,637	806	847	710	1459	353	549			
Odor	4.12	3.84	3.75	3.91	4.28	3.87	4	4.07	1.22	**	**
Flavor	3.57	3.55	3.5	3.54	3.64	3.58	3.48	3.62	1.18	*	NS

*Odor and flavor scores from 1 (like very much) to 7 (dislike very much).
†Different letters within main effects indicate significant differences (P<0.05).
‡**: P<0.01, *: P<0.05, NS: not significant.

Interaction between androstenone sensitivity and androstenone level in cooked pork for the odor scores of German and Spanish consumers

Androstenone level*	Androstenone sensitivity						Significance of the effect of sensitivity*
	Highly sensitive			Mildly sensitive/insensitive			
	n	LSM*	SE		LSM	SE	
Germany							
Low	231	4.36	0.1	1035	4.00	0.05	*
Medium	63	4.43	0.19	352	3.88	0.08	*
High	111	4.63	0.14	512	4.06	0.07	*
Spain							
Low	455	3.90 b	0.06	1006	3.85	0.04	NS
Medium	97	4.22 a	0.12	256	3.78	0.08	**
High	174	4.25 a	0.09	375	3.88	0.06	***

*Significance of difference between androstenone levels;
†LSM within the same column and country with different letters are significantly ($P<0.05$) different;
‡***: $P<0.001$, **: $P<0.01$, *: $P<0.05$ NS: not significant;
Source: From Weiler *et al.* (2000).

Table 19.3 Intensity of perception of the four substances.

Substance	Intensity of perception				
	Very strong	**Strong**	**Intermediate**	**Weak**	**No perception**
Skatole	21.3%	36.9%	30.3%	10.7%	0.8%
Indole	1.7%	3.3%	17.4%	50.4%	27.2%
3α-ole	0.0%	1.7%	16.7%	52.5%	29.1%
A-one	4.1%	17.4%	21.5%	33.9%	23.1%

Influence of sex on the sensitivity of perception for the four substances

		Sensitivity to perception		**% Anosmic**
Substance	**Sex**	**Very strong to intermediate**	**Weak to no perception**	**No perception**
Skatole	Males	88.4%	11.6%	0.0%
	Females	88.6%	11.4%	1.3%
Indole	Males	23.3%	76.7%	33.3%
	Females	22.8%	77.2%	24.1%
3α-ole	Males	18.6%	81.4%	32.6%
	Females	20.3%	79.7%	27.3%
A-one	Males	37.2%	62.8%	30.2%
	Females	46.9%	53.1%	19.2%

Influence of sensitivity of perception on pleasantness of the different scents (participants with "no perception" for a substance are excluded for the evaluation of that particular substance)

		Ranking		
Substance	**Intensity of perception**	**Pleasant**	**Neutral**	**Unpleasant**
Skatole	Weak perception	7.1%	35.7%	57.2%
	Very strong to intermediate perception	0.9%	0.9%	98.2%
	All (except no perception)	1.6%	4.9%	93.5%
Indole	Weak perception	22.2%	65.1%	12.7%
	Very strong to intermediate perception	11.1%	51.9%	37.0%
	All (except no perception)	18.9%	61.1%	20.0%
3α-ole	Weak perception	33.3%	48.5%	18.2%
	Very strong to intermediate perception	34.8%	21.7%	43.5%
	All (except no perception)	33.7%	41.5%	24.8%
A-one	Weak perception	17.1%	63.4%	19.5%
	Very strong to intermediate perception	0.0%	3.7%	96.3%
	All (except no perception)	7.4%	29.5%	63.1%

Sensitivity of perception for androstenone in the test panel of the consumer survey

Sex (*n*)	**Sensitive (very strong to intermediate)**	**Insensitive (weak to no perception)**
Males (223)	29.60%	70.40%
Females (257)	33.90%	66.10%

Source: From Weiler *et al.* (2000).

Table 19.4 Perception of boar taint.

Attribute	Female	Low	High	Attribute	Female	Low	High
Weighted least square (WLS) means* for the various attributes of ham							
Overall odor	6.7 a	6.6 a	6.9 a	Boar odor	1.6 ab	1.5 a	2.0 b
Overall flavor	7.1 ab	7.1 a	7.6 b	Boar flavor	1.8 a	1.4 a	2.9 b
Weighted least square (WLS) means* for the various attributes of bacon							
Overall odor	6.6 a	6.4 a	8.8 b	Boar odor	1.4 a	1.5 a	5.8 b
Overall flavor	7.8 a	7.8 a	9.3 b	Boar flavor	1.1 a	1.4 a	3.8 b
Weighted least square (WLS) means* for the various attributes of salami							
Overall odor	7.6 a	7.5 a	7.7 a	Boar odor	1.5 b	0.8 a	1.5 b
Overall flavor	7.3 a	7.6 a	7.3 a	Boar flavor	1.8 b	1.1 a	2.5 c
Weighted least square (WLS) means* for the various attributes of dry, oven-roasted pork							
Overall odor	6.1 a	6.2 a	7.9 b	Boar odor	2.5 a	3.0 a	5.6 b
Overall flavor	6.4 a	6.3 a	7.9 b	Boar flavor	2.4 a	2.7 a	5.0 b
Weighted least square (WLS) means* for the various attributes of stewed, oven cooked pork							
Overall odor	7.8 a	8.1 a	9.7 b	Boar odor	4.9 a	5.4 a	8.6 b
Overall flavor	6.5 a	6.8 a	8.0 b	Boar flavor	2.7 a	3.6 b	5.9 c
Weighted least square (WLS) means* for the various attributes of marinated, oven cooked pork							
Overall odor	8.0 a	7.7 a	9.0 b	Boar odor	1.0 a	1.1 a	3.6 b
Overall flavor	7.2 a	7.1 a	7.7 b	Boar flavor	0.9 a	1.0 a	3.2 b

*Means on the same line not followed by a common letter are significantly different at the 5% level.
Source: From McCauley *et al.* (1997).

products are cooked to an internal temperature of 68°C while androstenone contributes most to off odor when cooked to 80°C. Since the temperature levels fall into the range where most pork is cooked for sensory analysis, it is to be expected that both compounds would have the opportunity to contribute to off odors. McCauley *et al.* (1997) reported that although cookery and/or serving temperatures may influence intact male pork acceptability, the method of cookery nor methods of processing, except in the case of salami and hams, decreased perception of boar taint by trained professionals (Table 19.4).

Siret *et al.* (1997) also evaluated cookery methods' affects on boar odor from intact male products. They prepared products at commercial stabilization temperatures (250°C for 10 min) and by pan frying for 6 min. All products were served to consumers or trained French and German panelists. Table 19.5 indicates the results of the consumer test with the pan-frying method being the less discriminating method. Although products with high levels of boar taint were equally rejected by panelists regardless of cookery method, products with lower concentrations of either steroid were less detectable with the pan-frying method. In addition, they reported that skatole levels contribute a greater amount to the off odor than androstenone compounds. In a study comparing the effects of boar taint in cooked versus dry-cured products, Banon *et al.* (2003) found that although there were complex and intricate interactions among the various sensory parameters measured during boar taint analysis, that method of product preparation did not exclude boar taint occurrence (Table 19.6). Therefore,

Table 19.5 Consumer test results.

Scores for acceptability of pork according to skatole/androstenone classes (mean ± SD)

	Classes			
	Gilts	**LS/LA**	**LS/HA**	**HS/HA**
Oven procedure				
$F = 15.57$ $P < 0.001$	3.1 ± 1.2 a	2.7 ± 1.5 b	2.5 ± 1.5 b	1.9 ± 1.5 c
Hot plate procedure				
$F = 3.49$ $P = 0.016$	3.6 ± 1.2 a	3.5 ± 1.3 ab	3.3 ± 1.3 ab	3.1 ± 1.3 b

Scores for cooking-odor attributes in the French Sensory Study, according to skatole/androstenone classes (mean ± SD)

			Classes			
Attributes	**F**	**Probability level**	**Gilts**	**LS/LA**	**LS/HA**	**HS/HA**
Oven procedure						
Pig	13.92	<0.001	3.5 ± 2.2 a	3.7 ± 2.1 a	3.2 ± 2.1 a	1.7 ± 1.6 b
Urine	11.96	<0.001	2.7 ±2.5 a	2.2 ± 2.4 a	3.8 ± 2.7 b	4.7 ± 3.0 c
Manure	19.35	<0.001	2.7 ± 2.9 ab	2.3 ± 2.0 a	3.5 ± 2.2 b	5.5 ± 2.4 c
Naphthalene	10.14	<0.001	1.8 ± 2.2 a	1.5 ± 2.1 a	2.1 ± 2.5 a	3.3 ±2.7 b
Rancid	1.42	0.24	1.0 ± 1.8	0.8 ±1.7	1.1 ±1.0	1.3 ±2.3
Sweet	6.27	<0.001	2.2 ±2.0 ab	2.7 ± 2.1 a	1.7 ±1.9 bc	1.3 ±1.8 c
Sweat	12.62	<0.001	2.0 ± 2.2 a	1.9 ± 1.9 a	2.7 ± 2.0 a	3.9 ± 2.5 b
Abnormal	15.15	<0.001	4.7 ± 2.9 ab	4.3 ± 2.7 a	5.5 ±2.2 b	7.1 ± 2.2 c
Hot plate procedure						
Pig	4.77	0.003	3.1 ± 2.1 ab	3.1 ±1.9 ab	3.7 ± 2.1 a	2.4 ±1.8 b
Urine	5.37	0.002	3.0 ±2.5 ab	3.0 ± 2.7 ab	2.2 ± 2.0 a	4.1 ±2.5 b
Manure	7.59	<0.001	3.1 ± 2.9 a	2.9 ±2.9 a	2.2 ±2.2 a	4.2 ± 2.9 b
Naphthalene	2.52	0.06	1.1 ± 1.8	1.0 ± 1.6	0.6 ± 1.1	1.3 ± 1.7
Rancid	0.56	0.643	0.7 ± 1.3	0.6 ±1.1	0.5 ± 1.0	0.5 ± 0.8
Sweet	1.40	0.245	1.8 ± 1.9	1.7 ± 1.9	1.7 ± 1.9	1.2 ± 1.5
Sweat	5.19	0.002	2.3 ± 1.8 a	2.3 ± 1.9 a	1.9 ± 1.7 a	3.3 ± 2.4 b
Abnormal	9.34	<0.001	5.1 ± 2.7 b	4.8 ± 2.4 b	3.8 ± 2.4 a	6.2 ± 2.2 c

Source: From Siret *et al.* (1997).

although processing methods such as fry curing alters perception of various traits, the detection of the phenomenon still occurs. Mathews *et al.* (1997) however found no differences in eating quality of British-style sausages prepared from intact males versus females, although differences readily existed in chops prepared from the same animals (Table 19.7).

Babol *et al.* (2002) reported that in a study of Canadian consumers, the meat from intact males with low levels of androstenone and skatole was as acceptable as that from gilts initially and over time (Table 19.8). They suggested that products with a controlled level of boar taint could be readily utilized in the marketplace and that skatole analysis would be beneficial in avoiding products with excess levels. However, they cautioned that consumers might become more sensitive to boar taint over time.

Table 19.6 Androstenone and skatole fat content and boar taint scores for samples used in sensory evaluation (mean 6 SD).

		Backfat		Processed meat			
				Cooked		Dry cured	
	N	**Androstenone**	**Skatole**	**Odor**	**Flavor**	**Odor**	**Flavor**
Castrates	24	0.256 ± 0.097	0.029 ± 0.018	1.47 ± 0.36 a	1.15 ± 0.21 a	2.41 ± 0.63 a	1.80 ± 0.52 a
Entires	24						
LL	4	0.328 ± 0.129	0.024 ± 0.016	2.23 ± 0.21 a	1.83 ± 0.23 ab	3.01 ± 0.20 ab	2.20 ± 0.37 ab
LM	4	0.251 ± 0.069	0.123 ± 0.054	3.27 ± 0.23 b	1.27 ± 0.23 ab	3.75 ± 0.30 ab	2.04 ± 0.37 ab
ML	4	0.694 ± 0.116	0.033 ± 0.018	3.47 ± 0.12 b	2.53 ±0.12 cd	2.79 ± 0.57 ab	2.46 ± 0.43 ab
MM	4	0.569 ± 0.072	0.106 ± 0.029	3.66 ± 0.33 b	2.06 ± 0.24 bc	3.25 ± 0.20 ab	2.47 ± 0.27 ab
HL	4	1.551 ± 0.562	0.043 ± 0.027	4.60 ± 0.40 c	3.53 ± 0.90 d	3.40 ± 0.30 ab	2.48 ± 0.03 ab
HM	4	1.987 ± 0.614	0.131 ± 0.055	4.93 ± 0.04 c	3.53 ± 0.50 d	4.26 ± 0.09 b	3.84 ± 0.04 b

Fat samples for entire males: H, high; M, medium; L, low levels of AN/SK. Means with different letters are significantly different at *P* 0.05. AN/SK content expressed in μg g^{-1}. Scoring scale: 1, minimum; 5, maximum.
Source: From Banon *et al.* (2003).

Table 19.7 Trained sensory panel scores for loin chops and sausages from entire male and female pigs (24 Point Scale, 1 5 very low intensity, 24 5 very high intensity for attribute).

	Entire Male	Female	Significance	LSD
Lean				
Juiciness	15.6	15.7	ns	0.82
Tenderness	14	14	ns	0.82
Pork flavor	13.8	13.7	ns	0.60
Abnormal flavor	4.9	4.4	ns	0.62
Fat				
Pork flavor	11.3	11.4	ns	0.56
Boar flavor	2.8	2.4	*	0.37
Abnormal flavor	4	4	ns	0.60
Pork odor	10.7	11.3	*	0.60
Boar odor	4.2	2.8	***	0.68
Abnormal odor	4.3	4	ns	0.62
Overall acceptability	13.7	13.9	ns	0.76
Tenderness	18.3	18.6	ns	0.40
Texture (particle size)	12.8	12.5	ns	0.54
Firmness	13	13	ns	0.64
Juiciness	12.5	12.8	ns	0.46
Greasiness	8.9	9	ns	0.42
Breadiness	11.9	11.8	ns	0.52
Springiness	12.2	11.8	ns	0.60
Graininess	10.5	10.2	ns	0.42
Meatiness	15.9	16	ns	0.42
Pork flavor	15.1	15.4	ns	0.45
Sweetness	8.6	8.5	ns	0.46
Spiciness	10.7	10.7	ns	0.54
Saltiness	10.8	11.3	*	0.50
Herbiness	6	5.8	ns	0.39
Off flavor	3.2	2.7	ns	0.46
Overall acceptability	15.7	16.2	*	0.46

Source: From Mathews *et al.* (1997).

Table 19.8 Mean consumer sensory scores* and standard errors (SE) for meat samples evaluated in different weeks.†

Sensory attribute	Week 1		Week 2		Week 3		Week 4		*P*-Value
	Mean	SE	Mean	SE	Mean	SE	Mean	SE	
Cooking aroma (n = 240)‡	2.61	0.16	2.98	0.2	2.93	0.21	2.67	0.16	0.301
Flavor (n = 664)	2.16†	0.07	2.48*	0.1	2.53*	0.11	2.41*†	0.09	0.014
Texture (n = 664)	2.46†	0.08	2.70*†	0.1	2.89*	0.12	2.82*	0.13	0.001
Overall (n = 663)	2.25†	0.07	2.58*	0.1	2.73*	0.12	2.68*	0.11	0.001

*Sensory scale: 1, like strongly; 2, like moderately; 3, like slightly; 4, dislike slightly; 5, dislike moderately; 6, dislike strongly.

†Means with different letters within a row differ (P <0.05).

‡Total number of evaluations.

Source: From Babol *et al.* (2002).

19.4 CONCLUSIONS

In summary, although pork taint can be the result of improper dry curing or usage of meat from intact males, the cause in most countries outside the USA is due to the swine production methods that exclude the use of castration for various reasons. The meat from intact males is detectable by a large portion of the population regardless of cookery or processing method and even country of origin. Although both androstenone and skatole contribute to the off odors and off flavors, it appears that more people tend to object to lower levels of skatole present in the product.

REFERENCES

Agerhem, H. & Tornberg, E. (1995) A comparison of the off-flavour of meat from entire males cooked to two different internal end-point temperatures. In: *Proceedings EAAP Workshop on Production and Utilization of Meat from Entire Males*, Milton Keynes, U.K., September 27–29, 1995.

Ames, J.M. & MacLeod, G. (1985) Volatile components of a yeast extract composition. *Journal of Food Science*, **50**, 125–135.

Andersson, K., Hullberg, A., Mulmgren, L., Lundstrom, K., Squires, J. & Rydhmer, L. (1997) Relations between boar taint and puberty in entire male pigs. In: *Proceedings of the EAAP Working Group: Production and Utilization of Meat from Entire Male Pigs.* (eds M. Bonneau, K. Lundstrom & B. Malmfors), Stockholm, Sweden.

Babol, J., Squires, E.J. & Gullett, E.A. (2002) Factors affecting the level of boar taint in entire male pigs as assessed by consumer sensory panel. *Meat Science*, **61**, 33–40.

Banon, S., Costa, E., Gil, M.D. & Garrido, M.D. (2003) A comparative study of boar taint in cooked and dry-cured meat. *Meat Science*, **63**, 381–388.

Blanco, D., Arino, A., Conchello, P., Perez, C., Yanguella, J. & Herrera, A. (1997) Physiochemical characterization of "bonetaint" in Spanish dry-cured hams. *Journal of Food Protection*, **60**, 667–672.

Bonneau, M., Kempster, A.J., Claus, R., *et al.* (2000a) An international study on the importance of androstenone and skatole for boar taint: I. *Presentation of the programme and measurement of boar taint compounds with different analytical procedures. Meat Science*, **54**, 251–259.

Bonneau, M., Walstra, P., Claudi-Magnussen, C., *et al.* (2000b) An international study on the importance of androstenone and skatole for boar taint: IV. Simulation studies on consumer dissatisfaction with entire male pork and the effect of sorting carcasses on the slaughter line, main conclusions and recommendations. *Meat Science*, **54**, 285–295.

Cordoba, J.J., Antequera, T., Garcia, C., Ventanas, J., Lopez, C. & Asensio, M.A. (1994) Evolution of free amino acids and amines during ripening of Iberian cured ham. *Journal of the Agricultural and Food Chemistry*, **42**, 2296–2301.

Dijksterhuis, G.B., Engel, B., Walstra, P., *et al.* (2000) An international study on the importance of androstenone and skatole for boar taint: II. *Sensory evaluation by trained panels in seven European countries. Meat Science*, **54**, 261–269.

Garcia, C., Martin, A., Timon, M.L. & Cordoba, J.J. (2000) Microbial populations and volatile compounds in the "bone taint" spoilage of dry cured ham. *Letters in Applied Microbiology*, **30**, 61–66.

Giolitti, G., Cantoni, C.A., Bianchi, M.A. & Renon, P. (1971) Microbiology and chemical changes in raw hams of Italian type. *Journal of Applied Microbiology*, **34**, 51–61.

Griffith, N.M. & Patterson, R.L.S. (1970) Human olfactory response to 5a-androst-16ene-3one. *Journal of the Science of Food and Agriculture*, **21**, 4–6.

de Kock, H.L., van Heerden, S.M., Heinze, P.H., Dijksterhuis, G.B. & Minnaar, A. (2001) Reaction to boar odour by different South African consumer groups. *Meat Science*, **59**, 353–362.

Lerche, D. (1936) Geschlechtsgeruch bei Eberkastraten. *Zeitschrift fur Fleisch-und Milchhygiene*, **46**, 417–420.

Malmfors, B. & Lundstrom, K. (1983) Consumer reaction to boar meat—A review. *Livestock Production Science*, **10**, 187–196.

Mathews, K.R., Homer, D.B. & Leskanich, C.O. (1997) The effects of sex on the eating quality of British style sausages prepared from individual entire male and female pigs in: Boar taint in entire male pigs. In: *Proceedings of the EAAP Working Group: Production and Utilization of Meat from Entire Male Pigs* (eds M. Bonneau, K. Lundstrom & B. Malmfors), Stockholm, Sweden.

Mathews, K.R., Homer, D.B., Punter, P., *et al.* (2000) An international study on the importance of androstenone and skatole for boar taint: III. Consumer survey in seven European countries. Meat Science, **54**, 271–283.

McCauley, I., Hennessy, D.P., Boghossian, V., *et al.* (1997) Effect of cooking methods and processing pork on the perception of boar taint: Boar taint in entire male pigs. In: *Proceedings of the EAAP Working Group: Production and Utilization of Meat from Entire Male Pigs* (eds M. Bonneau, K. Lundstrom & B. Malmfors), Stockholm, Sweden.

Paterson, R.L.S. (1968) 5a-androst-16-ene-3-one: Compound responsible for taint in boar fat. *Journal of the Science of Food and Agriculture*, **19**, 31–38.

Siret, F., Beague, M.P., Fischer, K. & Chevillon, P. (1997) Consumer acceptability and characteristics of the cooking odour of pork with different anstrostenone and skatole contents: Comparison of two cooking procedures: Boar taint in entire male pigs. In: *Proceedings of the EAAP Working Group: Production and Utilization of Meat from Entire Male Pigs* (eds M. Bonneau, K. Lundstrom & B. Malmfors), Stockholm, Sweden.

Vold, E. (1970) Fleischproduktionseigenschaften bei Ebern und Kastraten IV. Organoleptische und gaschromatografische Untersuchungen Wassedampffluchtiger Stooffe des Ruckckenspeckes von Ebern. *Meldinger Nordlandbruckhoegskole*, **49**, 1–25.

Walstra, P., Claudi-Magnussen, C., Chevillon, P. & Cook, G.L. (1997) EU research program on boar taint: Androstenone and skatole levels according to seasons and counties: Boar taint in entire male pigs. In: *Proceedings of the EAAP Working Group: Production and Utilization of Meat from Entire Male Pigs* (eds M. Bonneau, K. Lundstrom & B. Malmfors), Stockholm, Sweden.

Weiler, U., Font I Furnols, M., Fischer, K., *et al.* (2000) Influence of differences in sensitivity of Spanish and German consumers to perceive androstenone on the acceptance of boar meat differing in skatole and androstenone concentrations. *Meat Science*, **54**, 297–304.

20 Microbiological and Sensory Properties of Fresh and Frozen Pork Products

Lisa McKee

Abstract: Reducing the microbial contamination of pork and other meats continues to be a challenge for the food industry. The presence of pathogens on pork can be a significant safety risk, particularly in light of the many reports of improper handling of meat by consumers. Growth of both pathogens and spoilage organisms on meat can also have a significant impact on organoleptic properties of pork and other meats and contribute to food waste. Methods used to reduce microbial loads on meats can also have significant impacts on texture, juiciness, appearance, odor, and other sensory properties. Continued research into newer technologies such as high-pressure processing may provide improved ways to produce pork products that are microbiologically stable and organoleptically acceptable.

Keywords: Pork contamination, pathogens, controlling pork microorganisms, washing, sanitizing, irradiation, high-pressure processing, pork meat packaging

20.1 INTRODUCTION

Pork has been used as human food for centuries. Evidence indicates hogs were domesticated for food in the Middle East about 7000 BC and arrived in the region that became the USA with Hernando de Soto in 1539 (USDA-FSIS 2003). During the nineteenth century, the addition of salt as a preservative resulted in the popular product "salt pork." The tradition of salt pork remains today in the form of cured products such as ham and lunchmeat, which account for about 60% of pork consumption at lunch. Fresh pork products, however, particularly in the form of chops, are more popular at dinner, accounting for about 35% of pork consumption at that meal (NPD 2003).

An increase in consumer knowledge about food safety has been attributed to greater media coverage of food-borne illness outbreaks (Woodburn & Raab 1997). Forty percent of respondents in that survey, however, either thought food could not be made safe to eat or could not cite a way to make food safe. Although consumers were reported to be knowledgeable about the relationship between cross-contamination and proper cooking and prevention of food-borne illness in a study by Fischer *et al.* (2007), 7 of the 25 participants allowed juices from raw meat to contaminate the final meal. One-third of the respondents in a survey by Altekruse *et al.* (1996) reported unsafe food preparation practices, including not washing their hands and cross-contaminating raw meat and other foods. Although improper hand washing is a major factor in food-borne illness, Strohbehn *et al.* (2008) reported compliance rates in food service workers for hand-washing frequency and procedures

Handbook of Meat, Poultry and Seafood Quality, Second Edition. Edited by Leo M. L. Nollet.

ranged from 5% in restaurants to 33% in assisted-living facilities. Improper food handling has frequently been reported to be a leading cause of food-borne illness worldwide (Scott 1996; Jay *et al.* 1999). The concerns about consumer food handling point to the continuing need to produce meat products that meet high standards for microbiological as well as sensory quality.

20.2 GENERAL MICROBIAL TYPES AND LOADS ON PORK

Contamination of pork carcasses with fecal and ingested materials during processing is unavoidable. The Pathogen Reduction, Hazard Analysis and Critical Control Point Systems (PR-HACCP) rule, published by the US Department of Agriculture Food Safety and Inspection Service in 1996, requires pork processors to create and implement HACCP plans and standard procedures for sanitation operations. Pork production plants must also evaluate products for generic *Escherichia coli* as a measure of process control adequacy and meet *Salmonella* standards set in the rule (USDA-FSIS 2000).

A variety of microorganisms, including pathogens such as *Salmonella*, *Campylobacter*, *Escherichia coli*, *Yersinia*, and *Listeria*, and spoilage organisms such as *Lactobacillus* and *Pseudomonas*, are commonly detected on pork and other meats. *Trichinella spiralis*, a parasitic roundworm, has also been associated with pork products. Rates of contamination vary widely and depend on a number of factors including area of the carcass evaluated, method of evaluation, and where in the process the evaluation is made. Hides, intestinal contents, personnel, and equipment are only a few of the numerous potential sources of contamination during pork processing.

20.3 COMMON PATHOGENS ON PORK

Trichinella spiralis is a parasitic roundworm that invades the tissues of both carnivorous and omnivorous animals. Humans become infected with the organism through consumption of inadequately cooked meat containing *Trichinella* larval cysts. Once in the gastrointestinal tract, the larvae invade the bowel wall, producing symptoms such as abdominal pain, vomiting, diarrhea or constipation, and fever. As the larvae migrate into muscle tissues, symptoms such as edema, headaches, rashes, and coughing may develop.

Trichinella has historically been associated with pork as a result of hogs fed garbage containing animal waste materials (Roy *et al.* 2003). Rates of trichinellosis have declined substantially over time due to laws such as the Federal Swine Health Protection Act, which prohibits the use of potentially contaminated garbage as swine feed. The *Trichinae* Herd Certification Program—a joint effort of the USDA, the National Pork Producers Council, and the pork-processing industry—is a preharvest documentation system that allows production facilities that follow the program stipulations to be certified as *Trichinella*-free (National Pork Board 2000). Despite such programs, pork was reported to be the source of trichinellosis in 22 of 72 cases in the USA between 1997 and 2001 (Roy *et al.* 2003) and 10 of 66 cases in the USA between 2002 and 2007 (Kennedy *et al.* 2009).

Although *Salmonella* is more often associated with poultry products, the pathogen is a hazard in all meats. Provisions in the PR-HACCP rule require all meat processing plants to meet a standard for *Salmonella* rates. For pork that performance standard is 8.7% or no more than 6 *Salmonella*-positive samples in a 55-sample set (USDA-FSIS 2000).

Prior to implementation of PR-HACCP, Epling *et al.* (1993) reported that 29% of 225 swine carcasses were *Salmonella* positive. Subsequent to PR-HACCP implementation, analysis of 1532 sponge samples collected during 1998 and 1999 from swine carcasses at 33 large US processing plants indicated a *Salmonella*-positive rate of 6.5% (USDA-FSIS 2000). Algino *et al.* (2009), however, reported a *Salmonella*-positive rate of 9.4% on prewash and 18.8% on chilled pork carcasses in small pork abattoirs in Wisconsin. Muth *et al.* (2009) reported 28 of 121 federally inspected pork slaughter plants exceeded half the *Salmonella* performance standard between 2003 and 2005.

Data generally indicate *Salmonella* is most often isolated prior to fabrication and/or refrigerated storage, and decreases in *Salmonella* levels are frequently noted as carcasses proceed through processing (Epling *et al.* 1993; Saide-Alboronoz *et al.* 1995). Duffy *et al.* (2001), however, detected *Salmonella* on 9.6% of retail pork samples in the USA. More recently, Kegode *et al.* (2008) reported a 0% detection rate for *Salmonella* in retail pork samples in the Fargo, North Dakota metropolitan area. Delhalle *et al.* (2009) reported *Salmonella* detection rates ranging from 0% to 50% in pork-processing plants but only 0.3%–4.3% in minced pork at the retail level in Belgium. Prendergast *et al.* (2008) found the prevalence of *Salmonella*, which ranged from 0% to 31.7%, in pork abattoirs in Ireland was dependent on the specific abattoir and the day of sampling. *Salmonella* spp. were detected in 13 of 500 pork samples collected in 2007 from butchers and supermarkets in Ireland by Prendergast *et al.* (2009). *Salmonella* typhimurium was the most common serotype detected in that study, with one strain, *Salmonella* typhimurium U310, recovered from an abattoir being 100% identical to a strain isolated one year later from a retail pork sample.

Escherichia coli (*E. coli*) is a gram-negative microorganism that has been recognized as a food-borne pathogen since 1971 but has probably been responsible for food-borne illnesses since at least the 1700s. Classified as a coliform, *E. coli* grows well on a wide variety of foods and in a variety of environmental conditions. Although five virulence groups are recognized for *E. coli*, the enterohemorrhagic strain O157:H7 is one of the most studied in foods. Disease syndromes associated with *E. coli* O157:H7 include hemolytic uremic syndrome and hemorrhagic colitis.

Seventy percent of pork samples from the Fargo, North Dakota metropolitan area, tested positive for *Escherichia coli* in a study by Kegode *et al.* (2008). Pork chops (16%), pork ribs (15%), pork roast (13%), and ground pork (12%) were found to have the highest levels of the bacterium. Wong *et al.* (2008) reported a prevalence rate of 1% for *E. coli* O157:H7 in domestic pig carcasses in New Zealand. The 1.8% prevalence rate for the microorganism in imported pig carcasses was attributed to pork from Australia with no *E. coli* O157:H7 detected in pork imported from the USA or Canada. Tamplin *et al.* (2001), however, detected *E. coli* biotype 1 in 30 of 100 pork carcasses processed in the USA under a HACCP-based inspection model project.

A variety of other pathogens have been detected in pork products. *Yersinia enterocolitica* was detected in 19.8% of whole muscle, store-packaged pork samples and in 11.5% of store-ground pork samples while *Campylobacter jejuni/coli* was detected in 1.3% of retail and 6.7% of processing plant pork samples by Duffy *et al.* (2001). Ohlendorf and Murano (2002) detected *Arcobacter* spp. in 20% of low-fat and 4% of high-fat samples in a study of 200 fresh pork samples collected from different US geographical regions and evaluated using different recovery methods. Although pork carcasses were not sampled directly, Chasseignaux *et al.* (2002) reported 37% of swab samples taken from environment and equipment sources during processing of pork in three production plants were positive for *Listeria monocytogenes*.

20.4 CONTROLLING MICROORGANISMS ON PORK PRODUCTS AND THE EFFECT ON SENSORY PROPERTIES TEMPERATURE

Although the effect of both hot and cold temperatures on microbial levels in pork products has been investigated, cold temperatures are by far the most frequent method of controlling microbial loads in meats. Chilling has been cited as a critical control point in pork processing although microbial loads have been reported to both increase (Bolton *et al.* 2002) and decrease (Gill & Bryant 1992; Carr *et al.* 1998) during chilling. The effects of the method of chilling as well as temperature of storage on microbial loads and sensory properties of pork have been investigated.

Freezing can be an effective means of reducing microbial loads/infectivity on foods. Kotula *et al.* (1990) predicted the thermal death time of *Trichinella spiralis* (*T. spiralis*) in pork to be 8 min at −20°C with an upper confidence level of 48 min. Medina-Larena *et al.* (2009) determined the effect of processing procedures including cold storage and freezing on the viability and infectivity of *Trichinella spiralis* ML. Mean level of *T. spiralis* ML in 50 g samples of pork prior to processing was 93 ± 15. Freezing at −20°C was associated with elimination of both viability and infectivity of the parasite after only 15 days of storage. Although no odor was detected in frozen samples throughout the 266-day storage period, color became increasingly darker red over time.

Greer and Murray (1991) noted a greater lag time for psychrotrophs, pseudomonads, *Brochothrix thermosphacta*, and enteric bacteria in fresh pork compared to vacuum-packaged pork frozen for 90 days at −30°C and then thawed at 1.7°C for 2 days. Bacterial generation time was not different between fresh and frozen/thawed samples. Although freezing was associated with only a slight darkening and loss of brightness, an interaction between muscle quality and freezing/thawing on appearance case-life of the pork was noted. While freezing/thawing significantly reduced the appearance case-life of the pork regardless of muscle quality, odor was not affected by freezing/thawing treatments.

Cryogenic chilling is another method of cooling that can affect both microbial loads and quality attributes of pork. Neither chilling in liquid nitrogen for 1 or 3 min nor chilling at 1°C for 24 h was effective in reducing mesophilic bacterial loads in pork and no differences in protein solubility or sarcomere length were noted (Jones *et al.* 1991). Liquid nitrogen was most effective at reducing *Pseudomonas* spp.; reductions in *Brochothrix thermosphacta* B2, *Escherichia coli* ATCC 11775, and *Salmonella typhimurium* ATCC 14028 due to liquid nitrogen treatments were also noted. Although the color of liquid nitrogen-treated muscle was slightly darker and muscle shear values were higher for carcasses treated with liquid nitrogen for 1 min, few significant differences in objective pork quality characteristics were found. Subjective evaluation of palatability characteristics such as initial tenderness, juiciness, and flavor desirability by a six-member sensory panel indicated little difference in all characteristics except sour off flavors that were higher in liquid nitrogen-treated carcasses from pigs with high glycolytic rates.

Lower storage temperatures are typically associated with lower microbial growth. Lee *et al.* (1985) found pork samples stored above 0°C had shelf lives of 14–28 days while those stored at −4°C had minimal changes in microbial and quality characteristics through 49 days. Pork stored at 7°C was consistently rated lower for overall appearance and had increased exudate, green discoloration, and off odors compared to samples stored at −4°C, 0°C, or 3°C. Liu *et al.* (2006) found *Brochothrix thermosphacta* and coliforms were the most common organisms associated with spoilage of sliced pork, regardless of temperature or atmospheric conditions. A six-member trained sensory panel detected increasing off odor spoilage in all samples as storage time increased, with higher off odor development associated with storage at 10°C compared to −2°C and 4°C. Flavor characteristics, however, did not differ between pork stored under different atmospheric conditions at either −2°C or 10°C.

20.5 WASHING AND SANITIZING

Washing and sanitizing are some of the most commonly used methods for reducing microbial loads on pork and other meats. A water spray of 90°C for 2 min reduced the total viable count on pork carcasses from 5.58 to 3.26 logs with minimal effects on color and odor characteristics in a study by Latha *et al.* (2009). Water was compared to several combination treatments as a decontamination method on lean and fat-covered pork tissue (Castelo *et al.* 2001). Reductions in coliform levels of 1.46 $\log_{10}$ colony-forming units (CFU)/square centimeters (cm^2) (15 s) to 2.52 $\log_{10}$ CFU/cm^2 (180 s) were noted when 15°C water was used to wash lean tissue contaminated with feces. Water plus lactic acid and three treatments using combinations of water, hot water, hot air, and lactic acid were more effective than water alone at reducing levels of aerobic bacteria, coliforms, *Escherichia coli*, lactic acid bacteria, and psychrotrophic bacteria. Combination treatments were associated with a decrease in redness and an increase in yellowness. Ground pork samples treated with water plus lactic acid were reported to be dark red with a mushy appearance. Brown cooked spots were noted on samples treated with any combination method.

The use of acids as decontaminating agents has also been investigated by others. Prasai *et al.* (1992) sprayed 1% lactic acid at 55°C onto pork carcasses immediately after dehairing, after evisceration, and at both points in the processing. Mean total aerobic counts were always lower in acid-treated carcasses, but the decreases were not statistically significant when compared to control carcass results. Overall appearance, lean color, fat color, surface discoloration, and off odor as evaluated by a four-member trained panel also did not differ between treated and untreated samples. van Netten *et al.* (1995), however, reported undesirable color changes in pork carcasses decontaminated with either 2% or 5% lactic acid. Hot (55°C) lactic acid solutions were more consistently effective in reducing *Salmonella typhimurium* contamination on pork carcasses than were cold solutions in that study.

Mendonca *et al.* (1989) dipped pork chops into 1% acetic acid, 1% acetic acid/1% lactic acid, 1.5% acetic acid/1.5% sodium acetate, 3% acetic acid/3% sodium ascorbate, 3% acetic acid/2% sodium chloride (NaCl), or sterile, distilled water. Chops were subsequently vacuum packaged and stored at 2–4°C for 6 weeks. All samples except those containing 3% acetic acid had microbial loads of 10^7–10^8 CFU/cm^2 after 3 weeks' storage and were considered spoiled. Samples treated with solutions containing 3% acetic acid had lower mesophilic and psychrotrophic counts than all other samples throughout the evaluation period. All treatments containing acetic acid, however, were effective at reducing *Enterobacteriaceae* levels. Treated samples were darker and tended to be less red and yellow than controls, but the presence of sodium ascorbate minimized these detrimental effects on pork color.

Fu *et al.* (1994) studied the microbial and quality characteristics of pork cuts from carcasses treated with 1.5% acetic acid, citric, or lactic acids. Acetic and citric acids were associated with a decrease in total aerobic plate and coliform counts during the first 14 days of storage at 2–4°C and all acid-treated samples had lower total aerobic plate counts compared to controls at 42 days of storage. Off odors were detected by sensory panelists at day 5 for citric acid, day 7 for acetic acid, and day 14 for lactic acid. All samples exhibited undesirable off odors by day 35. Surface color of chops changed from red 1 day after cutting to gray-pink

8–12 days after cutting. Color, however, as well as other sensory characteristics were not different between treatments.

Both acetic acid and citric acid were effective in reducing total bacterial growth in pork fillets after 28 days of storage in a study by Schirmer and Langsrud (2010). The 3% level was more effective at reducing microbial growth than the 1% level for both acids but acetic acid had greater antimicrobial activity. A significant white precipitate was observed in pork treated with 3% citric acid while acetic acid did not alter the appearance. Addition of up to 1.5% acetic acid and 5% citric acid in three marinades was reported to improve sensory qualities of the pork after cooking. Acetic acid was also found to be the most effective antimicrobial agent on pork loin chops by Lin and Chuang (2001), but the low pH of the chops was associated with an undesirable pale and watery appearance.

Huang *et al.* (2005) compared the shelf life characteristics of pork chops dipped in 500 ppm ascorbic acid, 250 ppm citric acid, or no acid dip, followed by simulated retail display in either modified atmosphere packaging (MAP) with gas exchange or vacuum-packaging/gas permeable overwrap. Psychrotrophic counts were lower for samples dipped in ascorbic acid compared to citric acid but all counts reached log 6 CFU/g or more at the end of day 7 which was considered to be the effective end of shelf life. Chops dipped in either ascorbic or citric acid were lighter and more yellow compared to no acid dip samples before packaging but citric acid was associated with less red color compared to ascorbic acid and no dip. Addition of ascorbic and citric acids was also associated with increased lipid stability compared to control samples when packaged in MAP.

20.6 SALTS

Pork longissimus muscle marinated with a combination of citric acid and sodium acid pyrophosphate had lower total aerobic plate counts at 14 and 21 days when stored at 4°C compared to those marinated with either sodium tripolyphosphate or sodium acid pyrophosphate (Cannon *et al.* 1993). The lower pH in pork from the citric acid treatment, however, was associated with lower juiciness, tenderness, and palatability compared with muscle from the higher pH treatments.

Another phosphate compound, trisodium phosphate (TSP), has been reported to be an effective antimicrobial agent in beef and poultry, but results have been less consistent in pork. No improvements in mean aerobic plate counts were found when pork carcasses and loins were dipped in, sprayed with, or scalded with TSP solutions in a study by Morris *et al.* (1997). Trained sensory panelists found few significant differences in discoloration, color uniformity, color, texture, or off odor. Lin and Chuang (2001), however, reported TSP-treated pork chops had lower total aerobic and psychrotrophic counts compared to control chops after 6 days of refrigerated storage. The TSP-treated chops were also reported to have an unacceptable dark, dry surface.

The addition of 2–3% sodium or potassium lactate (KL) has been reported to reduce total aerobic plate counts and extend refrigerated shelf life of pork products by 12–14 days (Brewer *et al.* 1991; Lamkey *et al.* 1991; O'Connor *et al.* 1993; Lin and Chuang 2001; Tan & Shelef 2002). The delay in microbial deterioration was associated with a delay in sour and off flavor development (Brewer *et al.* 1991). Protection of red colors and enhancement of juiciness and pork flavors have been consistently reported. Protection of fat color (white) during chub storage (Lamkey *et al.* 1991) and fat stability (Tan & Shelef 2002) with the addition of lactates have also been reported. Brewer *et al.* (1991) reported enhanced salty flavors in fresh pork sausage containing sodium lactate, but O'Connor *et al.* (1993) found the salty flavor of sodium was less intense from sodium lactate than from sodium chloride.

Latha *et al.* (2009) evaluated the bacteriological and sensory properties of inoculated pork carcasses sprayed with either a 5% potassium sorbate solution or a combination solution containing 5% NaCl + 2.5% each of sodium acetate, sodium citrate, sodium lactate, and potassium sorbate. Total viable counts were reduced by 0.96 log units by the potassium sorbate treatment and 1.31 log units by the combination treatment with minimal effects on color or odor characteristics. *Staphylococcus aureus*, *Escherichia coli*, *Salmonella typhimurium*, *Yersinia enterocolitica*, *Listeria monocytogenes*, *Serratia marcescens*, *Pseudomonas aeruginosa*, and *Proteus vulgaris* were more sensitive to the combination salt treatment than to potassium sorbate alone. Both the potassium sorbate (8 days) and combination (11 days) treatments increased shelf life of the pork compared to the control (4 days).

Fresh pork sausages prepared from different combinations of prerigor pork meat and fresh pork trim were formulated with 0% or 3% KL by Seyfert *et al.* (2006). Sausage patties were displayed at 4°C in either a lighted display case for 10 days or in an unlighted display case for 7 days followed by 3 days in the lighted display. The addition of 3% KL was associated with lower aerobic plate counts and lower lipid oxidation compared to samples without KL. Although a number of interactions were noted, the addition of KL was associated with greater redness and vividness and less visual discoloration regardless of storage treatment. Pork sausages formulated with KL had similar color characteristics after 10 days in lighted displays as those formulated without KL and stored in the dark for 7 days, indicating KL increased color stability even in less desirable lighting conditions.

20.7 SPICES, HERBS, AND OTHER ADDED SUBSTANCES

Martínez *et al.* (2006a, b) evaluated the quality characteristics and shelf life of fresh pork sausages formulated with a variety of spices and herbs, including black, white, red sweet, and hot cayenne peppers, rosemary, and red beet root (RBR) extract. All peppers were associated with lower psychrotrophic counts at 8 and 12 days of storage at 2°C ± 1°C compared to the control, but only 2% cayenne and 1% black peppers maintained the lower counts at day 16. Psychrotrophic loads increased in sausages formulated with rosemary or RBR and reached 8 log CFU/g by the end of the 16-day storage period. Higher levels of red sweet, cayenne, and black peppers were associated with lower CIE L* values while addition of red sweet and cayenne peppers was associated with increasing redness and yellowness values. A sharp decrease in redness was noted after 8 days of storage regardless of spice addition. Although RBR extract contributed to increased redness values, it did not protect sausages from discoloration. Red sweet pepper, cayenne pepper, and rosemary plus ascorbic acid were associated with the greatest inhibition of oxidation. Discoloration increased with storage in all sausages according to data from a six-member trained sensory panel, but samples containing all levels of black pepper, 2% red or cayenne peppers, and a combination of rosemary + ascorbic acid + sodium lactate maintained a score below 3 (the limit set for acceptability) through storage day 12. Samples treated with a combination of rosemary + ascorbic acid + sodium lactate + RBR extract did not reach a score of 3 at any point in the 16-day storage period. Samples containing 2% red or cayenne peppers and 1% black pepper maintained scores of 1 or less for off odor development throughout the 16-day storage period.

Martínez *et al.* (2006c) also investigated the effects of NaCl, rosemary extract, green tea powder, pu-erh tea infusion, and ascorbic acid in various combinations on fresh pork sausages packaged in an atmosphere of four parts O_2 and one part CO_2 and stored in the dark for 20 days. Aerobic psychrotroph counts, CIE L* and a* values, thiobarbituric acid reactive substance (TBARS), and off odor and discoloration as evaluated by a six-member trained sensory panel were determined at 4-day intervals during the storage period. Use of NaCl caused darkening and promoted lipid oxidation but protected sausages from discoloration. Green tea powder, pu-erh tea infusion, and borage meal, on the other hand, strongly inhibited lipid oxidation but were associated with significant color loss. The most effective substance evaluated was a combination of rosemary extract plus ascorbic acid. Lipid oxidation was delayed in sausages containing that combination and only a slight off odor was detected by sensory panelists at the 20th day of storage. Rosemary extract plus ascorbic acid was also associated with a delay in sausage discoloration and slower microbial growth, which resulted in a 4-day extension of the shelf life of the fresh pork sausages.

The antimicrobial activity of clove oil and extracts prepared from honeysuckle, Scutellaria, Forsythia suspensa, cinnamon, and rosemary in 75% ethanol in vacuum-packaged fresh pork was determined by Kong *et al.* (2007). Although all extracts were found to suppress growth, initial testing in inoculated agar indicated equal volume mixtures of Scutellaria + honeysuckle + Forsythia + cinnamon and cinnamon + rosemary + clove oil had the strongest antimicrobial activity against *Escherichia coli*, *Pseudomonas fluorescens*, and *Lactobacillus plantarum*. When applied to vacuum-packaged fresh pork, these extract combinations were associated with reductions of 1.81–2.32 $\log_{10}$ in microbial loads compared to the control over a 28-day storage period. Minimal differences in surface color and off odors were detected by sensory panelists between treated and untreated pork samples.

Shan *et al.* (2009) evaluated the effects of cinnamon stick, oregano, clove, pomegranate peel, and grape seed extracts on pH, color, TBARS, and growth of *Listeria monocytogenes*, *Staphylococcus aureus*, and *Salmonella enterica* in raw pork at 20°C. Pork treated with the spice and herb extracts contained 0.8–2.24 log CFU/g fewer microorganisms compared to the control after 9 days of storage when inoculated with a low level of each organism. Lower levels of *Staphylococcus aureus* and reductions in *Salmonella enterica* were found in all treated samples compared to the control. When the pork was inoculated with a high level of each microorganism, *Listeria monocytogenes* was 1.36 log CFU/g, *Staphylococcus aureus* was 0.96 log CFU/g, and *Salmonella enterica* was 1.75 CFU/g lower in spice/herb treated samples compared to the control. Clove and cinnamon extracts were the most effective antimicrobial agents. Spice/herb extracts were generally associated with darker, more red, and more yellow color characteristics compared to the control. Only slight changes in color were noted during storage for treated samples while the control became significantly darker, less red, and more yellow over time. TBARS of spice/herb treated pork samples increased 21%–28% during the 9 day storage period while the TBARS value for the control increased 3000%. As with antimicrobial activity, clove extract was a particularly effective antioxidant.

Birk and Knochel (2009) studied the effect of marination with red wine and yogurt on *Brochothrix thermosphacta*, *Carnobacterium maltaromaticum*, *Listeria monocytogenes*, and *Campylobacter jejuni* on pork meat. Submersion of pork meat in 42°C red wine for 15 min followed by storage in the wine at 4°C for 3 days reduced *Carnobacterium maltaromaticum* 1.5 logs, *Listeria monocytogenes* 2 logs, *Brochothrix thermosphacta* 3 logs, and *Campylobacter jejuni* 6 logs.

20.8 IRRADIATION

Irradiation treatment of pork and other meats as a means of microbial reduction was found safe by the USDA in 1997 (USDA-FSIS 1999). The final rule approving irradiation of meats was issued in 2000. The PR-HACCP rules require irradiation be done in accordance with an established HACCP plan with irradiation being designated as a critical control point. Although not currently used extensively in the USA, the effects of irradiation on microbial loads and sensory properties of pork continue to be studied.

An absorbed dose of 1.91 kGy or higher was identified by Thayer *et al.* (1993) as necessary to reduce microbial loads in vacuum-packaged fresh pork to undetectable levels. The levels of mesophiles, psychrotrophs, and anaerobes/facultative anaerobes were reduced by a 1 kGy irradiation treatment (Mattison *et al.* 1986; Ehioba *et al.* 1987), but a sublethal effect was noted by Ehioba *et al.* (1987) in vacuum-packaged pork treated with that irradiation level. An irradiation dose of 1.75 kGy reduced microbial loads of 10^6 cells/g by one to five logs depending on the microbial species (Grant & Patterson 1991a). Of the pathogens studied, *Clostridium perfringens* was the most resistant and *Yersinia enterocolitica* the least resistant to irradiation. Gram-negative organisms, particularly *Pseudomonas* and *Enterobacter*, accounted for 96% of isolates in fresh ground pork initially but only 76% of isolates in nonirradiated vacuum-packaged pork after 9 days at 5°C (Ehioba *et al.* 1988). Microflora in irradiated samples was primarily gram-positive immediately after treatment. *Lactobacillus* and coryneforms predominated after 9 and 12 days of storage at 5°C. *Lactobacillus* was also the predominant organism detected in irradiated pork stored under 25% or 50% carbon dioxide (CO_2) by Grant and Patterson (1991b), and Lebepe *et al.* (1990) noted lactobacilli seemed to be the microbial type least affected by irradiation. Thayer *et al.* (1993) detected *Staphylococcus*, *Micrococcus*, and yeasts in fresh pork irradiated with 0.57 kGy. *Listeria monocytogenes* at an inoculated level of 10^5–10^6 CFU/g was inactivated by an irradiation dose of 3.0 kGy in ground pork (Bari *et al.* 2006), but an irradiation dose of 4.7 kGy was needed to inactivate 10^6 CFU/g *Salmonella enteritidis* in vacuum-packaged pork meat (Sinanaglou *et al.* 2009). A dose of 2.5 kGy, however, was sufficient to inactivate the *Salmonella enteritidis* when the inoculation level was 10^3 CFU/g.

In contrast to studies with poultry, only limited effects on sensory properties of pork due to irradiation have been reported. Mattison *et al.* (1986), using triangle tests, reported short-term irradiation affects on pork sensory properties 24 h after treatment, but those effects disappeared after 2 weeks of storage at 4°C. Grant and Patterson (1991b), however, reported a distinct "irradiation odor" in pork chops. Although it did not change in intensity over storage time, the odor was described as more "sour" or "dairy" as storage progressed. Bari *et al.* (2006) noted sensory panelists were able to distinguish between sausages made from irradiated and nonirradiated ground pork but were divided as to whether irradiation adversely affected sausage quality. Panelists who preferred the radiated sausage noted it had more flavor than the nonirradiated product. On the other hand, panelists who preferred the nonirradiated sausage reported the irradiated sausage had a sharp, processed, chemical, or refrigerator taste. Lebepe *et al.* (1990) noted a higher red value in irradiated pork chops stored at 5°C under fluorescent light but inconsistent differences in lightness and yellowness. Sinanaglou *et al.* (2009), however, reported increasing redness and yellowness but no changes in lightness in pork irradiated at 2.5 and 4.7 kGy. A trained descriptive panel was able to detect differences in some characteristics of irradiated chilled (bitterness, sour, browned/roasted, pork identity, and sweet flavors) or frozen (browned/roasted flavors, juiciness, and toughness) pork chops, but results for many of the flavor characteristics evaluated were inconsistent (Luchsinger *et al.* 1996). Dose level, irradiation source, and packaging type did not influence fat-like flavor, juiciness, metallic flavor, toughness, or raw and cooked pork aromas in chilled pork chops nor bitterness, fat-like, pork identity, sour, or sweet flavors, or raw and cooked pork aromas in frozen chops. No differences in overall acceptance, meatiness, freshness, tenderness, or juiciness were detected by consumer panelists in chilled, vacuum-packaged pork chops irradiated at 2.5 kGy compared to nonirradiated controls (Luchsinger *et al.* 1996).

20.9 HIGH-PRESSURE PROCESSING

High-pressure processing has received increased attention in recent years as a food preservation method. The application of high pressure causes volume changes in the food or its constituent parts, resulting in protein denaturation, textural alterations, gelation, greater flavor and color retention, and enzyme modifications (Knorr 1993; Hugas *et al.* 2002). High-pressure processing can also effectively reduce microbial loads, but the results depend on many factors including amount of pressure applied, type of microorganism, presence of spores, temperature, and water activity.

Although hydrodynamic pressure was associated with a decrease in numbers of *Trichinella spiralis* recovered from pork, infectivity of the larvae was not eliminated by the treatment (Gamble *et al.* 1998). D-values of 2.17 min at 414 MPa and 25°C for *Listeria monocytogenes* and 1.48 min at 414 MPa and 2°C for *Salmonella typhimurium* were reported by Ananth *et al.* (1998). Psychrotrophs reached 5.71 $\log_{10}$ CFU/g after 33 days of storage at 4°C in pork loin treated with pressure at 25°C compared to 7.0 $\log_{10}$ CFU/g in untreated loins. Mesophiles exceeded 6.0 $\log_{10}$ CFU/g in both untreated and pressure-treated samples after 3 days when stored aerobically at 25°C and all samples were considered spoiled after 5 days of storage at 25°C. Murano *et al.* (1999) found a 10 $\log_{10}$ reduction in a resistant strain of *Listeria monocytogenes* in ground pork patties treated with 414 MPa at 50°C for 6 min. The addition of the mild heat to the pressure treatment reduced the D-value ranges from 1.89–4.17 to 0.37–0.63 min. Treated samples had a shelf life of 28 days compared to 5 days for control patties when stored at 4°C.

The use of high-pressure processing to reduce microbial loads may also result in changes in pork quality characteristics. Sensory panelists, however, could not distinguish between high-pressure plus heat-treated pork patties and controls after grilling (Murano *et al.* 1999). Ananth *et al.* (1998) also reported no differences in flavor, juiciness, firmness, color, texture, water-holding capacity, and moisture between pressure-treated (25°C) and control samples when the pork loins were cooked and evaluated by panelists using descriptive analysis.

20.10 PACKAGING MATERIALS AND ENVIRONMENTS

Packaging materials and environments can have significant affects on both microbiological and sensory qualities of pork products. Weakley *et al.* (1986) compared the effects of parchment wrapping versus vacuum packaging on quality characteristics of fresh pork loins that had been either hot or conventionally processed. Few differences in processing/packaging treatments were noted during storage at 2°C until day 14 when vacuum-packaged loins exhibited both lower total plate counts and lower off odor scores. Vacuum-packaged pork loins had less surface discoloration than parchment-wrapped loins at days 7 and 14. Higher surface discoloration was also associated with parchment wrapping in a study by Vrana *et al.* (1985). Fresh pork loins were first stored unwrapped, wrapped in parchment, or vacuum packaged. Chops cut from each type of stored loin were then displayed wrapped in either oxygen-permeable or high oxygen barrier film. Lean color scores differed little, but overall appearance was lower in chops from parchment-wrapped and vacuum-packaged loins compared to unwrapped loins. Chops from unwrapped loins also had less off odors and cooked off odors. The high off odor scores for vacuum-packaged chops were the result of microbial counts of 10^8 CFU/cm^2 or higher at 6–10 days of storage. Cooked off odors were described as putrid in unwrapped chops displayed in oxygen-permeable wrap, gaseous hydrogen sulfide in parchment-wrapped/high oxygen barrier packaged chops, and sour-acidic in vacuum-packaged/high oxygen barrier packaged chops. Microflora of chops displayed in high oxygen barrier wrap was primarily *Lactobacillus* and *Micrococcus* spp. while *Pseudomonas* spp. predominated in chops displayed in oxygen-permeable wrap. Although microbial loads for chops from unwrapped loins were also 10^7–10^8 CFU/cm^2 at 6–10 days of storage, lower off odor scores were apparently due to a shorter exposure time to high microbial loads.

Modification of the atmosphere within packaged pork products has also been investigated. One of the most common gases used in such modifications is CO_2. Greer *et al.* (1993) detected only lactic acid bacteria in pork loin stored under CO_2 at −1.5°C through 24 weeks, with a maximum of 10^7 CFU/cm^2 at 9 weeks. Lactic acid bacteria and pseudomonads were the primary spoilage organisms detected during retail display of the loins. When a five-member sensory panel evaluated the odor of displayed pork on a seven-point hedonic scale and appearance on a five-point acceptability scale, scores for both characteristics decreased as storage time in CO_2 increased. Linear regression indicated a case-life reduction of about 1 day for each 6 weeks of storage in CO_2. Although lactic acid bacteria dominated the microflora of both vacuum-packaged pork and pork stored under CO_2 at 3°C, spoilage was attributed to a substantial increase in *Brochothrix thermosphacta* by Gill and Harrison (1989). Color and texture were stable throughout the storage period for all samples. Sour-aromatic odors from *Brocothrix thermosphacta* and acid-dairy odors from lactic acid bacteria were unacceptably strong after 11 days of storage in vacuum-packaged samples and 39 days in CO_2-stored pork and were associated with microbial levels of 10^8 organisms/cm^2. Similar trends were noted for pork samples stored at −1.5°C, although storage life increased to 10 weeks or more.

Scholtz *et al.* (1992a) compared laboratory and commercial bulk packaging systems for pork using a 100% CO_2 atmosphere followed by aerobic retail display. Initial total microbial loads at day 0 of retail display were about 4 $\log_{10}$ CFU/cm^2 for all CO_2 bulk storage times. Counts increased as display time increased, reaching an unacceptable level of greater than 7 $\log_{10}$ CFU/cm^2 after 5 days of retail display. Anaerobic, lactic acid bacteria, and pseudomonad counts followed similar trends. Color of all samples as evaluated by a ten-member trained panel was pale to normal throughout bulk storage and retail display. The panel detected a fresh meat odor in all samples after 14 days of storage under CO_2. Odor became increasingly unacceptable, however, as retail display progressed. The microbiological and sensory properties of pork stored under 100% CO_2 were not different from chops stored in polyvinyl chloride (PVC), modified atmosphere (25% CO_2/75% O_2), or vacuum packages through 4 days of storage in a second study by Scholtz *et al.* (1992b). Interactions between packaging type and color, odor, overall acceptability, and microbial loads were noted, however, during longer storage times. Storage life of the 100% CO_2 chops was superior overall.

Several investigators have evaluated MAP environments containing combinations of CO_2 and CO. Wilkinson *et al.* (2006) compared the color and microbiological qualities of pork chops stored in modified atmospheres with 100% CO_2 or a mixture of 80% CO_2/19.6% N_2/0.4% CO. No differences were detected in total aerobic loads (2.98–3.82 $\log_{10}$ CFU/g) and total anaerobic loads (1.23–2.33 $\log_{10}$ CFU/g) between treatments during the 8 week storage period. Inclusion of CO in the environment was associated with a 32% increase in vividness and a shift toward a lighter red color. When the samples were exposed to O_2 on opening of the packages, blooming in the pork chops was associated with a lighter, more intense red color in all treatments but the color was more intense in samples treated with CO. Although CO was found to stabilize pork chop color, the lack of inhibition of pathogens was cited as a concern for use of the gas in packaging environments.

Microbiological, color, and sensory properties of pork loin stored at 4°C under different MAP conditions (vacuum, 100% CO_2, 99% CO_2 + 1% CO, 100% O_2, and 100% CO + vacuum) were evaluated by Viana *et al.* (2005). Psychrotroph loads were 10^8 CFU/g in samples in 100% O_2 and 10^7 CFU/g in all other pork samples after 20 days of storage while *Pseudomonas* growth was limited in all treatments except 100% O_2. The 99% CO_2 + 1% CO environment produced pork loin with the most desirable color characteristics and the highest consumer acceptance scores at both 24 h and 20 days of storage.

The effects of microbial inhibitors (KL plus sodium diacetate), packaging method (vacuum or MAP with 99.5–100% CO_2), and storage temperature (4°C and 10°C) alone and in combination on growth of *Listeria monocytogenes* and *Salmonella enterica* Typhimurium in pork chops and ham were studied by Michaelsen *et al.* (2006). The most significant factor effecting growth of both microorganisms at 4°C was the storage temperature. At 10°C, MAP, addition of antimicrobials, and the combination of MAP + antimicrobials prevented growth of *Salmonella* Typhimurium in pork chops and *Listeria monocytogenes* in ham but MAP did not enhance the activity of the antimicrobials. A 12-member trained sensory panel found raw pork chops stored under MAP treatments were more pink and less tender than those that were vacuum packaged. Addition of antimicrobials was associated with lower raw pork odor and increased vinegar odor. Juiciness, pork flavor, saltiness, and sourness were not different between treatments in cooked pork chops.

The presence of environmental oxygen can have significant effects on both microbiological and quality characteristics of pork. Presence of residual oxygen in the packaging atmosphere has generally been associated with lighter, less red, and more yellow pork compared to oxygen-free atmospheres (Lambert *et al.* 1992a; Sørheim *et al.* 1997). Increased discoloration was also detected by sensory panelists as the oxygen level increased in the storage atmosphere. Although strong off odors were associated with the presence of oxygen immediately after treatment, those odors weakened during storage (Lambert *et al.* 1992a). Microflora after modified atmosphere storage is often lactic acid bacteria, and few differences are generally reported regardless of whether or not oxygen is present (Lambert *et al.* 1992b; Sørheim *et al.* 1997).

Controlled atmospheres using combinations of CO_2, O_2, and N_2 have also been studied. The presence of O_2 has been associated with a greater increase in psychrotrophs compared to environments containing CO_2 or CO_2 plus N_2 at storage temperatures of both 2°C and 5°C (Spahl *et al.* 1981). *Pseudomonas* accounted for as much as 60% of microbial loads in the samples. Rejection of chops by a sensory panel was associated with objectionable odors before appearance became a determining factor. Although the 100% CO_2 environment resulted in the greatest extension of shelf life based on sensory results, combinations containing mixtures of CO_2 and N_2 were reported to be best for extending overall shelf life of pork chops. Holley *et al.* (1994) found pork stored in atmospheres of 100% CO_2 and 50% CO_2/50% N_2 had lower psychrotroph growth at both −1°C and 4°C compared to 100% N_2. Storage environment had no effect on L, a, or b values or expressible juice and little effect on tenderness.

20.11 CONCLUSION

Reducing the microbial contamination of pork and other meats continues to be a challenge for the food industry. The presence of pathogens on pork can be a significant safety risk, particularly in light of the many reports of improper handling of meat by consumers. Growth of both pathogens and spoilage organisms on meat can also have a significant impact on organoleptic properties of pork and other meats and contribute to food waste. Methods used to reduce microbial loads on meats can also have significant impacts on texture, juiciness, appearance, odor, and other sensory properties. Continued research into newer technologies such as high-pressure processing may provide improved ways to produce pork products that are microbiologically stable and organoleptically acceptable.

REFERENCES

Algino, R.J., Badtram, G.A., Ingham, B.H. & Ingham, S.C. (2009) Factors associated with *Salmonella* prevalence on pork carcasses in very small abattoirs in Wisconsin. *Journal of Food Protection*, **72** (4), 714–721.

Altekruse, S.F., Street, D.A., Fein, S.B. & Levy, A.S. (1996) Consumer knowledge of foodborne microbial hazards and food-handling practices. *Journal of Food Protection*, **59** (3), 287–294.

Ananth, V., Dickson, J.S., Olson, D.G. & Murano, E.A. (1998) Shelf life extension, safety and quality of fresh pork loin treated with high hydrostatic pressure. *Journal of Food Protection*, **61** (12), 1649–1656.

Bari, M.L., Hayakawa, F., Kawamoto, S., Todoriki, S. & Sommers, C. (2006) Irradiation inactivation of *Listeria monocytogenes* in low-fat ground pork at freezing and refrigeration temperatures. *Journal of Food Protection*, **69** (12), 2955–2960.

Birk, T. & Knochel, S. (2009) Fate of food-associated bacteria in pork as affected by marinade, temperature and ultrasound. *Journal of Food Protection*, **72** (3), 549–555.

Bolton, D.J., Pearce, R.A., Sheridan, J.J., Blair, I.S., McDowell, D.A. & Harrington, D. (2002) Washing and chilling as critical control points in pork slaughter hazard analysis and critical control point (HACCP) systems. *Journal of Applied Microbiology*, **92**, 893–902.

Brewer, M.S., McKeith, F., Martin, S.E., Dallmier, A.W. & Meyer, J. (1991) Sodium lactate effects on shelf-life, sensory and physical characteristics of fresh pork sausage. *Journal of Food Science*, **56** (5), 1176–1178.

Cannon, J.E., McKeith, F.K., Martin, S.E., Novakofski, J. & Carr, T.R. (1993) Acceptability and shelf-life of marinated fresh and precooked pork. *Journal of Food Science*, **58** (6), 1249–1253.

Carr, M.A., Thompson, L.D., Miller, M.F., Ramsey, C.B. & Kastner, C.S. (1998) Chilling and trimming effects on the microbial populations of pork carcasses. *Journal of Food Protection*, **61** (4), 487–489.

Castelo, M.M., Kang, D.H., Siragusa, G.R., Koohmaraie, M. & Berry, E.D. (2001) Evaluation of combination treatment processes for the microbial decontamination of pork trim. *Journal of Food Protection*, **64** (3), 335–342.

Chasseignaux, E., Gérault, P., Toquin, M.T., Salvat, G., Colin, P. & Ermel, G. (2002) Ecology of *Listeria monocytogenes* in the environment of raw poultry and raw pork meat processing plants. *FEMS Microbiology Letters*, **210**, 271–275.

Delhalle, L., Saegerman, C., Farnir, F., *et al.* (2009) *Salmonella* surveillance and control at post-harvest in the Belgian pork meat chain. *Food Microbiology*, **26**, 265–271.

Duffy, E.A., Belk, K.E., Sofos, J.N., Bellinger, G.R., Paper, A. & Smith, G.C. (2001) Extent of microbial contamination in United States pork retail products. *Journal of Food Protection*, **64** (2), 172–178.

Ehioba, R.M., Kraft, A.A., Molins, R.A., *et al.* (1987) Effect of low-dose (100 krad) gamma radiation on the microflora of vacuum-packaged ground pork with and without added sodium phosphates. *Journal of Food Science*, **52** (6), 1477–1480, 1505.

Ehioba, R.M., Kraft, A.A., Molins, R.A., *et al.* (1988) Identification of microbial isolates from vacuum-packaged ground pork irradiated at 1 kGy. *Journal of Food Science*, **53** (1), 278–279, 281.

Epling, L.K., Carpenter, J.A. & Blankenship, L.C. (1993) Prevalence of *Campylobacter* spp. and *Salmonella* spp. on pork carcasses and the reduction effected by spraying with lactic acid. *Journal of Food Protection*, **56** (6), 536–537.

Fischer, A.R.H., De Jong, A.E.I., Van Asselt, E.D., De Jonge, R., Frewer, L.J. & Nauta, M.J. (2007) Interdisciplinary investigation of microbial hazards during food preparation. *Risk Analysis*, **27** (4), 1065–1077.

Fu, A.H., Sebranek, J.G. & Murano, E.A. (1994) Microbial and quality characteristics of pork cuts from carcasses treated with sanitizing sprays. *Journal of Food Science*, **59** (2), 306–309.

Gamble, H.R., Solomon, M.B. & Long, J.B. (1998) Effects of hydrodynamic pressure on the viability of *Trichinella spiralis* in pork. *Journal of Food Protection*, **61** (5), 637–639.

Gill, C.O. & Bryant, J. (1992) The contamination of pork with spoilage bacteria during commercial dressing, chilling and cutting of pig carcasses. *International Journal of Food Microbiology*, **16**, 51–62.

Gill, C.O. & Harrison, J.C.L. (1989) The storage life of chilled pork packaged under carbon dioxide. *Meat Science*, **26**, 313–324.

Grant, I.R. & Patterson, M.F. (1991a) Effect of irradiation and modified atmosphere packaging on the microbiological safety of minced pork stored under temperature abuse conditions. *International Journal of Food Science & Technology*, **26**, 521–533.

Grant, I.R. & Patterson, M.F. (1991b) Effect of irradiation and modified atmosphere packaging on the microbiological and sensory quality of pork stored at refrigeration temperatures. *International Journal of Food Science & Technology*, **26**, 507–519.

Greer, G.G. & Murray, A.C. (1991) Freezing effects on quality, bacteriology and retail-case life of pork. *Journal of Food Science*, **56** (4), 891–894, 912.

Greer, G.G., Dilts, B.D. & Jeremiah, L.E. (1993) Bacteriology and retail case life of pork after storage in carbon dioxide. *Journal of Food Protection*, **56** (8), 689–693.

Holley, R.A., Delaquis, P., Rodrigue, N., Doyon, G., Gagnon, J. & Gariépy, C. (1994) Controlled-atmosphere storage of pork under carbon dioxide. *Journal of Food Protection*, **57** (12), 1088–1093.

Huang, N.Y., Ho, C.P. & McMillin, K.W. (2005) Retail shelf-life of pork dipped in organic acid before modified atmosphere or vacuum packaging. *Journal of Food Science*, **70** (8), M382–M387.

Hugas, M., Garriga, M. & Monfort, J.M. (2002) New mild technologies in meat processing: High pressure as a model technology. *Meat Science*, **62**, 359–371.

Jay, L.S., Comar, D. & Govenlock, L.D. (1999) A video study of Australian domestic food-handling practices. *Journal of Food Protection*, **62**, 1285–1296.

Jones, S.D.M., Greer, G.G., Jeremiah, L.E., Murray, A.C. & Robertson, W.M. (1991) Cryogenic chilling of pork carcasses: Effects on muscle quality, bacterial populations and palatability. *Meat Science*, **29**, 1–16.

Kegode, R.B., Doetkott, D.K., Khaitsa, M.L. & Wesley, I.V. (2008) Occurrence of *Campylobacter* species, *Salmonella* species and generic *Escherichia coli* in meat products from retail outlets in the Fargo metropolitan area. *Journal of Food Safety*, **28**, 111–125.

Kennedy, E.D., Hall, R.L., Montgomery, S.P., Pyburn, D.G. & Jones, J.L. (2009) Trichinellosis surveillance-United States, 2002–2007. *MMWR*, **58** (SS9), 1–8.

Knorr, D. (1993) Effects of high-hydrostatic-pressure processes on food safety and quality. *Food Technology*, **47** (6), 156–161.

Kong, B., Xiong, Y.L. & Wang, J. (2007) Antimicrobial activity of several herb and spice extracts in culture medium and in vacuum-packaged pork. *Journal of Food Protection*, **70** (3), 641–647.

Kotula, A.W., Sharar, A.K., Paroczay, E., Gamble, H.R., Murrell, K.D. & Douglass, L. (1990) Infectivity of *Trichinella spiralis* from frozen pork. *Journal of Food Protection*, **53** (7), 571–573.

Lambert, A.D., Smith, J.P. & Dodds, K.L. (1992a) Physical, chemical and sensory changes in irradiated fresh pork packaged in modified atmosphere. *Journal of Food Science*, **57** (6), 1294–1299.

Lambert, A.D., Smith, J.P., Dodds, K.L. & Charbonneau, R. (1992b) Microbiological changes and shelf life of MAP irradiated fresh pork. *Food Microbiology*, **9** (3), 231–244.

Lamkey, J.W., Leak, F.W., Tuley, W.B., Johnson, D.D. & West, R.L. (1991) Assessment of sodium lactate addition to fresh pork sausage. *Journal of Food Science*, **56** (1), 220–223.

Latha, C., Sherikar, A.T., Waskar, V.S., Dubal, Z.B. & Ahmed, S.N. (2009) Sanitizing effect of salts on experimentally inoculated organisms on pork carcasses. *Meat Science*, **83**, 796–799.

Lebepe, S., Molins, R.A., Charoen, S.P., Farrar IV, H. & Skowronski, R.P. (1990) Changes in microflora and other characteristics of vacuum-packaged pork loins irradiated at 3.0 kGy. *Journal of Food Science*, **55** (4), 918–924.

Lee, B.H., Simard, R.E., Laleye, L.C. & Holley, R.A. (1985) Effects of temperature and storage duration on the microflora, physicochemical and sensory changes of vacuum- or nitrogen-packed pork. *Meat Science*, **13**, 99–112.

Lin, K.W. & Chuang, C.H. (2001) Effectiveness of dipping with phosphate, lactate and acetic acid solutions on the quality and shelf-life of pork loin chops. *Journal of Food Science*, **66** (3), 494–499.

Liu, F., Yang, R.Q. & Li, Y.F. (2006) Correlations between growth parameters of spoilage micro-organisms and shelf-life of pork stored under air and modified atmosphere at −2, 4 and 10°C. *Food Microbiology*, **23**, 578–583.

Luchsinger, S.E., Kropf, D.H., Garcia Zepeda, C.M., Chambers IV, E., Hollingsworth, M.E., Hunt, M.C., Marsden, J.L., Kastner, C.L. & Kuecker, W.G. (1996) Sensory analysis and consumer acceptance of irradiated boneless pork chops. *Journal of Food Science*, **61** (6), 1261–1266.

Martínez, L., Cilla, I., Beltrán, J.A. & Roncalés, P. (2006a) Effect of *Capsicum annuum* (red sweet and cayenne) and *Piper nigrum* (black and white) pepper powders on the shelf life of fresh pork sausages packaged in modified atmosphere. *Journal of Food Science*, **71** (1), S48–S53.

Martínez, L., Cilla, I., Beltrán, J.A. & Roncalés, P. (2006b) Combined effect of modified atmosphere packaging and addition of rosemary (*Rosmarinus officinalis*), ascorbic acid, red beet root (*Beta vulgaris*), and sodium lactate and their mixtures on the stability of fresh pork sausages. *Journal of Agricultural and Food Chemistry*, **54**, 4674–4680.

Martínez, L., Cilla, I., Beltrán, J.A. & Roncalés, P. (2006c) Antioxidant effect of rosemary, borage, green tea, pu-erh tea and ascorbic acid on fresh pork sausages packaged in a modified atmosphere: Influence of the presence of sodium chloride. *Journal of the Science of Food and Agriculture*, **86**, 1298–1307.

Mattison, M.L., Kraft, A.A., Olson, D.G., Walker, H.W., Rust, R.E. & James, D.B. (1986) Effect of low dose irradiation of pork loins on the microflora, sensory characteristics and fat stability. *Journal of Food Science*, **51** (2), 284–287.

Medina-Larena, M.S., Ramirez-Ákvarez, A., Kühne, M., Gómez-Priego, A. & de-la-Rosa, J.L. (2009) Influence of different processing procedures on reproductive capacity of *Trichinella spiralis* in pork meat. *Tropical Animal Health and Production*, **41**, 437–442.

Mendonca, A.F., Molins, R.A., Kraft, A.A. & Walker, H.W. (1989) Microbiological, chemical and physical changes in fresh, vacuum-packaged pork treated with organic acids and salts. *Journal of Food Science*, **54** (1), 18–21.

Michaelsen, A.R., Sebranek, J.G. & Dickson, J.S. (2006) Effects of microbial inhibitors and modified atmosphere packaging on growth of *Listeria monocytogenes* and *Salmonella enterica* Typhimurium and on quality attributes of injected pork chops and sliced cured ham. *Journal of Food Protection*, **69** (11), 2671–2680.

Morris, C.A., Lucia, L.M., Savell, J.W. & Acuff, G.R. (1997) Trisodium phosphate treatment of pork carcasses. *Journal of Food Science*, **62** (2), 402–403, 405.

Murano, E.A., Murano, P.S., Brennan, R.E., Shenoy, K. & Moreira, R.G. (1999) Application of high hydrostatic pressure to eliminate *Listeria monocytogenes* from fresh pork sausage. *Journal of Food Protection*, **62**, 480–483.

Muth, M.K., Karns, S.A. & Fahimi, M. (2009) Analysis of *Salmonella* control performance in U.S. young chicken slaughter and pork slaughter establishments. *Journal of Food Protection*, **72** (1), 6–13.

National Pork Board. (2000) Quality assurance: A program of America's pork producers. Level III. National Pork Producers Council, Des Moines, Iowa. Accessed on January 2002 at www.aphis.usda.gov/vs/trichinella/

van Netten, P., Mossel, D.A.A. & Veld, J.H. (1995) Lactic acid decontamination of fresh pork carcasses: A pilot plant study. *International Journal of Food Microbiology*, **25** (1), 1–9.

NPD (2003) National eating trends in-home/away from home combined; 2001/2002. Accessed on May 2004 at www.porkfoodservice.com/documents/RAC-MEAL%20USAGE%202003.PPT

O'Connor, P.L., Brewer, M.S., McKeith, F.K., Novakofski, J.E. & Carr, T.R. (1993) Sodium lactate/sodium chloride effects on sensory characteristics and shelf-life of fresh ground pork. *Journal of Food Science*, **58** (5), 978–990.

Ohlendorf, D.S. & Murano, E.A. (2002) Prevalence of *Arcobacter* spp. in raw ground pork from several geographical regions according to various isolation methods. *Journal of Food Protection*, **65**, 1700–1705.

Prasai, R.K., Accuff, G.R., Lucia, L.M., Morgan, J.B., May, S.G. & Savell, J.W. (1992) Microbiological effects of acid decontamination of pork carcasses at various locations in processing. *Meat Science*, **32**, 413–423.

Prendergast, D.M., Duggan, S.J., Fanning, S., *et al.* (2008) Prevalence and numbers of *Salmonella* spp. and Enterobacteriaceae on pork cuts in abattoirs in the Republic of Ireland. *Journal of Applied Microbiology*, **105**, 1209–1219.

Prendergast, D.M., Butler, F., Cormican, M., *et al.* (2009) Prevalence, numbers and characteristics of *Salmonella* spp. on Irish retail pork. *International Journal of Food Microbiology*, **131** (2–3), 233–239.

Roy, S.L., Lopez, A.S. & Schantz, P.M. (2003) Trichinellosis surveillance-United States, 1997–2001. *MMWR*, **52** (SS06), 1–8.

Saide-Alboronoz, J.J., Knipe, C.L., Murano, E.A. & Beran, G.W. (1995) Contamination of pork carcasses during slaughter, fabrication and chilled storage. *Journal of Food Protection*, **58** (9), 993–997.

Schirmer, C. & Langsrud, S. (2010) Dissolving CO_2 headspace combined with organic acids prolongs the shelf-life of fresh pork. *Meat Science*, **85**, 280–284.

Scholtz, E., Krüger, J., Nortjé, G. & Naudé, R. (1992a) Centralized bulk pre-packaging of pork retail cuts. *International Journal of Food Science & Technology*, **27**, 391–398.

Scholtz, E.M., Jordaan, E., Krüger, J., Nortjé, G. & Naudé, R.T. (1992b) The influence of different centralised pre-packaging systems on the shelf-life of fresh pork. *Meat Science*, **32**, 11–29.

Scott, E. (1996) Foodborne disease and other hygiene issues in the home. *Journal of Applied Bacteriology*, **80**, 5–9.

Seyfert, M., Hunt, M.C., Grobbel, J.P., Ryan, S.M., Johnson, D.E. & Monderen, R.A. (2006) Potassium lactate and fresh-pork-sausage formulation effects on shelf life in lighted and unlighted display. *Journal of Food Science*, **71** (7), C390–C394.

Shan, B., Cai, Y.Z., Brooks, J.D. & Corke, H. (2009) Antibacterial and antioxidant effects of five spice and herb extracts as natural preservatives of raw pork. *Journal of the Science of Food and Agriculture*, **89**, 1879–1885.

Sinanaglou, V., Mantis, F., Sflomos, K., Konteles, S. & Batrinou, A. (2009) Effects of gamma radiation on microbiological status, fatty acid composition, and color of vacuum-packaged cold-stored fresh pork meat. *Journal of Food Protection*, **72** (3), 556–563.

Sørheim, O., Erlandsen, T., Nissen, H., Lea, P. & Høyem, T. (1997) Effects of modified atmosphere storage on colour and microbiological shelf life of normal and pale, soft and exudative pork. *Meat Science*, **47** (2), 147–155.

Spahl, A., Reineccius, G. & Tatini, S. (1981) Storage life of pork chops in CO_2-containing atmospheres. *Journal of Food Protection*, **44** (9), 670–673.

Strohbehn, C., Sneed, J., Paez, P. & Meyer, J. (2008) Hand washing frequencies and procedures used in retail food service. *Journal of Food Protection*, **71** (8), 1641–1650.

Tamplin, M.L., Feder, I., Palumbo, S.A., Oser, A., Yoder, L. & Luchansky, J.B. (2001) *Salmonella* spp. and *Escherichia coli* biotype I on swine carcasses processed under the hazard analysis and critical control point-based inspection models project. *Journal of Food Protection*, **64**, 1305–1308.

Tan, W. & Shelef, L.A. (2002) Effects of sodium chloride and lactates on chemical and microbiological changes in refrigerated and frozen fresh ground pork. *Meat Science*, **62**, 27–32.

Thayer, D.W., Boyd, G. & Jenkins, R.K. (1993) Low-dose gamma irradiation and refrigerated storage *in vacuo* affect microbial flora of fresh pork. *Journal of Food Science*, **58** (4), 717–719, 733.

USDA-FSIS (1999) USDA issues final rule on meat and poultry irradiation. U.S. Department of Agriculture Food Safety and Inspection Service. Accessed on June 2004 at www.fsis.usda.gov/os/background/irrad_final.htm

USDA-FSIS (2000) HACCP implementation: First year *Salmonella* test results. U.S. Department of Agriculture Food Safety and Inspection Service. Accessed on March 2004 at www.fsis.usda.gov/ophs/salmdata.htm

USDA-FSIS (2003) Safety of fresh pork…from farm to table. Accessed on December 2003 at www.fsis.usda.gov/OA/pubs/pork.htm

Viana, E.S., Gomide, L.A.M. & Vanetti, M.C.D. (2005) Effect of modified atmospheres on microbiological, color and sensory properties of refrigerated pork. *Meat Science*, **71**, 696–705.

Vrana, J.A., Savell, J.W., Dill, C.W., Smith, G.C., Ehlers, J.G. & Vanderzant, C. (1985) Retail appearance, odor and microbiological characteristics of pork loin chops packaged in different oxygen-barrier films as affected by pork loin storage treatment. *Journal of Food Protection*, **48** (6), 476–481.

Weakley, D.F., McKeith, F.K., Bechtel, P.J., Martin, S.E. & Thomas, D.L. (1986) Effects of packaging and processing procedures on the quality and shelf life of fresh pork loins. *Journal of Food Science*, **51** (2), 281–283.

Wilkinson, B.H.P., Janz, J.A.M., Morel, P.C.H., Purchas, R.W. & Hendriks, W.H. (2006) The effect of modified atmosphere packaging with carbon monoxide on the storage quality of master-packaged fresh pork. *Meat Science*, **73**, 605–610.

Wong, T.L., MacDiarmid, S. & Cook, R. (2008) *Salmonella*, *Escherichia coli* O157:H7 and *E. coli* biotype 1 in a pilot survey of imported and New Zealand pig meats. *Food Microbiology*, **26**, 177–182.

Woodburn, M.J. & Raab, C.A. (1997) Household food preparers' food-safety knowledge and practices following widely publicized outbreaks of foodborne illness. *Journal of Food Protection*, **60** (9), 1105–1109.

21 Shelf Life of Fresh and Frozen Pork

Sancho Bañon Arias

Abstract: Pork shelf life is limited by several factors that influence meat spoilage, such as meat properties, handling, packaging, chilling, display lighting, mechanical stresses, or the addition of preservatives. These factors will determine the predominant spoilage mechanism of retailed fresh pork: browning caused by metamyoglobin formation; putrefaction caused by aerobic spoilage bacteria; rancidity caused by lipid hyper-oxidization under high-oxygen MAP; or acidity caused by lactic acid bacteria under vacuum or free-oxygen MAP. Chilling technologies, retail lighting selections, display temperatures, packaging modifications, and product irradiation can offer color stability, microbial reductions, and delayed lipid oxidation that will extend the shelf life of pork and promote consumer satisfaction and repeat purchasing.

Keywords: Pork quality, pork retail case life, pork shelf life, vitamin E, preservatives, pork chilling, packaging, modified atmosphere packaging, display lighting, irradiation

21.1 INTRODUCTION

The length of time considered suitable for sale, use, or consumption is described as the shelf life of pork, or as the recommended time that pork can be stored, during which the defined quality of a specified proportion of the goods remains acceptable under specified conditions of distribution, storage, and display (Gyesley 1991). Shelf life is different from expiry date: the former relates to pork quality and the latter to pork safety. Pork that has passed its shelf life might still be safe, but quality is no longer guaranteed. The shelf life of fresh pork will be limited by factors such as color stability, bacterial spoilage, and lipid oxidation, while retailed frozen pork suffers dehydration and oxidation phenomena, involving mainly lipid and muscle proteins. Regardless of possible health risks, the shelf life of retailed pork ends when the appearance and odor of fresh pork is not acceptable, or when the subsequent eating quality is critically worsened by previous display. This chapter will update the information concerning pork shelf life available in the previous edition written by Carr (2007).

21.2 PORK QUALITY

Postmortem pH decline and temperature will play a decisive role in the early quality of pork (Aberle *et al.* 2001). When pigs are slaughtered for meat production, muscle metabolism soon changes to anaerobic glycolysis. The lack of oxygen transport from the lungs to the

Handbook of Meat, Poultry and Seafood Quality, Second Edition. Edited by Leo M. L. Nollet.

muscle by hemoglobin reduces the amount of oxygen transferred to the myoglobin pigment in muscle. Although more adenosine-5′-triphosphate (ATP) is produced through the aerobic pathway, some ATP is generated after death through the anaerobic pathway. However, the anaerobic pathway is also associated with the buildup of lactic acid. Without a functioning circulatory system, the lactic acid cannot be transported to the liver for conversion to glucose and glycogen as an energy source nor can it be taken to the heart to be converted to carbon dioxide and water. Once glycogen stores are used up and metabolic enzymes become inactive, lactic acid concentration increases and muscle pH drops. At the point of death, the pH of muscle is near 7.0. Normal pH in pork gradually decreases to 5.6 or 5.7 within 6–8 h postmortem, and a final pH of about 5.5 is reached near 24 h postmortem. The meat produced will be reddish-pink in color.

However, pork may suffer an abnormal pH decline, causing deviations in quality. Pale, soft, and exudative (PSE) pork may occur as a result of porcine stress syndrome or be promoted by external stress factors. Stress induces adrenaline release, which intensifies anaerobic glycolysis prior to death, resulting in the production of more lactic acid. This may lower the postmortem pH to beyond the normal range within the first hour postmortem to reach 5.4 or 5.5, and will produce an equal or slightly lower final pH than normal. This condition will result in meat that is paler in color compared to normal meat. The buildup of lactic acid—before body heat can be removed from the carcass by chilling—will promote muscle protein denaturation. As the pH of muscle drops, it approaches the critical pH known as the isoelectric point of muscle proteins (5.1–5.2), where the numbers of positive and negative reactive groups on the protein chains are equal. As muscle pH approaches the isoelectric point, fewer reactive groups are available, and the ability of the muscle to bind water diminishes. The resulting muscle is pale through the loss of color pigmentation, exudative from the loss of water and protein-binding capabilities, and soft in texture from the loss of protein stability. Dark, firm, and dry (DFD) pork results when little glycogen is available in muscle before death due to exercise or other handling factors. An inadequate amount of lactic acid will be produced through metabolism, and so pH decline will be lower than in a normal, reaching values higher than 6.0–6.2. The resulting meat color will be darker than that produced by the normal pH decline. In this case, the pH falls only slightly and the final pH is reached quickly. As the ultimate pH is far from the isoelectric point of proteins, this muscle will bind water very well and will hold its pigmentation, resulting in a dark color. The higher pH also supports or enhances microbial growth. Acid pork (Monin & Sellier 1985) can occur in muscle with a high glycolytic potential, where postmortem pH declines at a normal rate reaching final values close to 5.0. Therefore, the normal pH declines at high temperature resulting in a lower protein denaturation compared with PSE pork. However, the low ultimate pH reached will decrease the water-binding capacity of acid pork.

Fresh pork quality can be subjectively evaluated using pork quality standards as a reference. Color may be the most important factor that influences the appearance and attractiveness of fresh pork (Faustman & Cassens 1990). The wide production of early pig breeds has resulted in less pigmented meat, making it difficult to discriminate pork color. Fresh pork color began to be classified in the 1970s using the Japanese Pork Color Standards (Nakai *et al.* 1975). Fresh pork can also be assigned to five quality categories according to National Pork Producers Council (NPPC) standards (NPPC 1999): PSE (pale, soft, exudative), PFN (pale, firm, nonexudative), RSE (red, soft, exudative), RFN (red, firm, nonexudative), and DFD (dark, firm, dry). The NPPC standards range from 1 (pale pinkish gray to white) to 6 (dark purplish red). The visual assessment of pork quality can be helped by using a color machine vision, which improves the repeatability of measurements (Tan *et al.* 2000). Faucitano *et al.* (2010) proposed lightness, ultimate pH, and drip loss (expressed as filter paper wetness, "FPW") values to assess fresh pork quality; pork was assessed as pale ($L^* > 50$ CIE units), red ($L^* = 42–50$ CIE units), or dark ($L^* < 42$ CIE units); as exudative or nonexudative (pH < 6) and dry (pH ≥ 6); and as exudative (FPW ≥ 80 mg), nonexudative (FPW = 41–79 mg), or dry (FPW ≤ 40 mg). The use of objective measurements of firmness instead of pork standards has also been discussed. Rincker *et al.* (2007) found reasonable correlations between objective (texture profile analysis) and subjective (NPPC standards) firmness scores in fresh pork loins. This suggests that handheld or online measuring devices could be developed to standardized pork firmness as an alternative to NPPC standards.

Norman *et al.* (2003) found that 20.8% of consumers surveyed chose pale pork, 26.4% chose normal reddish-pink colored pork, and 52.8% chose dark pork. Buege (2001) also reported that normal RFN pork is the most desirable not only for its ability to hold water and provide juiciness once cooked, but also because color relates to appearance and these combine to serve as an important factor in consumers' meat-purchasing decisions. Brewer and McKeith (1999) found that when consumers were asked to evaluate pork loin slices in a retail setting, the pinkest PSE samples and all DFD samples were ranked highest by consumers in the intent to purchase category, which generally related to overall acceptability. Greer and Murray (1991) reported that loin chops that were frozen and then thawed produced a higher pH and darker color than nonfrozen, which reduced shelf life related to appearance compared with fresh pork.

Quality deviations seem to play a role in the shelf life of fresh pork. The general appearance and color scores decreased at a more rapid rate with storage time in PSE loin chops versus chops of normal color, resulting in a shorter shelf life for the PSE chops (Fox *et al.* 1980). Zhu and Brewer (1998) evaluated PSE, "normal," and DFD pork over 7 days at 4°C: PSE pork was lighter (higher L^*) than normal and DFD pork, but did not change over time. An increase in a^* occurred the first day for normal and DFD but not for PSE samples. A subsequent increase in hue angle and loss of redness occurred in PSE and normal, but not in DFD samples. Changes in hue angle were more rapid in PSE than in normal samples, while light exposure accelerated all changes except L^*. Faucitano *et al.* (2010) determined the shelf life of vacuum packed pork from five different quality classes. After 35 days of storage at 4°C, spoilage bacteria growth was highest in DFD pork, followed by RSE pork, whereas PFN, RFN, and PSE pork had lower and similar microbial loads.

21.3 RETAIL CASE LIFE OF PORK

The key to retail pork display is to present products in an attractive and salable form. This includes a fresh appearance and acceptable quality without abnormal traits, such as unattractive color, purge, or dehydration (Kropf 1998). Carr and Miller (1998) surveyed multiple retail stores across the USA to define the color that develops when pork becomes unacceptable, the average shelf life, styles of retail cases in use, types of lighting, and other information related to the retail display and shelf life of pork. The majority of refrigerated retail cases were triple level (50%), followed by single level (coffin style; 36.7%), and finally quad level (12.5%). The most popular description of unacceptable pork packages in the retail case as described by meat market managers was poor color (63.3% of responses). The poor color was most often gray (50%), grayish green (36.7%), or light green to brown (10%), but the color was seldom too dark (3.3%). The reason market managers removed pork products from display was poor color (43.3% of responses), followed closely by poor color and past sell-by date (36.7%).

Retailed fresh pork suffers color deterioration due to the oxidation of myoglobin pigment. An increase in metamyoglobin pigment is associated with hue angle increase and chroma reduction in lean meat, resulting in a gradual loss of redness and progressive browning of the fresh pork (Bekhit *et al.* 2003). Flavor deteriorates slower than color in retailed raw pork. The first sign is the loss of the typical odor and the serum- or blood-like taste of fresh meat. Several off-odors can appear during storage of pork. Depending of storage conditions, the most usual being putrid, acid, and rancid. Putrid odor is caused by the formation of volatiles from the degradation of proteins by aerobic bacteria, including *Brochothrix thermosphacta*, cold-tolerant *Enterobacteriaceae*, and *Pseudomonas*, among others. Rancid odor is caused by lipid hyper-oxidization under the high-O_2 atmospheres used to preserve meat redness. Lipid oxidation will occur in pork muscle in refrigerated and frozen storage (Eleftheriadou *et al.* 2010). Acid odor is caused by lactic acid formation due to microbial activity in pork packed under a vacuum or under free-O_2 atmosphere. Fresh pork shelf life can be mathematically modeled using growth parameters of microorganisms, such as the initial cell number (N_0), maximum cell number (N_{max}), maximum specific growth rate (μ_{max}), and lag-phase (λ). The shelf life of fresh pork is mainly correlated with the μmax and λ than by N_{max} (Liu *et al.* 2006).

The shelf life of retailed frozen pork cuts will depend on the freezing, packaging, and display conditions. In general, frozen meat can be consumed without risk to health after several months or even years of storage. However, frozen pork can suffer gradual dehydration and oxidation phenomena, mainly involving lipid and muscle proteins, resulting in freeze burns, discoloration, off-flavors, and poor texture. Raw meat aging lowers the sensorial quality of cooked meat, giving rise to rancidity and other off-flavors. Lipid oxidation reactions are the main determinants of the eating quality and shelf life of chilled-stored cooked meat (Jensen *et al.* 1998). Prior to consumption, retailed cooked meat can remain for hours or even days exposed to the oxidizing action of air and light, which alter the color, aroma, and flavor before microbial spoilage occurs. The first sensory signs of cooked meat spoilage include loss of odor and flavor. The terms “warmed-over flavor” (WOF) (Timms & Watts 1958) and “meat flavor deterioration” (MFD) (Spanier *et al.* 1992) have been coined to describe the overall increase in off-flavor notes and loss of desirable meat flavor. These phenomena become more noticeable when refrigerated cooked meat is reheated. Pentanal, hexanal, and 2-4-decadienal are the major compounds responsible for the off-flavor and off-odor. The flavor and closely related odors produced from these compounds are often characterized as “stale,” “cardboard-like,” “painty,” or “rancid” (Vega & Brewer 1994).

21.4 METHODS TO IMPROVE PORK SHELF LIFE

21.4.1 Vitamin E and other dietary strategies

Dietary supplementation with Vitamin E (alpha-tocopherol) is a wide practice in modern day pig production. Muscle vitamin E levels reflect the dietary intake (Jensen *et al.* 1998), since vitamin E is deposited in the tissue membranes where it can act more effectively as antioxidant (Cayuela *et al.* 2003). Dietary supplementation for 6–12 weeks before pig slaughter with 100–200 mg of alpha-tocopherol per kilogram of feed has consistently been shown to reduce lipid oxidation or rancidity in raw, frozen, and cooked pork under different retail conditions, although most studies have found no beneficial effects of vitamin E on fresh pork color and drip loss in different packaging conditions (Monahan *et al.* 1990; Cannon *et al.* 1995, 1996; Dirinck *et al.* 1996; Jensen *et al.* 1998; Phillips *et al.* 2001; Hoz *et al.* 2003; Boler *et al.* 2009). Some skepticism has also been expressed concerning the success of adding vitamin E to muscle postmortem, as the compound may be unable to successfully integrate into the cell membranes. However, Whang *et al.* (1986) indicated that lipid oxidation can be controlled in refrigerated and cooked pork patties by adding 200 mg/kg of vitamin E during processing.

Other dietary strategies have also been used to extend the shelf life of pork. Apple *et al.* (2007) reported that supplementing swine diets with 80 ppm manganese (as amino acid complex) may improve pork color during retail display without increasing the likelihood of lipid oxidation. Manganese is involved in the activity of calcium ATPase and sodium/potassium ATPase and may play an important role in lipid oxidation. Current strategies for improving the nutritional quality of pork, in particular the use of dietary fat sources rich in unsaturated fatty acids, will probably require effective strategies to prevent oxidation in retailed meat. The dietary use of plants or agro-industrial by-products rich in polyphenols offer great potential for prolonging the shelf life of fresh meat. Polyphenols are excellent electron and proton donors, and their intermediate radicals are quite stable due to electron delocation phenomena and due to the lack of positions attackable by O_2. Like vitamin E, dietary polyphenols are deposited in muscle where they act as efficient antioxidants (Djenane *et al.* 2005). The use of plants in pig diets should take into account the bioavailability of polyphenols, and the possible accumulation of active compounds from plants that can modify the meat flavor.

21.4.2 Preservatives

Sulfite (expressed as SO_2 units) is widely used as a preservative to prolong the shelf life of minced fresh pork. EU Directive 95/2/EC (1995) restricts the use of sulfite in raw meat products to a maximum of 450 mg SO_2 per kg of meat. The FAO–WHO (1986) established a maximum permissible daily ingestion of 0–0.7 mg SO_2 per kg of body weight and considered it especially important that SO_2 be minimized in food with high thiamine content, such as red meat. There is substantial information concerning the preservative action of sulfite on meat. SO_2 is very active against gram-negative bacteria, such as Enterobacteriaceae (Banks *et al.* 1985). Sulfite reduces the myoglobin heme group and is then oxidized to sulfate. This favors the formation of oxymyoglobin and deoxymyoglobin and reduces metamyoglobin, providing a fresh appearance to red meat (Wedzicha & Mountfort 1991). Sulfite can transform hydroperoxides into nonradical forms and also acts as a secondary lipid antioxidant (Günther *et al.* 1997). Reducing agents such as ascorbate could protect SO_2 against oxidation, enabling SO_2 to act for a longer duration. Ascorbate is added at doses of around 500 mg/kg to minced pork, alone or combined with lactate or citrate as antioxidant enhancers.

Several attempts have been made to extend the shelf life of pork cuts using different methods of spraying and soaking with different preservative solutions. When Lin and Chuang (2001) evaluated the shelf life of pork cuts dipped in different solutions, 10% sodium lactate solution provided better visual appearance and lower microbial counts, suggesting the possibility of extending the shelf life of pork to 6 days under refrigeration. However, 10% trisodium phosphate resulted in a dry and dark visual appearance on the meat surface, while 2% acetic acid was similar to PSE pork. Psychrotrophic counts decreased during retail display in pork chops dipped in 500 mg/kg ascorbic acid and 250 mg/kg of citric acid compared with untreated pork (Huang *et al.* 2005). A novel method combining the advantages of modified atmosphere packaging (MAP) and organic acids was successfully tested by Schirmer and Langsrud (2010). A CO_2 headspace (initial gas/product ratio 0.2/1.0) combined with a brine solution containing citric acid (3% w/w, pH 5), acetic acid (1% w/w, pH 5), or a combination of both prolonged the shelf life of fresh pork.

Herb extracts are also used to extend the shelf life of pork and are generally regarded as safe (GRAS). Zhang *et al.* (2009) evaluated the potential of 14 common spice extracts as natural antimicrobials in fresh pork retailed in MAP. The extracts of clove, rosemary, cassia bark, and licorice were seen to inhibit pathogenic and spoilage bacteria. The mixture of rosemary and licorice demonstrated great potential as a natural preservative when sprayed on fresh pork. However, Rojas and Brewer (2008) obtained only modest results using grape seed, rosemary, and oregano derivatives to delay the oxidation of pork patties vacuum packed and stored frozen for 4 months. Whatever the case, herbs and other natural preservatives that are not considered food additives can be considered suitable to extend the shelf life of pork, especially of refrigerated-minced products whose microbiological quality must be ensured.

21.4.3 Chilling

Pork chilling aims to prolong its shelf life, delaying spoilage and reducing weight loss through drip or evaporation. Carcass chilling is crucial for subsequent pork quality, but it is difficult to chill pig carcass by air convection systems, such as rooms or tunnels, because of the size and fat cover. The chilling methods can be classified as convectional (0°C–4°C, 90%–95% RH, and 0.25–0.5 m/s air speed) or rapid (from −1°C to −2°C, 90%–95% RH, and 2–5 m/s air speed). Rapid chilling quickly provided improvements in aspects that reduced the incidence of PSE meat (Crenwelge *et al.* 1984; Milligan *et al.* 1998; Springer *et al.* 2003). However, Weakley *et al.* (1986) reported no extension in shelf life when pork carcasses were rapidly chilled. Chilled pork cuts were subsequently exposed in display cabinets for retail. Refrigeration delays the activity of microorganisms and enzymes in fresh meat, although it does not paralyze meat spoilage. Meat temperature, film permeability to gases, and air flow in the cabinet must be adjusted to prevent ice formation or excessive dehydration. The temperature of the meat should be as low as possible without forming surface ice that can alter the appearance of packed cuts. Miller and Carr (2000) evaluated the development of undesirable color changes in pork displayed at various temperatures in different case types. Retail display cases were set so that the airflow temperature around the pork loin chops was at −1.1°C, 1.1°C, or 3.3°C. The highest temperature setting was discontinued 3 days into the "warm-up" period before data collection could begin because of a significant increase in product deterioration related to poor color. These authors found that single-level cases to an airflow temperature of −1.1°C produced the least amount of product temperature gain in the hour before, during, and in the hour after their defrost cycle. However, if the display cases were multilevel, a setting of 1.1°C produced the least product temperature gain in the same time frame. The color stability of MAP pork cuts can be improved by display at −1.5°C, compared with 2°C and 5 C (Jeremiah and Gibson 1997). Freezing practically paralyzes the activity of microorganisms and enzymes in meat, so that it can be kept for months or years. The standard temperature in frozen meat display systems is −18°C, although temperature variations may occur due to lighting or loss of coolness. Calkins *et al.* (1986) reported an increase from −16°C to −10°C in the surface temperature of the frozen pork chops under retail display conditions.

21.4.4 Packaging

Packaging is necessary to extend the shelf life of pork under retail display conditions, because otherwise the meat would be preserved for few days under refrigeration. MAP is the removal and/or replacement of the atmosphere surrounding the meat before sealing in vapor-barrier materials. Most of the shelf life properties of pork can be extended by using MAP. Current MAP options include air-permeable overwrapped trays in master packs, low O_2 formats of shrunk film vacuum packaging (VP), MAP with CO_2 and N_2 and their peelable barrier film derivatives, and high O_2 MAP. Anoxic forms of MAP without CO extend the shelf life of pork, although they do not provide the bloomed red meat color. VP can successfully reduce the amount of oxygen in contact with the meat's surface, considerably slowing down myoglobin oxidation that results in the development of browning, although the more purple appearance of retailed pork cuts provided by deoxymyoglobin is not as well accepted by consumers. VP inhibits the growth of aerobic bacteria, but can also promote the growth of anaerobic microbes with lipolytic and proteolytic activity or lactic acid producers, which can negatively affect meat odor. High O_2/CO_2 MAPs (filled with N_2 or not) were applied in fresh pork to enhance redness and to inhibit the growth of aerobic bacteria. Levels below 15% CO_2 did not inhibit microorganism growth satisfactorily, while levels above 40% resulted in package collapse as CO_2 was absorbed by the meat tissue (McMillin *et al.* 1999). Zhang and Sundar (2005) studied the effect of different levels of O_2 (5%–45%) in MAP (20% CO_2 and filled with N_2) on the shelf life of refrigerated pork cuts, proposing 45% O_2 to provide a better equilibrium between color deterioration, acceptability, and microbial counts, although higher O_2 levels were not examined. It is now fairly common to use O_2/CO_2 percentages ranging from 20/80 to 70/30 in MAPs for fresh pork (Cayuela *et al.* 2004; Šcetar *et al.* 2010). N_2 is mainly used as gas filler because it provides few advantages in red fresh meat over other gases used in red meat MAP. Pork loin slices displayed at 4°C in 100% CO_2, or 50% N_2/50% CO_2, could be held in retail display 7 days longer than if packaged in 100% N_2 (Holley *et al.* 1994). Low CO amounts have also been used to enhance pork color by carboxymyoglobin formation, providing cherry-red chromaticity. CO improved color stability and sensory panel scores for both injected and noninjected pork chops during 36 days of refrigerated storage (Krause *et al.* 2003). Higher acceptance scores were given by consumers and color was lighter for pork treated with 1% CO and then stored for 20 days compared with VP, 100% O_2, and 100% CO_2 (Viana *et al.* 2005). The color of master-packed fresh pork was enhanced by 0.4% CO, with no difference in plate counts or lipid oxidation (Wilkinson *et al.* 2006).

Some characteristics, particularly color, of fresh pork under MAP may change when the pack is opened, meat blooming resulting from contact with air oxygen. Jeremiah and Gibson (1997) examined the color stability of display-ready pork cuts stored in three different atmospheres (100% CO_2, 100% N_2, and 70% O_2/30% CO_2) under refrigeration for up to 28 days, and then subsequently displayed aerobically for up to 30h. Pork cuts with 100% CO_2 tended to maximize the retention of color stability, although 70% O_2/30% CO_2 MAP provided brighter and redder meat following storage and prior to subsequent aerobic display. Other research reported that pork loin stored in 100% O_2 for 20 days was rejected by consumers due to browning, while pork stored in VP after 1 h exposure to 1% CO/99% CO_2 or 100% CO_2 before VP were still found acceptable (Viana *et al.* 2005). Unlike fresh meat, the quality of frozen pork is quite stable under display retail conditions, although it has been reported that lipid oxidation gradually increases in frozen pork cuts aerobically stored at −18°C for up to 180 days (Eleftheriadou *et al.* 2010). Thus, the shelf life of displayed frozen pork is preferably extended by using anoxic MAPs, in particular, skin VP. In contrast, Park *et al.* (2007) reported no substantial differences as regards lipid oxidation and off-odor volatiles in pork loin cuts that were either vacuum or air packed and stored frozen at −10°C for up to 120 days.

Different films are used, depending on the MAP system. Lidding films are often polyvinylidene chloride coated polypropylene-polyethylene, polyvinylidene chloride coated polyethylene terephthalate-polyethylene, or polyamide-polyethylene. Flow wrap films may be polyamide-polyethylene, polyamide-ionomer, or polyamide-ethylene vinyl acetate-polyethylene (Mullan & McDowell 2003). A single layer or type of plastic generally does not have all of the needed properties for a food package application, so lamination, coating, or coextrusion is used to create layers of plastic with the desired properties (Jenkins & Harrington 1991). A common multilayer structure has a structural external material (polyethylene terephthalate, polyamide, or polypropylene), an intermediate material with gas-barrier properties (polyvinyl chloride, polyvinylidene chloride, ethylene vinyl acetate, or others), and an inner material that provides the seal (polyethylene, ethylene vinyl acetate, ionomer, or others). Modifications in packaging film have shown good success in extending pork shelf life. VP pork chops displayed in high oxygen-barrier film had less off-odor when displayed 3 or 4 days under refrigeration, but not longer, because off-odor limited extended shelf life even though the chops were visually acceptable for 10 days (Vrana *et al.* 1985). Microperforated polypropylene film (allowing water vapor to escape) delayed discoloration and gave a more desirable outer appearance to retailed pork cuts stored under refrigeration for up to 14 days (Lee *et al.* 2003). The use of gas exchange systems can extend the shelf life of retailed pork. Pork chops in MAP (80% N_2/20% CO_2) with gas exchange (CO_2 80%/O_2 20%) had a lighter and redder color, increased weight retention, decreased psychrotrophic counts, and increased lipid oxidation compared with conventional vacuum and overwrap packaging systems (Huang *et al.* 2005). Antifogging agents can be externally applied to the polymer surface by dipcoating or spraying, or blended into the polymer for migration to the surface. Commonly used agents to prevent film fogging are glycerol esters, polyglycerol esters, sorbitan esters and their ethoxylates, alcohol ethoxylates, and nonylphenol ethoxylates (Osswald *et al.* 2006). MAP applications may have thermoformed or preformed base trays made from different materials.

Active packaging (AP) is becoming increasingly significant to extend the shelf life of fresh meat. AP is a type of packaging where the product, its packaging, and the surrounding environment interact to extend food shelf life, improving the organoleptic properties and/or food security (Suppalku *et al.* 2003). Principal AP systems include those that involve oxygen scavenging, moisture absorption and control, carbon dioxide and ethanol generation, and antimicrobial migrating and nonmigrating systems. As examples, the microbial growth and drip loss were inhibited and a^* value maintained on chilled pork by combining a chitosan coating with spice extracts (Xia *et al.* 2007), whereas bovine gelatin reduced the purge and color deterioration in pork, but was not an effective barrier for lipid oxidation in high-O_2 MAP or VP under refrigeration (Antoniewski *et al.* 2007).

Estimating the shelf life of retailed pork carries some risk of error, since, as be seen, meat spoilage is affected by many factors. According to recent data reviewed by McMillin (2008) and Scetar *et al.* (2010), the ranges of shelf life for fresh pork cuts displayed at 4°C may be established at 5–7 days (air packaging), 12–16 days (high-O_2/CO_2 MAP), 21–35 days (CO MAP), and 28–60 days (VP). The shelf life for fresh minced pork displayed at 4°C would be about half that of pork cuts, depending on the mincing conditions. The shelf life of retailed frozen pork may be established in 4–6 months; however, pork filets or chops processed in good conditions (e.g., quickly frozen, vacuum packed, chilled at −18°C or less, and kept with some protection against light) can be eaten without risk to health after several months or longer of storage.

21.4.5 Display lighting

Lighting is essential for the marketing and presentation of retailed pork, both for traditional and for case-ready sales, although it may speed up meat discoloration, reducing the shelf life. Lighting systems with a color temperature of 2900–3750 K and display light intensity of 270–1615 lux are recommended for pork display (Kropf 1998). Light types include incandescent, fluorescent, and metal halide (mercury vapor or high-intensity sodium). The color of fresh pork has been described as pink, pink brown or dark pink, and brown or dull pink, under incandescent, fluorescent, and metal halide lights, respectively. Relative light scattering, collected with a fiber-optic probe connected to a photo diode array, demonstrated the reason to be a lack of redness in the commonly used fluorescent and metal halide light sources (Barbut 2001). Display lighting effects on the appearance or rate of discoloration of meat could result from a temperature elevation at the meat surface, photochemical effects and/or differences in color rendition because of different spectral energy distribution patterns. Kropf (1998) estimated a 0.56°C rise in temperature for every 107 lux of incandescent lighting in a display case with 2 m^2 per minute air velocity. This increase in surface temperature would not only increase deterioration in color by the oxidation of pigments, but could also increase microbial growth. Deluxe fluorescent lights, as well as other types, can be substituted as they release about one-fifth the heat of incandescent lights for equivalent lighting. Calkins *et al.* (1986) reported that fresh loin chops displayed for 5 days under cool flood incandescent lights became darker, had a lower oxymyoglobin percentage, and were likely to show greater microbial growth due to an increase in chop temperature. It appears that cool flood incandescent and deluxe cool white lights provide the most desirable color and that precautions against temperature elevation are needed. Ultraviolet wavelengths causing meat discoloration can be filtered by using packaging film and/or lamps designed to reduce ultraviolet emission (Bertelsen and Skibsted 1987; Martínez *et al.* 2007). Lamps that emit low light in the red area of the spectrum (or correspond to more light in the blue area such as a colder, bluer light emitting at 4200 K) can promote undesirable appearance of blue or less red surface in retailed meat, compared to a warm light with a high proportion of red (such as one emitting at 2600 K) (Kropf 1998).

21.4.6 Irradiation

Irradiation can be applied to reduce microbial loads of retailed pork. Lipid oxidation and the subsequent development of off-flavors after cooking may be the main limitation of irradiated meat. Minimal doses (1 kGy) can sufficiently increase the shelf life of pork loins and reduce spoilage microbes after 15 days of storage, depending on initial microbe charges (Dogbevi *et al.* 1999). A 1 kGy dose of irradiation can also increase the shelf life of VP minced pork by 2.5–3.5 days if stored at 5 °C (Ehioba *et al.* 1987). VP pork loin chops irradiated at 2.5 kGy cobalt6′ were redder and possessed more color and lipid oxidation stability than controls, although oxidative rancidity increased and color stability decreased in loin irradiated in aerobic packaging (Luchsinger *et al.* 1996a). Similarly, irradiated pork under oxygen-permeable film became less red, more yellow, more brown, and less uniform compared to nonirradiated controls (Nanke *et al.* 1999). Lambert *et al.* (1992) found that applications of irradiation (1.0 kGy) in an environment without oxygen extended the shelf life of pork loin chops from 9 to 26 days at 5°C. However, oxygen in the package headspace combined with irradiation adversely affected the physical, chemical, and sensory characteristics of pork. Combined treatments with organic acids (lactic, citric, and acetic at 2%) and irradiation (2 and 3 kGy) were more effective than irradiation alone for extending the shelf life of pork loins for 14 days of aerobic storage, providing good results as regards microbial quality and lipid oxidation (Kim *et al.* 2004). Cooking denatures antioxidants in meat and damages cell membranes, thus exposing the lipids in the cell membrane to environmental factors that promote oxidation, including irradiation doses (Ahn *et al.* 1998). However, Chen *et al.* (1999) reported no influence in lipid oxidation of cooked pork during storage if irradiated before cooking. Off-odor has been characterized as metallic, sulfide, wet dog, wet grain, or burnt (Huber *et al.* 1953). Luchsinger *et al.* (1996b) found no difference between irradiated and nonirradiated pork chops in consumer sensory evaluations, which included acceptance and freshness.

21.5 CONCLUSION

Pork shelf life is limited by several factors that influence meat spoilage, such as meat properties (microbial charge, pH, muscle pigments, water-binding capacity, and endogenous antioxidants), handling, packaging, chilling, display lighting, mechanical stresses, and the addition of preservatives. These factors will determine the predominant spoilage mechanism of retailed fresh pork: browning caused by metamyoglobin formation, putrefaction caused by aerobic spoilage bacteria, rancidity caused by lipid hyper-oxidization under high-oxygen MAP, or acidity caused by lactic acid bacteria under vacuum or free-oxygen MAP. Chilling technologies, retail lighting selections, display temperatures, packaging modifications, and product irradiation can offer color stability, microbial reductions, and delayed lipid oxidation that will extend the shelf life of pork and promote consumer satisfaction and repeat purchasing.

REFERENCES

Aberle, E.D., Forrest, J.C., Gerrard, D.E. & Mills, E.W. (2001) Conversion of muscle to meat and development of meat quality. In: *Principles of Meat Science* (eds E.D. Aberle, J.C. Forrest, D.E. Gerrard & E.W. Mills), pp. 84–85. Kendall/Hunt, Dubuque, Iowa.

Ahn, D.U., Olson, D.G., Jo, C., Chen, X., Wu, C. & Lee, J.I. (1998) Effect of muscle type, packaging, and irradiation on lipid oxidation, volatile production and color in raw pork patties. *Meat Science*, **49**, 27–39.

Antoniewski, M.N., Barringer, S.A., Knipe, C.L. & Zerby, H.N. (2007) Effect of a gelatin coating on the shelf life of fresh meat. *Journal of Food Science*, **72** (6), 382–387.

Apple, J.K., Roberts, W.J., Maxwell, C.V., Rakes, L.K., Jr., Friesen, K.G. & Fakler, T.M. (2007) Influence of dietary inclusion level of manganese on pork quality during retail display. *Meat Science*, **75** (4), 640–647.

Banks, J.G., Dalton, H.K., Nychas, G.J. & Board, R.G. (1985) Sulfite, an elective agent in the microbiological and chemical changes occurring in uncooked comminuted meat products. *Journal of Applied Biochemistry*, **7** (3), 161–179.

Barbut, S. (2001) Effect of illumination source on the appearance of fresh meat cuts. *Meat Science*, **59**, 187–191.

Bekhit, A.E.D., Geesink, G.H., Ilian, M.A., Morton, J.D., Sedcole, R. & Bickerstaffe, R. (2003) Particulate metmyoglobin reducing activity and its relationship with meat colour. *Journal of Agricultural and Food Chemistry*, **51**, 6026–6035.

Bertelsen, G. & Skibsted, L.H. (1987) Photooxidation of oxymyoglobin Wavelength dependence of quantum yields of relation to light discoloration of meat. *Meat Science*, **19**, 243–251.

Boler, D.D., Gabriel, S.R., Yang, H., *et al.* (2009) Effect of different dietary levels of natural-source vitamin E in grow-finish pigs on pork quality and shelf life. *Meat Science*, **83** (4), 723–730.

Brewer, M.S. & McKeith, F.K. (1999) Consumer-rated quality characteristics as related to purchase intent of fresh pork. *Journal of Food Science*, **64**, 171–174.

Buege, D.R. (2001) Pork facts: Variation in pork lean quality. Des Moines, Iowa, National Pork Producers Council, pp. 1–8. NPPC Publications, Washington, District of Columbia.

Calkins, C.R., Goll, S.J. & Mandigo, R.W. (1986) Retail display lighting type and fresh pork color. *Journal of Food Science*, **51**, 1141–1143.

Cannon, J.E., Morgan, J.B., Schmidt, G.R., *et al.* (1995) Vacuum packaged precooked pork from hogs fed supplemental vitamin E: Chemical, shelf-life and sensory properties. *Journal of Animal Science*, **60**, 1179–1182.

Cannon, J.E., Morgan, J.B., Schmidt, G.R., *et al.* (1996) Growth and fresh meat quality characteristics of pigs supplemented with vitamin E. *Journal of Animal Science*, **74**, 98–105.

Carr, M. (2007) Shelf life of fresh and frozen pork. In: *Handbook of Meat, Poultry and Seafood Quality* (ed. L. Nollet), pp. 417–426. Blackwell, Ames, Iowa.

Carr, M.A. & Miller, M.F. (1998) Improvement of retail case life of pork-phase I. Des Moines, Iowa, National Pork Producers Council, pp. 101–107. NPPC Publications, Washington, District of Columbia.

Cayuela, J.M., Garrido, M.D., Bañón, S. & Ros, J.M. (2003) Simultaneous HPLC analysis of alpha-tocopherol and cholesterol in fresh pig meat. *Journal of Agricultural and Food Chemistry*, **51**, 1120–1124.

Cayuela, J.M., Gil, M.D., Bañón, S. & Garrido, M.D. (2004) Effect of vacuum and modified atmosphere packaging on the quality of pork loin. *European Food Research and Technology*, **219**, 316–320.

Chen, X., Jo, C., Wu, C., Lee, J.I. & Ahn, D.U. (1999) Effect of natural antioxidants on lipid oxidation, volatiles and color changes of irradiated pork patties. *Journal of Food Science*, **64**, 16–29.

Crenwelge, D.D., Terrell, R.N., Dutson, T.R., Smith, G.C. & Carpenter, Z.L. (1984) Effect of chilling method and electrical stimulation on pork quality. *Journal of Animal Science*, **59**, 697–705.

Djenane, D., Montañés, L. & Roncales, P. (2005) New perspectives for natural preservation of meat. *Eurocarne*, **133**, 153–180.

Dirinck, P., De Winne, A., Casteels, M. & Frigg, M. (1996) Studies on vitamin E and meat quality.1. Effect of feeding high vitamin E levels on time related pork quality. *Journal of Agricultural and Food Chemistry*, **44**, 65–68.

Dogbevi, M.K., Vachon, C. & Lacroix, M. (1999) Physiochemical and microbiological changes in irradiated fresh pork loins. *Meat Science*, **51**, 349–354.

Ehioba, R.M., Kraft, A.A., Molins, R.A., *et al.* (1987) Effect of low-dose (100 krad) gamma radiation on the microflora of vacuum-packaged ground pork with and without added sodium phosphates. *Journal of Food Science*, **52**, 1477–1480.

Eleftheriadou, A., Samouris, G., Ioannidou, M. & Kasapidou, E. (2010) Changes in lipid oxidation and fatty acid composition in pork and poultry meat during refrigerated and frozen storage. *Archiv fur Lebensmittelhygiene*, **61**, 173–178.

EU No. 1831. (2003) Of the European Parliament and of the Council of 22 September 2003 on additives for use in animal nutrition. *Official Journal of the European Union*, **L268**, 29–43.

FAO/WHO. Food and Agriculture Organization/World Health Organization. (1986) Evaluation of certain additive and polluting agents in foods. 291 Report of Mixed Committee FAO/WHO of Food Additive Experts. Series of Technical Information N 733. FAO/WHO publications, Geneva, Italy.

Faucitano, L., Lelo, M.C., Ster, C., Lo Fiego, D.P., Methot, S. & Saucier, L. (2010) Shelf life of pork from five different quality classes. *Meat Science*, **84** (3), 466–469.

Faustman, C. & Cassens, R.G. (1990) The biochemical basis for discoloration in fresh meat: A review. *Journal of Muscle Foods*, **1**, 217–243.

Fox, J.D., Wolfram, S.A., Kemp, J.D. & Langlois, B.E. (1980) Physical, chemical, sensory, and microbiological properties and shelf-life of PSE and normal pork chops. *Journal of Food Science*, **45**, 786–790.

Greer, G.G. & Murray, A.C. (1991) Freezing effects on quality, bacteriology and retail-case life of pork. *Journal of Food Science*, **56**, 891–894.

Günther, A., König, T., Habicher, W.D. & Schwetlick, K. (1997) Antioxidant action of organic sulphites I. Esters of sulphurous acid as secondary antioxidants. *Polymer Degradation and Stability*, **55** (2), 209–216.

Gyesley, S.W. (1991) Total systems approach to predict shelf life of packaged foods. ASTM Special Technical Publications 1113-EB, Philadelphia, Pennsylvania.

Holley, R.A., Gariepy, C., Delaquis, P.J., Doyon, G. & Gagnon, J. (1994) Static, controlled (CO_2) atmosphere packaging of retail ready pork. *Journal of Food Science*, **59**, 1296–1301.

Hoz, L., Lopez-Bote, C.J., Cambero, M.I., *et al.* (2003) Effect of dietary linseed oil and alpha-tocopherol on pork tenderloin (Psoas major) muscle. *Meat Science*, **65** (3), 1039–1044.

Huang, N.Y., Ho, C.P. & McMillin, K.V. (2005) Retail shelf-life of pork dipped in organic acid before modified atmosphere or vacuum packaging. *Journal of Food Science*, **70** (8), 382–387.

Huber, W., Brasch, A. & Waly, A. (1953) Effect of processing conditions on organoleptic changes in foodstuffs sterilized with high intensity electrons. *Food Technology*, **1** (7), 109–115.

Jenkins, W.A. & Harrington, J.P. (1991) Packaging foods with plastics. In: *Technology and Packaging Applications* (eds W.A. Jenkins & J.P. Harrington). Technomic, Lancaster, Pennsylvania.

Jensen, C., Lauridsen, C. & Bertelsen, G. (1998) Dietary vitamin E: Quality and storage a stability of pork and poultry. *Trends in Food Science and Technology*, **9** (2), 62–72.

Jeremiah, L.E. & Gibson, L.L. (1997) The influence of storage and display conditions on the color stability of display-ready pork loin roasts. *Meat Science*, **47**, 1–16.

Kim, B.H., Jang, A., Lee, S.O., Min, J.S. & Lee, M. (2004) Combined effect of electron-beam (beta) irradiation and organic acids on shelf life of pork loins during cold storage. *Journal of Food Protection*, **67**, 168–171.

Krause, T.R., Sebranek, J.G., Rust, R.E. & Honeyman, M.S. (2003) Use of carbon monoxide packaging for improving the shelf life of pork. *Journal of Food Science*, **68** (8), 2596–2603.

Kropf, D.H. (1998) Pork facts-meat display lighting. Des Moines, Iowa, National Pork Producers Council, pp. 1–8. NPPC Publications, Washington, District of Columbia.

Lambert, A.D., Smith, J.P. & Dodds, K.L. (1992) Physical, chemical and sensory changes in irradiated fresh pork packaged in modified atmosphere. *Journal of Food Science*, **57**, 1294–1299.

Lee, K., Choi, W. & Yoon, C. (2003) Effects of microperforated film on the quality and shelf life improvements of pork loins during chilled storage. *Meat Science*, **66**, 77–82.

Lin, K.W. & Chuang, C.H. (2001) Effectiveness of dipping with phosphate, lactate and acetic acid solutions on the quality and shelf-life of pork loin chop. *Journal of Food Science*, **66** (3), 494–499.

Liu, F., Yang, R.Q. & Li, Y.F. (2006) Correlations between growth parameters of spoilage micro-organisms and shelf-life of pork stored under air and modified atmosphere at −2, 4 and 10 degrees C. *Food Microbiology*, **23** (6), 578–583.

Luchsinger, S.E., Kropf, D.H., Garcia Zepeda, C.M., *et al.* (1996a) Color and oxidative rancidity of gamma and electron beam-irradiated boneless pork chops. *Journal of Food Science*, **61**, 1000–1005.

Luchsinger, S.E., Kropf, D.H., Garcia Zepeda, C.M., *et al.* (1996b) Sensory analysis and consumer acceptance of irradiated boneless pork chops. *Journal of Food Science*, **61**, 1261–1266.

Martínez, L., Cilla, I., Beltrán, J.A. & Roncalés, P. (2007) Effect of illumination on the display life of fresh pork sausages packaged in modified atmosphere. Influence of the addition of rosemary, ascorbic acid and black pepper. *Meat Science*, **75**, 443–450.

Miller, M.F. & Carr, M.A. (2000) Improvement of retail case life of pork-phase II. Des Moines, Iowa, National Pork Producers Council, pp. 290–292. NPPC Publications, Washington, District of Columbia.

Milligan, S.D., Ramsey, C.B., Miller, M.F., Kaster, C.S. & Thompson, L.D. (1998) Resting of pigs and hot-fat trimming and accelerated chilling of carcasses to improve pork quality. *Journal of Animal Science*, **76**, 74–86.

McMillin, K.W. (2008) Where is MAP going? A review and future potential of modified atmosphere packaging for meat. *Meat Science*, **80**, 43–65.

Monahan, F.J., Buckley, D.J., Gray, J.I., *et al.* (1990) Effect of vitamin E on the stability of raw and cooked pork. *Meat Science*, **27**, 99–108.

Monin, G. & Sellier, P. (1985) Pork of low technological quality with a normal rate of muscle pH fall in the immediate post mortem period: The case of the Hampshire breed. *Meat Science*, **13** (1), 49–63.

Mullan, M. & McDowell, D. (2003) Modified atmosphere packaging. In: *Food Packaging Technology* (eds R. Coles, D. McDowell & M.J. Kirwan), pp. 303–339. Blackwell Publishing, London, U.K.

Nakai, H., Saito, F., Ikeda, T., Ando, S. & Komatsu, A. (1975) Standard models of pork colour. *Bulletin National of Institute of Animal Industry*, **29**, 69–75.

Nanke, K.E., Sebranek, J.G. & Olson, D.G. (1999) Color characteristics of irradiated aerobically packaged pork, beef, and turkey. *Journal of Food Science*, **64**, 272–278.

Norman, J.L., Berg, E.P., Heymann, H. & Lorenzen, C.L. (2003) Pork loin color relative to sensory and instrumental tenderness and consumer acceptance. *Meat Science*, **65**, 927–933.

NPPC. (1999) Pork quality standards. Des Moines, Iowa, National Pork Producers Council, p. 1. NPPC Publications, Washington, District of Columbia.

Osswald, T.A., Baur, E., Brinkmann, S., Oberbach, K. & Schmachtenberg, E. (2006) Polymer additives. In: *International Plastics Handbook. The Resource for Plastics Engineers* (eds T.A. Osswald, E. Baur, S. Brinkmann, K. Oberbach & E. Schmachtenberg), pp. 507–699, 708. Hanser, Munich, Germany.

Park, S.Y., Yoo, S.S., Uh, J.H., Eun, J.B., Lee, H.C., Kim, Y.J. & Chin, K.B. (2007) Evaluation of lipid oxidation and oxidative products as affected by pork meat cut, packaging method, and storage time during frozen storage (−10°C). *Food Chemistry and Toxicology*, **72** (2), 114–119.

Phillips, A.L., Faustman, C., Lynch, M.P., Govoni, K.E., Hoagland, T.A. & Zinn, S.A. (2001) Effect of dietary alpha-tocopherol supplementation on color and lipid stability in pork. *Meat Science*, **58**, 389–393.

Rincker, P.J., Killefer, J. & McKeith, F.K. (2007) Evaluating an objective method to measure fresh pork loin firmness. *Meat Science*, **77**, 213–219.

Rojas, M.C. & Brewer, M.S. (2008) Effect of natural antioxidants on oxidative stability of frozen, vacuum-packaged beef and pork. *Journal of Food Quality*, **31** (2), 173–188.

Scetar, M., Kurek, M. & Galic, K. (2010) Trends in meat and meat products packaging—A review. *Croatian Journal of Food Science and Technology*, **2** (1), 32–48.

Schirmer, B.C. & Langsrud, S.A. (2010) Dissolving CO_2 headspace combined with organic acids prolongs the shelf-life of fresh pork. *Meat Science*, **85** (2), 280–284.

Scholtz, E.M., Jordaan, E., Kruger, J., Nortje, G.L. & Naude, R.T. (1992) The influence of different centralized pre-packaging systems on the shelf-life of fresh pork. *Meat Science*, **32**, 11–29.

Sorheim, O., Erlandsen, T., Nissen, H., Lea, P. & Hoyem, T. (1997) Effects of modified atmosphere storage on colour and microbiological shelf life of normal and pale, soft, and exudative pork. *Meat Science*, **47**, 147–155.

Spanier, A.M., Vercellotti, J.R. & James Jr., C. (1992) Correlation of sensory, instrumental and chemical attributes of beef as influenced by meat structure and oxygen exclusion. *Journal of Food Science*, **57** (1), 10–15.

Springer, M.P., Carr, M.A., Ramsey, C.B. & Miller, M.F. (2003) Accelerated chilling of carcasses to improve pork quality. *Journal of Animal Science*, **81**, 1464–1472.

Suppalku, P., Miltz, J., Sonneveld, K. & Bigger, S.W. (2003) Antimicrobial properties of basil and its possible application in food packaging. *Journal of Agricultural and Food Chemistry*, **51**, 3197–3207.

Tan, F.J., Morgan, M.T., Ludas, L.I., Forrest, J.C. & Gerrard, D.E. (2000) Assessment of fresh pork color with color machine vision. *Journal of Animal Science*, **78**, 3078–3085.

Timms, M.J. & Watts, B.M. (1958) Protection of cooked meats with phosphates. *Food Technology*, **12**, 240–243.

Vega, J.D. & Brewer, M.S. (1994) Detectable odor thresholds of selected lipid oxidation compounds at various temperatures in a gelatin model system. *Journal of Food Lipids*, **1**, 229–245.

Viana, E.S., Gomide, L.A.M. & Vanetti, M.C.D. (2005) Effect of modified atmospheres on microbiological, color and sensory properties of refrigerated pork. *Meat Science*, **71**, 696–705.

Vrana, J.A., Savell, J.W., Dill, C.W., Smith, G.C., Ehlers, J.G. & Vanderzant, C. (1985) Retail appearance, odor and microbiological characteristics of pork loin chops packaged in different oxygen-barrier films as affected by loin storage treatment. *Journal of Food Protection*, **48**, 476–481.

Weakley, D.F., McKeith, F.K., Bechtel, P.J., Martin, S.E. & Thomas, D.L. (1986) Effects of different chilling methods on hot processed vacuum packaged pork. *Journal of Food Science*, **51**, 757–765.

Wedzicha, B.L. & Mountfort, K.A. (1991) Reactivity of sulphur dioxide in comminuted meat. *Food Chemistry*, **39** (3), 281–297.

Whang, K., Aberle, E.D., Judge, M.D. & Peng, I.C. (1986) Antioxidative activity of alpha-tocopherol in cooked and uncooked ground pork. *Meat Science*, **17**, 235–249.

Wilkinson, B.H.P., Janz, J.A.M., Morel, P.C.H., Purchas, R.W. & Hendriks, W.H. (2006) The effect of modified atmosphere packaging with carbon monoxide on the storage quality of master-packaged fresh pork. *Meat Science*, **73** (4), 605–610.

Xia, X., Kong, B., Xiong, Y.L. & Meng, X. (2007) Extending the shelf-life of chilled pork by combination of chitosan coating with spice extracts. In: *Proceedings of 53rd International Congress of Meat Science and Technology*, pp. 457–458, Beijing, China.

Zhang, M. & Sundar, S. (2005) Effect of oxygen concentration on the shelf-life of fresh pork packed in a modified atmosphere. *Packaging Technology and Science*, **18** (4), 217–222.

Zhang, H.Y., Kong, B.H., Xiong, Y.L.L. & Xu Sun, X. (2009) Antimicrobial activities of spice extracts against pathogenic and spoilage bacteria in modified atmosphere packaged fresh pork and vacuum packaged ham slices stored at 4 degrees C. *Meat Science*, **81**, 686–692.

Zhu, L.G. & Brewer, M.S. (1998) Discoloration of fresh pork as related to muscle and display conditions. *Journal of Food Science*, **63**, 763–767.

22 Packaging of (Fresh and Frozen) Pork

Maurice O'Sullivan and Joseph P. Kerry

Abstract: The primary objective of packaging of fresh and processed meats is the limitation and retardation of the growth of spoilage and pathogenic organisms, followed by maintaining the optimal sensory characteristics within the shelf-life specification of the product.

Generally, there are four different packaging formats for pork products: high oxygen modified atmosphere packs (high O_2 MAP), low oxygen modified atmosphere packs (low O_2 MAP), controlled atmosphere packs (CAP), and vacuum packaging (VP). Different products are suited to different formats. The first three formats are used for refrigerated retail display whereas VP is used for both refrigerated and frozen storage. All these methods will extend the shelf life of pork products, but will also affect sensory quality to an extent.

Lipid oxidation is the primary cause of rancidity during frozen storage of meat and meat products and pork-based products are much more susceptible to rancidity development than other types of meat products. Various other factors can affect the preservation of fresh and frozen pork products.

Keywords: Modified atmosphere packaging—high O_2 packaged pork, controlled atmosphere packs, low O_2 modified atmosphere packaging pork, vacuum packaging, frozen pork, antimicrobial agents, vitamin E, active packaging

22.1 INTRODUCTION

The primary objective of packaging of fresh and processed meats is the limitation and retardation of the growth of spoilage and pathogenic organisms, followed by maintaining the optimal sensory characteristics within the shelf-life specification of the product (O'Sullivan & Kerry 2011). Also, packaging protects the product from the incursion of foreign matter; physical damage during transport, storage, and retail display; as well as presents the product in an aesthetically appealing fashion to the consumer.

One of the most important quality traits of fresh pork in the decision to purchase by the consumer is the appearance or color of the meat (Risvik 1994). Color perception plays a major role in consumer evaluation of meat quality (Lanari *et al.* 1995). Consumers need first to be entirely satisfied with the sensory properties of products, before other quality dimensions become relevant (Chambers & Bowers 1993). Unfortunately, fresh meat color is short-lived and surface discoloration that occurs during frozen storage is considered a sign of unwholesomeness and product deterioration (Faustman & Cassens 1990).

Handbook of Meat, Poultry and Seafood Quality, Second Edition. Edited by Leo M. L. Nollet.

Generally there are four different packaging formats for pork products; high oxygen modified atmosphere packs (high O_2 MAP); low oxygen modified atmosphere packs (low O_2 MAP); controlled atmosphere packs (CAP) (Gill & Gill 2005); and vacuum packaging (VP). Different products are suited to different formats, that is, pork chops in MAP; cooked pork products in low O_2 MAP or CAP; and pork fillet, rashers, or bacon products in VP. The first three formats are used for refrigerated retail display, whereas VP is used for both refrigerated and frozen storage. All these methods will not only extend the shelf life of pork products but also affect sensory quality to an extent which will also be discussed in this chapter. Packaging pork in high O_2 MAP will result in the formation of off-flavors from the reaction of the relatively highly polyunsaturated fat tissue in pork muscle to the O_2.

Lipid oxidation is the primary cause of rancidity during frozen storage of meat and meat products, and pork-based products are much more susceptible to rancidity development than other types of meat products (Buckley *et al.* 1989). Lipid oxidation is an autocatalytic chain reaction that occurs in the presence of O_2 which results in the development of rancid off-flavors. These off-flavors increase over the storage time of pork meats packed in O_2 environments. Also, various other factors can affect the preservation of fresh and frozen pork products. Feeding of materials high in polyunsaturated fatty acids (PUFAs) will result in an increase in these compounds in the resulting meat postslaughter. Feeding PUFA (n-3) to pigs improves the quality of pork in terms of human nutrition; it is well documented that these compounds are considered a source of "good fats." These good fats have a role in reducing the risk of coronary heart disease in humans. However, feeding PUFA also increases this susceptibility to lipid oxidation which occurs at the double bonds of these fatty acids. Free radicals are generated which lead not only to the production of off-flavors and off-odors but also reduce the nutritive value of the meat (Sheard *et al.* 2000). However, this can be counteracted by the incorporation of antioxidants such as α-tocopherol by dietary means or of other natural antioxidants such as rosemary extract by exogenous addition to the muscle tissue, which inhibits lipid oxidation and thus prolongs the sensory quality of the meat product.

Active packaging is defined as packaging, which "changes the condition of the packed food to extend shelf-life or to improve safety or sensory properties, while maintaining the quality of packaged food" (Ahvenainen 2003). This will also be discussed along with the above topics in this chapter.

22.2 SENSORY QUALITY OF FRESH AND FROZEN PORK

The packaging of fresh and frozen pork will ultimately affect the sensory quality of the product with respect to color, odor/flavor, and texture changes. The three sensory properties by which consumers most readily judge meat quality are: appearance, texture, and flavor (Liu *et al.* 1995). Pork retail shelf life is largely limited by a change in color that usually occurs before microbial spoilage. Discoloration and spoilage begin with surface exposure to oxygen and limit distribution of fresh pork cuts over long distances (Viana *et al.* 2005). At the point of sale, color and color stability are the most important attributes of meat quality, and various approaches have been used to meet consumer expectations that an attractive, bright-red color indicates long shelf life and good eating quality (Hood & Mead 1993).

One of the principal flavor defects to occur in pork products includes warmed-over flavor or WOF. Tims and Watts (1958) were among the first workers to recognize WOF as a sensory challenge to meat products. In the last few decades, there has been a rapid development of fast-food facilities and widespread use of precooked frozen meals (Dethmers & Rock 1975).

WOF is an objectionable flavor which becomes most noticeable when refrigerated cooked meat is reheated. It is a rapidly occurring phenomenon, in contrast to the slowly developing oxidative rancidity which results from long-term storage of frozen raw meat or storage of vegetable oils at ambient temperature (St. Angelo & Bailey 1987). Jensen *et al.* (1997a) found that in cooked pork meat a significant improvement in oxidative stability, as measured by thiobarbituric reactive substances (TBARS), an index of lipid oxidation, during chill storage was achieved by supplementation with 700 mg α-tocopheryl acetate/kg feed (Jensen *et al.* 1997a). O'Sullivan *et al.* (2003c) found that cooked *M. psoas major* was more susceptible to WOF development as determined by sensory profiling compared to *M. longissimus dorsi* for all treatments (control, vitamin E, iron/vitamin E, and iron). Development of WOF in *M. longissimus dorsi* was highest in the iron-supplemented groups followed by the control and vitamin E-supplemented groups. In subsequent sections, we will discuss the determination of packaging-induced off-flavors, such as lipid oxidation in MAP, and how this can be reduced through vacuum packaging, low O_2 packaging, product freezing, and the utilization of antioxidants such as vitamin E and rosemary extract.

Texture is also an important sensory quality parameter for pork and pork products. As will be discussed in the subsequent sections, MAP can increase sensory-evaluated toughness of pork when stored in high O_2 atmospheres (Lund *et al.* 2007).

22.3 MAP—HIGH O_2 PACKAGED PORK

MAP is one of the most popular refrigerated packaging formats for pork products such as pork loin chops (Fig. 22.1). It involves placing a retail cut of meat in a bag or pouch with low gas permeability, evacuation of the air, replacement of the air by a single gas or a gas mixture, and finally clipping or heat sealing the bag or pouch (Seideman & Durland 1984; Zakrys *et al.* 2009). Typically, fresh red meats are stored in MAP containing 80% O_2:20% CO_2 (Georgala & Davidson 1970) and cooked meats are stored in 70% N_2:30% CO_2 (Smiddy *et al.* 2002). MAP usually contains mixtures of two or three gases: O_2 (to enhance color stability), CO_2 (to inhibit microbiological growth), and N_2 (to maintain pack shape) (Sørheim *et al.* 1999; Jakobsen & Bertelsen 2000). Most aerobic bacteria are inhibited by CO_2 concentrations of 10% or higher, while obligate anaerobic bacteria are inhibited by the presence of even small amounts of oxygen (Ray 2001).

Fig. 22.1 Modified atmosphere packaged (MAP) loin chops.

The color of pork is extended by storage under MAP conditions (Lanari *et al.* 1995; Kerry *et al.* 2000). The high oxygen concentration allows for formation of oxymyoglobin, creating the more desirable pink/red color of fresh pork (Gill 1991). Also, different muscles exhibit different rates of oxymyoglobin oxidation under controlled conditions (Ledward 1985). The vast majority of meat products have been and continue to be offered in high oxygen pack formats (approximately 80% O_2) in order to maintain bloom, with at least 20% CO_2 to prevent selective microbial growth (Eilert 2005). Discoloration in retail meats during display conditions may occur as a combined function of muscle pigment oxidation (oxymyoglobin to metmyoglobin) and lipid oxidation in membrane phospholipids (Sherbeck *et al.* 1995). In the USA, there is a high demand for fresh "bone in" pork chops; however, color deterioration of both muscle and bone limits the marketing options (Lanari *et al.* 1995). Also, the color stability of meat products depends on variables such as muscle type, pH, storage temperature, oxygen availability, and lighting type and intensity during display (Andersen & Skibsted 1991). Discoloration in retail meats during display conditions may occur as a combined function of muscle pigment oxidation (oxymyoglobin to metmyoglobin) and lipid oxidation in membrane phospholipids (Sherbeck *et al.* 1995).

The oxidation of PUFAs causes the rapid development of meat rancidity (Kanner 1994). Packaging under high oxygen atmospheres also increases lipid oxidation in pork (Lund *et al.* 2007) and may impact negatively on the oxidative stability of muscle lipids and lead to development of undesirable flavors (Rhee & Ziprin 1987; Estevez & Cava 2004). The breakdown products of lipid oxidation have been associated with the development of off-flavors and off-odors and loss of color in meat (Faustman & Cassens 1989). Feeding PUFA also increases this susceptibility to lipid oxidation which occurs at the double bonds of these fatty acids. Free radicals are generated which lead not only to the production of off-flavors and off-odors but also reduce the nutritive value of the meat (Sheard *et al.* 2000).

Lund *et al.* (2007) investigated the effect of MAP (70% O_2/30% CO_2) and skin packaging (no oxygen) on protein oxidation and texture of pork *M. longissimus dorsi* muscle during storage for 14 days at 4°C. Results reported by Lund *et al.* (2007) showed that a significant effect of packaging atmosphere on tenderness and juiciness of pork LD slices was determined through sensory analysis (using nine assessors, all nonsmokers recruited from the professional sensory panel at the Danish Meat Research Institute). They found that the high oxygen atmosphere resulted in reduced tenderness and juiciness of samples. Additionally, their SDS–PAGE (sodium dodecyl sulfate polyacrylamide gel electrophoresis) data revealed cross-linking of myosin heavy chain through disulfide bonding, and the content of protein thiols was reduced indicating protein oxidation.

Overall, high O_2 MAP will prolong the color quality of packaged pork products, but may negatively affect other sensory quality variables such as flavor, aroma, and texture.

22.4 CAP AND LOW O_2 MAP PORK

Pork products can also be packed in the refrigerated CAP and low O_2 packaging formats. These use combinations of 70–80% nitrogen and 20–30% CO_2 in the packaging headspace and include pork products such as sausages (Fig. 22.2) and cooked sliced ham (Fig. 22.3). Again the CO_2 is present to prevent the growth of spoilage bacteria and the N_2 assists in maintaining the pack shape. Jeremiah and Gibson (1997) reported that pork packaged with 100% CO_2 has great color stability. CAP is essentially O_2-free MAP and has also been used commercially for shipment of chilled lamb to distant markets (Gill 1990). High concentrations of CO_2 in packaged meat products can result in the formation of CO_2 taint in the resulting cooked

Fig. 22.2 Controlled atmosphere packaged (CAP) sausages.

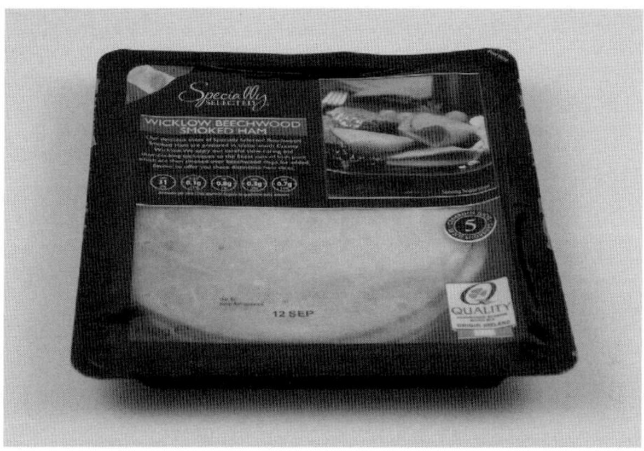

Fig. 22.3 Controlled atmosphere packaged (CAP) sliced ham.

meat. This CO_2 taint is an off-flavor that can be tasted by the consumer after cooking the meat (Nattress & Jeremiah 2000; O'Sullivan *et al.* 2011). However, this taint effect can be removed by opening CAP packs up to 30 min prior to cooking which allows dissolved CO_2 in the meat to dissipate (O'Sullivan *et al.* 2011).

Low O_2 MAP can be used to pack fresh meats, such as pork, but the major drawback is that high CO_2 concentrations are associated with a certain degree of darkening as a result of metmyoglobin formation especially when low O_2 concentrations are present (Gill & Jones, 1996). On the contrary, Sørheim *et al.* (1997b) reported that CO_2 has no effect upon meat color. In low O_2 MAP, meat color discoloration can be prevented by the inclusion of low levels of carbon monoxide (CO) in the gas mixture (Luno *et al.* 2000). MAP with low concentrations of CO and high concentrations of CO_2 has been shown to improve beef and pork color and other meat quality traits (Clark *et al.* 1976; Sørheim *et al.* 1999, 2001; Luno *et al.* 2000; Jayasingh *et al.* 2002). CO is a colorless, odorless, and tasteless gas, and may also be used for prolonging meat color integrity (Sørheim *et al.* 1997a). CO reacts with myoglobin to form carboxymyoglobin (COMb), which is more resistant to oxidation than oxymyoglobin, owing to the stronger binding of CO to the iron–porphyrin site on the myoglobin molecule (Wolfe 1980). CO in MAP systems counteracts the discoloration that results from a high carbon dioxide concentration; even in the presence of oxygen, it can enhance myoglobin reduction (Lanier *et al.* 1978; Luno *et al.* 2000). Studies with beef and pork confirmed that the desired cherry red color is retained when a low concentration of CO is introduced into a low-oxygen MAP system (Sørheim *et al.* 1999; Jayasingh *et al.* 2002).

22.5 VACUUM PACKAGING OF PORK

VP extends the storage life of chilled meats by maintaining an O_2-deficient environment within the pack (Bell *et al.* 1995) and is used for both refrigerated (VP pork fillet) and frozen products (VP pork belly). VP involves the evacuation of air from the packs prior to sealing and is used extensively for the packaging of pork products such as pork steak or fillet, bacon rashers (Fig. 22.4), whole hams (Fig. 22.5), and cured or barbecue ribs to mention just a few products. VP was one of the earliest forms of MAP methods developed commercially and still is extensively used for products such as primal cuts of fresh red meat and cured meats (Parry 1993). According to Holley *et al.* (2004), the commercial storage life attainable with

Fig. 22.4 Vacuum-packed rashers.

Fig. 22.5 Vacuum-packed bacon in a heat shrinkable film.

vacuum-packaged pork is about 8 weeks, because the psychotropic lactic acid bacteria (LAB) which predominate in the flora of vacuum-packaged pork will usually number ≥6 log cfu cm^{-2}, which is the maximum number for acceptance of the products by some customers, irrespective of whether or not the product is overtly spoiled (Holley *et al.* 2004). Respiration of the meat in vacuum packs will quickly consume any residual O_2, replacing it with CO_2, which eventually increases to 10–20% within the package (Taylor 1985; Parry 1993; Gill 1996), and the amount of O_2 remaining in the pack at the time of closure must be very small if the product is to be effectively preserved, as the capacity of the muscle tissue for removing O_2 is limited (Gill & Gill 2005).

Low O_2 permeability of the packaging film causes a change in meat color from red to purple due to the conversion of oxymyoglobin to deoxymyoglobin. From the consumers' perspective, these forms of myoglobin have been considered as unacceptable meat colors (Parry 1993; Allen *et al.* 1996). Generally vacuum-packaged meat is unsuitable for the retail market, except for pork steak or pork fillet which are extensively sold in this packaging format in the retail sector in Ireland. Also, extended storage of meat in vacuum packs results in the accumulation of drip, which is very unappealing to consumers (Jeremiah *et al.* 1992; Parry 1993; Payne *et al.* 1997). However, this can be counteracted by using vacuum skin packaging (VSP), which uses a film that fits very tightly to the meat surface, leaving little space for the accumulation of drip (Hood & Mead 1993).

Pork has a shorter shelf life than beef which may be due to a lower hygienic status during the production of pork. Also, the more rapid depletion of glycogen and glucose in pork compared to that in beef is of importance in dictating the shorter shelf life of pork. Blixt and Borch (2002) found that vacuum-packed, cold-stored minced pork and beef, of the same initial bacterial loads, showed differences in rates of spoilage and bacterial growth and indicated that factors other than the initial bacterial loading were of importance in determining the shelf life of vacuum-packed meat. They also present that differences were not related to whether the meat was pork or beef, but rather to other intrinsic factors; pH and the concentrations of fat and L-lactate were the three most important factors. Also, a low initial pH and fat content, and a high initial L-lactate concentration, positively affected the sensory shelf life of the meat.

22.6 FROZEN PORK

A good frozen food package must withstand low temperatures and sometimes high temperatures (microwave, boil in bag products). It must be nontoxic and not impart odors or flavors to the food, must provide a barrier to the transmission of water vapor, and must be water resistant (Paine & Paine 1992).

Freezing is one of the oldest, most common, and effective methods of preserving foods. Freezing slows down the metabolic rate of microorganisms, thus retarding their growth. Additionally, freezing reduces the availability of free water, which is required for microbial growth, and also concentrates dissolved compounds in any remaining water, further reducing the availability of free water required for microbial growth. The microbial loading of meat prior to freezing is a very important factor in preservation and food safety as freezing will only slightly reduce loadings. While microbial growth is stopped by freezing, not all microorganisms are destroyed. Spores and preformed toxins will survive and damaged bacterial cells may still be viable after a long resuscitation period (Paine & Paine 1992). The more rapid freezing takes place, the smaller the ice crystals form between muscle tissue cells. Slow freezing allows ice crystals to grow which rupture cells resulting in excessive drip loss on thawing.

The storage temperature of frozen meat might fluctuate over an often unknown temperature range during transport, distribution, storage, and retail display (Hansen *et al.* 2004). Ulrich (1981) reported that there was little evidence of any reduction in the shelf life in relation to temperature fluctuations at temperatures of −18°C or lower.

Lipid oxidation is the primary cause of rancidity during frozen storage of meat and meat products, and pork-based products are much more susceptible to rancidity development than other types of meat products (Buckley *et al.* 1989). During frozen storage, the oxidation processes are slowed down but are not completely arrested; some lipid soluble radicals may even be more stable at the lower temperatures and thereby propagate oxidation (Kanner 1994). Hansen *et al.* (2004) measured the oxidative stability of pork patties during frozen storage, using TBARS, at −10°C, −23°C, and −40°C. The pork patties were made from pork loin (low fat 1.8%) or pork belly (high fat 22.7%). These authors found that lower storage temperatures resulted in less lipid oxidation, with temperatures fluctuating between −40°C and −23°C, whereas temperatures fluctuating between −23°C and −10°C resulted in intermediate oxidation in samples stored at constant low or high temperatures of the fluctuation interval. These authors also found that the level of α-tocopherol in samples was unaffected by the extent of oxidation in the frozen samples (Hansen *et al.* 2004). It is thought that α-tocopherol present in the meat prior to freezing will inhibit lipid oxidation further during freezing. However, there is little information on the effects of vitamin E on the quality of processed pork products, particularly the effect it has on oxidative rancidity during frozen storage (Channon & Trout 2002). Whang *et al.* (1986) showed that vitamin E slowed the rate of oxidation in cooked postrigor pork stored at −20°C for 14 days. Vitamin E was found to be highly stable in raw pork during frozen storage for up to 10 months at −25°C (Jensen *et al.* 1998). Ice is formed in high-moisture foods at temperatures below 0°C.

VP is commonly used in conjunction with freezing of muscle foods. The exclusion of O_2 in conjunction with freezing will further prolong storage, inhibit lipid oxidation, and prevent the deterioration of sensory quality.

22.7 ANTIMICROBIAL AGENTS AND PACKAGED PORK

Modified atmosphere packaging (MAP) using a high carbon dioxide (CO_2) environment is an effective means of prolonging microbial shelf life of meat during extended storage (Sørheim *et al.* 1999). However, shelf life of MAP pork may be further extended by the incorporation of other hurdle effects to inhibit microbial growth. MAP pork loin chops in retail display can have a retail display life of up to 14 days. However, there is a significant amount of shelf-life variation among individual packages (Livingston *et al.* 2004). In order to optimize shelf life, sensory quality, and microbiological safety of foods using MAP, the packaging system applied must be highly product specific (Church & Parsons 1995). The initial microbial loading of the meat is one of the most important factors affecting shelf life in MAP pork products. The microflora Brochothrix thermosphacta, cold-tolerant Enterobacteriaceae, LAB, and pseudomonads will occur in meat stored under aerobic conditions (Lambropoulou *et al.* 1996). However, in the majority of modified atmosphere and vacuum-packaged meat products, the bacterial flora is dominated by the presence of LAB (Korkeala & Björkroth 1997; Samelis *et al.* 2000; Stolzenbach *et al.* 2009). Livingston *et al.* (2004) found that enhanced-MAP chops from loins with higher ultimate pH values (>5.65) had greater color stability, but were more susceptible to microbial growth over time than loins with lower ultimate pH. They also found that a lower ultimate pH (<5.65) group had reduced microbial growth but discolored more rapidly over time.

The metabolic activity of LAB results in spoilage appearing as "sour" off-flavors and off-odors, the presence of milky exudates which are slimy, and swelling of the pack, sometimes accompanied by product greening (Samelis *et al.* 1998, 2000). A CO_2 concentration of 20–30% is required to inhibit growth effects (Stiles 1991; Patsias *et al.* 2006). Reducing the level of microbial contamination of pork has been shown to extend shelf life (Jeremiah & Gibson 1995). Bacterial counts have been shown to be higher on pig carcasses than on beef carcasses (Johanson *et al.* 1983). Zakrys-Waliwander *et al.* (2011) compared experimental and commercial MAP beef steaks and found that microbial counts did not exceed recommended legal ranges over the storage period for these treatments, but the numbers of LAB were the highest for commercially packed samples (O_2Com) in comparison to laboratory-packaged samples ($O_2$80, $O_2$70, and $O_2$50). This had a knock-on effect in that the commercially packaged samples eventually spoiled due to the occurrence of off-flavors and off-odors and subsequently were the least acceptable to consumers. Thus, initial hygiene during production will have an effect from a microbial quality perspective which in turn can affect subsequent sensory quality (Zakrys-Waliwander *et al.* 2011).

Additionally, there are a variety of approaches to shelf-life extension, including enhancement with antimicrobial, organic acid solutions, which have been evaluated in an effort to create a more desirable, longer lasting product (Jensen *et al.* 2002). Lactic acid, which does not have undesirable odor-forming effect even at 5%-level application in lamb carcasses, is one of the widely used acid decontaminants in meat industry (Smulders 1987; Shrestha & Min 2006). Shrestha and Min (2006) found that lactic acid application in fresh pork reduced the bacterial (4–6% lactic acid) load, but there was a significant loss of color and overall acceptability scores. Firstenberg-Eden (1981) found that lactic acid used to decontaminate carcasses 1–2 days after slaughter is better than delaying for more than 2 days after slaughter. Active packaging can also be used to enhance the effectiveness of MAP and is discussed in a later section of this chapter.

22.8 VITAMIN E AND EXTENDING THE SENSORY SHELF LIFE OF PORK

As discussed above, O_2 present in packaged pork, either overwrapped or MAP, will ultimately have a negative effect on the product quality of the product. Off-flavors can be produced which can have a negative effect on consumer acceptance. However, the incorporation of antioxidants such as α-tocopherol into the cell membrane will stabilize the membrane lipids and consequently enhance the quality of meat storage. α-Tocopherol or vitamin E is an antioxidant that retards the onset of lipid oxidation. Skeletal muscle is particularly susceptible to oxidative reactions since it contains high concentrations of pro-oxidants (transition metals, haem-containing proteins, i.e., myoglobin, hemoglobin) and lipid membrane which contain higher percentages of PUFAs (Kanner 1994). Also, pork-based products are much more susceptible to rancidity development than other types of meat products (Buckley *et al.* 1989). This is because pork contains higher concentrations of PUFAs compared to beef and lamb. Oxidation occurs at the double bonds of these fatty acids which generate free radicals which lead to the production of off-flavors and off-odors (Sheard *et al.* 2000).

Dietary vitamin E supplementation results in elevated concentrations of α-tocopherol in cell membranes, especially in the mitochondria and the microsomes resulting in a significantly lower susceptibility to lipid oxidation in the membranes (Monahan *et al.* 1990).

The level of accumulation of vitamin E in muscle tissue will depend on the level of supplementation and the type of muscle involved. Accumulation of α-tocopherol in muscles has been found to take place in a dose- and duration-dependent manner. In general, the greater the amount of vitamin E fed and/or the duration of supplementation period, the higher the tissue concentration of α-tocopherol (Arnold *et al.* 1993). O'Sullivan *et al.* (2002b) showed that porcine *M. psoas major* had significantly higher levels of vitamin E (μg/g of meat) compared to porcine *M. longissimus dorsi. M. longissimus dorsi* is a white muscle and contains less phospholipids compared to the red muscle *M. psoas major* (Lawrie 1991). This explains the greater uptake of vitamin E in *M. psoas major* after vitamin E supplementation. Similar findings were presented by O'Sullivan *et al.* (1997), who reported that increasing the level of α-tocopheryl acetate supplementation from 0 to 160 mg/kg in pigs significantly ($p < 0.05$) increased α–tocopherol levels in muscle tissue. The highest levels were found in the muscles of the thoracic limb, followed by the neck and thorax, pelvic limb, and back (O'Sullivan *et al.* 1997). No single factor has, however, been identified to account for the large variation in skeletal muscle concentrations of α–tocopherol in the same animal (Liu *et al.* 1995). Variation in lipid content, fiber-type distribution, and metabolic characteristics, including uptake and excretion of tocopherols, of the individual muscles together seem to determine the steady-state level (Buckley *et al.* 1995). Oxidation studies with chicken (Jensen *et al.* 1995), turkeys (Marusich *et al.* 1975), pigs (O'Sullivan *et al.* 1997), cattle (Faustman *et al.* 1989), veal (Shorland *et al.* 1981), and fish (Frigg *et al.* 1990) have all demonstrated reduced lipid oxidation in muscles and adipose tissue from the supplemented animals compared to the same muscles from the nonsupplemented animals.

A number of workers have demonstrated that oxymyoglobin stability is greater in beef that contains higher concentrations of α-tocopherol (Arnold *et al.* 1993; Chan *et al.* 1996; Lanari *et al.* 1996). Similarly, the color stability of packaged pork chops can be increased by vitamin E supplementation (Faustman & Cassens 1990; Morrissey *et al.* 1994; Lanari *et al.* 1995; O'Sullivan *et al.* 1998; O'Sullivan *et al.* 2002a). As discussed previously, the three sensory properties by which consumers most readily judge meat quality are: appearance, texture, and flavor (Liu *et al.* 1995). At the point of sale, color and color stability are the most important attributes of meat quality and various approaches have been used to meet consumer expectations that an attractive, bright-red color indicates long shelf life and good eating quality (Hood & Mead 1993). It appears that α-tocopherol protects oxymyoglobin indirectly by delaying oxidation of phospholipids (Yin *et al.* 1993). However, O'Sullivan *et al.* (2002a) showed that in a sensory color stability experiment on pork *M. longissimus dorsi* muscle, all treatments appeared initially red, but became less red and browner, the greater the duration of time in the refrigerated display cabinet at 5°C, with samples on day 5 appearing the brownest. Vitamin E treated samples appeared redder and less brown compared to the other samples on successive days in the display cabinet followed by the control, iron/vitamin E, and iron treatment groups. In another study, O'Sullivan *et al.* (2003a, b) found that minced porcine *M. longissimus dorsi* was more stable to oxidative deterioration compared to *M. psoas major* for all the experimental treatments as determined by sensory and instrumental assessment (Minolta colorimeter, L, a, b values). Zanardi *et al.* (1998) used a ten-member panel, trained to judge the appearance of the brown color of metmyoglobin, and carried out color stability evaluations on pork chops from animals supplemented with either vitamin E, oleic acid (sunflower oil), or copper. Panelists evaluated the samples according to a 5-point scale of brown varying from 1 (very light) to 5 (very deep). These authors found that the sensory scores did not show significant differences in the rate of brown appearance in oxygen-permeable packed chops. Increased color stability was observed under protective atmosphere (80% O_2 and 20% CO_2) for groups supplemented with 100 and 200 ppm vitamin E. Dietary supplementation with vitamin E increases the concentration of α-tocopherol in muscle and reduces the susceptibility of the muscle to lipid oxidation (Morrissey *et al.* 1998).

A potential mechanism by which α-tocopherol retards oxymyoglobin pigment oxidation has been proposed by Lynch *et al.* (1998). They demonstrated that natural reductants such as NADH, glutathione, and NADPH are capable of reducing metmyoglobin in the presence of a partially purified extract of beef liver containing cytochrome b_5. They further postulated that α-tocopherol may directly reduce cytochrome b_5-mediated metmyoglobin reduction. The *in vitro* reduction observed was only 15%, which may indicate this color stabilizing mechanism mediated by α-tocopherol plays only a minor role in meat.

The effect of dietary α-tocopherol supplementation in increasing the oxidative stability of meats are particularly pronounced in products exposed to increased oxidative stress such as comminuted meats and meats held in frozen storage (Liu *et al.* 1995; Jensen *et al.* 1997b). Precooked meats are highly susceptible to lipid oxidation resulting in WOF (Liu *et al.* 1995; Faustman *et al.* 1989). In cooked pork meat, significant improvement in oxidative stability, as measured by TBARS during chill storage, was achieved by supplementation with 700 mg α-tocopheryl acetate/kg feed (Jensen *et al.* 1997a). O'Sullivan *et al.* (2003c) found that cooked *M. psoas major* was more susceptible to WOF development as determined by sensory profiling compared to *M. longissimus dorsi* for all treatments (control, vitamin E, iron/vitamin E, and iron). Development of WOF in *M. longissimus dorsi* was highest in an iron-supplemented group followed by the control and vitamin E–supplemented groups. The development of WOF in *M. psoas major* was less well defined between dietary treatments because the level of vitamin E uptake in *M. psoas major* was probably sufficient to protect the muscles from oxidative attack. Also, these authors found that the sensory analysis could predict the chemical measurements vitamin E, iron, and TBARS, and vice versa (O'Sullivan *et al.* 2003c).

22.9 OTHER NATURAL ANTIOXIDANTS—HERBS AND SPICES

Herbs and spices, and the oleoresins and essential oils extracted from them, are widely recognized as powerful agents for the preservation of food quality (Janz *et al.* 2007). Concerns regarding the safety and toxicity of synthetic antioxidants have resulted in much research on natural antioxidants derived from plant sources (O'Grady *et al.* 2006). Synthetic antioxidants have been commonly used by the meat industry to reduce lipid oxidation, but in recent years consumers and food manufacturers have been opting for meat products with all natural or clean labels. This prompted a surge in the use of natural antioxidants due to the increased limitations on the use of synthetic antioxidants (Hayes *et al.* 2009). Tea catechins (TC) are a major group of polyphenolic flavonoids found in green tea (O'Grady *et al.* 2006). The principal catechins found in green tea (Camellia sinensis) are (-)epicatechin (-EC), (-)-epigallocatechin (-EGC), (-)-epicatechin gallate (-ECG), and (-)-epigallocatechin gallate (-EGCG) (Tang *et al.* 2001). TC are polyphenolic antioxidants which possess a range of health-promoting properties (Dreosti 1996; Katiyar & Mukhtar 1997). McCarthy *et al.* (2001) evaluated the antioxidant activities (TBARS determination) of TC at levels of 0%, 0.01%, 0.05%, 0.10%, 0.25%, 0.50%, and 1% in fresh and previously frozen pork patties. They reported that optimum working concentrations of TC ranged from 0.25% to 1% (2 500–10 000 mg/kg), and that optimum concentration of TC was 0.25% (2500 mg/kg). Also, Jo *et al.* (2003) reported that raw pork patties with added 0.1% (1000 mg/kg meat) green tea extract powder had a higher Hunter a value than that of the control during chilled storage, and that sensory color of cooked pork patties was scored better with green tea extract powder than the control.

Rosemary (*Rosmarinus officinalis* L.) extracts contain antioxidant compounds, the most active being phenolic diterpenes such as carnosol, carnosic acid, rosmanol, epirosmanol isorosmanol, methyl carnosate, and other phenolic acids, such as rosmarinic acid (Frankel *et al.* 1996). Fernández-López *et al.* (2003) showed that rosemary and hyssop added to cooked pork may be useful as additives for meat processing to prevent lipid oxidation and discoloration, thus increasing shelf life. McCarthy *et al.* (2001) evaluated the antioxidant activities of aloe vera, fenugreek, ginseng, mustard, rosemary, sage, soya protein, TC, and whey protein concentrate (35% protein) in pork patties prepared from both fresh and previously frozen (−20°C) pork. They identified TC, rosemary, and sage as being effective antioxidants in previously frozen pork patties (McCarthy *et al.* 2001). Nissen *et al.* (2004) found that rosemary extract in a sensory acceptable level was found to yield efficient protection as demonstrated both by sensory evaluation and chemical analyses and thus displayed potential for maintaining sensory eating quality in processed pork products (Nissen *et al.* 2004). Sebranek *et al.* (2005) showed that rosemary extract provided significant protection of color and freshness for pork sausage products (Sebranek *et al.* 2005).

Carpenter *et al.* (2007) in a study on pork meat found that that grape-seed extract (GSE) was effective at reducing biochemical markers of oxidation but had no effect on sensory scores or color in raw and cooked pork. Additionally, O'Grady *et al.* (2008) displayed that antioxidant activity of GSE and bear berry (BB) was observed in cooked pork patties demonstrating the thermal stability of GSE and BB (O'Grady *et al.* 2008).

Hayes *et al.* (2011) evaluated fresh and cooked pork sausages, stored in aerobic or MAP, with the addition of sesamol, ellagic acid, and olive leaf extract. These authors found that these compounds improved lipid stability in raw and cooked pork sausages. These results demonstrate the potential of natural antioxidants, lutein, sesamol, ellagic acid, and olive leaf extract to the meat industry in the development of novel healthy meat products with improved shelf life (Hayes *et al.* 2011).

22.10 ANIMAL DIET AND SHELF-LIFE STABILITY

Pig meat contains high levels of unsaturated fatty acids (UFA) relative to ruminant meats (Enser *et al.* 1996) and is more susceptible to oxidative deterioration of lipids and myoglobin. All meat products can exhibit lipid oxidation during frozen storage but pork-based products are particularly susceptible to oxidation due to their high PUFA levels compared to similar lamb and beef products. The effect of even small changes in PUFA levels in muscle can have substantial effects on rancidity development. PUFAs are highly susceptible to oxidation due to the presence of labile double bonds (Channon & Trout 2002). The intramuscular fatty acid composition of the monogastric animals, and in particular the triacylglycerols, is a reflection of the dietary fatty acids, while in ruminants the biohydrogenation in the rumen (i.e., saturation of the dietary UFA) is responsible for the smaller variations in intramuscular fatty acid composition (Raes *et al.* 2004). As the degree of unsaturation of fatty acids increases, susceptibility to oxidation and rancidity development increases disproportionally (Channon & Trout 2002). Feeding n-3 PUFAs to pigs to improve the quality of pig meat in human nutrition increases this susceptibility to oxidation further (Sheard *et al.* 2000). Boselli *et al.* (2008) describes why swine feeding has been formulated with a higher content of natural sources of UFA, such as n-3 series or conjugated linoleic acid (CLA), due to human health concerns (Boselli *et al.* 2008). Diets enriched with vegetable oils (such as sunflower oil, soybean oil, or corn oil) that contain an elevated UFA percentage should result in healthier products for consumers (Mitchaothai *et al.* 2007). Also, beneficial effects of n-3 fatty acids have been shown in the secondary prevention of coronary heart disease, hypertension, type 2 diabetes, and in some patients with renal disease, rheumatoid arthritis, ulcerative colitis, Crohn's disease, and chronic obstructive pulmonary disease (Simopoulos 1999). Among the fatty acids, it is the omega-3 PUFAs which possess the most potent immunomodulatory activities, and among the omega-3 PUFAs, those from fish oil eicosapentaenoic acid (EPA) and docosahexaenoic acid (DHA) are more biologically potent (Simopoulos 2002).

However, the high level of incorporation of PUFAs into the phospholipid membranes further increases the potential for oxidation since the membranes themselves are highly susceptible to oxidation due to their high surface area and close proximity to pro-oxidant compounds in the cell (Channon & Trout 2002).

As described earlier, dietary vitamin E supplementation results in elevated concentrations of the antioxidant α-tocopherol in cell membranes (mitochondria and the microsomes), resulting in a significantly lower susceptibility to lipid oxidation in the meat of both pigment and lipids (Monahan *et al.* 1990). Feeding dietary α-tocopherol to pigs in parallel with PUFAs will counteract to a certain extend the propensity for lipid oxidation to occur. Cardenia *et al.* (2011) found that dietary supplementation with high-oleic sunflower oil (HOSO) and/or α-tocopheryl acetate used was an appropriate supplementation strategy to produce pork meat with a suitable oxidative stability (Cardenia *et al.* 2011).

22.11 ACTIVE PACKAGING

Active packaging has been defined as "a type of packaging that changes the condition of the packaging to extend shelf-life or improve safety or sensory properties while maintaining the quality of the food" (Vermeiren *et al.* 1999; Quintavalla & Vicini 2002). Ahvenainen (2003) defined active packaging as packaging that "changes the condition of the packed food to extend shelf-life or to improve safety or sensory properties, while maintaining the quality of packaged food" (Ahvenainen 2003). Packaging may be termed active when it performs some desired role in food preservation other than providing an inert barrier to external conditions (Hutton 2003). Rooney (1995) defined an active package as a material that "performs a role other than an inert barrier to the outside environment." Looking to the consumers' demand for chemical preservative-free foods, food manufacturers are now using naturally occurring antimicrobials to sterilize and/or extend the shelf life of foods (Han 2005). Active packaging systems can include oxygen scavengers, carbon dioxide scavengers and emitters, moisture control agents, and antimicrobial packaging technologies (Kerry *et al.* 2006). The aim of active packaging is to increase the display life of contained products, while maintaining their quality, safety, and sensory properties, without direct addition of the active agents to the product (Camo *et al.* 2008).

The most prevalent form of active packaging in the meat industry is based on oxygen scavenging. The scavengers are usually made from iron powders which are combined with acids or salts, or both, and humectants to promote oxidation of the iron (Gill & McGinnis 1995). Oxygen scavengers operate on the principle of binding free oxygen present in packages after sealing to make it unavailable for the growth of aerobic microorganisms. Additionally, this also prevents lipid oxidation and retards the formation of oxidized off-flavors in meat. Oxygen scavenging technologies utilize one or more of the following concepts: iron powder oxidation, ascorbic acid oxidation, photosensitive dye oxidation, enzymatic oxidation (e.g., glucose oxidase and alcohol oxidase), UFA (e.g., oleic or linolenic acid), rice extract, or immobilized yeast on a solid substrate (Floros *et al.* 1997).

Carbon dioxide emitters function on the basis of suppressing microbial growth. The release of carbon dioxide operates similar to O_2 scavenging. Since the permeability of carbon dioxide is 3–5 times higher than that of oxygen in most plastic films, it must be continuously produced to maintain the desired concentration within the package (Ozdemir & Floros 2004).

Carbon dioxide scavengers or absorbers (sachets), consisting of either calcium hydroxide and sodium hydroxide, or potassium hydroxide, calcium oxide, and silica gel, may be used to remove carbon dioxide during storage in order to prevent bursting of the package (Kerry *et al.* 2006).

Antimicrobial packaging is a promising form of active food packaging, in particular for meat products (Quintavalla & Vicini 2002). Active packaging refers to the incorporation of additives into packaging systems with the aim of maintaining or extending meat product quality and shelf life. Antimicrobial packaging includes systems such as adding a sachet into the package, dispersing bioactive agents in the packaging, coating bioactive agents on the surface of the packaging material, or utilizing antimicrobial macromolecules with film-forming properties or edible matrices (Coma 2008). Antimicrobial food-packaging materials have to extend the lag phase and reduce the growth phase of microorganisms in order to extend shelf life and to maintain product quality and safety (Han 2000). Several categories of antimicrobials have been tested for antimicrobial packaging applications: organic acids, fungicides, bacteriocins, proteins, inorganic gases, species, silver substitute zeolite, and enzymes (Han 2000; Cutter & Sumner 2002; Suppakul *et al.* 2003). Among the functional substances which are thought most suitable for their incorporation in the package wall are phytochemicals, vitamins, nanofibers, and prebiotics (Lopez-Rubio *et al.* 2006). Also, antimicrobial substances incorporated into packaging materials can control microbial contamination by reducing the growth rate and maximum growth population and/or extending the lag phase of the target microorganism, or by inactivating microorganisms by contact (Quintavalla & Vicini 2002).

22.12 FUTURE TRENDS

Consumer demand for more environmentally friendly packaging and more natural products will also create increased demand for packaging from biodegradable and renewable resources (Cutter 2006). As discussed above, vacuum or MAP materials can extend the shelf life of muscle food, including pork, products by inhibiting microbiological growth and, in the case of MAP, prolonging color quality which is a key purchase characteristic as far as the consumer is concerned. These packaging materials are principally made from plastics including polyethylene. However, a variety of bio-based materials have been shown to prevent drip loss, reduce lipid oxidation, and improve flavor attributes, as well as enhance the handling properties, color retention, and microbial stability of foods (Cutter 2006). These materials can also be made from renewable resources and include polysaccharides such as alginates, carrageenans, chitosan/chitin, pectin, and starch. Proteins such as casein, whey, collagen, gelatin, corn, soy, and wheat can also be used, as well as lipids including fats, waxes, and oils (Cutter & Sumner 2002; Cutter 2006). The materials used for food packaging today consist of a variety of petroleum-derived plastic materials (Weber *et al.* 2002). The volatility of oil prices has a direct effect on the cost of traditional petrochemical-based packaging materials. Also, the environmental considerations of disposing of traditional packaging after use has become center stage in recent years with respect to green solutions to modern living. The increased costs of petroleum will continue to drive the demands for bio-based packaging materials (Cutter 2006; Marsh & Bugusu 2007; O'Sullivan & Kerry 2009).

As mentioned in the above section, active packaging has the advantage of maintaining the preservative effects of various compounds (antimicrobial, antifungal, or antioxidant), but without being in direct contact with the pork-based muscle food product. Thus, the shelf life of the product is extended and the clean-label status of the product maintained. This is an important development, considering the consumer drive toward clean labeling of food products and the desire to limit the use of food additives (O'Sullivan & Kerry 2009).

ACKNOWLEDGMENT

This work was funded by the Irish Food Industry Research Measure (FIRM) as part of the project titled "Understanding the physico-chemistry of the meat matrix and its potential as a delivery system for added bioactive compounds."

REFERENCES

Ahvenainen, R. (2003) Active and intelligent packaging: An introduction. In: *Novel Food Packaging Techniques* (ed. R. Ahvenainen), pp. 5–21. Woodhead Publishing Limited, Cambridge, U.K.

Allen, P., Doherty, A.M., Buckley, D.J., Kerry, J., O'Grady, M.N. & Monahan, F.J. (1996) Effect of oxygen scavengers and vitamin E supplementation on colour stability of MAP beef. In: *42nd International Congress of Meat Science and Technology*, Lillehammer, Norway, pp. 88–89.

Andersen, H.J. & Skibsted, L.H. (1991) Oxidative stability of frozen pork patties-effect of light and added salt. *Journal of Food Science*, **56**, 1182–1184.

Arnold, R.N., Arp, S.C., Schleller, K.K., Williams, S.N. & Schaefer, D.M. (1993) The equilibration and subcellular distribution of vitamin E relative to myoglobin and lipid oxidation displayed in beef. *Journal of Animal Science*, **71**, 105–118.

Bell, R.G., Penny, N. & Moorhead, S.M. (1995) Growth of the psychotropic pathogens *Aeromonas hydrophila*, *Listeria monocytogenes* and *Yersinia enterocolitica* in smoked blue cod *Parapercis colias* packed under vacuum or carbon dioxide. *International Journal of Food Science and Technology*, **30**, 515–521.

Blixt, Y. & Borch, E. (2002) Comparison of shelf life of vacuum-packed pork and beef. *Meat Science*, **60**, 371–378.

Boselli, E., Pacetti, D., Lucci, P., Di Lecce, P. & Frega, N. (2008) Supplementation with high-oleic sunflower oil and α-tocopheryl acetate: Effects on meat pork lipids. *European Journal of Food Science and Technology*, **110**, 381–391.

Buckley, D.J., Gray, J.I., Asghar, A., *et al.* (1989) Effects of dietary antioxidants and oxidised oil on membranol lipid stability pork product quality. *Journal of Food Science*, **54**, 1193–1197.

Buckley, D.J., Morrissey, P.A. & Gray, J.I. (1995) Influence of dietary vitamin E on the oxidative stability and quality of pig meat. *Journal of Animal Science*, **73**, 3122–3130.

Camo, J., Antonio Beltrán, J. & Roncalés, P. (2008) Extension of the display life of lamb with an antioxidant active packaging. *Meat Science*, **80**, 1086–1091.

Cardenia, V., Rodriguez-Estrada, M.T., Cumella, F., Sardi, L., Della Casa, G. & Lercker, G. (2011) Oxidative stability of pork meat lipids as related to high-oleic sunflower oil and vitamin E diet supplementation and storage conditions. *Meat Science*, **88**, 271–279.

Carpenter, R., O'Grady, M.N., O'Callaghan, Y.C., O'Brien, N.M. & Kerry, J.P. (2007) Evaluation of the antioxidant potential of grape seed and bearberry extracts in raw and cooked pork. *Meat Science*, **76**, 604–610.

Chambers IV, E. & Bowers, J. (1993) Consumer perception of sensory quality in muscle foods: Sensory characteristics of meat influence consumer decisions. *Food Technology*, **47**, 116–120.

Chan, W.K.M., Hakkarainen, K., Faustman, C., Schaefer, D.M., Scheller, K.K. & Liu, Q. (1996) Dietary vitamin E effect on colour stability and sensory assessment of spoilage in three beef muscles. *Meat Science*, **42**, 387–399.

Channon, H.A. & Trout, G.R. (2002) Effect of tocopherol concentration on rancidity development during frozen storage of a cured and an uncured processed pork product. *Meat Science*, **62**, 9–17.

Church, I.J. & Parsons, A.L. (1995) Modified atmosphere packaging technology: Review. *Journal of the Science of Food and Agriculture*, **67**, 143–152.

Clark, D.S., Lentz, C.P. & Roth, L.A. (1976) Use of carbon monoxide for extending shelf-life of prepackaged fresh beef. *Canadian Institute of Food Science and Technology*, **9** (3), 114–117.

Coma, V. (2008) Bioactive packaging technologies for extended shelf life of meat-based products. *Meat Science*, **78**, 90–103.

Cutter, C.N. (2006) Opportunities for bio-based packaging technologies to improve the quality and safety of fresh and further processed muscle foods. *Meat Science*, **74**, 131–142.

Cutter, C.N. & Sumner, S.S. (2002) Application of edible coatings on muscle foods. In: *Protein-Based Films and Coatings* (ed. A. Gennadios), pp. 467–484. CRC Press, Boca Raton, Florida.

Dethmers, A.E. & Rock, H. (1975) Effect of added sodium nitrite on sensory quality and nitrosamine formation in thuringer sausage. *Journal of Food Science*, **40**, 491–495.

Dreosti, I.E. (1996) Bioactive ingredients: antioxidants and polyphenols in tea. *Nutrition Reviews*, **54** (11), S51–S58.

Eilert, S.J. (2005) New packaging technologies for the 21st century. *Meat Science*, **71**, 122–127.

Enser, M., Hallett, K., Hewitt, B., Fursey, G.A.J. & Wood, J.D. (1996) Fatty acid content and composition of English beef, lamb and pork at retail. *Meat Science*, **42**, 443–456.

Estevez, M. & Cava, R. (2004) Lipid and protein oxidation, release of iron from heme molecule and colour deterioration during refrigerated storage of liver pate. *Meat Science*, **68**, 551–558.

Faustman, C. & Cassens, R.G. (1989) Strategies for improving fresh meat colour. In: *Proceedings of the 35th International Congress of Meat Science and Technology*, Copenhagen, Denmark, pp. 446–453.

Faustman, C. & Cassens, R.G. (1990) Influence of aerobic metmyoglobin reducing capacity on colour stability of beef. *Journal of Food Science*, **55**, 1279–1283.

Faustman, C., Cassens, R.G., Schaefer, D.M., Buege, D.R., Williams, S.N. & Scheller, K.K. (1989) Improvement of pigment and lipid stability in Holstein steer beef by dietary supplementation with vitamin E. *Journal of Food Science*, **54**, 858–862.

Fernández-López, J., Sevilla, L., Sayas-Barberá, E., Navarro, C., Marín, F. & Pérez-Alvarez, J.A. (2003) Evaluation of the antioxidant potential of hyssop (*Hyssopus officinalis* L.) and rosemary (*Rosmarinus officinalis* L.) extracts in cooked pork meat. *Journal of Food Science*, **68**, 660–664.

Firstenberg-Eden, R. (1981) Attachment of bacteria to meat surfaces: A review. *Journal of Food Protection*, **44**, 602–607.

Floros, J.D., Dock, L.L. & Han, J.H. (1997) Active packaging technologies and applications. *Food Cosmetics and Drug Packaging*, **20**, 10–17.

Frankel, E.N., Huang, S.W., Aeschbach, R. & Prior, E. (1996) Antioxidant activity of a rosemary extract and its constituents, carnosic acid, carnosol, and rosmarinic acid, in bulk oil and oil-in-water emulsion. *Journal of Agricultural and Food Chemistry*, **44**, 131–135.

Frigg, M., Prabucki, A.L. & Ruhdel, E.U. (1990) Effect of dietary vitamin E levels on oxidative stability of trout fillets. *Aquaculture*, **84**, 145–158.

Georgala, D.L. & Davidson, C.M. (1970) Food package. *British Patent,* **1,** 199–998.

Gill, C.O. (1990) Controlled atmosphere packaging of chilled meat. *Food Control*, **1**, 74–78.

Gill, C.O. (1991) Extending the storage life of raw meat. I. Preservative atmospheres, Western Canada Research Group on extended storage of meat and meat products. Technical Bulletin Number 1, Department of Applied Microbiology and Food Science, University of Saskatchewan, Saskatoon, Saskatchewan, Canada.

Gill, C.O. (1996) Extending the storage life of raw chilled meats, *Meat Science*, **43** (Suppl.), S99–S109.

Gill, A.O. & Gill, C.O. (2005) CH13. Preservative packaging for fresh meats, poultry and fin fish. In: *Innovations in Food Packaging* (ed. J.H. Han), pp. 204–220. Elsevier/Academic Press, San Diego, California.

Gill, C.O. & Jones, T. (1996) The display life of retail packaged pork chops after their storage in master packs under atmospheres of N_2, CO_2, or $O_2 + CO_2$. *Meat Science*, **42** (2), 203–213.

Gill, C.O. & McGinnis, J.C. (1995) The use of oxygen scavengers to prevent the transient discolouration of ground beef packaged under controlled, oxygen-depleted atmospheres. *Meat Science*, **41**, 19–27.

Han, J.H. (2000) Antimicrobial food packaging. *Food Technology*, **54** (3), 56–65.

Han, J.H. (2005) Antimicrobial packaging systems. In: *Innovations in Food Packaging* (ed. J.H. Han), pp. 81–107. Elsevier Academic Press, Amsterdam, the Netherlands.

Hansen, E., Lauridsen, L., Skibsted, L.H., Moawad, R.K. and Andersen, M.L. (2004) Oxidative stability of frozen pork patties: Effect of fluctuating temperature on lipid oxidation. *Meat Science*, **68**, 185–191.

Hayes, J.E., Stepanyan, B., Allen, P., O'Grady, M.N., O'Brien, N.M. & Kerry, J.P. (2009) The effect of lutein, sesamol, ellagic acid and olive leaf extract on lipid oxidation and oxymyoglobin oxidation in bovine and porcine muscle model systems. *Meat Science*, **83**, 201–208.

Hayes, J.E., Stepanyan, V., Allen, P., O'Grady, M.N. & Kerry, J.P. (2011) Evaluation of the effects of selected plant-derived nutraceuticals on the quality and shelf-life stability of raw and cooked pork sausages. *LWT—Food Science and Technology*, **44**, 164–172.

Holley, R.A., Peirson, M.D., Lam, J. & Tan, K.B. (2004) Microbial profiles of commercial, vacuum-packaged, fresh pork of normal or short storage life. *International Journal of Food Microbiology*, **97**, 53–62.

Hood, D.E. & Mead, G.C. (1993) Modified atmosphere storage of fresh meat and poultry. In: *Principles and Applications of Modified Atmosphere Packing of Food* (ed. R.T. Parry), pp. 269–298. Blackie Academic and Professional, London, U.K.

Hutton, T. (2003) *Food Packaging: An introduction. Key Topics in Food Science and Technology—Number 7*, p. 108. Campden and Chorleywood Food Research Association Group, Chipping Campden, U.K.

Jakobsen, M. & Bertelsen, G. (2000) Colour stability and lipid oxidation of fresh beef. Development of a response surface model for predicting the effects of temperature, storage time, and modified atmosphere composition. *Meat Science*, **54**, 49–57.

Janz, J.A.M., Morel, P.C.H., Wilkonson, B.H.P. & Purchas, R.W. (2007) Preliminary investigation of the effects of low-level dietary inclusion of fragrant essential oils and oleoresins on pig performance and pork quality. *Meat Science*, **75**, 350–355.

Jayasingh, P., Cornforth, D.P., Brennand, C.P., Carpenter, C.E. & Whittier, D.R. (2002) Sensory evaluation of ground beef stored in high-oxygen modified atmosphere packaging. *Journal of Food Science*, **67** (9), 3493–3496.

Jensen, C., Skibsted, L.H., Jakobsen, K. & Bertelsen, G. (1995) Supplementation of broiler diets with all-rac-α- or a mixture of natural source RRR-α-γ-δ-tocopheryl acetate. 2. Effect of oxidative stability of raw and pre-cooked broiler meat products. *Poultry Science*, **74**, 2048–2056.

Jensen, C., Guidera, J., Skovgaard, I.M., *et al.* (1997a) Effects of dietary α-tocopheryl acetate supplementation on α-tocopherol deposition in porcine *m. psoas major* and *m. longissimus dorsi* and on drip loss, colour stability and oxidative stability of pork meat. *Meat Science*, **45**, 491–500.

Jensen, C., Skibsted, L.H. & Bertelsen, G. (1997b) Influence of dietary fat source and vitamin E on quality of frozen pork meat products. In: *Proceedings of 43rd International Congress of Meat Science and Technology*, Auckland, New Zealand, pp. 402–402.

Jensen, C., Skibsted, L.H. & Bertelsen, G. (1998) Oxidative stability of frozen-stored raw pork chops, chill-stored pre-frozen raw pork chops, and frozen-stored pre-cooked sausages in relation to dietary CuSO4, rapeseed oil and vitamin E. *European Food Research and Technology*, **207**, 363–368.

Jensen, J.M., Robbins, K.L., Ryan, K.J., Homco-Ryan, C., McKeith, F.K. & Brewer, M.S. (2002) Effects of lactic and acetic acid salts on quality characteristics of enhanced pork during retail display. *Meat Science*, **63**, 501–508.

Jeremiah, L.E. & Gibson, L.L. (1995) Flavour changes accompanying nominally anoxic spoilage of pork with commercial and reduced levels of microbial contamination. *Journal of Muscle Foods*, **6**, 341–358.

Jeremiah, L.E. & Gibson, L.L. (1997) The influence of controlled atmosphere storage on the flavor and texture profiles of display ready pork cuts. *Food Research International*, **30**, 117–129.

Jeremiah, L.E., Gill, C.O. & Penney, N. (1992) The effect on pork storage life of oxygen contamination in nominally anoxic packagings. *Journal of Muscle Foods*, **3**, 263–281.

Jo, C., Son, J.H., Son, C.B. & Byun, M.W. (2003) Functional properties of raw and cooked pork patties with added irradiated, freeze-dried green tea leaf extract powder during storage at 4°C. *Meat Science*, **64**, 13–17.

Johanson, L., Underdal, B., Grosland, K., Whelehan, O.P. & Roberts, T.A. (1983) A survey of the hygienic quality of beef and pork carcasses in Norway. *Acta Vetermarla Seandmawca*, **24**, 1–13.

Kanner, J. (1994) Oxidative processes in meat and meat products: Quality implications. *Meat Science*, **36**, 169–189.

Katiyar, S.K. & Mukhtar, H. (1997) Tea antioxidants in cancer chemoprevention. *Journal of Cellular Biochemistry, Supplement*, **27**, 59–67.

Kerry, J.P., O'Sullivan, M.G., Buckley, D.J., Lynch, P.B. & Morrissey, P.A. (2000) The effects of dietary α-tocopheryl acetate supplementation and modified atmosphere packaging (MAP) on the quality of lamb patties. *Meat Science*, **56**, 61–66.

Kerry, J.P., O'Grady, M.N. & Hogan, S.A. (2006) Past, current and potential utilisation of active and intelligent packaging systems for meat and muscle-based products: A review. *Meat Science*, **74**, 113–130.

Korkeala, H. & Björkroth, K.J. (1997) Spoilage and contamination of vacuum-packaged cooked sausages: A review. *Journal of Food Protection*, **60**, 724–731.

Lambropoulou, K.A., Drosinos, E.H. & Nychas, G.J.E. (1996) The effect of glucose supplementation on the spoilage microflora and chemical composition of minced beef stored aerobically or under a modified atmosphere at 4°C. *International Journal of Food Microbiology*, **30**, 281–291.

Lanari, M.C., Schaefer, D.M. & Scheller, K.K. (1995) Dietary vitamin E supplementation and discoloration of pork bone and muscle following modified atmosphere packaging. *Meat Science*, **41**, 237–250.

Lanari, M.C., Schaefer, D.M., Liu, Q. & Cassens, R.G. (1996) Kinetics of pigment oxidation in beef from steers supplemented with vitamin E. *Journal of Food Science*, **61**, 884–889.

Lanier, T.C., Carpenter, J.A., Toledo, R.T. & Reagan, J.O. (1978) Metmyoglobin reduction in beef systems as affected by aerobic, anaerobic and carbon monoxide-containing environments. *Journal of Food Science*, **43** (6), 1788–1792.

Lawrie, R.A. (ed.) (1991) Chemical and biochemical constitution of muscle. In: *Meat Science*, 5th edn., pp 71–72. Pergamon Press, Oxford, U.K.

Ledward, D.A. (1985) Post-slaughter influences on the formation of metmyoglobin in beef muscles. *Meat Science*, **15**, 149–171.

Liu, Q., Lanari, M.C. & Schaefer, D.M. (1995) A review of dietary vitamin E supplementation for improvement of beef quality. *Journal of Animal Science*, **73**, 3131–3140.

Livingston, M., Brewer, M.S., Killifer, J., Bidner, B. & McKeith, F. (2004) Shelf life characteristics of enhanced modified atmosphere packaged pork. *Meat Science*, **68**, 115–122.

Lopez-Rubio, A., Gavara, R. & Lagaron, J.M. (2006) Bioactive packaging: Turning foods into healthier foods through biomaterials. *Trends in Food Science & Technology*, **17**, 567–575.

Lund, M.N., Lametsch, R., Hviid, M.S., Jensen, O.N. & Skibsted, L.H. (2007) High oxygen packaging atmosphere influences protein oxidation and tenderness of porcine longissimus dorsi during chill storage. *Meat Science*, **77**, 295–303.

Luno, M., Roncales, P., Djenane, D. & Beltran, J.A. (2000) Beef shelf life in low O_2 and high CO_2 atmospheres containing different low CO concentrations. *Meat Science*, **55** (4), 413–419.

Lynch, M.P., Faustman, C., Chan, W.K.M., Kerry, J.P. & Buckley, D.J. (1998) A potential mechanism by which α-tocopherol maintains oxymyoglobin pigment through cytochrome b_5 mediated reduction. *Meat Science*, **3**, 333–342.

Marsh, K. & Bugusu, B. (2007) Food packaging roles, materials, and environmental issues. *Food Science*, **72** (3), 39–55.

Marusich, W.L., De Ritter, E., Ogrinz, E.F., Keating, J., Mitrovic, M. & Bunnell, R.H. (1975) Effect of supplemental vitamin E in control of rancidity in poultry meat. *Poultry Science*, **54**, 831–844.

McCarthy, T.L., Kerry, J.P., Kerry, J.F., Lynch, P.B. & Buckley, D.J. (2001) Assessment of the antioxidant potential of natural food and plant extracts in fresh and previously frozen pork patties. *Meat Science*, **57**, 177–184.

Mitchaothai, J., Yuangklang, C., Wittayakun, S., Vasupen, K., Wongsutthavas, S., Srenanul, P., *et al.* (2007) Effect of dietary fat type on meat quality and fatty acid composition of various tissues in growing-finishing swine. *Meat Science*, **76**, 95–101.

Monahan, F.J., Buckley, D.J., Gray, J.I., Morrissey, P.A., Asghar, A., Hanrahan, T.J. & Lynch, P.B. (1990) Effect of dietary vitamin E on the stability of raw and cooked pork meat. *Meat Science*, **27**, 99–108.

Morrissey, P.A., Buckley, D.J., Sheehy, P.J.A. & Monaghan, F.J. (1994) Vitamin E and meat quality. *Proceedings of the Nutrition Society*, **53**, 289–295.

Morrissey, P.A., Sheehy, P.J.A., Galvin, K., Kerry, J.P. & Buckley, D.J. (1998) Lipid stability in meat and meat products. *Meat Science*, **49**, 73–86.

Nattress, F.M. & Jeremiah, L.E. (2000) Bacterial mediated off-flavours in retail-ready beef after storage in controlled atmospheres. *Food Research International*, **33**, 743–748.

Nissen, L.R., Byrne, D.V., Bertelsen, G. & Skibsted, L.H. (2004) The antioxidative activity of plant extracts in cooked pork patties as evaluated by descriptive sensory profiling and chemical analysis. *Meat Science*, **68**, 485–495.

O'Grady, M.N., Maher, M., Troy, D.J., Moloney, A.P. & Kerry, J.P. (2006) An assessment of dietary supplementation with tea catechins and rosemary extract on the quality of fresh beef. *Meat Science*, **73**, 132–143.

O'Grady, M.N., Carpenter, R., Lynch, P.B., O'Brien, N.M. & Kerry, J.P. (2008) Addition of grape seed extract and bearberry to porcine diets: Influence on quality attributes of raw and cooked pork. *Meat Science*, **78** (2008), 438–446.

O'Sullivan, M.G. & Kerry, J.P. (2009) CH 13, Meat packaging. In: *Handbook of Meat Processing* (ed. F. Toldrá), pp. 211–230. John Wiley & Sons, Chichester, U.K.

O'Sullivan, M.G. & Kerry, J.P. (2011) Sensory quality of fresh and processed meats. In: *Improving the Sensory and Nutritional Quality of Fresh and Processed Meats* (eds J.P. Kerry & D.A. Ledward). Woodhead Publishing Limited, Cambridge, U.K.

O'Sullivan, M.G., Kerry, J.P., Buckley, D.J., Lynch, P.B. & Morrissey, P.A. (1997) The distribution of dietary vitamin E in the muscle of the porcine carcass. *Meat Science*, **45**, 297–305.

O'Sullivan, M.G., Kerry, J.P., Buckley, D.J., Lynch, P.B. & Morrissey, P.A. (1998) The effect of dietary vitamin E supplementation on quality aspects of porcine muscles. *Irish Journal of Agriculture and Food Research*, **37**, 227–235.

O'Sullivan, M.G., Byrne, D.V., Stagsted, J., Andersen, H.J. & Martens, M. (2002a) Sensory colour assessment of fresh meat from pigs supplemented with iron and vitamin E. *Meat Science*, **60**, 253–265.

O'Sullivan, M.G., Byrne, D.V. & Martens, M. (2002b) Data analytical methodologies in the development of a vocabulary for evaluation of meat quality. *Journal of Sensory Studies*, **17**, 539–558.

O'Sullivan, M.G., Byrne, D.V. & Martens, M. (2003a) Evaluation of pork colour: Sensory colour assessment using trained and untrained sensory panellists. *Meat Science*, **63**, 119–129.

O'Sullivan, M.G., Byrne, D.V., Martens, H., Gidskehaug, L.H, Andersen, H.J. & Martens, M. (2003b) Evaluation of pork meat colour: Prediction of visual sensory quality of meat from instrumental and computer vision methods of colour analysis. *Meat Science*, **65**, 909–918.

O'Sullivan, M.G., Byrne, D.V., Nielsen, J.H., Andersen, H.J. & Martens, M. (2003c) Sensory and chemical assessment of pork supplemented with iron and vitamin E. *Meat Science*, **64**, 175–189.

O'Sullivan, M.G., Cruz, M. & Kerry, J.P. (2011) Evaluation of carbon dioxide flavour taint in modified atmosphere packed beef steaks. *LWT—Food Science and Technology*, **44**, 2193–2198.

Ozdemir, M. & Floros, J.D. (2004) Active food packaging technologies. *Critical Reviews in Food Science and Nutrition*, **44**, 185–193.

Paine, F.A. & Paine, H.Y. (1992) A Handbook of Food Packaging. Ch9 Frozen Foods. Chapman & Hall, London, U.K.

Parry, R.T. (1993) Introduction. In: *Principles and Applications of Modified Atmosphere Packaging of Food* (ed. R.T. Parry). Blackie Academic and Professional, New York.

Patsias, A., Chouliara, I., Badeka, I.N., Savvaidis, M.G. & Kontominas, M.G. (2006) Shelf-life of chilled precooked chicken product stored in air and under modified atmospheres: Microbiological, chemical, sensory attributes. *Food Microbiology*, **23**, 423–429.

Payne, S.R., Durham, C.J., Scott, S.M., Penney, N., Bell, R.G. & Devine, C.E. (1997) The effects of rigor temperature, electrical stimulation, storage duration and packaging systems on drip loss in beef. In: *Proceedings of the 43rd International Congress of Meat Science and Technology*, Auckland, New Zealand, (Gl-22), pp. 592–593.

Quintavalla, S. & Vicini, L. (2002) Antimicrobial food packaging in meat industry. *Meat Science*, **62**, 373–380.

Raes, K., De Smet, S. & Demeyer, D. (2004) Effect of dietary fatty acids on incorporation of long chain polyunsaturated fatty acids and conjugated linoleic acid in lamb, beef and pork meat: A review. *Animal Feed Science and Technology*, **113**, 199–221.

Ray, B. (2001) *Fundamental Food Microbiology*, 2nd edn., pp. 65–69. CRC Press LLC, Boca Raton, Florida.

Rhee, K.I. & Ziprin, Y.A. (1987) Lipid oxidation in retail beef, pork and chicken muscles as affected by concentrations of heme pigments and nonheme iron and microsomal enzymic lipid peroxidation activity. *Journal of Food Biochemistry*, **11**, 1–15.

Risvik, E. (1994) Sensory properties and preferences. *Meat Science*, **36**, 67–77.

Rooney, M.L. (1995) Overview of active food packaging. In: *Active Food Packaging* (ed. M.L. Rooney), pp. 1–37. Blackie Academic and Professional, London, U.K.

Samelis, J., Kakouri, A., Georgiadou, K.G. & Metaxopoulos, J. (1998) Evaluation of the extent and type of bacterial contamination at different stages of processing of cooked ham. *Journal of Applied Microbiology*, **84**, 649–660.

Samelis, J., Kakouri, A. & Rementzis, J. (2000) Selective effect of the product type and the packaging conditions on the species of lactic acid bacteria dominating the spoilage microbial association of cooked meats at 4°C. *Food Microbiology*, **17**, 329–340.

Sebranek, J.G., Sewalt, V.J.H., Robbins, K.L. & Houser, T.A. (2005) Comparison of a natural rosemary extract and BHA/BHT for relative antioxidant effectiveness in pork sausage. *Meat Science*, **69**, 289–296.

Seideman, S.C. & Durland, P.R. (1984) The utilization of modified atmosphere packaging for fresh meat: A review. *Journal of Food Quality*, **6**, 239–252.

Sheard, P.R., Enser, M., Wood, J.D., Nute, G.R., Gill, B.P. & Richardson, R.I. (2000) Shelf life and quality of pork and pork products with raised n-3 PUFA. *Meat Science*, **55**, 213–221.

Sherbeck, J.A., Wulf, D.M., Morgan, J.B., Tatum, J.D., Smith, G.C. & Williams, S.N. (1995) Dietary supplementation of vitamin E to feedlot cattle affects retail display properties. *Journal of Food Science*, **60**, 250–252.

Shorland, F.B., Igene, J.O., Pearson, A.M., Thomas, J.W., McGuffey, R.K. & Aldridge, A.E. (1981) Effects of dietary fat and vitamin E on the lipid composition and stability of veal during frozen storage. *Journal of Agriculture and Food Chemistry*, **29**, 863–871.

Shrestha, S. & Min, Z. (2006) Effect of lactic acid pretreatment on the quality of fresh pork packed in modified atmosphere. *Journal of Food Engineering*, **72**, 254–260.

Simopoulos, A.P. (1999) Essential fatty acids in health and chronic disease. *American Journal of Clinical Nutrition*, **70**, 3, 560S–569S.

Simopoulos, A.P. (2002) Omega-3 fatty acids in inflammation and autoimmune diseases. *Journal of the American College of Nutrition*, **21**, 6, 495–505.

Smiddy, M., Papkovskaia, N., Papkovsky, D.B. & Kerry, J.P. (2002) Use of oxygen sensors for the non-destructive measurement of the oxygen content in modified atmosphere and vacuum packs of cooked chicken patties: Impact of oxygen content on lipid oxidation. *Food Research International*, **35**, 577–584.

Smulders, F.J.M. (1987) Prospectives for microbial decontamination of meat and poultry by organic acids with special reference to lactic acid. In: *Elimination of Pathogenic Organisms from Meat and Poultry* (ed. F.J.M. Smulders), pp. 319–341. Elsevier, Amsterdam, the Netherlands.

Sørheim, O., Aune, T. & Nesbakken, T. (1997a) Technological, hygienic and toxicological aspects of carbon monoxide used in modified-atmosphere packaging of meat. *Trends in Food Science & Technology*, **8** (9), 307–312.

Sørheim, O., Tore, A. & Truls, N. (1997b) Technological, hygienic and toxicological aspects of carbon monoxide used in modified-atmosphere packaging of meat. *Trends in Food Science & Technology*, **8**, 307–312.

Sørheim, O., Nissen, H. & Nesbakken, T. (1999) The storage life of beef and pork packaged in an atmosphere with low carbon monoxide and high carbon dioxide. *Meat Science*, **52**, 157–164.

Sørheim, O., Nissen, H. & Nesbakken, T. (2001) Use of carbon monoxide in retail meat packaging. In: *Proceedings of the 54th Reciprocal Meat Conference*, Indianapolis, Indiana, pp. 47–51.

St. Angelo, A.J. & Bailey, M.E. (1987) *Warmed-Over Flavour of Meat*, pp. vii–viii. Academic Press, Orlando, Florida.

Stiles, M.E. (1991) Scientific principles of controlled/modified atmosphere packaging. In: *Modified Atmosphere Packaging of Foods* (eds B. Ooraikul & M.E. Stiles), pp. 18–25 and 118–147. Ellis Horwood Ltd., Chichester, England.

Stolzenbach, S., Leisner, J.J. & Byrne, D.V. (2009) Sensory shelf life determination of a processed meat product 'rullepøse' and microbial metabolites as potential indicators. *Meat Science*, **83**, 285–292.

Suppakul, P., Miltz, J., Sonneveld, K. & Bigger, S.W. (2003) Active packaging technologies with an emphasis on antimicrobial packaging and its applications. *Journal of Food Science*, **68** (2), 408–420.

Tang, S.Z., Kerry, J.P., Sheehan, D., Buckley, D.J. & Morrissey, P.A. (2001) Antioxidative effect of dietary tea catechins on lipid oxidation of long-term frozen stored chicken meat. *Meat Science*, **57**, 331–336.

Taylor, A.A. (1985) Packaging fresh meat. In: *Developments in Meat Science* (ed. R.A. Lawrie), Vol. 3. Elsevier Applied Science Publishers, Essex, England.

Tims, M.J. & Watts, B.M. (1958) Protection of cooked meats with phosphates. *Food Technology*, **12**, 240–243.

Ulrich, R. (1981) Variations de temperature et qualite des produits surgeles. *Revue Generale du Froid*, **71**, 371–389.

Vermeiren, L., Devlieghere, F., van Beest, M., de Kruijf, N. & Debevere, J. (1999) Developments in the active packaging of foods. *Trends in Food Science and Technology*, **10**, 77–86.

Viana, E.S., Gomide, L.A.M. & Vanetti, M.C.D. (2005) Effect of modified atmospheres on microbiological, colour and sensory properties of refrigerated pork. *Meat Science*, **71**, 696–705.

Weber, C.J., Haugaard, V., Festersen, R. & Bertelsen, G. (2002) Production and applications of biobased packaging materials for the food industry. *Food Additives and Contaminants*, **19**, 172–177.

Whang, K., Aberle, E.D., Judge, M.D. & Peng, I.C. (1986) Antioxidative activity of alpha-tocopherol in cooked and uncooked ground pork. *Meat Science*, **17**, 235–249.

Wolfe, S.K. (1980) Use of CO- and CO_2 enriched atmospheres for meats, fish, and produce. *Food Technology*, **34** (3), 55–63.

Yin, M.C., Faustman, C., Riesen, J.W. & Williams, S.N. (1993) The effects of α-tocopherol and ascorbate upon oxymyoglobin and phospholipid oxidation. *Journal of Food Science*, **58**, 1273–1276.

Zakrys, P.I., O'Sullivan, M.G., Allen, P. & Kerry, J.P. (2009) Consumer acceptability and physiochemical characteristics of modified atmosphere packed beef steaks. *Meat Science*, **81**, 720–725.

Zakrys-Waliwander, P.I., O'Sullivan, M.G., Walshe, H., Allen, P. & Kerry, J.P. (2011) Sensory comparison of commercial low and high oxygen modified atmosphere packed sirloin beef steaks. *Meat Science*, **88**, 198–202.

Zanardi, E., Novelli, E., Nanni, N., *et al.* (1998) Oxidative stability and dietary treatment with vitamin E, oleic acid, copper of fresh and cooked pork chops. *Meat science*, **49**, 309–320.

Part Five
Poultry Quality

23 Poultry Meat Flavor

Paul L. Dawson and Nick Spinelli

Abstract: Poultry meat flavors result from interactions between proteins, nucleic acids, lipids, and other components found in the meat. Cooking promotes the development of freshly cooked flavors that are derived via amino and fatty acid reactions to form volatile and nonvolatile compounds. The Maillard reaction or Maillard-like reactions occur during heating, especially when high heat and dehydrating types of cooking are used. Once poultry meat is cooked, oxidative reactions quickly initiate: first, within the phospholipids fraction, and then within hours result in the development of detectable oxidized off-flavors. These oxidized off-flavors have been described as "warmed-over flavors" being formed from the scission of fatty acids into shorter chain carbon compounds such as aldehydes, ketones, alcohols, and acids. Since unsaturated fatty acids oxidize at a faster rate than saturated fatty acids, poultry meat's higher degr ee of unsaturation compared to other meats facilitates faster oxidized flavor development. Certain strategies have been employed to slow the development of off-flavors including feeding the animal high levels of antioxidants prior to slaughter or adding antioxidants prior to cooking. Exclusion of oxygen from direct contact with the meat and lowering the storage temperature also slow the rate of oxidation and off-flavor development. Poultry meat flavor also varies due to muscle types within animals (i.e., breast vs. leg). Dark chicken meat has a slightly higher fat content than lighter meat; however, the difference in color is due to pigment content.

Keywords: Poultry meat flavor, cooked flavors, sulfur compounds, carbonyl compounds, thiobarbituric acid reactive substances, oxidation-derived flavors, lipid oxidation, warmed-over flavor, smoked meat flavors, cured meat flavors

23.1 INTRODUCTION

Poultry meat flavors result from interactions between proteins, nucleic acids, lipids, and other components found in the meat. Cooking promotes the development of freshly cooked flavors that are derived via amino and fatty acid reactions to form volatile and nonvolatile compounds. The Maillard reaction or Maillard-like reactions occur during heating especially when high heat and dehydrating types of cooking are used. Once poultry meat is cooked, oxidative reactions quickly initiate: first, within the phospholipids fraction, and then within hours result in the development of detectable oxidized off-flavors. These oxidized off-flavors have been described as "warmed-over flavors" (WOF) being formed from the scission of fatty acids into shorter chain carbon compounds such as aldehydes, ketones, alcohols, and

Handbook of Meat, Poultry and Seafood Quality, Second Edition. Edited by Leo M. L. Nollet.

acids. Since unsaturated fatty acids oxidize at a faster rate than saturated fatty acids, poultry meat's higher degree of unsaturation compared to other meats facilitates faster oxidized flavor development. Certain strategies have been employed to slow the development of off-flavors including feeding the animal high levels of antioxidants prior to slaughter or adding antioxidants prior to cooking. Exclusion of oxygen from direct contact with the meat and lowering the storage temperature also slow the rate of oxidation and off-flavor development. Poultry meat flavor also varies due to muscle types within animals (i.e., breast vs. leg). Dark chicken meat has a slightly higher fat content than lighter meat; however, the difference in color is due to pigment content. This chapter will discuss specific flavors of poultry meat resulting from cooking, oxidation after cooking, and flavor differences due to animal diet or species.

23.2 COOKED FLAVORS

There are several excellent reviews of literature on poultry flavors (Wilson & Katz 1972; Steverink 1981; Ramaswamy & Richards 1982; Chen & Ho 1994). These studies discussed previous research and their own studies on both the water-soluble and lipid portions of the flavor of poultry. Most used chicken broth to focus on the nonlipid portion of poultry flavor. The separation of chicken flavor into individual compounds evolved with the advent and development of gas chromatography (GC) technology (James & Martin 1952). The complexity of meat and poultry flavor was mentioned in 1937 (Howe and Barbella) and in 1948 (Crocker). Crocker (1948) published the first report attempting to link chemical composition and flavor. Bouthilet (1949, 1950, 1951a, b) published the earliest in-depth work on poultry meat flavor and concluded that chicken flavor consists of at least two main fractions: a sulfur and a fatty acid fraction. This researcher further concluded that meat flavor was due to the muscle tissue and not primarily linked to the lipid fraction since the main meat flavor was extracted with a polar solvent. This conclusion was verified by Pippen *et al.* (1954, 1969) when the meat, bone, skin, and composite fractions were compared for flavor and the meat fraction yielded the most meaty flavor. Using carbonyl compounds found by Pippen *et al.* (1958) and Pippen and Nonaka (1960), Lineweaver *et al.* (1962) reported the first use of GC to characterize chicken flavor volatiles. Pippen and Nonaka (1963) found that cooking chicken in air yielded more complex and larger volatile compound fractions compared to chicken cooked in a nitrogen gas environment. This led to the assumption that oxygen greatly affects the development of cooked chicken flavor. These researchers also identified hexanal and 2,4-decadienal as prominent volatiles in cooked chicken flavor. Minor *et al.* (1965b, c) used GC to elucidate alcohols, amines, esters, mercaptans, and sulfides in addition to ammonia, hydrogen sulfide, and other 15 or so carbonyl compounds as chicken flavor components. Minor *et al.* (1965b, c) listed 29 and 25 volatiles from cooked meat and water slurries from leg and breast muscle, respectively. These researchers also confirmed the observation by Bouthilet (1949, 1950) that chicken flavor comprises two major components: a "meaty" sulfur component and a "chickeny" carbonyl component and that sulfur amino acids and lipid components are the precursors of these components, respectively. For instance, Grey and Shrimpton (1967a, b) identified 20 and then 27 volatile components of chicken breast muscle. The use of Kovat's indices facilitated the identification of volatile compounds by different researchers. Before the development of high-resolution GC and mass spectroscopy (MS), less than 40 compounds were identified. Nonaka *et al.* (1967) separated 227 compounds using these technologies but were only able to identify 62 due to interference and lack of separation. Hobson-Frohock (1970) identified about 30 new chicken flavor compounds by sequentially removing classes of compounds. Wilson and Katz (1972) added eleven new chicken flavor compounds (mostly unsaturated alcohols) listing a total of 178, while Janney *et al.* (1974) identified 27 compounds from fried chicken including butylated hydroxy toluene, a common food antioxidant. Wilson and Katz (1972) divided chicken flavor compounds into categories of hydrocarbons, aromatic hydrocarbons, furans, alcohols, esters, ketones, ethers, fatty acids, aldehydes, sulfur compounds, inorganic compounds, amino acids/peptides, sugars, amines, lactones, and miscellaneous compounds. Harkes and Begemann (1974) found 11 new unsaturated aldehydes and Horvat (1976) identified 30 previously unknown compounds from chicken, bringing the total of identified compounds to 236 from raw, cooked, and fried chicken. Tang *et al.* (1983) determined the flavor components of fried chicken and found three peaks containing various cyclic compounds were related to the unique fried flavor. These researchers concluded that of the 130 identified volatiles found in fried chicken flavor, they were likely to be formed from either (1) chicken constituents during frying, (2) flour constituents during frying, (3) oxidation of the frying oil itself, (4) and/or interaction between the decomposition products of chicken or flour and the frying oil. By 1984, over 300 compounds had been identified as components of cooked chicken flavor (Maarse, 1984). Further work by Noleau and Toulemonde (1986) using a Likens–Nickerson apparatus listed 193 components from roasted chicken flavor. The list included hydrocarbons (27), alcohols (17), aldehydes (47), ketones (33), acids (14), esters (6), bases (22), sulfur compounds (10), halogens (3), acetals (2), nitriles (1), phenols (5), and furans (10).

The chemical reactions that create many of the chicken and meat flavors during cooking can be attributed to Maillard reactions. These are a series of complex reactions initiated with heating foods containing protein (amine groups) and compounds with reducing groups (sugars, oxidized lipids). The type and intensity of chicken flavor developing from cooking can be influenced by many factors including the amount of Maillard flavor precursors found in the raw meat. Alaini and Farmer (2005a) reported that variation occurs both between different commercial sources and between individual chickens from the same source. Coefficients of variation higher than 30% were found for inosine 5′-monophosphate, guanosine 5′-monophosphate, and inosine. These levels were similar to those previously determined for reducing sugars and their phosphates. There were concentration ranges of threefold and higher, which may have the potential to affect odor and flavor formation. In contrast, thiamine and amino acids (both protein and nonprotein) most often vary less than twofold.

23.3 SULFUR COMPOUNDS IN FLAVOR

Sulfur compounds formed during cooking are important flavor compounds yielding a chicken flavor. Since meat lipids do not contain sulfur, the sulfur is derived from the protein fraction, namely the sulfur amino acids (cysteine, cystine, methionine) or other sulfur-containing compounds found in meat such as glutathione. Bouthilet (1951a, b) found the presence of ammonia and hydrogen sulfide in cooked chicken and claimed that sulfur compounds were important to chicken flavor. The presence of sulfides in cooked chicken volatiles was verified by Pippen and Eyring (1957) when they concluded sulfur compounds were part of the chickeny flavor and that while ammonia was present it did not contribute to the chicken flavor. In addition to hydrogen sulfide, Mecchi *et al.* (1964) found trithiane in cooked chicken volatiles. Minor *et al.* (1965c) detected thirteen sulfur compounds in chicken volatiles. Several researchers reported the release of substantial quantities of hydrogen sulfide during cooking of chicken (Mecchi *et al.* 1964; Klose 1965; Minor *et al.* 1965b). Pippen and Mecchi (1969) reported hydrogen sulfide concentrations of 180–730 ppb in simmered, roasted, and fried chicken while only 35 ppb chicken broth. The human threshold level of detection of hydrogen sulfide in broth is 10 ppb, thus is likely to contribute strongly to chicken flavor. While storage, freezing, and reheating can reduce hydrogen sulfide to below threshold levels (Pippen & Mecchi 1969), possible interaction with carbonyl compounds may give a more stable chicken flavor. The possible interaction of hydrogen sulfide with carbonyls was demonstrated by Pippen and Mecchi (1969) who blew hydrogen sulfide gas over chicken fat and found the resulting fat high in fixed sulfur and quite odorous. Several researchers have found that sulfur compounds decompose during cooking (Balance 1961; Mecchi *et al.* 1964) and that compounds released during cooking may dissolve in the fat yielding a characteristic flavor. For example, trithiane can be formed in cooked chicken from a reaction between hydrogen sulfide and formaldehyde (Mecchi *et al.* 1964). Swoboda (1970) detected dimethyltrisulfide at 10 ppb in roasted chicken leg meat and concluded it was a strong flavor component. A variety of sulfur compounds in chicken volatiles have been subsequently reported by various researchers (Hobson-Frohock 1970; Janney *et al.* 1974; Horvat 1976). Farbood and MacNeil (1979) traced the volatile sulfur compounds produced from methionine using radioactive labeling and determined that 90.7% were trapped in the protein fraction of the chicken meat slurry and 8.9% held with the lipid fraction. Bouthilet (1951b) concluded glutathione was the precursor for hydrogen sulfide. Pippen (1967) stated that poultry meat contains several sulfur-containing compounds including cystine, cysteine, methionine, taurine, and glutathione and any of these could be the source of volatile sulfur compounds. Biotin, thiamine, and coenzyme A are other sulfur-containing compounds found naturally in poultry meat, but there is little evidence that these contribute significantly to sulfur flavor volatiles. Hydrogen sulfide was produced at 200–1000 ppb during oven cooking of chicken at 70°C–125°C (Parr & Levett 1969). These researchers further stated that cysteine and cystine in protein provided 90% of the sulfur for hydrogen sulfide production and that cysteine in glutathione was a minor contributor. Mecchi *et al.* (1964) also concluded that cysteine was the major sulfur contributor for hydrogen sulfide and that leg muscle protein produced hydrogen sulfide 80% faster than did whole leg muscle. Leg meat nonprotein nitrogen did not produce hydrogen sulfide until after 2.5 h of boiling and the glutathione was the precursor for less than 9% of the hydrogen sulfide formed during boiling. The formation of sulfur compounds was shown to occur from the primary and secondary reactions of cysteine degradation products (Boelens *et al.* 1975). One example is the formation of thiozoles, oxazoles, imadazoles, oxothio ketones, mercapto ketones, disulfide ketones, and 2,5-dialklthiophenes from the reaction of diketones with ethyl alcohol, hydrogen sulfide, and ammonia (Boelens *et al.* 1975).

Gasser and Grosch (1990) identified 2-methyl-3-furanthiol as an important compound contributing to the meaty flavor of chicken broth as well as for beef (Gasser & Grosch 1988) and canned tuna (Withycombe & Mussinan 1988). The furan ring structure can be formed by the cyclization of a 5-carbon reductone. Reductones are common intermediates of Maillard-type reactions and their formation is enhanced by heating. The first and fifth carbons of the reductone link via the oxygen to the furan ring structure (Fig. 23.1). A related Maillard pathway, the Strecker degradation of methionine, can lead to the formation of methional and dimethyl disulfide (Pippen & Mecchi 1969). In fact, the Strecker degradation has been identified as the likely pathway for the production of carbonyl flavor compounds involving amino acids (Schonberg & Maubacher 1952). Ambler (1929) and Schonberg and Maubacher (1952)

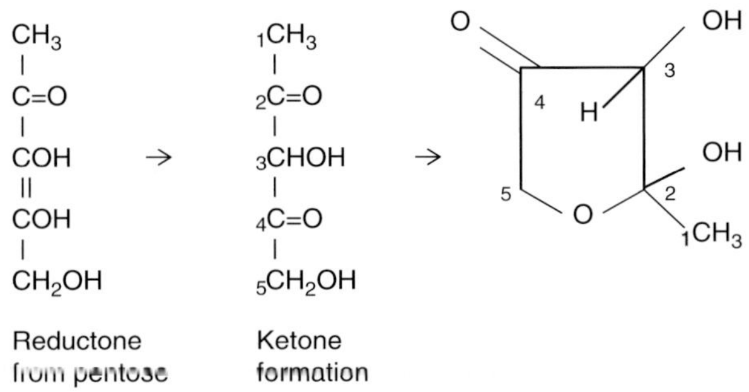

Fig. 23.1 Pathway for formation of methyl-furan.

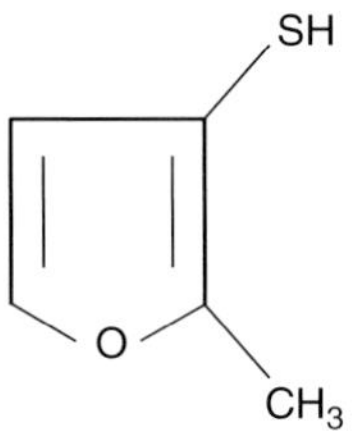

Fig. 23.2 Structure of 2-methyl-3-furanthiol.

demonstrated that acetaldehyde could be formed via the Strecker degradation of alanine, aspargine, or aspartic acid. The presence of thiols and sulfides linked to alkyl groups suggest that sulfur-containing amino acids undergo breakdown by more complex mechanisms than the Strecker degradation (Pippen 1967).

2-methyl-3-furanthiol (Fig. 23.2) and a similarly structured dimer have been created by heating thiamine and cysteine hydrochloride with hydrolyzed vegetable protein (Evers *et al.* 1976). These meat flavor compounds have also been detected in volatiles from heated yeast extract (Ames & MacLeod 1985) and heated thiamine (van der Linde *et al.* 1979; Hartman *et al.* 1984; Reineccius and Leardon 1985), verifying that thiamine is one likely source of freshly cooked meat flavor. In fact, Giacino (1968) patented a method to produce chicken flavor by heating thiamine with sulfur-containing polypeptides and alkanones followed by addition of hexanal and diacetyl. Simple sugars (i.e., ribose) and nucleic acids (i.e., inosine monophosphate [IMP]) when reacted with cysteine or glutathione yielded 2-methyl-3-furanthiol. The pathway to forming 2-methyl-3-furanthiol and its dimer bis(2-methyl-3-furanthiol)disulfide can occur when heating liberates hydrogen sulfide from either cysteine or glutathione which in turn reacts with a 5-carbon ketone from IMP or ribose (Fig. 23.3). Alaini and Farmer (2005b) studied thiamine, inosine 5′-monophosphate, ribose, ribose-5-phosphate, glucose, and glucose-6-phosphate as possible chicken flavor precursors and reported that sensory panelists detected an increase in flavors and aromas associated with chicken when ribose levels were increased two- to fourfold. GC odor assessment by GC MS of volatiles found that ribose increased odors described as "roasted" and "chicken" and that the changes in odor due to additional ribose are probably caused by elevated concentrations of compounds such as 2-furanmethanethiol, 2-methyl-3-furanthiol, and 3-methylthiopropanal. Glutathione is a major source of hydrogen sulfide in the initial stages of heating (Ohloff *et al.* 1985), while cysteine is a major contributor of hydrogen sulfide during long-term cooking (Mecchi *et al.* 1964).

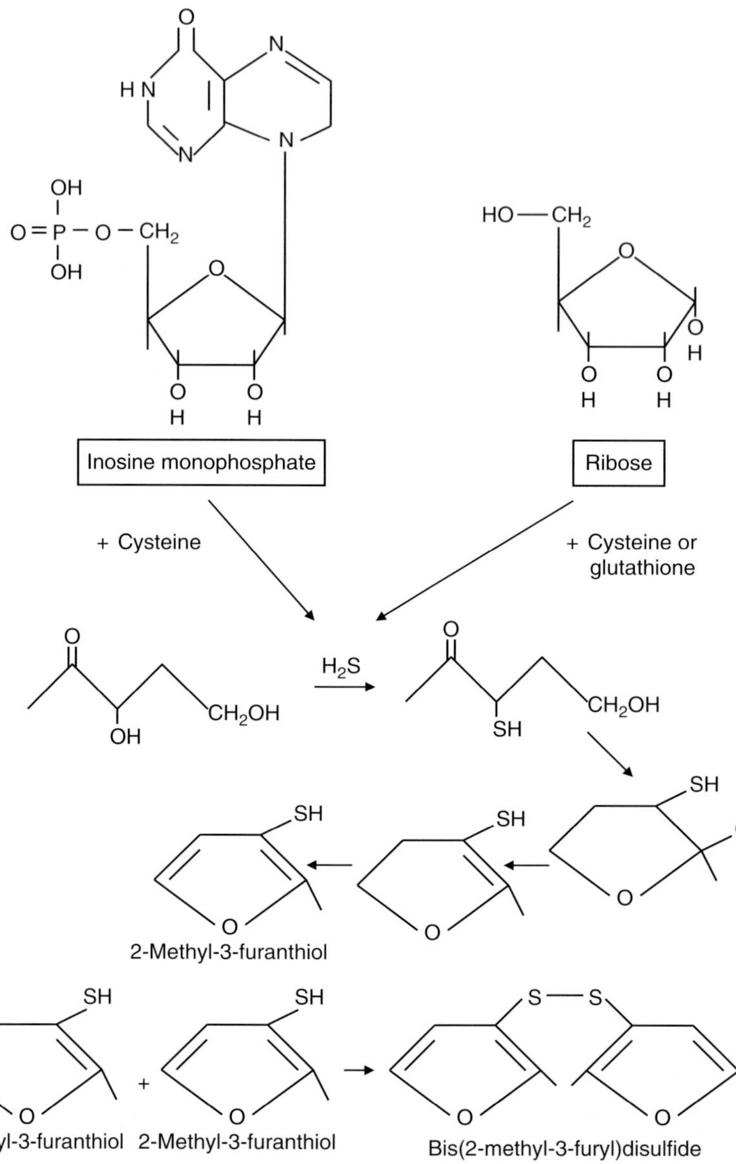

Fig. 23.3 Pathway to formation of 2- methyl-3-furanthiol and dimer from IMP and ribose with cysteine and glutathione.

23.4 CARBONYL COMPOUNDS IN FLAVOR

Carbonyls are a diverse group of compounds that include ketones, aldehydes, and acids. Compounds containing a carbonyl functional group have been considered as a major source of poultry meat flavors (Lineweaver & Pippen 1961; Minor *et al.* 1965c). The carbonyl flavor compounds are derived from oxidative reactions of lipids which are promoted by cooking and storage. Thus, the type and amount of carbonyls produced from poultry meat are affected by type of heating and length of storage. The development of oxidized carbonyls can be somewhat controlled by the addition of antioxidants and by exclusion of oxygen via packaging. The single-digit ppb threshold values of some carbonyls (Lea & Swoboda 1958; Meijboom 1964) make them potent flavor compounds. Meijboom (1964) associated the flavor perception of carbonyls to their chemical structures, degree of saturation, and chain length. For example, 2-trans-alkenals had higher threshold values than their saturated counterparts, and aldehydes with even number of carbons had a higher threshold value than ones with an odd number of carbons. Specific odor volatiles in turkey-skin aroma were linked to several flavor/aroma descriptors by expert sensory panelists (Table 23.1) (MacNeil & Dimick 1970a). The conversion of acetoin to diacetyl was reported to create a freshly cooked chicken aroma (Pippen *et al.* 1960) since acetoin (Fig. 23.4a) presence did not contribute to chicken broth odor while diacetyl (Fig. 23.4b) made a significant contribution to freshly cooked chicken flavor. Narasimhan *et al.* (1993) reported isolating 1,2-dibutylcyclopentane, 2,6-bis(1,1,-dimethylethyl)-4-methylphenol, and 1,12-dodecadiol as unique chicken flavor compounds compared to beef and pork.

Carbonyl compounds are also a source of oxidized off-flavors or rancidity of chicken. Nutter *et al.* (1943) found raw, frozen turkey fat became rancid in 3 weeks while chicken fat was stable up to 4 months based on measurements of fatty acid content, acetyl value, iodine value, and thiocyanogen numbers. Fat stability also varies according to the part of the animal from which the fat was derived (Pool *et al.* 1950), with back fat, neck fat, and visceral fat sequentially decreasing in stability (most to least) (Klose *et al.* 1951). The formation and content of specific aldehydes (malondialdehyde and hexanal) have been employed to measure the degree of meat oxidation. The thiobarbituric acid (TBA) value (also known as thiobarbituric acid reactive substances (TBARS)) and the malondialdehyde value (MDA) are chemical tests measuring the presence of malondialdehyde in meat and other substances including human blood. Harris and Lindsay (1972) reported that off-flavor development in cooked chicken was linked to malondialdehyde formation using the TBA test. Furthermore, a trained sensory panel rejection of stored poultry skin was directly related to high levels of 2-enal and 2,4-dienal carbonyls (MacNeil & Dimick 1970b). By capturing volatiles in air bubbled through a heated mixture of chicken meat and water, Pippen *et al.* (1958) identified diacetyl, acetone, (C_2–C_6, C_8, C_9) saturated aldehydes, (C_5–C_7, C_{10}, C_{11}) 2-alkenals, and 2,4-heptadienal. Subsequently,

Table 23.1 Response of expert sensory panelists to specific volatile carbonyls generated from turkey skin.

Turkey-skin volatile	Expert sensory panel descriptor
Methyl ketone	Oily, minty
Alkanal	Meaty, turkey-like
Alk-2-enal	Strong, oxidized, broth-like
Alk-2,4-dienal	Strong, painty, nutmeg-like, spicy

Sources: Adapted from MacNeil and Dimick (1970a).

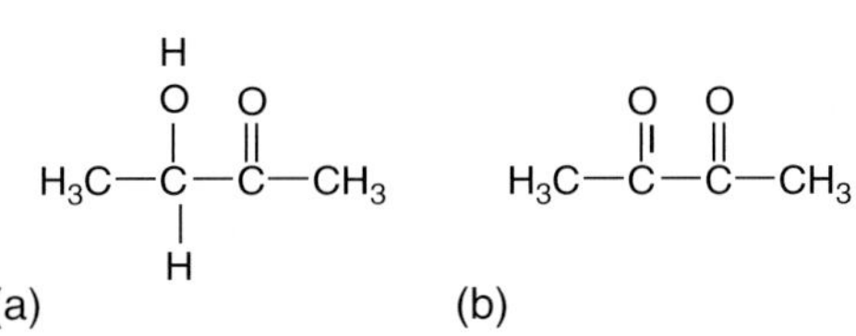

Fig. 23.4 Structure of acetion (a) and diacetyl (b).

Pippen *et al.* (1960) identified many more carbonyls in a steam distillate from fresh chicken including 2-alkanones and 1,2-dicarbonyls with the predominant compounds being diacetyl, acetaldehyde, n-hexanal, and 2,4-decadienal. In a later study, Pippen and Nanoka (1963) reported that when chicken and turkey meat was cooked in a nitrogen atmosphere, fewer carbonyls and smaller amounts of those produced were detected compared to turkey meat cooked in an air atmosphere. Furthermore, rancid chicken meat contained the same volatiles but in larger quantities than fresh chicken meat. Rao *et al.* (1976) found the addition of polyphosphates to poultry meat prior to cooking could reduce the amount of carbonyls (mostly dicarbonyls, methyl ketones, 2,4-dienals) produced.

23.5 PRODUCTION/PROCESSING FACTORS

Various production and processing factors can affect the flavor of poultry in addition to the complexity of flavor in culinary applications. Production factors include animal diet, age, gender, and genetics, while muscle type, chilling method, thermal processing, irradiation, storage, antioxidants, and microbiology are some processing factors affecting poultry flavor. While animal genetics has some effect on the flavor of meat, it is minimal when compared to other production and processing factors. Hanson *et al.* (1960) demonstrated this by making over 600 comparisons between "slow-growing" 1930s genetic strains and "faster-growing" 1950s broilers and reporting no difference on flavor.

23.5.1 Microbiological effects on flavor

The microbiological content and makeup of meat can have a significant effect on flavor during storage (Carlin *et al.* 1957; Adamcic & Clark 1970; Freeman *et al.* 1976). In fact, Harris *et al.* (1968, 1970) reported significant flavor differences between germ-free and conventional chicken. Grey and Shrimpton (1967a) matched flavor compounds found in chicken breast muscle with ones found in the animal cecum (part of the digestive tract). Klinger and Basker (1977) found that chickens fed 1,3, b-p-chlorobenzylidine aminoguanidine hydrochloride (an anticoccidial agent) from birth to slaughter had a longer storage life and better flavor than controls. Sheldon (1979) further demonstrated that chicken meat flavor could be altered by changing the intestinal flora and resulting microbial fatty acid production via antibiotic administration during growth.

Several studies have confirmed that poultry meat spoilage is largely due to Gram-negative, psychrotrophic bacteria (McMeekin 1975, 1977; Daud *et al.* 1979). Hawke (1966) established that bacteria could produce methyl ketones from lipids. Aerobic lipophilic bacteria with the capacity to hydrolyze lipids including *Pseudomonas* spp., *Achromobacter* spp., and *Micrococci* spp. have all been associated with poultry meat (Frazier 1967; Jay 2000). Microbial enzymes may also degrade lipids (Smith & Alford 1968) resulting in carbonyl off-flavor compounds. Off-flavor due to bacteria has been linked to chicken breast and leg meat (McKeekin 1975) and to chicken skin (Daud *et al.* 1979). McKeekin and Patterson (1975) linked hydrogen sulfide-producing bacteria to chicken-processing plants and Freeman *et al.* (1976) associated spoilage odor volatiles to refrigerated chicken. *Pseudomonas* spp. has been identified as a major spoilage strain linked to poultry meat and several researchers have isolated volatile compounds from pure *Pseudomonas* spp. cultures. For example, Freeman *et al.* (1976) detected dimethyl sulfide, dimethyl disulfide, and methyl thiolacetate from a *P. fluorescens* culture grown on previously irradiated chicken meat. In this study, dimethyl disulfide was the only sulfur compound detected from sterile chicken meat when inoculated with *P. putida*. Lee *et al.* (1979) reported that acetone, 2-butanone, methyl thiolacetate, dimethyl disulfide, toluene, dimethyl trisulfide, and 2-nonanone were found when *P. putida* was grown on either standard microbiological growth media or chicken breast meat and that acetone, toluene, methyl thiolacetate, dimethyl disulfide, dimethyl trisulfide, 2-nonanone, and 1-undecance were detected for *P. fluorescens* under the same conditions. Pittard *et al.* (1982) found that 12 of the same volatiles out of 25 total compounds were produced by four different fluorescent *Pseudomonas* strains and that only 4 compounds out of 25 were unique to one strain.

23.5.2 Animal dietary effects on flavor

Animal diet has a strong impact on poultry flavor. Objectively, animal dietary effect on meat flavor is dramatic when the flavor of wild game birds is compared to the flavor of birds grown under controlled dietary conditions. The effect of diet on flavor was documented as early as 1935 by Maw who reported that poultry fed a corn diet were more flavorful than those fed a diet consisting of barley, oats, and wheat. The addition of dried chicken feces (Lewis 1955) or dairy products (Weisberg 1956) to broiler feed was reported to produce a more flavorful meat.

Considerable work has been reported on off-flavor, especially related to "fishy" off-flavors. The transfer of fishy off-flavor from dietary fish meal or fish oil to poultry meat was observed many years ago (Cruickshank 1939; Murphy 1939). Asmundson *et al.* (1938) found that 25% fish meal in turkey diets did not result in a meat with fishy taste but that 2–5% fish oil added to turkey diets did result in meat with a fishy flavor. The fishy flavor in poultry meat has been attributed to the inclusion of highly unsaturated fatty acids (three or more double bonds). These highly unsaturated fatty acids are metabolized and deposited in the fatty tissue of the animal that when cooked produce fishy flavor volatiles (Klose *et al.* 1953). Klose *et al.* (1951) found that feeding linseed oil (highly unsaturated fat) to turkeys resulted in cooked meat with a fishy off-flavor. Crawford *et al.* (1975) further reported that fish off-flavor in turkey meat could be induced by feeding turkeys highly unsaturated fatty acids and later it was determined that these off-flavors developed during cooking (Crawford & Kretsch 1979). Crawford and Kretsch (1976) identified 71 unique volatile compounds in turkeys fed with tuna oil and that 21 of those were likely to be the cause of fishy off-flavor. Since the fishy flavors were a result of the oxidative breakdown products of unsaturated fatty acids, fishy flavor in poultry fed highly unsaturated fatty acids could be reduced by the addition of antioxidants (α-tocopherol) (Crawford *et al.* 1975). Sheldon *et al.* (1997) studied the effects of various dietary vitamin E levels on the oxidative stability, flavor, color, and volatile compounds of refrigerated and frozen turkey meat. Turkeys were fed 5, 10, or 25 times the National Research Council recommended amounts, which are 12 and 10 IU/kg body weight from 0 to 8 and 9 to 18 weeks of age, respectively. Oxidation rate was inversely proportional to dietary vitamin E as indicated by TBA values and sensory panels.

Animal di*et al*terations other than degree of unsaturated fatty acids or antioxidants in the diet of poultry have little effect on meat flavor. Variations in poultry rations using 12.5% raw peanuts (Offiong 1976), 45% field beans (Grey *et al.* 1972), or 38.2% dried poultry waste (Cunningham 1976) had no significant effect on cooked meat flavor. Furthermore, Carlson *et al.* (1962) found no difference in the flavor of meat from turkeys fed varying levels of fat. Thus, the percentage of highly unsaturated fat in poultry feed is the only dietary factor playing a significant role in poultry meat flavor. Long-chain unsaturated fatty acids such as those found in fish have been linked to improved cardiovascular health and addition of these to meat has been accomplished through animal diets. Van Elswyk *et al.* (1995) compared the incorporation of docosohexanoic acid (DHA) into chicken meat by adding the fatty acid to the diet as menhaden oil or as natural marine algae. Meat from broilers fed the algae source of DHA was more acceptable and stable than meat from birds receiving feed supplemented with menhaden oil.

23.5.3 Effect of animal gender and age on flavor

There are conflicting reports as to the effect of gender on poultry meat flavor, and the age of the animal sometimes confounds these results. For instance, Sweetman and MacKellar (1954) found that meat from male birds had stronger flavor compared to meat from female birds, while Carlson *et al.* (1962) reported no flavor difference in meat from male and female turkeys. Fry *et al.* (1958) found chicken broth flavor was affected by the gender of the meat source; for chickens 10–14 weeks old, however, no measurable difference in this meat was found when baked. Fry *et al.* (1958) also reported flavor differences in broth and baked chicken due to hormone levels in the meat from 6- to 14-week-old birds. MacNeil and Dimick (1970a) reported higher carbonyl levels in meat from male turkeys compared to female turkeys.

Meat from older chickens was found to be more flavorful (Sweetman & MacKellar 1954; Vail *et al.* 1967). This was supported by Minor *et al.* (1965a) who reported meat from older chickens had higher concentrations of flavor compounds compared to meat from younger birds. Between 9 week-, 19 week-, and 28-month-old chickens, Wells *et al.* (1962) reported that cooked, freeze-dried meat from the 28-month-old chickens was most flavorful. The generation conclusion is that meat from older chickens and turkeys tends to taste more flavorful.

23.6 OXIDATION-DERIVED FLAVORS

Off-flavors via lipid oxidation are a common concern in poultry products due to the relatively high degree of unsaturation of the fatty acids compared to other meats. Turkey meat is even less stable than chicken (Wilson *et al.* 1976; Einerson & Reineccius 1977), partly due to the lower levels of natural antioxidants in turkey lipids (Nutter *et al.* 1943), even when animals are fed the same amount of tocopherol (Mecchi *et al.* 1956). Criddle and Morgan (1951) reported that tocopherol content in turkey meat could be greatly increased by feeding tocopherol during growth. WOF was a phrase introduced by Tims and Watts (1958) to describe the rapid onset of rancidity in cooked meats detected when reheated after refrigerated storage. The main components in meat that promote the onset of oxidation leading to rancidity are heme, nonheme iron, unsaturated fatty acids, and phospholipids. St. Angelo *et al.* (1988) identified volatile compounds responsible for WOF as hexanal (grassy), propanal (alcoholic), pentanal (pungent), 2,3-octadione, nonanal (soapy), and 2-pentylfuran. Johnson and Civille (1986) and later Byrne *et al.* (2002) found a decrease in the meaty and sweet notes on cooked refrigerated chicken meat and an increase in WOF notes such as cardboard, linseed oil, rancid, and sulfur/rubber. Hexanal has been identified as an indicator of oxidized chicken meat using GC (Goodridge *et al.* 2003). Byrne *et al.* (2002) reported compounds as products of different reactions including 3-methylbutanal (malty), 2-methylbutanal (roasted corn), and 3-methylthiopropanal from the Strecker degradation; dimethly trisulfide (garlic) and dimethyl tetrasulfide (cabbage) from thermal sulfur amino acid degradation. Ruenger *et al.* (1978) reported that WOF in cooked turkey was detected equally by sensory panelists in dark and light meat contradictory to Wilson *et al.* (1976) who reported that dark meat developed WOF to a greater degree than light meat. Ruenger *et al.* (1978) further found that three compounds increased in reheated samples. Two of the compounds were identified as heptaldehyde (flavor characterized as harsh, pungent, and unpleasant fatty) and n-nona-3,6-dienal (dienal flavors characterized as fishy).

Mechanically deboned chicken meat (MDCM) is particularly susceptible to rancidity (Dimick *et al.* 1972) due to mixing of the meat with bone marrow and cellular constituents during deboning, in particular the heme pigments (Froning & Johnson 1973). Mixing of meat tissue with air at this point will also contribute to accelerated oxidation (Dawson *et al.* 1990). Oxidation of MDCM results in off-flavors that will carry through to the final product, making oxidized meat less valuable.

Poultry by-product meal is used in pet foods and is derived from the rendering of parts of the animal not typically used for human consumption (head, viscera, legs, feathers). These parts are prepared into a dry meal via the rendering process and are an inexpensive source of protein. Greenberg (1981) characterized the flavor volatiles from poultry by-product meal and concluded that the majority of the flavor volatiles in the meal were derived from lipid oxidation. These compounds are formed from the 14% fat contained in the final dried meal, and the formation of these compounds is promoted by the heat and oxidative environment used to render the meal safe for use as feed.

23.7 SMOKED MEAT FLAVORS

Smoking of foods was historically used to preserve meats by absorption of aromatics from the smoke into the meat. Smoked poultry meats are desired more for their flavor than for the preservation effect in modern times. Of the more than 300 compounds found in smoke, only a few have been found in smoked meats, reflecting the interaction between the smoke and the meat (Hamm 1977; Daun 1979; Clifford *et al.* 1980). In addition to the hydrocarbons, acids, ketones, and aldehydes found in cooked meat flavor, phenols are found in smoked meats and are responsible for smoked meat flavor (Sink 1979; Clifford *et al.* 1980; Shiau & Chai 1985; Hassan 1988). A specialty smoked meat item is duck and the subcutaneous fat covered fillet known in French cuisine as "magret." The unique flavor developed in this product is a result of smoking, lipid oxidation, and the compounds forming from compound interactions. Lesimple *et al.* (1995) identified 62 volatiles, 34 of which were related to smoke and 7 that had not been previously reported in smoked duck meat. The smoked meat contained 15 phenolic compounds, none of which were found in nonsmoked duck meat. A comparison of water-boiled and Cantonese-style roasted duck flavors revealed that roasting produced heterocyclic compounds such as pyrazines, pyridines, and thiazoles (Wu and Liou 1992), all of which are Maillard reaction products. Wu and Liou (1992) also reported that indole was a component of both water-boiled and roasted duck flavor and that indole is not found in most other meat flavors. Indole is responsible for floral notes at low concentrations (1000–0.2 ppm) and acts as a positive flavor promoter dependent on other flavor compounds at concentrations below 0.21 ppm (Arctander 1969). Indole is also a primary component of certain essential oils in flowers from certain plants (lemon, bitter orange, coffee, etc.) and in some teas (Furia & Bellanca 1979).

23.8 CULINARY ASPECTS OF COOKED POULTRY MEAT

The application of heat to food in various ways is a crude definition of cooking. Meat is cooked to render it safe, easier to digest, and more flavorful (McGee 1984). Cooking can be divided into three categories as dry heat, moist heat, and a combination of dry and moist heat

Table 23.2 Categories of different types of cooking.

Dry heat	Moist heat	Combination
Grilling or broiling	Steaming	Stewing
Barbecuing	Poaching	Braising
Roasting or baking	Simmering	
Poeleing	Boiling	
Sauteing		
Pan-frying		
Deep-fat frying		

with several different types of cooking possible within each category (Table 23.2). The dry heat methods produce a variety of results (Culinary Institute of America 1996). For example, grilling and broiling results in highly flavored exteriors with moist interiors, while roasted meats develop a rich roasted aroma, well-developed color, and tender texture. Sautéed foods develop a specific degree of browning, while pan-fried and deep-fried items have a crispy exterior. In general, dry heat does not tenderize meat to the degree moist heat does, thus is not normally used for tough meat cuts. Moist heat cooking can be used to tenderize meat and is combined with a flavorful liquid in the preparation. The combination of methods is sometimes called "slow cooking" with a preliminary searing step (dry heat) or conversely a blanching step (moist heat). The combination of the two methods develops complex flavor profiles not possible with single methods. Imafidon and Spanier (1994) linked the five basic meat tastes of sweet, salty, acid, bitter, and umami to acids, peptides, and salts: sweet (certain amino acids); bitter and umami (peptides and hypoxanthine); acid (lactic, inosinic, orthophosphoric, and succinic acids); sweet and salty (glucose and sodium salts of glutamic and aspartic acids).

The flavor of chicken allows for other flavors to develop and blend together. Chicken is used in numerous cultures\cuisines. It is extremely versatile due to the subtle profile it has and the manner in which flavors, from simple to complex, can be developed with it. Chicken can be prepared in many ways, including baking, broiling, boiling, roasting, frying, sautéing, braising, barbecuing, and stewing. Each of these means of preparation provides a different flavor profile, yet it is easily enhanced by any number of flavor combinations in any of the cooking techniques. Therefore, the cooked flavor of chicken is greatly influenced by numerous factors.

The cooked flavor of chicken, when not influenced by a cooking method or seasoning, is very pleasant, unlike poultry with more dominant flavors, such as duck, turkey, goose, and Cornish game hens. This cooked flavor can best be experienced when boiling fresh chicken with its bones. Meat that is held in refrigeration for too many days or thawed from the freezer has a discernable different flavor. Professional kitchens will often intensify the cooked poultry flavor notes. One method is to reduce poultry liquids, otherwise referred to as stocks and broths. These liquids are further enhanced by creating a thickening agent that is made from heated poultry fat.

The use of particular herbs and vegetables aid the common poultry flavor in the European and North American cultures. The herbs used are rosemary, thyme, marjoram, parsley, and sage, plus the addition of the vegetables consisting of carrots, onions, leeks, and celery.

23.9 CURED MEAT FLAVORS

According to Egyptian records, meat curing originated from the addition of impure salt to preserve meat. The salt contained nitrates from the soil which was converted to nitrite by bacteria and natural meat enzymes. In 1925, the USDA limited the level of nitrite to 200 ppm in the final product, and then in 1978 this level was lowered to 156 ppm before cooking. The regulations have further limited the level of sodium nitrite in some cured products to 120 ppm with 550 ppm ascorbate or erythorbate to reduce any residual nitrite. Sodium nitrate is allowed only for dry-cured or fermented meat products. Curing agents are added to meat one of three ways: directly as with frankfurters, externally on the surface as with bacon, or injected into the meat in a solution as with corned beef. Most cured products are cooked after the addition of nitrite functioning to inhibit outgrowth and toxin production by *Clostridium botulinum*, to develop cured meat color, and to create a unique cured meat flavor.

REFERENCES

Adamcic, M. & Clark, D.S. (1970) Bacteria-induced biochemical changes in chicken skin stored at 5 C. *Journal of Food Science*, **35**, 103–106.

Alaini, M. & Farmer, L.J. (2005a) Precursors of chicken flavor. I. Determination of some flavor precursors in chicken muscle. *Journal of Agricultural and Food Chemistry*, **53**, 6067–6072.

Alaini, M. & Farmer, L.J. (2005b) Precursors of chicken flavor. II. Identification of key flavor precursors using sensory methods. *Journal of Agricultural and Food Chemistry*, **53**, 6455–6452.

Ambler, J.A. (1929) The reaction between amino acids and glucose. *Indian Engineering Chemical Industry*, **21**, 47–51.

Ames, J.M. & MacLeod, G. (1985) Volatile components of a yeast extract composition. *Journal Food Science*, **50**, 125–131.

Arctander, S. (1969) Perfume and Flavor Chemicals (ed. S. Arctander). S. Arctander, Elizabeth, New Jersey.

Asmundson, V.S., Jukes, T.H., Fyler, H.M. & Maxwell, M.L. (1938) The effect of certain fish meals and fish oils in the ration on the flavor of turkey. *Poultry Science*, **17**, 147.

Balance, P.E. (1961) Production of volatile compounds related to the flavor of foods from the Strecker degradation of DL-methionine. *Journal of the Science of Food and Agriculture*, **12**, 532–535.

Boelens, H., Van der Linde, L.M., De Vallois, P.J., Van Dort, J.M. & Takken, H.J. (1975) Organic sulfur compounds as flavor compounds: Reaction products of carbonyl compounds, H_2S and NH_3. Aroma Research Proceedings of the International Symposium (eds P.H. Marse & P.J. Groeman), p. 75. Center Agricultural Publications & Document Publisher, Wageningen, the Netherlands.

Bouthilet, R.J. (1949) A note on the nature of a flavor constituent from poultry meat. *Food Technology*, **3**, 118–119.

Bouthilet, R.J. (1950) Chicken flavor: Separation and concentration of its volatile components from broth. *Food Research*, **15**, 322–325.

Bouthilet, R.J. (1951a) Chicken flavor: The fractionation of the volatile constituents. *Food Research*, **16**, 137–141.

Bouthilet, R.J. (1951b) Chicken flavor: The source of the meat flavor component. *Food Research*, **16**, 201–204.

Byrne, D.V., Bredie, W.L.P., Mottram, D.S. & Martens, M. (2002) Development of a sensory vocabulary for warmed-over-flavor: Part II. *In chicken meat. Journal of Sensory Studies*, **14**, 67–78.

Carlin, A.F., Holl, B.E. & Walker, H.M. (1957) Correlation between flavor and number of micro-organisms associated with eviscerated chicken treated with chlortetracycline. *Food Technology*, **11**, 573–577.

Carlson, C.W., Adams, A.W., Wilcox, R.A., Gastler, G.F. & Burrill, L.M. (1962) Dietary energy, sex, strain and storage influencing composition and/or palatability of broad breasted bronze turkeys. *Poultry Science*, **41**, 150–160.

Chen, C.W. & Ho, C.T (1994) The flavour of poultry meat. In: *Flavor of Meat, Meat Products and Seafoods* (ed. F. Shahidi), 2nd edn., pp. 84–98. Blackie Academic & Professional, Chapman & Hall, New York.

Clifford, M.N., Tang, S.L. & Eyo, A.A. (1980) Smoking of foods. *Process Biochemistry*, **15**, 8,10,11,17,26.

Crawford, L. & Kretsch, M.J. (1976) GC-MS identification of the volatile components extracted from roasted turkeys from a basal diet supplemented with tuna oil: Some comments on fishy flavor. *Journal of Food Science*, **41**, 1470–1478.

Crawford, L. & Kretsch, M.J. (1979) The effects of cooking in air or nitrogen on the development of fishy flavor in the breast meat of turkeys fed tuna oil with or without α-tocopherol supplement or injection. *Fishery Bulletin*, **74**, 89–91.

Crawford, L., Kretsch, M.J., Peterson, D.W. & Lilyblade, A.L. (1975) The remedial and preventative effect of dietary α-tocopherol on the development of fish flavor in turkey meat. *Journal of Food Science*, **40**, 751–755.

Criddle, J.E. & Morgan, A.F. (1951) Effect of tocopherol feeding on the composition of turkey tissues. In: *Proceedings of the Society of Experimental Biology*, **78**, 41–45.

Crocker, E.C. (1948) Flavor of meat. *Food Research*, **13**, 179–183.

Cruickshank, E.M. (1939) The effect of cod liver oil and fish meal on the flavor of poultry products. In: *Proceedings of the 7th World Poultry Congress,* Cleveland, Ohio. Waverly Press, Baltimore, Maryland.

Culinary Institute of America. (1996) *The New Professional Chef*, 6th edn. John Wiley & Sons Incorporated, New York, pp. 302–352.

Cunningham, F.E. (1976) Using dehydrated poultry waste in poultry rations—A review. *Agricultural Environment*, **3** (1), 69–79.

Daud, A.B., McMeekin, T.A. & Thomas, C.J. (1979) Spoilage association of chicken skin. *Applied and Environmental Microbiology*, **37**, 399–401.

Daun, H. (1979) Interaction of wood smoke components and food. *Food Technology*, 33, 66,68,70,71,83.

Dawson, P.L., Sheldon, B.W., Ball, H.R. & Larick, D.K. (1990) Changes in the neutral lipid and phospholipid fractions of mechanically deboned chicken meat due to washing, cooking and storage. *Poultry Science*, **69**, 166–175.

Dimick, P.S., MacNeil, J.H. & Grunden, L.E. (1972) Poultry product quality: Carbonyl composition and organoleptic evaluation of mechanically deboned poultry meat. *Journal of Food Science*, **37**, 544–546.

Einerson, M.A. & Reineccius, E.A. (1977) Inhibition of warmed over flavor in retorted turkey by antioxidants formed during processing. *Journal of Food Processing and Preservation*, **11**, 279–283.

Evers, W.J., Heinsohn, H.H., Mayers, B.J. & Sanderson, A. (1976) Furans substituted at the three position with sulfur. In: *Phenolic, Sulfur and Nitrogen Compounds in Food Flavors* (eds G. Charalambous & I. Katz), pp. 184–193. American Chemical Society, Washington, District of Columbia.

Farbood, M.I. & MacNeil, J.H. (1979) Retention and stability of volatile sulfur containing compounds in chicken meat. *Journal of Food Science*, **44**, 652–655.

Frazier, W.C. (1967) *Food Microbiology*, 2nd edn. McGraw-Hill Book Company, New York.

Freeman, L.R., Silverman, G.J., Angelini, P., Merritt, C., Jr. & Esselen, W.B. (1976) Volatiles produced by microorganisms isolated from refrigerated chicken at spoilage. *Applied and Environmental Microbiology*, **32**, 222–231.

Froning, G.W. & Johnson, F. (1973) Improving the quality of mechanically deboned fowl meat by centrifugation. *Journal of Food Science*, **38**, 279–281.

Fry, J.L., Bennet, G. & Stadelman, W.J. (1958) The effect of age, sex and harmonization on the flavor of chicken meat. *Poultry Science*, **37**, 331–335.

Furia, T.E. & Bellanca, N. (1979) *Fenaroli's Handbook of Flavor Ingredients*, 2nd edn. CRC Press, Cleveland, Ohio.

Gasser, U. & Grosch, W. (1988) Identification of volatile flavour compounds with high aroma values from cooked beef. *Z Lebensm Unters Forsch*, **186**, 489–494.

Gasser, U. & Grosch, W. (1990) Primary odorants of chicken broth. *Z Lebensm Unters Forsch*, **190**, 3–8.

Giacino, C. (1968) Poultry flavor composition and process. US Patent 3,394,017.

Goodridge, C.F., Beaudry, R.M., Pestka, J.J. & Smith, D.M. (2003) Solid-phase microextraction-gas chromatography for quantifying headspace hexanal above freeze-dried chicken myofibrils. *Journal of Agricultural and Food Chemistry*, **51**, 4185–4190.

Greenberg, M. (1981) Characterization of poultry byproduct meal flavor volatiles. *Journal of Agricultural and Food Chemistry*, **29**, 831–834.

Grey, T.C. & Shrimpton, D.H. (1967a) Volatile components of raw chicken breast muscle. *British Poultry Science*, **8**, 23–33.

Grey, T.C. & Shrimpton, D.H. (1967b) Volatile components in the breast muscle of chickens of different ages. *British Poultry Science*, **8**, 35–41.

Grey, T.C., Griffiths, N.M., Wilson, B.J. & Bolton, W. (1972) The effect of field beans on growth and flavor of broilers. *British Poultry Science*, **13**, 7.

Hamm, R. (1977) Analysis of smoke and smoked foods. *Pure Applied Chemistry*, **49**, 1655–1666.

Hanson, H.L., Campbell, A.A., Kraft, A.A., Gilan, G.L. & Harkin, A.M. (1960) The flavor of modern and old type chickens. *Poultry Science*, **38**, 1071–1076.

Harkes, P.D. & Begemann, W.J. (1974) Identification of some previously unknown aldehydes in cooked chicken. *Journal of American Oil Chemists Society*, **51**, 356–359.

Harris, N.D. & Lindsay, R.C. (1972) Flavor changes in reheated chicken. *Journal of Food Science*, **37**, 19–22.

Harris, N.D., Strong, D.H. & Sunde, M.L. (1968) Intestinal flora and chicken flavor. *Journal of Food Science*, **33**, 543–547.

Harris, N.D., Strong, D.H. & Sunde, M.L. (1970) Comparison of headspace volatiles of germfree-reared and conventionally-reared chickens. *Poultry Science*, **49**, 1597–1599.

Hartman, G.J., Carlin, J.T., Scheide, J.D. & Ho, C.T. (1984) Volatile products formed from the thermal degradation of thiamine at high and low moisture levels. *Journal of Agriculture and Food Chemistry*, **32**, 1015–1018.

Hassan, I.M. (1988). Processing of smoked common carp fish and its relation to some chemical, physical and organoleptic properties. *Food Chemistry*, **27**, 95–106.

Hawke, J.C. (1966) The formation and metabolism of methyl ketones and related compounds. *Journal of Dairy Research*, **33**, 225–228.

Hobson-Frohock, A. (1970) Use of molecular sieve 5A in the analysis of the flavour volatiles of cooked chicken. *Journal of Agriculture and Food Chemistry*, **21**, 152–155.

Horvat, R.J. (1976) Identification of some volatile compounds in cooked chicken. *Journal of Agriculture and Food Chemistry*, **24**, 953–958.

Howe, P.E. & Barbella, N.G. (1937) The flavor of meat and meat products. *Food Research*, **2**, 197–202.

Imafidon, I. & Spanier, A.M. (1994) Unraveling the secret of meat flavor. *Trends in Food Science and Technology*, **5**, 315–321.

James, A.T. & Martin, A.J.P. (1952) Gas-liquid partition chromatography: The separation and micro estimation of volatile fatty acids from formic acid to dodecanoic acid. *Biochemistry Journal*, **50**, 679–692.

Janney, C.G., Kale, K.K., Jr. & Higman, H.C. (1974) The identification of fried chicken volatiles by gas chromatographic and mass spectral analysis. *Poultry Science*, **53**, 1758–1761.

Jay, J. (2000) Microorganisms in food. In: Modern Food Microbiology (ed. J. Jay), 6th edn., pp. 57–60. Aspen Publishers, Inc., Gaithersburg, Maryland.

Johnson, P.B. & Civille, G.V. (1986) A standardized lexicon of meat WOF descriptors. *Journal of Sensory Science*, **1**, 99–104.

Klinger, I. & Baker, D. (1977) The effect of feed containing cycostat on the taste of broiler chickens. *Refuah Veterinary*, **34**, 21–25.

Klose, A.A. (1965) Direct olfactory demonstration of major fractions of chicken aroma. *25th Annual Meeting of Institute of Food Technologists*, Abstracts, 25, 97.

Klose, A.A., Mecchi, E.P., Hanson, H.L. & Lineweaver, H. (1951) The role of dietary fat in the quality of fresh and frozen storage turkeys. *Journal of American Oil Chemists Society*, **28**, 162–165.

Klose, A.A., Hanson, H.L., Mecchi, E.P., Anderson, J.H., Streeter, I.V. & Lineweaver, H. (1953) Quality and stability of turkeys as a function of dietary fat. *Poultry Science*, **32**, 82–88.

Lea, C.H. & Swoboda, P.A.T. (1958) The flavor of aliphatic aldehydes. *Chemical Industry*, **4**, 1289–1290.

Lee, M.L., Smith, D.L. & Freeman, L.R. (1979) High-resolution gas chromatographic profiles of volatile compounds produced by microorganisms at refrigerated temperatures. *Applied and Environmental Microbiology*, **37**, 85–90.

Lesimple, S., Torres, L., Mitjavila, S., Fernandez, Y. & Durand, L. (1995) Volatile compounds in processed duck fillet. *Journal of Food Science*, **60**, 615–618.

Lewis, R.W. (1955) The diet's effect on flavor of chicken. MSc thesis # 13,027. Kansas State College dissertation abstract 15, 1827.

van der Linde, L.M., van Dort, J.M., de Valois, P., Boelens, H. & de Rijke, D. (1979) Volatile components from thermally degraded thiamin. In: *Progress in Flavour Research* (eds D.G. Land & H.E. Nurten), pp. 214–219. Applied Science, London, U.K.

Lineweaver, H. & Pippen, E.L. (1961) Chicken flavor. In: Proceedings of the Flavor Chemistry Symposium, pp. 21–33. Campbell Soups Co., Candem, New Jersey.

Lineweaver, H., Pippen, E.L. & Nonaka, M. (1962) Chromatography of chicken flavor volatiles. In: *12th World's Poultry Congress Proceedings*, pp. 405–408.

Maarse, H. (1984) Volatiles components in food-Qualitative data. TNO, Supplement 1. *Chicken (heated)*, 1, 57.1–57.5.

MacNeil, J.H. & Dimick, P.S. (1970a) Poultry product quality. 1. Compositional changes during cooking of turkey roasts. *Journal of Food Science*, **35**, 184–186.

MacNeil, J.H. & Dimick, P.S. (1970b) Poultry product quality. 3. Organoleptic evaluation of cooked chicken and turkey skin fractions as affected by storage time and temperature. *Journal of Food Science*, **35**, 191–194.

Maw, W.A. (1935) How quality in meat is affected by fat distribution in the carcass. *U.S. Poultry Magazine*, **41**, 32.

McGee, H. (1984) *Food and Cooking. The Science and Love of the Kitchen*, pp. 87–122. Charles Scribner's & Sons, New York.

McMeekin, T.A. (1975) Spoilage association with chicken breast muscle. *Applied Microbiology*, **29**, 44–47.

McMeekin, T.A. (1977) Spoilage association with chicken leg muscle. *Applied and Environmental Microbiology*, **33**, 1244–1246.

McMeekin, T.A. & Patterson, J.T. (1975) Characterization of hydrogen sulfide producing bacteria isolated from meat and poultry plants. *Applied Microbiology*, **29**, 165.

Mecchi, E.P., Pool, M.F., Behman, G.A., Hamachi, M. & Klose, A.A. (1956) The role of tocopherol content in the comparative stability of chicken and turkey fat. *Poultry Science*, **35**, 1238–1246.

Mecchi, E.P., Pippen, E.L. & Lineweaver, H. (1964) Origin of hydrogen sulfide in heated chicken muscle. *Journal of Food Science*, **29**, 393–399.

Meijboom, P.W. (1964) Relationship between molecular structure and flavor perceptibility of aliphatic aldehydes. *Journal of American Oil Chemists Society*, **41**, 326–329.

Minor, L.J., Pearson, A.M., Dawson, L.E. & Scheigert, B.S. (1965a) Gas chromatographic analysis of volatile constituents from cooked chicken carcasses of old and young chickens. *Poultry Science*, **44**, 535–543.

Minor, L.J., Pearson, A.M., Dawson, L.E. & Scheigert, B.S. (1965b) Chicken flavor and aroma: Separation and identification of carbonyl and sulfur compounds in the volatile fraction of cooked chicken. *Journal of Agricultural and Food Chemistry*, **13**, 298–300.

Minor, L.J., Pearson, A.M., Dawson, L.E. & Scheigert, B.S. (1965c) Chicken flavor: The identification of some chemical components and the importance of sulfur compounds in the cooked volatile fraction. *Journal of Food Science*, **30**, 686–696.

Murphy, R.R. (1939) Flavor of turkey meat as affected by feeding fish meal and fish oil. In: *Proceedings of the 7th World's Poultry Congress*, Cleveland, Ohio. Waverly Press, Baltimore, Maryland.

Narasimhan, R., Runbin, L.J. & Diosady, L.L. (1993) Studies on meat flavor. IV. Fractionation, characterization, and quantitation of volatiles from uncured and cured beef and chicken. *Journal of Agricultural and Food Chemistry*, **41**, 939–945.

Noleau, I. & Toulemonde, B. (1986) Quantitative study of roasted chicken flavor. *Lebensn-Wisss u Technology*, **19**, 122–125.

Nonaka, M., Black, D.R. & Pippen, E.L. (1967) Gas chromatographic and mass spectral analysis of cooked chicken meat volatiles. *Journal of Agriculture and Food Chemistry*, **15**, 713–717.

Nutter, M.K., Lockhart, R.E. & Harris, R.S. (1943) The chemical composition of depot fats in chickens and turkeys. *Oil and Soap*, **20**, 231–233.

Offiong, S.A. (1976) The effect of rations containing raw peanuts on the eating quality of broilers. *Journal of the Science of Food and Agriculture*, **27**, 771–775.

Ohloff, G., Flament, I. & Pickenhagen, W. (1985) Flavour chemistry. *Food Review International*, **1**, 99–148.

Parr, L.J. & Levett, G. (1969) Hydrogen sulfide in cooked meat. *Journal of Food Technology*, **4**, 283–285.

Pippen, E.L. (1967). Poultry flavor. In: *The Chemistry and Physiology of Flavors* (ed. H.W. Schultz). AVI Publishing Co. Inc., Westport, Connecticut.

Pippen, E.L. & Eyring, E.J. (1957) Characteristics of volatiles nitrogen and volatile sulfur fractions of cooked chicken and their relation to flavor. *Food Technology*, **11**, 53–56.

Pippen, E.L. & Mecchi, E.P. (1969) Hydrogen sulfide, a direct and potentially indirect contributor to cooked chicken aroma. *Journal of Food Science*, **34**, 443–446.

Pippen, E.L. & Nonaka, M. (1960) Volatile carbonyl compounds of cooked chicken. II. Compounds volatilized with steam during cooking. *Food Research*, **25**, 764–769.

Pippen, E.L. & Nonaka, M. (1963) Gas chromatography of chicken and turkey volatiles: The effect of temperature, oxygen and type of tissue on the composition of the volatile fraction. *Journal of Food Science*, **28**, 334–341.

Pippen, E.L., Campbell, A.A. & Streeter, I.V. (1954) Origin of chicken flavor. *Journal of Agricultural and Food Chemistry*, **2**, 364–367.

Pippen, E.L., Nonaka, M., Jones, F.T. & Stitt, F. (1958) Volatile carbonyl compounds of cooked chicken I. Compounds obtained by air entrainment. *Food Research*, **23**, 103–113.

Pippen, E.L., Erying, E.J. & Nonaka, M. (1960) Occurrence and flavor significance of acetion in aqueous extracts of chicken. *Poultry Science*, **39**, 922–924.

Pippen, E.L., Mecchi, E.P. & Nonaka, M. (1969) Origin and nature of aroma in fat of cooked poultry. *Journal of Food Science*, **34**, 436–442.

Pittard, B.T., Freeman, L.R., Later, D.W. & Lee, M.L. (1982) Identification of volatile organic compounds produced by fluorescent Pseudomonads on chicken breast muscle. *Applied and Environmental Microbiology*, **43**, 1504–1506.

Pool, M.F., Hanson, H.L. & Klose, A.A. (1950) Effect of prefreezing hold time and antioxidant spray on storage stability of frozen eviscerated turkeys. *Poultry Science*, **29**, 347–350.

Ramaswamy, H.S. & Richards, J.F. (1982) Flavor of poultry meat- a review. *Canadian Institute Food Science Technology Journal*, **15**, 7–18.

Rao, C.S., Day, E.J., Chen, T.C. & Miniyard, J.P. (1976) Effects of polyphosphates on carbonyl volatiles of poultry meat. *Journal of Food Science*, **41**, 241–243.

Reineccius, G.A. & Leardon, R. (1985) The use of charcoal traps and microwave desorption for the analysis of headspace volatiles above heated thiamine solutions. In: *Topics in Flavour Research* (eds R.G. Berger, S. Nitz & P. Schreir), pp. 125–136. Eichhorn, Marzling-Hangenhan.

Ruenger, E.L., Reineccius, G.A. & Thompson, D.R. (1978) Flavor compounds related to the warmed-over flavor of turkey. *Journal of Food Science*, **43**, 1198–1200.

Saint Angelo, A., Vercellotti, J.R., Dupuy, H.P. & Spanier, A.M. (1988) Assessment of beef flavor quality: A multidisciplinary approach. *Food Technology*, **42**, 133–138.

Schonberg, A. & Maubacher, R. (1952) The Strecker degradation of α-amino acids. *Chemical Research*, **50**, 261–264.

Sheldon, B.W. (1979) Feed and intestinal microflora as factors in the flavor and quality of broiler meat. Thesis. Virginia Polytechnic Institute and State University.

Sheldon, B.W., Curtis, P.A., Dawson, P.L. & Ferket, P.R. (1997) Effect of dietary vitamin E on the oxidative stability, flavor, color, and volatile profiles of refrigerated and frozen turkey breast meat. *Poultry Science*, **76**, 634–641.

Shiau, C.Y. & Chai, T. (1985). Smoked dogfish processing and its refrigerated storage stability. *Journal of Food Science*, **50**, 1348–1350.

Sink, J.D. (1979) Effect of smoke processing on muscle food product characteristics. *Food Technology*, **33**, 72,74,78,80,83.

Smith, J.L. & Alford, J.A. (1968) Action of microorganisms on the peroxides and carbonyls of rancid fat. *Journal of Food Science*, **33**, 93–97.

Steverink, A.T.G. (1981) Review of literature on the chemical constituents of chicken meat flavor and suggestion for future research. In: *Quality of Poultry Meat: Proceedings of the Fifth European Symposium on Quality of Poultry Meat*, pp. 356–371. Spelderholt Institute for Poultry Research, Ministry of Agriculture and Fisheries, 7361 DA, Beekbergen, the Netherlands.

Sweetman, M.D. & MacKellar, I. (1954) *Food Selection and Preparation*, 4th edn. John Wiley & Sons, Inc., New York.

Swoboda, P.A.T. (1970) Dimethyltrisulfide: A constituent of roast chicken aroma. *160th Annual American Chemical Society Meeting*, Abstracts, 160, AGFD80.

Tang, J., Jin, Q.Z., Shen, G.-H., Ho, C.-T. & Chang, S.S. (1983) Isolation and identification of volatile compounds from fried chicken. *Journal of Agricultural and Food Chemistry*, **31**, 1287–1292.

Tims, M.J. & Watts, B.M. (1958) Protection of cooked meats with phosphates. *Food Technology*, **12**, 240–243.

Vail, G.E., Griswol, R.M., Justin, M.M. & Rust, L.O. (1967) *Foods*, 5th edn. Houghton Mifflin Co., Boston, Massachusetts.

Van Elswyk, M.E., Mooney, J.A., Hirschler, E.M. & Sams, A.R. (1995) Lipid stability and flavor quality of stored breast meat from broilers fed marine algae. In: *Institute of Food Technologists Annual Conference Proceedings*, p. 184.

Weisberg, S.M. (1956) The effect of milk products on the eating quality of broilers. Presented at *Texas Nutrition Conference*, Texas A & M, College Station, Texas. Abstract, 31, 45.

Wells, G.H., May, K.N. & Powers, J.J. (1962) Taste panel and shear press evaluation of tenderness of freeze dried chicken as affected by age and preslaughter feeding of ions. *Food Technology*, **16**, 137.

Wilson, R.A. & Katz, I. (1972) Review of literature on chicken flavor and report of isolation of several new chicken flavor components from aqueous cooked chicken broth. *Journal of Agricultural and Food Chemistry*, **20**, 741–747.

Wilson, B.R., Pearson, A.M. & Shorland, F.B. (1976) Effect of total lipids and phospholipids on warmed over flavor in red and white muscle from several species as measured by thiobarbituric analysis. *Journal of Agricultural and Food Chemistry*, **24**, 7–11.

Withycombe, D.A. & Mussinan, C.J. (1988). Identification of 2-methyl-3-furanthiol in the steam distillate from canned tuna fish. *Journal of Food Science*, **53**, 658–659.

Wu, C.M. & Liou, S.E. (1992) Volatile components of water-boiled duck meat and Cantonese style roasted duck. *Journal of Agricultural and Food Chemistry*, **40**, 838–841.

24 Poultry Quality and Tainting

María de Lourdes Pérez-Chabela and Alfonso Totosaus

Abstract: A safe production chain for poultry products is necessary to avoid epidemiological and toxicological human illness. Substances that can cause tainting in poultry are naturally present in nature, but greater risks are associated with those produced by anthropogenic activities often used to increase poultry yield. Although complete avoidance of tainting substances is unlikely, drugs to combat poultry illness and growth promoters are two contaminants that can be avoided with good manufacturing practices from farm to processing. There are four main categories of contaminants that contribute to poultry taint: naturally present, man-caused (anthropogenic), secondary origin, and microbiological.

Keywords: Poultry tainting, heavy metals, mycotoxins, pesticides, dioxins, perchloroethylene, organometallic compounds, nitrosamines, benzopyrenes, antibiotics, growth promoters, biogenic amines, PAHs, H5N1 avian influenza

24.1 INTRODUCTION

Meat quality is a generic term used to describe the perceptions of meat product properties such as color, freshness, and texture. Quality includes attributes such as carcass composition and conformation, contamination, health issues associated with meat consumption, and production issues such as animal welfare and environmental impact. Consumer evaluation of eating properties is the major determinant of meat quality, with tenderness, juiciness, and flavor of meat being the most important elements (Maltin *et al.* 2003).

Residual undesirable substances present in food products are a concern for consumers and governments due to the human health risks they imply. Birds can ingest a large quantity of tainting substances from air, water, feed, or medications that can be dangerous to human health. Some substances are quickly metabolized and therefore difficult to trace in poultry tissue, while others are slowly excreted and will be present in varying amounts in different tissues (Pérez-Chabela 2006). Poisoning appears to be an uncommon cause of illness in animals compared to other clinical problems such as infectious diseases, trauma, or neoplasia. One reason for this may be a lack of information on the most common toxicants affecting poultry, resulting in little information available to guide and facilitate diagnosis. This leads to limited dissemination of toxicoepidemiological data, with little or no availability to veterinarians in regions or countries where livestock and poultry are most frequently reported

Handbook of Meat, Poultry and Seafood Quality, Second Edition. Edited by Leo M. L. Nollet.

to contain mycotoxins, pesticides, metals, or contaminants of industrial origin such as dioxins (Guitart *et al.* 2010).

Food safety incidents involving chemical substances have occurred in the past. Examples are animal consumption of unapproved genetically modified food crops, intoxications with botanical products containing unintentionally admixed toxic herbs, residues of unapproved antibiotics and contaminants, and adverse effects of chemical and biological pesticides of natural origin. Various activities aim at the early identification of such emerging chemical risks. The establishment of databases would be beneficial, for example, for contaminants potentially present within products, since this proactive reconnaissance can facilitate the identification of products latent with hazardous substances in food and feed (Kleter *et al.* 2009).

Environmental contamination of poultry products has not been extensively reported but such contamination deserves attention. The regulation of the levels of polychlorinated biphenyls (PCBs), heavy metals such as lead and cadmium, and organochlorine pesticides (OCPs) in animals destined for human consumption is necessary. Programs to monitor levels of such chemicals in animals are also essential to provide factual information concerning both the present state and past trends in environmental behavior (Ruiter 1985). Extending such monitoring to components in the environment as a whole and not solely in food and feedstuffs can help establish patterns that can be used to insure the safety of the food supply.

This chapter deals with the main substances that taint poultry and cause detrimental effects on the quality of poultry meat. There are four main categories of contaminants that contribute to poultry taint: naturally present, man-caused (anthropogenic), secondary origin, and microbiological (Fig. 24.1).

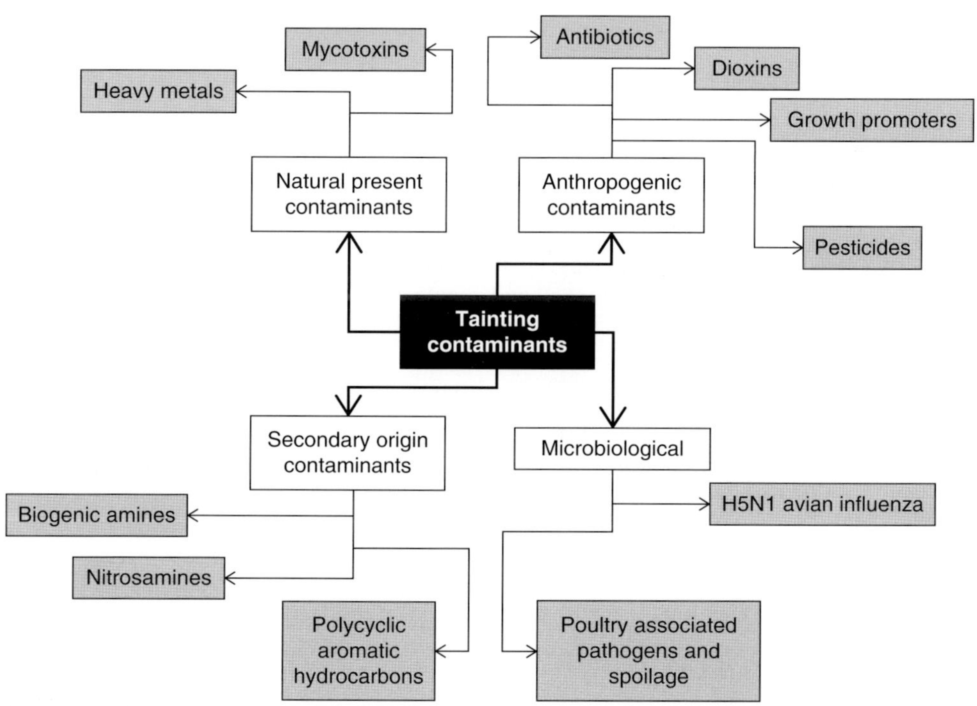

Fig. 24.1 Tainting contaminants.

24.2 POULTRY QUALITY

Quality is determined by a composite of characteristics that differentiate individual units of a product. The group of characteristics that pertain to a particular product are significant in determining the degree of acceptability of the product to a consumer. Differences in meat quality may be due to strain, age, nutritional regime, climate, or genetics.

Dransfield and Sosnicki (1999) studied the relationship between muscle growth and the quality of poultry meat. Muscle fiber cross-sectional area increased with age and fast-growing chickens had larger diameter fibers than slow-growing lines. This increase was also associated with an increase in the number of giant fibers that typically have cross-sectional areas three to five times larger than normal. A higher proportion of glycolytic fibers and a lower proteolytic potential in the muscles were also noted. Duval *et al.* (2008) evaluated the genetic parameters of several breast meat quality traits and their genetic relationships to muscle characteristics in a heavy line of broilers. The study results confirmed that genetic selection could be used to improve meat characteristics without impairing profitability, since no genetic conflict was detected between the quality and quantity of meat.

The two most important quality attributes for poultry meat are appearance and texture. Appearance is critical for both initial consumer selection of the product as well as for final satisfaction. Appearance quality attributes include skin color, meat color, and appearance defects such as bruises and hemorrhages. Texture is the single most important sensory property affecting final quality assessment. Modern production practices produce highly uniform young birds in which the major problems associated with meat texture are the result of processing errors or early deboning (Fletcher 2002). Maltin *et al.* (2003) studied factors that affect the texture of poultry meat. Postmortem factors such as sarcomere length, temperature, pH, and proteolysis were found to have major impacts on tenderness. Intrinsic properties of muscle, including fiber type, composition, and collagen content, together with factors related to breed, genotype, growth rate, and nutrition, may also play a part in determining tenderness.

24.3 POULTRY TAINTING OR CONTAMINATION

Generally, foods of animal origin play an important role in determining the exposure of humans to contaminants of both biological and chemical origin. A potentially large number of chemicals could be considered, but several of them deserve particular attention due to their occurrence and intake scenarios reflecting the differences existing in the economical, environmental, social, and ecological contexts. The environmental persistence of organic pollutants—defined as chemical substances that persist in the environment, bioaccumulate through the food web, and pose a risk of causing adverse effects to both human health and the environment—is generally correlated to their chemical stability. These substances are highly resistant to biological and chemical degradation, resulting in potentially long-term exposure. Although several factors may play a role, the bioaccumulation magnitude of organic pollutants is primarily related to the solubility of these substances in lipids. Greater lipid solubility strengthens the tendency of the pollutants to be concentrated in the fatty tissue of living organisms resulting in a higher concentration in food-producing animals than those present in the environment and/or feedstuffs (Brambilla *et al.* 2009).

Under normal conditions, chemical substances like polychlorinated dibenzodioxins (PCDDs), polychlorinated dibenzofurans (PCDFs), PCBs, polycyclic aromatic hydrocarbons

(PAHs), and heavy metals such as cadmium, mercury, arsenic, and lead are widespread in the environment. This gives rise to contamination of cereals and plants that are employed as animal feeds. Although their use is carefully controlled, pesticides, including insecticides, herbicides, and fungicides, can also be found in low levels in animal feed. Commercial compound feed, fodder, forage, or natural grass/straw used as animal feed is, therefore, the main exposure route for animals to environmental contaminants and pesticides. The chemicals are typically found in the target organ(s), but residues as well as metabolites from the metabolism of the chemicals produce concurrent meat contamination (Gilbert 2005).

The term "residue" involves all chemical compounds or substances that can be present in the food chain and can be potentially dangerous to human health. These residues can be present in meat-based foods and can be divided into three primary groups (Pérez-Chabela 2006):

Naturally present residues: Substances present in important amounts in nature as part of the food chain due to modern agricultural methods and environmental pollution. Examples include heavy metals and mycotoxins.

Man-caused residues: Substances caused directly by man that are very stable in the environment. Examples include agriculture chemicals such as pesticides, DDT, lindane, and hexachlorobenzene, industrial chemicals such as organochlorines, dioxins, perchloroethylene, and organometallic compounds, and medicines and growth promoter substances employed in overdoses.

Secondary origin residues: Compounds generated due to production processes added in excessive amounts causing a residue effect. Examples include nitrosamines and benzopyrenes.

There are a number of routes for contaminants to get into the food chain and efforts over the past few years have focused on minimizing environmental contamination, particularly if contaminants are known to bioaccumulate (Gilbert 2005).

24.3.1 Naturally present contaminants

24.3.1.1 Heavy metals

The presence of metals in poultry tissue is due to feed. Potassium is the most abundant metal found in poultry liver, kidney, and spleen, followed by phosphorus and sodium. Zinc, magnesium, and calcium are found in lower concentrations (Fig. 24.2). Heavy metals are also present at levels consistent with the presence of toxic metals in feed, soil, and water. Chickens, like other animals, ultimately depend on plants to synthesize inorganic material into organic compounds they can use as food (Demirbaş 1999). The fact that toxic metals are present in high concentrations in the feeds is of particular importance in relation to the standards for lead and cadmium as toxic metals.

Less lead accumulated in the gizzard and femoral muscle than in the kidney and liver of ducks (Jeng *et al.* 1997). Drinking water containing a mixture of arsenic (0.8 ppm), benzene (1.3 ppm), cadmium (5.1 ppm), lead (6.7 ppm), and trichloroethylene (0.65 ppm) resulted in lowered reproductive performance as indicated by decreased egg production, decreased egg weight, increased embryonic mortality, and decreased hatchability in broilers (Vodela *et al.* 1997). However, an individual chemical cannot be held responsible for altering reproductive performance because multichemical exposure can cause additive, synergistic, or antagonistic effects, affecting the sensitivity of the chicken's reproductive system to adverse effects of contaminants commonly found in groundwater.

The U.S. Department of Agriculture Food Safety and Inspection Service (FSIS) conducted an exploratory assessment in 2003–2004 to determine the occurrence and levels of cadmium

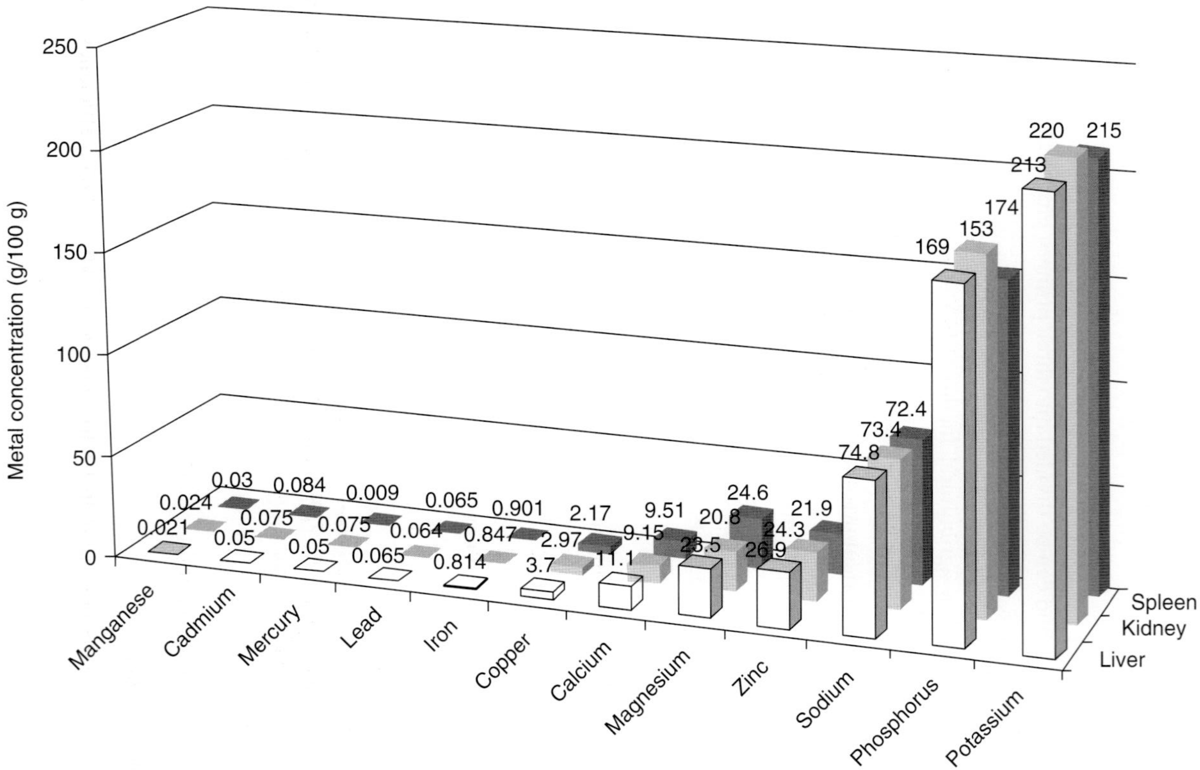

Fig. 24.2 Presence of metals in poultry tissues.

and lead in randomly collected samples of kidney, liver, and muscle tissues of mature chicken, boars/stags, dairy cows, and heifers. The data generated were qualitatively compared to data that FSIS gathered in 1985–1986. In each production class tested, levels of cadmium and lead were higher in kidney and liver samples than in muscle samples. None of the muscle samples contained cadmium or lead levels exceeding the maximum residue levels established by other countries or international organizations. The incidence of cadmium and lead in the animals did not increase between 1985 and 2004, indicating that the risk of human exposure to heavy metals in meat and poultry is low (Pagán-Rodriguez *et al.* 2007).

24.3.1.2 Mycotoxins

Mycotoxins are toxic substances elaborated by fungi that constitute a heterogeneous group of secondary metabolites with potent pharmacological and toxic effects in humans and animals. These molecules are produced during mold development on plants in the field or during storage and can be found as natural contaminants of cereals, fruits, nuts, grains, forage, and compound foods intended for human or animal consumption. The most important mycotoxins are produced by molds in the *Aspergillus*, *Penicillium*, and *Fusarium* genus. Mycotoxins are suspected in several pathological syndromes in humans including ochratoxin A and Balkan endemic nephropathy, esophageal cancer, and fumonisin B1. Toxicity can be classified as hepatotoxic (aflatoxin) and immunotoxic (trichothecenes, fumonisins). Other mycotoxins have estrogenic potential (zearalenone) and several are considered carcinogenic or suspected of having carcinogenic properties (Bailly & Guerre 2009).

Aflatoxin B1 is rapidly metabolized in the liver and muscles of broiler chicks but no regulatory limits exist for levels of aflatoxin in poultry feed. Secondary exposure to aflatoxins through consumption of chicken liver and meat derived from poultry-fed aflatoxin-contaminated feed may pose a risk to consumer health (Hussain *et al.* 2010). The mycotoxin zearalenone, one of the most widely distributed fusariotoxins, is common in maize and maize products, soybeans, and various cereal grains and represents a major component of human food and animal feed. Zearalenone is metabolized to alpha-zearalenol (α-ZOL) and beta-zearalenol (β-ZOL). The efficiency in the production of alpha-reduced metabolites appears of particular interest in risk assessment as these metabolites constitute activated forms whereas beta-reduced metabolites are less estrogenic than zearalenone. Kolf-Clauw *et al.* (2008) found that chickens, hens, and quail presented the highest β-ZOL concentrations, followed by duck, Guinea-fowl, and goose, with quail having lower α-ZOL/β-ZOL ratios. Variations in the zearalenone metabolites may be explained by the various isoforms of dehydrogenases in the different species.

24.3.2 Anthropogenic contaminants

24.3.2.1 Antibiotics

Antibiotics are used in the poultry industry to enhance growth and feed efficiency and reduce disease. Antibiotic usage has facilitated the efficient production of poultry, allowing consumers to purchase high quality poultry meat and eggs at a reasonable cost. Antibiotic usage has also enhanced the health and well-being of poultry by reducing the incidence of disease. Although these uses benefit all involved, consumer perceptions are that edible poultry tissues are contaminated with harmful concentrations of drug residues (Donoghue 2003).

The term "antibiotic" should strictly apply to a range of compounds that are of biological origin from three main sources: (1) compounds produced metabolically from molds of the filamentous *Penicillium* species (benzylpenicillin, cephalosporin); (2) compounds extracted from cultures of specific bacteria such as several *Streptomyces* species (streptomycin, gentamicin, tetracycline, spiramycin); and (3) semisynthetic substances that are additionally modified by chemical synthesis (florfenicol, amoxicillin, cephalexin). Other compounds with antibiotic effects like lincosamides and cephamycins are considered subfamilies of antibiotics. The terms "antibacterial" and "antimicrobial" apply to a broader set of compounds including not only the natural and semisynthetic antibiotics, but also several other classes of molecules having antibacterial properties. These compounds are produced by chemical synthesis and are classified as nitrofurans, nitroimidazoles, quinolones, and sulfonamides. Veterinary drugs with some antibacterial activity include carbadox, dapsone, malachite green, olaquindox, novobiocin, and virginiamycin (Verdon 2011).

Eighteen days of withdrawal are needed before birds medicated with levamisole are slaughtered if their tissues are to be safe for human consumption (El-Kholy & Kemppainen 2005). The longer withdrawal time is needed due to the larger recommended dose of the drug. The occurrence of chloramphenicol (CAP) in broiler chicken liver, kidney, and tissues was detected by Tajik *et al.* (2010) despite the prohibition on CAP application in food animals. One hundred and sixty broiler chicken carcasses were collected from three provinces in Iran. Four-plate test (FPT), ELISA, and high pressure liquid chromatography (HPLC) were used to qualify and quantify contamination of samples with CAP. The results of FPT revealed that up to 17.5% of the samples were contaminated with the antibiotic. The ELISA assay showed that out of 28 positive samples in FPT, 22 liver, 21 kidney, and 14 muscle samples were positive for CAP. ELISA analyses indicated a minimum level of 0.54 ng/g and a maximum level of 155.2 ng/g in kidney and liver. HPLC analyses confirmed the ELISA findings although the level of contamination was lower than that of ELISA analyses. These data showed that despite the prohibition of CAP application in food animals such as poultry, detectable CAP residues are present in chickens indicating illegal use of this antibiotic. Pena *et al.* (2010) developed and validated a sensitive, accurate, and robust method for determining fluoroquinolone levels in poultry meat based on liquid chromatography with spectrofluorimetric detection. Of the samples analyzed, 44.2% of chickens and 37.8% of turkeys were contaminated. The levels detected were less than 114.2 μg/kg in chicken muscle and less than 87.6 μg/kg in turkey muscle. One positive chicken sample was contaminated with enrofloxacin at levels higher than the maximum residue levels.

24.3.2.2 Dioxins

Polychlorinated dibenzo-p-dioxins (PCDDs) are resistant global pollutants. The dioxin-like PCB congeners are classified according to the number of chlorines in the ortho position: non-ortho-chloro, mono-ortho-chloro, and di-ortho-chloro. These compounds accumulate in the food chain because of their chemical stability and lipophilicity. Food has been widely recognized as the main source of human intake for this kind of toxic chemical.

Polychlorinated naphthalenes (PCNs) are chlorinated PAHs, a potentially vast group of little-known environmental contaminants of anthropogenic origin. PCNs are the most studied subgroup of these compounds and some of the 75 congeners have recognized toxic, bioaccumulative, and persistence properties. These characteristics coupled with the similarity in structural planar configurations to some PCBs and dioxins can bestow a dioxin-like mode of toxic action. PCNs are an industrial chemical produced over most of the last century although manufacture is currently banned in some countries (Fernandes *et al.* 2010).

A statistical survey of dioxins and dioxin-like compounds in domestic meat and poultry was carried out in 2007–2008 (Huwe *et al.* 2009). Seventeen toxic polychlorinated dibenzo-p-dioxins and dibenzofurans (PCDD/F) and four non-ortho-polychlorinated diphenyls (no-PCB) were measured in turkey and chicken samples. The results of the survey showed that the sum of PCDD/F and non-PCB toxic equivalencies (sum-TEQ) ranged from none detected to 4.5 pg/g of lipid. In comparison to data from 2002 to 2003, there was a decrease in the TEQ trend, although data from the two time periods were difficult to compare due to the different laboratories and techniques employed. In general, however, a decline in both PCDD/F and PCB was noted. One exception was increasing PCB levels in turkeys, which was likely due to a localized contamination.

A simple and fast method based on an automated cleanup system for simultaneous analysis of PCBs, PCDDs, and PCDFs was used for a surveillance program of food samples produced in Spain in 2000 (Eljarrat *et al.* 2002). All the calculated TEQPCDD/F (total toxicity equivalent) values were below the limits proposed in the draft of the European Community regulation. However, when the TEQPCB was taken into account, the muscle samples exceeded the World Health Organization limits of 3 pg/g in the whole product.

24.3.2.3 Growth promoters

Meat quality can be affected by the use of growth promoters. Such substances may increase connective tissue production and collagen cross-links giving a tougher meat. Inhibition of muscle proteases responsible for protein breakdown in postmortem meat, acceleration of lipolysis, and breakdown of triacylglycerols may also occur. Growth promoters have also been associated with reductions in marbling and tenderness, with a negative effect on palatability (Reig & Toldrá 2011).

Growth promoters are substances administered through the feed or drinking water that are employed in farm animals for therapeutic and prophylactic purposes. These substances promote growth rate and improve efficiency of feed utilization. Their use contributes to an increase in protein deposition that is usually linked to fat utilization, which means a reduction in the fat content in the carcass and an increase in meat leanness. Anabolic promoters like estradiol, progesterone, and testosterone are allowed in many countries, but the use of growth promoters has been officially banned in the European Union since 1988 due to concerns about harmful effects on consumers. The fraudulent practice of using combinations of low amounts of several substances such as β-agonists (clenbuterol), corticosteroids (dexamethasone), and anabolic steroids—mixtures known as cocktails—can have a synergistic effect and exert growth promotion but is also more difficult to detect analytically (Reig & Toldrá 2011).

24.3.2.4 Pesticides

For the past several decades, pesticides have been used in the public health sector to control disease vectors and in agriculture to control/eradicate crop pests. OCPs, effective against a variety of insects, have been extensively used around the world. Their low volatility combined with their extreme stability and probably indiscriminate use in the past has led to their high persistence both in the environment after application and in organisms after exposure. OCPs bioaccumulate in the food chain and tend to concentrate in fatty tissues due to their lipophilic nature. Exposure to OCPs also occurs via low level food contamination (Barriada-Pereira *et al.* 2003).

Organophosphorus pesticides (OPPs) are less persistent than OCPs and are frequently preferred because of their efficacious, safe, and cost-effective control of a wide range of pests. They can be absorbed by all routes including inhalation, ingestions, and dermal absorption. The awareness that OPPs may also concentrate along the food chain has led to the establishment of low maximum residue limits (MRLs) in meat by the European Union (Basheer *et al.* 2002). Both OCPs and OPPs are known to induce or aggravate human health problems such as cancer, immune system disorders, and the disruption of hormonal functions.

Garrido *et al.* (2006) developed and validated a multiresidue method for the simultaneous determination of about 45 pesticides, mainly OCPs and OPPs, by gas chromatography/mass spectroscopy (GC/MS) in chicken, pork, and lamb. Only three OCPs (endosulfan α, endosulfan sulfate, and dichloran) were detected in three different lamb samples while endosulfan α was detected in only one pork sample. Pesticide residues were not found in any of the 10 chicken samples analyzed. Ahmad *et al.* (2010) investigated the extent of contamination of eggs, chicken, and meat in Jordan with aldrin, DDTs, dieldrin, endosulfan isomers, endrin, HCHs, heptachlor, heptachlor epoxide, and HCB residues. Results indicated that 28% of eggs, 20% of chicken, and 49% of meat samples examined were contaminated with OCP residues. HCHs and DDTs were the most prominently noted compounds present, indicating meat may contribute substantially to the intake of OCPs through food consumption.

24.3.3 Secondary origin residues

24.3.3.1 Biogenic amines

Biogenic amines are nitrogenous compounds of a basic nature that show biological activity. They are synthesized and degraded by animal, plant, and microbial metabolisms and consequently are found in a wide variety of food products. The biogenic amines most commonly found in foods are grouped according to their chemical structure as (1) aromatic monoamines (tyramine and phenylalanine), (2) heterocyclic amines (histamine and tryptamine), (3) aliphatic diamines (cadaverine and putrescine), and (4) aliphatic polyamines (agmatine, spermidine, and spermine). Classically, biogenic amines are defined as "biogenic" or "endogenous or natural" depending on their synthesis. Endogenous or natural amines are formed as a result of the intracellular metabolic processes of animals, plants, and microorganisms (Vidal-Carou *et al.* 2011).

The content of heterocyclic amines in heat-treated poultry products varies with the conditions and methods of cooking (Skog & Solyakov 2002). The most abundant heterocyclic amine is 2-amino-1-methyl-6-phenylimidazo[4,5-b]pyridine (>40 ng/g) followed by 1-methyl-9 H-pyrido[3,4-b]indole, 9 H-pyrido[3,4-b]indole (<10 ng/g), 2-amino-3,8-dimethylimidazo[4,5-f]quinoxaline (<3 ng/g), and 2-amino-3,-trimethylimidazo[4,5-f] quinoxaline (<3 ng/g). Other heterocyclic amines, including 2-amino-3-methylimidazo [4,5-f]quinoline, 2-amino-3-methylimidazo[4,5-f]quinoxaline, 3-amino-1,4-dimethyl-5 H-pyrido[4,3-b]indole, 3-amino-1-methyl-5 H-pyrido[4,3-b]indole, 2-amino-9 H-pyrido[2,3-b] indole, and 2-amino-3-methyl-9 H-pyrido[2,3-b]indole, are generally found at low levels (<1 ng/g).

24.3.3.2 Nitrosamines

Color and appearance are the first and most important characteristics for consumers in cured cooked meat products. The color of the meat product is the result of naturally present pigments or added colorants. The primary natural meat pigment is myoglobin, which reacts with nitrous oxide to form nitrosyl myoglobin, a stable pink pigment responsible for the pink color of cured meats. Adding curing agents and nitrite also contributes typical flavor, inhibits *Clostridium* spp., and provides antioxidant capacity (Bázan-Lugo 2008).

N-nitrosamines are readily absorbed from the gastrointestinal tract without bioaccumulation, requiring metabolic activation to exhibit their mutagenic and carcinogenic action. The formation of nitrosamines in meat and meat products is a complex process and several factors/substances could influence nitrosation reactions. The nitrosamine concentration in meat products depends on the residual nitrite concentration, presence of nitrosation catalysts and inhibitors, cooking method, cooking temperature and time, and storage conditions. The presence of microorganisms, which are able to reduce nitrate to nitrite and promote the degradation of proteins to amines and amino acids, may also play a role. In general, the concentration of nitrosamines in meat products currently lies at levels lower than 30 μg/kg (Rath & Reyes-Reyes 2011).

24.3.3.3 Polycyclic aromatic hydrocarbons

A prolonged period of smoking is required to produce a rich deep-brown color and a strong aroma profile in meats. Called farmhouse smoked, this process often occurs under uncontrolled technological conditions and without legislative regulation. Meat products produced by this smoking method contain far higher levels of PAH, one of the most important groups of chemicals that are actually harmful to human health. PAH are formed during the thermal decomposition of wood, especially under limited oxygen access, in the range of 500–900°C. PAH are characterized by two or more condensed aromatic rings in a molecular structure and have a strong lipophilic character. Some PAH are able to interact in organisms with enzymes such as aryl hydrocarbon hydroxylases to form PAH dihydrodiol derivatives, reactive products believed to be ultimate carcinogens that are able to form covalently bonded adducts with proteins and nucleic acids. Moreover, PAH content in smoked foods can be affected not only by environmental factors, but also by diffusion processes from plastic packaging materials (Šimko 2011).

24.3.4 Microbial contamination

Microbial safety and quality of poultry meat is a concern for producers, retailers, and consumers. One group of microorganisms associated with poultry products are those that are generally harmless to human health but that multiply during chill storage due to their psychrotrophic nature, resulting in decreased shelf life. The organisms that predominate at spoilage when poultry meat is stored aerobically under chill conditions are invariably *Pseudomonas* spp. accompanied by lower numbers of Gram-negative bacteria such as *Shewanella putrefaciens*, *Acinetobacter*, *Moraxella*, and *Psychrobacter*.

Illness due to contamination of poultry meat with foodborne pathogens remains an important public health issue. Improper handling, undercooking, cross-contamination, and improper postcooking storage may all lead to contamination of poultry products with pathogens. The use of antimicrobials in poultry production, whether for prophylactic, therapeutic, or performance-enhancing purposes, contributes to the development of resistant pathogens causing serious consequences for the treatment of human illness from these organisms. *Salmonella* spp. and *Campylobacter* spp. are the predominant foodborne pathogens associated with poultry and are frequently implicated in human illness from this source. Other pathogens reportedly found on poultry products include *Clostridium perfringens*, *Escherichia coli* O157:H7, *Listeria monocytogenes*, *Arcobacter* spp., and *Helicobacter* spp. Hazard analysis and critical control points (HACCP) programs are a universally applied approach to food safety control in the poultry industry while quantitative risk analysis (QRA) is being applied to microbial hazards found in these products (Mead 2004).

Poor economic returns received by the egg and poultry industries has resulted in intensive production methods being commonly used, which has in turn resulted in increased susceptibility of poultry to pathogen colonization. An international pandemic of human salmonellosis has occurred over the last 20 years from eggs contaminated with *Salmonella enteritidis*. *Campylobacter jejuni* and *Campylobacter coli* are the two main species associated with human infections, accounting for over 95% of campylobacteriosis cases (King & Adams 2008). *Campylobacter* spp. and most if not all of the common zoonotic *Salmonella* spp. are essentially commensal in chickens. They usually form part of the gut flora and contaminate chicken carcasses by fecal spillage during slaughter. Chickens mount antibody responses to both pathogens, which indicates something other than commensalism. Such immune responses, however, do not always result in the clearance of the pathogens.

Suboptimal environments, poor diets, and/or psychological or physical stresses increase the susceptibility of poultry being infected by these pathogens (Humphrey 2006).

24.3.5 H5N1 avian influenza

A new highly pathogenic strain of avian influenza, designated H5N1, emerged in the poultry markets of Hong Kong in 1997 and subsequently reemerged in Vietnam in 2003. Since then, the virus has spread rapidly across the globe and is likely to be endemic in poultry in many parts of the world. At present, its transmission in humans remains limited but ongoing human exposure potentially increases the risk that a pandemic virus strain will emerge that is easily transmissible to humans. Particular concern has arisen over the continued human infection with the highly pathogenic avian influenza (HPAI) of the H5N1 subtype (Mumford *et al.* 2007).

Garske *et al.* (2007) analyzed data published from four outbreaks of HPAI in commercial poultry in industrialized countries to explore the different intervention measures implemented during these outbreaks. Results indicated that current controls might not be enough to eradicate the disease, but effectiveness is dependent on the particular situation. Close monitoring of outbreaks was recommended. Catolli *et al.* (2009) reviewed and updated the information concerning the evolution and molecular epidemiology of HPAI/H5N1 with particular reference to the European continent. Results showed that genetically indistinguishable HPAI/H5N1 viruses belonging to the main clade 2.2 were introduced in Europe in 2005–2006 and co-circulated. Although genetic diversity of clade 2.2 can be traced back to the first outbreak at Quinhai Lake in April 2005, the authors highlight the significant role of the Russian territory, and perhaps other Eastern European countries, as the source of the majority of the early viruses spreading into the rest of Europe.

Li *et al.* (2010) analyzed 51 representative H5N1 avian influenza viruses isolated from domestic poultry, wild birds, and humans in China in 2004–2009. Twenty-one genotypes were detected based on whole-genome sequences. The viruses have caused outbreaks in poultry in more than 60 countries. The H5N1 influenza viruses that emerged in chickens in northern China in 2006 have antigenically drifted and could not be protected efficiently by the GS/GD/1/96 virus-derived vaccines in China demonstrating the complexity of the influenza viruses. Clearly, there is a critical need for the continued surveillance of poultry and for regularly updated control measures.

24.4 CONCLUSIONS

A safe production chain for poultry products is necessary to avoid epidemiological and toxicological human illness. Substances that can cause tainting in poultry are naturally present in nature, but greater risks are associated with those produced by anthropogenic activities often used to increase poultry yield. Although complete avoidance of tainting substances is unlikely, drugs to combat poultry illness and growth promoters are two contaminants that can be avoided with good manufacturing practices from farm to processing.

REFERENCES

Ahmad, R., Salem, M. & Estaitieh, H. (2010) Occurrence of organochlorine pesticide residues in eggs, chicken and meat in Jordan. *Chemosphere*, **78**, 667–671.

Bailly, J.-D. & Guerre, P. (2009) Mycotoxins in meat and processed meat products. In: *Safety of Meat and Processed Meat* (ed. F. Toldrá), pp. 83–124. Springer, New York.

Barriada-Pereira, M., Concha-Grana, E., González-Castro, M.J., *et al.* (2003) Micro-wave assisted extraction versus Soxhlet extraction in the analysis of 21 organochlorine pesticides in plants. *Journal of Chromatography A*, **1008**, 115–122.

Basheer, C., Lee, H.K. & Obbard, J.P. (2002) Determination of organochlorine pesticides in seawater using liquid-phase hollow fibre membrane microextraction and gas chromatography-mass spectrometry. *Journal of Chromatography A*, **968**, 191–199.

Bázan-Lugo, E. (2008) Nitritos y Nitratos: Su uso, control y alternatives en embutidos cárnicos. *Nacameh*, **2**, 160–187.

Brambilla, G., Iamiceli, A. & di Domenico, A. (2009) Priority environmental chemical contaminants in meat. In: *Safety of Meat and Processed Meat* (ed. F. Toldrá), pp. 391–424. Springer, New York.

Catolli, G., Fusaro, A., Monne, I. & Capua, I. (2009) H5N1 virus evolution in Europe—An updated overview. *Viruses*, **1**, 1351–1363.

Demirbaş, A. (1999) Proximate and heavy metal composition in chicken meat and tissues. *Food Chemistry*, **67**, 27–31.

Donoghue, D.J. (2003) Antibiotic residues in poultry tissues and eggs: Human health concerns? *Poultry Science*, **82**, 618–621.

Dransfield, E. & Sosnicki, A.A. (1999) Relationship between muscle growth and poultry meat quality. *Poultry Science*, **78**, 743–746.

Duval, E.L., Debut, M., Berri, C.M., *et al.* (2008) Chicken meat quality: Genetic variability and relationship with growth and muscle characteristics. *BMC Genetics*, **9**, 53.

Eljarrat, E., Monjonell, A., Caixach, J. & Rivera, J. (2002) Toxic potency of polychlorinated dibenzo-p-dioxins, polychlorinated dibenzofurans, and polychlorinated biphenyls in food samples from Catalunia (Spain). *Journal of Agricultural and Food Chemistry*, **50**, 1161–1167.

El-Kholy, H. & Kemppainen, B.W. (2005) Levamisole residues in chicken tissues and eggs. *Poultry Science*, **84**, 9–13.

Fernandes, A., Mortimer, D., Gem, M., *et al.* (2010) Polychlorinated naphtalenes (PCNs): Congener specific analysis, occurrence in food, and dietary exposure in the UK. *Environmental Science and Technology*, **44**, 3533–3538.

Fletcher, D.L. (2002) Poultry meat quality. *World Poultry Science Journal*, **58**, 131–145.

Garrido, F.A., Martinez, V.J.L., Cruz, S.A.D., González, R.M.J. & Plaza, B.P. (2006) Multiresidue analysis of organochlorine and organophosphorus pesticides in muscle of chicken, pork and lamb by gas chromatography-triple quadruple mass spectrometry. *Analytica Chimica Acta*, **558**, 42–52.

Garske, T., Clarke, P. & Ghani, A.C. (2007) The transmissibility of highly pathogenic avian influenza in commercial poultry in industrialised countries. *Plos One*, **2**, e349.

Gilbert, J. (2005) Environmental contaminants and pesticides in animal feed and meat. In: *Improving the Safety of Fresh Meat* (ed. J. Sofos), Woodhead Publishing Co, Cambridge.

Guitart, R., Croubels, S., Caloni, F., et al. (2010) Animal poisoning in Europe. Part 1: Farm livestock poultry. *Veterinary Journal*, **183**, 249–254.

Humphrey, T. (2006) Are happy chickens safer chickens? Poultry welfare and disease susceptibility. *British Poultry Science*, **47**, 379–391.

Hussain, Z., Khan, M.Z., Khan, A., *et al.* (2010) Residues of aflatoxin B1 in broiler meat: Effect of age and dietary aflatoxin B1 levels. *Food and Chemical Toxicology*, **48**, 3304–3307.

Huwe, J., Pagan-Rodriguez, D., Abdelmajid, N., et al. (2009) Survey of polychlorinated dibenzo-p-dioxins, polychlorinated dibenzofurans, and non-ortho-polychlorinated biphenyls in U.S. meat and poultry 2007–2008: Effect of new toxic equivalency factors on toxic equivalency levels, patterns and temporal trends. *Journal of Agricultural and Food Chemistry*, **57**, 11194–11200.

Jeng, S.L., Lee, S.J., Liu, Y.F., Yang, S.C. & Liou, P.P. (1997) Effect of lead ingestion on concentrations of lead in tissues and eggs of laying Tsaiya ducks in Taiwan. *Poultry Science*, **76**, 13–16.

King, S. & Adams, M.C. (2008) Incidence of campylobacter in processed poultry: Is it a concern for human health? *Journal of Food Safety*, **28**, 376–388.

Kleter, G.A., Groot, M.J., Poelman, M., Kok, E.J. & Marvin, H.J.P. (2009) Timely awareness and prevention of emerging chemical and biochemical risks in foods: Proposal for a strategy based on experience with recent cases. *Food and Chemical Toxicology*, **47**, 992–1008.

Kolf-Clauw, M., Ayouni, F., Tardieu, D. & Guerre, P. (2008) Variations in zearalenone activation in avian food species. *Food and Chemical Toxicology*, **46**, 1467–1473.

Li, Y., Shi, J., Zhong, G., *et al.* (2010) Continued evolution of H5N1 influenza viruses in wild birds, domestic poultry and humans in China from 2004 to 2009. *Journal of Virology*, **84**, 8389–8397.

Maltin, C., Balcerzak, D., Tilley, R. & Delday, M. (2003) Determinants of meat quality: Tenderness. *Proceedings of the Nutrition Society*, **62**, 337–347.

Mead, G.C. (2004) Microbiological quality of poultry meat: A review. *Brazilian Journal of Poultry Science*, **6**, 135–142.

Mumford, E., Bishop, J., Hendrickx, S., Embarek, P.B. & Perdue, M. (2007) Avian influenza H5N1: Risks at the human–animal interface. *Food and Nutrition Bulletin*, **28**, S357–S363.

Pagán-Rodriguez, D., O'Keefe, M., Deyrup, C., Zervos, P., Walker, H. & Thaler, A. (2007) Cadmium and lead residue control in a hazard analysis and critical control points (HACCP) environment. *Journal of Agricultural and Food Chemistry*, **55**, S1638–S1642.

Pena, A., Silva, L.J.G., Pereira, A., Meisel, L. & Lino, C.M. (2010) Determination of fluoroquinolone residues in poultry muscle in Portugal. *Analytical and Bioanalytical Chemistry*, **397**, 2615–2621.

Pérez-Chabela, M.L. (2006) Residuos quimicos. In: *Ciencia y Technologia de Carnes* (eds Y.H. Hui, I. Guerrero & M.R. Rosmini), pp. 371–391. Limusa Noriega Editores, Mexico City.

Rath, S. & Reyes-Reyes, F.G. (2011) Nitrosamines. In: *Safety Analysis of Foods of Animal Origin* (eds L.M.L. Nollet & F. Toldrá), pp. 421–440. CRC Press, Boca Raton, Florida.

Reig, M. & Toldrá, F. (2011) Growth promoters. In: *Safety Analysis of Foods of Animal Origin* (eds L.M.L. Nollet & F. Toldrá), pp. 229–247. CRC Press, Boca Raton, Florida.

Ruiter, A. (1985) Contaminates in meat and meat products. In: *Developments in Meat Science* (ed. R. Lawrie) Vol. **3**, pp. 185–220. Elsevier Applied Science Publishers, London.

Šimko, P. (2011) Polycyclic aromatic hydrocarbons. In: *Safety Analysis of Foods of Animal Origin* (eds L.M.L. Nollet & F. Toldrá), pp. 441–460. CRC Press, Boca Raton.

Skog, K. & Solyakov, A. (2002) Heterocyclic amines in poultry products: A literature review. *Food and Chemical Toxicology*, **40**, 1213–1221.

Tajik, H. Malekinejad, H., Razavi-Rouhani, S.M., Pajouhi, M.R., Mahmoudi, R. & Haghnazari, A. (2010) Chloramphenicol residues in chicken liver, kidney and muscle: A comparison among the antibacterial residues monitoring methods of four plate test, ELISA and HPLC. *Food and Chemical Toxicology*, **48**, 2464–2468.

Verdon, E. (2011) Antibiotic residues in muscle tissues of edible animal products. In: *Safety Analysis of Foods of Animal Origin* (eds L.M.L. Nollet & F. Toldrá), pp. 250–347. CRC Press, Boca Raton, Florida.

Vidal-Carou, M.C., Latorre-Moratalla, M.L. & Bover-Cid, S. (2011) Biogenic amines. In: *Safety Analysis of Foods of Animal Origin*. (eds L.M.L. Nollet & F. Toldrá), pp. 399–420. CRC Press, Boca Raton.

Vodela, J.K., Lenz, S.D., Renden, J.A., McElhenney, W.H. & Kemppainen, B.W. (1997) Drinking water contaminants (arsenic, cadmium, lead, benzene, and trichloroethylene). 2. Effects on reproductive performance, egg quality, and embryo toxicity in broiler breeders. *Poultry Science*, **76**, 1493–1500.

25 Microbiological and Sensory Properties of Fresh and Frozen Poultry

Lisa McKee

Abstract: Microbial contamination of poultry products remains a problem for the food industry. Pathogens such as *Campylobacter*, *Salmonella*, and *Listeria* and spoilage organisms such as *Pseudomonas* can play significant roles in the safety and shelf life of poultry products. Although washing and sanitizing, treatment with organic and inorganic compounds, modified atmosphere packaging/storage, irradiation, and other processes have been investigated as potential methods for reducing microbial loads on poultry, the effect of such methods on color, flavor, texture, and other sensory properties is often undesirable. New technologies such as hydrodynamic pressure and ozone treatment are only now being investigated and may provide a means in the future to produce safer poultry products with acceptable organoleptic properties.

Keywords: Poultry microbial loads, poultry spoilage organisms, vacuum packaging, modified atmosphere packaging, irradiation

25.1 INTRODUCTION

Poultry consumption has risen dramatically in the USA over the past 25 years (see Table 25.1). Although the term poultry can include everything from chicken and turkey to duck and geese, approximately 98% of poultry consumption is in the form of chicken and turkey (National Chicken Council 2002).

As consumption of poultry has increased, so has concern over consumer handling of such products. Numerous factors are involved in outbreaks of poultry-borne illness, including cross-contamination (Brown *et al.* 1988; Bryan & Doyle 1995; MMWR 1998; Jiménez *et al.* 2005; Moffatt *et al.* 2006); inadequate cooking (Bryan 1980; Bean & Griffin 1990; MMWR 2008); preparation of food well in advance of consumption (Worsfold & Griffith 1997); improper handling of food (Altekruse *et al.* 1996; Shiferaw *et al.* 2000); and temperature abuse during transport, preparation, or storage (Worsfold & Griffith 1997). A particular problem that is consistently reported is consumers not washing their hands or cutting boards after handling raw poultry. Although a variety of educational strategies have been designed to improve consumer food safety knowledge and safe food-handling skills, poultry-related food-borne illness problems persist (see Table 25.2).

Handbook of Meat, Poultry and Seafood Quality, Second Edition. Edited by Leo M. L. Nollet.

Table 25.1 US per capita consumption of poultry and other meats.

Year	Pounds per capita per year			
	Chicken	**Turkey**	**Beef**	**Pork**
1980	32.7	8.1	72.1	52.1
1990	42.4	13.8	63.9	46.4
2001	52.4	13.8	63.1	46.9
2007	86.5	17.3	65.0	50.5

Sources: Compiled from Economic Research Service/USDA data and American Meat Institute (2009).

Table 25.2 Selected food-borne illness outbreaks associated with poultry.

Year	Potential source	Organism	Place(s)
1982	Poultry giblet gravy	*Salmonella enteritidis*	Maine
1983	Chicken liver paté	*Salmonella heidelberg*	Maine
1986	Improperly thawed and cooked chicken	*Salmonella*	Oklahoma
1988	Turkey franks	*Listeria monocytogenes*	Oklahoma
1990	Improperly thawed and cooked turkey	*Salmonella reading*	Connecticut
1995	Chicken salad	*Cryptosporidium parvum*	Minnesota
1996	Lettuce cross-contaminated with raw chicken	*Campylobacter jejuni*	Oklahoma
1999	Chicken	*Campylobacter jejuni*	England
2000	Chicken salad sandwiches	*Group A rotavirus*	District of Columbia
2000	Deli turkey meat	*Listeria monocytogenes*	New York, Georgia, Connecticut, Ohio, Michigan, Tennessee, California, Utah, Pennsylvania, Wisconsin
2002	Deli turkey meat	*Listeria monocytogenes*	Pennsylvania, New York, Delaware, Maryland, New Jersey, Michigan, Connecticut, Massachusetts
2003	Custard cross-contaminated with raw chicken	*Campylobacter jejuni*	Spain
2005	Chicken salad	*Campylobacter jejuni*	Denmark
2005	Rolls cross-contaminated with raw chicken	*Salmonella typhimurium*	Australia

Sources: Compiled from *Morbidity and Mortality Weekly Reports* with the exception of England 1999 (Pearson *et al.* 2000), Spain 2003 (Jiménez *et al.* 2005), Denmark 2005 (Mazick *et al.* 2006), and Australia 2005 (Moffatt *et al.* 2006).

25.2 GENERAL MICROBIAL TYPES AND LOADS ON POULTRY

Raw poultry is often highly contaminated. The United States Department of Agriculture (USDA) Food Safety and Inspection Service (FSIS) recognizes the challenge in reducing the microbial contamination of such products. The Pathogen Reduction, Hazard Analysis and

Critical Control Point (PR-HACCP) Systems rule, published in 1996, requires poultry plants to develop and implement HACCP plans and sanitation standard operating procedures. All plants must test for generic *Escherichia coli* as an indicator of fecal contamination and must ensure *Salmonella* rates are below standards set in the rule (Schlosser *et al.* 2000; USDA-FSIS 2000b).

A number of microorganisms can be found in raw poultry meat. Lahellec *et al.* (1975) isolated 5920 strains of psychrotrophic bacteria from chicken carcasses in three processing plants, including *Pseudomonas*, *Achromobacter*, *Flavobacterium*, enterobacteriaceae, and *Corynebacterium*. Pathogens such as *Salmonella* and *Campylobacter* are also frequently reported contaminants on poultry. Loads vary widely, ranging from less than $\log_{10}$ 1.0 colony-forming units (CFU)/square centimeter (cm^2) to greater than $\log_{10}$ 7.0 CFU/cm^2, but are reported to most commonly be on the order of $\log_{10}$ 4.0 CFU/cm^2 (National Advisory Committee on Microbiological Criteria for Foods 1997). Sources of poultry microorganisms also range widely and include everything from farm management practices and transportation to processing operations such as defeathering, scalding, and chilling.

25.2.1 Common pathogens on poultry

Poultry products are often thought to be a significant source of *Salmonella*. There are thousands of recognized serovars of this rod-shaped, gram-negative, facultative anaerobe. The bacteria have been reported to grow in a wide range of temperatures and pH levels, are typically inhibited by 3%–4% sodium chloride (NaCl), and generally require an a_w of at least 0.93. *Salmonella* infections resulting in enterocolitis/gastroenteritis produce symptoms ranging from abdominal pain and diarrhea to nausea and vomiting. Other syndromes associated with salmonellosis include enteric (typhoid) fever and chronic conditions such as arthritis and Reiter's syndrome.

Although many consumers commonly associate *Salmonella* with poultry products, detection rates for the organism in such products is often low. In a baseline study conducted prior to implementation of the PR-HACCP rule, Schlosser *et al.* (2000) reported an average rate of 15.9% in chicken carcasses and 11.9% in turkey carcasses for the four most frequently detected *Salmonella* species. Only 10.9% of the 5697 rinse samples taken from broiler carcasses collected from 124 large processing plants during the first year of HACCP implementation were *Salmonella* positive (USDA-FSIS 2000a). Logue *et al.* (2003) reported an overall incidence rate of 16.7% in turkey carcasses collected monthly for one year from US Midwestern processing plants. During the first quarter of 2010, USDA-FSIS (2010) reported only 7% of broilers, 0% of turkeys, 13.1% of ground chicken, and 7.9% of ground turkey samples were positive for *Salmonella*.

Salmonella detection rates for poultry products in countries other than the USA have varied. Dufrenne *et al.* (2001) reported 68% of fresh and frozen chicken and chicken products collected from retail outlets in the Netherlands contained <10 *Salmonella* per carcass, while Straver *et al.* (2007) reported only 8.6% of retail chicken breast fillets collected from five retail outlets in the Netherlands were contaminated with >10 *Salmonella* per fillet. Soultos *et al.* (2003) detected *Salmonella* in only 3 of 205 retail chicken samples in Northern Ireland. Arsenault *et al.* (2007), however, reported a *Salmonella*-positive rate of 31.2% for turkey carcasses slaughtered in Quebec, Canada, while Bohaychuk *et al.* (2006) found 30% of chicken legs collected from retail outlets in Edmonton, Alberta, Canada, were positive for *Salmonella*.

A variety of *Salmonella* serovars have been isolated from poultry products. *Salmonella heidelberg* and *Salmonella hadar* were two of the most common species detected in both chicken and turkey carcasses by Schlosser *et al.* (2000). Zhao *et al.* (2008) recovered 298 *Salmonella enterica* serovar Heidelberg isolates from retail meats, of which 59.7% were from ground turkey and 36.9% were from chicken breasts. Altekruse *et al.* (2006) detected *Salmonella enteritidis* in 280 of 51 327 broiler rinses (0.5%) between 2000 and 2005. The proportion of isolates that were *Salmonella enteritidis*, percentage of positive rinses, establishments testing positive, and the proportion of *Salmonella enteritidis*-positive establishments with multiple positive rinses all increased during the 6-year study period. *Salmonella infantis* was isolated from 100% of carcass and retail poultry meat samples in Hungary by Nógrády *et al.* (2008). Rasschaert *et al.* (2008) also isolated *Salmonella infantis*, as well as *Salmonella agona*, *Salmonella hadar*, *Salmonella typhimurium*, and other serotypes, from neck skins of broilers after slaughter.

Campylobacter spp. were not identified as a significant human pathogen until 1977. The organisms are gram-negative, asporogenous rods that may be aerobic, anaerobic, or microaerophilic depending on the strain. Low pH levels, temperatures less than 30°C, high oxygen levels, and drying inhibit growth of *Campylobacter*. *Campylobacter* infections cause fever, abdominal pain, and diarrhea. Incidence is generally highest in infants, but a potential high risk of campylobacteriosis has also been reported for 18–30 year olds (Bryan & Doyle 1995; Rosenquist *et al.* 2003).

Detection rates for *Campylobacter* spp. in poultry are typically much higher than those for *Salmonella* and are frequently reported to be greater than 50% of samples evaluated (Park & Stankiewicz 1981; Rayes *et al.* 1983; Moore *et al.* 2002; Pezzotti *et al.* 2003; Cui *et al.* 2005; Rodrigo *et al.* 2005; Bohaychuk *et al.* 2006). Meldrum *et al.* (2006) reported overall positive *Campylobacter* rates of 68.6%–73.1% in whole retail raw chicken samples evaluated between 2001 and 2004 in the United Kingdom, while Huat *et al.* (2010) reported *Campylobacter* contamination rates of 84% for chickens in conventional slaughterhouses, 94% in processing plants, and 92% in supermarket samples. Populations of up to 10^7 organisms per uneviscerated chicken carcass (Hood *et al.* 1988) and 10^6 organisms per eviscerated chicken carcass (Stern *et al.* 1985; Tokumaru *et al.* 1991; Waldroup *et al.* 1992; Jorgensen *et al.* 2002) have also been reported.

Campylobacter levels less than 50% have also been reported. Habib *et al.* (2008) reported an overall *Campylobacter*-positive rate of 48.02% for a variety of chicken products, including wings, breasts, legs, sausages, burgers, and ground/minced meat, taken from 11 processing plants in Belgium. The mean *Campylobacter* level was 1.68 $\log_{10}$ CFU/g. Arsenault *et al.* (2007) reported an average *Campylobacter*-positive prevalence rate of 36.9% in turkeys slaughtered in Quebec, Canada, while Dufrenne *et al.* (2001) found about 38% of fresh chicken carcasses and products in the Netherlands were contaminated with <100 *Campylobacter* per sample and 34% were contaminated with >1000 *Campylobacter*.

Campylobacter is reportedly responsible for about 2 million food-borne infections each year (Mead *et al.* 1999; Samuel *et al.* 2004). Wingstrand *et al.* (2006) reported that the primary risk factor for campylobacteriosis in Denmark between October 2000 and September 2001 was consumption of fresh, unfrozen chicken. Other case-controlled studies have implicated handling raw poultry, consumption of chicken and turkey, consumption of undercooked poultry, and cross-contamination of poultry with salads and ready-to-eat foods as sources of *Campylobacter* illnesses (Hopkins & Scott 1983; Hopkins *et al.* 1984; Harris *et al.* 1986; Deming *et al.* 1987).

Escherichia coli (*E. coli*) was first isolated in 1885 by Escherich but was not established as a food-borne pathogen until 1971. The organism, which is typically used as an indicator of fecal contamination, is a gram-negative, asporogeneous rod that ferments lactose to produce acid and gas. *E. coli* is subdivided into five virulence groups, including enteroaggregative, enterohemorrhagic, enteroinvasive, enteropathogenic, and enterotoxigenic. The enterohemorrhagic *Escherichia coli* O157:H7 strain is one of the most studied members of this food-borne pathogen.

Testing for generic *Escherichia coli* is required by the PR-HACCP systems rule for all poultry slaughter plants as an indication of proper processing and sanitation procedures (USDA-FSIS 2000b). Detection rates for the organism in poultry have varied widely, with reported levels ranging from 11.9% to 71% in turkey (Zhao *et al.* 2001; Kegode *et al.* 2008) and 38.7% to nearly 100% in chicken (Zhao *et al.* 2001; Northcutt *et al.* 2003; Forward *et al.* 2004; Kegode *et al.* 2008). *Escherichia coli* was the most frequently detected bacteria in a study by Jiménez *et al.* (2003). The organism was identified in 50% of whole chicken carcasses without visible fecal contamination and 59.7% of carcasses with visible contamination.

Listeria is widely distributed in nature and has been detected in foods ranging from milk and soft cheese to poultry and raw vegetables. Growth of the rod-shaped, gram-positive organism is best at a neutral pH and is enhanced by glucose. *Listeria* prefers colder environments and is able to grow in the presence of 10% NaCl. Pregnant women and those with compromised immune systems are particularly susceptible to listeriosis. Meningitis, sepsis, and stillbirths are commonly associated with the illness.

Although *Listeria* is often associated with dairy products such as milk and soft cheeses, the bacteria has frequently been detected on poultry products. Studies by Genigeorgis *et al.* (1989, 1990) indicated rates of 40.6% in chicken and 45.0% in turkey for overall *Listeria*. Forty-eight percent of retail raw chicken samples collected from outlets in Northern Ireland were positive for the organism in a study by Soultos *et al.* (2003). Several researchers (Antunes *et al.* 2002; Barbalho *et al.* 2005) have reported incidence rates of 100% for *Listeria* in poultry products.

Listeria monocytogenes is often the most prevalent type of *Listeria* detected and/or analyzed in poultry products. *Listeria monocytogenes* was the most frequently isolated species (18%) in retail poultry samples in Northern Ireland (Soultos *et al.* 2003) and in poultry carcasses (41%) in Portugal (Antunes *et al.* 2002). Miettinen *et al.* (2001) detected *L. monocytogenes* at rates ranging from 1% to 19% on the air chiller, skin-removing machine, and the conveyor belt leading to the packaging area in poultry-processing plants and abattoirs and in 62% of raw broiler pieces purchased from retail outlets in Finland. Gudbjörnsdóttir *et al.* (2004) reported *L. monocytogenes* rates of 20.6%–24.1% in poultry plants and 22% on raw poultry products in a study conducted in Finland, Iceland, Norway, and Sweden. Seventy percent of raw broiler legs were positive for *L. monocytogenes* in a study conducted in Estonia (Praakle-Amin *et al.* 2006). The organism was more prevalent in broiler legs originating in Estonia (88%) than in legs imported from Denmark, Finland, Hungary, Sweden, and the USA (53%).

25.2.2 Common spoilage organisms on poultry

As with most meat processing, poultry slaughter involves a shift from a mesophilic environment at the start of processing to a psychrotrophic environment at the end of the production cycle. Spoilage of poultry is therefore most often associated with psychrotrophic microorganisms. Storage and packaging conditions also influence the microorganisms associated with spoilage in poultry products.

Pseudomonas spp. are the most common organisms associated with spoilage of refrigerated meats such as poultry stored under aerobic conditions. In addition to being psychrotrophic, the pseudomonads are able to grow at pH levels of 5.5–7.0 and are highly oxidative, which gives them the ability to use nitrogen compounds as an energy source. These characteristics give pseudomonads a competitive advantage in spoilage of poultry and other meats. Levels of *Pseudomonas* spp. in poultry prior to treatment have been reported to be in the 3.0–4.0 $\log_{10}$ CFU/g range (Chouliara *et al.* 2007, 2008; Patsias *et al.* 2008; Economou *et al.* 2009).

Lactobacillus spp. are often associated with spoilage of refrigerated poultry stored under microaerophilic or anaerobic conditions. These gram-positive rods are common on many foods but are generally present in very low levels on freshly slaughtered poultry. Cold temperatures and vacuum packaging, however, encourage the growth of *Lactobacillus*.

Yeast loads have been reported to increase in both chicken and turkey during cold storage (Ismail *et al.* 2000; Chouliara *et al.* 2008), and yeasts were the fifth most common organism detected in a study of 5920 psychrotrophic organisms isolated from chickens (Lahellec *et al.* 1975). Proteolytic and/or lipolytic yeasts such as *Candida zeylanoides* and *Yarrowia lipolytica* are some of the most common species isolated from fresh and spoiled poultry (Diriye *et al.* 1993; Viljoen *et al.* 1998).

25.3 MICROBIAL LOADS AND SENSORY PROPERTIES OF FRESH AND FROZEN POULTRY

Spoilage of fresh and frozen poultry meat is associated with changes in the organoleptic properties of the products. The psychrotrophic microorganisms responsible for spoilage of poultry meat are typically present at low levels immediately after slaughter but flourish rapidly during cold storage to produce changes in the aroma, appearance, and texture of the meat.

Off odors are typically one of the first organoleptic characteristics to be detected in spoiling poultry and become noticeable when microbial levels reach 10^6–10^8 CFU/cm^2. Pseudomonads are by far the most commonly reported microorganisms associated with the off odors of spoiled poultry. Most of the microbial load in chicken used in a study by Chen *et al.* (1991) was found to be pseudomonads. When rated on a scale of 1 = absent or excellent to 9 = extremely poor, aroma scores of both raw and simmered chicken were positively correlated with microbial loads. Intensity of aroma increased as aroma became less desirable. Controlled spoilage at 3°C resulted in odors described subjectively as "sulfur," "dishrag," "ammonia," "wet dog," "skunk," "dirty socks," "rancid fish," and "canned corn" (Russell *et al.* 1995). These odors were attributed primarily to *Shewanella putrefaciens* A, B, and D; *Pseudomonas fluorescens* A, B, and D; and *Pseudomonas fragi*. Subjective detection of sulfur-like odors associated with pseudomonads such as *Pseudomonas fluorescens* and *Pseudomonas putida* has also been reported by other researchers (Viehweg *et al.* 1989; Russell 1997).

Off odors detected subjectively in spoiling poultry are associated with sulfur-containing compounds. Dimethyl sulfide, dimethyl disulfide, and propylene sulfide are some of the compounds frequently detected in spoiled poultry (Pittard *et al.* 1982; Bowman *et al.* 1983). Lovestead and Bruno (2010) identified dimethyl disulfide, dimethyl trisulfide, phenyl sulfide, and allyl methyl sulfide as well as several other compounds as potential markers for poultry spoilage using a purge-and-trap method. Although sulfur-containing compounds are typically associated with organisms such as *Pseudomonas fluorescens* and *Pseudomonas putida*, the specific strain of bacteria, the storage conditions, and the substrate will influence the exact compounds produced during low temperature spoilage of poultry.

Sliminess in poultry typically occurs after off odors are apparent, and has been associated with microbial levels of 10^6–10^9 CFU/cm^2. Vareltzis *et al.* (1997) noted slime formation under the wings and between the leg and breast of whole chicken carcasses after 5 days of storage in oxygen-permeable packaging at 4°C. *Pseudomonas* populations on the carcasses were reported to be 10^6 CFU/cm^2 at the time slime was detected. A positive correlation between subjective raw sliminess and microbial loads was reported by Chen *et al.* (1991). The slime smear test was positively correlated to both sensory evaluation of sliminess and to pseudomonad, total aerobic plate, total aerobic psychrotrophic, and fluorescent pseudomonad counts.

25.4 METHODS FOR REDUCING MICROBIAL LOADS AND SENSORY PROPERTIES OF POULTRY

25.4.1 Atmosphere modifications

Since it is primarily a surface phenomenon, poultry spoilage is most commonly the result of growth of aerobic bacteria. Modification of the environment around the meat can therefore influence the microflora as well as subsequent spoilage characteristics. Vacuum packaging, carbon dioxide, and nitrogen are some of the most common atmosphere modifications made to increase the shelf life of poultry.

Vacuum packaging involves evacuation of air from the package. The lack of oxygen inhibits growth of gram-negative psychrotrophs such as *Pseudomonas* but results in the growth of facultative anaerobes. Lactic acid bacteria, in particular *Lactobacillus* spp., have been reported to be the predominant spoilage microorganisms in poultry packaged under vacuum (Bailey *et al.* 1979; Thomas *et al.* 1984; Sawaya *et al.* 1993). Although vacuum packaging undoubtedly extends the shelf life of poultry, the effects of such atmospheric modification are dependent on several factors.

Storage temperature has a significant effect on the shelf life of vacuum-packaged poultry. Although a panel of five to six judges rated odor as unacceptable in vacuum-packaged chicken carcasses after 17 days of storage at 4°C, color, texture, and overall appearance were acceptable through 20 days of storage (Sawaya *et al.* 1993). Vacuum-packaged carcasses stored at 7°C were rated unacceptable for all characteristics after 17 days and those stored at 9°C were unacceptable at 10 days. Carcasses stored in conventional packages at the same temperatures were judged unacceptable 3–4 days sooner than vacuum-packaged samples. Vacuum-packaged whole broilers were free of slime through 13 days of storage at about 5°C, but objectionable ammonia-like odors were noted by judges at day 9 in a study by Thomas *et al.* (1984). A less appealing appearance was also noted in vacuum-packaged samples due to yellowing of the skin. Lactic acid odors and greening were reported by Patterson *et al.* (1984) in vacuum-packaged chicken breast portions after 42 days and in leg and thigh portions after 35 days when stored at 1°C. However, off odors were detected after only 28 days in breasts and 15 days in legs and thighs when stored at 4°C–5°C.

Interactions between storage temperature and treatments applied to poultry prior to vacuum packaging also affect the shelf life of the products. Potassium sorbate has generally been found to reduce microbial growth and increase shelf life when vacuum-packaged poultry is stored at temperatures of 1°C–2°C (McMeekin *et al.* 1984; Patterson *et al.* 1984). The positive effects of the sorbate, however, are negated when storage temperatures increase. An untrained taste panel could not detect flavor changes in sorbate-dipped, vacuum-packaged breast fillets through 6 weeks of storage at 2°C nor could the panel differentiate between sorbate-dipped fillets and control, vacuum-packed, or untreated fillets (McMeekin *et al.* 1984). Lactic acid also increased the shelf life of vacuum-packaged chicken legs and thighs stored at 1°C, but the skin and meat of the pieces were described as "gray and unattractive," while those dipped in the potassium sorbate were described as having "an attractive white appearance" (Patterson *et al.* 1984).

Flushing packages with gases such as CO_2 and nitrogen generally results in an increase in shelf life of poultry products by extending the lag phase of aerobic microorganisms. Facultative and true anaerobic microorganisms, however, often find such environments acceptable and are commonly found in high levels in spoiled products stored under such atmospheric modifications. The shift in microflora from aerobic bacteria such as *Pseudomonas* to facultative/anaerobic bacteria such as *Lactobacillus* results in changes in the organoleptic properties and spoilage characteristics.

As with other treatments, shelf life of CO_2-treated poultry is temperature dependent. An environment of 65% CO_2 resulted in lower microbial loads and a slightly longer shelf life (19 days) than environments of either 20% CO_2 (18 days) or air (14 days) when chicken samples were stored at 2°C (Bailey *et al.* 1979). Spoilage was defined as "strong off odor" and was associated with microbial loads of 10^6 CFU/cm^2 with *Lactobacillus* comprising more than 90% of the microorganisms in the spoiled chicken. Thomas *et al.* (1984) noted off odors in CO_2-treated whole chickens at 20 days and cut-up broilers at 17 days when stored at about 5°C. Microbial loads were reported to be 10^6 for mesophiles and 10^4 for lactobacilli at the time of spoilage. Breast portions had a shelf life of 42 days and leg/thigh portions a shelf life of 35 days when treated with either CO_2 or nitrogen and stored at 1°C (Patterson *et al.* 1984). Microbial loads were 10^7 CFU/cm^2 at the end of storage for both breasts and legs/thighs. *Listeria monocytogenes* growth was arrested by storage at 1°C, regardless of atmosphere, and inhibited by environments containing CO_2 when chicken breasts were held at 6°C (Hart *et al.* 1991). Pseudomonads predominated in breasts stored aerobically and were associated with strong, fruity odors. *Carnobacterium* and *Lactobacillus* spp. predominated in breasts stored under CO_2. The odor of spoiled poultry stored under CO_2 is often described as acid or sour (Bailey *et al.* 1979; Patterson *et al.* 1984; Hart *et al.* 1991). Hart *et al.* (1991) also reported cheese odors. Such odors are typically attributed to the presence of lactobacilli. Patterson *et al.* (1984) also noted sulfide odors in gas-flushed packages of legs/thighs after 15 days of storage at 4°C. Fecal odors detected in breast portions at 42 days when stored at 1°C and 28 days when stored at 4°C were associated with increases in *Enterobacteriaceae*. Chicken thighs treated with a combination of potassium sorbate and CO_2 were described as slightly sour at a total microbial load of 10^7 CFU/cm^2, while thighs treated with sorbate and vacuum packaging were described as putrid at the same microbial level (Elliott *et al.* 1985).

Nitrogen, alone or in combination with CO_2, has also been used to modify the atmosphere around poultry products. Chouliara *et al.* (2007) evaluated the shelf life of chicken breast meat packaged under atmospheres of 30% CO_2/70% N_2 and 70% CO_2/30% N_2. Taste and odor intensities were evaluated on a scale of 0 (least acceptable) to 5 (most acceptable) with a score of 3.5 set as the lower limit of acceptability. Both taste and odor were scored 3.5 after 6 days when packaged in air. Total viable counts for the air-packaged samples at day 6 were 7.19 $\log_{10}$ CFU/g, *Pseudomonas* spp. loads were 6.28 $\log_{10}$ CFU/g, and lactic acid bacteria populations were 6.41 $\log_{10}$ CFU/g. Chicken packaged under 30% CO_2/70% N_2 was scored 3.5 at 10–11 days for odor and at 5–6 days for taste, while samples stored under 70% CO_2/30% N_2 reached the 3.5 lower acceptability limit at 11–12 days for odor and 7–8 days for taste. Although all samples reached the 3.5 lower acceptability limit at 6–7 days for taste, microbial loads for the samples stored under both modified atmospheres were lower for all types of microorganisms evaluated compared to air-packaged samples.

Modified atmospheres also affect poultry color. Turkey meat stored under 100% CO_2 plus an O_2 scavenger had the most stable red values, the lowest myoglobin oxidation rates, and the lowest microbial loads compared to turkey stored under 100% O_2, 100% N_2, or 25% CO_2/9% N_2/66% O_2 (Sante *et al.* 1994). Yellowness of the CO_2-treated turkey also increased slightly over 21 days of storage at 3°C–4°C. A 100% CO_2 concentration has also been reported to result in a whitening of chicken skin (Elliott *et al.* 1985). A higher level of redness and more appealing color characteristics were reported for ground chicken and turkey meats packaged under a modified atmosphere of 20% CO_2/80% N_2 but coliform and *Escherichia coli* levels were lower in samples packaged in a modified atmosphere containing 62% CO_2/8% O_2/30% N_2 (Saucier *et al.* 2000). Ellis *et al.* (2006) used a combination of modified atmosphere packaging and slow or fast release chlorine dioxide sachets to extend the shelf life of chicken breasts. Total plate counts increased for all treatments after 6–9 days but those containing chlorine dioxide sachets had lower counts compared to samples without the sachets. Use of the sachets was associated with a significantly improved odor but discoloration of the chicken was evident, particularly in areas in close proximity to the sachets.

25.4.2 Use of organic and inorganic compounds

The antimicrobial activity of acids such as lactic, acetic, and citric is associated with both a decrease in pH and the presence of undissociated molecules (Ouattara *et al.* 1997). Weak organic acids tend to be more effective than strong acids, but concentration, the acid dissociation constant, and the type of acid all affect the antimicrobial activity. The use of acids can also have significant effects on spoilage characteristics and organoleptic properties of the poultry.

Acetic acid had greater antimicrobial activity than either lactic or citric acids in a study by Kim and Marshall (2000). Chicken wings immersed in a 1% acetic acid solution for 10 min and stored at 4°C had lower aerobic plate counts for 12 days compared to wings treated with 1% solutions of either lactic or citric acids. Untrained sensory panelists were unable to detect color differences among the treatments even though Hunter color values indicated wings immersed in acetic acid were lighter, less red, and less yellow than untreated wings. Significant reductions in microbial loads and no differences in skin appearance in whole broiler carcasses treated with 1% acetic acid (Dickens & Whittemore 1997) and either 0.5% acetic acid or 0.25% lactic acid (Sakhare *et al.* 1999) have also been reported. Odor of acid-treated wings (Kim & Marshall 2000) was always rated higher by panelists compared to controls with those treated with acetic acid having the most acceptable odor over the 12-day storage period. Off odors were also suppressed by 1% acetic acid in a study by Jiménez *et al.* (1999). Chicken breasts in that study reportedly had a slightly acidic and pleasant odor with no slime formation after 21 days of storage. All bacterial genera evaluated in the study, including pseudomonads and enterobacteria, were reduced by the acetic acid treatment.

Broiler carcasses treated with food-grade vinegar (0.6% acetic acid) darkened/yellowed and the feather follicles puckered according to Dickens *et al.* (1994). No differences in Warner–Bratzler shear values and sensory triangle tests were detected for treated and untreated muscle cooked by either boiling in sealed bags or oven roasting in foil boats. Although acetic acid was not effective in reducing total aerobic counts, it did significantly reduce the *Enterobacteriaceae* counts.

Lactic acid was more effective than citric acid (CA) for decontaminating chicken skin (Deumier 2004). Although total viable counts, *Enterobacteriaceae*, total coliforms, and *Escherichia coli* levels were reduced by the acid treatments, increasing acid concentrations resulted in yellowing and greening of the chicken samples. Gulmez *et al.* (2006) reported lactic acid was superior to water or sumac extract in decontamination of both spoilage and pathogenic microorganisms on chicken wings but resulted in an undesirable pale color and acid odor in the samples. Color of chicken legs was also adversely affected by treatment with 2% and 5% lactic acid in a study by González-Fandos and Dominguez (2006). *Listeria monocytogenes* levels were reduced but not eliminated and lag phase of microbial growth was extended by the lactic acid treatments. Chicken legs treated with a 10% lactic acid buffered system were judged acceptable for smell and visual appearance by 10 trained judges after 12 days' storage at 6°C (Zeitoun & Debevere 1990). Untreated legs stored for 6 days at 6°C were judged unacceptable even though aerobic psychrotrophic loads were similar to those of the treated samples at 12 days' storage. Hydrogen sulfide and other sulfide-like flavors were detected after eight days' storage at 6°C in cooked legs that had been treated with 2% lactic acid. van der Marel *et al.* (1988) also reported typical sulfide-like odors in chicken legs treated with 5% lactic acid after 9 days of storage at 6°C. No other adverse sensory effects were reported due to the lactic acid. Sinhamahapatra *et al.* (2004) reported a reduction in total microbial loads of 1.36 log/cm^2 with no detectable effects on appearance, smell, tenderness, or overall acceptability in dressed broilers treated with 2% lactic acid.

Anang *et al.* (2006) evaluated oxalic acid dips at concentrations ranging from 0% to 2% for 10–30 min on microbial and quality characteristics of chicken breasts. *Pseudomonas* spp. populations on treated breasts were 3.88 log CFU/g compared to 8.83 log CFU/g for untreated breasts after 14 days' storage at 4°C. Similar reductions were noted for *Enterobacteriaceae* counts. Breasts treated with oxalic acid were darker, more red, and more yellow compared to untreated samples.

Spoilage bacteria, including total aerobic loads, *Brochothrix thermosphacta*, *Pseudomonas*, and coliform counts, were always lower for chicken carcasses treated with sodium tripolyphosphate compared to controls (Vareltzis *et al.* 1997). Sensory evaluation using a trained panel found treated carcasses had a putrid odor after 7 days of storage at 4°C, while untreated carcasses were putrid after 4 days. Judges detected slime formation after 5 days in control carcasses, but slime was only beginning to form on treated carcasses after 8 days at 4°C. Rodriguez de Ledesma *et al.* (1996) used trisodium phosphate (TSP) and hot water alone and in combination to decontaminate chicken wings. A combination of dipping in 10% TSP followed by 95°C hot water resulted in reductions of 99.67% for *Salmonella typhimurium*, 96.68% for *Listeria monocytogenes*, and 94.88% for *Staphylococcus aureus* when stored at 4°C. Organoleptic evaluation by a six-person panel indicated wings treated with the TSP/hot water combination were preferred at 6 days' storage at 4°C based on odor and appearance. del Rio *et al.* (2008) compared microbial and sensory properties of chicken legs dipped in a 12% solution of (TSP) with those dipped in water, 1200 ppm acidified sodium chlorite (ASC), 2% CA, and 220 ppm peroxyacids (PA). TSP, ASC, and CA had the strongest antimicrobial activity, with TSP particularly effective at reducing mesophilic bacteria, *Micrococcaceae*, pseudomonads, and lactic acid bacteria. Only samples treated with TSP, ASC, or CA were scored greater than 6 (the lowest limit of acceptability) for color, smell, and overall acceptability at the end of 5 days' storage at 3°C by an untrained sensory panel. ASC was also reported to be an effective antimicrobial treatment on air-chilled broiler carcasses with minimal effects on oxidative, visual, and flavor characteristics by Schneider *et al.* (2002).

Russell (2008) evaluated the effectiveness of an acidic, copper sulfate-based commercial sanitizer on microbiological quality of chicken broiler carcasses. When added to the scalder during processing, the sanitizer was able to eliminate *Salmonella typhimurium, Listeria monocytogenes*, *Staphylococcus aureus*, *Pseudomonas fluorescens*, and *Shewanella putrefaciens*, and reduced *Escherichia coli* by 4.9 $\log_{10}$. Broiler carcasses dipped in the sanitizer after processing had significantly lower psychrotrophic bacterial loads from days 6 through 14 of storage and improved odor scores from days 8 through 14 of storage. An extension of 4 days was estimated for shelf life of broiler carcasses treated with the sanitizer.

25.4.3 Irradiation

The use of irradiation in doses of 1.5–3.0 kilogray (kGy) to eliminate bacteria on poultry was approved by the U.S. Food and Drug Administration (FDA) in 1992 (Federal Register 1992). Studies have shown that such treatment is effective in eliminating pathogens and extending the refrigerated shelf life of poultry. Factors such as irradiation dose, storage temperature, and poultry characteristics also affect the organoleptic properties of the irradiated poultry.

An irradiation dose of as little as 1.0 kGy was effective in eliminating coliforms, generic *Escherichia coli*, *Salmonella*, *Campylobacter*, and psychrotrophs on boneless, skinless chicken breasts and reduced total aerobic bacteria from $\log_{10}$ 4.6 to $\log_{10}$ 2.23 (Lewis *et al.* 2002). A consumer taste panel found no differences in appearance, texture, flavor, color desirability, and overall acceptability between irradiated and nonirradiated breast fillets stored at 0°C at 0 days' storage, but irradiated samples had lower texture and flavor scores at 14 days' storage. Texture, flavor, and overall acceptability of the irradiated breasts were even less desirable at 28 days' storage. Objective measurements also indicated lipid oxidation increased and breasts became redder with increasing irradiation level and storage time.

Viable *Listeria monocytogenes* was still present in chicken breasts cooked to an internal temperature of 71.1°C under vacuum (sous vide) but was undetectable in samples treated with a combination of sous vide and an irradiation dose of 3.1 kGy (Shamsuzzaman *et al.* 1995). Evaluation of odor, flavor, and texture indicated nonirradiated sous vide breasts spoiled at 16 days when stored at 8°C, while those treated with an irradiation dose of 1.0 kGy spoiled at 23 days. Breasts treated with 2.0 and 3.0 kGy were acceptable through day 35. No significant differences in odor, flavor, or texture were associated with any of the irradiation doses evaluated.

Abu-Tarboush *et al.* (1997) also reported few differences in the sensory properties of raw and cooked chicken breasts and thighs taken from whole carcasses irradiated with doses of 2.5, 5.0, 7.5, and 10.0 kGy. Panelists rejected samples treated with 2.5 kGy, however, at day 21 when stored at 4°C. As has been noted in other studies, a pink color in raw samples and elimination by cooking of some of the volatile compounds associated with odor deterioration were reported for irradiated samples. Although panelists were able to detect specific problems in raw meat, including rancidity in thighs and odor deterioration in breasts, that resulted in rejection, color, appearance, odor, juiciness, taste, and tenderness of cooked samples were acceptable for samples irradiated with doses above 2.5 kGy through 21 days of storage at 4°C.

A radiation dose of 2.5 kGy reduced the aerobic microbial load on chicken breast halves packaged on polystyrene trays overwrapped with polyvinyl chloride (PVC) film by two log cycles compared to untreated controls (Lescano *et al.* 1991). Although the aerobic count reached 10^6 CFU/g on the 19th day of storage at 2°C, *Escherichia coli*, enterococci, and presumptive *Salmonella* were not detected in irradiated breasts. Lynch *et al.* (1991) reported negligible microorganism growth in chill-stored turkey breast fillets irradiated at 2.5 kGy.

In contrast to other studies, a strong irradiation odor was detected in the raw treated meat and was related to an unpleasant odor detected by trained sensory panelists by both Lescano *et al.* (1991) and Lynch *et al.* (1991). Odors were described as sour, rancid, manure, bad meat, and putrid when turkey breast fillets were stored in oxygen-impermeable films (Lynch *et al.* 1991). A slight pink color developed in irradiated chicken breasts over a 22-day storage period (Lescano *et al.* 1991), while an intense pink color was reportedly associated with irradiation of turkey breast fillets (Lynch *et al.* 1991). Lescano *et al.* (1991) also reported tenderness of cooked, irradiated chicken was lower than that of untreated samples and was related to a decrease in water-holding capacity. Flavor and overall acceptability of irradiated samples were similar to controls through 22 days' storage. Both rancidity (by peroxide value) and free fatty acids decreased in the chicken samples irradiated with 2.5 kGy.

Irradiation was more effective than modified atmospheres containing 30% CO_2/70% N_2 and 70% CO_2/30% N_2 in reducing levels of *Pseudomonas*, lactic acid bacteria, and *Enterobacteriaceae* on chicken breast meat (Chouliara *et al.* 2008). A combination of 70% CO_2/30% N_2 plus irradiation at 4 kGy resulted in a shelf-life extension of 12 days compared to air-packaged samples when stored at 4°C. Irradiation was associated with a very low level of lipid oxidation and no changes in L^* or b^* values, but an increase in redness was noted in particular for samples treated with 4 kGy. The lower acceptability sensory rating for odor of 3.5 was reached at 12 days in samples irradiated at 2.0 kGy and 15–16 days for those treated at 4.0 kGy compared to 6 days for air-packaged samples. For taste, the lower acceptability level was reached at 6 days for air packaging, 8–9 days for the 2.0 kGy dose level, and 15–16 days for the 4.0 kGy level of irradiation. An "irradiation odor" was noted for irradiated samples during the first days of storage but was found to dissipate as storage time increased.

As in other studies, Gomes *et al.* (2006) noted an increase in redness in Cornish hens irradiated at dosages of 2, 3, and 7 kGy. In contrast to other studies, however, irradiation was associated with toughening of the chicken meat. Although lipid oxidation as determined by malonaldehyde content increased as irradiation dosage and storage time increased, sensory panelists did not detect oxidation in the samples. Nonirradiated carcasses reached a total viable microorganism count of 6 log CFU/50 cm^2 in less than 14 days at 4°C, while those irradiated at 2 and 3 kGy did not reach that microbial level until day 21 of storage. Microbial loads for Cornish hens irradiated at 7 kGy did not exceed 2.5 CFU/50 cm^2 at any point during the 21-day storage period.

Several compounds have been investigated as a means of reducing the increase in lipid oxidation associated with irradiation. Jayathilakan *et al.* (2009) found 2% lactic acid controlled the development of lipid oxidation in hurdle processed chicken legs through 6 months' storage at 5°C. A 3 log reduction in total microbial loads, a 2 log reduction in yeast/mold loads, and the absence of pathogens were reported for legs treated with 2 kGy. Chicken legs treated with a combination of 2 kGy irradiation plus 2% lactic acid had an overall acceptability score of 7.84 on a 9-point hedonic scale after 6 months' storage at 5°C. Rababah *et al.* (2010) found a combination of green tea and grape seed extracts at 6000 ppm effectively minimized lipid oxidation in chicken breasts over 9 months' storage at −20°C. The plant extracts also decreased the redness, hardness, and maximum shear force associated with irradiation of the samples.

25.5 CONCLUSION

Microbial contamination of poultry products remains a problem for the food industry. Pathogens such as *Campylobacter*, *Salmonella*, and *Listeria* and spoilage organisms such as *Pseudomonas* can play significant roles in the safety and shelf life of poultry products. Although washing and sanitizing, treatment with organic and inorganic compounds, modified atmosphere packaging/storage, irradiation, and other processes have been investigated as potential methods for reducing microbial loads on poultry, the effect of such methods on color, flavor, texture, and other sensory properties is often undesirable. New technologies such as hydrodynamic pressure and ozone treatment are only now being investigated and may provide a means in the future to produce safer poultry products with acceptable organoleptic properties.

REFERENCES

Abu-Tarboush, H.M., Al-Kahtani, H.A., Abou-Arab, A.A., Bajaber, A.S. & El-Mojaddidi, M.A. (1997) Sensory and microbial quality of chicken as affected by irradiation and postirradiation storage at 4°C. *Journal of Food Protection*, **60** (7), 761–770.

Altekruse, S.F., Street, D.A., Fein, S.B. & Levy, A.S. (1996) Consumer knowledge of foodborne microbial hazards and food-handling practices. *Journal of Food Protection*, **59** (3), 287–294.

Altekruse, S.F., Bauer, N., Chanlongbutra, A., *et al.* (2006) *Salmonella* enteritidis in broiler chickens, United States, 2000–2005. *Emerging Infectious Disease*, **12** (12), 1848–1852.

American Meat Institute. (2009) U.S. meat and poultry production and consumption: An overview. AMI Fact Sheet. Accessed June 2010 at www.MeatAMI.com

Anang, D.M., Rusul, G., Radu, S., Bakar, J. & Beuchat, L.R. (2006) Inhibitory effect of oxalic acid on bacterial spoilage of raw chilled chicken. *Journal of Food Protection*, **69** (8), 1913–1919.

Antunes, P., Réu, C., Sousa, J.C., Pestana, N. & Peixe, L. (2002) Incidence and susceptibility to antimicrobial agents of *Listeria* spp. and *Listeria monocytogenes* isolated from poultry carcasses in Porto, Portugal. *Journal of Food Protection*, **65** (12), 1888–1893.

Arsenault, J., Letellier, A., Quessy, S., Morin, J.P. & Boulianne, M. (2007) Prevalence and risk factors for *Salmonella* and *Campylobacter* spp. carcass contamination in turkeys slaughtered in Quebec, Canada. *Journal of Food Protection*, **70** (6), 1350–1359.

Bailey, J.S., Reagan, J.O., Carpenter, J.A., Schuler, G.A. & Thomson, J.E. (1979) Types of bacteria and shelf-life of evacuated carbon dioxide-injected and ice-packed broilers. *Journal of Food Protection*, **42** (3), 218–221.

Barbalho, T.C.F., Almeida, P.F., Almeida, R.C.C. & Hofer, E. (2005) Prevalence of *Listeria* spp. at a poultry processing plant in Brazil and a phage test for rapid confirmation of suspect colonies. *Food Control*, **16**, 211–216.

Bean, N.H. & Griffin, P.M. (1990) Foodborne disease outbreaks in the United States, 1973–1987. Pathogens, vehicles, and trends. *Journal of Food Protection*, **53**, 804–817.

Bohaychuk, V.M., Gensler, G.E., King, R.K., Manninen, K.I., Sorensen, O., Wu, J.T., Stiles, M.E. & McMullen, L.M. (2006) Occurrence of pathogens in raw and ready-to-eat meat and poultry products collected from the retail marketplace in Edmonton, Alberta, Canada. *Journal of Food Protection*, **69** (9), 2176–2182.

Bowman, E.V., Freeman, L.R., Later, D.W. & Lee, M.L. (1983) Comparison of volatiles produced by selected pseudomonads on chicken skin. *Journal of Food Science*, **48**, 1358–1359.

Brown, P., Kidd, D., Riordan, T. & Barrell, R.A. (1988) An outbreak of food-borne *Campylobacter jejuni* infection and the possible role of cross-contamination. *Journal of Infection*, **17**, 171–176.

Bryan, F.L. (1980) Foodborne diseases in the United States associated with meat and poultry. *Journal of Food Protection*, **43**, 140–150.

Bryan, F.L. & Doyle, M.P. (1995) Health risks and consequences of *Salmonella* and *Campylobacter jejuni* in raw poultry. *Journal of Food Protection*, **58** (3), 326–344.

Chen, Y., Woodburn, M.J. & Kelsey, M.W. (1991) Microbial and sensory quality of refrigerated market fryers. *Journal of Food Protection*, **54** (9), 704–710.

Chouliara, E., Badeka, A., Savvaidis, I. & Kontominas, M.G. (2008) Combined effect of irradiation and modified atmosphere packaging on shelf-life extension of chicken breast meat: Microbiological, chemical and sensory changes. *European Food Research and Technology*, **226**, 877–888.

Chouliara, E., Karatapanis, A., Savvaidis, I.N. & Kontominas, M.G. (2007) Combined effect of oregano essential oil and modified atmosphere packaging on shelf-life extension of fresh chicken breast meat stored at 4°C. *Food Microbiology*, **24**, 607–617.

Cui, S., Ge, B., Zheng, J. & Meng, J. (2005) Prevalence and antimicrobial resistance of *Campylobacter* spp. and *Salmonella* serovars in organic chickens from Maryland retail stores. *Applied and Environmental Microbiology*, **71** (7), 4108–4111.

Deming, M.S., Tauxe, R.V., Blake, P.A., *et al.* (1987) *Campylobacter enteritis* at a university: Transmission from eating chicken and from cats. *American Journal of Epidemiology*, **126**, 526–534.

Deumier, F. (2004) Pulsed-vacuum immersion of chicken meat and skin in acid solutions. Effects on mass transfers, colour and microbial quality. *International Journal of Food Science and Technology*, **39**, 277–286.

Dickens, J.A. & Whittemore, A.D. (1997) Effects of acetic acid and hydrogen peroxide application during defeathering on the microbiological quality of broiler carcasses prior to evisceration. *Poultry Science*, **76**, 657–660.

Dickens, J.A., Lyon, B.G., Whittemore, A.D. & Lyon, C.E. (1994) The effect of acetic acid dip on carcass appearance, microbiological quality and cooked breast meat texture and flavor. *Poultry Science*, **73**, 576–581.

Diriye, F.U., Scorzetti, G. & Martini, A. (1993) Methods for the separation of yeast cells from the surfaces of processed frozen foods. *International Journal of Food Microbiology*, **19**, 27–37.

Dufrenne, J., Ritmeester, W., Delfgou-van Asch, E. van Leusden, F. & de Jonge, R. (2001) Quantification of the contamination of chicken and chicken products in the Netherlands with *Salmonella* and *Campylobacter*. *Journal of Food Protection*, **64** (4), 538–541.

Economou, T., Pournis, N., Ntzimani, A. & Savvaidis, I.N. (2009) Nisin-EDTA treatments and modified atmosphere packaging to increase fresh chicken meat shelf-life. *Food Chemistry*, **114**, 1470–1476.

Elliott, P.H., Tomlins, R.I. & Gray, R.J.H. (1985) Control of microbial spoilage on fresh poultry using a combination of potassium sorbate/carbon dioxide packaging system. *Journal of Food Science*, **50**, 1360–1363.

Ellis, M., Cooksey, K., Dawson, P., Han, I. & Vergano, P. (2006) Quality of fresh chicken breasts using a combination of modified atmosphere packaging and chlorine dioxide sachets. *Journal of Food Protection*, **69** (8), 1991–1996.

Federal Register. (1992) Irradiation of poultry products; Final rule. Food Safety and Inspection Service. *Federal Register*, **57**, 43588–43600.

Forward, K.R., Matheson, K.M., Hiltz, M., Musgrave, H. & Poppe, C. (2004) Recovery of cephalosporin-resistant *Escherichia coli* and *Salmonella* from pork, beef and chicken marketed in Nova Scotia. *Canadian Journal of Infectious Diseases & Medical Microbiology*, **15** (4), 226–230.

Genigeorgis, C.A., Dutulescu, D. & Fernandez-Garayzabal, J. (1989) Prevalence of *Listeria* spp. in poultry meat at the supermarket and slaughterhouse level. *Journal of Food Protection*, **52** (9), 618–624.

Genigeorgis, C.A., Oanca, P. & Dutulescu, D. (1990) Prevalence of *Listeria* spp. in turkey meat at the supermarket and slaughterhouse level. *Journal of Food Protection*, **53** (4), 282–288.

Gomes, C., Da Silva, P.F., Castell-Perez, M.E. & Moreira, R.G. (2006) Quality and microbial population of Cornish game hen carcasses as affected by electron beam irradiation. *Journal of Food Science*, **71** (7), E327–E336.

González-Fandos, E. & Dominguez, J.L. (2006) Efficacy of lactic acid against *Listeria monocytogenes* attached to poultry skin during refrigerated storage. *Journal of Applied Microbiology*, **101**, 1331–1339.

Gudbjörnsdóttir, B., Suihko, M.L., Gustavsson, P., *et al.* (2004) The incidence of *Listeria monocytogenes* in meat, poultry and seafood plants in the Nordic countries. *Food Microbiology*, **21**, 217–225.

Gulmez, M., Oral, N. & Vatansever, L. (2006) The effects of water extract of sumac (*Rhus coriaria* L.) and lactic acid on decontamination and shelf life of raw broiler wings. *Poultry Science*, **85**, 1466–1471.

Habib, I., Sampers, I., Uyttendaele, M., Berkvens, D. & De Zutter, L. (2008) Baseline data from a Belgium-wide survey of *Campylobacter* species contamination in chicken meat preparations and considerations for a reliable monitoring program. *Applied and Environmental Microbiology*, **74** (17), 5483–5489.

Harris, N.V., Thomson, D., Martin, D.C. & Nolan, C.M. (1986) A survey of *Campylobacter* and other bacterial contaminants of pre-market chicken and retail poultry and meats, King County, Washington. *American Journal of Public Health*, **76**, 401–406.

Hart, C.D., Mead, G.C. & Norris, A.P. (1991) Effects of gaseous environment and temperature on the storage behaviour of *Listeria monocytogenes* on chicken breast meat. *Journal of Applied Bacteriology*, **70**, 40–46.

Hood, A.M., Pearson, A.D. & Shahamat, M. (1988) The extent of surface contamination of retailed chicken with *Campylobacter jejuni* serogroups. *Epidemiology and Infection*, **100**, 17–25.

Hopkins, R.S. & Scott, A.S. (1983) Handling raw chicken as a source for sporadic *Campylobacter jejuni* infections. *Journal of Infectious Diseases*, **148**, 770.

Hopkins, R.S., Olmsted, R. & Istre, G.R. (1984) *Campylobacter jejuni* infection in Colorado: Identified risk factors. *American Journal of Public Health*, **74**, 249–250.

Huat, J.T.Y., Aziz, S.A., Abu, J., *et al.* (2010) Thermophilic *Campylobacter* spp. occurrence on chickens at farm, slaughter house and retail. *International Journal of Poultry Science*, **9** (2), 134–138.

Ismail, S.A.S., Deak, T., Abd El-Rahman, H.A., Yassien, M.A.M. & Beuchat, L.R. (2000) Presence and changes in populations of yeasts on raw and processed poultry products stored at refrigeration temperature. *International Journal of Food Microbiology*, **62**, 113–121.

Jayathilakan, K., Sultana, K., Radhakrishna, K. & Bawa, A.S. (2009) Effect of lactic acid and irradiation on the shelf stability characteristics of hurdle processed chicken legs. *International Journal of Poultry Science*, **8** (7), 665–670.

Jiménez, S.M., Salsi, M.S., Tiburzi, M.C., Rafaghelli, R.C. & Pirovani, M.E. (1999) Combined use of acetic acid treatment and modified atmosphere packaging for extending the shelf-life of chilled chicken breast portions. *Journal of Applied Microbiology*, **87**, 339–344.

Jiménez, S.M., Tiburzi, M.C., Salsi, M.S., Pirovani, M.E. & Moguilevsky, M.A. (2003) The role of visible faecal material as a vehicle for generic *Escherichia coli*, coliform, and other enterobacteria contaminating poultry carcasses during slaughtering. *Journal of Applied Microbiology*, **95**, 451–456.

Jiménez, M., Soler, P., Venanzi, J.D., Canté, P., Varela, C. & Martinez-Navarro, F. (2005) An outbreak of *Campylobacter jejuni* enteritis in a school of Madrid, Spain. *Eurosurveillance*, **10** (4–6), 118–121.

Jorgensen, F., Bailey, R., Williams, S., *et al.* (2002) Prevalence and numbers of *Salmonella* and *Campylobacter* spp. on raw, whole chickens in relation to sampling methods. *International Journal of Food Microbiology*, **76** (1/2), 151–164.

Kegode, R.B., Doetkott, D.K., Khaitsa, M.L. & Wesley, I.V. (2008) Occurrence of *Campylobacter* species, *Salmonella* species and generic *Escherichia coli* in meat products from retail outlets in the Fargo metropolitan area. *Journal of Food Safety*, **28**, 111–125.

Kim, C.R. & Marshall, D.L. (2000) Quality evaluation of refrigerated chicken wings treated with organic acids. *Journal of Food Quality*, **23** (3), 327–335.

Lahellec, C., Meurier, C., Bennejean, G. & Catsaras, M. (1975) A study of 5,920 strains of psychrotrophic bacteria isolated from chicken. *Journal of Applied Bacteriology*, **38**, 89–97.

Lescano, G., Narvaiz, P., Kairiyama, E. & Kaupert, N. (1991) Effect of chicken breast irradiation on microbiological, chemical and organoleptic quality. *Lebensmittel-Wissenschaft und-Technologie*, **24**, 130–134.

Lewis, S.J., Velasquez, A., Cuppett, S.L. & McKee, S.R. (2002) Effect of electron beam irradiation on poultry meat safety and quality. *Poultry Science*, **81**, 896–903.

Logue, C.M., Sherwood, J.S., Olah, P.A., Elijah, L.M. & Dockter, M.R. (2003) The incidence of antimicrobial-resistant *Salmonella* spp. on freshly processed poultry from US Midwestern processing plants. *Journal of Applied Microbiology*, **94**, 16–24.

Lovestead, T.M. & Bruno, T.J. (2010) Detection of poultry spoilage markers from headspace analysis with cryoadsorption on a short alumina PLOT column. *Food Chemistry*, **121** (4), 1274–1282.

Lynch, J.A., Macfie, H.J.H. & Mead, G.C. (1991) Effect of irradiation and packaging type on sensory quality of chill-stored turkey breast fillets. *International Journal of Food Science and Technology*, **26**, 653–668.

van der Marel, G.M., Van Logtestijn, J.G. & Mossel, D.A.A. (1988) Bacteriological quality of broiler carcasses as affected by in-plant lactic acid decontamination. *International Journal of Food Microbiology*, **6**, 31–42.

Mazick, A., Ethelberg, S., Møller Nielsen, E., Mølbak, K. & Lisby, M. (2006) An outbreak of *Campylobacter jejuni* associated with consumption of chicken, Copenhagen, 2005. *Eurosurveillance*, **11** (4–6), 137–139.

McMeekin, T.A., Pennington, P.I. & Thomas, C.J. (1984) Effect of potassium sorbate on the microbiology of vacuum-packed poultry. *Journal of Food Safety*, **6**, 261–270.

Mead, P.S., Slutsker, L., Dietz, V., *et al.* (1999) Food-related illness and death in the United States. *Emerging Infectious Diseases*, **5**, 607–625.

Miettinen, M.K., Palmu, L., Björkroth, K.J. & Korkeala, H. (2001) Prevalence of *Listeria monocytogenes* in broilers at the abattoir, processing plant, and retail level. *Journal of Food Protection*, **64** (7), 994–999.

Meldrum, R.J., Smith, R.M.M. & Wilson, I.G. (2006) Three-year surveillance program examining the prevalence of *Campylobacter* and *Salmonella* in whole retail raw chicken. *Journal of Food Protection*, **69** (4), 928–931.

MMWR. (1998) Outbreak of *Campylobacter enteritis* associated with cross-contamination of food—Oklahoma, 1996. *MMWR*, **47** (07), 129–131.

MMWR. (2008) Multistate outbreak of *Salmonella* infections associated with frozen pot pies—United States, 2007. *MMWR*, **57** (47), 1277–1280.

Moffatt, C.R.M., Combs, B.G., Mwanri, L., *et al.* (2006) An outbreak of *Salmonella* typhimurium phage type 64 gastroenteritis linked to catered luncheons in Adelaide, South Australia, June 2005. *Communicable Diseases Intelligence*, **30**, 443–448.

Moore, J.E., Wilson, T.S., Wareing, D.R.A., Humphrey, T.J. & Murphy, P.G. (2002) Prevalence of thermophilic *Campylobacter* spp. in ready-to-eat foods and raw poultry in Northern Ireland. *Journal of Food Protection*, **65** (8), 1326–1328.

National Advisory Committee on Microbiological Criteria for Foods. (1997) Generic HACCP application in broiler slaughter and processing. *Journal of Food Protection*, **60** (5), 579–604.

National Chicken Council. (2002) Statistics and research. National Chicken Council. Accessed January 2004 at www.nationalchickencouncil.com/statistics/stat_detail.cfm?id=8

Nógrády, N., Kardos, G., Bistyák, A., *et al.* (2008) Prevalence and characterization of *Salmonella infantis* isolates originating from different points of the broiler chicken-human food chain in Hungary. *International Journal of Food Microbiology*, **127**, 162–167.

Northcutt, J.K., Berrang, M.E., Dickens, J.A., Fletcher, D.L. & Cox, N.A. (2003) Effect of broiler age, feed withdrawal and transportation on level of coliforms, *Campylobacter*, *Escherichia coli* and *Salmonella* on carcasses before and after immersion chilling. *Poultry Science*, **82**, 169–173.

Ouattara, B., Simard, R.E., Holley, R.A., Piette, G.J.P. & Begin, A. (1997) Inhibitory effect of organic acids upon meat spoilage bacteria. *Journal of Food Protection*, **60**, 246–253.

Park, C.E. & Stankiewicz, Z.K. (1981) Incidence of *Campylobacter jejuni* in fresh eviscerated whole market chickens. *Canadian Journal of Microbiology*, **27**, 841–842.

Patsias, A., Badeka, A.V., Savvaidis, I.N. & Kontominas, M.G. (2008) Combined effect of freeze chilling and MAP on quality parameters of raw chicken fillets. *Food Microbiology*, **25**, 575–581.

Patterson, J.T., Gillespie, C.W. & Hough, B. (1984) Aspects of the microbiology of vacuum- and gas-packaged chicken, including pre-treatments with lactic acid and potassium sorbate. *British Poultry Science*, **25**, 457–465.

Pearson, A.D., Greenwood, M.H., Donaldson, J., *et al.* (2000) Continuous source outbreak of campylobacteriosis traced to chicken. *Journal of Food Protection*, **63** (3), 309–314.

Pezzotti, G., Serafin, A., Luzzi, I., Mioni, R., Milan, M. & Perin, R. (2003) Occurrence and resistance to antibiotics of *Campylobacter jejuni* and *Campylobacter coli* in animals and meat in northeastern Italy. *International Journal of Food Microbiology*, **82** (3), 281–287.

Pittard, B.T., Freeman, L.R., Later, D.W. & Lee, M.L. (1982) Identification of volatile organic compounds produced by fluorescent pseudomonads on chicken breast muscle. *Applied and Environmental Microbiology*, **43** (6), 1504–1506.

Praakle-Amin, K., Hänninen, M.L. & Korkeala, H. (2006) Prevalence and genetic characterization of *Listeria monocytogenes* in retail broiler meat in Estonia. *Journal of Food Protection*, **69** (2), 436–440.

Rababah, T., Over, K., Hettiarachchy, N.S., Horax, R., *et al.* (2010) Infusion of plant extracts during processing to preserve quality attributes of irradiated chicken breasts over 9 months storage at −20°C. *Journal of Food Processing and Preservation*, **34**, 287–307.

Rasschaert, G., Houf, K., Godard, C., Wildemauwe, C., Pastuszczak-Frak, M. & De Zutter, L. (2008) Contamination of carcasses with *Salmonella* during poultry slaughter. *Journal of Food Protection*, **71** (1), 146–152.

Rayes, H.M., Genigeorgis, C.A. & Farver, T.B. (1983) Prevalence of *Campylobacter jejuni* on turkey wings at the supermarket level. *Journal of Food Protection*, **46** (4), 292–294.

del Rio, E., Panizo-Morán, M., Prieto, M., Alonso-Calleja, C. & Capita, R. (2008) Effect of various chemical decontamination treatments on natural microflora and sensory characteristics of poultry. *International Journal of Food Microbiology*, **115** (3), 268–280.

Rodrigo, S., Adesiyun, A., Asgarali, Z. & Swanston, W. (2005) Prevalence of *Campylobacter* spp. on chickens from selected retail processors in Trinidad. *Food Microbiology*, **22**, 125–131.

Rodriguez de Ledesma, A.M., Riemann, H.P. & Farver, T.B. (1996) Short-time treatment with alkali and/or hot water to remove common pathogenic and spoilage bacteria from chicken wing skin. *Journal of Food Protection*, **59** (7), 746–750.

Rosenquist, H., Nielsen, N.L., Sommer, H.M., Norrung, B. & Christensen, B.B. (2003) Quantitative risk assessment of human campylobacteriosis associated with thermophilic *Campylobacter* species in chicken. *International Journal of Food Microbiology*, **83** (1), 87–103.

Russell, S.M. (1997) A rapid microbiological method for enumeration of *Pseudomonas fluorescens* from broiler chicken carcasses. *Journal of Food Protection*, **60** (4), 385–390.

Russell, S.M. (2008) The effect of an acidic, copper sulfate-based commercial sanitizer on indicator, pathogenic, and spoilage bacteria associated with broiler chicken carcasses when applied at various intervention points during poultry processing. *Poultry Science*, **87**, 1435–1440.

Russell, S.M., Fletcher, D.L. & Cox, N.A. (1995) Spoilage bacteria in fresh broiler chicken carcasses. *Poultry Science*, **74** (12), 2041–2047.

Sakhare, P.Z., Sachindra, N.M., Yashoda, K.P. & Narasimha Rao, D. (1999) Efficacy of intermittent decontamination treatments during processing in reducing the microbial load on broiler chicken carcasses. *Food Control*, **10**, 189–194.

Samuel, M.C., Vugia, D.J., Shallow, S., *et al.* (2004) Epidemiology of sporadic *Campylobacter* infection in the United States and declining trend in incidence, FoodNet 1996–1999. *Clinical Infectious Diseases*, **38** (Suppl 3), S165–S174.

Sante, V., Renerre, M. & Lacourt, A. (1994) Effect of modified atmosphere packaging on color stability and on microbiology of turkey breast meat. *Journal of Food Quality*, **17**, 177–195.

Saucier, L., Gendron, C. & Gariépy, C. (2000) Shelf life of ground poultry meat stored under modified atmosphere. *Poultry Science*, **79**, 1851–1856.

Sawaya, W.N., Abu-Ruwaida, A.S., Hussain, A.J., Khalafawi, M.S. & Dashti, B.H. (1993) Shelf-life of vacuum-packaged eviscerated broiler carcasses under simulated market storage conditions. *Journal of Food Safety*, **13**, 305–321.

Schlosser, W., Hogue, A., Ebel, E., *et al.* (2000) Analysis of *Salmonella* serotypes from selected carcasses and raw ground products sampled prior to implementation of the Pathogen reduction, hazard analysis and critical control point final rule in the US. *International Journal of Food Microbiology*, **58**, 107–111.

Schneider, K.R., Kere Kemp, G. & Aldrich, M.L. (2002) Acidified sodium chlorite antimicrobial treatment of air chilled broiler carcasses. *Dairy, Food and Environmental Sanitation*, **22** (2), 102–108.

Shamsuzzaman, K., Lucht, L. & Chuaqui-Offermanns, N. (1995) Effects of combined electron-beam irradiation and sous-vide treatments on microbiological and other qualities of chicken breast meat. *Journal of Food Protection*, **58** (5), 497–501.

Shiferaw, B., Yang, S., Cieslak, P., *et al.* (2000) Prevalence of high-risk food consumption and food-handling practices among adults: A multistate survey, 1996 to 1997. *Journal of Food Protection*, **63** (11), 1538–1543.

Sinhamahapatra, M., Biswas, S., Das, A.K. & Bhattacharyya, D. (2004) Comparative study of different surface decontaminants on chicken quality. *British Poultry Science*, **45** (5), 624–630.

Soultos, N., Koidis, P. & Madden, R.H. (2003) Presence of *Listeria* and *Salmonella* spp. in retail chicken in Northern Ireland. *Letters in Applied Microbiology*, **37**, 421–423.

Stern, N.J., Rothenberg, F.J. & Stone, J.M. (1985) Enumeration and reduction of *Campylobacter jejuni* in poultry and red meats. *Journal of Food Protection*, **48**, 606–610.

Straver, J.M., Janssen, A.F.W., Linnemann, A.R., van Boekel, M.A.J.S., Beumer, R.R. & Swietering, M.H. (2007) Number of *Salmonella* on chicken breast filet at retail level and its implications for public health risk. *Journal of Food Protection*, **70** (9), 2045–2055.

Thomas, Y.O., Kraft, A.A., Rust, R.E. & Hotchkiss, D.K. (1984) Effect of carbon dioxide flushing and packaging methods on the microbiology of packaged chicken. *Journal of Food Science*, **49**, 1367–1371.

Tokumaru, M., Konuma, H., Umesako, M., Konno, S. & Shinagawa, K. (1991) Rates of detection of *Salmonella* and *Campylobacter* in meats in response to sample size and the infection level of each species. *International Journal of Food Microbiology*, **13**, 41–46.

USDA-FSIS. (2000a) HACCP implementation: First year *Salmonella* test results. Accessed March 2004 at www.fsis.usda.gov/ophs/salmdata.htm

USDA-FSIS. (2000b) Microbiological testing program for meat and poultry. Accessed March 2004 at www.fsis.usda.gov/OA/background/microtest.htm

USDA-FSIS. (2010) Quarterly progress report on *Salmonella* testing of selected raw meat and poultry products: Preliminary results, January–March 2010. Accessed December 2010 at www.fsis.usda.gov/science/Q1_2010_Salmonella_Testing/index.asp

Vareltzis, K., Soultos, N., Koidis, P., Ambrosiadis, J. & Genigeorgis, C. (1997) Antimicrobial effects of sodium tripolyphosphate against bacteria attached to the surface of chicken carcasses. *Lebensmittel Wissenschaft und Technologie*, **30** (7), 665–669.

Viehweg, S.H., Schmitt, R.E. & Schmidt-Lorenz, W. (1989) Microbial spoilage of refrigerated fresh broilers. Part VII. Production of off odours from poultry skin by bacterial isolates. *Lebensmittel Wissenschaft und Technologie*, **22**, 356–367.

Viljoen, B.C., Geornaras, I., Lamprecht, A. & von Holy, A. (1998) Yeast populations associated with processed poultry. *International Journal of Food Microbiology*, **15**, 113–117.

Waldroup, A.L., Rathgeber, B.M., Forsythe, R.H. & Smoot, L. (1992) Effects of six modifications on the incidence and levels of spoilage and pathogenic organisms on commercially processed post chill broilers. *Journal of Applied Poultry Research*, **1**, 226–234.

Wingstrand, A., Neimann, J., Engberg, J., *et al.* (2006) Fresh chicken as main risk factor for campylobacteriosis, Denmark. *Emerging Infectious Diseases*, **12** (2), 280–284.

Worsfold, D. & Griffith, C.J. (1997) Assessment of the standard of consumer food safety behavior. *Journal of Food Protection*, **60** (4), 399–406.

Zeitoun, A.A.M. & Debevere, J.M. (1990) The effect of treatment with buffered lactic acid on microbial decontamination and on shelf life of poultry. *International Journal of Food Microbiology*, **11**, 305–312.

Zhao, C., Ge, B., De Villena, J., Sudler, R., *et al.* (2001) Prevalence of *Campylobacter* spp., *Escherichia coli* and *Salmonella* serovars in retail chicken, turkey, pork and beef from the Greater Washington, DC area. *Applied and Environmental Microbiology*, **67** (12), 5431–5436.

Zhao, S., White, D.G., Friedman, S.L., *et al.* (2008) Antimicrobial resistance in *Salmonella enterica* serovar Heidelberg isolates from retail meats, including poultry, from 2002 to 2006. *Applied and Environmental Microbiology*, **74** (21), 6656–6662.

26 Quality Indicators in Poultry Products

Lisa McKee, Elizabeth Cobb, and Sarah Padilla

Abstract: Food manufacturers try to maintain the quality of their products through the use of quality control/quality assurance (QC/QA) personnel who are responsible for everything from training workers to answering consumer complaints. One of the primary functions of QC/QA departments is the testing and evaluation of products to maintain the standards and specifications designated for those products. In poultry products, organoleptic properties such as color, flavor, aroma, and texture as well as microbiological characteristics are some of the indicators of quality. A variety of techniques, including both instrumental and sensory methods, are used to determine the quality of poultry products.

Gas chromatography (GC), alone or in combination with mass spectrometry (GC/MS), has been used in many studies of poultry volatile components.

Keywords: Gas chromatography, mass spectrometry, poultry meat quality indicators, poultry meat color, poultry meat flavor, panelists, sensory evaluation, olfactometry, high-pressure/performance liquid chromatography, thiobarbituric acid reactive substances (TBARs) test, poultry texture, poultry microbial characteristics, polymerase chain reaction

26.1 INTRODUCTION

Quality is a complex term with a variety of meanings. Consistency, safety, conforming to specifications, usefulness for the designated purpose, and a lack of defects all contribute to a product being termed "high quality." Consumers are the final arbiters of quality and consumer satisfaction may be the ultimate definition of quality. What defines "quality" for one person, however, may mean rejection by another consumer.

Food manufacturers try to maintain the quality of their products through the use of quality control/quality assurance (QC/QA) personnel who are responsible for everything from training workers to answering consumer complaints. One of the primary functions of QA/QC departments is the testing and evaluation of products to maintain the standards and specifications designated for those products. In poultry products, organoleptic properties such as color, flavor, aroma, and texture as well as microbiological characteristics are some of the indicators of quality. A variety of techniques, including both instrumental and sensory methods, are used to determine the quality of poultry products.

Handbook of Meat, Poultry and Seafood Quality, Second Edition. Edited by Leo M. L. Nollet.

26.2 COLOR

Color is the perception that arises from the detection of light waves after they have interacted with a food. Such interaction can result in refraction, reflection, transmission, or absorption of the light by the food. These interactions, as well as factors such as the chemical/physical make-up of the food, the characteristics of the light source, the surrounding environment, and the spectral sensitivity of the viewer, all play a role in the description of "color." Since it is frequently the first characteristic used by consumers in determining quality of poultry and other meats, measurement of color is of primary importance to the poultry industry. While both instrumental and sensory methods of evaluation are used to determine the color characteristics of poultry products, instrumental evaluation of color is often used more extensively.

Humans perceive color by way of the rods and cones located on the retina of the eye. The rods, which generate only achromatic (black and white) information, are capable of operating under low light intensities. The cones, which require higher levels of illumination, are responsible for the chromatic information. Color-sensitive pigments in the cones respond to red, green, and blue light. Electrical nerve impulses generated by the cones travel to the brain which then responds in terms of red-green and blue-yellow pairs.

Color difference meters are the most common instrumental method of measuring color in foods and are designed to duplicate the response of the human eye to light. Such instruments generate a beam of light which strikes the surface of a food and is reflected back into the meter through a series of three filters which duplicate the red-green-blue response. A photocell interprets the information mathematically to produce tristimulus color values. Color systems such as the Hunter L, a, b, the CIELAB $L^*a^*b^*$, and the L, C, h color space are used to describe the resulting color characteristics of the food.

As with any food, instrumental measurement of poultry meat color is dependent on a number of factors. Sandusky and Heath (1996) examined the effects of sample thickness (0.5, 1.0, and 1.5 cm), background color (white, pink, green, and gray), and illuminant (D-65, A, and F) on color measurement of broiler muscle. Light was able to pass through the 0.5-cm thick samples to such an extent that reflection by the background prevented measurement of the true color of the muscle. Reflection was greatest with the white and pink backgrounds. Muscle color was also affected by the predominant wavelength of the illuminant used. Bianchi and Fletcher (2002) also investigated the influence of meat thickness and background color on the lightness (L^*), redness (a^*), and yellowness (b^*) of turkey and chicken breast meat. Redness and yellowness decreased as thickness of turkey sample increased from 1 to 2 cm, while lower L^* values were detected in turkey when thickness increased from 1 to 3 cm. All three color values decreased as broiler meat thickness increased from 1 to 2 cm. One-centimeter-thick broiler breast meat samples were more red on white paper and yellow polystyrene backgrounds and less red on the black plastic background, while 1-cm-thick turkey breast meat samples were more red on the white paper background and less red and yellow on the black plastic background. Color characteristics were not different for either type of poultry meat when sample thickness was greater than 1 cm. Results of these studies point to the importance of considering all potential variables when measuring poultry meat color.

Color in poultry is related to a variety of factors, including genetics. Although no differences in L^*, a^*, or b^* values were detected between slow-growing and fast-growing turkey strains (Werner *et al.* 2008), Berri *et al.* (2005) found increased wing flapping duration resulted in lower L^* and b^* values and higher a^* values in slow-growing broilers but only increased breast a^* values in a heavy line of broilers. Ermellinata di Rovigo broilers, an Italian dual-purpose breed, were found to have greater L values compared to the Italian Robusta Maculata breed and the hybrid genotypes Hy Line White 36 and Hy Line Brown in a study by Rizzi *et al.* (2007). Bianchi *et al.* (2006) reported few differences in L^*, a^*, and b^* values of breast meat from Ross 508 and Cobb 500 broilers, but H^* was lower for Ross 508 (45.12) compared to Cobb 500 (48.52). Hunter L values did not differ between inbred Leghorn, inbred Fayoumi, commercial, F_5 broiler-inbred Leghorn cross, or F_5 broiler-inbred Fayoumi cross groups, but breasts from the inbred Leghorn line were a more pure and intense red color compared to other genetic types in a study by Lonergan *et al.* (2003). Although specific genotypes were not identified, Wilkins *et al.* (2000) found large variations in the color of breast meat taken from broilers representing 23 different flocks. Wide distributions for L^* (45.0–67.3), a^* (–0.40 to 7.5), and b^* (–3.63 to 6.91) were noted, with meat from two flocks identified as free range having lighter and less red color characteristics compared to the other samples.

Color has frequently been used as an indicator of quality and functional properties in poultry meat. Qiao *et al.* (2002a) investigated the relationship between color and composition of raw broiler breast meat. Breast fillets were separated into color groups defined as lighter than normal ($L^* > 53$), normal ($48 < L^* < 51$), and darker than normal ($L^* < 46$). No differences between color groups were detected for moisture, total lipid, iron, or residual glycogen contents but percent protein was significantly lower in the light color group (22.44%) compared to the normal (22.96%) and dark (23.27%) groups. Qiao *et al.* (2001) found L^* values were negatively correlated with emulsification capacity while a^* values were positively correlated with water-holding capacity. Breast fillets classified as lighter than normal ($L^* > 53$) were reported to have high moisture, low emulsification capacity, and low water-holding capacity. Breast meat classified as dark absorbed significantly less marinade compared to light and normal samples but had a greater cooked meat yield compared to light fillets in a study by Qiao *et al.* (2002b). Allen *et al.* (1998), however, reported dark fillets picked up significantly more marinade compared to light fillets. Dark fillets also had higher levels of bound moisture and lower drip and cook losses.

Fletcher (1999) found a highly significant correlation between color and pH of broiler breast meat. Results indicated pH values decreased as L^* values increased. Similar results have been reported by Boulianne and King (1995, 1998), Allen *et al.* (1998), Wilkins *et al.* (2000), Qiao *et al.* (2001), Woelfel *et al.* (2002), and Petracci *et al.* (2004). The pale, low-pH meat is often associated with undesirable functional properties including lower water-holding capacity, higher drip loss, and softer texture. Allen *et al.* (1997), however, found lighter broiler breast fillets had a longer shelf life compared to darker fillets. Although this defect is frequently referred to as pale, soft, exudative (PSE), Smith and Northcutt (2009) suggested the PSE label may be misleading since a standard definition of "pale" and "normal" has not been used in research. Those authors suggested "pale chicken muscle" or "pale poultry muscle syndrome" be used to describe poultry with these undesirable conditions.

Another defect in poultry meat related to color is pinking. In this case, white poultry meat displays a pink color even when cooked appropriately. A number of factors have been associated with pinking defect in poultry, including genetics, environmental stressors, processing methods, chemical composition, and cooking methods. Holownia *et al.* (2003, 2004) investigated the color of chicken breast meat subdivided into normal, lighter than normal, and darker than normal color groups and subjected to a variety of treatments to induce pinking. A subjective threshold of $a^* = 3.8$ was established through experimentation and used to judge pink discoloration. Treatments (sodium chloride, sodium tripolyphosphate [STP], sodium erythorbate, and sodium nitrate) alone and in various combinations were injected into the fillets, which were then tumbled and stored overnight at 4–5°C. Half the samples were evaluated raw while the other half was cooked before evaluation. Pinking occurred in all eight treatments containing sodium nitrate as well as in normal color and darker than normal fillets injected with a combination of sodium chloride, STP, and sodium erythorbate. Simulation of pinking was most effective in dark samples and least effective in light samples. The most critical factor associated with pinking was the initial lightness (L^*) of the raw muscle. Data collected were used to develop a logistic regression model to predict pinking. Nitrosopigment, pH, and reducing conditions were the most important factors in the model which was found to have a predictive ability of greater than 90%.

26.3 AROMA

Aroma is the perception that arises when the olfactory receptors come in contact with airborne or volatile molecules. These volatile compounds come from external sources drawn into the body by normal inspiration through the nose or as the result of aromatics from foods in the mouth. This physiological connection between the human senses of smell and taste can lead to confusion in the characterization of these food properties. Much of what is described by humans as flavor is in fact mediated by the sense of smell. While taste comprises only sweet, sour, salty, and bitter, aroma is a complex property involving hundreds of chemical compounds.

The olfactory receptors are located on the roof of the nasal cavity. This remote location is thought to protect the organs from damage but results in only a minute portion of the aroma compounds flowing through the system actually reaching the receptors. Millions of cilia located on the nasal epithelium and a complex network of nerves leading into the brain help to enhance the weak aroma signals. The ability to detect food aromas depends on a variety of factors, including sensitivity of the evaluator, concentration of the aroma compounds, the combinations of chemicals found in a food, adaptation, and air flow.

Raw meat has little aroma. The characteristic aromas associated with meat are the result of a variety of reactions occurring during cooking, including the Maillard reaction, lipid oxidation, and degradation of thiamin. These reactions can produce hundreds of volatile substances such as pyrazines, aldehydes, alcohols, piperidines, and sulfur-containing compounds. Processing techniques such as irradiation and vacuum packaging as well as growth of microorganisms are also associated with odor generation in meat products.

A variety of chemical analysis methods have been used to determine the volatile composition of poultry products. Chromatography is a frequently used method of separation in which a mobile phase, typically gas or liquid, carries the sample through a column packed with material holding a stationary phase. The partitioning of the sample between the mobile and stationary phases results in different elution rates for the chemical compounds in the sample. A detector translates the results into peaks that are recorded on a chart; retention times and peak heights can then be used to identify types and quantities of volatiles present.

Aliani and Farmer (2005a) used high-performance liquid chromatography (HPLC) to analyze odor and flavor precursors in raw chicken breasts and thighs. Compounds determined included inosine 5′-monophosphate disodium salt (IMP), inosine, hypoxanthine, adenosine 5′-monophosphate (AMP), threonine, proline, glycine, cysteine, methionine, tyrosine, histidine, and thiamine. Significant differences were found in IMP, inosine, AMP, hypoxanthine, and contents of amino acids in the protein fraction between breasts and thighs, between sources of chicken, and between individual chickens from a single source. These differences would have substantial influence on odor and flavor generation due to reactions such as the Maillard reaction.

Gas chromatography (GC), alone or in combination with mass spectrometry (GC-MS), has been used in many studies of poultry volatile components. Wettasinghe *et al.* (2000) used GC-MS to investigate the volatiles produced by roasting poultry processing by-products such as backbones, breastbones, spent bones, and skin. Total volatiles ranged from 2030 ppb (backbones) to 4049 ppb (skin) and were composed of aldehydes, ketones, alcohols, hydrocarbons, esters, and nitrogen- and sulfur-containing compounds. Although 50–135 different volatiles were detected in the by-products, hexanal (an aldehyde associated with aromas from oxidation of lipids); 2-methyl-2-buten-1-ol (an alcohol possibly associated with mushroom or metallic notes); and heterocyclic nitrogen/sulfur-containing compounds such as piperidine, tetrahydropyridazine, and benzothiazole (responsible for characteristics such as meaty and roast) were found in the greatest quantities in the by-products.

Ramarathnam *et al.* (1993) used GC-MS to characterize and quantify volatiles from uncured and cured chicken. A nitrogen purge-and-trap technique was used in the study to collect the volatiles. Of the 52 aroma compounds identified in chicken, 21 were hydrocarbons, 11 were carbonyls, 4 were alcohols, 6 were phenols, 5 were esters, and the remaining 6 were heterocyclic compounds. Seven of the aroma compounds were unique to that species. Three compounds (1,2-dibutylcyclopentane, 2,6-bis(1,1-dimethyl-ethyl)-4-methylphenol, and 1,12-dodecanediol) were said to be newly identified.

Irradiation has often been associated with off-odor development in poultry products. Kwon *et al.* (2008) used GC-MS to determine the volatiles in chicken thighs, beef loins, and pork loins irradiated before and after cooking. Chicken thighs contained the greatest number of volatile compounds and had the greatest increase in volatiles after irradiation. New (1-butene, 1-pentene, 1-hexene, 1-heptene, dimethyl disulfide, and dimethyl trisulfide) and increased (butane, dimethyl sulfide, hexane, and heptane) levels of volatiles were detected in all three meats due to irradiation. Sulfur-containing compounds were primarily associated with irradiation while increased lipid oxidation was associated with cooking. Irradiation off-odor due to sulfur-containing compounds diminished during storage for 6 months in aerobic conditions.

Rababah *et al.* (2006) used GC to determine the volatile components in raw and cooked irradiated chicken breast meat infused with antioxidant green tea and grape-seed extracts (GSE). Hexanal was reported to be present in the highest levels followed by pentanal. Other compounds detected included acetylaldehyde, propanol, propanal, butanal, and nonanal. Irradiation as well as cooking was associated with increased lipid oxidation and levels of hexanal and pentanal. The combination of green tea and GSE was the most effective in reducing lipid oxidation as well as levels of hexanal and pentanal.

GC-MS has been used to study the effects of antioxidants on production of off-odor volatiles and lipid oxidation in irradiated turkey products in several studies (Lee and Ahn 2003; Lee *et al.* 2003). Dimethyl disulfide, the primary volatile associated with irradiation off-odor in poultry, increased in both turkey homogenates and turkey breast patties due to irradiation. Hexanal, a chemical produced during lipid oxidation, increased in both irradiated and nonirradiated turkey meat. Addition of antioxidants, including sesamol, extracts of far-infrared-treated rice hulls, rosemary, tocopherol, gallate, and trolox, as well as combinations of gallate + tocopherol and sesamol + tocopherol, were effective in reducing levels of volatiles. Storage in aerobic conditions was associated with decreases in off-odor volatiles.

Arnold and Senter (1998) used GC-MS in combination with solid-phase microextraction (SPME) and digital aroma technology (electronic nose) to analyze the headspace volatiles of representative pathogenic bacteria isolated from broiler carcasses when grown on defined media. Alcohols, particularly ethanol, were the predominant compounds detected from all microbial species except *Pseudomonas aeruginosa* and *Acinetobacter calcoaceticus*. Production of 3-methyl butanal was characteristic of *Listeria monocytogenes* while indole characterized *Escherichia coli* growth. Although the electronic nose was reported to have potential for headspace analysis, improvements in the reproducibility and sensitivity of the method were reported to be necessary before extensive use of the technique is possible.

Pignoli *et al.* (2009) used GC in combination with headspace solid-phase microextraction (HS-SPME) to study the suitability of saturated aldehydes as a marker of lipid oxidation. Changes in headspace concentrations of propanal and hexanal were determined using a model system of turkey muscle washed with sodium phosphate buffer. Results were compared to a traditional thiobarbituric acid-reactive substance (TBAR) test. TBARs and concentration of propanal did not change over time when an antioxidant was added to samples just prior to analysis but hexanal concentrations unexpectedly decreased over time. The authors hypothesized that the reduced volatility of hexanal was due to increased interaction of the chemical with turkey muscle. TBARs, however, was found to be positively correlated with hexanal levels, leading the authors to conclude TBARs was a more sensitive measure of lipid oxidation than HS-SPME-GC under the conditions of the study.

Although instrumental methods such as chromatography are useful in analyzing aroma characteristics of poultry products, such methods do not necessarily duplicate human evaluation of foods. Sensory evaluation is a measurement method that uses humans as the "instrument" in analysis of foods. Test methods are typically categorized into three classes. Discrimination tests involve determination of any differences between samples. Panelists for discrimination tests are generally oriented to the terminology and test method and screened for acuity in the sensory property(ies) of interest; they may also be trained. Descriptive tests require panelists to quantify perceived intensities of organoleptic properties and describe those properties using advanced terminology. In addition to the orientation and acuity testing, panelists for descriptive tests are often highly trained. Untrained consumers are used as panelists in affective tests. These tests determine how well products are liked or which products are preferred. A variety of sensory techniques have been used in the aroma evaluation of poultry products.

Sensory profiles of chicken, turkey, and 13 other types of meat were created by Rødbotten *et al.* (2004) using a sensory panel composed of 11 well-trained assessors. Panelists developed a list of nonspecies specific attributes as well as the definition of each attribute. Nonstructured, continuous 15-cm intensity line scales were used to evaluate individual attributes. Results were analyzed using univariate analysis of variance and principal component analysis. When all attributes were included, color explained 68% of the first component variation. Chicken and turkey were the most pale in color. When color was removed from the analysis, odor and flavor components explained 66% of the variation in the first component. Chicken and turkey had two of the lowest odor intensity ratings of the 17 animal groups evaluated.

Aliani and Farmer (2005b) used trained and experienced but untrained panelists and several discrimination and descriptive methods to determine the importance of potential precursors to odor and flavor characteristics of chicken. Natural and added precursors of Maillard and other flavor-forming reactions, including IMP, ribose, cysteine + ribose, glucose, and thiamine, were used in the experiments. Ribose was reported to be the most important compound associated with chicken aroma. Descriptive evaluation indicated addition of ribose was associated with increased nutty and woody aromas and decreased rancid and bloody aromas while cysteine + ribose was associated with increased chicken aroma. When assessed by GC-MS, ribose was found to be responsible for roasted and chicken odor descriptors probably as a result of increased concentrations of 2-furanmethanethiol, 2-methyl-3-furanthiol, and 3-methylthiopropanal.

A six-member descriptive panel was used by Brannan (2009) to evaluate the effect of GSE on quality characteristics of ground chicken during refrigerated storage. Panelists were trained for 20 h with the final 3–4 h being used to achieve consensus on the odor and flavor descriptors as well as the definitions and references used for each descriptor. Odor attributes evaluated included chicken brothy, fishy, sulfury, musty, and rancid. Higher musty odor was detected in control patties, precooked patties, and after 12 days of refrigerated storage. Control patties were also determined to have a greater rancid odor compared to GSE-treated patties.

An eight-member trained sensory panel was used by Yan *et al.* (2006) to study the turkey aroma and irradiation off-odor of raw and cooked breast patties from turkeys fed diets containing vitamin E (VE), selenium (Se), conjugated linoleic acids (CLA), VE + Se, VE + CLA, Se + CLA, and VE + Se + CLA. Panelists detected an intense turkey aroma in raw meat packaged aerobically and an intense irradiation off-odor in raw meat when vacuum packaged. Turkey aroma scores were higher for raw meat from dietary treatments containing CLA, while meat from treatments containing VE had lower turkey aroma when compared to the control. Irradiation off-odor in both raw and cooked turkey meat, however, was lowered by VE and increased by CLA. Irradiation and packaging had no effect on turkey meat aroma and irradiation off-odor in cooked meat. Dietary supplementation of VE and VE + Se improved consumer acceptance of irradiated raw turkey.

26.4 FLAVOR

Poultry flavor is derived from a multifaceted network of factors, including all aspects of both preslaughter and postmortem (PM) influences (Diosady *et al.* 1991). The flavor of poultry is molded by a series of circumstances that begin during the embryonic stage and continue to the grocery store. The type of animal feed, preslaughter stressors, slaughtering procedures, and application of PM antioxidants and preservatives contribute to the final chemical composition and flavor of poultry meat. The exact amino acid content, fatty acid composition, fat content, sugar composition and content, and the animal's breed also influence the end product flavor (Dugo *et al.* 2008).

Cooking poultry meat allows various chemical reactions to occur that utilize meat compounds to produce volatile compounds and specific flavors (Aliani & Farmer 2005b). Major chemical reactions include the Maillard reaction, Strecker degradation, and the oxidation of lipids (Dugo *et al.* 2008). The specific cooking method and the final meat temperature affect the development of flavors by means of these chemical reactions (Dugo *et al.* 2008).

The analysis of poultry flavor relies heavily on sensory evaluation due to the inability of analytical instruments to fully reciprocate the human ability to translate food-induced chemical signals into flavor and aroma descriptions. Flavor is a complex phenomenon which relies on the interaction of human sensory abilities, such as aroma and flavor perception (Cook *et al.* 2005). Instrumental methods such as GC and HPLC, however, are used to analyze chemical components which can be related to human flavor perceptions.

26.4.1 Sensory evaluation of poultry flavor

Many studies on poultry flavor have relied on trained sensory panels. Such panels vary in numbers and levels of experience in general and study-specific sensory techniques. Common methods of selecting panel members include administering basic taste tests, administering triangle tests, and using British Standards Institution models to select ideal candidates (Franqueira de Toledo *et al.* 2005; Patsias *et al.* 2006; Sandercock *et al.* 2009). Once panelists have been selected, additional training is typically needed to familiarize evaluators with the attributes of the poultry products under study and the analysis methods. Training sessions may include definitions of terms, general product information, product attribute descriptions, and scorecard explanations (Kayaardi *et al.* 2005; Zhuang *et al.* 2009). The additional training may be as few as one to as many as 36 sessions or more. The sensory analysis methods vary among studies and may be designed by the coordinator or constructed based on the panelists' feedback from trial sessions.

A 0–15-point intensity scale was used by Zhuang *et al.* (2009) to analyze the effects of dry-air chilling, immersion chilling, no chilling, and muscle types on flavor characteristics of chicken. Panelists analyzed nine flavor attributes: chicken, brothy, barnyard/wet feathers, bloody/serumy/metallic, cardboardy, sweet, salty, sour, and bitter. Results indicated there were no significant differences in flavor attributes among the different chilling methods or specific muscle types.

Vara-Ubol and Bowers (2002) used a 0–15-point intensity scale to study the inhibitory effects of STP with or without NaCl on oxidative flavor development in ground turkey breast. Turkey meat with 1% NaCl had stronger meaty flavors and reduced staleness characteristics compared to turkey samples without NaCl after 2 days of refrigerated (4°C) storage but sensory scores for staleness were almost the same for samples with and without NaCl after 4 days of storage. The 0.5% level of STP was associated with a slightly increased metallic aftertaste and a greater ability to reduce oxidative flavor changes such as loss of meaty and turkey flavors and staleness compared to the 0.3% level.

A 9-point scale was used by a trained sensory panel to rate acceptability of flavor and odor characteristics of chicken packaged in air, vacuum, a modified atmosphere of 30% CO_2/65% N_2/5% O_2 (M1), or a modified atmosphere of 65% CO_2/30% N_2/5% O_2 (M2) in a study by Balamatsia *et al.* (2007). A rating of 0 was designated as unacceptable, 6 was defined as the first off-taste/off-odor and no longer acceptable, and 9 was excellent. Chicken packaged in air had the least acceptable stability overall. Air and vacuum packaged chicken samples received a score of 6 on day 7 of storage while the M1 packaged chicken received a score of 6 on day 10. The M2 packaged chicken was the most acceptable sample over the storage period, receiving a score of 7 after 15 days of storage.

Patsias *et al.* (2006) also used a 9-point scale to rate odor and flavor acceptability of precooked chicken stored under air or modified atmospheres (M1 = 30% CO_2/70% N_2; M2 = 60% CO_2/40% N_2; and M3 = 90% CO_2/10% N_2). A score of 1 was defined as "extreme foreign flavor or dislike intensely" while 9 was designated as "no foreign flavor or like extremely." A score of 6 marked the lower level of acceptability. Air-packaged chicken was rated as more acceptable compared to any of the modified atmospheres up to 12 days of storage but received a score of 6 before day 16. The M2 and M3 treatments maintained the highest acceptability ratings overall, receiving a score of 7 after 20 days of storage.

Sandercock *et al.* (2009) studied the effects of genetic differences of different chicken lines on carcass and muscle yield, meat quality, and sensory scores. Panelists used 8-point scales to evaluate chicken flavor and abnormal flavor intensities and an 8-point hedonic scale to rate flavor and overall liking. Treatments involved 37 different lines of commercial broiler, layer, and traditional chickens slaughtered at 6 and 10 weeks of age. At 6 weeks of age, the broiler chicken samples were scored higher for chicken flavor and overall liking than the layer and traditional samples. At 10 weeks of age, both the broiler and layer samples were scored similarly and higher for chicken flavor intensity compared to the traditional samples.

Line scales were used by Franqueira de Toledo *et al.* (2005) to study the effects of no irradiation and irradiation with Co^{60} at 2, 4, 6, and 8 kGy doses on seven chicken flavor categories: typical, bitter, salty, metallic, intensity, smoked, and strange. Samples were evaluated after irradiation treatment and again after frozen storage. Panel members collaborated with a coordinator to develop terminology used for the line scales. Before frozen storage, nonirradiated samples and those irradiated with 8 kGy were scored lowest for bitterness. The 8 kGy samples were scored highest for saltiness while the 4 kGy sample was scored lowest for that attribute. The nonirradiated sample was scored lowest for metallic flavor and highest for smoke flavor. After frozen storage, the sample irradiated with 6 kGy was rated highest for typical flavor and lowest for bitterness. The sample irradiated with 2 kGy and the nonirradiated sample were scored highest for metallic flavor. The nonirradiated sample was also scored highest for bitterness and strange flavor characteristics. Overall, the flavor of fresh irradiated chicken was negatively impacted by irradiation while freezing tended to improve the flavor characteristics.

Flavor of breast meat from organic, free range, and corn-fed chickens purchased from retail sources was evaluated by Jahan *et al.* (2005) using conventional profiling and free-choice profiling sensory techniques. Flavor analysis included general attributes and personal vocabularies such as roasted, oily, pungent, sweet, buttery, and bland. Generalized Procrustes analysis of free-choice profiling indicated two groups existed: one group could distinguish flavor differences while the other could not. Organic chicken was scored lower for buttery and salty flavors but higher for oily flavor compared to free-range chicken. Overall, the authors found most panelists could not easily discriminate flavor attributes among the different treatments.

Aliani and Farmer (2005b) used a variety of sensory methods with trained panels in two experiments to identify flavor precursors leading to important flavor compounds in cooked chicken. Ribose was found to be one of the most important precursors, corresponding to stronger roasted chicken flavors and less off-flavors. Desirable flavors in chicken breast were increased when two to four times the natural concentration of ribose was added to the chicken breasts. However, panelists noted no significant flavor differences when additional ribose was added to chicken legs.

Consumer sensory panels offer a way to determine the average person's opinion about the general characteristics of a specific food product. Panelists may come from various backgrounds, such as university students and employees, local citizens, and/or friends (Ballard *et al.* 2007; Battula *et al.* 2008). The sensory evaluation forms are typically straightforward to eliminate the need for training or prior experience. Five-, eight-, and nine-point hedonic scales are commonly used to evaluate specific categories such as flavor and overall acceptability of products.

A 5-point just-about-right (JAR) scale, with 1 meaning too low, 5 meaning too high, and 3 meaning JAR, was used by Saha *et al.* (2009) to determine overall flavor, chicken flavor, and saltiness of broiler breast fillets marinated with different amounts of salt. Panelists also used a 9-point hedonic scale to rate flavor and overall impression. Average hedonic scores showed marinated chicken had a more desirable flavor compared to the nonmarinated sample. Based on average JAR ratings, the treatments with 0.5% and 0.75% salt marinades had a more ideal overall flavor, chicken flavor, and saltiness compared to the higher salt levels (1.0% and 1.25%), which tended to be rated as too salty.

The effects of adding garlic bulbs (2% and 4%) or garlic husks (2% and 4%) to chicken feed on sensory characteristics of the chicken meat was studied by Kim *et al.* (2009). Consumer panelists used a 5-point hedonic scale to rate chicken thigh flavor. Analysis showed chicken meat from broilers fed feed containing garlic bulbs or husks had better flavor than that from broilers fed traditional feed, with the 4% garlic husk addition associated with a slightly higher preference compared to other treatments. The preference for the flavor of meat from broilers fed either garlic bulbs or husks was reportedly due to lower fat content and lipid oxidation products detected in those samples compared to meat from broilers fed traditional feed when TBARs and fatty acid composition analyses were performed.

An 8-point hedonic scale was used by consumer panelists in a study by Waskar *et al.* (2009) to rate flavor of chicken meat from broilers fed feed containing a herbal toxin binder compared to meat from broilers fed traditional feed. Scores for flavor and overall palatability indicated no preference for either treatment regardless of whether the sample was the breast fillet or the tender. However, feed containing the herbal toxin binder was associated with improved appearance and carcass yield.

Sensory panelists used a 9-point hedonic scale to evaluate the effects of electrical and vacuum stunning under low atmospheric pressure as well as PM deboning times of 0.75, 2, and 4 h on chicken breast flavor in a study by Battula *et al.* (2008). Stunning method did not affect flavor or overall acceptability scores but both scores increased as deboning time increased. A 9-point hedonic scale was also used in a study by Gorsuch and Alvarado (2010). Consumer panelists in the study rated overall flavor of pale broiler breast fillets marinated with STP, high pH (11.9) phosphate, high pH STP, agglomerated phosphate, or nonagglomerated phosphate. Consumer scores indicated no preference for any of the treatments, with all treatments being scored between "neither like nor dislike" and "like slightly."

26.4.2 Instrumental evaluation of poultry flavor

Vara-Ubol and Bowers (2002) used GC to identify levels of hexanal compounds as an indicator of oxidative flavor changes in ground turkey with and without added antioxidant or salt. Detection of off-flavors by trained sensory panelists was correlated with increasing levels of hexanal. Addition of 0.5% STP inhibited increases in hexanal during storage. Although α-tocopherol alone was reported to have no effect on hexanal values, the combination of 0.3% STP plus 0.03% α-tocopherol was more effective at inhibiting increases in hexanal than STP alone.

Rababah *et al.* (2006) used GC to determine levels of hexanal and pentanal, compounds associated with fat oxidation and off-flavors, in irradiated chicken infused with plant extracts. Cooking and irradiation were associated with increased total volatile levels but infusion with antioxidant extracts led to lower hexanal levels. A combination of green tea and GSE (6000 ppm) was the most effective in reducing total hexanal and pentanal levels.

GC is often combined with other instrumental analysis methods to improve the accuracy and detection of flavor compounds in foods. MS ionizes gases produced in GC analyses, causing fragmentation of substances into specific mass to charge ratios which can then be separated and identified using a mass spectrum graph. Olfactometry (O) utilizes a human or mechanical "sniffer" to smell the effluent from the olfactometer outlet. The "sniffer" then provides a description and often an intensity rating of each individual component passing from the GC column to the detector. The various peaks on the chromatogram graph are then correlated to the "sniffer" descriptions.

Zhu *et al.* (2004) used GC-MS to identify compounds in irradiated turkey breast rolls which were then correlated to sensory evaluation data. Levels of acetaldehyde increased as irradiation dose increased from 0 to 2.0 kGy. This compound was thought to correspond to the higher metallic flavor described by panelists at higher irradiation levels. Higher levels of dimethyl disulfide and other sulfur-based compounds detected in the GC-MS analysis also correlated to stronger sulfur flavors and aromas reported by sensory panelists in product irradiated with 2.0 kGy.

Liu *et al.* (2007) used GC-MS to identify volatiles, later correlated with flavor compounds, in Chinese Nanjing marinated duck. Steam distillation, supercritical fluid extraction (SDE), and purge and trap (P&T), as well as SPME were used to extract volatiles. P&T and SPME were found to be better suited to extraction of volatiles with low masses while SDE was better suited to extraction of heavy masses. Fat degradation and lipid oxidation products were the primary components detected in the duck and were associated with flavor and aroma compounds described as nutty, meaty, green, and sulfur-like.

Aliani and Farmer (2005b) combined GC-MS, GC-O, and sensory evaluation to characterize specific compounds responsible for odor and flavor in chicken. Treatments included a control, addition of D-ribose, and addition of D-ribose-5-phosphate disodium salt (R5P). Human detectors were used in the GC-O procedure to note odor types and intensities of eluted volatiles while GC-MS was used to identify the specific components described by the detectors during GC-O. Results of the GC-O and GC-MS analyses were then compared to sensory evaluation ratings for flavor and aroma of the treatments. Higher levels of added ribose and R5P were associated with increased sensory descriptions of savory, potato, and roasted flavors and tended to decrease undesirable flavor traits.

Rather than relying on gravity as in GC, HPLC employs pressure to force a liquid through a column. The use of pressure allows more complete separation of particles and is particularly useful for separation of smaller particles since a column with smaller retention holes can be used. HPLC also provides an alternative to GC when analyzing less volatile compounds.

Flavor precursors of chicken breasts and legs, including nucleotides, inosine, hypoxanthine (Hx), and thiamine, were analyzed using HPLC by Aliani and Farmer (2005a). Results were compared to sensory evaluations. Many of the flavor compounds differed significantly among chicken samples from different commercial sources and among samples purchased from the same source. However, IMP disodium salt usually occurred in the highest levels in commercial chicken muscle followed by inosine with Hx occurring at lower levels. Thiamine levels differed little between samples. Based on sensory data, reducing sugars, nucleotides, and similar compounds were thought to aid in chicken flavor development during cooking.

In another study by Aliani and Farmer (2002), HPLC was used to analyze the reducing sugar and phosphorylated reducing sugar contents of chicken. Higher levels of glucose and ribose were found in chicken breast muscle compared to leg muscle. Significant differences in ribose, glucose, ribose phosphate, and glucose phosphate concentrations were also detected between the different chicken samples.

26.4.3 Chemical evaluation of poultry flavor

One of the most common methods of analyzing poultry flavor characteristics is the TBARs test. In this method, levels of lipid peroxidation compounds, primarily malondialdehydes, are measured as an indication of lipid degradation and the resulting loss of desirable flavor characteristics. Increased levels of TBARs correspond to higher levels of lipid oxidation and off-flavors.

Rababah *et al.* (2006) used TBARs to determine the levels of lipid peroxidation in irradiated and nonirradiated chicken samples infused with green tea and GSE. GSE was associated with slightly lower TBARs compared to green tea extracts but the combination of the two antioxidants was much more effective than either extract alone. TBARs for irradiated samples infused with 6000 ppm of the extract combination were lower than those infused with 3000 ppm.

TBARs was also used by Jo *et al.* (2006) to analyze the benefits of *Prunus mume* extract versus rosemary extract on reducing lipid oxidation in cooked chicken breast after 0, 1, and 3 days of storage. Rosemary extract was more effective in preventing lipid degradation in all samples. The antioxidant effects of rosemary were thought to be related to the high levels of phenolic compounds while antioxidant effects of *Prunus mume* were related to flavonoids and other compounds.

Patsias *et al.* (2006) used TBARs values to track changes in lipid deterioration in precooked chicken stored under different modified atmosphere packaging (MAP) conditions over a

20-day period. TBARs values decreased in chicken stored in air or under 30% CO_2/70% N_2, 60% CO_2/40% N_2, or 90% CO_2/10% N_2 through day 8. After day 8, there was little change in the TBARs values for chicken stored under any of the MAP conditions but significant increases in TBARs were noted for air-packaged chicken samples.

Kim *et al.* (2009) used TBARs to evaluate the effects of feeding chicken garlic bulbs or garlic husks on lipid oxidation. Addition of either the bulbs or husks was associated with lower TBARs in thigh meat with no significant ($P > 0.05$) differences found between the two garlic products. Four percent garlic bulbs or husks was associated with slightly lower TBARs compared to 2%.

TBARs values of PSE versus normal broiler fillets were compared by Gorsuch and Alvarado (2010). PSE fillets marinated in NaCl, STP, high pH (11.9) phosphate, high pH STP, agglomerated phosphate, and nonagglomerated phosphate had similar TBARs values compared to normal fillets. Since PSE is associated with lower water-holding capacity which in turn causes accelerated lipid degradation rates, the researchers suggested the marinades provided a buffer against oxidation in the PSE fillets.

26.5 TEXTURE

Texture is the rheological and structural properties of a food as perceived by humans through mechanical, tactile, visual, and/or auditory means. Unlike color and flavor, which are often used by consumers to assess food safety, texture is typically used to determine food quality. In poultry and other meats, textural characteristics such as tenderness, juiciness, chewiness, and fibrousness are primary factors associated with the quality of the products (Braxton *et al.* 1996). The importance of texture to consumers has driven the need for research on methods to measure the texture properties of poultry products.

In the past, texture was evaluated solely using groups of trained sensory panelists. Sensory testing is ideal when evaluating textural properties of poultry since every person experiences food breakdown differently within the mouth, but such testing can be inefficient, time consuming, and costly. Sensory evaluation requires recruitment of panelists, in-depth training of descriptive panels, multiple evaluation sessions, and potentially large quantities of sample. Variability between human subjects as well as the complexity of the mastication process further complicate the sensory evaluation of meats. This has led to development of instrumental methods of texture measurement.

26.5.1 Instrumental analysis of poultry texture

Analytical testing methods are used for their reliability, ease of use, cost-effectiveness, and the ability to implement QC during testing (Xiong *et al.* 2006). Analytical techniques used in the past to study texture include gnathosonics (sounds measured during mastication), ultra-high-speed cinematography (mandibular movement and velocity measurements), laryngophone or adaphotodetector (monitoring of swallowing), recordings from synchro transmitters, and pressure transducers (food and tooth pressure measurements) (Lee *et al.* 2009). Today, one of the most common instrumental methods used to measure meat texture is shear analysis.

Instrumental methods which measure shear force, including the Warner–Bratzler shear force system, the Allo–Kramer shear system, and the razor blade shear test, operate by measuring characteristics that are directly related to the physical components of meat products. In these methods, samples are cut perpendicularly to the muscle fiber orientation. Shear testing is most commonly used when attempting to predict tenderness in meat samples and is designed to simulate a human taking the initial bite of a meat sample rather than the entire mastication process.

The Warner–Bratzler shear force system is one of the most popular mechanical methods of analyzing meat texture and has been used frequently to evaluate the effects of deboning and aging on poultry tenderness. Zhuang and Savage (2009) studied the effects of deboning time and cold storage on the texture of early deboned broiler breast fillets. Treatments included deboning at 2 h PM with no cold storage, 24 h PM with no cold storage, 2 h PM and stored at 3°C for 7 days, 2 h PM and stored at −20°C for 7 days, and 2 h PM and stored at −20°C for 6 days , followed by 1 day at 3°C. Warner–Bratzler shear analysis was performed on samples which had been vacuum bagged and cooked at 85°C to an internal temperature of 78°C using three 1.9 cm strips from each breast. Breast fillets deboned at 2 h PM with no cold storage had the highest mean shear value (107 N) while those deboned at 24 h PM with no cold storage had the lowest mean shear value (37 N). Overall, cold storage resulted in reduced average Warner–Bratzler shear values. Warner–Bratzler shear data indicated aging chicken carcasses for 24 h at 0°C prior to deboning produced meat with lower shear values than hot-boning carcasses in a study by Seabra *et al.* (2001). An interaction between broiler age and postchill aging on Warner–Bratzler shear was reported by Northcutt *et al.* (2001). Breast fillets harvested at less than 2 h postchill aging were more tender when taken from broilers slaughtered at 42 or 44 days of age compared to those harvested from birds 49 or 51 days of age.

The affects of marination with various compounds on tenderness of poultry meat has also been studied using Warner–Bratzler shear analysis. Use of STP on chicken forequarters resulted in 60% greater shear values compared to controls when no postchill aging was used but had no effect on shear values at 120 min postchill aging (Young & Lyon 1997b). Polyphosphates were reported to enhance tenderness of skinless deboned chicken breast meat, however, when used in a still-marinating process (Lemos *et al.* 1999). Optimal conditions to produce the most tender breast meat were 8–14 h marination in polyphosphate concentrations between 0.5% and 2.0%. Marination of peri-rigor chicken breasts in calcium at 50, 100, 150, and 200 mM $CaCl_2$ improved meat tenderness according to Warner–Bratzler shear values, but cooking losses increased as calcium concentration increased (Young & Lyon 1997a). Shear force decreased in spent hen meat treatment with ginger extract compared to controls in a study by Naveena and Mendiratta (2001). Tenderness increased as concentration of ginger extract increased and was effective in both prechill and postchill phases.

Another method of texture analysis is the Allo–Kramer shear method. In this test, a meat sample is placed in a compression cell and sheared with multiple blades during a single stroke. Texture of spent hen meat treated with $CaCl_2$ was evaluated in a series of studies by Nurmahmudi and Sams (1997a, b) and Nurmahmudi *et al.* (1997) using the Allo–Kramer method. In the initial study, a combination of injection with 0.3 M $CaCl_2$ and tumbling resulted in the greatest tenderness. Injection to a 10% level of 0.3 M $CaCl_2$ at 24h PM was a more effective tenderizer than injection at less than 30 min PM. Biochemical evaluation indicated $CaCl_2$ produced tenderization through ionic strength and calcium-specific actions.

Cavitt *et al.* (2005) compared the use of Allo–Kramer shear, razor blade shear, and needle puncture for measuring texture in poultry meat. Both Allo–Kramer and razor blade shear values indicated significant differences in tenderness of breast fillets deboned less than 1.5 h PM and more than 6 h PM, but the two methods were only moderately correlated ($r = 0.72$). Razor blade shear, however, was a better predictor than the Allo–Kramer shear of descriptive sensory panel ratings for initial hardness, chewdown hardness, cohesiveness, cohesiveness of mass, and number of chews to swallow in a study by Cavitt *et al.* (2004).

The razor blade shear test has two subtests: the Meullenet–Owens razor shear (MORS) and the blunt Meullenet–Owens razor shear (BMORS). Lee *et al.* (2008a) evaluated the effectiveness and reliability of the MORS and BMORS for predicting tenderness in broiler breast meat. Tenderness of breast fillets which had been frozen, thawed overnight, and then cooked in a convection oven to an internal temperature of 76°C was evaluated using several different texture analyzers. BMORS was better able to detect toughness in breast fillets deboned at 0.25, 2.5, 4.0, and 6.0 h but both MORS and BMORS were found to be reliable methods for determining poultry meat texture.

26.5.2 Sensory evaluation of poultry texture

While instrumental analysis can provide rapid, reliable, and economical data on texture of poultry products, such data often only partially correlates with human sensory evaluation of textural properties. Humans are able to provide information on a wide range of texture characteristics such as mouthfeel, juiciness, toothpacking, and wetness as well as textural changes during mastication that cannot be analyzed instrumentally. Both untrained consumer panels and trained descriptive panels have been used to provide information on textural properties of poultry products.

Sensory tenderness was correlated to instrumental shear value ranges by Lyon and Lyon (1991). Shear values from four instruments, including the Warner–Bratzler and the Allo–Kramer, were correlated to six consumer tenderness categories. Standards were shear value <3.62 kg = very tender, 3.62–6.61 kg = moderately to slightly tender, 6.62–9.60 kg = slightly tender to slightly tough, 9.61–12.60 kg = slightly to moderately tough, and >12.60 kg = very tough. The relationship between descriptive profiles and shear values of deboned broilers was also determined by Lyon and Lyon (1997). Shear values from both Warner–Bratzler and Allo–Kramer were highly correlated to mechanical properties such as springiness, chewiness, cohesiveness, hardness, and fibrousness as well as chewdown characteristics such as chew count, bolus size, and ease of swallow. Xiong *et al.* (2006) used both a trained 7-member descriptive panel and a 74-member consumer panel to evaluate the texture characteristics of chicken breasts deboned at PM times ranging from 0.25 to 24 h. Results indicated deboning time significantly affected instrumental shear properties as well as hardness, cohesiveness, texture/tenderness acceptance, and JAR tenderness ratings as determined by the consumer panel. Razor blade shear was more highly correlated to sensory tenderness as determined by a trained descriptive panel but consumer tenderness ratings were predicted equally well by Warner–Bratzler, Allo–Kramer, and razor blade shear tests.

Lee *et al.* (2008b) evaluated the ability of a laser air puff and shape profile method to predict broiler breast texture as determined by a seven-member descriptive sensory panel using a 15-point numerical scale with assigned intensities. As has been reported in other studies, sensory results indicated initial hardness, hardness of mass, cohesiveness of mass, and number of chews decreased as PM deboning time increased. A similar trend was noted for the results from MORS and BMORS analyses. Air puff and shape profile was only 83% accurate in predicting sensory tenderness.

26.6 MICROBIAL CHARACTERISTICS

As with most meats, raw poultry is frequently highly contaminated with microorganisms. Pathogens such as *Salmonella*, *Campylobacter*, *Escherichia coli*, and *Listeria monocytogenes* and spoilage organisms including *Pseudomonas* and *Lactobacillus* have been reported on poultry products. Microbial levels range widely, from less than 1.0 colony forming unit (CFU)/square centimeter (cm^2) to greater than $\log_{10}$ 7.0 CFU/cm^2. Changes in organoleptic properties, including off-odors and sliminess, are typically associated with proliferation of microorganism. Detection, enumeration, and identification of the microorganisms present on poultry are therefore primary indicators of product quality.

Methods used for recovery of microorganisms from poultry meat can have a significant effect on subsequent enumeration and identification of microbial loads. Sarlin *et al.* (1998) and Gill and Badoni (2005) compared whole carcass rinses, skin excision, and swabbing for recovery of bacteria from broiler carcasses during commercial processing. Results indicated swabbing was the least effective method of sampling regardless of whether the swabs were cotton or cellulose acetate sponges. Skin excision and whole carcass rinsing produced similar results for numbers of *Salmonella*, *Escherichia coli*, coliforms, and total aerobic bacteria detected. The combined use of rinsing plus excision has been reported to be even more effective in recovering bacteria such as *Campylobacter* (Jørgensen *et al.* 2002; Moran *et al.* 2009).

Time of sampling may also be an important factor in levels of microorganisms detected on poultry. Whyte *et al.* (2004) investigated the effect of both time of day and point of sampling during processing on levels of microorganisms detected on broiler carcasses. Levels of total aerobic bacteria, *Enterobacteriaceae*, *Escherichia coli*, and *Staphylococcus aureus* were lower in samples taken in the morning compared to those taken during the evening. Total aerobic bacteria, *Enterobacteriaceae*, and *E. coli* loads in scald tank water were lower in morning samples taken before processing start-up than in evening samples after 42 475 carcasses had been processed. Levels of *Listeria* increased as processing progressed regardless of whether samples were taken in the morning or evening.

Enumeration and/or identification of microorganisms collected from poultry products have traditionally been accomplished through the use of culture techniques. Although nonspecific agar such as standard plate agar is used for total aerobic loads, many culture techniques are developed for use with specific microorganisms. These techniques often include an enrichment step to improve recovery of the bacteria under study. The plating and enrichment media used can have significant effects on the levels of bacteria detected.

Rall *et al.* (2005) evaluated three enrichment broths (selenite cystine [SC], tetrathionate brilliant green [TBG], and Rappaport Vassiliadis [RV]) and five plating media (Salmonella–Shigella agar, brilliant green agar, xylose lysine desoxycholate agar, Rambach agar, and CHROMagar Salmonella agar) for detection of *Salmonella* in poultry carcasses collected from supermarkets, butcher shops, and hospital kitchens. A total of 29 poultry carcasses were positive for *Salmonella* when results from all plating media were combined. Based on those results, RV broth and CHROMagar were reported to be the most effective media for isolation of *Salmonella*.

Yuste *et al.* (2003) compared tryptone soy agar (TSA); six selective media (xylose lysine desoxycholate agar, Eosin methylene blue agar, Cefsulodin-Irgasan-novoblocin medium, Baird-Parker medium, Oxford Listeria medium, and modified Oxford Listeria medium); and a thin-layer method (each selective agar overlaid with TSA) for isolation of five pathogens in chicken breast meat. Although no significant differences were reported for counts on the different media, an enrichment step using brain heart infusion was necessary for growth of *Salmonella* Typhimurium on TSA. The thin-layer method was reported to be more effective than selective agars alone for differentiation and enumeration of the pathogens under study.

Traditional microbial culture methods are often labor intensive and time consuming. A variety of rapid and molecular techniques have been developed to detect, differentiate, and enumerate microorganisms in poultry products. Bohaychuk *et al.* (2005) compared enzyme-linked immunosorbent assay (ELISA), polymerase chain reaction assay (PCR), and lateral flow immunoprecipitation (LFI) with the conventional culture method for the detection of *Salmonella*, *Campylobacter*, *Listeria*, and *Escherichia coli* O157:H7 in raw and processed poultry. PCR was reported to be highly sensitive for the detection of all target pathogens except *Listeria*. ELISA was found to have sensitivity and specificity rates of 100% for *Salmonella*. LFI tests were statistically different from the culture method results for *Salmonella* and *Listeria* due to false-positive results but the method was still reported to be suitable for screening poultry for pathogens.

A variety of PCR techniques have been used for the detection and identification of *Salmonella* (Croci *et al.* 2004; Catarame *et al.* 2006; Balakrishna *et al.* 2010; Wang & Mustapha 2010), *Campylobacter* (Katzav *et al.* 2008), *Yersinia enterocolitica* (Balakrishna *et al.* 2010), *Listeria* (Antunes *et al.* 2002), and other pathogens on poultry. Regardless of the specific method used, PCR has been reported to have comparable or better sensitivity and specificity compared to traditional culture methods. Results are typically more rapid with PCR and the method often requires less time for enrichment before enumeration and differentiation can be completed.

Valdivieso-Garcia *et al.* (2007) used enrichment in Rosef's enrichment broth followed by hydrophobic grid membrane filter on semisolid medium (HGMF-SSM) to isolate thermophilic *Campylobacter*. The technique was compared to a standard method using selective antibiotics. In an initial study, pure cultures of campylobacters were isolated after 16–18 h versus 48 h for the standard culture method. HGMF-SSM was more effective in isolation of *Campylobacter coli* due to the lack of antibiotics. Extensive contamination of retail chicken and turkey samples was detected using the method, with 59.6% of chicken and 45.8% of turkey samples contaminated with one or more species of *Campylobacter*. HGMF-SSM was reported to be quicker, cheaper, less labor intensive, and more sensitive for isolation of *Campylobacter* compared to a standard antibiotic-based culture method.

26.7 CONCLUSION

While the term quality can have a variety of meanings, quality evaluation of poultry products most often depends on organoleptic properties such as color, aroma, flavor, and texture as well as the microbiological characteristics of the products. Food scientists have developed a number of instrumental techniques for evaluating quality characteristics of poultry. Although such techniques can provide a wealth of information about chicken, turkey, and other products, it is ultimately the consumer evaluation that determines poultry quality.

REFERENCES

Aliani, M. & Farmer, L.J. (2002) Postcolumn derivatization method for determination of reducing and phosphorylated sugars in chicken by high performance liquid chromatography. *Journal of Agricultural and Food Chemistry*, **50** (10), 2760–2766.

Aliani, M. & Farmer, L.J. (2005a) Precursors of chicken flavor. I. Determination of some flavor precursors in chicken muscle. *Journal of Agricultural and Food Chemistry*, **53** (15), 6067–6072.

Aliani, M. & Farmer, L.J. (2005b) Precursors of chicken flavor. II. Identification of key flavor precursors using sensory methods. *Journal of Agricultural and Food Chemistry*, **53** (16), 6455–6462.

Allen, C.D., Russell, S.M. & Fletcher, D.L. (1997) The relationship of broiler breast meat color and pH to shelf-life and odor development. *Poultry Science*, **76**, 1042–1046.

Allen, C.D., Fletcher, D.L., Northcutt, J.K. & Russell, S.M. (1998) The relationship of broiler breast color to meat quality and shelf-life. *Poultry Science*, **77**, 361–366.

Antunes, P., Réu, C., Sousa, J.C., Pestana, N. & Peixe, L. (2002) Incidence and susceptibility to antimicrobial agents of *Listeria* spp. and *Listeria monocytogenes* isolated from poultry carcasses in Porto, Portugal. *Journal of Food Protection*, **65** (12), 1888–1893.

Arnold, J.W. & Senter, S.D. (1998) Use of digital aroma technology and SPME GC-MS to compare volatile compounds produced by bacteria isolated from processed poultry. *Journal of the Science of Food and Agriculture*, **78**, 343–348.

Balakrishna, K., Murali, H.S. & Batra, H.V. (2010) A novel multiplex polymerase chain reaction for simultaneous detection of *Yersinia enterocolitica*, *Staphylococcus aureus*, *Aeromonas* and *Salmonella* from chicken meat and milk samples. *Journal of Food Safety*, **30**, 263–275.

Balamatsia, C.C., Patsias, A., Kontominas, M.G. & Savvaidis, I.N. (2007) Possible role of volatile amines as quality-indicating metabolites in modified atmosphere-packaged chicken fillets: Correlation with microbiological and sensory attributes. *Food Chemistry*, **104**, 1622–1628.

Ballard, T.S., Mallikarjunan, P. & Duncan, S.E. (2007) Consumer assessment of crispness of pressure fried chicken nuggets using nitrogen gas. *Journal of Foodservice*, **18**, 1–6.

Battula, V., Schilling, M.W., Vizzier-Thaxton, Y., Behrends, J.M., Williams, J.B. & Schmidt, T.B. (2008) The effects of low-atmosphere stunning and deboning time on broiler breast meat quality. *Poultry Science*, **87**, 1202–1210.

Berri, C., Debut, M., Santé-Lhoutellier, V., *et al.* (2005) Variations in chicken breast meat quality: Implications of struggle and muscle glycogen content at death. *British Poultry Science*, **46** (5), 572–579.

Bianchi, M. & Fletcher, D.L. (2002) Effects of broiler breast meat thickness and background on color measurements. *Poultry Science*, **81**, 1766–1769.

Bianchi, M., Petracci, M. & Cavani, C. (2006) The influence of genotype, market live weight, transportation, and holding conditions prior to slaughter on broiler breast meat color. *Poultry Science*, **85**, 123–128.

Bohaychuk, V.M., Gensler, G.E., King, R.K., Wu, J.T. & McMullen, L.M. (2005) Evaluation of detection methods for screening meat and poultry products for the presence of foodborne pathogens. *Journal of Food Protection*, **68** (12), 2637–2647.

Boulianne, M. & King, A.J. (1995) Biochemical and color characteristics of skinless boneless pale chicken breast. *Poultry Science*, **74**, 1693–1698.

Boulianne, M. & King, A.J. (1998) Meat color and biochemical characteristics of unacceptable dark-colored broiler chicken carcasses. *Journal of Food Science*, **63** (5), 759–762.

Brannan, R.G. (2009) Effect of grape seed extract on descriptive sensory analysis of ground chicken during refrigerated storage. *Meat Science*, **81**, 589–595.

Braxton, D., Dauchel, C. & Brown, W.E. (1996) Association between chewing efficiency and mastication patterns for meat, and influence on tenderness perception. *Food Quality and Preference*, **7**, 217–223.

Catarame, T.M.G., O'Hanlon, K.A., McDowell, D.A., Blair, I.S. & Duffy, G. (2006) Comparison of a real-time polymerase chain reaction assay with a culture method for the detection of *Salmonella* in retail meat samples. *Journal of Food Safety*, **26**, 1–15.

Cavitt, L.C., Youm, G.W., Meullenet, J.E., Owens, C.M. & Xiong, R. (2004) Prediction of poultry meat tenderness using razor blade shear, Allo-Kramer shear, and sarcomere length. *Journal of Food Science*, **69** (1), SNQ11–SNQ15.

Cavitt, L.C., Meullenet, J.F., Gandhapuneni, R.K., Youm, G.W. & Owens, C.M. (2005) Rigor development and meat quality of large and small broilers and the use of Allo-Kramer shear, needle puncture and razor blade shear to measure tenderness. *Poultry Science*, **84**, 113–118.

Cook, D., Hollowood, T., Linforth, R. & Taylor, A. (2005) Correlating instrumental measurements of texture and flavor release with human perception. *International Journal of Food Science & Technology*, **40**, 631–641.

Croci, L., Delibato, E., Volpe, G., De Medici, D. & Palleschi, G. (2004) Comparison of PCR, electrochemical enzyme-linked immunosorbent assays, and the standard culture method for detecting *Salmonella* in meat products. *Applied and Environmental Microbiology*, **70** (3), 1393–1396.

Diosady, L., Ramarathnam, N. & Rubin, L. (1991) Studies on meat flavor. 2. A quantitative investigation of the volatile carbonyls and hydrocarbons in uncured and cured beef and chicken. *Journal of Agricultural and Food Chemistry*, **39** (10), 1839–1847.

Dugo, G., Dugo, P., Mondello, L. & Zellner, B. (2008) Gas chromatography-olfactometry in food flavor analysis. *Journal of Chromatography A*, **1186**, 123–143.

Fletcher, D.L. (1999) Broiler breast meat color variation, pH, and texture. *Poultry Science*, **78**, 1323–1327.

Franqueira de Toledo, T.C., Canniatti-Brazaca, S.G., Spoto, M.H.F. & Arthur, V. (2005) Sensory evaluation of chicken breast under gamma irradiation at commercial doses. *Journal of Food Science*, **70** (1), S8–S12.

Gill, C.O. & Badoni, M. (2005) Recovery of bacteria from poultry carcasses by rinsing, swabbing or excision of skin. *Food Microbiology*, **22**, 101–107.

Gorsuch, V. & Alvarado, C.Z. (2010) Postrigor tumble marination strategies for improving color and water-holding capacity in normal and pale broiler breast fillets. *Poultry Science*, **89**, 1002–1008.

Holownia, K., Chinnan, M.S., Reynolds, A.E. & Koehler, P.E. (2003) Evaluation of induced color changes in chicken breast meat during simulation of pink color defect. *Poultry Science*, **82**, 1049–1059.

Holownia, K., Chinnan, M.S., Reynolds, A.E. & Davis, J.W. (2004) Relating induced in situ conditions of raw chicken breast meat to pinking. *Poultry Science*, **83**, 109–118.

Jahan, K., Paterson, A. & Piggott, J. (2005) Sensory quality in retailed organic, free range and corn-fed chicken breast. *Food Research International*, **38**, 495–503.

Jo, S.C., Nam, K.C., Min, B.R., *et al.* (2006) Antioxidant activity of *Prunus mume* extract in cooked chicken breast meat. *International Journal of Food Science & Technology*, **41**, 15–19.

Jørgensen, F., Bailey, R., Williams, S., *et al.* (2002) Prevalence and numbers of *Salmonella* and *Campylobacter* spp. on raw, whole chickens in relation to sampling methods. *International Journal of Food Microbiology*, **76**, 151–164.

Katzav, M., Isohanni, P., Lund, M., Hakkinen, M. & Lyhs, U. (2008) PCR assay for the detection of *Campylobacter* in marinated and non-marinated poultry products. *Food Microbiology*, **25**, 908–914.

Kayaardi, S., Kundakci, A., Kayacier, A. & Gok, V. (2005) Sensory and chemical analysis of doner kebab made from turkey meat. *Journal of Muscle Foods*, **17**, 165–173.

Kim, Y.J., Jin, S.K. & Yang, H.S. (2009) Processing, products and food safety: Effect of dietary garlic bulb and husk on the physiochemical properties of chicken meat. *Poultry Science*, **88**, 398–405.

Kwon, J.H., Kwon, Y., Nam, K.C., Lee, E.J. & Ahn, D.U. (2008) Effect of electron-beam irradiation before and after cooking on the chemical properties of beef, pork, and chicken. *Meat Science*, **80**, 903–909.

Lee, E.J. & Ahn, D.U. (2003) Odor volatiles and lipid oxidation in irradiated turkey breast meat and meat homogenates. *Journal of Food Science*, **68** (5), 1631–1638.

Lee, S.C., Kim, L.H., Nam, K.C. & Ahn, D.U. (2003) Antioxidant properties of far infrared-treated rice hull extract in irradiated raw and cooked turkey breast. *Journal of Food Science*, **68** (6), 1904–1909.

Lee, Y.S., Owens, C.M. & Meullenet, J.F. (2008a) The Meullenet-Owens razor shear (MORS) for predicting poultry meat tenderness: Its application and optimization. *Journal of Texture Studies*, **39**, 655–672.

Lee, Y.S., Owens, C.M. & Meullenet, J.F. (2008b) A novel laser air puff and shape profile method for predicting tenderness of broiler breast meat. *Poultry Science*, **87**, 1451–1457.

Lee, Y.S., Owens, C.M. & Meullenet, J.F. (2009) Tenderness perception of poultry *major pectoralis* muscle during mastication. *Journal of Food Science*, **74**, 413–422.

Lemos, A.L.S.C., Nunes, D.R.M. & Viana, A.G. (1999) Optimization of the still-marinating process of chicken parts. *Meat Science*, **52**, 227–234.

Liu, Y., Xu, X.L. & Zhou, G.H. (2007) Comparative study of volatile compounds in traditional Chinese Nanjing marinated duck by different extraction techniques. *International Journal of Food Science & Technology*, **42**, 543–550.

Lonergan, S.M., Deeb, N., Fedler, C.A. & Lamont, S.J. (2003) Breast meat quality and composition in unique chicken populations. *Poultry Science*, **82**, 1990–1994.

Lyon, B.G. & Lyon, C.E. (1991) Research note: Shear value ranges by instron Warner-Bratzler and single-blade Allo-Kramer devices that correspond to sensory tenderness. *Poultry Science*, **70**, 188–191.

Lyon, B.G. & Lyon, C.E. (1997) Sensory descriptive profile relationships to shear values of deboned poultry. *Journal of Food Science*, **62** (4), 885–888, 897.

Moran, L., Kelly, C. & Madden, R.H. (2009) Factors affecting the recovery of *Campylobacter* spp. from retail packs of raw, fresh chicken using ISO 10272-1:2006. *Letters in Applied Microbiology*, **48**, 628–632.

Naveena, B.M. & Mendiratta, S.K. (2001) Tenderisation of spent hen meat using ginger extract. *British Poultry Science*, **42**, 344–349.

Northcutt, J.K., Buhr, R.J., Young, L.L., Lyon, C.E. & Ware, G.O. (2001) Influence of age and postchill carcass aging duration on chicken breast fillet quality. *Poultry Science*, **80**, 808–812.

Nurmahmudi & Sams, A.R. (1997a) Tenderizing spent fowl meat with calcium chloride. 1. Effects of delivery method and tumbling. *Poultry Science*, **76**, 534–537.

Nurmahmudi & Sams, A.R. (1997b) Tenderizing spent fowl meat with calcium chloride. 3. Biochemical characteristics of tenderized breast meat. *Poultry Science*, **76**, 543–547.

Nurmahmudi, Veeramuthu, G.J. & Sams, A.R. (1997) Tenderizing spent fowl meat with calcium chloride. 2. The role of delayed application and ionic strength. *Poultry Science*, **76**, 538–542.

Patsias, A., Chouliara, I., Badeka, A., Savvaidis, I.N. & Kontominas, M.G. (2006) Shelf-life of a chilled precooked chicken product stored in air and under modified atmospheres: Microbiological, chemical, sensory attributes. *Food Microbiology*, **23**, 423–429.

Petracci, M., Betti, M., Bianchi, M. & Cavani, C. (2004) Color variation and characterization of broiler breast meat during processing in Italy. *Poultry Science*, **83**, 2086–2092.

Pignoli, G., Bou, R., Rodriguez-Estrada, M.T. & Decker, E.A. (2009) Suitability of saturated aldehydes as lipid oxidation markers in washed turkey meat. *Meat Science*, **83**, 412–416.

Qiao, M., Fletcher, D.L., Smith, D.P. & Northcutt, J.K. (2001) The effect of broiler breast meat color on pH, moisture, water-holding capacity, and emulsification capacity. *Poultry Science*, **80**, 676–680.

Qiao, M., Fletcher, D.L., Smith, D.P. & Northcutt, J.K. (2002a) The relationship between raw broiler breast meat color and composition. *Poultry Science*, **81**, 422–427.

Qiao, M., Fletcher, D.L., Smith, D.P. & Northcutt, J.K. (2002b) Effects of raw broiler breast meat color variation on marination and cooked meat quality. *Poultry Science*, **81**, 276–280.

Rababah, T., Hettiarachchy, N.S., Horax, R., Cho, M.J., Davis, B. & Dickson, J. (2006) Thiobarbituric acid reactive substances and volatile compounds in chicken breast meat infused with plant extracts and subjected to electron beam irradiation. *Poultry Science*, **85**, 1107–1113.

Rall, V.L.M., Rall, R., Aragon, L.C. & da Silva, M.G. (2005) Evaluation of three enrichment broths and five plating media for *Salmonella* detection in poultry. *Brazilian Journal of Microbiology*, **36**, 147–150.

Ramarathnam, N., Rubin, L.J. & Diosady, L.L. (1993) Studies on meat flavor. 4. Fractionation, characterization, and quantitation of volatiles from uncured and cured beef and chicken. *Journal of Agricultural and Food Chemistry*, **41**, 939–945.

Rizzi, C., Marangon, A. & Chiericato, G.M. (2007) Effect of genotype on slaughtering performance and meat physical and sensory characteristics of organic laying hens. *Poultry Science*, **86**, 128–135.

Rødbotten, M., Kubberød, E., Lea, P. & Ueland, Ø. (2004) A sensory map of the meat universe. Sensory profiles of meat from 15 species. *Meat Science*, **68**, 137–144.

Saha, A., Lee, Y., Meullenet, J.F. & Owens, C.M. (2009) Consumer acceptance of broiler fillets marinated with varying levels of salt. *Poultry Science*, **88**, 415–423.

Sandercock, D.A., Nute, G.R. & Hocking, P.M. (2009) Quantifying the effects of genetic selection and genetic variation for body size, carcass composition, and meat quality in the domestic fowl (*Gallus domesticus*). *Poultry Science*, **88**, 923–931.

Sandusky, C.L. & Heath, J.L. (1996) Effect of background color, sample thickness and illuminant on the measurement of broiler meat color. *Poultry Science*, **75**, 1437–1442.

Sarlin, L.L., Barnhart, E.T., Caldwell, D.J., *et al.* (1998) Evaluation of alternative sampling methods for *Salmonella* critical control point determination at broiler processing. *Poultry Science*, **77**, 1253–1257.

Seabra, L.M., Zapata, J.F., Fuentes, M.F., Aguiar, C.M., Freitas, E.R. & Rodrigues, M.C. (2001) Effect of deboning time, muscle tensioning, and calcium chloride marination on texture characteristics of chicken breast meat. *Poultry Science*, **80**, 109–112.

Smith, D.P. & Northcutt, J.K. (2009) PSE syndrome in poultry symposium. Pale poultry muscle syndrome. *Poultry Science*, **88**, 1493–1496.

Valdivieso-Garcia, A., Harris, K., Riche, E., *et al.* (2007) Novel *Campylobacter* isolation method using hydrophobic grid membrane filter and semisolid medium. *Journal of Food Protection*, **70** (2), 355–362.

Vara-Ubol, S. & Bowers, J.A. (2002) Inhibition of oxidative flavor changes in meat by α-tocopherol in combination with sodium tripolyphosphate. *Journal of Food Science*, **67** (4), 1300–1307.

Wang, L. & Mustapha, A. (2010) EMA-real-time PCR as a reliable method for detection of viable *Salmonella* in chicken and eggs. *Journal of Food Science*, **75** (3), M134–M139.

Waskar, V.S., Devangare, A.A., Gosavi, P.P., Ravikanth, K., Maini, S. and Rekhe, D.S. (2009) Meat quality attributes of broilers supplemented with herbal toxin binder product. *Veterinary World*, **2** (7), 274–277.

Werner, C., Riegel, J. & Wicke, M. (2008) Slaughter performance of four different turkey strains, with special focus on the muscle fiber structure and the meat quality of the breast muscle. *Poultry Science*, **87**, 1849–1859.

Wettasinghe, M., Vasanthan, T., Temelli, F. & Swallow, K. (2000) Volatiles from roasted byproducts of the poultry processing industry. *Journal of Agricultural and Food Chemistry*, **46**, 3485–3492.

Whyte, P., McGill, K., Monahan, C. and Collins, J.D. (2004) The effect of sampling time on the levels of micro-organisms recovered from broiler carcasses in a commercial slaughter plant. *Food Microbiology*, **21**, 59–65.

Wilkins, L.J., Brown, S.N., Phillips, A.J. & Warriss, P.D. (2000) Variation in the colour of broiler breast fillets in the UK. *British Poultry Science*, **41**, 308–312.

Woelfel, R.L., Owens, C.M., Hirschler, E.M., Martinez-Dawson, R. & Sams, A.R. (2002) The characterization and incidence of pale, soft, and exudative broiler meat in a commercial processing plant. *Poultry Science*, **81**, 579–584.

Xiong, R., Cavitt, L.C., Meullenet, J.F. & Owens, C.M. (2006) Comparison of Allo-Kramer, Warner-Bratzler and razor blade shears for predicting sensory tenderness of broiler breast meat. *Journal of Text Studies*, **37**, 179–199.

Yan, H.J., Lee, E.J., Nam, K.C., Min, B.R. & Ahn, D.U. (2006) Effects of dietary functional ingredients and packaging methods on sensory characteristics and consumer acceptance of irradiated turkey breast meat. *Poultry Science*, **85**, 1482–1489.

Young, L.L. & Lyon, C.E. (1997a) Effect of calcium marination on biochemical and textural properties of peri-rigor chicken breast meat. *Poultry Science*, **76**, 197–201.

Young, L.L. & Lyon, C.E. (1997b) Effect of postchill aging and sodium tripolyphosphate on moisture binding properties, color, and Warner-Bratzler shear values of chicken breast meat. *Poultry Science*, **76**, 1587–1590.

Yuste, J., Capellas, M., Pla, R., Llorens, S., Fung, D.Y.C. & Mor-Mur, M. (2003) Use of conventional media and thin agar layer method for recovery of foodborne pathogens from pressure-treated poultry products. *Journal of Food Science*, **68** (7), 2321–2324.

Zhu, M.J., Mendonca, A., Lee, E.J. & Ahn, D.U. (2004) Influence of irradiation and storage on the quality of ready-to-eat turkey breast rolls. *Poultry Science*, **83**, 1462–1466.

Zhuang, H. & Savage, E.M. (2009) Variation and Pearson correlation coefficients of Warner-Bratzler shear force measurements within broiler breast fillets. *Poultry Science*, **88**, 214–220.

Zhuang, H., Savage, E.M., Smith, D.P. & Berrang, M.E. (2009) Effect of dry-air chilling on sensory descriptive profile of cooked broiler breast meat deboned four hours after the initiation of chilling. *Poultry Science*, **88**, 1282–1291.

27 Shelf Life of Fresh and Frozen Poultry

María de Lourdes Pérez-Chabela

Abstract: The perishable characteristics of poultry make necessary alternatives that improve shelf life with the minimum physicochemical and sensory changes necessary. Frozen storage has most often been employed in poultry transportation. Methods such as packaging that help prevent contamination with pathogens and spoilage flora have also been used to maintain an acceptable poultry meat or product during storage. Recent technologies including the use of natural compounds such as antimicrobials and antioxidants and application of hydrostatic pressure are being increasingly used to improve the shelf life of fresh and frozen poultry products.

Keyowrds: Poultry shelf life; poultry color change; oxidative quality; protein denaturation; pale, soft, and exudative meat; irradiation; high hydrostatic pressure; vacuum packaging; modified atmosphere packaging; active packaging; poultry microbiological contamination

27.1 INTRODUCTION

The shelf life of a food can be defined as the time the food is safe to consume and/or has an acceptable quality to consumers (Fu & Labuza 1993). Poultry meat is one of the most popular muscle foods in the USA, where the per capita meat consumption of chicken and turkey has increased dramatically since the 1970s. Chicken consumption rose from 40.2 lb per person in 1970 to 86.5 lb per person in 2007. Turkey consumption climbed from 8.1 lb per person in 1970 to 17.3 lb per person in 2007 (American Meat Institute 2009). Overall, per capita consumption of red meat and poultry has not changed significantly but when beef, pork, and chicken are examined separately, beef appears to be losing market share to chicken. Consumer's concerns with regard to eating less beef are related to cholesterol, calorie content, artificial ingredients, convenience, characteristics of how beef is displayed in the store, and price (Resurrección 2003). Other authors reported a differentiation between beef and poultry related to taste, healthiness, and convenience, and by process characteristics like organic production and animal welfare (Grunert *et al.* 2004). Whereas straightforward meat lovers focus mainly on taste as the decisive criterion, indifferent consumers are strongly oriented to price (Verbeke & Vackier 2004). Besides, cholesterol has been implicated in diseases associated with modern life, especially in developed countries. These include various cancers and especially coronary heart diseases (Wood *et al.* 2004).

Extending the shelf life of poultry products is a major concern for the poultry industry. The shelf life of poultry depends on several factors, particularly initial bacterial loads, storage temperature, and the gaseous environment around the product (Mead 1990).

Handbook of Meat, Poultry and Seafood Quality, Second Edition. Edited by Leo M. L. Nollet.

Teratanavat and Hooker (2004) examined the characteristics of US meat and poultry recalls between 1994 and 2002 and highlighted the public health implications of these trends. Most recalls involve serious health consequences including biological hazards. This may be explained by improving inspection methods for detecting microbial pathogens, greater consumer awareness, and/or better surveillance of food-borne illness.

Food shelf life is not totally dependent on time but also on environmental conditions such as temperature, humidity, light, and oxygen. The prime factors, temperature and humidity, may increase deterioration as they rise or slow the process as they decrease. However, their impact depends on how widely they vary and on the product itself. Although more difficult to measure in certain cases, sensory quality is the most important characteristic for consumers and processors alike, and in some cases with foods with long shelf life, this attribute may also be an indicator of nutrient quality. More than 75% of further processed chicken products are frozen for distribution. Many of these products are thawed prior to preparation, where ice crystals alter muscle cell integrity and enable fluid losses that vary with the nature of crystallization.

27.2 DETERIORATION MODES OF FRESH AND FROZEN POULTRY

The quality deterioration and potential hazards are usually exaggerated or complicated by a fluctuating time–temperature environment during storage. On the other hand, the shelf life of a frozen food can be extended through ingredient selection, process modification, and change of package or storage conditions. Microbes usually are not a problem since they cannot grow at freezing temperatures unless subjected to extensive temperature abuse above the freezing point. For any specific frozen product, the mode that determines its shelf life depends on the product characteristics, prefreezing treatment, freezing process, packaging film and process, and storage conditions (Fu & Labuza 1993). The main modes of fresh and frozen poultry meat deterioration are shown in Table 27.1.

27.2.1 Microorganisms

Microbiological food safety standards have been around for a long time, but reasons for their development and the differences between countries and jurisdictions lead to ambivalence by industry and trading countries. There is an underlying assumption that there is a need to protect the public by reducing the contamination level of pathogens in raw meats and poultry even if they are not completely eliminated. But there is little apparent connection between public health goals and standards or guidelines except in a general way for reducing or limiting contamination (Todd 2003).

Alvarez-Astorga *et al.* (2002) studied the microbiological quality of retail chicken parts reporting that quality was generally regarded as unacceptable when psychrotrophs,

Table 27.1 Mode of deterioration of fresh and freezing poultry.

Food product	Mode of deterioration	Critical environmental	Average shelf life
Fresh poultry	Pathogen growth, microbial decay, rancidity	Oxygen, temperature, light	2–7 days at refrigerated temperature
Frozen poultry	Protein denaturation, color change	Temperature	6–12 months

Escherichia coli (*E. coli*), and *Staphylococcus aureus* counts were higher than the maximum limits established in the guidelines for poultry meat. Feed withdrawal may have increased prechill carcass counts for *E. coli* and *Campylobacter*, particularly in older broilers, but with no effect on postchill counts when chlorine was used (Northcutt *et al.* 2003). Nayak and Kenney (2002) reported that approximately 25% of *Salmonella* serotypes isolated from poultry were antibiotic resistant, identifying *S. heidelberg* as the most frequent antibiotic resistance strain. Other factors like cooking methods and feed can also affect the presence of pathogens in poultry. Wang *et al.* (2004) studied the *sous vide* treatment to prolong the shelf life of chicken wings and concluded that this treatment retarded the microorganism growth for 7 weeks. On other hand, Jones and Richardson (2004) suggested that *Salmonella* contamination rates are related to poultry feed as a result of mill management practices. *Campylobacter*, a major food-borne pathogen found in poultry products, remains a serious problem for poultry processors and has been recognized as an important cause of human illness. Case control studies have estimated that 50–70% of *Campylobacter* contamination in human food is due to poultry and poultry products. Short-time exposure of *Escherichia coli* and *Campylobacter jejuni* in chicken meat to microwaves had no significant effect on their numbers or growth (Goksoy *et al.* 2000). Frozen chicken nuggets and strips have recently been identified as a cause of salmonellosis due to undercooking. *Salmonella* can survive frozen storage when inoculated in frozen processed chicken products, since cell populations remained relatively constant at −20°C during a 16-week storage period (Domínguez & Schafner 2009). Polymerase chain reaction/Rappaport–Vassailiadis (PCR-RV) is a rapid, time-saving technique applicable to detect *Salmonella* directly from chicken samples (Moussa *et al.* 2010). Other authors (Maziero & de Oliveira 2010) studied the survivability of *Campylobacter jejuni* stored for 7 days at 4°C and for 28 days at −20°C. After refrigeration, 53.3% of the analyzed samples tested positive, but after storage at −20°C, 36.6% of the samples were positive, showing that this microorganism can survive under the tested storage conditions. The population of yeast in fresh poultry meat during storage at refrigeration temperature was studied and results suggested that yeast, particularly *Yarrowia lipolytica*, may play a more prominent role than previously recognized in the spoilage of fresh and processed poultry products stored at 5°C (Ismail *et al.* 2000).

Microbial contamination is a serious problem and many decontaminant treatments have been proposed (Table 27.2). Washing systems including combinations of inside/outside carcass washers and homemade cabinet washers provide minimal reductions in *Campylobacter* populations found on poultry in processing plants (Bashor *et al.* 2004). The recurrent use of antibiotic substances represents a potential hazard for the poultry industry. *Salmonella* and other food-borne pathogens could acquire antibiotic resistance by random chromosomal mutations, mutation of existing genes, and through specific mechanisms such as transduction, transformation, and conjugation.

27.2.2 Color change

Color, appearance, and texture are important factors that consumers will consider before making a decision to buy poultry. The effect of stunning and decapitation is an important factor for meat quality (McNeal *et al.* 2003). Rodbotten *et al.* (2004) investigated if meat from different species could be described and related to each other by sensory analysis. Sixty-eight percent of the sensory variation was contained in the first component, which was dominated by color attributes. Qiao *et al.* (2002) reported that extreme variations in breast meat color were related to differences in muscle pH. Instrumental shear force measurement and CIE-Lab color parameters or sensory evaluation techniques can provide reliable information about the

Table 27.2 Mean poultry decontamination treatments.

Treatment	Examples and general mode of action
Chemical	Phenols (caustic chemical compounds that denature proteins) Alcohols (denature proteins and disrupt lipids) Halogens (iodine, chlorine, bromine, and fluorine have both disinfectant and antiseptic properties) Oxidizing agents (hydrogen peroxide that releases oxygen radicals, which are toxic to many microbes, especially anaerobes) Surfactants (disrupt cellular membranes)
Physical	Temperature (high temperature to destroy vegetative forms; low temperatures to slow down microbial growth) High pressure (4000–9000 atm) or ultrasonic energy (direct cavitational damage to microbial cell membranes) Irradiation and UV light (cellular DNA damage)
Microbiological	Antimicrobial compounds produced by other microorganisms. Competence for nutrients and inhibition of competitive flora (several lactic acid bacteria producing organic acids and/or bacteriocins)

Source: Adapted from Capita et al. 2002

quality of poultry meat. However, many of these techniques are destructive, time consuming, and unsuitable for online application. Visible near-infrared spectroscopy could form the basis for such techniques due to speed, ease of use, and lesser interference from moisture or color of meat samples (Liu *et al.* 2004). When poultry meat is frozen and thawed, an accentuated L value can be achieved, usually indicating a poor quality because of the soft character and extensive water loss, which is similar to the pale, soft, and exudative (PSE) condition in swine white muscle (Galobart & Moran 2004). Variation in the color of raw poultry meat, very light or very dark, has been reported as PSE.

27.2.3 Oxidative quality

Lipid oxidation is considered to be one of the major problems in the meat industry, due to the resultant flavor deterioration and loss of nutritional value (Ahn *et al.* 1992). Changes in quality are manifested by adverse changes in flavor, color, texture, nutritive value, and the possible production of toxic compounds (Gray *et al.* 1996). Lipid oxidation is a complex process whereby unsaturated fatty acids react with the molecular oxygen via free radicals and form peroxides or other primary products of oxidation. Secondary oxidation products, such as aldehydes, ketones, and esters, are responsible for the increased deterioration and rancid flavor during frozen storage. Antioxidant activity depends on the concentration of active compounds present in the samples available to scavenge the free radicals formed during the storage period. Supplementation with antioxidants could be an alternative method to prevent oxidative degradation of the meat during frozen storage when vacuum packaging is not practical (Mielnik *et al.* 2003). In a sensory evaluation study, descriptive profiling indicated that a warmed-over flavor was described by an increase of rancid and sulfur/rubber sensory notes and a concurrent decrease of chicken meaty characteristics (Byrne *et al.* 2002).

Russell *et al.* (2003) studied the effect of supplementation with tocopheryl acetate on the quality of frozen duck meat. This was done because poultry meat, particularly that of duck, has relatively high levels of unsaturated fatty acids and low levels of antioxidants. Ducks consume twice as much feed as broilers during growth. Other authors (Batifoulier *et al.* 2002) reported that the supplementation of turkeys with a-tocopheryl acetate increased the vitamin E content of microsomal membranes and had a protective effect on lipid oxidation. The supplementation with vitamin E significantly protected free thiols from oxidation but had only a small effect on carbonyl group formation. Narciso-Gaytán *et al.* (2010) evaluated the diet effect of supplementation with a low (33 ppm) or high (200–400 ppm) level of vitamin E on lipid oxidation stability of fresh and cooked chicken meat. The results showed that lipid oxidation development in both fresh chicken parts and cooked meat patties was influenced by the interaction of either dietary lipid source of vitamin E level with storage time. The high level of vitamin E supplementation showed lower malonaldehyde values than the control treatment. The lipid oxidation stability of chicken meat was influenced by the lipid source and vitamin E level included in the diet upon storage time and processing of the meat. Gheisari and Motamedi (2010) investigated the effects of NaCl and KCl at varying ionic strengths on catalase and glutathione peroxidase (GSH-Px) activities and lipid oxidation in ground cattle, camel, and chicken muscle during refrigerated storage. Results indicated that negative correlation between lipid oxidation and GSH-Px activity and the accelerated lipid oxidation in salted meat may be partly related to a decrease in GSH-Px activity. Chicken samples had lower enzyme activities and thiobarbituric acid reactive substances (TBARS) content than cattle and camel. Beltran *et al.* (2003) reported that salt and mechanical processing had a greater prooxidant effect on pressurized samples.

27.2.4 Protein denaturation

Protein denaturation is a consequence of the meat microstructure disruption, caused by changes in temperature and/or humidity (Molina-Garcia *et al.* 2004). Van Laack and Lane (2000) studied the susceptibility to denaturation of myofibrillar protein from chicken muscles and compared it with denaturation of myofibrillar proteins from pork. They utilized white muscles and red muscles. Protein solubility as an indicator of denaturation was assessed after 0-, 10-, 20-, and 60-min incubation. Results indicated that proteins in chicken white myofibrils seem more resistant to denaturation than those from the chicken red myofibrils and those from pork myofibrils. Yoon (2002) observed no significant texture toughening in frozen chicken breast after 10 months of storage at –20°C, suggesting that toughening is not a determinant factor in quality loss of frozen chicken breast.

27.2.5 Pale, soft, and exudative meat

The occurrence of PSE meat in poultry has begun to gain large-scale industry interest during the past few years because of its economic importance. Overall, the poultry grading system around the world is based on aesthetic attributes, which puts emphasis on factors such as conformation, presence/absence of bruises, skin tears, and missing parts. Based on this system, no consideration is given to functional properties of the meat such as water-holding capacity, texture, and integrity (Barbut 1997). The occurrence of PSE poultry meat is believed to be the result of accelerated postmortem glycolysis while the carcass is still warm. Allen *et al.* (1997) determined whether a relationship exists between broiler breast meat color, pH, and rate of microbial spoilage during storage. Data suggested that darker broiler breast meat fillets have a shorter shelf life than lighter breast fillets. The shorter shelf life may be due to differences in pH. High muscle pH produced conditions that made dark-colored fillets more susceptible to bacterial spoilage than light-colored fillets when held at the same refrigerated storage conditions. Swatland (2008) studied why pH causes paleness or darkness in chicken breast muscle. The experiment results indicated that pH-dependent light scattering in chicken meat exhibits the same features as in pork and beef. This pH-related scattering has a major effect on the translucency of poultry meat causing low-pH meat to appear pale and high-pH to appear dark. Uniformity of meat coloration matched to the requirements of a particular market is extremely important in the poultry industry. Kissel *et al.* (2009) evaluated the emulsion properties of broiler chicken PSE meat in the production of mortadella. Although the technological functional properties of broiler PSE meat are weak, they were strengthened by the addition of the other mortadella ingredients. As a result, this abnormal meat can be used as a fresh raw material in the production of a broiler breast meat emulsion. In Europe, the consumer's preference for more traditional products coupled with the EU policy to limit additive use increases the importance of raw meat quality used for the preparation of further processed products. This makes European processing industries more vulnerable to the PSE-like condition. Data from several studies have indicated that the incidence of poultry PSE-like meat in Europe is around 10% (Petracci *et al.* 2009).

Barbut (2009) suggested the poultry industry can reduce or eliminate the PSE problem using both short- and long-term approaches. In the short term, attention should be given to factors currently known to trigger the formation of PSE in susceptible birds (reducing stress during transportation, selection or separation of PSE meat). In the long term, more attention should be given to genetics, selection, and finding a reliable marker. The genetic basis of PSE in poultry is still not fully understood, but it is likely that one or more mutations predispose the birds to develop PSE conditions.

27.3 METHODS TO PROLONG SHELF LIFE

27.3.1 Irradiation

The chronology of food irradiation events in poultry started in 1976, when the US Army contracted with commercial companies to study the wholesomeness of irradiated ham, pork, and chicken. In 1990, the Food and Drug Administration (FDA) approved the irradiation of poultry to control *Salmonella* and other food-borne bacteria. Maximum permitted dosage was set at 3.0 kGy. In 1992, the United States Department of Agriculture (USDA) approved irradiation for poultry to control *Salmonella* and other food-borne bacteria. Although irradiation is the best method to ensure the microbiological safety of raw meat, it causes a few radiolytic meat quality defects. Irradiation of pork and poultry meat accelerates lipid oxidation (Katusin-Razem *et al.* 1992). Color changes in irradiated fresh meat occur because of the susceptibility of the myoglobin molecule, especially the iron, to alterations in the chemical environment and to energy input. The potential for iron electrons to exist in various states makes the environment adjacent to the iron atom particularly vulnerable to the presence of electron-donating compounds and high-energy inputs (irradiation). Generation of stable red pigments or brown pigments that become red over time appears to be due to binding of irradiation-generated reactive oxygen species or gasses that become ligands bound by iron under altered reducing conditions. Generation of green pigments appears to be due to breakdown of the porphyrin integrity and/or formation of sulfmyoglobin. Maintenance of ideal meat color during irradiation can be enhanced by many combinations of preslaughter feeding of antioxidants to livestock, optimizing the condition of the meat prior to irradiation, addition of antioxidants to the meat, gas atmosphere, packaging, and temperature control (Brewer 2004).

Gomes *et al.* (2003) reported that irradiated mechanically deboned chicken meat showed higher values for a* (redness) than the nonirradiated samples from the fourth day under refrigeration. Considering the sensory analysis, color, psychrotrophic bacterial counts, and TBARS analyses as a whole, the samples irradiated with doses of 0.0, 3.0, and 4.0 kGy were acceptable under refrigerated storage for 4, 10, and 6 days, respectively. Nam and Ahn (2003b) reported that the redness of meat decreased during a 7-day storage period, but irradiated meat maintained a redder color than nonirradiated samples. Irradiated meat also produced more sulfur volatiles and aldehydes than nonirradiated meats. Liu *et al.* (2003) reported that oxymyoglobin increased as a result of irradiation. Zhu *et al.* (2003) studied the effect of irradiation on the quality of turkey ham during storage. Their results showed that up to 2 kGy of irradiation had limited effects on the color and oxidation of vacuum-packaged commercial turkey ham. However, irradiation had a significant influence on the odor and flavor of the turkey ham. Both sensory panelists' and volatiles' analyses showed that there were significant changes in sulfur-related odor/flavor in ready-to-eat (RTE) turkey products by irradiation. Irradiation of fresh chicken pieces with a dose of 3 kGy appeared to be able to extend the microbial shelf life by a factor of 2 (Mahrour *et al.* 1998). When the chicken was marinated and irradiated at 3 kGy or when irradiated at 5 kGy without marinating, the microbial shelf life was extended by a factor of 7–8. *Salmonella* was found in samples irradiated at 5 kGy under vacuum (Mahrour *et al.* 1998). Lewis *et al.* (2002) utilized e-beam irradiation to improve poultry product safety and shelf life. Skinless chicken breasts were subjected to an electron beam irradiation dose of 0.8 kGy and stored under aerobic or vacuum-packaged conditions for 42 days at 4°C. Results showed that the 0.8 kGy dose was effective in reducing population levels of total aerobic bacteria and psychrotrophs, eliminating *Salmonella* and *Campylobacter*, increasing product redness, and had no major effects on appearance, texture, flavor, and overall acceptability of the breast fillets. Shelf life of the breast fillets increased from 14 to 42 days due to the irradiation treatment. Teets and Were (2008) studied the antioxidant effects of electron beam irradiation on raw minced chicken breast during refrigerated and frozen storage. They reported decreased formation of lipid oxidation products during both refrigerated and frozen storage of the minced chicken breast.

27.3.2 Natural compounds

Natural compounds such as essential oils, chitosan, nisin, and lysozyme have been investigated to replace chemical preservatives and to obtain "green label" products. Various spices and essential oils have preservative properties and have been used to extend the storage life of meat products (Zhou *et al.* 2010). For example, oregano supplements added to chicken meat protected against stress-induced increases in TBARS in different muscles with no effect on the water-holding capacity (Young *et al.* 2003). Rosemary extract had an antioxidant effect, ethylenediamine tetraacetic acid (EDTA) strongly inhibited oxidation, and hexametaphosphate also showed antioxidant ability when added to chicken slurries (Beltran *et al.* 2004).

Phenolic compounds are ubiquitous in plants. When plant foods are consumed, these phytochemicals contribute to the intake of natural antioxidants in the human diets. Agro-industrial by-products are good sources of phenolic compounds and have been explored as a source of natural antioxidants. While the use of naturally occurring phenolic compounds as food antioxidants is particularly interesting, practical aspects that need to be considered include extraction efficiency, availability of sufficient raw material, and toxicity and other safety considerations (Balasundram *et al.* 2006). The supplementation of broiler diets with plant-derived extracts rich in natural antioxidants improved the oxidative stability of meat (Smet *et al.* 2008). The combination of natural antimicrobial treatments (rosemary or oregano oil) with lysozyme and EDTA on semicooked coated chicken meat stored in vacuum packages (4°C for 18 days) was effective against the growth of Gram-negative and Gram-positive bacteria and to a lesser extent against yeasts, although the oregano treatment was not as pleasant as the rosemary treatment (Ntzimani *et al.* 2010). Pettersen *et al.* (2004) investigated the effects of antioxidants in packaging material under different packaging atmospheres. Natural antioxidants like α-tocopherol used in the manufacture of polyethylene layers decreased TBARS values and hexanal content in mechanically deboned turkey meat.

27.3.3 High hydrostatic pressure

Derived from material science, high-pressure technology (100–1000 MPa) is of increasing interest to biological and food systems. High hydrostatic pressure is a nonthermal technology of primary interest because it can inactivate product-spoiling microorganisms and enzymes at low temperatures without changing the sensory or nutritional characteristics of the product (Cheftel & Culioli 1997). The pressure applied is isostatically transmitted (Pascal's law and Le Chatelier's principle) inside the pressure vessel; food is compressed independently of the product size and geometry because the transmission of pressure to the core is not mass/time dependent and thus the process is minimized. A process of 500–600 MPa may take 2.5 min to charge, 5 min for pressurization, and 3 min to discharge (Aymerich *et al.* 2008).

Patterson *et al.* (2010) studied the effect of high pressure on the microbiological quality of cooked chicken and identified those organisms that dominated during storage for up to 35 days. *Weissella viridescens* was found to be the dominant microorganism in the pressure-treated meat, constituting 100% of the microflora identified. The selection of a pressure-resistant organism such as this *Weissella* spp. could be advantageous in extending the shelf life and improving the microbiological safety of products as it could give protection in addition to the pressure treatment itself.

Generally, it is assumed that the fresh flavor is retained in pressurized food products, since small flavor molecules are not affected by high pressure. Schindler *et al.* (2010) studied changes in the aroma profiles of chicken breast high-pressure treated at 400 and 600 MPa, which is in the vicinity of the critical pressure for lipid oxidation. They confirmed that pressurization does not result in severe changes to the aroma profiles. The practical implementation of the high-pressure technology to produce high quality sensorially stable meat products is promising.

27.3.4 Packaging

Packaging protects products against deteriorative effects, which include discoloration, off-flavor and off-odor development, nutrient loss, texture changes, pathogenicity, and other measurable factors. Packaging materials used for meat products are usually plastics, in which polymers with good oxygen barrier properties are incorporated with polymers with good humidity barrier and sealing properties such as polyethylene and polypropylene. While permeability varies with the particular film type used, fresh meat packaging is only minimally permeable to moisture and so surface desiccation is prevented. Packaging options are vacuum packaging, modified atmosphere packaging, and active packaging (McMillin *et al.* 1999).

27.3.4.1 Vacuum packaging

Nam and Ahn (2003a) studied the effects of combinations of aerobic and anaerobic packaging on color, lipid oxidation, and volatile production to establish a modified packaging method to control quality changes in irradiated raw turkey meat. Lipid oxidation was the major problem with aerobically packaged irradiated turkey breast while retention of volatiles such as dimethyl sulfide, dimethyl disulfide, and dimethyl trisulfide associated with irradiation off-odor was the primary problem in vacuum-packaged turkey. Although irradiation was associated with increased a values, exposure to aerobic conditions reduced the redness characteristics.

27.3.4.2 *Modified atmosphere packaging*

The applicability of time–temperature indicators for quality control of modified atmosphere packaging (MAP) broiler chicken cuts was evaluated at various constant and variable temperature conditions (Smolander *et al.* 2004). It was found that microbiological shelf life could be considerably improved when the cold-chain was carefully maintained. Temperature had a critical effect on the amount of enterobacteriaceae, proteolytic bacteria, hydrogen sulfide-producing bacteria, and clostridia, the microbial groups most likely to have an effect on the sensory quality. The results also indicated that time–temperature indicators seemed to be useful tools for evaluation of the quality of broiler chicken cuts. Quality changes in MAP broiler chicken cuts during cold storage were monitored using four different methods in a study by Vainionpaa *et al.* (2004). The comparison of the sensitivities of the quality-indicating methods was performed using principal component analysis. According to this study, the microbiological analysis and the time–temperature indicators gave the same result and were more critical than either the quality-indicating metabolites or sensory evaluations. Pexara *et al.* (2002) evaluated shelf life by comparing products stored in vacuum and modified atmospheres and concluded that the use of MAP in these tests did not extend nor reduce the product shelf life in comparison to vacuum packaging. Patsias *et al.* (2006) evaluated the effect of MAP on shelf-life extension of a precooked chicken meat product stored at 4°C using microbiological, physicochemical, and sensory analysis. They utilized three gas mixtures: M1: 30% CO_2/70% N_2, M2: 60% CO_2/40% N_2, and M3: 90% CO_2/10% N_2. The results showed that the limit of sensory acceptability was reached for the aerobically stored and M1 gas mixture chicken samples at days 16 and 20 of storage, respectively. The use of MAP resulted in an extension of shelf life of precooked chicken by about 4 days with the M1 gas mixture and by more than 6 days with the M2 and M3 gas mixtures. Precooked chicken meat was better preserved under the M2 and M3 mixtures, maintaining desirable odor/taste attributes even on the final day of storage. Fraqueza and Barreto (2009) evaluated the effect of MAP environments containing argon on the shelf life of turkey meat stored at 0°C. The use of 50% argon as the packaging gas improved microbiological quality as compared to aerobic packaging, but provided no additional protective effect on the lipid oxidation of turkey meat.

27.3.4.3 *Active packaging*

Active packaging involves the inclusion of agents either in or on the packaging to enhance some aspect of the packaging system performance. Examples of active packaging include systems with oxygen scavenging abilities, moisture absorption systems, and packaging systems that control generation of compounds such as ethanol or carbon dioxide (Coma 2008). Much of the work on active packaging has remained at the research stage but commercialization has taken place in the USA and Japan. Of particular importance is the use of antimicrobial packaging systems.

Antimicrobial packaging systems can take on a variety of forms including the following (Cooksey 2010, Coma 2008):

1. Antimicrobial infused sachets connected to the package that releases volatile bioactive substances during storage;
2. Packaging materials in which the antimicrobial agent is directly incorporated into the film;
3. Packaging materials coated with a carrier matrix infused with the antimicrobial agent;
4. Use of film-forming polymers with inherent antimicrobial properties; and
5. Application of edible coatings with bioactive properties directly onto the surface of foods.

The potential for survival and growth of psychrotrophic microaerophilic pathogens in vacuum-packaged or MAP products stored under refrigeration has led to increased interest in the utilization of bioactive packaging systems. Coma (2008) reviewed a number of such systems and their applications to shelf-life extension of meats such as poultry.

27.4 CONCLUSIONS

The perishable characteristics of poultry make necessary alternatives that improve shelf life with the minimum physicochemical and sensory changes necessary. Frozen storage has most often been employed in poultry transportation. Methods such as packaging that help prevent contamination with pathogens and spoilage flora have also been used to maintain an acceptable poultry meat or product during storage. Recent technologies including the use of natural compounds such as antimicrobials and antioxidants and application of hydrostatic pressure are being increasingly used to improve the shelf life of fresh and frozen poultry products.

REFERENCES

Ahn, D.U., Wolfe, F.H., Sim, J.S. & Kim, D.H. (1992) Packaging cooked turkey meat patties while hot reduces lipid oxidation. *Journal of Food Science*, **57**, 1075–1078.

Allen, C.D., Russell, S.M. & Fletcher, D.L. (1997) The relationship of broiler breast meat color and pH to shelf-life and odor development. *Poultry Science*, **76**, 1042–1046.

Alvarez-Astorga, M., Capita, R., Alonso-Calleja, C., Moreno, B. & Garcia-Fernandez, M.C. (2002) Microbiological quality of retail chicken by-products in Spain. *Meat Science*, **62**, 45–50.

American Meat Institute (2009) U.S. meat and poultry production & consumption: An overview. Available at http://www.meatami.com/ht/a/GetDocumentAction/i/48781 (accessed on January 31, 2010).

Aymerich, T., Picouet, P.A. & Monfort, J.M. (2008) Decontamination technologies for meat products. *Meat Science*, **78**, 114–129.

Balasundram, N., Sundram, K. & Samman, S. (2006) Phenolic compounds in plants and agri-industrial by-products: Antioxidant activity, occurrence and potential uses. *Food Chemistry*, **99**, 191–203.

Barbut, S. (1997) Problem of pale soft exudative meat in broiler chickens. *Poultry Science*, **38**, 355–358.

Barbut, S. (2009) Pale, soft, and exudative poultry meat—Reviewing ways to manage at the processing plant. *Poultry Science*, **88**, 1506–1512.

Bashor, M.P., Curtis, P.A., Keener, K.M., Sheldon, B.W., Kathariou, S. & Osborne, J.A. (2004) Effects of carcass washers on Campylobacter contamination in large broiler processing plants. *Poultry Science*, **83**, 1232–1239.

Batifoulier, F., Mercier, Y., Gatellier, P. & Renerre, M. (2002) Influence of vitamin E on lipid and protein oxidation induced by H_2O_2-activated MetMb in microsomal membranes from turkey muscle. *Meat Science*, **61**, 389–395.

Beltran, E., Pla, R., Yuste, J. & Mor-Mur, M. (2004) Use of antioxidants to minimize rancidity in pressurized and cogged chicken slurries. *Meat Science*, **66**, 719–725.

Beltran, E., Pla, R., Yuste, J. & Mor-Mur, M. (2003) Lipid oxidation of pressurized and cooked chicken: Role of sodium chloride and mechanical processing on TBARS and hexanal values. *Meat Science*, **64**, 19–25.

Brewer, S. (2004) Irradiation effects on meat color—A review. *Meat Science*, **68**, 1–17.

Byrne, D.V., Bredie, W.L.P., Mottram, D.S. & Martens, M. (2002) Sensory and chemical investigations on the effect of oven cooking on warmed-over flavour development in chicken meat. *Meat Science*, **61**, 127–139.

Capita, R., Alonso-Calleja, C., Garcia-Fernandez, M.C. & Moreno, B. (2002) Review: Trisodium phosphate (TSP) treatment for decontamination of poultry. *Food Science and Technology International*, **8**, 11–24.

Cheftel, J.C. & Culioli, J. (1997) Effects of high pressure on meat: A review. *Meat Science*, **46**, 211–236.

Coma, V. (2008) Bioactive packaging technologies for extended shelf life of meat-based products. *Meat Science*, **78**, 90–103.

Cooksey, K. (2010) Active packaging and the shelf life of foods. In: *Food Packaging and Shelf Life. A Practical Guide* (ed. G.L. Robertson), pp. 367–381. CRC Press, Boca Raton, Florida.

Domínguez, S.A. & Schafner, D.W. (2009) Survival of Salmonella in processed chicken products during frozen storage. *Journal of Food Protection*, **27**, 2088–2092.

Fraqueza, M.J. & Barreto, A.S. (2009) The effect on turkey meat shelf life of modified-atmosphere packaging with an argon mixture. *Poultry Science*, **88**, 1991–1998.

Fu, B. & Labuza, T.P. (1993) Shelf life prediction: theory and application. *Food Control*, **4**(3), 125–133.

Galobart, J. & Moran, Jr., E.T. (2004) Refrigeration and freeze-thaw effects on broiler fillets having extreme L* values. *Poultry Science*, **83**, 1433–1439.

Gheisari, H.R. & Motamedi, H. (2010) Chloride salt type/ionic strength and refrigeration effects on antioxidant enzymes and lipid oxidation in cattle, camel and chicken meat. *Meat Science*, **86**, 377–383.

Goksoy, E.O., James, C. & Corry, J.E.L. (2000) The effect of short-time microwave exposures on inoculated pathogens on chicken and the shelf-life on uninoculated chicken meat. *Journal of Food Engineering*, **45**, 153–160.

Gomes, H.A., Da Silva, E.N., Cardello, H.M.A.B. & Cipolli, K.M.V.A. (2003) Effect of gamma radiation on refrigerated mechanically deboned chicken meat quality. *Meat Science*, **65**, 919–926.

Gray, J.I., Gomaa, E.A. & Buckley, D.J. (1996) Oxidative quality and shelf life of meats. *Meat Science*, **43**, 111–123.

Grunert, K.G., Bredahl, L. & Brunso, K. (2004) Consumer perception of meat quality and implications for product development in the meat sector—A review. *Meat Science*, **66**, 259–272.

Ismail, S.A.S., Deak, T., Abd El-Rahman, H.A., Yassien, M.A.M. & Beuchat, L.R. (2000) Presence and changes in populations of yeast on raw and processed poultry products stored at refrigeration temperature. *International Journal of Food Microbiology*, **62**, 113–121.

Jones, F.T. & Richardson, K.E. (2004) Salmonella in commercially manufactured feeds. *Poultry Science*, **83**, 384–391.

Katusin-Razem, B., Mihaljevic, K.W. & Razem, D. (1992) Time-dependent post irradiation oxidative chemical changes in dehydrated egg products. *Journal of Agricultural and Food Chemistry*, **40**, 1948–1952.

Kissel, C., Soares, A.L., Rossa, A. & Shimokomaki, M. (2009) Functional properties of PSE (pale, soft, exudative) broiler meat in the production of mortadella. *Brazilian Archives of Biology and Technology*, **52**, 213–217.

Lewis, S.J., Velásquez, A., Cuppett, S.L. & McKee, S.R. (2002) Effect of electron beam irradiation on poultry meat safety and quality. *Poultry Science*, **81**, 896–903.

Liu, Y., Fan, X., Chen, Y.-R. & Thayer, D.W. (2003) Changes in structure and color characteristics of irradiated chicken breast as a function of dosage and storage time. *Meat Science*, **63**, 301–307.

Liu, Y., Lyon, B.G., Windham, W.R., Lyon, C.E. & Savage, E.M. (2004) Prediction of physical, color, and sensory characteristics of broiler breast by visible near infrared reflectance spectroscopy. *Poultry Science*, **83**, 1467–1474.

Mahrour, A., Lacroix, S., Nketsa-Tabiri, J., Calderon, N. & Gagnon, M. (1998) Antimicrobial properties of natural substances in irradiated fresh poultry. *Radiation Physics and Chemistry*, **52**, 81–84.

Maziero, M.T. & de Oliveira, T.C.R.M. (2010) Effect of refrigeration and frozen storage on the *Campylobacter jejuni* recovery from naturally contaminated broiler carcasses. *Brazilian Journal of Microbiology*, **41**, 501–505.

McMillin, K.W., Huang, N.Y., Ho, C.P. & Smith, B.S. (1999). Quality and shelf life of meat in case-ready modified atmosphere packaging. In: *Quality Attributes of Muscle Foods* (eds Y.L. Xiong, F. Shahidi & C.T. Ho), pp. 73–93. ACS Symposium Series, New York.

McNeal, W.D., Fletcher, D.L. & Buhr, R.J. (2003) Effects of stunning and decapitation on broiler activity during bleeding, blood loss, carcass and breast meat quality. *Poultry Science*, **82**, 163–168.

Mead, G.C. (1990) Standardized method for determining the microbiological condition of processed poultry in relation to potential shelf-life. *Poultry Science*, **46**, 14–18.

Mielnik, M.B., Aaby, K. & Skrede, G. (2003) Commercial antioxidants control lipid oxidation in mechanically deboned turkey meat. *Meat Science*, **65**, 1147–1155.

Molina-Garcia, A.D., Otero, L., Martino, M.N., *et al.* (2004) Ice VI freezing of meat: Supercooling and ultrastructural studies. *Meat Science*, **66**, 709–718.

Moussa, I.M., Gassen, M.A., Al-Doss, A.A., Sadik, M.W.A. & Mawgood, A.A.L. (2010) Using molecular techniques for rapid detection of Salmonella serovars in frozen chicken and chicken products collected from Riyadh, Saudi Arabia. *African Journal of Biotechnology*, **9**, 612–619.

Nam, K.C. & Ahn, D.U. (2003a) Combination of aerobic and vacuum packaging to control lipid oxidation and off-odor volatiles of irradiated raw turkey breast. *Meat Science*, **63**, 389–395.

Nam, K.C. & Ahn, D.U. (2003b) Use of double packaging and antioxidant combinations to improve color, lipid oxidation and volatiles of irradiated raw and cooked turkey breast patties. *Poultry Science*, **82**, 850–857.

Narciso-Gaytán, C., Shin, D., Sams, A.R., *et al.* (2010) Dietary lipid source and vitamin E effects on lipid oxidation stability of refrigerated fresh and cooked chicken meat. *Poultry Science*, **89**, 2726–2734.

Nayak, R. & Kenney, P.B. (2002) Screening of Salmonella isolated from a turkey production facility for antibiotic resistance. *Poultry Science*, **81**, 1496–1500.

Northcutt, J.K., Berrang, M.E., Dickens, J.A., Fletcher, D.L. & Cox, N.A. (2003) Effect of broiler age, feed withdrawal, and transportation on levels of coliforms, Campylobacter, *Escherichia coli* and *Salmonella* on carcasses before and after immersion chilling. *Poultry Science*, **82**, 169–173.

Ntzimani, A.G., Giatrakou, V.I. & Savvaidis, I.N. (2010) Combined natural antimicrobial treatments (EDTA, lysozyme, rosemary and oregano oil) on semi cooked coated chicken meat stored in vacuum packages at 4°C: Microbiological and sensory evaluation. *Innovative Food Science and Emerging Technologies*, **11**, 187–196.

Patsias, A., Chouliara, I., Badeka, A., Savvaidis, I.N. & Kontominas, M.G. (2006). Shelf-life of a chilled precooked chicken product in air and under modified atmospheres: Microbiological, chemical, sensory attributes. *Food Microbiology*, **23**, 423–429.

Patterson, M.F., McKay, A.M., Connolly, M. & Linton, M. (2010) Effect of high pressure on the microbiological quality of cooked chicken during storage at normal and abuse refrigeration temperatures. *Food Microbiology*, **27**, 266–273.

Petracci, M., Bianchi, M. & Cavani, C. (2009) The European perspective on pale, soft, exudative conditions in poultry. *Poultry Science*, **88**, 1518–1523.

Pettersen, M.K., Mielnik, M.B., Eie, T., Skrede, G. & Nilsson, A. (2004) Lipid oxidation in frozen, mechanically deboned turkey meat as affected by packaging parameters and storage conditions. *Poultry Science*, **83**, 1240–1248.

Pexara, E.S., Metaxopoulos, J. & Drosinos, E.H. (2002) Evaluation of shelf life of cured, cooked, sliced turkeys fillets and cooked pork sausages stored under vacuum and modified atmospheres at +4 and +10°C. *Meat Science*, **62**, 33–43.

Qiao, M., Fletcher, D.L., Smith, D.P. & Northcutt, J.K. (2002) Effects of raw breast meat color variation on cooked meat quality. *Poultry Science*, **81**, 276–280.

Resurrección, A.V.A. (2003) Sensory aspects of consumer choices for meat and meat products. *Meat Science*, **66**, 11–20.

Rodbotten, M., Kubberod, E., Lea, P. & Ueland, O. (2004) A sensory map of the meat universe. Sensory profile of meat from 15 species. *Meat Science*, **68**, 137–144.

Russell, E.A., Lynch, A., Galvin, K., Lynch, P.B. & Kerry, J.P. (2003) Quality of raw, frozen and cooked duck meat as affected by dietary fat and tocopherol acetate supplementation. *International Journal of Poultry Sciences*, **2**(5), 324–334.

Schindler, S., Krings, U., Berger, R.G. & Orlien, V. (2010) Aroma development in high pressure treated beef and chicken meat compared to raw and heat treated. *Meat Science*, **86**, 317–323.

Smet, K., Raes, K., Huyhebaert, G., Haak, L., Arnouts, S. & de Smet, S. (2008) Lipid and protein oxidation of broiler meat as influenced by dietary natural antioxidant supplementation. *Poultry Science*, **87**, 1682–1688.

Smolander, M., Alakomi, H.-L., Ritvanen, T., Vainionpaa, J. & Ahvenainen, R. (2004) Monitoring of the quality of modified atmosphere packaged broiler chicken cuts stored in different temperature conditions, A. Time-temperature indicators as quality-indicating tools. *Food Control*, **15**, 217–229.

Swatland, H.J. (2008) How pH causes paleness or darkness in chicken breast meat. *Meat Science*, **80**, 396–400.

Teets, A.S. & Were, L.M. (2008) Inhibition of lipid oxidation in refrigerated and frozen salted raw minced chicken breast with electron beam irradiated almond skin powder. *Meat Science*, **80**, 1326–1332.

Teratanavat, R. & Hooker, N.H. (2004) Understanding the characteristics of US meat and poultry recalls: 1994–2002. *Food Control*, **15**, 359–367.

Todd, E.C.D. (2003) Microbiological safety standards and public health goals to reduce foodborne disease. *Meat Science*, **66**, 33–43.

Vainionpaa, J., Smolander, M., Alakomi, H.-L., *et al.* (2004) Comparison of different analytical methods in the monitoring of the quality of modified atmosphere packaged broiler chicken cuts using principal component analysis. *Journal of Food Engineering*, **65**, 273–280.

Van Laack, R.L.J.M. & Lane, J.L. (2000) Denaturation of myofibrillar proteins from chicken as affected by pH, temperature, and adenosine triphosphate concentration. *Poultry Science*, **79**, 105–109.

Verbeke, W. & Vackier, I. (2004) Profile and effects of consumer involvement in fresh meat. *Meat Science*, **67**, 159–168.

Wang, S.H., Chang, M.H. & Chen, T.C. (2004) Shelf-life microbiological profiler of chicken wing products following sous vide treatment. *International Journal of Poultry Sciences*, **3**(5), 326–332.

Wood, J.D., Richardson, R.I., Nute, G.R., *et al.* (2004) Effects of fatty acids on meat quality: A review. *Meat Science*, **66**, 21–32.

Yoon, K.S. (2002) Texture and microstructure properties of frozen chicken breast pretreated with salt and phosphate solutions. *Poultry Science*, **81**, 1910–1915.

Young, J.F., Stagsted, J., Jensen, S.K., Karlsson, A.H. & Henckel, P. (2003) Ascorbic acid, a-tocopherol and oregano supplements reduce stress-induced deterioration of chicken meat quality. *Poultry Science*, **82**, 1343–1351.

Zhou, G.H., Xu, X.L. & Liu, Y. (2010) Preservation technologies for fresh meat—A review. *Meat Science*, **86**, 119–128.

Zhu, M.J., Lee, E.J., Mendonca, A. & Ahn, D.U. (2003) Effect of irradiation on the quality of turkey ham during storage. *Meat Science*, **66**, 63–68.

28 Packaging of Fresh and Frozen Poultry

Alfonso Totosaus and V. Kuri

Abstract: Packaging has been developed as a way to protect food from the environment during handling, from transportation to preparation. The development of frozen packaging is significant and food packaging has continued to evolve since that time as an important method of preserving foods, particularly high value-added products such as poultry and other muscle foods. Today, food packaging aims to maintain quality characteristics such as physicochemical and microbiological parameters as well as aesthetic and retailing functions. Packaging should slow down or reduce deteriorative changes such as enzyme-mediated catalytic activity, lipid oxidation, or color changes in fresh or frozen foods. In addition, packaging should also reduce or inhibit the growth of spoilage and pathogenic microorganisms. The combination of packaging technologies with cold or frozen storage helps to maintain an adequate shelf life. The use of other preservative technologies could help to improve and extend the shelf life of poultry meat and products. Recent developments and trends in packaging include active packaging, the use of biodegradable materials, and convenience and multifunctional packaging.

Keywords: Poultry packaging functions, modified atmosphere packaging, vacuum packaging, irradiation, active packaging, recyclable packaging, biodegradable films

28.1 INTRODUCTION

Packaging has been developed as a way to protect food from the environment during handling, from transportation to preparation. Packaging was used in prehistorical markets. The Aztecs, for example, packaged foodstuffs in leafs for sale and transportation. Later, as suitable technologies became available, packaging and the thermal conservation of foods was developed by Nicholas Appert and played a role in feeding Napoleon's army. The further development of frozen packaging by Birdseye was significant and food packaging has continued to evolve since that time as an important method of preserving foods, particularly high value-added products such as poultry and other muscle foods. Today, food packaging aims to maintain quality characteristics such as physicochemical and microbiological parameters as well as aesthetic and retailing functions.

Packaging should slow down or reduce deteriorative changes such as enzyme-mediated catalytic activity, lipid oxidation, or color changes in fresh or frozen foods. In addition, packaging should also reduce or inhibit the growth of spoilage and pathogenic microorganisms. The combination of packaging technologies with cold or frozen storage helps to maintain an adequate shelf life. The use of other preservative technologies could help to improve and

Handbook of Meat, Poultry and Seafood Quality, Second Edition. Edited by Leo M. L. Nollet.

extend the shelf life of poultry meat and products. Recent developments and trends in packaging include active packaging, the use of biodegradable materials, and convenience and multifunctional packaging.

28.2 PACKAGING FUNCTIONS

Generally, food packaging seeks to maintain and protect the quality and hygiene during storage and transport. The choice of suitable packaging materials and in particular the suitability of processing conditions should ensure that the impact of undesirable influences on the contents is minimized as much as possible. Regarding frozen packaging, there are three main factors that affect product quality of any given frozen food (Zaritzky 2008): (1) the initial quality of the original foodstuff, (2) the processing and packaging of the product, and (3) the temperature and duration of storage. The shelf life of foodstuffs depends on process characteristics including raw materials, ingredients, formulations, prefreezing and freezing processes, packaging, and storage conditions. During storage, the material and type of packaging influence the frozen shelf life. Wrapping a product in a tightly fitting pack with low water and oxygen permeability (such as a vacuum pack) can more than double the storage life. Waterproof packing also helps to prevent freeze burn defects, while the tight packing helps to prevent ice buildup in the pack between the meat and the film. Rancidity occurs in unwrapped meats because the surface dries, allowing oxygen to reach subcutaneous fat. Without wrapping, freeze burns may occur, causing unsightly marks, extreme toughening, and the development of rancidity in the affected area. Packaging can be effective in some cases in reducing discoloration by eliminating or minimizing oxygen penetration into the meat. Therefore, products kept in dark or opaque packages would be expected to retain their color longer than those exposed to light (James 2008). Alternative applications include an absorbent pad in conventional foam trays with a polyvinyl chloride overwrap that may or may not control microbial growth, but maintains an aesthetic appeal by absorbing all visible moisture released from the meat during storage (Charles *et al.* 2006). Fig. 28.1 shows different types of commonly used packaging for fresh and frozen poultry. Most fresh poultry (legs, thighs, or breasts) are packed in a tray with or without a moisture absorption pad and wrapped with a stretchable film such as Saran or PVC. This packaging system is also employed in frozen cuts such as deboned breasts. Laminated preformed packaging is also employed for fresh products. Whole birds are packaged in thermo shrinkable films to be frozen. Poultry meat for further processing is usually frozen in air-blast freezers as carcasses or boned-in and boned-out portions in cartons weighing up to 25 kg (James 2008). Alternatives to corrugated cardboard boxes (with high heat transfer resistance) used commercially for poultry freezing in continuous convective air-blast tunnels are perforated metal boxes that allow shorter freezing times (45% less) and spatially homogeneous freezing (Santos *et al.* 2008).

28.3 CONVENIENCE AND PACKAGING

Market differentiation of poultry products is achieved through packaging development and product innovation. One of the observed trends is for multifunctional packaging. This involves the use or reuse of the packaging material for the preparation of the meal. Examples include polymer bags used for marinating or basting and cooking trays designed as case-ready trays that maintain the product attributes during the retail and storage product and are then used as

Fig. 28.1 Different types of poultry packaging: Clockwise from top left, tray and film; skin tight; heat sealed tray, modified atmosphere packaging (MAP) with pad; mold formed MAP, tray and film, fresh.

disposable baking oven trays. Convenience is one of the drivers for these developments, as often novel packaging systems include the preparation of fresh or frozen meats and the addition of ingredients such as dressings, stuffing, sauces, or marinades. In a more advanced version of such convenient packages, vegetables for casseroles or stir fry dishes are included as part of the package (Fig. 28.2).

28.4 ATMOSPHERE MODIFICATION

Modification of the packaging environment represents one of the most widely applied methods, in combination with a chill chain, to extend muscle-based foods such as poultry. The primary principle of modified atmosphere (MA) packaging is the exclusion of oxygen. Oxygen limits the shelf life of meat by causing oxidative rancidity and by enhancing the growth of spoilage microorganisms. By using a barrier film or modifying the gaseous environment surrounding the meat, the enclosure of a food product in gas-barrier materials provides an environment that has been changed or modified to inhibit the action of spoilage microorganisms. This results either in the maintenance of the quality of a perishable food during its natural life or in the actual extension of the shelf life. The product is generally stored in an atmosphere containing an appropriate composition of nitrogen, carbon dioxide, and oxygen (Hood & Mead 1993; Balasubramaniam & Chinnan 1997; Rao & Sachindra 2002). The proliferation of microorganisms present in poultry depends on environmental conditions. Gaseous composition and temperature are two of the main factors associated with the foodstuffs storage environment.

Fig. 28.2 Convenience and multifunctional packaging: Clockwise from top left, fresh whole birds on film, on heat sealed tray, on tray, prepared with stuffing; divided tray with vegetables on one side, and fresh prepared portions on the other side; individually wrapped fresh portions, for home freezing in resealable bag; on oven ready tray, with sauce, fresh; cardboard box detail about baking system, and baking bag with frozen portion with flavored butter; fresh duck portion basted in sauce on oven tray, with cardboard sleeve.

The control of the environment by packaging atmosphere, based on the control of gas composition plus temperature or other factors, will retard spoilage and extend shelf life. *Pseudomonas* spp., *Enterobacteriaceae*, and *Brochothrix thermosphacta* are commonly found in poultry meat, but they are inhibited by high concentrations of CO_2 or vacuum packaging (VP). This selective effect on the microbial flora determines the storage life and the type of spoilage that eventually occurs. Under some conditions, microbial growth may stop and then restart after the adaptation of the microorganisms to a new set of environmental conditions. In a similar way, the degree of inhibition increases as temperature decreases (Rao & Sachindra 2002).

28.4.1 Gas mixtures

As implied by the definition, the basic concept of MA packaging is the replacement of air surrounding the food in the package by a mixture of atmospheric gases different in proportion from that in air. Oxygen, nitrogen, and carbon dioxide are the three main gases used commercially, although trace gases such as carbon monoxide, nitrous oxide, and even sulfur dioxide are also used for MA packaging of foods. These gases can be applied individually or in combination in order to obtain the cumulative effect of these gases, which have different purposes in food preservation (Rao & Sachindra 2002).

The shelf life of poultry packaged in MA is dependent on several factors such as gas composition, storage temperature, degree of initial carcass contamination, film permeability

Table 28.1 Gas mixtures employed and their effect on poultry modified atmosphere packaging.

Gas composition (%)				
CO_2	N_2	O_2	Main effect	References
90	10	0	Maintains desirable odor/taste	Patsias *et al.* (2006)
65	30	5	Lower *Pseudomonas* counts	Balamatsia *et al.* (2007)
65	30	5	With nisin-EDTA-like antimicrobial treatments extended refrigerated storage	Economou *et al.* (2009)
70	30		Irradiation of 4 kGy resulted in shelf-life extension (lower viable count and TBA values)	Chouliara *et al.* (2008)
80	20		With nisin-EDTA shelf life was extended	Cosby *et al.* (1999)
62	30	8	Lowers coliforms and *E. coli*	Saucier *et al.* (2000)
60	40	0	Maintains desirable odor/taste attributes, decreases TVC and lipid oxidation	Patsias *et al.* (2006)
50	0	0	Plus 50% argon inhibits total psychrotrophic counts, total anaerobic counts, and *Brochothrix thermosphacta*	Fraqueza & Barreto (2009)
30	65	5	Reduction of aerobic spoilage microbial flora	Balamatsia *et al.* (2007)
30	70	0	Plus freeze chilling reduced *Pseudomonas*, inhibited yeasts and molds, reduced drip loss, no major effect on color	Patsias *et al.* (2008)
30	70	0	Plus 2000 ppm of ozone no sensory deterioration	Al-Haddad *et al.* (2005)
30	70	0	Reduced microbial populations and TBA values	Chouliara *et al.* (2007)
30	70	0	Lowers putrescine and tyramine content	Patsias *et al.* (2006a, b)
25	75		Chloride dioxide reduced plate counts and *S. typhimurium*, affecting color	Ellis *et al.* (2006)
20	80	0	Maintains color, reduces coliforms and *Escherichia coli* counts	Saucier *et al.* (2000)
20	80	0	Plus rosemary oleoresin lowers oxidation	Keokamnerd *et al.* (2008)

to oxygen and carbon dioxide, and headspace volume in the package. Table 28.1 shows the main gas mixtures employed and their effect on poultry MA packaging. Packaging atmospheres used with poultry are designed to prevent changes in coloration and inhibit coliforms and *E. coli*. These goals have been effectively achieved with the use of high CO_2 atmospheres (Saucier *et al.* 2000; Patsias *et al.* 2006a; Balamatsia *et al.* 2007). Other gases like argon have also been employed with an efficient effect on delaying flora development (Fraqueza & Barreto 2009). It has also been reported that ozone used at a 2000 ppm level reduces the salmonellae contamination of chicken carcasses and improves their shelf life (Al-Haddad *et al.* 2005). The addition of other compounds such as rosemary or oregano essential oils can help improve lipid rancidity by reducing TBA reactive substances and hexanal, promote a more stable color, and result in more acceptable sensory scores (Chouliara *et al.* 2007; Keokamnerd *et al.* 2008).

Nonetheless, most of the problem in poultry packaging is related to spoilage and the putrefaction compounds formed. An electronic nose has been developed to determinate the deteriorated quality from poultry packages, consistent with *Enterobacteriaceae* and hydrogen sulfide-producing bacteria (Rajamäki *et al.* 2006). The biogenic amine values for tryptamine, histamine, and tyramine are well correlated with a detriment in sensory data (Ntzimani *et al.* 2006; Patsias *et al.* 2006b). The use of antimicrobial treatments like nisin-EDTA has resulted in an organoleptic extension of MA packaging (Cosby *et al.* 1999; Economou *et al.* 2009).

MA packaging in combination with freeze chilling could decrease populations of *Pseudomonas* spp., yeasts, and molds, promote lower basic nitrogen values, and decrease drip loss with no major effect on color (Patsias *et al.* 2008).

28.5 VACUUM PACKAGING

VP could also be considered as a type of MA packaging since removing the air from the environment itself is a modification of the atmosphere (Hood & Mead 1993; Rao & Sachindra 2002). VP involves placing a product in a film of low-oxygen permeability, removing air from the package, and applying a hermetic seal. The method is effective in inhibiting bacterial growth at low temperatures, resulting in reduced spoilage and extended product shelf life (Balasubramaniam & Chinnan 1997).

VP with shrinkable film results in a second skin formed around the exact contours of the birds when shrunk. Products are then either frozen in brine or blast freezers, according to the processor facilities. The tight vacuum-shrunk bag protects the birds against freezer burn, water absorption is insignificant, and the oxygen barrier is enough to prevent rancidity (Lee 2006). The addition of antimicrobial compounds such as EDTA-nisin or EDTA-lysozyme has been proposed to improve the shelf life of refrigerated vacuum-packaged poultry products (Cosby *et al.* 1999; Ntzimani *et al.* 2010). Natural antioxidants such as honey or essential oils from rosemary or oregano may provide protection against rancidity (Antony *et al.* 2006; Ntzimani *et al.* 2010).

28.6 IRRADIATION

Irradiating fresh meat, even at low doses, can result in off-odors and off-flavors. The methods used to decrease the detrimental effects of irradiation include packaging with oxygen exclusion (VP), replacement of atmospheric oxygen with inert gases (nitrogen), addition of protective agents (antioxidants), and postirradiation repackaging or double packaging in an oxygen-permeable film to allow the flavor to return to near-normal levels (Brewer 2009). In addition to decreasing oxidation, gas-atmosphere packaging and temperature control could help to maintain an ideal meat color during irradiation (Brewer 2004).

Packaging in appropriate materials contributes to the safety of irradiated foods. Materials like nitrocellulose-coated cellophane, polyethylene films, Saran, or nylon are approved as packaging materials for radiation preservation of foods (Elias 1985). Recent studies have been conducted on the application of radiation to poultry products. Irradiation and aerobic packaging promoted the production of aldehydes (propanal and hexanal) related to lipid oxidation in turkey breast and thigh meats (Nam & Ahn 2003c). On the other hand, the characteristic irradiation off-odor volatiles such as dimethyl sulfide, dimethyl disulfide, and dimethyl trisulfide were a concern for VP (Nam & Ahn 2003a), although VP was better than aerobic packaging in preventing lipid oxidation and oxidation-dependent volatile production (Nam *et al.* 2002). The use of double packaging alone was not enough to reduce the pink color of irradiated raw turkey meat, where lipid oxidation and the irradiation of off-odor S-volatile compounds like methanethiol, dimethyl sulfide, dimethyl disulfide, and dimethyl trisulfide (Nam & Ahn 2003c) also occurred, but the combined use of antioxidants (gallate and α-tocopherol) and double packaging would be a useful method to control the oxidative quality changes of irradiated raw and cooked turkey breast (Nam & Ahn 2003b). Other treatments such as marinating with thyme or rosemary reduced the total microbial count and controlled microbial proliferation during storage at 4°C of fresh chicken legs (Mahrour *et al.* 1998). Irradiated packaging of precooked vacuum-frozen turkey breast muscles maintained a stable color, although ionizing radiation produced characteristic off-odor volatiles (dimethyl disulfide and methylthioethane) and lipid oxidation products (Nam *et al.* 2002). Alternatives like MA packaging in combination with irradiation could extend the shelf life by reducing populations of total viable microorganisms, *Pseudomonas* spp., lactic acid bacteria, yeasts, *Brochothrix thermosphacta*, and *Enterobacteriaceae*, and help lower TBA values (Chouliara *et al.* 2008).

28.7 ACTIVE PACKAGING

These packaging systems change the conditions of the packages to extend shelf life or improve safety or sensory properties while maintaining the quality of the food. Removal of O_2, moisture control, ethylene or CO_2 release, addition of preservatives like ethanol, and antimicrobial action are some of the techniques employed in active packaging systems (Rao & Sachindra 2002). This innovative concept in packaging technology can be defined as a system in which the poultry meat, the packaging material, and the environment interact to extend the shelf life, ensuring the innocuousness and the sensory properties (Suppakul *et al.* 2003). There are two types of active packaging systems: one termed "active" that progressively modifies the packaging environment by active compounds that remove or scavenge oxygen and/or generate ethanol or carbon dioxide, and another named "antimicrobial" packaging that releases antimicrobial compounds onto the food surface.

Antimicrobial packaging controls or prevents the growth of nondesirable bacteria including pathogens and those responsible for food deterioration (Bouncore *et al.* 2003). Ming *et al.* (1997) demonstrated the possibility of reducing the risk but not controlling the growth of *L. monocytogenes* by the application of pediocin to the film's surface. Polymeric films control the diffusion of low molecular weight compounds from outside to inside or in the opposite direction (Del Nobile *et al.* 2003). MA packaging may cause a change in the physicochemical parameters and the development of pathogenic microorganisms, so it is necessary to combine MA with other preservation techniques, like the use of natural antimicrobial compounds. The antimicrobial compounds can be incorporated in several ways: by addition into the packaging material, by coating onto the surface of the packaging film, or by slow release during the storage from a sachet contained within the package. Essential oils, enzymes, bacteriocins, organic acids, and quitosan are commonly employed (Mastromatteo *et al.* 2010). Allyl isothiocyanate has been employed as an active compound against *L. monocytogenes* and *S. typhimurium*, both bacteria associated with poultry (Shin *et al.* 2010). The use of sachets or pads containing volatile antimicrobial agents from oregano (*Origanum onites*) essential oil extended the shelf life of overwrap packed fresh chicken drumsticks (Oral *et al.* 2009).

28.8 RECYCLABLE AND BIODEGRADABLE PACKAGING

The use of food packaging is a socioeconomic indicator of the increased spending ability of the population or the gross domestic product, as well as of regional (rural as opposed to urban) food availability (Brody *et al.* 2008). Food packaging has been influenced by notable changes in food distribution, including globalization of the food supply and consumer trends for more natural, disposable, potentially biodegradable, as well as recyclable food packaging materials

(Ruban 2009). The primary purpose of food packaging must continue to be the safety, wholesomeness, and quality of the food, but the impact of packaging waste on the environment can be minimized by prudently selecting materials and reviewing expectations of packaging in terms of environmental impact (Marsh & Bugusu 2007). The importance of VP or MA is increasing as meat exportation becomes more prevalent. However, there are concerns regarding the increased use of plastic-based packaging, which will be linked to significant packaging waste disposal problems in the future (Cutter & Sumner 2002).

However, while some materials used in food packaging (aluminum, glass, plastics, paper, and carton) can be easily recovered, reused, or recycled, most packaging materials in use by the meat industry require different types of recovery and disposal systems that may not always be available. One factor to consider is safety, since thawing juices support microbial growth. Biodegradable materials could be a better option for frozen muscle-based foods; such materials would respond to sustainability demands and expectations in the modern food industry while meeting the technological requirements for hygiene, performance, and shelf-life extension. Additionally, future research will also be required to determine the edible nature of packaging films and the consumer attitudes to such films (Ruban 2009) in the meat and poultry industry.

28.9 BIODEGRADABLE FILMS

Biodegradable films are gaining attention for their potential to reduce the pollution produced by synthetic materials that are difficult to recycle. Because reusing or recycling used packaging from meat, poultry, or seafood may involve an operation of washing before reuse or disposal, the use of biodegradable materials for packaging frozen meat and similar products seems to be more feasible.

Biodegradable or edible films can also be used as packaging materials, acting as "delivery ingredient systems" in the application or release of preweighed ingredients during processing. Such a method would potentially avoid errors in the measurement, and handling of these compounds (Kim & Ustonol 2001). Alteri *et al.* (2004) proposed the design of oxygen scavenger films using aerobic microorganisms as the "active compounds."

Even if biodegradable packaging is not practical on a broad basis, the advantages are significant for certain applications, since the litter argument for biodegradable plastics has merit to the extent that biodegradable plastics will tend to break down and become less obtrusive after disposal (Marsh & Bugusu 2007). The functions of biodegradable materials are important to control and modify their mechanical and barrier properties, which primarily depend on the structure of the polymeric packaging material (Siracusa *et al.* 2008). Even though a standard worldwide definition for biodegradable plastics has not been established, the general consensus relates the degradability of a material to a specific disposal environment, and to a specific standard test method that simulates this environment in a time period that determines its classification. In this view, according to De Vlieger (2003), these materials had to perform as follows:

a. Demonstrate inherent biodegradability at a rate and degree similar to that of natural biodegradable polymers;
b. Disintegrate during active composting, so that there are no visible, distinguishable pieces found on the screens; and
c. Have no ecotoxicity nor impact the ability of the resultant compost to support microbial and plant growth.

Table 28.2 Some biodegradable polymers employed in food packaging.

Material	Properties
Starch-based polymer	Combination of polymers from the synthesis of oil-based monomer employing a different percentage (10%, 50%, or 90%) of starch as additive (starch, consumed by microbial action, accelerates the disintegration or fragmentation of polymer chain by producing pores in the materials)
Polysaccharides (other than starch)	Good oxygen barrier properties, similar to plastics, preventing dehydratation, oxidative rancidity, and surface browning, although their hydrophilic nature makes them poor barriers for water vapor. Most reported: alginate, carrageenan, cellulose ether, pectin, agar and chitin/chitosan
Lipids	Larding or enrobing muscle foods with waxes, oils, or fats with emulsifiers and optional seasoning and/or preservative agents (like lactic acid–glycerol lactopalmite). Fats as edible coatings maintain the quality of the coated product during extended storage by acting as excellent oxygen and moisture barriers
Proteins	Collagen, gelatin, and whey protein films create a barrier against lipid oxidation, color, microbial growth, and improving sensory attributes
Poly(lactic acid) (PLA)	Recyclable and compostable polymer, with high transparency, high molecular weight, and water solubility resistance. Resulted of controlled depolymerization of the lactic acid monomer (obtained by fermentation)

Biodegradable films and protective coatings have been used for centuries to prevent quality loss such as shrinkage, oxidative off-flavors, microbial contamination, and discoloration in meat and poultry products (Ustonol 2009). Freezing procedures and frozen storage can compromise the meat quality attributes, but coating with edible materials can also improve the freezing procedure, providing an economically favorable method of meat storage (Yu *et al.* 2008). Biodegradable films and coatings have numerous advantages including protecting meat products during frozen storage, dissolving during thawing and cooking, preserving quality, reducing moisture loss and freezer burn, maintaining color integrity, reducing lipid oxidation and discoloration, enhancing product appearance in retail packages by eliminating drip, sealing in volatile flavors, functioning as carriers of food additives such as antimicrobial and antioxidant agents, and reducing oil uptake by battered and breaded products during frying (Genadios *et al.* 1997).

Biopolymers or edible materials have been developed from various sources for application to muscle foods, including exopolysaccharides or fermentation by-products. These polymers may be divided into three main categories based on their origin and production (Siracusa *et al.* 2008; Ruban 2009):

1. Polymers directly extracted or removed from biomass, such as proteins, lipids, or polysaccharides, for example, polysaccharides such as starch and cellulose, chitosan/chitin, and proteins such as casein and gluten;
2. Polymers produced by classical chemical synthesis including aliphatic copolymers, aliphatic polyesters, polylactide aliphatic copolymers from renewable bio-based monomers such as polylactic acid (PLA), or oil-based monomers such as polycaprolactones; and
3. Polymers produced by microorganisms or bacteria like polyhydroxyalkanoates.

Although these materials fulfill the environmental concerns, some shortcomings on performance include limitations on thermal resistance, barrier, and mechanical properties and high cost (Siracusa *et al.* 2008). Proposed alternatives are the combination of two or more materials to improve mechanical properties, gas exchange, adherence to surfaces, and/or moisture barrier properties or applications such as emulsions or bilayers (Ustonol 2009). Additionally, plasticizers can be used to modify film mechanical properties (Ruban 2009). Table 28.2 lists some examples of biodegradable polymers employed in packaging.

As packaging technologies evolve, retailers and consumers demand innovative packaging, convenience, and high performance to maintain and, in some cases, to modify and improve the quality attributes of meats. An emerging concern of sectors of the market for sustainable and lean technologies is likely to result in the development and use of lean and green packaging and research is underway to develop suitable materials.

REFERENCES

Al-Haddad, K.S.H., Al-Qassemi, R.A.S. & Robinson, R.K. (2005) The use of gaseous ozone and gas packaging to control populations of *Salmonella infantis* and *Pseudomonas aeruginosa* on the skin of chicken portions. *Food Control*, **16**, 405–410.

Alteri, C., Sinigaglia, M., Corbo, M.R., Buonocore, G.G., Falcone, P. & Del Mobile, M.A. (2004) Use of entrapped microorganisms as biological oxygen scavengers in food packaging applications. *LWT—Food Science and Technology*, **37**, 9–15.

Antony, S., Rieck, J.R., Acton, J.C., Han, I.Y., Halpin, E.L. & Dawson, P.L. (2006) Effect of dry honey on the shelf life of packaged turkey slices. *Poultry Science*, **85**, 1811–1820.

Balamatsia, C.C., Patsias, A., Kontominas, M.G. & Savvaidis, I.N. (2007) Possible role of volatile amines as quality-indicating metabolites in modified atmosphere-packaged chicken fillets: Correlation with microbiological and sensory attributes. *Food Chemistry*, **104**, 1622–1628.

Balasubramaniam, V.M. & Chinnan, M.S. (1997) Role of packaging in quality preservation of frozen foods. In: *Quality in Frozen Food* (eds M.C. Erickson & Y.-C. Hung), pp. 296–309. Chapman & Hall, New York.

Bouncore, G.G., Del Nobile, M.A., Panizza, A., Boue, S., Battaglia, B. & Nicolais, L. (2003) Modeling the lysozyme release kinetics from antimicrobial films intended for food packaging applications. *Journal of Food Science*, **68**, 1365–1370.

Brewer, S. (2004) Irradiation effects on meat color—A review. *Meat Science*, **68**, 1–17.

Brewer, M.S. (2009) Irradiation effects on meat flavor: A review. *Meat Science*, **81**, 1–14.

Brody, A.L., Bugusu, B., Han, J.H., Sand, C.K. & McHugh, T.H. (2008) Innovative food packaging solutions. *Journal of Food Science*, **73**, R107–R116.

Charles, N., Williams, S.K. & Rodrick, G.E. (2006) Effects of packaging systems on the natural microflora and acceptability of chicken breast meat. *Poultry Science*, **85**, 1798–1801.

Chouliara, E., Karatapanis, A., Savvaidis, I.N. & Kontominas, M.G. (2007) Combined effect of oregano essential oil and modified atmosphere packaging on shelf-life extension of fresh chicken breast meat stored at 4°C. *Food Microbiology*, **24**, 607–617.

Chouliara, E., Badeka, A., Savvaidis, I. & Kontominas, M.G. (2008) Combined effect of irradiation and modified atmosphere packaging on shelf-life extension of chicken breast meat: Microbiological, chemical and sensory changes. *European Food Research and Technology*, **226**, 877–888.

Cosby, D.E., Harrison, M.A. & Toledo, R.T. (1999) Vacuum or modified atmosphere packaging and EDTA-nisin treatment to increase poultry shelf life. *Journal of Applied Poultry Research*, **8**, 185–190.

Cutter, C.N. & Sumner, S.S. (2002) Application of edible coating on muscle foods. In: *Protein-Based Films and Coatings* (ed. A. Genadios). CRC Press, Boca Raton, Florida.

De Vlieger, J.J. (2003) Green plastics for food packaging. In: *Novel Food Packaging Techniques* (ed. R. Ahvenainen). Woodhead Publishing Ltd, Cambridge.

Del Nobile, M.A., Bouncore, G.G., Actieri, C., Battaglia, G. & Nicolais, L. (2003) Modeling the water barrier properties of nylon film intended for food packaging applications. *Journal of Food Science*, **68**, 1334–1340.

Economou, T., Pournis, N., Ntzimani, A. & Savvaidis, I.N. (2009) Nisin–EDTA treatments and modified atmosphere packaging to increase fresh chicken meat shelf-life. *Food Chemistry*, **114**, 1470–1476.

Elias, P.S. (1985) Irradiation preservation of meat and meat products. In: *Developments on Meat Science-3* (ed. R.A. Lawrie), pp. 115–183. Elsevier Applied Science Publishers, London.

Ellis, M., Cooksey, K., Dawson, P., Han, I. & Vergano, P. (2006) Quality of fresh chicken breasts using a combination of modified atmosphere packaging and chlorine dioxide sachets. *Journal of Food Protection* **69** (8), 1991–1996.

Fraqueza, M.J. & Barreto, A.S. (2009) The effect on turkey meat shelf life of modified-atmosphere packaging with an argon mixture. *Poultry Science*, **88**, 1991–1998.

Genadios, A., Hanna, M.A. & Kurth, L.B. (1997) Application of edible coating on meats, poultry and seafood: A review. *LWT—Food Science and Technology*, **30**, 337–350.

Hood, D.E. & Mead, G.C. (1993) Modified atmosphere packaging of fresh meat and poultry. In: *Principles and Applications of Modified Atmosphere Packaging of Foods* (ed. R.T. Parry), pp. 269–298. Blackie Academic & Professional, London.

James, S. (2008) Freezing meat. In: *Frozen Food Science and Technology* (ed. J.A. Evans). Blackwell, Oxford.

Keokamnerd, T., Acton, J.C., Han, I.Y. & Dawson, P.L. (2008) Effect of commercial rosemary oleoresin preparations on ground chicken thigh meat quality packaged in a high-oxygen atmosphere. *Poultry Science*, **87**, 170–179.

Kim, S.-J. & Ustonol, Z. (2001) Thermal properties, heat sealability and attributes of whey protein isolate lipid emulsion edible films. *Journal of Food Science*, **66**, 985–990.

Lee, K.H. (2006) Plastic packaging of frozen foods. In: *Handbook of Frozen Food Processing and Packaging* (ed. D.W. Sun), pp. 641–651. CRC Press, Boca Raton, Florida.

Mahrour, A., Laeroix, M., Nketsa-Tabiri, J., Calderon, N. & Gagnon, M. (1998) Antimicrobial properties of natural substances in irradiated fresh poultry. *Radiation Physics and Chemistry*, **52**, 81–84.

Marsh, K. & Bugusu, B. (2007) Food packaging—Roles, materials, and environmental issues. *Journal of Food Science*, **72**, R39–R55.

Mastromatteo, M., Conte, A. & Del Nobile, M.A. (2010) Combined use of modified atmosphere packaging and natural compounds for food preservation. *Food Engineering Reviews*, **2**, 28–38.

Ming, X.T., Weber, G.H., Ayres, J.W. & Sandine, W.E. (1997) Bacteriocins applied to food packaging materials to inhibit *Listeria monocytogenes* on meats. *Journal of Food Science*, **62**, 413–415.

Nam, K.C. & Ahn, D.U. (2003a) Combination of aerobic and vacuum packaging to control lipid oxidation and off-odor volatiles of irradiated raw turkey breast. *Meat Science*, **63**, 389–395.

Nam, K.C. & Ahn, D.U. (2003b) Double-packaging is effective in reducing lipid oxidation and off-odor volatiles of irradiated raw turkey meat. *Poultry Science*, **82**, 1468–1474.

Nam, K.C. & Ahn, D.U. (2003c) Use of double packaging and antioxidant combinations to improve color, lipid oxidation, and volatiles of irradiated raw and cooked turkey breast patties. *Poultry Science*, **82**, 850–857.

Nam, K.C., Kim, Y.H., Du, M. & Ahn, D.U. (2002) Off-odor volatiles and pink color development in precooked, irradiated turkey breast during frozen storage. *Poultry Science*, **81**, 269–275.

Ntzimani, A.G., Paleologos, E.K., Savvaidis, I.N. & Kontominas, M.G. (2006) Formation of biogenic amines and relation to microbial flora and sensory changes in smoked turkey breast fillets stored under various packaging conditions at 4°C. *Food Microbiology*, **25**, 509–517.

Ntzimani, A.G., Giatrakou, V.I. & Savvaidis, I.N. (2010) Combined natural antimicrobial treatments (EDTA, lysozyme, rosemary and oregano oil) on semi cooked coated chicken meat stored in vacuum packages at 4°C: Microbiological and sensory evaluation. *Innovative Food Science and Emerging Technologies*, **11**, 187–196.

Oral, N., Vatansever, L., Sezer, Ç., *et al.* (2009) Effect of absorbent pads containing oregano essential oil on the shelf life extension of overwrap packed chicken drumsticks stored at four degrees Celsius. *Poultry Science*, **88**, 1459–1465.

Patsias, A., Chouliara, I., Paleologos, E.K., Savvaidis, I. & Kontominas, M.G. (2006a) Relation of biogenic amines to microbial and sensory changes of precooked chicken meat stored aerobically and under modified atmosphere packaging at 4°C. *European Food Research and Technology*, **223**, 683–689.

Patsias, A., Chouliara, I., Badeka, A., Savvaidis, I.N. & Kontominas, M.G. (2006b) Shelf-life of a chilled precooked chicken product stored in air and under modified atmospheres: Microbiological, chemical, sensory attributes. *Food Microbiology*, **23**, 423–429.

Patsias, A., Badeka, A.V., Savvaidis, I.N. & Kontominas, M.G. (2008) Combined effect of freeze chilling and MAP on quality parameters of raw chicken fillets. *Food Microbiology*, **25**, 575–581.

Rajamäki, T., Alakomi, H.-L., Ritvanenb, T., Skyttä, E., Smolander, M. & Ahvenainen, R. (2006) Application of an electronic nose for quality assessment of modified atmosphere packaged poultry meat. *Food Control*, **17**, 5–13.

Rao, D.N. & Sachindra, N.M. (2002) Modified atmosphere and vacuum packaging of meat and poultry products. *Food Reviews International*, **18**, 263–293.

Ruban, S.W. (2009) Biobased packaging—Application in meat industry. *Vet World*, **2**, 79–82.

Santos, C.A., Laurindo, J.B., Silveira, V. Jr. & Hense, H. (2008) Influence of secondary packing on the freezing time of chicken meat in air blast freezing tunnels. *Ciência e Tecnologia de Alimentos*, **28**, 252–258.

Saucier, L., Gendron, C. & Gariépy, C. (2000) Shelf life of ground poultry meat stored under modified atmosphere. *Poultry Science*, **79**, 1851–1856.

Shin, J., Harte, B., Ryser, E. & Selke, S. (2010) Active packaging of fresh chicken breast, with allyl isothiocyanate (AITC) in combination with modified atmosphere packaging (MAP) to control the growth of pathogens. *Journal of Food Science*, **75**, M65–M71.

Siracusa, V., Rocculi, P., Romani, S. & Rosa, M.D. (2008) Biodegradable polymers for food packaging: A review. *Trends in Food Science & Technology*, **19**, 634–643.

Suppakul, P., Miltz, J., Sonneveld, K. & Bigger, S.W. (2003) Active packaging technologies with an emphasis on antimicrobial packaging and its applications. *Journal of Food Science*, **68**, 408–420.

Ustonol, Z. (2009) Edible films and coatings for meat and poultry. In: *Edible Films and Coatings for Food Applications* (eds M.E. Embuscado & K.C. Huber), pp. 244–268. Springer, New York.

Yu, X.L., Li, X.B., Xu, X.L. & Zhou, G.H. (2008) Coating with sodium alginate and its effects on the functional properties and structure of frozen pork. *Journal of Muscle Foods*, **19**, 333–351.

Zaritzky, N.E. (2008) Frozen storage. In: *Frozen Food Science and Technology* (ed. J.A. Evans). Blackwell, Oxford.

Part Six
Seafood Quality

29 Quality Index Methods

Grethe Hyldig, Allan Bremner, Emilia Martinsdóttir, and Rian Schelvis

Abstract: Recent activity has been in the development of quality index method (QIM) schemes suited to individual fish species. Earlier schemes did not take into account the differences among species. To do that, it is necessary to develop one scheme for each species, and the aim when developing QIM for various species is, also, to have the QI increase linearly with storage period of the fish expressed in equivalent days in ice. To develop any new scheme, QIM takes into account the inherent differences among fish species. It is necessary to have some specific knowledge about the fish species, to have on hand two tested and trained sensory panels, a facility to conduct storage experiments under standardized conditions, and to be able to make a statistical validation of the developed QIM scheme.

Keywords: Quality index methods, principal component analysis, sensory panelist, fish shelf life, quality grading relation EU system, consumer quality index method

29.1 INTRODUCTION

The quality index method (QIM), as it has come to be known, was born out of necessity. During the late 1970s and early 1980s, the Tasmanian Food Research Unit (TFRU) of the Commonwealth Scientific and Industrial Research Organisation (CSIRO), Australia, was conducting a series of trials on different postcatch handling and storage methods for fish and on different packaging and storage methods for fish and for fillets. These trials involved comparisons of the product and it was evident that no satisfactory method existed that allowed differentiation between the various materials over the course of the storage period. Furthermore, no method existed that clearly evaluated the nature of the starting material so that comparisons could be drawn between the different trials, since start material that was exactly similar could not always be obtained. This is a situation that is faced daily by fish buyers and agents as well as by experimenters seeking to determine storage properties.

Although changes occurring in stored fish had been recorded many times, the majority of the schemes used were descriptive and comparisons were difficult. Furthermore, the TFRU trials were (naturally) being conducted on species from the Southern Hemisphere, which had scarcely been fished previously on a commercial basis and about which very little was known. There was thus no body of standard information about the nature of the changes and no experienced personnel to train assessors. In Australia, no standard schemes were in use at the time and those available, for example, the Torry schemes (Shewan *et al.* 1953) and the

Handbook of Meat, Poultry and Seafood Quality, Second Edition. Edited by Leo M. L. Nollet.

European Union (EU) grading scheme (Anonymous 1996), had been developed on Northern Hemisphere species most of which were unrelated to those in Australia.

At that time, the concept of the relative rate of spoilage had been published (Olley & Ratkowsky 1973a, b). This states that most protein foods, such as fish, meats, and milk, all spoil at similar rates, expressed as a ratio to the rate they spoil at the reference temperature of zero degrees Celsius (0°C). This means that spoilage at a variety of temperatures can be reexpressed in terms equivalent days of storage at 0°C and that the integrated effects of storage at different temperatures can be taken into account. Because much recently caught fish is chilled in ice, this provides a uniform method for expressing spoilage and shelf life. It also follows from the nature of this relationship that fish spoil four times faster at 10°C than at 0°C, and twice as fast at 4°C as at 0°C. No schemes proposed up to that point had considered this aspect. It was therefore obvious that a new approach was necessary that could be used on a variety of species, did not require extensive training, was robust, easy to understand, and capable of integrating the effects of time and temperature.

When a scheme is developed, consideration must be given to why it is required, what advantages it has over existing schemes, its accuracy and precision, whether it fits a need, what it will be used for, the circumstances in which it will be used, its robustness in different hands, its adaptability to changing circumstances, its potential in meeting future requirements, its communication value, its ease of use, its cost, its likelihood of adoption, its consistency with known theory, and thus its predictive capacities.

It is fair to say that the developed scheme meets all these criteria and is probably the only scheme yet devised that does so, or in which all these criteria have been considered in its development.

The terminology used in the scheme has been deliberately chosen to avoid conveying the impression that the scheme measures quality itself or freshness or even spoilage since these are vague concepts about which different opinions are common (Bremner 2000; Bremner & Sakaguchi 2000). What the scheme measures is the degree and rate of change in important criteria and in the sum total of these changes, which can be interpreted into equivalent days of storage and remaining shelf life. These are very factual, very practical considerations. The later coining of the term quality index method (QIM) shows that the result obtained in the scheme can be used as an index of what the material may be like for an appropriate end use. This may be a judgment of its ability to withstand a process, or a selection of it for a particular grade of product, or it may foreshadow what the product is anticipated to be like when it is cooked and eaten by the consumer. There are now also QIM schemes developed for cod and plaice fillet (Larsen *et al.* 1998–1999; Bonilla *et al.* 2007) and for frozen cod (Hyldig & Nielsen 1997; Warm *et al.* 1998; Herrero *et al.* 2003). The QIM scheme for frozen cod measures the storage history. It has already been stated that the total score can be used to assess storage history and to estimate remaining shelf life, but this is not its only use. The output score is a small number, generally below a value of 20, and this makes labeling and assessment of changes very easy. Although schemes can be readily expressed and constructed in different languages, a numerical value as an output is universal and can be understood by all. The score can be used to judge handling and storage procedures, efficiencies of personnel, and duration of transport, and it is particularly useful in troubleshooting.

The original scheme was a general scheme used with slight modifications to assess some southern temperate species (Statham & Bremner 1983, 1985; Statham *et al.* 1985; Fitzgerald & Bremner 1998) and some tropical species (Bremner *et al.* 1984). With suitable modification, scampi (Bremner 1988) and shrimp (Chinivasagam *et al.* 1995) were also assessed using the same general principles.

The technique was adopted at the Danish Institute for Fisheries Research, Department for Seafood Research, and during the late 1980s, a number of more specific schemes suited to Northern Hemisphere species were developed in a Nordic project (Jespersen and Helbo 1990; Larsen *et al.* 1992). In the Nordic collaboration project between fish research institutes in Iceland, Norway, Denmark, and the Faroe Islands, QIM schemes were developed for some of the most important fish species in the Nordic countries. These were published in Danish (Anonymous 1992a, b). The name quality index method was coined as a concise term for the method.

An EU-funded Concerted Action (CA) project on fish freshness recommended QIM as a general reference method (Martinsdóttir 1997). Later, several EU-funded projects have had the objective to develop QIM schemes for the various fish species and to study the possibility of incorporating QIM into normal practice, specifications, and standards.

On the basis of these projects, a major development has recently been announced in the formation of the QIM-Eurofish Foundation (www.qimeurofish.com), which is an alliance among the three major fish research institutes in Holland, Iceland, and Denmark that have been most concerned with the development and exploitation of QIM. The aim of this alliance is not merely to promote the use of QIM but to safeguard the manner in which it is used. In this respect, training and audit of schemes has been done on a needs basis. Needs from the industry are based on new technological developments such as auctioning through the internet (i.e., PEFA.com) and remote trade. Borders of countries have no trade barriers any more, and these days it is common practice to buy fish sight unseen in different European countries. The industry therefore needs a reliable, universal, and useful tool to assess the quality of fish. In practice, the EU scheme (Anonymous 1996) is found to be inaccurate and not practical and is mainly used for detecting fish that are unfit for human consumption at the point of landing. For quality grading, the current EU scheme is not adequate. In a grading scheme, sensory attributes of three to six different criteria may be used. Each of them has to be evaluated according to the specifications of the grade standards being used, which often consists of three to six subdescriptions for each of the different grades. Thereafter, a final grade is determined. It often happens that the sensory characteristics of the criterion being evaluated do not agree with all of the other subdescriptions. This will increase the time used to evaluate the fish and can cause confusion to the assessor (Botta 1995). The EU scheme is not species related and the sensory characteristics among the various fish species are very different. QIM Eurofish will continue to stimulate the implementation of QIM within the fish sector and its recognition and adoption by official bodies like the European Commission (Luten & Martinsdóttir 1997). Currently, there are QIM manuals in 11 European languages (Danish, Dutch, English, French, German, Greek, Icelandic, Italian, Norwegian, Portuguese, and Spanish) covering 13 QIM schemes for commercially important species (Martinsdóttir *et al.* 2001, 2004).

29.1.1 Principle

The scheme is based on the proposition that evaluators cannot judge degrees of perfection but can very readily detect deviations or changes from it. A simple illustration of this would be a crack in an otherwise perfect wall. Thus, defects in the product were allotted demerit points, which were summed to a total to provide an overall evaluation. The higher the number of demerit points, the more defects the product had. This approach was derived from the understanding that during storage of fish, changes occur in it that are readily detectable and often measurable. This is also in keeping with the fact that the vast majority of chemical, biochemical, and microbiological tests on fish products start from either zero or a low value and increase with both temperature and period of storage.

Setting the start point at zero is a fundamentally different view from many other evaluative schemes in which a perfect product starts at a high score, often arbitrarily chosen at 10 or 100, and deviations from it are subtracted to result in a lower total. The approach where the start value is a high score requires a priori knowledge of the stages of change in order to set the scaling system and, although it is commonly used in many fields, it is counterintuitive. Only its familiarity makes it seem comfortable to users. It is valid to score a product on a single criterion but differentiation among products is difficult unless accuracy and precision are high and the scale is sufficiently large. These factors are rarely ever the case when evaluating fish, and if a single characteristic such as the shape of the eyes was used, there are few descriptive categories available and judges would have reservations about differentiating among the descriptors flat and slightly flat.

Conversely, scoring large numbers of criteria is laborious, tedious, time consuming, and hence prone to error due to operator fatigue; many may be irrelevant and may not result in any better differentiation than a smaller number of criteria. In addition, the scoring allotted to each criterion was such that no single criterion could dominate and that the score values were easy to judge. Using the example for shape of eyes, for fish species such as cod, only three scores maybe necessary: 0 = convex, 1 = flat, 2 = sunken. All of these are very easy for an untrained assessor to judge. A fourth score 3 = very sunken may be included in a QIM scheme if it is necessary (see Table 29.1).

The resultant scheme became a list of attributes, each of which was scored on a restricted scale (0–3), and the scores were then added to provide a total, a quality index (QI).

With further experience, the attributes and the scores were adjusted slightly to result in a scoring system that gave a straight-line relationship with period of storage of the fish (Bremner 1985; Larsen *et al.* 1992; Huidobro *et al.* 2000; Sveinsdóttir *et al.* 2002, 2003; Nielsen *et al.* 2003; Hyldig & Nielsen 2004; Hyldig & Green-Petersen 2005). Further trials and theoretical investigations underpinned the validity of the scheme and demonstrated that it was capable of integrating the effects of time and temperature during storage (Bremner *et al.* 1987; Sveinsdóttir *et al.* 2002, 2003). The slope of the line is the rate of demerit point accumulation per day of storage, and a simple calculation can indicate the equivalent of the number of days at 0°C that the product has been stored. If the decision had been made at which value the product should no longer be sold, or where it crosses some arbitrary set boundary between product grades, then the remaining shelf life can be calculated for the appropriate end use.

29.1.2 One scheme for each species

The recent activity has been in the development of QIM schemes suited to individual species. As already mentioned, some of the problems among earlier schemes, such as the EU scheme (Anonymous 1996), are that they do not take into account the differences among species. To do that, it is necessary to develop one scheme for each species, and the aim when developing QIM for various species is, also, to have the QI increase linearly with storage period of the fish expressed in equivalent days in ice. QIM does take the inherent differences among fish species into account and therefore it is necessary to develop QIM schemes for each fish species. To illustrate this, parts of different QIM schemes are shown in Tables 29.1 and 29.2. Table 29.1 shows the part with the eyes from the QIM schemes for cod (*Gadus morhua*) and salmon (*Salmo salar*). There are three quality parameters concerning the eyes in the scheme for cod (corona, form of the eyes, and the pupil) and only two in the scheme for salmon (pupil and form of the eyes).

Table 29.2 shows that there are large differences in appearances between plaice and salmon. The texture is not in the scheme for plaice, because here texture is not easy to measure and it does not change much during storage.

Table 29.1 The part of the QIM schemes with the quality parameter for the eyes for cod (*Gadus morhua*) and salmon (*Salmo salar*).

Quality parameters		Description	Score
		Cod (Gadus morhua)	
Eyes	*Cornea*	Clear	0
		Opalescent	1
		Milky	2
	Form	Convex	0
		Flat, slightly sunken	1
		Sunken, concave	2
	Color of pupil	Black	0
		Opaque	1
		Gray	2
		Clear	0
		Salmon (Salmo salar)	
Eyes	*Pupils*	Clear and black, metal shiny	0
		Dark gray	1
		Mat, gray	2
		Convex	0
	Form	Flat	1
		Sunken	2

Source: From Martinsdóttir *et al.* (2004).

Table 29.2 The part of the QIM scheme with the quality parameter for the appearance for plaice (*Pleuronectes platessa*) and salmon (*Salmo salar*).

Quality parameters		Description	Score
		Salmon (Salmo salar)	
Skin: Salmon	*Color/appearance*	Pearl-shiny all over the skin	0
		The skin is less pearl-shiny	1
		The fish is yellowish, mainly near the abdomen	2
	Mucus	Clear, not clotted	0
		Milky, clotted	1
		Yellow and clotted	2
	Odor	Fresh seaweedy, neutral	0
		Cucumber, metal, hay	1
		Sour, dish cloth	2
		Rotten	3
	Texture	In rigor	0
		Finger mark disappears rapidly	1
		Finger leaves mark over 3 s	2
		Plaice (Pleuronectes platessa)	
Skin: Plaice	*Skin (both dark and white side)*	Fresh, bright, metallic, no discoloration	0
		Bright, but without shine	1
		Matt, rather dull, slight green/blue or purple discoloration	2
		Dull, green/blue, purple discoloration	3
	Mucus	Clear, not clotted	0
		Slightly clotted and milky	1
		Clotted and slightly yellow	2
		Yellow and clotted	3

Source: From Martinsdóttir et al. (2004).

29.2 DEVELOPMENT OF SCHEMES

As described in the Introduction, there are several considerations to take into account to develop any new scheme. It is necessary to have some specific knowledge about the fish species, to have on hand two tested and trained sensory panels, a facility to conduct storage experiments under standardized conditions, and to be able to make a statistical validation of the developed QIM scheme. In the following paragraphs, this standardized development is described in detail.

29.2.1 The raw material

The selection of fish species is based on practical considerations and economic value. The fishing gear used to catch them and the fishing grounds where they are caught correspond with this selection and are fixed for QIM schemes. The handling of the fish should be according to good manufacturing practice (GMP) (Codex 1976). GMP can mean various standards or technical specifications, such as that the fish is gutted at sea (if gutting is the normal procedure) and washed. The fish is directly cooled down to 0°C in melting ice or equivalent cooling media. The fish is stored in fish boxes, with sufficient ice, which may be replenished, during the whole storage period. The storage trials begin with homogeneous batches of fish, preferably from one haul, with known history such as date of catch, storage condition, etc. For the complete development of a new QIM scheme, at least three storage trials are needed and the experiment must begin at the time of catch and continue till after the end of shelf life. The QIM schemes developed with these batches are valid for these conditions.

29.2.2 The draft scheme

For the development of a draft QIM scheme, the first storage experiment is needed. In this stage, the fish is described in detail during the complete ice storage until the end of an expected shelf life. This is done at fixed time intervals of 12–48 h, depending on the expected shelf life of the species. The shorter the expected shelf life is, the shorter the time intervals required. A group of a maximum of five experienced panelists, including a panel leader, is put together. This group is selected, besides the usual criteria for sensory panelists, for their good use of vocabulary and knowledge of the fish species. The panel leader is experienced in controlling the discussion and enables everyone to express their own meaning. Approximately, five fishes are used at the time, and they are discarded after each session. During the sessions, the fishes are placed on chill plates or ice. Further sessions are continued until the fish is completely spoiled. All attributes are listed and described in detail. The attributes to be assessed vary per species; for example, for flat fish species, the appearance of both sides might be relevant which is, obviously, not the case for round fish species. The appearance for shrimp will have a completely different description than for cod. All descriptions are written down, preferably in terms the whole group agrees on. A description is written on how the assessment is done—for example, the assessment of the texture attribute for cod is done by pressing the finger (firmly but not too hard) on the spine muscle and observing how fast the flesh recovers.

After this first storage trial, the panel leader selects those attributes that change over the storage period, the descriptions are grouped together per attribute, and major changes are scored with "demerit" points from 0 to 3. If, for example, the maximum of three demerit points are scored within the first 5 days, but it is generally known that then the shelf life is, in total, about 14 days, the description per demerit point needs to be changed in such a way that the scoring covers more of the complete shelf life. This results in the draft QIM scheme that will be used in the next storage trial.

29.2.3 Testing the draft scheme

The testing of the draft QIM scheme is done with a second batch of fish during a new storage trial. For this test, a new panel of approximately ten panelists (experts in QIM) is organized, and the panelists who participated in the development of the scheme are included.

During the storage trial at fixed time intervals, again depending on the shelf life but less frequently than in the first storage trial, five fishes are assessed individually by using the draft QIM scheme. Each expert scores every attribute and comments are written down. After each session, the scores, comments, and questions on how to assess the attributes are discussed among the panelists. The results are analyzed and the attributes are selected for being discriminative and for ease of assessment. Statistical analyses used to assist these decisions are principal component analyses (PCA) or calculation of linear regression lines per attribute (Sveinsdóttir *et al.* 2002, 2003). These decisions on attributes and scores are of major importance for meeting several of the above-mentioned considerations like the balance between the strength of the scheme, reliability of the results, and practical use of the scheme, and cannot be made without knowledge on the fish species, spoilage pattern, practical importance of the sensory attributes, and QIM principles in general.

At the same time when the draft scheme is tested, the end of shelf life needs to be determined. The shelf life is defined as the number of days that whole, fresh (gutted) fish can be stored in ice until it becomes unfit for human consumption. It needs to be emphasized that the estimated shelf life is based upon optimal catching and storage conditions (Sveinsdóttir *et al.* 2002, 2003; Martinsdóttir *et al.* 2001).

Spoilage due to microbial activity is the main limitation of the shelf life. Another cause of spoilage may be rancidity, especially in fatty fish species. The flesh of newly caught fish is free of bacteria. However, considerable amounts of bacteria may be in the viscera, on the gills, and on the skin, which can give contamination during processing. When the fish is stored whole in ice, the deterioration caused by bacteria is minimal for the first days of storage and then it will increase.

In general, when fish is stored in ice, the flavor and odor compounds that characterize newly caught fish decrease and disappear in the first few days during storage, and the fish flesh becomes almost flavorless and odorless for a while. Then an increase in bad-smelling sulfur and nitrogenous volatiles will result in rejection of the fish for human consumption. This can be measured by descriptive sensory profiling or by using the Torry scale. A tested and trained sensory panel evaluates cooked samples from the storage experiment (Hyldig *et al.* 2008). The sensory panel must be a second panel of different composition in order to overcome bias. The descriptive sensory profiling and the Torry scale are described in the previous chapter. From the results of the descriptive sensory analysis, the shelf life is defined.

For future use of the QIM scheme, such as training of QIM inspectors and illustration of the descriptions of the different attributes, pictures are of utmost importance. During the development of QIM, pictures need to be taken of all the changes of the different attributes. These pictures are preferably taken by a professional photographer in order to be clear and with the right use of flashlights. In practice, the importance of the quality of the pictures and the difficulties in taking them are highly underestimated. This can result in low-quality pictures that are not suitable for printing in reference manuals.

29.2.4 Calibration curve

The next step in the development of a QIM scheme is the so-called calibration curve. As mentioned in the Introduction of this chapter, the QIM resulted in a scoring system that gave a linear relationship with storage period and the slope of the line is the rate of demerit point accumulation per day of storage. A simple calculation can indicate the equivalent of the number of days at 0°C that the product has been stored (i.e., predicted storage time). To calculate the calibration curve, a simple linear regression analysis is performed. The data for this calculation are obtained from a third, more complex storage trial. Before this third trial, the QIM panel is trained for the newly developed QIM scheme. The QIM panel will assess according to good sensory practice. They will evaluate unknown coded samples in random order. This implies that, during the storage trial, several new batches, which have been stored for different equivalent days in ice, are randomly allocated so that fish of different QI are assessed in one session. From each batch, five fishes are assessed within each trial and a maximum of five assessments are performed per fish. If more than five panelists assess the fish, a double batch is needed. The storage trial is finished when the fish is spoiled, meaning some days after the determined end of shelf life, in order to get a complete calibration curve for the QIM scheme.

The average results are used to calculate the linear regression. The equation is calculated as well as the variation around the curve. The intercept at the *y*-axis is of major importance and should be close to zero. This follows the QIM principle of "deviations from a perfect product." The correlation coefficient between the QI and the days in ice should be close to 1.0. With this result, the calibration curves give reliable prediction of the storage time. The calculated standard error of prediction (SEP) results in the reliability of the QIM assessments. Sveinsdóttir *et al.* (2002, 2003) published results with the assessment of three salmon per batch. The reliability was 2 days, and with assessment of five salmon per batch the reliability was 1.5 days (Sveinsdóttir *et al.* 2002, 2003).

The combination of the calibration curve and the determined end of shelf life makes it possible to calculate the remaining shelf life for the appropriate end use. Fig. 29.1 shows a batch of fish reaching a sum of ten points—corresponding to 12 days in ice—and having a remaining shelf life of 8 days in ice.

29.2.5 Validity of the QIM scheme

To finalize the QIM scheme, the foregoing steps need to be combined and will result in a QIM scheme that is more than only a table with attributes, descriptions, and scores. The scheme is developed under standardized conditions that need to be described in order to know the validity of the scheme under different circumstances.

A fish caught by handline/longline will generally show fewer signs of deterioration than one caught by trawling, and its starting characteristics when stored in ice will be different. A trawled fish stored in refrigerated seawater will have a different appearance than those stored in ice (Nielsen & Hyldig 2004). In both these instances, the QIM scores at the start of storage will be different, but later during the storage period, any differences may be insignificant. Either a general scheme that does not differentiate on the basis of catch or storage medium should be used or particular schemes should be developed that allow for this difference in properties. It depends on circumstances, anticipated storage life, and the use of the scheme. Huidobro *et al.* (2001) reported that the QIM of washed sea bream developed similarly to that of unwashed sea bream for up to 10 days' storage in ice. After that, less slime and lower trimethylamine (TMA) formation in the washed fish retarded the normal increase in score, in comparison to unwashed

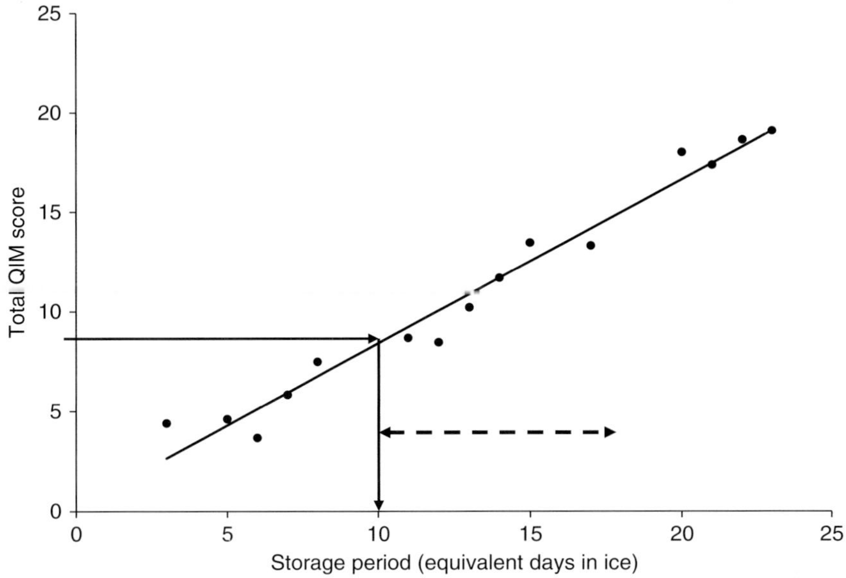

Fig. 29.1 QIM—linear calibration curve for salmon ($y=0.692x+1.57$; $R^2=0.953$). When a batch of fish reaches a sum of ten points it corresponds to 12 days in ice and the remaining shelf life is 8 days in ice.

fish. This illustrates another instance where the process may influence the score. But anticipated shelf life should also be taken into account since these were aquacultured sea bream destined for sale as fresh (i.e., unfrozen) fish. Clearly, there would be a failure in the market and distribution system if such a prime product was not sold in much less than 10 days, and a scheme should not necessarily be required to cover all treatment options beyond this time.

To a large extent, all changes in stored fish are governed by time and temperature with the proviso, as we have seen above, that species and catch and handling practices, and maybe seasonal factors, are similar. If the commercial fish handling system maintains fish at a steady temperature of 0°C, then any lack of ability to integrate is not of concern. However, it is a matter of common and recorded observation that temperature in the fish chain is not always well controlled. It may be that some schemes provide an incorrect assessment depending on how long the fish has been at elevated temperatures. However, most other instruments and schemes have not been tested in this regard either. The Torrymeter for example cannot integrate time and temperature at the correct rate (Bremner *et al.* 1987). The original scheme, on which QIM is based, was shown to be capable of integrating the effects of time and temperature (Bremner *et al.* 1984). Experiments at the Danish Institute for Fisheries Research in 2003 were set up to check whether the developed QIM schemes could integrate the time/temperature effects. The results from experiments with cod and plaice showed that if the fish have been stored at elevated temperatures (between 0°C and 5°C) and then stored at 0°C, the development in QI would follow the calibration curve for the fish species.

Some tropical reef fish have inordinately long shelf lives of more than 3 weeks when they are iced soon after catch and kept cold during storage (Bremner *et al.* 1984). This seems to be due to the absence of spoilage psychrotrophic bacteria, unless they are introduced, and to a stable level of inosine monophosphate (IMP) in the flesh. Consequently, the flesh remains acceptable in flavor for considerable periods of time when virtually all of the external indicators that would be used in a QIM scheme are at or near their maxima. However, it is still obvious that the fish have been stored for a considerable period and their remaining commercial

Table 29.3 QIM scheme for whole farmed salmon (*Salmo salar*) containing description for each parameter and the given scores in succession from 0 to 3.

Quality parameters		Description	Score
Skin	*Color/appearance*	Pearl-shiny all over the skin	0
		The skin is less pearl-shiny	1
		The fish is yellowish, mainly near the abdomen	2
	Mucus	Clear, not clotted	0
		Milky, clotted	1
		Yellow and clotted	2
	Odor	Fresh seaweedy, neutral	0
		Cucumber, metal, hay	1
		Sour, dish cloth	2
		Rotten	3
	Texture	In rigor	0
		Finger mark disappears rapidly	1
		Finger leaves mark over 3 s	2
Eyes	*Pupils*	Clear and black, metal shiny	0
		Dark gray	1
		Mat, gray	2
	Form	Convex	0
		Flat	1
		Sunken	2
Gills	*Color/appearance*	Red/dark brown	0
		Light red, pink/hazel	1
		Gray-brown, brown, gray, green	2
	Mucus	Transparent	0
		Milky, clotted	1
		Brown, clotted	2
	Odor	Fresh, seaweed	0
		Metal, cucumber	1
		Sour, moldy	2
		Rotten	3
Abdomen	*Blood in abdomen*	Blood red/not present	0
		Blood more brown, yellowish	1
	Odor	Neutral	0
		Cucumber, melon	1
		Sour, reminds of fermentation	2
		Rotten/rotten kale	3
Quality index (0–24)			

life must be extremely limited. The use of a QIM approach for these species is still valid but care must be taken in formulating a scheme in which the maximum QIM score does not undershoot shelf life by too great a margin.

To validate the new developed QIM scheme, a storage experiment at another location, season, or catching ground can be conducted (Sveinsdóttir *et al.* 2002, 2003).

29.2.6 QIM schemes for fish kept in ice

In a reference manual for the fish industry (Martinsdóttir *et al.* 2001, 2004), QIM schemes for the following fish species are published: brill (*Rhombus laevis*), cod (*Gadus morhua*), deep-water shrimp (*Pandalus borealis*), farmed salmon (*Salmo salar*), whole fjord shrimp (*Pandalus borealis*), haddock (*Melanogrammus aeglefinus*), herring (*Clupea harengus*), peeled shrimp

Table 29.4 QIM scheme for whole cod (*Gadus morhua*) containing description for each parameter and the given scores in succession from 0 to 3.

Quality parameter		**Description**	**Score**
Appearance	*Skin*	Bright, iridescent pigmentation	0
		Rather dull, becoming discolored	1
		Dull	2
	Stiffness	In rigor	0
		Firm, elastic	1
		Soft	2
		Very soft	3
Eyes	*Cornea*	Clear	0
		Opalescent	1
		Milky	2
	Form	Convex	0
		Flat, slightly sunken	1
		Sunken, concave	2
	Color of pupil	Black	0
		Opaque	1
		Gray	2
Gills	*Color*	Bright	0
		Less colored, becoming discolored	1
		Discolored, brown spots	2
		Brown, discolored	3
	Smell	Fresh, seaweedy, metallic	0
		Neutral, grassy, musty	1
		Yeast, bread, beer, sour milk	2
		Acetic acid, sulfuric, very sour	3
	Mucus	Clear	0
		Milky	1
		Milky, dark, opaque	2
Blood	*Color*	Red	0
		Dark red	1
		Brown	2
Fillets	*Color*	Translucent, bluish	0
		Waxy, milky	1
		Opaque, yellow, brown spots	2
Quality index (0–23)			

(*Pandalus borealis*), plaice (*Pleuronectes platessa*), pollock (*Pollachius virens*), redfish (*Sebastes mentella/marinus*), sole (*Solea vulgaris*), and turbot (*Scophthalmus maximus*).

Tables 29.3–29.5 show the schemes for salmon, cod, and plaice.

Moreover Andrade *et al.* (1997) have published schemes for Atlantic mackerel (*Scomber scombrus*), horse mackerel (*Trachurus trachurus*) and European sardine (*Sardina pilchardus*); Barbosa and Vaz-Pires (2004) a QIM- scheme for common octopus (Octopus vulgaris); Huidobro *et al.* (2000) for raw gilthead Seabream (*Sparus aurata*); Baixas-Nogueras *et al.* (2003) for Mediterranean Hake (*Merluccius merluccius*); Pons-Sánchez-Cascado *et al.* (2006) for Mediterranean anchovies (*Engraulis encrasicholus*); and Herrero *et al.* (2003) for frozen Hake (*M. capensis and M. paradoxus*). Furthermore, QIM schemes have been developed for cod fillets (*Gadus morhua*) (Bonilla *et al.* 2007), flounder (*Paralichthys patagonicus*) (Massa *et al.* 2005), air and modified atmosphere–packed maatjes herring (*Clupea harengus*) (Lyhs & Schelvis-Smit 2005), cuttlefish (*Sepia officinalis*) and broadtail shortfin squid (*Illex coindetii*) (Vaz-Pires & Seixas 2006), farmed Atlantic halibut (*Hippoglossus hippoglossus L.*)

Table 29.5 QIM scheme for whole plaice (*Pleuronectes platessa*) containing description for each parameter and the given scores in succession from 0 to 3.

Quality parameter		Description	Score
Appearance	*Skin(both dark and white side)*	Fresh, bright, metallic, no discoloration	0
		Bright, but without shine	1
		Matt, rather dull, slight green/blue or purple discoloration	2
		Dull, green/blue, purple discoloration	3
	Mucus	Clear, not clotted	0
		Slightly clotted and milky	1
		Clotted and slightly yellow	2
		Yellow and clotted	3
Eyes	*Form*	Convex	0
		Convex but slightly sunken	1
		Flat or swollen (like a balloon)	2
		Flat, sunken in the middle	3
	Brightness	Clear, black shining pupil	0
		Rather matt, black pupil	1
		Matt, opaque pupil	2
		Milky, gray pupil	3
Gills	*Odor*	Fresh oil, sea weedy, metallic, peppery	0
		Neutral, oily, grassy, slightly musty	1
		Musty, bread, beer, malt, slightly rancid	2
		Rancid, sour, rotten, sulfurous	3
	Color	Bright, light red	0
		Slightly discolored, especially at the end of the gill filaments	1
		Discolored	2
		Yellowish, brown, gray	3
	Mucus	No mucus	0
		Clear	1
		Yellowish, slightly clotted	2
		Yellow, brown, clotted	3
Flesh, fillets	*Color*	Fresh, translucent, bluish	0
		Waxy, milky	1
		Dull, slight discolored, yellowish	2
		Opaque, discolored, yellow, brown	3
Quality index (0–24)			

(Guillerm-Regost *et al.* 2006), tub gunard (*Chelidonichthys lucernus*) (Bekaert 2006), European cuttlefish (*Sepia officinalis*, L.) (Sykes *et al.* 2009), and trout (*Onchorhynchus mykiss*) (Wünnenburg & Oehlenschläger 2008).

Collaborative work has resulted in development of the Australian Quality Index Manual (Bremner *et al.* 2009) that contains schemes for 14 species: Atlantic Salmon (*Salmo salar*), cooked Black Tiger Prawns (Penaeus monodon), Goldband Snapper (*Pristopomoides multidens*), Sea Mullet (*Mugil cephalus*), Snapper (*Pagrus auratus*), Tiger Flathead (*Neoplatycephalus richardsoni*), Barramundi (*Lates calcarifer*), cooked King Prawns (*Melicertus plebejus*), Pink Ling (*Genypterus blacodes*), Redthroat Emperor (*Lethrinus miriatus*), Saddletail Snapper (*Lutjanus malabaricus*), Scaly Mackerel (*Sardinella limuru*), Silver Warehou (*Seriolella punctata*), and Yellowtail Kingfish (*Seriola lalandi*).

Badiani *et al.* 2009 have published a practical guide with QIM schemes for European Hake (*Merluccius merluccius*), Chub Mackerel (*Scomber japonicus*), and Horse Mackerel (*Trachurus spp.*).

29.3 TRAINING OF QIM ASSESSORS FOR THE INDUSTRY

Training QIM assessors for the industry implies both training for being a sensory panelist, including the standard sensory procedures, as well as training for implementation of QIM in practice. The sampling system, methods, and procedures for sensory evaluation must be very well defined to serve its purpose in quality management.

QIM is an objective method and, compared to other sensory methods, it is easy to work with, since it includes instructions and easily understood illustrational material.

29.3.1 Test, selection, and numbers of assessors

QIM assessors must be selected on their ability to evaluate appearance, color, odor, and texture. Assessors must also be healthy and possess normally sensitive taste and odor senses (ISO 8586-1 1993; ISO 11035 1994). Personal characteristics are also very important such as conscientiousness and accuracy, and they must be able to work in a group without disturbing the other assessors with noise, talk, and making faces. Depending on the regular duties of the individual, he or she must be readily available.

For a company, it is necessary to have a panel leader and a group of tested and trained assessors. The assessors for the QIM evaluation are then picked out from the group depending on availability, but it must be emphasized that all assessors in the sensory group are used frequently. A company could have a panel leader and six tested and trained assessors for the sensory group, even though all do not participate in all sensory evaluation sessions. Table 29.6 shows an example with four assessors in each session.

29.3.2 Standardized training of QIM assessors

QIM is well suited to train assessors and monitor performance of the panel. The QIM sessions must take place without any disturbance among the assessors. Assessors should know the nature and limits of the sense organs and learn how to recognize and evaluate appearance, taste, odor, and texture of fish after different periods of storage.

The training of the sensory panel should begin by describing the procedures of the sensory evaluation, what is expected of the assessors, etc. The nature and limits of the sense organs are described, such as the importance of breathing deeply and resting between samples during odor evaluation. The schemes intended for use must be carefully explained. The general descriptions of the parameters are shown in Table 29.7. It should be emphasized to the assessors that they must not let their hedonic personal judgment interfere with the evaluation.

Table 29.6 Distribution of assessors in a 3-week period, with a panel leader and four assessors in each session.

		Tested and trained assessors					
		A	B	C	D	E	F
Week 1	Panel leader		X		X	X	X
Week 2	Panel leader	X		X		X	X
Week 3	Panel leader	X	X	X	X		

Table 29.7 General description of parameters.

Parameter	Description
Appearance	
It is important that the fish do not lie for such a long time that the skin dries out	
Skin	The whole fish is inspected for the appearance of the skin and fins
Mucus	The appearance of mucus on the skin is assessed. Mucus can be difficult to find on fish such as salmon, but it is often located around the dorsal fin
Odor	The odor of the skin is assessed by smelling the spine. If the fish has been lying more than 15 min on the table, it should be turned over and smelled on the other side
Texture	
Texture/firmness: The texture is assessed by pressing a finger (firmly, but not too hard) on the spine muscle and observing if/how fast the flesh recovers. Only fish in rigor is given a score of 0. Prerigor fish is soft/very soft and therefore given a high score, but if it is known that it is a prerigor fish, the texture should be 0	
Belly	The consistency of the belly is assessed by pinching it between fingers or by stroking it with the fingertips.
Eyes	
Avoid touching the eyes with your fingers. If one eye is damaged, assess the other one	
Eyes where the cornea is swollen are often difficult to assess, but the membrane may be stung or cut for easier assessment of the eye	
Cornea	Color and clearness of the cornea is assessed
Form	The form of the eyes is assessed by looking at the eye directly or from the side
Gills	
The gills are assessed by lifting the opercula. If the gills have been cut on one side of the fish, assess the gills that have not been cut. Avoid touching the gills since the appearance and mucus of gills can easily be destroyed	
Gill color	The color of the gills is assessed
Gill odor	Odor of the gills is assessed by lifting the opercula and smelling by the gill bow
Mucus in gills	Color and appearance of the mucus is assessed
Viscera	
Fish kept in ice with the viscera (ungutted) must be opened. The appearance of the viscera is assessed	
Color of blood in abdomen	
Usually, remains of blood in abdomen are visible in gutted fish. Blood may also be assessed in the cut wound (near the gills); if no remaining blood is left in the abdomen the score should be 0	
Odor in abdomen	
Odor in the abdomen is assessed by smelling inside the abdomen	
Fillets/cut surface	
Color of fillets is assessed by the cut surface at the flaps or by assessing the fillets. Some fish such as redfish must be filleted from one side to be able to see the fillets and viscera	

For training, three to four samples of fish of different known storage periods in ice and treatment are used. The storage time of the fish is introduced to the assessors before they evaluate the fish and they are asked if they can agree on the scores that should be given for each sample. The samples are number coded. All panelists should become very familiar with fish of all freshness stages (i.e., not only raw material that is on the borderline of production). Training results should be evaluated. Average and standard deviations of each sample are calculated and a comparison is made between the assessors, that is, by performing statistical analysis (e.g., analysis of variance). The ability of the assessors can be examined during repeated evaluation of the same samples. Repetition of the training will show the capabilities of the assessors. Regular training of the sensory panel should be done and performance of the assessors monitored. It is also important to keep the assessors motivated. Finally, the assessors must be able to perform QIM in a fast and accurate way and be in agreement with other QIM panels in proficiency tests.

29.4 APPLICATION OF QIM IN PRACTICE

The application of QIM in practice must be tailor-made. The desire for standard sensory evaluation procedures must be balanced with practical purpose and possibilities. These standard considerations must be taken into account when QIM is to be implemented in a particular company. An experienced consultant is likely to be needed.

29.4.1 Test room or testing areas

Sensory evaluation of whole fish is generally carried out by trained assessors in the reception or processing halls of fish factories or at auction sites. In quality control procedures, special facilities or rooms are preferred for sensory evaluation, but it is not always possible. In industry and auctions, the testing area should be located giving consideration to what is practical.

- Noise level should be kept to a minimum. During a session, the assessors must be able to work without any interruption.
- Lighting is very important. It is preferable that the light is either real daylight according to ISO standard (ISO 8589 1988), but as a minimum be an intensity of 600–1500 lux/square meter (m2).
- The area must be free of any foreign odor. At a minimum, there must not be any waste or other matter, or operation with a strong smell nearby.
- No eating, drinking, or smoking shall be allowed in the testing area.
- The testing area must be easy to clean and disinfect. Regular cleaning and disinfecting should take place. It must be ensured that the cleaning agents used do not leave odors in the testing area.
- The temperature should be kept low and constant.

29.4.2 Sampling

The aim of nonbiased sampling is to obtain a representative random sample from a lot. It is vital that the sample is selected randomly to ensure that it is representative. The number of fish to be sampled is determined by the accepted uncertainty, the characteristic of the lot, and economy (NMKL procedure No. 6 1998; Codex XOT 13-1969).

From a defined homogeneous lot, preferably three to five fish (ten for small fish species) should be assessed according to QIM schemes. A homogeneous lot of fish should be assessed (i.e., from the same catching day). Number the boxes in a standard way, for example, always from left to right and from top to bottom and generate three to ten random numbers. Take one fish out of each of the three to ten numbered boxes as decided. Make sure the fishes are taken from different places in the boxes (not always from the top layer). Evaluate all three to ten fish using the QIM schemes as provided.

29.4.3 Preparation

The panel leader prepares the evaluation by giving the fish samples three-digit codes and places the fish in random order on the table. They are kept cool either by placing them on a cooling plate or on ice. To avoid bias, samples should always be coded with two- to three-digit numbers that provide no information about the samples. The samples must be kept cool under evaluation and the panelists should not see the samples being placed. The boxes where the samples have been taken from must also be removed from the testing area, because this might enhance expectation error. Order of presentation should be random and the order should be balanced. The panelists must be told in which order they should evaluate the samples.

29.4.4 Evaluation

Hunger or satiation can influence the performance of the assessors. The assessors must not eat or smoke for an hour before the sensory evaluation. The assessors must be quiet and concentrate during the evaluation. Trained assessors can evaluate 40 fish with QIM in 20 min, and the method is nondestructive.

During continuous assessment of odor, assessors become insensitive to odors after some time. People become desensitized to odors as the receptors in the olfactory senses become saturated. Therefore, it is necessary to rest and breathe fresh air between samples when evaluating odor. Also, by taking a deep breath, the airflow through the olfactory senses increases and the odor becomes easier to detect.

When applying the QIM schemes, the outer appearance of the fish, eyes, gills, and texture are evaluated. The odor of gills is evaluated, and for some species, the odor and mucus of the skin is also evaluated. The color of blood and fillets (or the cut surface at the flaps) are evaluated in gutted fish. All attributes are to be assessed in the same order for each fish. For some fish species that are not gutted, such as redfish, dissolution of viscera is evaluated as well.

The assessor must evaluate all the parameters involved in the scheme (he/she cannot determine which parameters are most important). The assessors write down the scores given. For (quality) control purposes, it is important to write down the information about the batch, date of assessment, and name of the assessors prior to assessing the fish. To make the scheme uniform and easy to use and to ensure all criteria were scored, it was programmed into a handheld computer (Branch & Vail 1985; Helbo 1990; Jónsdóttir *et al.* 1999) and a prototype dedicated handheld device was developed (Bremner *et al.* 1987). An Icelandic company has developed software that can be connected to a handheld computer. The software includes both QIM schemes and photographs of fish at different spoilage stage (Luten 2000).

29.5 RESULTS OF QIM ASSESSMENTS

The scores for all the characteristics are summarized to give an overall sensory score, the so-called QI. If a score for one of the parameters is missing, it is not possible to calculate the total sum and thereby the QI for the assessed fish. If this situation is an incident (e.g., damaged eyes makes it impossible to assess this attribute), the best way to deal with it is to leave that particular fish out and take another sample. If it occurs on a regular basis (e.g., because of washing procedures at a company the mucus is always removed), the scheme should be adapted and a new calibration curve constructed in which the particular attribute (e.g., mucus) is removed.

Having the QI on an electronic basis means the data can be rapidly communicated from a boat, the quayside, or an auction, and that it can be used in management systems to plan

Table 29.8 The estimated shelf life of some fish species.

Species	Estimated shelf life in ice (days)
Brill (*Rhombus laevis*)	14
Cod (*Gadus morhua*)	15
Deep-water shrimp (*Pandalus borealis*)	6
Farmed salmon (*Salmo salar*)	20
Fjord shrimp (*Pandalus borealis*)	6
Haddock (*Melanogrammus aeglefinus*)	15
Herring (*Clupea harengus*)	8
Peeled shrimp (*Pandalus borealis*)	6*
Plaice (*Pleuronectes platessa*)	13
Pollock (*Pollachius virens*)	18
Redfish (*Sebastes mentella/marinus*)	18
Sole (*Solea vulgaris*)	15
Turbot (*Scophthalmus maximus*)	15

*The storage life before peeling.

supply and production and to allocate product to different grades or to end uses according to production or market requirements. The full assessment data can be stored in databases. This ability to transmit a meaningful quality index along with identity and traceability information over the Internet represents a major advantage and a progressive step in electronic marketing of fish products. It further enhances opportunities for quality chain management to ensure product of known properties is handled correctly along the supply chain.

29.5.1 Prediction of remaining shelf life of fish stored in ice

As the quality index increases linearly with storage time in ice, the information may be used in production management. From the QI results, an estimate can be calculated for the remaining shelf life, which equals the shelf life minus the predicted storage time. In the following equations, the calibration curves for cod, salmon, and plaice are shown.

Cod: $QI = 1.20 \times \text{days in ice} - 0.04$ ($R^2 = 0.966$)
Salmon: $QI = 0.692 \times \text{days in ice} - 1.57$ ($R^2 = 0.953$)
Plaice: $QI = 1.28 \times \text{days in ice}$ ($R^2 = 0.89$)

It is emphasized that the remaining shelf life should be used with some precaution due to the uncertainty in the estimation. Various factors can affect the remaining shelf life. It depends on the handling of the fish. Rapid cooling after the catch and an uninterrupted cold storage, different fishing gear, and bleeding and gutting methods are important. The season and catching ground can also have an effect. In the literature, several storage studies are reported and the estimated shelf life of different species is recorded (Howgate 1985; Martinsdóttir & Blomsterberg 1987; Magnússon *et al.* 1990; Rehbein *et al.* 1994; Larsen *et al.* 1998–1999; Martinsdóttir *et al.* 2000; Sveinsdóttir *et al.* 2002). These are summarized in Table 29.8.

29.5.2 Quality grading relation EU system

When industry professionals use the QIM as quality grading scheme, they need to refer the scores to the current EU classifications of E, A, and B. For this purpose, QIM Eurofish advises the following relation between the QI scores and EU classification.

Table 29.9 The relation between the EU classification and the QI score for four fish species.

EU classification	Plaice (*Pleuronectes platessa*)	Cod (*Gadus morhua*)	Sole (*Solea vulgaris*)	Turbot (*Scophthalmus maximus*)
	QI score	QI score	QI score	QI score
E	0–5	0–4	0–5	0– 5
A	6–16	5–13	6–19	6–19
B	17–21	14–16	20–27	20–26
Rejected	>22	>17	>28	>27

In case the average score of the assessed fish samples (three to five) per batch is between the mentioned QIM ranges, the results should be rounded off to the nearby QIM range.

Table 29.9 clearly shows that A, after the EU classification, is too broad and not detailed enough to use in electronic auctions and in production management. Several studies have shown that the QIM proved to be more reliable in assessing sensory changes of different fish species as compared to the EU grading scheme (Triqui & Bouchriti 2003).

29.6 C-QIM

The QIM schemes have been developed from the viewpoint of the industry and from technical research, and the obvious question arose as to how this relates to the consumer. This question can now be answered since a first version of a so-called consumer (C) QIM (C-QIM) has been developed (Warm *et al.* 2000; Larsen *et al.* 2003). This work used an external consumer panel, which developed and tested its own vocabulary in comparison with the standard laboratory terms for five species. The important characteristics for both consumer and sensory scientist are appearance, odor, and texture. These selected key parameters describe intensity and give an overall evaluation of fish quality, and thereby fulfills the need for a method that does not mix analytical and hedonic terms.

C-QIM is not an acceptance test, but a tool for decision-making for the consumer buying fish in a market or at a fishmonger (Nielsen *et al.* 2002). The work with developing the first C-QIM was based on five fish species only, and therefore further work with developing a C-QIM is ongoing at the Danish Institute for Fisheries Research, Department for Seafood Research (Hyldig & Larsen 2003). Luten *et al.* (2011) have recently developed new apps for iphone (see http://www.qim-eurofish.com/).

29.7 FUTURE USE OF QIM IN THE WHOLE FISHERY CHAIN

A list of where and which sensory methods can be used in the whole fishery chain from catch to consumer is provided in Fig. 29.2. It has been shown that the QIM can be used at each key stage. This is a major advance and a validation of the general applicability of the QIM approach.

To establish an optimal communication of sensory quality in the seafood processing chain, Green-Petersen *et al.* 2010 developed the seafood sensory quality model (SSQM). The SSQM can be used to communicate the sensory quality of seafood, and make it possible to share the understanding of the sensory quality. The SSQM makes it achievable to document sensory quality in different test points and to relate it to every step in the chain. Not only results from

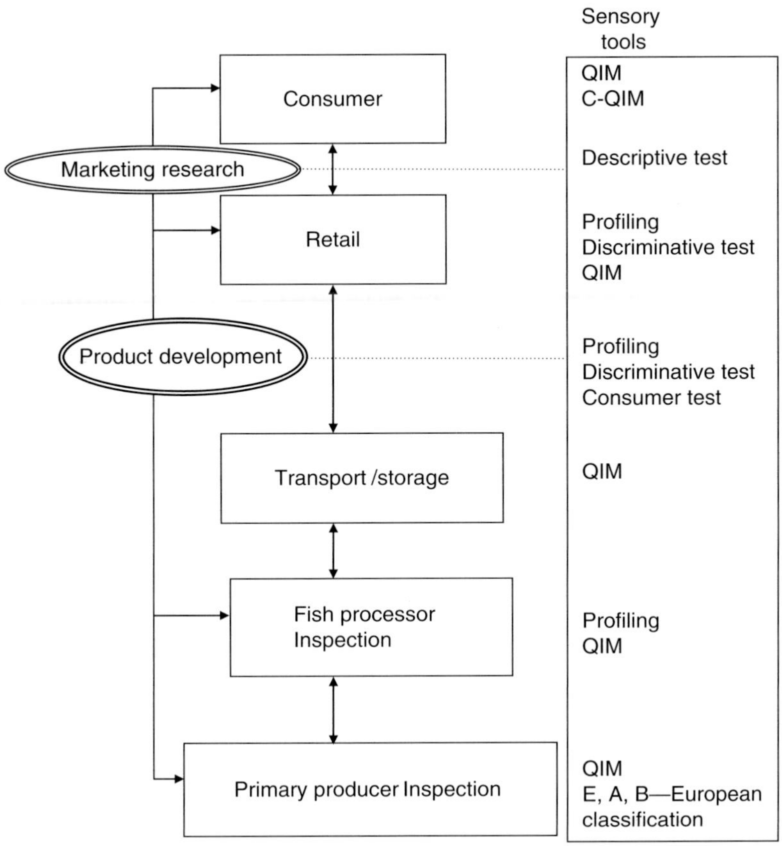

Fig. 29.2 Sensory methods to be used in the fishery chain.

sensory evaluation but also other information with an effect on the sensory quality can be included. Additionally, the SSQM is valuable in relation to deciding which sensory characteristics should be measured in the different test point in relation to product decision and product development. The use of SSQM can stimulate improvements in production of seafood of enhanced quality to meet consumers' preferences.

REFERENCES

Andrade, A., Nunes, M.L. & Batista, I. (1997) Freshness quality grading of small pelagic species by sensory analysis. In: Methods to Determine the Freshness of Fish in Research and Industry. Proceedings of the Final Meeting of the Concerted Action "Evaluation of Fish Freshness" AIR 3CT942283, Nantes Conference (eds G. Ólafsdóttir, *et al.*), November 12–14, 1997, Paris, International Institute of Refrigeration, pp. 333–338.

Anonymous. (1992a) Kvalitetsnormer på fisk. Manual for sensoriske bedømmelser af mørksej, rødfisk, sild og torsk. Lyngby. Danish Institute for Fisheries Research. Department of Seafood Research.

Anonymous. (1992b) Kvalitetsnormer på fisk. Slutrapport fase II. Lyngby. Danish Institute for Fisheries Research. Department of Seafood Research.

Anonymous. (1996) Council regulation (EC) No. 2406/96 of November 26, 1996, laying down common marketing standards for certain fishery products. *Official Journal of the European Communities*, **L334**, 1–14.

Badiani, A., Adinolfi, F., Bonaldo, A., *et al.* (2009) Freshness assessment of european hake, chub mackerel and horse mackerel, libreria bonomo editrice, Bologna, Italy, www.libreriabonomo.com

Baixas-Nogueras, S., Bover-Cid, S., Veciana- Nogues, T., Nunes, M.L. & Vidal-Carou, M.C. (2003) Development of a quality index method (QIM) to evaluate freshness in Mediterranean hake (*Merluccius merluccius*). *Journal of Food Science*, **68** (3), 1067–1071.

Barbosa, A. & Vaz-Pires, P. (2004) Quality index method (QIM): Development of a sensorial scheme for common octopus (*Octopus vulgaris*). *Food Control*, **15**, 161–168.

Bekaert, K. (2006) Development of quality index method scheme to evaluate freshness of tub gunard (*Chelidonichthys lucernus*). In: Seafood from fish to Dish (eds J.B. Luten, C. Jacobsen, K. Bekaert, A. Sæbö & J. Oehlenschläger), pp. 289–296. Academic Publishers, Wageningen, the Netherlands.

Bonilla, A.C., Sveinsdottir, K., Martinsdottir, E. (2007) Development of quality index method (QIM) scheme for fresh cod (*Gadus morhua*) fillets and application in shelf life study. *Food Control*, **18**, 352–358.

Botta, J.R. (1995) *Evaluation of Seafood Freshness Quality. Food Science and Technology*. VHC Publishers, Inc. New York.

Branch, A.C. & Vail, A.M.A. (1985) Bringing fish into the computer age. *Food Technology in Australia*, **37**, 352–355.

Bremner, A. (1985) A convenient easy-to-use system for estimating the quality of chilled seafood. In: Proceedings of the Fish Processing Conference (eds D.N. Scott & C. Summers), Nelson, New Zealand, April 23–25, 1985. *Fish Processing Bulletin* **7**, 59–703.

Bremner, H.A. (1988) Chill-storage trials on three species of scampi. *Lebensmittel-Wissenschaft und-Technologie*, **21**, 275–283.

Bremner, H.A. (2000) Toward practical definitions of quality for food science. *Critical Review in Food Science and Nutrition*, **40**, 83–90.

Bremner, H.A. & Sakaguchi, M. (2000) A critical look at whether 'freshness' can be determined. *Journal of Aquatic Food Product Technology*, **9**, 5–25.

Bremner, H.A., Statham, J.A. & Sykes, S.J. (1984) Tropical species from the north-west shelf of Australia. Sensory assessment and acceptability of fish stored on ice. In: Proceedings of the Sixth Session IPFC Working Party on Fish Technology and Marketing. Melbourne, VIC, Australia, 1984, FAO Fish Report 317 Suppl, pp. 41–53.

Bremner, H.A., Olley, J. & Vail, A.M.A. (1987) Estimating time-temperature effects by a rapid systematic sensory method. In: Advances in Food Research: Seafood Quality Determination, pp. 413–435. Elsevier Press, Amsterdam, the Netherlands.

Bremner, A., Poole, S., Kow, F., Weerasinghe, J. & Williams, H. (2009) Development of a Quality Index for Australian Seafoods (ed. M. Boulter). Fisheries Research and Development Corporation, Canberra, Australian Capital Territory, Australia.

Chinivasagam, H.N., Bremner, H.A., Thrower, S.J. & Nottingham, S.M. (1995) Spoilage pattern of five species of Australian Prawns: Deterioration is influenced by environment of capture and mode of storage. *Journal of Aquatic Food Product Technology*, **5**, 25–50.

Codex standards for methods of analysis and sampling, Sampling plans for prepackaged foods (AQL 6.5), XOT 13–1969, Rome, Italy, FAO/WHO Codex. Alimentarius.

Codex. (1976) Recommended international code of practice for fresh fish CAC/RCP 9.

Fitzgerald, C.H. & Bremner, H.A. (1998) The oxidative stability of chilled and frozen pilchards used as feed for captive southern bluefin tuna. *Journal of Aquatic Food Product Technology*, **7**, 27–44.

Green-Petersen, D., Nielsen, J. & Hyldig, G. (2012) A model for communication of sensory quality in the seafood processing chain. *Critical Reviews in Food Science and Nutrition*, **52** (5), 443–447.

Guillerm-Regost, C., Haugen, T., Nortvedt, R., *et al.* (2006) Quality characterization of farmed Atlantic halibut during ice storage. *Journal of Food Science*, **71** (2), 83–90.

Helbo, J. (1990) Information technology and production management in the white fish industry. Industrial PhD thesis, Danish Institute for Fisheries Research and The Royal Veterinary and Agricultural University, Copenhagen, Denmark.

Herrero, A.M., Huidobro, A.& Careche, M. (2003) Development of quality index method for frozen hake (M. capensis & M. paradoxus). *Journal of Food Science*, **68**, 1086–1092.

Howgate, P. (1985) Approaches to the definition and measurement of storage life of chilled and frozen fish. Torry Research Station, Aberdeen, U.K.

Huidobro, A., Pastor, A.& Tejada, M. (2000) Quality index method developed for raw gilthead seabream (*Sparus aurata*). *Journal of Food Science*, **65**, 1202–1205.

Huidobro, A., Pastor, A., Lopez-Caballero, M.E.& Tejada, M. (2001) Washing effect on the quality index method (QIM) developed for raw gilthead seabream (*Sparus auratus*). *European Food Research and Technology*, **212**, 408–412.

Hyldig, G. & Larsen, E. (2003) C-QIM—A tool for consumer evaluation. In: Poster at First Joint Trans Atlantic Fisheries Technology Conference (TAFT), 33rd WEFTA meeting, Reykjavik, Iceland, June 10–14, 2003.

Hyldig, G. & Nielsen, J. (1997) A rapid sensory method for quality management. In: Methods to Determine the Freshness of Fish in Research and Industry. *Proceedings of the Final Meeting of the Concerted Action "Evaluation of Fish Freshness" AIR3CT942283, Nantes Conference* (eds G. Ólafsdóttir, *et al.*), November 12–14, 1997, Paris, France, International Institute of Refrigeration, pp. 297–305.

Hyldig, G. & Nielsen, J. (2004) QIM—A tool for determination of fish freshness. In: Seafood Quality and Safety. Advances in the New Millennium (eds F. Shahidi & B.K. Simpson), pp. 81–89. ScienceTech Publishing Company, St John's, NL, Canada.

Hyldig, G. & Green-Petersen, D.M.B. (2005) Quality index method—An objective tool for determination of sensory quality. *Aquatic Food Product Technology*, **13**, 71–80.

Hyldig, G., Carlehög, M., Martinsdóttir, E., Sveinsdóttir, K. & Lilleberg, L. (2008) Guide for sensory analysis of fish and shellfish. NMKL Procedure No. 21. www.nmkl.org

ISO 8586-1. (1993) Sensory Analysis—General Guidance for the Selection, Training and Monitoring of Assessors. Part 1: Selected Assessors. International Standard, 1st edn.

ISO 11035. (1994) Sensory Analysis—Identification and Selection of Descriptors for Establishing a Sensory Profile by a Multi-Dimensional Approach. International Standard, 1st edn.

ISO 8589. (1988) Sensory Analysis—General Guidance for the Design of Test Rooms. International Standard, 1st edn.

Jespersen, C.M. & Helbo, J. (1991) Sensorisk kvalitetsindeks til torsk og sild. In: Kvalitet og Fiskeforædling. Report to the Nordic Council of Ministres, Chopenhagen, Denmark.

Jónsdóttir, S.M., Hyldig, G., Nielsen, J., Bleechmore, T. & Silberg, S. (1999) Rapid PC based sensory methods. *INFOFISH International*, **2**, 54–56.

Larsen, E.P., Heldbo, J., Jespersen, C.M. & Nielsen, J. (1992) Development of a standard for quality assessment on fish for human consumption. In: Quality Assurance in the Fish Industry (eds H.H. Huss, M. Jacobsen & J. Liston), pp. 351–358. Elsevier, Amsterdam, the Netherlands.

Larsen, E.P., Hyldig, G. & Frederiksen, M. (1998–1999) "Fersk fisk—Kvalitet i detailhandlen" DFU Rapport udgivet af Danmarks Fiskeriundersøgelser, Afdeling for Fiskeindustriel Forskning, Lyngby.

Larsen, E., Hyldig, G., Dalgaard, P., Bech, A.C. & Østerberg, C. (2003) Consumers and experts responses to fresh cod fillets. In Quality of Fish from Catch to Consumer: Labelling, Monitoring and Traceability (eds J.B. Luten, J. Oehlenschlager & G Ólafsdóttir), pp. 345–360. Wageningen Academic Publisher, Wageningen, the Netherlands.

Luten, J.B. (2000) Development and implementation of a computerised sensory system (QIM) for evaluating fish freshness. CRAFT FAIR CT97 9063. Final Report for the period from 01–01–98 to 31–03–00, RIVO The Netherlands Institute for Fisheries Research, Wageningen, the Netherlands, p. 18.

Luten, J.B. & Martinsdóttir, E. (1997) QIM-A European tool for fish freshness evaluation in the fishery Chain. In: Methods to Determine the Freshness of Fish in Research and Industry. Proceedings of the Final Meeting of the Concerted Action "Evaluation of Fish Freshness" AIR3CT942283, Nantes Conference (eds G. Ólafsdóttir, *et al.*), November 12–14, Paris, France, International Institute of Refrigeration, pp. 287–296.

Lyhs, U. & Schelvis-Smit, R. (2005) Development of a quality index method (QIM) for Maatjes Herring stored in air and under modified atmosphere. *Journal of Aquatic Food Product Technology*, **14** (2), 63–76.

Magnússon, H., Martinsdóttir, E. & Steinórsson, P. (1990) Áhrif frystingar og frystigeymslu orsks eftir íðingu. 26. Icelandic Fisheries Laboratories REPORT no. 26.

Martinsdóttir, E. (1997) Sensory evaluation in research of fish freshness. In: Methods to Determine the Freshness of Fish in Research and Industry. *Proceedings of the Final Meeting of the Concerted Action "Evaluation of Fish Freshness" AIR3CT942283, Nantes Conference* (eds G. Ólafsdóttir, *et al.*), November 12–14, 1997, Paris, France, International Institute of Refrigeration, pp. 306–312.

Martinsdóttir, E, & F Blomsterberg. (1987) Sjálfvirk ferskleikamæling með RT-gæðaflokkara 12. RIT Icelandic Fisheries Laboratories.

Martinsdóttir, E., Sveinsdóttir, K. & Ólafsdóttir, G. (2000) Development and implementation of a computerised sensory system (QIM) for fish freshness. Icelandic Fisheries Laboratories Project Report 11.

Martinsdóttir, E., Sveinsdóttir, K., Luten, J., Schelvis-Smit, R. & Hyldig, G. (2001, 2004) Sensory evaluation of fish freshness. A reference manual for the fish industry. QIM-Eurofish. Available from www.qimeurofish.com.

Massa, A.E., Palacios, D.L., Paredi, M.E. & Crupkin, M. (2005) Postmortem changes in quality indices of ice-stored flounder (*Paralichthys patagonicus*). *Journal of Food Biochemistry*, **29**, 570–590.

Nielsen, D. & Hyldig, G. (2004) Influence of handling procedures and biological factors on the QIM evaluation of whole herring (*Clupea harengus* L.). *Food Research International*, **37** (10), 975–983.

Nielsen, J., Hyldig, G. & Larsen, E. (2002) Eating quality of fish—A review. *Journal of Aquatic Food Product Technology*, **11**, 125–141.

Nielsen, D., Green, D., Bolton, G., Hodson, R. & Nielsen, J. (2003) Hybrid striped bass—A QIM scheme for a major U.S. aquaculture species. In: Poster at First Joint Trans Atlantic Fisheries Technology Conference (TAFT), 33rd WEFTA Meeting, June 10–14, 2003, Reykjavik, Iceland.

NMKL Procedure NR. 6. (1998) Generelle retningslinier for kvalitetssikring af sensoriske laboratorier. Nordisk Metodikkomité for Levnedsmidler.

Olley, J. & Ratkowsky, D.A. (1973a) Temperature function integration and its importance in the storage and distribution of flesh foods above the freezing point. *Food Technology, August* **25**, pp. 66–73.

Olley, J. & Ratkowsky, D.A. (1973b) The role of temperature function integration in monitoring of fish spoilage. *Food Technology in New Zealand*, **8**, 13, 14, 17.

Pons-Sánchez-Cascado, S., Vidal-Carou, M.C., Nunes, M.L. & Veciana-Nogués, M.T. (2006) Sensory analysis to assess the freshness of Mediterranean anchovies (*Engraulis encrasicholus*) stored in ice. *Food Control*, **17**, 564–569.

Rehbein, H., Martinsdóttir, E., Blomsterberg, F., Valdimarsson, G. & Öehlenschläger, J. (1994) Shelf life of ice-stored redfish, *Sebastes marinus* and *S. mentella. International Journal of Food Science & Technology*, **29**, 303–314.

Shewan, J.M., Macintosh, R.G., Tucker, C.G. & Ehrenberg, A.S.C. (1953) The development of a numerical scoring system for the sensory assessment of the spoilage of wet white fish stored in ice. *Journal of the Science of Food and Agriculture*, **4**, 283–298.

Statham, J.A. & Bremner, H.A. (1983) Effect of potassium sorbate on spoilage of blue grenadier (*Macruronus novaezelandiae*) as assessed by microbiology and sensory profiles. *Journal of Food Protection*, **46**, 1084–1091.

Statham, J.A. & Bremner, H.A. (1985) Acceptability of trevalla (Hyperoglyphe porosa Richardson) after storage in carbon dioxide. *Food Technology in Australia*, **37**, 212–215.

Statham, J.A., Bremner, H.A. & Quarmby, A.R. (1985) Storage of morwong (Nemadactylus macropterus Block and Schneider) in combinations of polyphosphate, potassium sorbate and carbon dioxide at 4°C. *Journal of Food Science*, **50**, 1580–1584 and 1587.

Sveinsdóttir, K., Martinsdóttir, E., Hyldig, G., Jørgensen, B. & Kristbergsson, K. (2002) Application of quality index method (QIM) scheme in shelf life study of farmed Atlantic salmon (*Salmo salar*). *Journal of Food Science*, **67** (4), 1570–1579.

Sveinsdóttir, K., Hyldig, G., Martinsdóttir, E., Jørgensen, B. & Kristbergsson, K. (2003) Development of quality index method (QIM) scheme for farmed Atlantic salmon (*Salmo salar*). *Food Quality and Preferences*, **14**, 237–245.

Sykes, A.V., Oliveira, A.R., Domingues, P.M., Cardoso, C.M., Andrade, J.P., Nunes, M.L. (2009) Assessment of European cuttlefish (*Sepia officinalis*, L.) nutritional value and freshness under ice storage using a developed quality index method (QIM) and biochemical methods, *LWT—Food Science and Technology*, **42**, 424–432.

Triqui, R. & Bouchriti, N. (2003) Freshness assessment of Moroccan Sardine (*Scardina pilchardus*): Comparison of overall sensory changes to instrumentally determined volatiles. *Journal of Agricultural and Food Chemistry*, **51**, 7540–7546.

Vaz-Pires, P. & Seixas, P. (2006) Development of new quality index method (QIM) schemes for cuttlefish (*Sepia officinalis*) and broadtail shortfin squid (*Illex coindetii*). *Food Control*, **17** (12), 942–949.

Warm, K., Bøknæs, N. & Nielsen, J. (1998) Development of quality index methods for evaluation of frozen cod (*Gadus morhua*) and cod fillets. *Journal of Aquatic Food Product Technology*, **7**, 45–59.

Warm, K., Nielsen, J., Hyldig, G. & Martens, M. (2000) Sensory quality of five fish species. *Journal of Food Quality*, **23**, 583–602.

Wünnenburg, A. & Oehlenschläger, J. (2008) Untersuchungen zur saisonalen Abhängigkeit der Haltbarkeit von Zuchtforellen (Onchorhynchus mykiss) während der Eislagerung mittels der Qualitäts-Index-Method (QIM) an Ganzfisch und Sensorik gegarter Filetproben. *Archiv für Lebensmittelhygiene*, **59**, 221–226.

30 Sensory Quality of Fish

Grethe Hyldig

Abstract: Sensory analysis plays an important role in quality control and quality assurance in the fish sector. Both objective and subjective sensory testing are used. The objective tests include discriminative (triangle test, forced choice) and descriptive (profiling, structured scaling) sensory tests. Both groups of tests are analytical measurements of the intrinsic quality of the product, whereas affective (subjective test) methods are used for consumer testing and measure the attitude and emotional response of the consumer toward the product. Tainting and sensory quality of fish depend on many factors such as feed, stress, or storage.

Keywords: Fish sensory quality, QIM, sensory profiling, assessors, fish tainting, fish storage

30.1 INTRODUCTION

Fish are different from all other food commodities regarding the method of harvesting, fragility of the product during transport to processing sites and further in the chain, temperature dependency, and the variety of species. This sensitivity makes it necessary to monitor the product quality in the whole chain, and sensory analysis is ideal for this monitoring. Sensory assessment of fish and fish products has therefore for years played a natural part of the seafood chain.

Fish are generally defined as aquatic vertebrates that use gills to obtain oxygen from water with fins consisting of a variable number of skeletal elements called fin rays. They are the most numerous of the vertebrates, with at least 20 000 known species, and more than half of the species are found in the marine environment (Thurman & Webber 1984). Each species is identified by a scientific name which has two parts—the genus and the specific epithet (binominal nomenclature). As an example, the scientific (species) name of the common cod is *Gadus morhua*.

30.1.1 The anatomy of fish

Fish (vertebrae) have a backbone composed of segments (vertebrae)—and a cranium covering the brain. The backbone runs from the head to the tail. The vertebrae are extended dorsally to form neural spines, and in the trunk region they have lateral processes that bear ribs. The ribs are cartilaginous or bony structures in the connective tissue between the muscle segments. Usually, there are also a corresponding number of false ribs or "pin bones" extending more or

Handbook of Meat, Poultry and Seafood Quality, Second Edition. Edited by Leo M. L. Nollet.

less horizontally into the muscle tissue. These bones have to be removed if a bone-free fillet is the goal. Fish have muscle cells running in parallel and connected to sheaths of connective tissue (myocommata) anchored to the skeleton and the skin. The bundles of parallel muscle cells are called myotomes. The myocommata run in an oblique pattern perpendicular to the long axis of the fish, from the skin to the spine. This anatomy is ideally suited for the flexing muscle movements necessary for swimming through the water. All muscle cells extend the full length between two myocommata, and run parallel with the longitudinal direction of the fish. The muscle mass on each side of the fish makes up the fillet, of which the upper part is termed the "dorsal muscle" and the lower part the "ventral muscle." During growth, the size of each muscle cell increases and not the number of muscle cells. Also, the proportion of connective tissue increases with age.

Most fish become sexually mature when they reach a size characteristic of the species and this is not necessarily directly correlated with age. In general, this critical size is reached earlier in males than in females. As the growth rate decreases after the fish has reached maturity, it is, therefore, often an economic advantage to produce female fish in aquaculture. Every year, mature fish use energy to build up the gonads (the roe and milk). This gonadal development causes a depletion of the protein and lipid reserves of the fish since it takes place during a period of low- or no-food intake. The length of the spawning season varies greatly between species.

30.1.2 Chemical composition

The variation in the chemical composition of fish is closely related to feed intake, migratory swimming, and sexual changes in connection with spawning. Fish will have starvation periods for natural or physiological reasons (such as migration and spawning) or due to external factors such as shortage of food. Fish raised in aquaculture may also show variation in chemical composition, but in this case several factors can be controlled, and the chemical composition may be predicted. To a certain extent, the fish farmer is able to design the fish by selecting the farming conditions. It has been reported that factors such as feed composition, environment, fish size, and genetic traits all have an impact on the composition and quality (Reinitz *et al.* 1979; Einen & Skrede 1998; Nortvedt & Tuene 1998; Refsgaard *et al.* 2000; Baron *et al.* 2009).

There is not only a difference between species, but also a considerable variation between individuals. These variations must be taken into account when setting up experiments using sensory analysis to characterize fish and fish products. For example, the lipid content in farmed salmon (*Salmo salar*) can differ from 10% to 19% from the same farm. The variation in sensory profile of rainbow trout (*Oncorhynchus mykiss*), belonging to the same aquaculture production batch and handled the same way, was explored by using objective sensory profiling on heat-treated minced fillets. Different groups of fish were sampled three different times during a production day. The results showed significant differences in the sensory profiles of individual fish within all three groups as well as significant differences between the groups. The results showed that significant differences in the sensory profiles of individual fish from the same aquaculture production batch may occur. Furthermore, the results also showed sensory differences between groups of samples taken at different times during a production day. A significant negative correlation between lipid content and firm texture was observed, but in general, the chemical and physical measurements could not explain the differences in the sensory profiling or in the mechanical texture measurements (Green-Petersen & Hyldig 2010). Also, for herring there is a wide variation in lipid content within catches. A single catch can contain herring (*Clupea harengus*) with lipid content ranging from 1% to 25%. This variation is due to heterogeneity caused by mixing between stocks (Nielsen *et al.* 2005b).

The lipid content of fillets from lean fish is low and stable, whereas the lipid content in fillets from fatty species varies considerably. In many species, most of the muscle is white or has a light color, but many fish species have a certain amount of dark tissue of a brown or reddish color. The dark muscle is located just under the skin along the body side. A typical pelagic fish as herring will contain approximately 25% of dark muscle required for prolonged aerobic muscle activity. The proportion of dark muscle varies with the activity of the fish: the more active the fish, the larger dark muscle. There are many differences in the chemical composition of the two muscle types, for example, higher levels of lipids and myoglobin in the dark muscle. From a technological point of view, the high lipid content of dark muscle is important and can give high intensities of desirable sensory attributes, but also a shorter shelf life due to lipid oxidation resulting in rancid and sour attributes among others. The demersal species as cod has a white flaky and tender muscle; textural changes as toughness and chewiness are here the most noticeable problem during storage. The white muscle is used for quick attacks or escapes—not long destinations. Even though the lipid content is low in lean fish, the lipid will oxidized during frozen storage. When fish such as cod go rancid, they will often taste somewhat like wet cardboard (cold-storage flavor).

A possible method for discriminating between lean and fatty fish species is to term fish that store lipids only in the liver as lean, and fish storing lipids in fat cells distributed in other body tissues as fatty fish. Typical lean species are the bottom-dwelling ground fish like cod fish and flatfish species. Fatty species include the pelagics like herring, mackerel, and sprat. Some species store lipids in limited parts of their body tissues only, or in lower quantities than typical fatty species, and are consequently termed semifatty species (e.g., barracuda, mullet, and shark).

Although the protein fraction is rather constant in most species, variations have been observed such as protein reduction occurring in salmon during the long-spawning migrations (Ando *et al.* 1985; Ando & Hatano 1986) and in Baltic cod during the spawning season (Borresen 1992).

30.2 SENSORY ANALYSIS OF FISH

Sensory analysis plays an important role in quality control and quality assurance in the fish sector.

Both objective and subjective sensory testing are used. The objective tests include discriminative (triangle test, forced choice) and descriptive (profiling, structured scaling) sensory tests. Both groups of tests are analytical measurements of the intrinsic quality of the product, whereas affective (subjective test) methods are used for consumer testing and measure the attitude and emotional response of the consumer toward the product. All these methods are used in the seafood chain.

Several conditions are unique for the sensory analysis of seafood. When selecting and training judges for sensory analysis, it is very important to be aware of that some people cannot taste rancid flavor, iodine, and geosmin and some have a very low response to cold-storage flavor and rancidity. Also, some people are allergic or hypersensitive to different fish-proteins, shellfish, or histamine.

The interpretation of the stimulus and response must be trained very carefully in order to receive objective responses. It is very easy, for example, to give an objective answer to the question: is the fish in rigor (completely stiff), but more training is needed if the assessor has to decide whether the fish is post- or prerigor.

30.2.1 Methods

30.2.1.1 EU scheme and quality index method

When fish is landed in Europe today, the most commonly used method for the quality assessment of raw fish in the inspection service is the EU scheme (Anon 1996). In the EU scheme, according to the Council Regulation (EC) No 2406/96 November 26, 1996 (Anon 1996), there are three quality levels, namely, E (Extra), A, and B, where E is the highest quality and below B is the level where fish are discarded for human consumption. The EU scheme does not take into account the differences between species as it only uses general parameters and there are also problems with mixing subjective and objective sensory in the scheme. Several studies have shown that the quality index method (QIM) proved to be more reliable in assessing sensory changes of different fish species as compared to the EU grading scheme (Triqui & Bouchriti 2003).

The Torry scale is used in some countries by fish industries and buyers of fish products. The more objective and descriptive sensory methods as sensory profiling and QIM are described in other chapters in this book.

30.2.1.2 Sensory profiling

A descriptive sensory analysis such as profiling can be used to characterize *postmortem changes* of all kinds of seafood. Descriptive testing provides quantitative data and can both be very simple and useful for the assessment of a single attribute, or more complex and give a total characterization of sensory quality. Objective descriptive sensory requires a tested and trained panel together with a test room (ISO 8589 1988; ISO 8586-1 1993). In the sensory method, profiling, a series of attributes are developed to describe the sensory characteristics describing appearance, smell, taste, and texture (see the previous chapter). The panel must be trained in identification of the attributes and furthermore it must be trained in using a scale for describing the intensity of the attributes. In a session, samples will be served in random order, and between the samples there will be served something that can be used to clean the mouth. Several conditions are unique for the objective sensory profile of seafood and will be described in the following.

30.2.1.3 Sampling and preparation of fish samples

Sampling for sensory analysis must be as representative as possible, and therefore consideration must be given not only to how and where the sample is cut, but also to the effect of the chosen method for heat treatment. The preparation must have a minimal sensory impact of the "innate" characteristics of the samples. Different types of seafood products demand different sample preparation. The optimal amount of seafood prepared for each assessor is a portion size of 30–100 g. All samples must be marked individually with three-digit or four-digit codes. The samples must be placed in individual containers, for instance, porcelain bowls covered with porcelain lids.

For fatty fish species, it is important to consider if the samples should be with or without skin. As the dark muscle is right under the skin and if the fillets are skinned, some of the dark muscle will be removed. This can have influence on the rancid flavor due to the high lipid content in the dark muscle. If the samples for replicates for each assessor are taken from the same position of each fillet, it is possible to have a real replicate and thereby identify if there are any differences between assessors or between fillets from different parts of the fish. If the texture is of no importance, it is possible to make minced fish samples to eliminate the differences between individuals.

30.2.1.4 Sampling and preparation of heat-treated samples

Several methods can be used for heating fish samples. Heating should be done gently because of the delicate muscle structure of fish. For all methods, the samples should be heated to an internal core temperature of 70°C.

Furthermore, it is very important that all material used for sample preparation is tested to guarantee no transferring of off-flavor to the sample. As an example, some aluminum foils give the samples a metal odor and taste. As a standard, all samples must be marked individually with three-digit or four-digit codes and cooked without salt or other spices.

30.2.1.5 Preparation of smoked and marinated fish products

When profiling smoked samples, it is necessary to decide the assessment temperature and the cut of the sample. Smoked salmon can be cut into ½cm-thick slices at a right angle to the filet surface. This gives a part of the dorsal muscle and also a part of the belly flap. The samples are served in petri dishes at room temperature (Einen & Skrede 1998; Rørå *et al.* 1998; Rasmussen *et al.* 2000; Cardinal *et al.* 2004).

It is difficult to assess marinated fish due to the intensive flavor. Nielsen *et al.* 2005a–c suggested the following: the left fillets of herring (*Clupea harengus*) were marinated in plastic containers in a solution of 7% acetic acid and 16% salt (1.5 parts herring to 1 part brine) and stored at 2°C for a minimum of 4 weeks. Before sensory analysis, the samples were dewatered in a 1% salt solution at 2°C for 2 days (1 part herring to 1 part brine). The middle part of the fillet was cut into two pieces (approximately 10–15 g each) and served in replicate to the same assessor. The samples were served (skin side facing up) in porcelain trays with lids.

30.2.1.6 Sessions

It is important to serve the heated samples warm and in containers with lids. In this way, the assessors get the full impression of the odor just after opening the lid. When samples are cooked in plastic bags, odor attributes are assessed immediately after opening the vacuum bags, while flavor and texture are evaluated after removing the samples from the bag (Bjerkeng *et al.* 1997; Nortvedt & Tuene 1998). All samples must be served to each assessor in randomized order to minimize possible carry-over effects between samples, and each assessor evaluates samples in replicate.

In storage experiments, extra samples can be included in each sensory session in the randomized order to prevent the assessors from guessing the sensory scores of samples due to the relative time of chilled storage. The degree of freshness of the extra samples should differ between sensory sessions, and data from these samples must not be included in the study.

The number of samples for each session depends on the number of attributes and type of sample. For smoked samples, the assessors get worn out after four samples, but for cooked samples without any spices added, the assessors can evaluate six samples within a session. If there are many attributes, it is necessary to cut down the number of samples in each session. It is important to give the assessors a short break (1–3 min) between each sample.

Depending on the samples, different cleaning items can be used to dissipate residual flavors and particles between evaluations. Tap water can be used if it is not too bitter and served at room temperature together with unsalted crackers or crisp bread. The tap water can be filtered through a domestic water filter to remove any extraneous flavors (Farmer *et al.* 2000), but distilled water and unsalted crackers can also be used (Cardello *et al.* 1983). To clean the palate between smoked samples, more than water is needed, for example, diluted apple juice combined with neutral bread can be used (Bech *et al.* 1997; Rasmussen *et al.* 2000) or mineral water and bread (Cardinal *et al.* 2004). Depending of the off-flavor or taint, some acidic solution can be used in combination with a longer break between samples. Bett *et al.* (2000) used unsalted soda crackers and citric acid (0.03%) solution as "palate cleansers."

30.2.1.7 Scales

Different types of scales can be used, but the most common is the line scale and the 9-point category scale. For assessing color, it is helpful if standards are used, such as the natural color system NCS® or for smoked salmon the *Salmo*Fan™ lineal from Roche.

30.2.1.8 Line scale with or without anchor point

The most commonly used scale is the 15 cm unstructured line scale, with two anchor points, "little" and "much," of attribute intensity, for example, placed 1.5 and 13.5 cm from the end point of the scale (Papadopoulos & Finne 1986; Hong *et al.* 1996; Warm *et al.* 2000; Bøkness *et al.* 2002; Sveinsdóttir *et al.* 2003; Nielsen *et al.* 2005a, b; Green-Petersen *et al.* 2006).

An intensity scale from 0 to 100 is also suggested (Schubring & Oehlenschläger 1997; Rasmussen *et al.* 2000; Ginés *et al.* 2004). Here, 0 is defined as a minimum mark or nothing of the attribute and 100 as the maximum mark of the attribute. Others use a nonstructured continuous line scale of different length: 8 cm (Stohr *et al.* 2001), 10 cm (Bech *et al.* 1997; Morita *et al.* 2001; Richards & Hultin 2001; Cardinal *et al.* 2004), or 15 cm (Rørå *et al.* 1998; Morita *et al.* 2003).

30.2.1.9 Category scale

Category scale is most often used for moderately trained assessors. It can also be used for simple profiling where only one flavor is assessed.

McGill *et al.* (1984) used an intensity scale ranging from 0 = absent to 5 = very strong and Cardello *et al.* (1983) used a 7-point category scale of intensity, of which three of the points had verbal labels, that is, 1 = slight, 4 = moderate, and 7 = extreme.

Bett *et al.* (2000) used a simple category scale for detecting the off-flavor muddy/earthy. They instructed the panelist to rate each sample as 0 if no muddy/earthy flavor was detected, 1 for slight or marginal, and 2 for obvious.

30.2.1.10 Training of assessors

Training is essential as it provides all assessors with a common vocabulary, thus decreasing the risk of different quantitative and qualitative interpretations of descriptors (ISO 8586-1, ISO 11035). Furthermore, continuous training during long-term projects, preferably using reference material, can reduce the risk of drifting. If the training is interrupted or stopped, the assessors might forget descriptor meanings and/or rating levels with time. Just as any other instrument in the laboratory, the performance of a sensory panel has to be evaluated and calibrated in order to give reliable results. If this is not the case, the risk of drawing wrong conclusions from the experiment is high. In a study of marinated herring, Nielsen *et al.* (2005c) used a reference material that was served both as mince and fillet portions. The sensory evaluations took place in five trials over a time period of 19 months. It was, therefore, important to prevent the "drifting" of profiling assessors and all sessions began with the evaluation of the reference. To exclude variation between fillets, 7–10 fillets were minced together and then served in 10–15 g portions in porcelain trays. The assessors evaluated the mince and "calibrated" their evaluations according to a completed list of assessments of odor and flavor of the herring mince performed by an in-house expert panel. The reference material was also served as a coded (unknown) fillet sample in order to check the performance of the panel. The samples were treated in all ways like the "real" sample material. The sensory properties were related to assessors and trials by partial least squares regression (PLSR) using The Unscrambler®, version 7.6 SR-1 (Camo ASA, Norway) (Martens & Næs, 1989). Both Discriminant-PLSR (DPLSR) and Anova-PLSR (APLSR) models were calculated (Martens & Martens, 1999; Martens & Martens, 2001). Data from the sensory profiling were corrected for the "level effect" (i.e., assessors using different parts of the line scale) by the method described by Thybo and Martens (2000).

By serving the same product as a reference before each session, assessors can easily recapitulate the descriptors and can recalibrate their evaluations to the same scale. The additional use of coded reference samples served together with the actual samples makes it possible to monitor the performance of the panel. Multivariate data analysis allows for quick calculations, and results are easily interpreted with the visual layout. No drifting was found during the evaluation period, and this is most likely an effect of the continuous use of the reference (Nielsen *et al.* 2005c).

30.2.1.11 Test room or testing areas

There are international and national standards and guidelines for the design and construction of sensory assessment rooms (ISO 8589, 1988; ISO 8586-1, 1993; ISO 11035, 1994; Meilgaard *et al.* 2006; Hyldig *et al.* 2008). The sensory evaluation of whole fish is generally carried out by trained assessors in the reception or processing halls of fish factories or at auction sites. In quality control procedures, special facilities or rooms are preferred for sensory evaluation, but it is not always possible. In industry and auctions, the testing area should adhere to the ISO 8589 standard as closely as possible.

30.2.1.12 Sensory attributes

The criteria for selection of sensory attributes are that they can discriminate between samples, are relevant for the specific seafood, must be nonredundant, and must be cognitively clear to the assessors.

In Table 30.1, there is a list of sensory attributes divided into groups and detail levels.

Table 30.1 Sensory attributes.

Appearance	Surface	Brightness	Shininess
			Translucent
			Waxy
		Spotted	Blood stain
			White stripes
			Coagulated protein
	Moist	Juicy appearance	
		Watery appearance	
	Fatty aspect	Oily	
		Fat droplets in water	
	Color	Light	Whiteness
			Cream
		Brownish	Beige
			Brown
			Gray
		Reddish	Orange
			Peach
			Pink
			Red
			Red-brown
		Yellowish	Yellow-green
			Yellow-brown
	Color characteristics	Darkness	
		Paleness	
		Hue	
		Homogeneity of color	
		Color intensity	
		Discoloration	
	Texture appearance	Flaky	
		Separation	
		Coarse-grained	
		Crumbly crumbliness	
Odor	Aquatic	Marine	Anchovy
			Herring
			Tuna
			Salmon
		Shellfish	Iodine
		Seaweed	
	Oily	Fresh	Fish oil
			Linseed oil
			Nutty
		Oxidized	Painty
			Rancid
			Train oil
	Milky	Warm milk	
		Fresh milk	
		Buttery	
	Sweetness	Sweet	
		Sickly-sweet	
	Sourish	Acidic	Vinegar
		Sour	
	Green	Cucumber	
		Grass	
		Hay	
	Product	Boiled	Boiled potato
			Boiled corn
		Smoked	Cold ashes
			Smoke odor
			Wood fire

Table 30.1 *(Continued)*

Appearance	Surface	Brightness	Shininess
		Processed	Canned fish
			Grilled
			Fried chicken
			Meatiness
	Fermented	Maltiness	
		Yeasty	
	Earthy-like	Earthy	
		Mold	
		Mushroom	
	Frozen storage	Cardboardy	
	Tainted fish	Fishy	Amine
		Muddy	
		Musty	
		Medicine	
		Plastic	
		Rubber	
		Metallic	
	Putrefaction	Boiled egg	
		Cabbage	
		Farmyard	
		Rotten seaweed	
		Wet dog	
Flavor	Sweetness	Sweet	
		Sickly-sweet	
	Bitterness	Bitter	
	Saltiness	Salty	
	Sourish	Acidic	Vinegar
			Citrus fruit
		Sour	Sour milk
			Sour dishcloth
	Aquatic	Marine	Flatfish
			Herring
			Tuna
			Salmon
		Shellfish	Iodine
	Oily	Fresh	Fish oil
			Linseed oil
			Nutty
		Oxidized	Painty
			Rancid
			Soapy
	Milky	Warm milk	
		Fresh milk	
		Buttery	
		Cheese	
		Creamy	
	Green	Cucumber	
		Grass	
		Hay	
	Product	Boiled	Boiled potato
			Boiled corn
		Smoked	Cold ashes
			Wood fire
			Smokiness
			Acid smokiness

(Continued)

Table 30.1 *(Continued)*

Appearance	Surface	Brightness	Shininess
		Processed	Chicken-like
			Liver-like/gall
			Meaty
			Salty fish
	Fermented	Malty	
		Yeasty	
	Earthy-like	Earthy	
		Mold	
		Mushroom	
	Frozen storage	Cardboardy	
	Tainted fish	Fishy	Amine
		Muddy	
		Musty	
		Medicine	
		Plastic	
		Rubber	
		Metallic	
		Chemical-metallic	
	Putrefaction	Boiled egg	
		Boiled cabbage	
		Farmyard	
		Rotten seaweed	
		Wet dog	
	Astringent	Astringent	
Texture	Resistance	Firmness	
		Hardness	
		Softness	
		Chewiness	
		Pastiness	
	Meaty	Tenderness	
		Toughness	
		Fibrousness	
	Cohesiveness	Adhesiveness	
		Stickiness	
		Flakiness	
	Resilience	Elasticity	
		Gumminess	
		Springiness	
	Brittle	Brittleness	
		Crispiness	
		Crunchy texture	
	Roughness	Coarseness	
		Graininess	
	Juicy	Dryness	
		Juiciness	
		Moistness	
		Moisture release	
		Mouth dryness	
		Watery	
	Fatty mouth feel	Fattiness	
		Greasiness	
		Oily mouth coating	
		Oily	

30.3 TAINTING

Occasionally, fish with off-flavors are caught, and in certain localities this is a fairly common phenomenon. Several of these off-flavors can be attributed to feeding on different contaminants or organisms. The planktonic mollusk, *Spiratella helicina*, gives rise to an off-flavor described as "mineral oil" or "petrol." It is caused by dimethyl-ß-propiothetin, which is converted to dimethylsulfide in the fish (Connell 1975). The larvae of *Mytilus* spp. cause a bitter taste in herring.

An iodine like flavor is found in some fish and shrimp species in the marine environment. This is caused by volatile bromophenolic compounds, and it has been suggested that the compounds are formed by marine algae, sponges, and Bryozoa and become distributed through the food chain (Anthoni *et al.* 1990).

Oil taint may be found in fish flesh in areas of the world where off-shore exploitation of oil is intensive or in areas where large oil spills occur. The fraction of the crude oil that is soluble in water is responsible for the off-flavors. This is caused by the accumulation of various hydrocarbon compounds, where particularly the aromatic compounds are strong flavorants (Martinsen *et al.* 1992).

A very well-known off-flavor is the muddy-earthy taint in many freshwater fish (Johnsen & Bett 1996; Howgate 2004). The flavor is mainly caused by two compounds: geosmin and 2-methyl-iso-borneol (MIB), which are also part of the chemical profile of wine with cork flavor.

The farm-raised catfish industry considers environmental off-flavors associated with blooms of blue-green algae and other microbes to be its most significant problem. Algal blooms can produce geosmin and 2-MIB, which impart a muddy, musty flavor in seafood. Microbial production of the muddy metabolites geosmin and 2-MIB can be a problem in freshwater fish such as trout, bream, and pike. Based on the values presented in the literature (Howgate 2004), the estimated odor detection threshold in water at a normal ambient temperature of natural geosmin is 0.015 and 0.035 $\mu g\ L^{-1}$ for MIB. Sensory evaluation is the accepted method for monitoring off-flavors in seafood, where the detection of unknown compounds or compounds at low concentrations is necessary (e.g., 1 ppb for geosmin and MIB) (Bett & Dionigi, 1997). In research, the most appropriate method would be a descriptive method to determine the intensity of the off-flavor, but in industry a more practical method is needed to discern off-flavored fish from acceptable fish. Extensive training is necessary for assessors to evaluate specific taints (Codex Standards, 1999).

The content of MIB and geosmin were measured both by chemical and sensory analysis in a study (Zimba *et al.* 2011) with smoked rainbow trout product from a commercial fish farm; the samples were frozen at –20°C until analysis. The levels of MIB were at least one order of magnitude greater than geosmin levels in each respective fillet; MIB was the dominant odor component. The MIB concentrations ranged from 4.8 to 19.7 $\mu g\ kg^{-1}$ while geosmin ranged from 0.27 to 0.59 $\mu g\ kg^{-1}$. The intensity of the off-flavor from MIB (moldy/fusty) and geosmin (muddy) was significantly higher than other odors in most samples, even though the intensity of smoke and wood tar was relatively high; the odors and flavors from MIB and geosmin were easily detectable and identifiable.

30.4 SENSORY QUALITY

Examples of sensory characterization of seafood.

30.4.1 Species

There can be large variations in the sensory characteristics of different fish species, but also similarities. Cardello *et al.* (1983) characterized 17 fish species by sensory profiling, and by using the cluster analysis of the data they found that the 17 species could be grouped in three major groups. The first group is characterized by primarily low-fat, low flavor intensity, white-fleshed fish and consists of tilefish (*Lopholatilus chamaeleonticeps*), pollock (*Pollachius virens*), haddock (*Melanogrammus aeglefinus*), wolfish (*Anarhichas lupus*), Atlantic cod (*Gadus morhua*), cusk (*Brosme brosme*), white hake (*Urophycis tenuis*), whiting (*Merluccius bilinearis*), blackback flounder (*Pseudopleuronectes americanus*), Atlantic halibut (*Hippoglossus hippoglossus*), monkfish (*Lophius americanus*), and grouper (*Mycteroperca microlepis*). The second major group consists of the high fat, high flavor intensity, dark-fleshed fish, including bluefish (*Pomatomus saltatrix*), Atlantic mackerel (*Scomber scombrus*), weakfish (*Cynoscion regalis*), and striped bass (*Morone saxatilis*). The third group consists solely of swordfish (*Xiphias gladius*), but none of the sensory parameters described the swordfish very good. The swordfish was included in another study by Morita *et al.* (2003). In this study, different saltwater fish, migratory coastal fish, coastal bottom fish, pelagic fish, deep-sea fish, freshwater fish, anadromous fish, and brackish water fish, in total 16 species, were sensory characterized. By multivariate data analysis, four groups were observed in a biplot using PCA (principal component analysis). Red sea bream snapper (*Pagrus major*), common Japanese conger (*Conger myriaster*), Japanese eel (*Anguilla japonica*), and freshwater species, that is, loach (*Misgurnus anguillicaudatus*), pond smelt (*Hypomesus nipponensis*), and carp (*Cyprinus carpio*), were grouped around the flavor "green." Migratory coastal fish species, that is, sardine (*Sardinops melanosticta*), banded blue-sprat (*Spratelloides gracilis*), and chub mackerel (*Scomber japonicus*), were located close to "fish oil," "grilled fish," "sea breeze," and "fishy." Swordfish (*Xiphias gladius*), sablefish (*Anoplopoma fimbria*), and chum salmon (*Oncorhynchus keta*) were located close to "fried chicken." Slime flounder (*Microstmus achne*), pacific cod (*Gadus macrocephalus*), bluefin tuna (*Thunnus thynnus*), and yellowfin goby (*Acanthogobius flavimanus*) were grouped around "cooked fish," "sweet," "canned tuna," and "roasted soy sauce."

30.4.2 Starvation

The length of starvation before slaughtering can have considerable influence on the sensory attributes. In farmed Atlantic salmon (*Salmo salar*), fresh flavor was significantly reduced in groups starved from 30 to 86 days, whereas no significant differences were found among groups starved for 0–30 days. The sensory test was performed on day 13–16 postmortem fish (Einen & Thomassen 1998).

30.4.3 Handling of fish

Rigor mortis begins immediately or shortly after death if the fish is starved and the glycogen reserves are depleted, or if the fish is stressed. The method used for stunning and killing the fish also influences the onset of *rigor*. Stunning and killing by hypothermia (the fish is killed in iced water) give the fastest onset of *rigor*, while a blow on the head gives a delay of up to 18 h (Azam *et al.* 1990; Proctor *et al.* 1992). The technological significance of *rigor mortis* is of major importance when the fish is filleted before or in *rigor*. In *rigor*, the fish body will be completely stiff; the filleting yield will drop significantly, and rough handling can cause gaping in the fillets. If the fillets are removed from the bone *prerigor*, the muscle can contract freely and the fillets will shorten following the onset of *rigor*. Dark muscle may shrink up to 52% of the original length and white muscle up to 15% (Buttkus 1963). If the fish is cooked *prerigor*, the texture will be very soft and pasty. In contrast, the texture is tough but not dry when the fish is cooked in *rigor*. *Postrigor*, the flesh will become firm, succulent, and elastic.

In some fisheries, bleeding of the fish is very important as a uniform white fillet is desirable. Bleeding is affected more by time onboard prior to bleeding/gutting than by the actual bleeding/gutting procedure. The best bleeding is obtained if live fish are handled, but it is of major importance to cut the fish before it enters *rigor mortis* since it is the muscle contractions that force the blood out of the tissues. However, it should be pointed out that the effect of bleeding must be weighted against the advantages of having a fast and effective handling procedure, resulting in rapid chilling of the catch. A practical solution could be handling small quantities of fish at the same time. Discoloration of the fillet may also be a result of rough handling during catch and catch handling while the fish is still alive. Physical rough handling in the net (long trawling time, very large catches) or on the deck (fishermen stepping on the fish or throwing boxes, containers, and other items on top of the fish) may cause bruises, rupture of blood vessels, and blood oozing into the muscle tissue (hematoma). Heavy pressure on dead fish, when the blood is clotted (e.g., overloading of fish boxes) does not cause discoloration, but the fish may suffer a serious weight loss.

30.4.4 Farmed fish

As demonstrated above, the chemical composition of the different fish species will show variations depending on the season, migratory behavior, sexual maturation, feeding cycles, etc. These factors are observed in wild, free-living fish in the open sea and inland waters. Fish raised in aquaculture may also show variation in chemical composition, but in this case several factors can be controlled, and the chemical composition may be predicted. To a certain extent, the fish farmer is able to design the fish by selecting the farming conditions. It has been reported that factors such as feed composition, environment, fish size, and genetic traits all have an impact on the composition and quality (Reinitz *et al.* 1979; Einen & Skrede 1998; Nortvedt & Tuene 1998). In salmon aquaculture, astaxanthin is included in the feed to compensate for the red color from crustaceans normally found in the natural feed of the wild fish. The red color of the flesh is one of the most important intrinsic quality criteria for salmon.

30.4.4.1 Feed

The content of the feed can influence not only the growth rate, but also the sensory characteristics of the fish. For the Atlantic halibut (*Hippoglossus hippoglossus*) fed with a diet containing 20% or 39% fat, the sensory test showed that larger fish (2.1–2.7 kg) were characterized by a fresher and more acidic flavor and a more juicy consistency, compared to smaller fish (1.4 kg). The growth of the 20 halibut used in the sensory test showed that high growth was associated with small fish that achieved higher scores for off-flavor and rancid flavor than the larger halibut with lower growth rate. This does not imply that a high growth rate is a contributing factor to the off-flavor or the rancid flavor of the small fish, but shows that these characteristics are typical for the small fish (Nortvedt & Tuene 1998). The odor for turbot (*Psetta maxima*) fed with three experimental diets (with 9% (w/w) added lipid—fish oil, soybean oil, or linseed oil) were described as having a fatty fish note, even though the lipid content of turbot muscle was very low. Grass and hay notes were also found to characterize the odor of flesh (Sérot *et al.* 2001).

Baron *et al.* (2009) conducted a study with rainbow trout (*Oncorhynchus mykiss*) fed a diet containing either fish oil or rapeseed oil and with or without 200 mg/kg carotenoid (either astaxanthin or canthaxanthin). A total of six diets were obtained. The fish were slaughtered and stored in polyethylene bags individually as butterfly fillets for 22 months at −20°C. The fish fed fish oil had a grainier, more fibrous, and firmer texture compared to fish fed vegetable oil. In contrast, fish fed vegetable oil were juicier. In general, they also had less rancid odor and flavor compared to the fish fed fish oil. Fish fed vegetable oil had cooked potato, sweet, and mushroom flavors, whereas fish fed fish oil had more sourish flavors. For the fish fed fish oil, the highest rancid odor and flavor were observed when astaxanthin was present in the feed. In contrast, the lower score for rancidity was observed when canthaxanthin was present in the feed. The picture was more complex and difficult to describe for fish fed vegetable oil, but fish receiving the no-pigment feed had a lower score for both rancid odor and flavor than both samples fed carotenoids.

The long-term effects of feeding high-energy, low-fishmeal feeds on growth and flesh characteristics were investigated in ~1–4.5 kg Atlantic salmon (Salmo salar) by Johnsen *et al.* (2011), in a 9 month growth trial. Feeds were high-, medium-, and low-fishmeal (20%, 15%, or 10% fishmeal of total feed ingredients). The sensory panel did not detect any significant differences between fish from different treatments. The sensory profile was characterized by high juiciness, color tone and intensity, whiteness, medium sea odor and flavor, metallic odor and taste, sour odor, bitter taste, coarseness and fatness, and by low intensity of feed odor and flavor, vegetable oil odor and flavor, rancid flavor, salty taste, and firmness. From a human nutritious point of view, it is important that the lipid in the fatty fish contains a high amount of long-chain n-3 polyunsaturated fatty acids, and this must be taken into consideration when designing the feed for farmed fish.

30.4.4.2 Stress

In a study made by Sveinsdóttir *et al.* (2010a) they introduced two levels of stress during handling of the farmed fish. Two sample groups of farmed cod (*G. morhua*) were used for the experiment. Cod were farmed in the same cage and half of the batch was individually harvested and immediately stunned by a blow on the head, followed by gill cut and bleeding in ice water (group with low stress). A conventional method was simulated for the remaining fish. Here, the fish were crowded for 30 min prior to identical procedures of slaughter (group with high stress). The fish were stored in ice for 2 days prior to filleting, trimming, and packing interleafed in plastic prior to freezing at −25°C. The fillets were stored frozen at −25°C for 2–3 weeks, and thawed overnight at 0°C–4°C prior to sensory evaluation. Both groups were mainly characterized with sweet, shellfish/algae-like meat, and vanilla/boiled milk odors. The odor of boiled potatoes was also distinguishable. The appearance was homogenous; the color white and the flesh had some white precipitation. Metallic, sweet, and meaty flavors were predominant in the samples, with a hint of salty flavor. The texture of the groups was very meaty, but less soft, juicy, and tender. The texture was slightly mushy, clammy, and rubbery. The sensory differences between the two groups were not very large, but the main differences were in texture. The low stress group had a more meaty and rubbery texture, but the high stress group was more tender. In addition, a marginal significance was found for flaky, juicy, and clammy texture, with the low stress group being slightly less flaky and juicy, but more clammy. In a study by Jessen and Hyldig (unpublished results), it was also found that the texture was affected by the stress level of trout. The results showed that unstressed trout were juicier and less fibrous than stressed trout and that the unstressed trout remained juicy during the whole ice storage period whereas the stressed trout became even more dry and fibrous during ice storage. The stress levels of the fish affect the sensory quality of the fish product and, therefore, the handling of the fish is important for keeping the quality of the fish.

30.4.5 Storage of fish

The spoilage rate and shelf life of fish is affected by many parameters and fish spoil at different rates. In general, it can be stated that larger fish spoil more slowly than smaller fish, lean fish keep longer than fatty fish under aerobic storage, and bony fish are edible longer than cartilaginous fish. Several factors probably contribute to these differences and whereas some are clear, many are still on the level of hypotheses. Rough handling will result in a faster spoilage rate. This is due to the physical damage to the fish, resulting in easy access for enzymes and spoilage bacteria. The surface/volume ratio of larger fish is lower than that of smaller fish, and, as bacteria are found on the outside, this is probably the reason for the longer shelf life of the former. This is true within a species but may not be universally.

The skin of the fatty pelagic fish is often very thin, and this may contribute to the faster spoilage rate. This allows enzymes and bacteria to penetrate more quickly. On the contrary, the thick skin of flatfish and the antibacterial compounds found in the slime of these fish may also contribute to the keepability of flatfish. As described earlier, the slime of flatfish contains bacteriolytic enzymes, antibodies, and various other antibacterial substances (Murray & Fletcher 1976; Hjelmland *et al.* 1983). Although large differences exist in the content of trimethylamine oxide (TMAO), this does not seem to affect the shelf life of aerobically stored fish but rather the chemical spoilage profile of the species.

30.4.5.1 Ice storage

During ice storage of fish, the intensity of the sensory attributes will change. The sensory profiling of ice storage farmed salmon showed that attributes characterizing the salmon (*Salmo salar*) on day 1 were seaweed, cucumber and sourish odor, and sweetish, sourish, fish oil, and mushroom flavor, and after 22–24 days of storage the salmon were characterized by rancid, sour, amine odor, and rancid flavor. The sensory attributes used in another storage experiment with farmed salmon stored in ice were grouped into "positive sensory parameters" and "negative sensory parameters." Samples from every second storage day were analyzed. The changes in the sensory attributes indicate that the salmon was approaching the end of the acceptable flavor after 20–21 days, when the salmon was characterized by an increasing intensity of sour, amine, and rancid odor and flavor. All positive attributes had a high intensity and were very characteristic for the salmon at the beginning of the storage time, but after 21–22 days of storage they were hardly detectable. It was concluded from the result of the sensory profiling that the shelf life for salmon, where the fish is no longer fit for human consumption, was 20 days in ice (Sveinsdóttir *et al.* 2003).

30.4.5.2 Storage in modified atmosphere packing

In storage studies with modified atmosphere packing (MAP), the development of sensory attributes are different than for ice storage of the same fish species. Hong *et al.* (1996) found that the odor of Atlantic mackerel (*Scomber scombrus* L.) during MAP storage changed from day 0 (seaweed, fishy, and rancid) to day 7 of storage (seaweed, cucumber-like, sour, fishy, painty, and rancid) and again from day 14 of storage (seaweed, sour, fishy, rancid) to day 21 of storage (seaweed, sour, fishy, metallic, and rancid).

30.4.5.3 Frozen storage

The rapid development of rancidity in stored muscle from fish such as mackerel, herring, capelin, and bluefish is often attributed to the fact that these species contain high levels of lipid. However, studies by Richards and Hultin (2001) indicate that blood-mediated oxidation of washed cod (*Gadus morhua*) lipids required $\leq 0.1\%$ phospholipid to cause rancidity. They also found that with long-term, frozen storage of mackerel (*Scomber scombrus*) muscle, sensory quality deterioration occurred in spite of no measurable change in fatty acid composition.

30.4.6 Fish product available for the consumers

Consumers' attitude toward the sensory quality of fish is very dependent of their ability to evaluate the quality and their knowledge of seafood. Juhl and Poulsen (2000) showed in a study in Denmark that many consumers are not able to use intrinsic cues for evaluating the quality of fresh fish. In their study, two segments were distinguished and the highly involved "fish lovers" were able to determine the freshness of the fish from its odor and appearance, while the less involved "traditional fish eaters" not were able to use the intrinsic cues in the same way. Brunsø *et al.* (2009) found that the most important motives for the consumption of fish in Spain and Belgium are that fish is perceived as very healthy food and as having a good taste and that these drivers for consumption did not change across the different segments with low versus high consumption levels of fish. They suggest a possible segmentation pattern of consumer groups that had a cross-cultural pattern from analysis of the data from focus group interviews conducted in Spain and Belgium. The consumers in the "skilled and demanding fish consumers" segment were the most demanding with regard to fish quality, and they consumed fish four to six times a week, they loved to cook, and they were very skilled and knowledgeable about how to prepare the fish. They were experts in evaluating the freshness of the fish and did not mind to pay a premium price in order to get the best possible quality. The segment with the "skilled, but willing to compromise consumers" consumed fish approximately twice a week. They had certain quality demands, but they were not willing to spend so much time on either the shopping or cooking of fish as the first segment, and were not willing to pay whatever price it took to get the best quality. The segment of "the unskilled light users" consisted mainly of consumers who consumed fish once a month or even less frequently. The segment was characterized primarily by the fact that they felt highly unskilled in evaluating the quality of the fish. They wanted fresh fish, but their best way of judging the freshness was by looking at the rotation in the shop and the price.

In the following, two studies of the sensory quality of fish products available for the consumers are presented.

Green-Petersen *et al.* (2006) gave an overview of both the sensory properties and the differences between the most common salmonid products available on the Danish market. Twelve salmon samples that differed in storage condition, packaging, and species were used in the experiment. All the samples were obtained from local shops or companies and bought as consumer products. There were nine samples of *Salmo salar*, two were MAP, four were storage in ice, and three were frozen. There were also two frozen samples of *Oncorhynchus kisutch* and one frozen sample of *Oncorhynchus keta*. The four samples stored in ice had a very similar sensory profile described by the attributes sea/seaweed odor, the flavors fresh fish oil, sweet, and mushroom, and the texture juicy and oily. Even though the ice storage samples represented different fish farms, different batches, and different storage times in ice (7 or 16 days), they did not observe any clear sensory differences that were in agreement with Sveinsdottir *et al.* (2003). The MAP samples stored for 5 days and the ice storage samples had a relatively similar sensory profile. Whereas the MAP that was stored for 7 days was rather different, it was rancid and sour. For the frozen samples, the results show that a longer freezing time gave a more soft texture, an increase in discolor, and a decrease in sea/seaweed odor. However, they found that freezing only had little influence on flavor. The results indicate that, in general, the sensory properties of *Oncorhynchus keta* and *Salmo salar* are fairly similar. Furthermore, the results indicate that the sensory properties of *Oncorhynchus kisutch* are very different from the sensory properties of *Oncorhynchus keta* and *Salmo salar*. The two samples of *Oncorhynchus kisutch* had a high intensity of firm texture, rancid flavor, and discoloration, combined with a low intensity of juicy and oily texture, sea/seaweed odor, and sweet, fresh fish oil, and mushroom flavor.

Sveinsdóttir *et al.* (2010b) investigated the sensory characteristics of different common cod products available to consumers. They used ten different samples of wild and farmed cod (*Gadus morhua*) fillets, stored under various conditions. The products were chosen based on which cod products are available on the market for consumers, but collected directly from fish producers/retailers. Wild and farmed cod stored in ice, wild cod MAP, and wild frozen cod were evaluated after a short and an extended storage period. The wild cod was caught with a long line, stored whole in ice after catch, and filleted after 3 up to 5 days. The fillets were stored at 0–1°C on ice in plastic boxes. Frozen cod fillets originated from a cod trawler and were processed by a fish producer. The farmed cod was slaughtered, kept whole in ice, filleted after 1–6 days, and then filleted and stored at 0–1°C on ice in plastic boxes. They found that the sample groups (farmed fresh, wild fresh, wild frozen, and wild MAP) with a short storage time had a characteristic sweet odor, metallic and sweet flavor, light and even color, and juicy texture. No signs of frozen storage, tablecloth, trimethylamine (TMA), sour, sulfur, or putrid odors were evident, nor were sour, frozen storage, pungent, TMA, or putrid flavors detected in the samples after short storage time. The fresh cod stored for 10 days was mainly described with attributes characteristic for cod at the end of shelf life, such as tablecloth, TMA sour and putrid odor, sour, pungent, and TMA and putrid flavor. After 9 days of storage, the MAP cod fillets were less described with freshness characteristics, such as sweet and metallic flavor, but more with dark and heterogeneous appearance. Hints of TMA odor and flavor were detected in this sample. This indicated that the MAP cod fillets were approaching the end of shelf life after 9 days. The farmed cod samples were clearly different from the wild cod samples. The three farmed cod samples (ice storage for 1, 9, and 12 days) were significantly more described with a meat flavor, with a more meaty, rubbery, and clammy texture and less mushy and tender texture compared with most of the wild cod samples. After 9 and 12 days storage, the farmed cod were also the least soft and juicy samples. Overall, the farmed cod samples appeared to retain the attributes characteristic for fresh cod after a short storage time for a longer duration than the wild cod of similar storage time. However, the farmed cod received higher scores for sour odor and flavor. Newly frozen cod fillets (frozen for 1 month) had very similar sensory characteristics as fresh cod fillets after short storage time. After extended frozen storage (6 months), the sweet flavor reduced, but other odor and flavor attributes characteristic for freshness did not reduce significantly. However, the fillets had a darker appearance, white precipitation, more odor of boiled potatoes, frozen storage flavor, and more clammy but less soft, juicy, and tender texture.

REFERENCES

Ando, S. & Hatano, M. (1986) Biochemical characteristics of chum salmon muscle during spawning migration. *Bulletin of the Japanese Society of Fisheries*, **52**, 1229–1235.

Ando, S., Hatano, M. & Zama, K. (1985) Deterioration of chum salmon (*Oncorhynchus keta*) muscle during spawning migration—1. Changes in proximate composition of chum salmon muscle during spawning migration. *Comparative Physiology and Biochemistry*, **80B**, 303–307.

Anon. (1996) Council Regulation (EC) No. 2406/96 of 26. November 1996 laying down common marketing standards for certain fishery products. *Official Journal of the European Communities*, **L334**, 1–14.

Anthoni, U., Borresen, T., Christophersen, C., Gram, L. & Nielsen, P.H. (1990) Is trimethylamine oxide a reliable indicator for the marine origin of fishes. *Comparative Physiology and Biochemistry*, **97B**, 569–571.

Azam, K., Mackie, L.M. & Smith, J. (1990) Effect of stunning methods on the time of onset, duration and resolution of rigor in rainbow trout (*salmo gardineri*) as measured by visual observation and analysis for lactic acid, nucleotide-degration products and glycogen. In: *Chilling and Freezing of New Fish Products. Science et Technique du Froid*. Proceedings of the meating of commission C2 1.1.F.-I.I.R, Aberdeen, U.K., pp. 351–358.

Baron, C.P., Hyldig, G. & Jacobsen, C. (2009) Does feed composition affect oxidation of rainbow trout (*Oncorhynchus mykiss*) during frozen storage? *Journal of Agricultural and Food Chemistry*, **57**, 4185–4194.

Bech, A.C., Kristensen, K., Juhla, H.J. & Poulsen, C.S. (1997) Development of farmed smoked eel in accordance with consumer demands. In: *Seafood from Producer to Consumer, Integrated Approach to Quality* (eds J.B. Luten, T. Børresen & J. Oehlenschläger), pp 3–19. Elsevier Science B.V., Amsterdam, the Netherlands.

Bett, K.L. (1997) Flavour-quality control in freshwater aquaculture. *The Progressive Fish-Culturist*, **59**, 149–154.

Bett, K.L., Ingram, D.A., Grimm, C.C., Vinyard, B.T., Boyette, K.D.C. & Dionigi, C.P. (2000) Alteration of the sensory perception of the muddy/earthy odorant 2-methylisoborneol in channel catfish *Ictalurus punctatus* fillet tissues by addition of seasonings. *Journal of Sensory Studies*, **15**, 459–472.

Bjerkeng, B., Refstie, S., Fjalestad, K.T., Storebakken, T., Røbotten, M. & Roem, A.J. (1997) Quality parameters of the flesh of Atlantic salmon (*Salmo salar*) as affected by dietary fat content and full-fat soybean meal as a partial substitute for fish meal in the diet. *Aquaculture*, **157**, 297–309.

Bøkness, N., Jensen, K.N., Guldager, H.S., Østerberg, C., Nielsen, J. & Dalgaard, P. (2002) Thawed chilled Barents Sea cod fillets in modified atmosphere packaging-appæication of multivariate data analysis to select key parameters in good manufacturing practice. *Lebensmittel-Wissenschaft und-Technologie*, **35**, 436–443.

Borresen, T. (1992) Quality aspects of wild and reared fish. In: *Proceedings of an International Conference on Quality Assurance in the Fish Industry* (eds H.H. Huss, M. Jacobsen & J. Liston), pp. 1–17. Copenhagen, Denmark, August 1991. Elsevier, Amsterdam, the Netherlands.

Brunsø, K., Verbeke, W., Olsen, S.O. & Jeppesen, L.F. (2009) Motives, barriers and quality evaluation in fish consumption situations. *British Food Journal*, **111** (7), 699–716.

Buttkus, H.J. (1963) Red and white muscle of fish in relation to rigor mortis. *Journal of the Fisheries Board of Canada*, **20**, 45–58.

Cardello, A.V., Sawyer, M.F., Prell, P., Maller, O. & Kapsalis, J. (1983) Sensory methodology for the classification of fish according to edibility characteristics. *Lebensmittel-Wissenschaft und-Technologie*, **16**, 190–194.

Cardinal, M., Gunnlaugsdottir, H., Bjoernevik, M., Ouisse, A., Vallet, J.L. & Leroi, F. (2004) Sensory characteristic of cold-smoked Atlantic salmon (*Salmo salar*) from European market and relationships with chemical, physical and microbiological measurements. *Food Research International*, **37**, 181–193.

Codex Standards. (1999) Codex standards for fish and fishery product, Guidelines for the sensory evaluation of fish and shellfish in laboratories, CAC-GL 31-1999 Rome, FAO/WHO Codex Alimentarius.

Connell, J.J. (1975) *Control of Fish Quality*. Fishing News (Books) Ltd. Farnham, Surry, U.K.

Einen, O. & Skrede, G. (1998) Quality characteristics in raw and smoked fillets of Atlantic salmon, *Salmo salar*, fed high-energy diets. *Aquaculture Nutrition*, **4**, 99–108.

Einen, O. & Thomassen, M.S. (1998) Starvation prior to slaughter in Atlantic salmon (*Salmo salar*) II. White muscle composition and evaluation of freshness, texture and colour characteristics in raw and cooked fillets. *Aquaculture*, **169**, 37–53.

Farmer, L.J., McConnell, J.M. & Kilpatrick, D.J. (2000) Sensory characteristic of farmed and wild Atlantic salmon. *Aquaculture*, **187**, 105–125.

Ginés, R., Valdimarsdottir, T., Sveinsdottir, K. & Thorarensen, H. (2004) Effect of rearing temperature and strain on sensory characteristics, texture, colour and fat of Arctic charr (*Salvelinus alpinus*). *Food Quality and Preferences*, **15**, 177–185.

Green-Petersen, D.M.B. & Hyldig, G. (2010) Variation in sensory profile between individual rainbow trout (*Oncorhynchus mykiss*) from the same production batch. *Journal of Food Science*, **75** (9), S499–S505.

Green-Petersen, D.M.B., Nielsen, J. & Hyldig, G. (2006) Sensory profiling of the most common Salmon products on the Danish market. *Journal of Sensory Studies*, **21**, 415–427.

Hjelmland, K., Christine, M. & Raa, J. (1983) Skin mucous protease from rainbow trout (*Salmo gairdneri*, Richardson). *Biological significance. Journal of Fish Biology*, **23**, 13–22.

Hong, L.C., Leblanc, E.L., Hawrysh, Z.J. & Hardin, R.T. (1996) Quality of Atlantic mackerel (*Scomber scombrus* L.) fillets during modified atmosphere storage. *Journal of Food Science*, **61**, 646–651.

Howgate, P. (2004) Tainting of farmed fish by geosmin and 2-methyl-iso-borneol: A review of sensory aspects and of uptake/depuration. *Aquaculture*, **234**, 155–181.

Hyldig, G., Carlehög, M., Martinsdóttir, E., Sveinsdóttir, K. & Lilleberg, L. (2008) *Guide for Sensory Analysis of Fish and Shellfish*. NMKL Procedure No. 21, pp. 1–27. Nordic Committee on Food Analysis. Norway.

ISO 8589. (1988) Sensory analysis—General guidance for the design of test rooms. Reference number ISO 8589, 1988(E). International Organization for Standardization, Geneva, Switzerland.

ISO 8586-1. (1993) Sensory analysis—General guidance for the selection, training and monitoring of assessors—Part 1: Selected assessors. International Organization for Standardization, Geneva, Switzerland.

ISO 11035. (1994) Sensory analysis—Identification and selection of descriptors for establishing a sensory profile by a multidimensional approach. International Organization for Standardization, Geneva, Switzerland.

Johnsen, P.B. & Bett, K.L. (1996) Sensory evaluation of the off-flavours Geosmin and 2-methylisoborneol (MIB) in farm raised catfish. *Journal of Applied Aquaculture*, **6** (2), 21–37.

Johnsen, C.A., Hagen, Ø. & Bendiksen, E.Å. (2011) Long-term effects of high-energy, low-fishmeal feeds on growth and flesh characteristics of Atlantic salmon (*Salmo salar* L.). *Aquaculture*, **312**, 109–116.

Juhl, H.J. & Poulsen, C.S. (2000) Antecedents and effects of consumer involvement in fish as a product group. *Appetite*, **34**, 261–267.

Martens, H. & Martens, M. (1999) Modified Jack-knife estimation of parameter uncertainty in bilinear modelling by partial least squares regression (PLSR). *Food Quality and Preference*, **11**, 5–16.

Martens, M. & Martens, H. (2001) *Multivariate Analysis of Quality. An Introduction*. John Wiley & Sons, England.

Martens, H. & Næs, T. (1989) *Multivariate Calibration*. John Wiley & Sons Ltd, Chichester, England.

Martinsen, C., Lauby, B., Nevissi, A. & Brannon, E. (1992) The influence of curde oil and dispersant on the sensory characteristics of steelhead (*Oncorhychus mykiss*) in marine waters. *Journal of Aquatic Food Product Technology*, **1**, 37–51.

McGill, A.S., Howgate, P., Thomson, A.B., Smith, G., Ritchie, A. & Hardy, R. (1984) The flavour of with fish and the relationship between the chemical analyses and sensory data. *Progress in flavour research. Development in Food Science*, **10**, 149–164.

Meilgaard, M., Civille, G.V. & Carr, B.T. (2006) *Sensory Evaluation Techniques*, 4th edn. CRC Press, Taylor & Francis Group, Boca Raton, Florida.

Morita, K., Kubota, K. & Aishima, T. (2001) Sensory characteristics and volatile components in aroma of boiled prawns prepared according to experimental designs. *Food Research International* **34**, 473–481.

Morita, K., Kubota, K. & Aishima, T. (2003) Comparison of aroma characteristics of 16 fish species by sensory evaluation and gas chromatographic analysis. *Journal of the Science of Food and Agriculture*, **83**, 289–297.

Murray, C.K. & Fletcher, T.C. (1976) The immunohistochemical location of lysozyme in plaice (*Pleuronectes platessa* L.) tissues. *Journal of Fish Biology*, **9**, 329–334.

Nielsen, D., Hyldig, G., Nielsen, J. & Nielsen, H.H. (2005a) Sensory properties of marinated herring (*Clupea harengus*) processed from raw material from commercial landings. *Journal of the Science of Food and Agriculture*, **85** (1), 127–134.

Nielsen, D., Hyldig, G., Nielsen, H.H. & Nielsen, J. (2005b) Lipid content in herring (*Clupea harengus* L)—Influence of biological factors and comparison of different methods of analyses: Solvent extraction, Fatmeter, NIR & NMR. *Lebensmittel-Wissenschaft und-Technologie*, **38**, 537–548.

Nielsen, D., Hyldig, G. & Sørensen, R. (2005c) An effective way to minimize drifting and monitor the performance of a sensory panel during long-term projects—A case study from a project on herring quality. *Journal of Sensory Studies*, **20**, 35–47.

Nortvedt, R. & Tuene, S. (1998) Body composition and sensory assessment of three weight groups of Atlantic halibut (*Hippoglossus hippoglossus*) fed three pellet sizes and three dietary fat levels. *Aquaculture*, **161**, 295–313.

Papadopoulos, L.S. & Finne, G. (1986) Effect of environmental salinity on sensory characteristics of penaeid shrimp. *Journal of Food Science*, **51**, 812–814.

Proctor, M.R.M., Ryan, J.A. & McLoughlin, J.V. (1992) The effects of stunning and slaughter methods on changes in skeletal muscle and quality of farmed fish. In: *Proceedings from TNO, International Conference: Upgrading and Utilization of Fishery Products*. the Netherlands.

Rasmussen, R.S., Ostenfeld, T.H., Rønsholdt, B. & McLean, E. (2000) Manipulation of end-product quality of rainbow trout with finishing diets. *Aquaculture Nutrition*, **6**, 17–23.

Refsgaard, H.H.F., Brockhoff, P. & Jensen, B. (2000) Free polyunsaturated fatty acids cause taste deterioration of Salmon during frozen storage. *Journal of Agriculture and Food Chemistry*, **48**, (8), 3280–3285.

Reinitz, G.L., Orme, L.E. & Hitzel, F.N. (1979) Variations of body composition and growth among strains of rainbow trout (*Slamo gairdneri*). *Transactions of the American Fisheries Society*, **108**, 204–207.

Richards, M. & Hultin, H.O. (2001) Rancidity development in a fish model system as affected by phospholipids. *Journal of Food Lipids*, **8**, 215–230.

Rørå, A.M.B., Kvåle, A., Mørkøre, T., Rørvik, K.A., Stein, S.S. & Thomassen, M.S. (1998) Process yield, colour and sensory quality of smoked Atlantic salmon (*Salmo salar*) in relation to raw material characteristics. *Food Research International*, **31**, 601–609.

Schubring, R. & Oehlenschläger, J. (1997) Comparison of the ripening process in salted Baltic and North Sea herring as measured by instrumental and sensory methods. *Zeitschrift fur Lebensmitteluntersuchung und Forschung A*, **205**, 89–92.

Sérot, T., Regot, C., Prost, C., Robin, J. & Arzel,. J. (2001) Effect of dietary lipid sources on odour-active compounds in muscle of turbot (*Psetta maxima*). *Journal of the Science of Food and Agriculture*, **81**, 1339–1346.

Stohr, V., Joffraud, J.J., Cardinal, M. & Leroi, F. (2001) Spoilage potential and sensory profile associated with bacteria isolated from cold-smoked salmon. *Food Research International*, **34**, 797–806.

Sveinsdóttir, K., Hyldig, G., Martinsdóttir, E., Jørgensen, B. & Kristbergsson, K. (2003) Development of quality index method (QIM) scheme for farmed Atlantic Salmon (*Salmo salar*). *Food Quality and Preferences*, **14**, 237–245.

Sveinsdóttir, K., Martinsdóttir, E., Thórsdóttir, F., Schlvis, R., Kole, A. & Thórsdóttir, I. (2010a) Evaluation of farmed cod products by a trained sensory panel and consumers in different test settings. *Journal of Sensory Studies*, **25**, 280–293.

Sveinsdóttir, K., Martinsdóttir, E., Hyldig, G. & Sigurgísladóttir, S. (2010b) Sensory characteristics of different cod products. *Journal of Sensory Studies*, **25**, 294–314

Thurman, H.V. & Webber, H.H. (1984) *Marine Biology*. Charles E Merrill Publishing CA Bell and Howell Co., Columbus, Ohio.

Thybo, A.K. & Martens, M. (2000) Analysis of sensory assessors in texture profiling of potatoes by multivariate modelling. *Food Quality and Preference*, **11**, 283–288.

Triqui, R. & Bouchriti, N. (2003) Freshness assessment of Moroccan Sardine (*Scardina pilchardus*): Comparison of overall sensory changes to instrumentally determined volatiles. *Journal of Agricultural and Food Chemistry*, **51**, 7540–7546.

Warm, K., Nielsen, J. & Hyldig, G. (2000) Sensory quality criteria for five fish species. *Journal of Food Quality*, **23**, 583–601.

Zimba, P.V., Schrader, K.K., Hyldig, G., Strabel, B.W. & Jørgensen, N.O.G. (2011) Evaluation of geosmin and 2-methylisoborneol off-flavour in smoked rainbow trout fillets by instrumental and sensory analyses. *Aquaculture Research*, **43**, 149–153.

31 Quality of Frozen Fish

Alex Augusto Gonçalves, Jette Nielsen,
and Flemming Jessen

Abstract Freezing is the most widely used preservation method for fish. The quality of seafood depends on the freezing rate. Different factors affect the quality of frozen fish. These parameters are biological or depend on the conditions of catching and processing fish.

The freeze or cold chain must be kept unbroken during storage and transport. Attention must be paid to retail display equipment and temperature monitoring.

The frozen state of fish has an impact on microorganisms. Freezing also influences biochemical attributes: fat and fat decomposition, oxidative rancidity, and protein changes. Sensory attributes are altered by freezing. How can these sensory changes be monitored? Fish populations belonging to the same fish species may have differences in freezing quality. Finally, freezing also has an impact on physical attributes.

Prefreezing factors and postfreezing treatments influence or protect the storage of frozen fish.

Keywords: Frozen fish quality, practical storage life (PSL), high-quality life (HQL), freeze chain, QIM, freezing process, fish rigor mortis, fat decomposition, oxidative rancidity, fish glazing

31.1 INTRODUCTION

Seafood quality, including safety, is a major concern facing the seafood industry today and is also one of the most abused words in food science, especially in fisheries and aquaculture research. A number of surveys have shown that consumer awareness about the quality of food is increasing. Fish and fish products are in the forefront of food safety and quality improvement because they are among the most internationally traded food commodities (Gonçalves & Blaha 2010).

Freezing is the most widely used preservation method for fish and it accounted for a 49.8% share of total processed fish for human consumption and 20.5% of total fish production in 2008 (FAO 2010). Nevertheless, the eating quality of fresh fish (not frozen before being sold on the market) is considered superior to that of frozen fish. It is, however, possible to produce frozen fish of high eating quality by quickly freezing the fish and subsequently storing it at low and stable temperatures. In this way, the physical and chemical processes causing the quality of the fish to deteriorate may be reduced. Quality deterioration is seen, for example, in the development of ice crystals, oxidation of fat, and degradation of muscle protein. The effect of freezing and frozen storage has also been studied for a few years and is described in a number of reviews (Sikorsky *et al.* 1976; Shenouda 1980; Haard 1992; Love 1992; Mackie 1993;

Handbook of Meat, Poultry and Seafood Quality, Second Edition. Edited by Leo M. L. Nollet.

Table 31.1 Storage time for PSL and storage time for HQL stated in months for lean fish (cod), big fatty fish (salmon), and small fatty fish (herring).

	Shelf life in months			
	−18°C		−30°C	
Fish species	**PSL**	**HQL**	**PSL**	**HQL**
Lean fish (e.g., cod)	7	3	12	6
Big fatty fish (e.g., salmon)	7	3	18	6
Small fatty fish (e.g., herring)	5	2	10	5

Sikorsky & Kolakowska 1994). Part of this chapter has also been published in Danish by Jessen and Nielsen (2002).

Frozen fish is easily stored and distributed. Under ideal circumstances (low and stable storage temperatures), some fish species may retain a fair eating quality for over a year. Shelf life can be assessed either in terms of practical storage life (PSL) or high-quality life (HQL). PSL is defined as the time the product can be in cold storage before it loses its characteristic properties or becomes unsuitable for consumption. PSL is often determined between trade partners, and no legislative rules apply to this area. HQL is a target for how long the product can be in cold storage before taste panels are able to discern a clear difference from the original quality of the fish. HQL is normally two to three times shorter than PSL. PSL, moreover, is what is eventually declared on the product. The shelf life of fish in cold storage depends on time, temperature, and the species of fish (Table 31.1).

Frozen fish can be marketed as super chilled, frozen thawed, freeze chilled, or refreshed with and without modified atmosphere packaging (Bøknæs *et al.* 2000; Olafsdottir *et al.* 2006).

Super chilling (partial freezing) extends shelf life by cooling the fish down to 1–2°C below the freezing point (Magnussen *et al.* 2008). The surface is thereby frozen and protects the fish by inhibition of microbial activity. Super chilling has been described as early as 1920 but has mostly been used to ease the slicing process of, for example, smoked salmon. Super chilling can be used to protect against temperature rises in poor cold chains for a short period. If the cold chain is out of control for a longer time, microstructural changes of the tissue are caused by ice crystallization and recrystallization causing extended drip loss resulting in a dry and tough product.

A frozen thawed, refreshed, and freeze chilled fish is a fish that has previously been frozen and thawed either before packaging and distribution or thawed during distribution. The quality of freeze chilled fish is dependent on the quality of the frozen fish and the thawing process. The process enables the manufacturing of bulk products to be released into the chill chain on demand (Fagan *et al.* 2003; Jensen *et al.* 2010).

31.2 THE FREEZING PROCESS—FROM WATER TO ICE

31.2.1 Seafood quality and freezing rate

Fish muscle contains a large amount of water, often about 80%, and on cooling below the freezing point of the tissue fluids (*c.* −1.5°C), water in the fluid is converted into ice (Fig. 31.1). As ice forms, proteins and solutes become concentrated in the nonfrozen fraction. The freezing point of the nonfrozen fraction is depressed as the solute concentration increases. Ultimately,

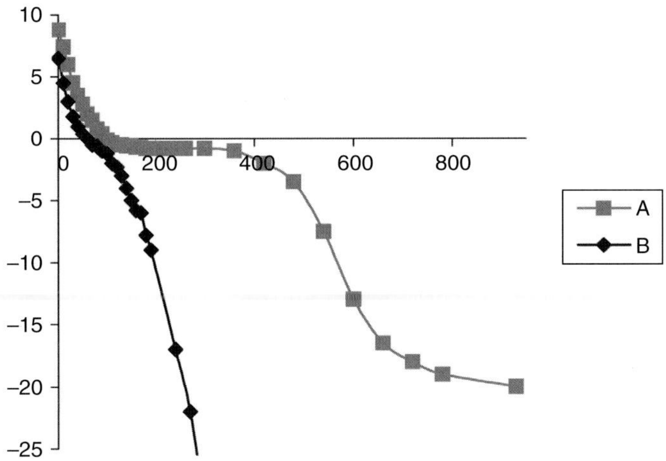

Fig. 31.1 Temperature profiles during freezing of whole cod. (A) Freezing at −20°C in still air (home freezer). (B) Freezing at −45°C in air blast freezer (industrial freezer). The temperature was measured in the center of the fish (core temperature). It is apparent from the figure that the fish can be brought through the critical temperature interval, from 0°C to 5°C, when quick freezing takes place at a low temperature and when a fan at the same time makes sure that the heat from the fish is removed quickly from the surface.

Table 31.2 Variation of percentage of water that remains unfrozen for cod fillet.

Temperature (°C)	% Unfrozen water
−1	92
−2	48
−3	33
−4	27
−5	21
−10	16
−20	11

Source: From Hedges and Nielsen (2000).

the freezing point of the nonfrozen fraction equals that of the freezing temperature. Once this point is reached, equilibrium is established between the ice phase and the nonfrozen fraction. The amount of nonfrozen fraction depends upon the composition of the fish product, and on the temperature at which the product is frozen or stored. When about 75% of the water is frozen, the temperature once more starts to drop rapidly. The lower the temperature (at −30°C), the more ice is formed (about 90% of water turns into ice) and the more concentrated the solutes become. Conversely, on increasing the temperature the amount of nonfrozen water increases (Gonçalves & Blaha 2010).

Table 31.2 shows the amount of unfrozen water in cod fillet as a function of temperature. As a consequence of forming a highly concentrated nonfrozen phase, potential reactants (e.g., enzymes and substrates) become far more concentrated than in the living tissue. Reactions are, potentially, accelerated. However, lowering the temperature decreases the rates of these reactions. Therefore, two antagonistic effects occur on cooling: an increase in reaction rate due to freeze concentration and a decrease in reaction rate due to a lowering of the temperature. In addition, in some foods, reaction rates may become diffusion controlled (Hedges & Nielsen 2000).

The actual freezer design and operation must compromise between quality and cost criteria. Unfortunately, the value of quality is difficult to quantify. Quality is usually considered to be related to freezing rate due to the latter's effect on ice crystal location and size and weight loss, but the relationship is ill-defined and very product specific. For most foods, as long as the freezing rate is "reasonably" fast, quality requirements become secondary to cost issues. Exceptions are usually frost-sensitive or valuable foods. Given that many of the advantages of fast freezing can be lost during subsequent storage, often too much emphasis is given to achieving the highest freezing rate in the process freezer. It is probably more critical that complete product freezing is achieved in the process freezer because rates of temperature reduction in frozen storage are much lower and can lead to both significant loss of product quality and increased storage freezer refrigeration costs (Cleland & Valentas 1997).

It is now well known that rapid freezing is essential for good quality. This is defined as taking no more than 2h to pass from 0°C to −5°C (from 32°F to 23°F) in the thickest part of the fish, and freezing at this rate or faster is now normal good commercial practice. Most fish begin to freeze at about −1°C (30°F), but the multiplication of the putrefactive bacteria is only completely arrested at about −9°C (15°F). Although bacterial spoilage is then suspended and there is some reduction in bacterial numbers, not all the bacteria are destroyed by freezing (Ranken *et al.* 1997).

Freezing rate affects the rate of tenderizing after thawing but not the ultimate tenderness. Freezing at −10°C more than doubles the rate; freezing in liquid nitrogen almost trebles the rate. Freezing is known to cause structural damage by ice crystal formation. It seems likely that ice crystals, particularly small intracellular ice crystals formed by very fast freezing rates, enhance the rate of conditioning probably by release of enzymes. Repeated freeze-thaw cycles using relatively low freezing rates does not seem to cause any enhanced tenderizing (James 2002).

For any study where attempts are made to vary the freezing rate, it is important to remember that a range of time–temperature profiles are likely to exist across a sample, and this situation will be exacerbated as the sample size is increased. Therefore, it should not be assumed that changing the external freezing temperature changes the overall rate at which a sample was frozen. However, a study by Doong (1998) indicated that the ice crystal size could be manipulated by changing freezing conditions, and that the ice crystal size correlated with the rate of change of, for example, the water-holding capacity (WHC) of the fillet on subsequent frozen storage. Also, Gonçalves and Ribeiro (2009) suggested that deep or more rapid freezing was more beneficial than slower freezing processes as it was more likely to produce smaller, less damaging ice crystals. A number of papers have also reported distinct advantages in freezing fish using liquid nitrogen (Lee 1982; Gonçalves & Ribeiro 2008a, 2009).

31.2.2 Formation of ice crystals

Freezing involves different factors in the conversion of water into ice: thermodynamic factors that define the position of the system under equilibrium conditions, and kinetic factors that describe the rates at which equilibrium might be approached. The freezing process includes two main stages: (1) the formation of ice crystals (nucleation) and (2) subsequent increase in crystal size (growth). The freezing point of a food may be described as *the temperature at which a minute crystal of ice exists in equilibrium with the surrounding water* (Fellows 2000; Zaritzky 2000).

The rate of ice crystal growth is controlled by the rate of heat transfer for the majority of the freezing plateau. The time taken for the temperature of a food to pass through the critical

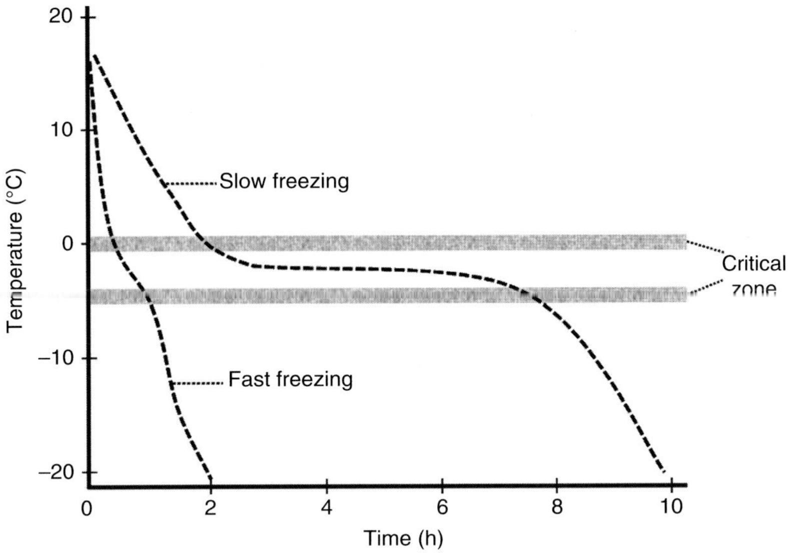

Fig. 31.2 Freezing: temperature changes of food through the critical zone.
Source: Adapted from Fellows (2000).

zone (Fig. 31.2), therefore, determines both the number and the size of ice crystals. The rate of mass transfer (of water molecules moving to the growing crystal and of solutes moving away from the crystal) does not control the rate of crystal growth except toward the end of the freezing period when solutes become more concentrated (Fellows 2000).

Ice formation involves a series of physicochemical modifications that decrease the food quality. The principal physical changes in frozen foods are freeze-cracking, moisture migration, recrystallization of ice, and drip loss during thawing (Zaritzky 2000), where crystals are formed and their size depends on whether the fish is frozen before (pre), during (in), or after (post) rigor mortis (stiffening after death), as well as on how speedy the freezing process is.

In prerigor fish, the water is only inside the muscle cells, and when the fish is frozen, the ice crystals form inside the cells. The ice crystals are mainly small. Large ice crystals form only if the freezing process is slow (e.g., in still air above −18°C). In in-rigor and prerigor fish, a small amount of water, however, remains outside the cells and thus the speed of freezing is significant for the formation of ice crystals. If the freezing is quick, small ice crystals form both inside and outside the cells. However, if freezing is slow, ice crystals first form outside the cells, resulting in an increase in salt concentration. The high salt concentration outside the cells extracts more water from the cells. When this happens, the ice crystals, which have already formed inside the cells, grow in size rather than allowing for the formation of new ice crystals. In in-rigor and postrigor fish, the speed at which the fish are frozen can, therefore, determine how large the ice crystals are and where in the muscle they form. This ranges from many small ice crystals both inside and outside the cells to only very few large ice crystals outside the cells.

Ice crystals form not only during freezing but also during cold storage. Here, not many new crystals are formed, but a recrystallization process takes place in which the surface of crystals, small ice crystals in particular, melts. This melting water subsequently freezes to the larger ice crystals, thereby leading to a greater number of large crystals at the expense of smaller ones. Recrystallization is especially rapid at fluctuating storage temperatures, even when the temperature fluctuates by only a few degrees.

Ice crystals consist of pure water and, therefore, substances that are normally dissolved in the muscle water have less water to be dissolved in as more water is converted into ice. These substances are primarily salts and enzymes, the concentrations of which become very high in the remaining "nonfreezable water." Many of these substances take part in processes that deteriorate the quality of frozen fish. High concentrations of salt, for instance, damage the proteins of the frozen fish. The speed of chemical reactions slows in step with the lowering of the temperature, whereas it accelerates as the concentration of the substances in the reaction increases. Thus, there still are chemical reactions in the "nonfreezable water" at the very low temperatures present in frozen fish due to the very high concentration levels. Reactions take place even at −30°C; however, they are very slow.

31.2.3 Rigor mortis

Immediately after death, the fish muscles are completely lax and the fish feels soft and elastic. When kept in ice, the fish goes into rigor mortis (stiffen) within a few hours. This is because the chemical energy present in the live fish is used up, causing a large part of the proteins in the muscle to bind into a network, which stiffens and hardens the muscle and thus the whole fish. After a couple of days in ice, the fish softens and becomes elastic as certain proteins in the network are broken down by enzymes.

Apart from the physical formation of ice leading to higher concentrations of substances, which in turn speeds up the destructive chemical processes, the ice crystals can directly and physically damage the structure of the muscle. Here, large ice crystals may cause much more damage than the smaller ones. Damage to membranes of muscle cells is particularly serious. This has to do with the fact that some of the cell salts and protein-disintegrating enzymes are in areas confined by membranes. When the membranes are ruptured, the salts are allowed to penetrate into other areas of the cell and increase the chemical processes there.

31.3 FACTORS THAT AFFECT FROZEN FISH QUALITY

31.3.1 Biological parameters

The quality of the frozen product is closely related to the freshness of the raw fish: the fresher the fish before freezing, the better it keeps in frozen storage. Whole fish 3–4 days old in ice show some signs of deterioration on thawing, and the staler the fish the more noticeable are the deteriorative changes that occur on storage. Much fish is now frozen and cold stored pending further processing; this development has been accelerated by the development of freezing at sea on board the fishing vessel (Ranken *et al.* 1997).

31.3.2 Conditions of catching and processing

The conditions of catching and processing are of paramount importance for the deterioration of frozen fish. According to Rehbein (2002), the quality is affected by:

1. *Method of catching, length of trawling time*: affected caloric, moisture, and protein contents, as well as some sensory attributes like muscle firmness.
2. *Stunning and killing procedure*: relationship between killing methods and quality—the more a fish struggles during catch and killing, the faster its pH falls after death. As low pH (~6.0) results in the denaturation of muscle proteins during the frozen storage of fish, the texture may become firm and dry if a stressed fish is processed.

3. *Rigor mortis*: fish can be processed either pre- or postrigor for high-quality products.
4. *Bleeding*: Bleeding the fish carefully prior to processing is beneficial to the quality for two reasons: color and appearance are improved by bleeding, as the presence of bloodstains leads to the devaluation of thawed fillets.
5. *Single and double freezing*: Frozen fillets are prepared directly after the catch on board freezer-trawlers or from fish that has been frozen at sea, transported to a land-based factory, thawed, and filleted. Depending on fish species and processing conditions, small or large differences in quality have been found between single and double-frozen fillets.
6. *Glazing and coating*: glazing protects frozen fillets against freezer burn and lipid oxidation, and edible coatings can reduce the rate of moisture and oxygen transfer between the fish flesh core and the surrounding atmosphere. Battered and breaded fish sticks or fillet portions have a longer shelf life than untreated fillet portions.
7. *Type of product*: the reason for the lower quality and stability is due to the contamination of muscle by blood, kidney, and other tissues that are rich in enzymes and hemeproteins.
8. *Storage conditions*: temperature and time of frozen storage have a large influence on shelf life of frozen stored fish and should be as low as economically possible, but at least less than −20°C.
9. *Freezing and thawing conditions*: it is well known that the speed of freezing has a great effect on the properties of the product. In industry, it is common practice to thaw fish in air or water. More sophisticated techniques, like vacuum thawing, or thawing by the use of microwave, dielectric, or electrical resistance heating, are faster but more expensive.

31.4 THE FREEZE (COLD) CHAIN

31.4.1 Cold storage

The route from catch to consumer is long (Fig. 31.3). Frozen fish products are usually processed in one of two ways: the fish is either frozen aboard the ship or on land. Aboard the ship, the fish is either frozen whole or as fillets. The frozen fillet is often used directly for sale, whereas the whole fish is thawed on land, processed, and then frozen again (double freezing). There is no tradition for double freezing in Danish waters, where the fish is already chilled (stored in ice) and brought on land and subsequently processed and frozen.

Industrial freezing of fish products is done at various stages of production. On board trawlers, whole fish are frozen in vertical plate freezers and fillets in horizontal plate freezers as interleaved fillets or standardized blocks (Bøknæs *et al.* 2001). On shore facilities, the fish are either frozen in blocks for further processing as fish fingers and portions or in retail sizes. The modern fish industry is equipped with quick freezing facilities such as air blast freezers and plate freezers. Fluidized bed and cryogenic freezers are used for individual quick freezing of shrimps and other shellfish. Quick freezing means that the time to pass from 0°C to −5°C must not normally exceed 5–10h and the warmest part of the product must be the temperature of the following frozen storage room. Due to the flow in production, it is preferable that the total freezing period is no more than 1–2h. Slow freezing over several days (fish stacked in insulated containers in silent air in a frozen store) can be avoided if the modern equipment is used.

Fish is particularly unstable during frozen storage. There are several reasons for this instability, some of which are more evident in certain fish species than in others. High-fat fish, for example, tend to have a short shelf life in frozen storage due to lipid oxidation, which produces a variety of off-flavors (Reid 1998). Because of the instability of frozen

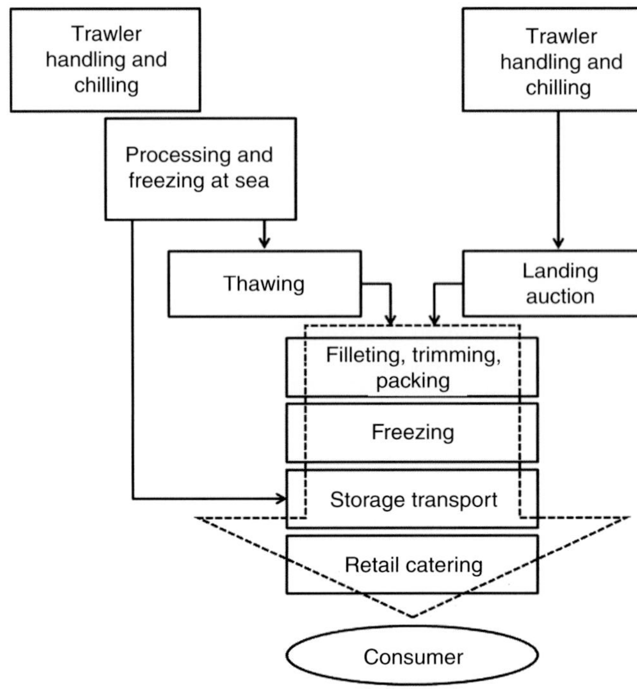

Fig. 31.3 The freeze chain. The fish is either frozen aboard ship or on land.

Table 31.3 Storage stability of frozen fish in days.

Species	−23°C (−10°F)	−18°C (0°F)	−12°C (10°F)
Trout	300	260	110
Cod fillet	300	210	90
Ocean perch	300	220	120
Mackerel	110	80	50
Halibut	350	260	170

Source: From Reid (1998).

fish, it is often recommended that very low storage temperatures be employed. Table 31.3 illustrates the relative frozen storage stability of several species of fish, typical lean fish, fatty fish, and gadoid fish.

The normally recommended air temperature for the storage of frozen foods is −18°C to −20°C. For frozen fish and a few other sensitive foods, the temperature should be much lower, and −30°C is recommended. Even at this temperature, fish do not keep indefinitely. Microbial action ceases below about −10°C, but chemical reactions leading to irreversible changes in odor, flavor, and appearance will slowly continue (Ranken *et al.* 1997). When fish are stored at temperatures above −30°C, deterioration is more rapid and storage life is reduced. Temperatures in distribution vehicles, secondary or distribution cold stores, frozen food rooms, and display cabinets, are all likely to be above −30°C. In some retail display cabinets, they may even be above the recommended level of −18°C. It is important to remember that cold stores are designed to hold frozen goods at the specified temperatures, not to freeze the goods down to those temperatures.

According to the Code of Practice for Fish and Fishery Products (FAO/WHO 2009), the frozen storage facility should be capable of maintaining the temperature of fish at or lower than −18°C, and with minimal temperature fluctuations. Severe fluctuations in storage temperature (more than 3°C) have to be avoided (especially for quick-frozen coated fish products). For frozen fish stored in a trawler, the same control should be done.

It has been suggested that the storage temperature on lean demersal fish should be between −24°C and −30°C. For oily fish, storage temperatures of −30°C to −70°C have been recommended in order to reduce the rate of oxidative changes. However, low-temperature storage is not currently employed, as the cost to the consumer for the final product is too high (Hedges 2002).

31.4.2 Thawing

Ideally frozen fish should be thawed immediately after removal from cold storage. The simplest methods are thawing in still air by simply leaving the frozen product out overnight at room temperature, or thawing whole fish in water. If fillets are thawed in water, they may become waterlogged and lose much of their flavor. For air thawing, the air temperature should not exceed 20°C to minimize bacterial growth on the surface, but very slow thawing must be avoided since bacterial growth may nonetheless cause the outer layers to spoil before the center is fully thawed. Thawing may be accelerated by mechanical means—air blast, heated water, or vacuum methods—or electrically by dielectric resistance or microwave heating. The electrical methods are usually restricted to applications where the size and shape of the units to be thawed are constant, in other words to frozen fish blocks (Ranken *et al.* 1997).

31.4.3 Frozen fish transport

Frozen fish delivered to a destination where they are to be sold immediately are likely to be consumed within a few hours and no harm is done if they are partially thawed on arrival at their destination. The frozen fish may in fact be carried in uninsulated containers depending on how long the journey takes. Enclosed vehicles, however, should be used or at least a cover provided to protect the fish from direct sunlight. An insulated vehicle will be required for long journeys depending on the initial temperature of the fish, whether the vehicle is fully or partly loaded, the size of the load, the insulation quality and thickness, the degree of air ingress, and the local climatic conditions. A local trial will ascertain the maximum range attainable (Johnston *et al.* 1994).

Frozen fish that are to be transferred to other cold stores must be transported in an insulated vehicle preferably with some form of refrigeration equipment to maintain the air space at a temperature of approximately −20°C. The following lists refrigeration methods that may be used: (1) Mechanical refrigeration using either wall coolers or forced convection coolers blowing air throughout the storage space. In some cases, a jacketed system for distributing the air is employed. This is the most common system. (2) Rechargeable eutectic plates. (3) Solid or liquid carbon dioxide or liquid nitrogen can be used with a total loss system (Johnston *et al.* 1994; Gonçalves & Blaha 2010).

During all the transportation steps of the end product, deep-frozen conditions should be maintained at −18°C (maximum fluctuation ±3°C) until the final destination of the product is reached. Cleanliness and suitability of the transport vehicle to carry frozen food products should be examined. Use of temperature-recording devices with the shipment is recommended (Johnston *et al.* 1994; FAO/WHO 2009).

Vehicles should be designed and constructed (1) such that walls, floors, and ceilings, where appropriate, are made of a suitable corrosion-resistant material with smooth, nonabsorbent surfaces. Floors should be adequately drained, where appropriate, (2) with chilling equipment to maintain chilled fish or shellfish during transportation to a temperature as close as possible to 0°C or, for frozen fish, shellfish, and their products, to maintain a temperature of −18°C or less (except for brine frozen fish intended for canning, which may be transported at −9°C or less); (3) so that live fish and shellfish are transported at temperatures tolerable for the species; (4) to provide the fish or shellfish with protection against contamination, exposure to extreme temperatures, and the drying effects of the sun or wind; and (5) to permit the free flow of chilled (Johnston *et al.* 1994; FAO/WHO 2009).

31.4.4 Retail display equipment

Retail display is one of the "weak" links in the frozen food cold chain, mainly due to the contrasting purposes that retail display cabinets have to serve, that is, persuading the customer to buy the product, while at the same time preserving it adequately. For merchandizing purposes, the product must be clearly visible and easy to reach in order to tempt potential customers; however, these features tend toward product temperature fluctuation (i.e., in excess of the recommended −18°C). This is known to be the primary cause of a loss of quality and safety in frozen foods. The best way to protect the product from temperature fluctuations is to keep it as far as possible from the shop environment, and from all possible heat sources (ambient air, lighting, etc.); but this means keeping it out of sight of customers. Matching these two conflicting needs is the main technological challenge facing display cabinet manufacturers (Cortella 2000; Gonçalves & Blaha 2010).

The main characteristics of a good frozen food retail display cabinet can be summarized as follows: (1) it has to guarantee good product temperature control, whatever the external ambient conditions; complying with the standards is crucial in this matter. (2) It has to prove an efficient seller, so the foodstuff must be visible and easily accessible to the customer. (3) It has to be cost-effective, not only in terms of the initial investment, but also in running costs; therefore, its energy consumption is extremely important, but so is easy access for loading, since this reduces the staff hours required to restock the shelves (Gonçalves & Blaha 2010).

While these are the main features of a good frozen food retail display cabinet, nobody—from the manufacturer to the shop managers to their employees—must forget that it is designed neither to freeze food, nor to reduce its temperature: Its purpose is simply to maintain the frozen food at the right temperature. Naturally, the environment where the cabinet is installed plays a major part in establishing how it will behave: Air conditioning and exposure to warm air streams and lighting must be carefully evaluated to guarantee best use of the cabinet (Cortella 2000).

31.4.5 Temperature monitoring

Refrigeration equipment is built to function for long periods without attention; however, there are many events apart from breakdown that can affect temperature control. The defrost cycles need attention to ensure they are at the correct frequency, and loading of food into refrigerated systems is often crucial to its operation and proper air flow. Air temperature monitoring can indicate whether refrigerated equipment is functioning correctly and is being operated correctly, even though it may be more difficult to extrapolate food temperatures. In some circumstances, air temperature monitoring is not possible and product temperature or product simulant temperature is required (Woolfe 2000).

There are an enormous number of different temperature monitoring systems available commercially, from a simple thermometer to a fully computerized system linked to a local refrigeration system or even central control system. The choice of system will depend on exactly the amount of detail the operator requires and the cost at which this information is provided. If the monitoring system is to provide detailed information on the operation of a system linked with other reactive management systems, then obviously a more elaborate and complex system is required. This may include a large number of sensors to enable a very complete picture of the temperature distribution within a refrigerated system to be obtained. It may also include other information such as defrost cycles, compressor and expansion valve pressures, door openings, and energy consumption, and may be linked to an alarm system (and even telephone) stock keeping and batch codes of products. On the other hand, if monitoring is being carried out only to ensure that food is being kept within certain temperatures as a critical control point, then the amount of information that is collected may be reduced (Gonçalves & Blaha 2010).

Temperature monitoring has been discussed in terms of displaying temperature readings of the surrounding air or of the food or simulated food itself. However, it is possible to use a physicochemical mechanism and a resulting color change to display (a) a current temperature, (b) the crossing of a threshold temperature, or (c) an integration of the temperature and the time of exposure to temperature after activation. Such devices are called temperature indicators (TIs) in the first two cases or time–temperature indicators (TTIs) in the last case (Selman 1990; Selman & Ballantyne 1988; Woolfe 2000).

The main causes of loss of storage life are fluctuating temperatures and the type of packaging used. Other factors, including type of raw material, prefreezing treatments, and processing conditions, have a great contribution on frozen seafood quality. Temperature fluctuation has a cumulative effect on food quality and the proportion of PSL or HQL lost can be found by integrating losses over time. Time–temperature tolerance (TTT) and product-processing-packaging (PPP) concepts are used to monitor and control the effects of temperature fluctuations on frozen food quality during production, distribution, and storage (Bogh-Sorensen 1984; Olsson 1984; Evans & James 1993; Ranken *et al.* 1997; Fellows 2000; Woolfe 2000).

Colored indicators are being developed to (1) show the temperature of food (e.g., liquid crystal coatings that change color with storage temperature); (2) indicate temperature abuse (e.g., wax melts and releases a colored dye when an unacceptable increase in temperature occurs); (3) integrate the time–temperature combination that a food has received after packaging; and (4) give an indication of the remaining shelf life (Fig. 31.4) (Gonçalves & Blaha 2010).

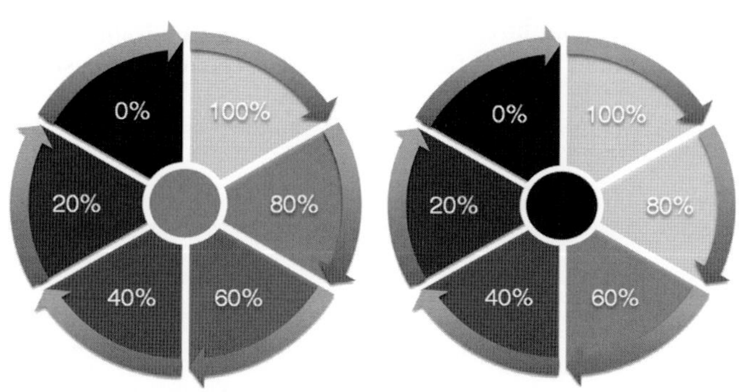

Fig. 31.4 Time–temperature integrator (indicator based on the color change of the inner circle: the percentages show the equivalence of color with the remaining life).

The indicators are normally integrated onto a packaging material that can be attached to the food packaging or the outside of the surrounding or bulk packaging, and can follow the food throughout the chill chain. The type of information that can be provided is one or more of the following: (1) reject or accept on the basis of a color change, (2) temperature abuse above a threshold temperature, (3) partial time–temperature history above a threshold temperature, and (4) full time–temperature history linked to shelf life. Over 100 patents have been filed on processes that could be used as a basis for indicators. These include changes with temperature based on melting-point temperature, enzyme reaction, polymerization, electrochemical corrosion, and liquid crystals. The result of the change is usually a color difference, which can be represented as a static change or moving band. Available temperature and time–temperature devices have been reviewed (Selman 1990; Selman & Ballantyne 1988; Woolfe 2000; Gonçalves & Blaha 2010).

31.5 INDICATORS OF DETERIORATION IN FROZEN FISH

For many years, attempts have been made to determine the shelf life of frozen fish by a number of chemical, biochemical, and physical methods; however, till date the gold standard is sensory assessment by a well-trained panel. Some sensory characteristics of spoilage of frozen fish, together with the reactions they are based on, are summarized in Table 31.4 (Rehbein 2002).

During chilling (i.e., temperatures between 0°C and 5°C), the quality of the fish changes due to oxidation, especially of its fat (oxidative rancidity), self-digestion brought on by the fish's own enzymes (autolysis), and bacterial growth. Normally, the waste products of the bacteria are what cause the fish to smell bad, and bacterial decomposition is what causes its

Table 31.4 Sensory description of thawed cooked fish flesh, and underlying chemical or physical reaction.

Aspect of quality	Underlying chemical or physical reaction
Appearance	
Dry	Protein denaturation
Gaping	Breakdown of connective tissue
Freezer burn	Sublimation of ice
Yellowish	Lipid oxidation; formation of FA
Odor	
Cold-store odor (cardboard)	Formation of carbonyls by lipid oxidation
Sour	Formation of carbonyls by lipid oxidation
Rancid	Formation of carbonyls by lipid oxidation
Amine	TMAO degradation into DMA and TMA
Flavor	
Cold-store flavor	Formation of carbonyls by lipid oxidation
Sour	Formation of carbonyls by lipid oxidation
Rancid	Formation of carbonyls by lipid oxidation
Soapy	Lipolysis
Amine	TMAO degradation into DMA and TMA
Texture	
Dry	Protein denaturation, loss of muscle structure
Firm, tough	Reaction between FA and protein
Soft, mushy	Proteolysis

Source: From Rehbein (2002).

putrefaction. Freezing and cold storage hinders bacterial growth whereas oxidation and autolysis continue, albeit these processes are slowed down when the temperature is below the freezing point. The quality of frozen fish moreover deteriorates because of changes in the structure of proteins (protein denaturation).

31.5.1 The impact of freezing on microorganisms' growth

Frozen foods have an excellent safety record and freezing has never been reported to be the cause of food poisoning. Temperature is one of the major factors affecting microbiological growth. The great advantage of freezing is that microorganisms do not grow in foods when the temperature is −10°C or colder. However, it should not be overlooked that although freezing kills some microorganisms, it does not eliminate pathogenic microorganisms nor microbial toxins present in the food product prior to freezing.

In favorable environments, growth of bacteria normally proceeds as shown in Fig. 31.5, where the logarithm to the number of bacteria is the vertical axis and time is the horizontal axis. First, there is a period of adjustment, the lag phase, where the bacteria are more or less inactive, that is, they do not multiply. The length of the lag time depends on many factors, for example, temperature, pH, inhibitors in the substrate (the food), etc. (Bogh-Sorensen 2000; James 2002). After the lag phase, growth begins and reaches a phase of exponential growth. In the exponential growth phase, the number of bacteria may rise very quickly, often expressed by the generation time, that is, the time required to double the number of bacteria. The generation time varies between bacteria, and depends on several factors, in particular the temperature. The two remaining phases are of less interest (James 2002).

Inactivation of microorganisms caused by freezing and thawing may take place in three ways: (1) When food is cooled so that vegetative microorganisms are kept at temperatures below their minimum for growth, some loss of viability can be expected. (2) Inactivation of microorganisms takes place during the freezing process. The response of microorganisms to freezing is mainly studied in thawed cultures; therefore, the thawing method is a very important aspect that must always be taken into consideration. (3) Finally, inactivation of microorganisms may take place during storage, depending on storage time and temperature.

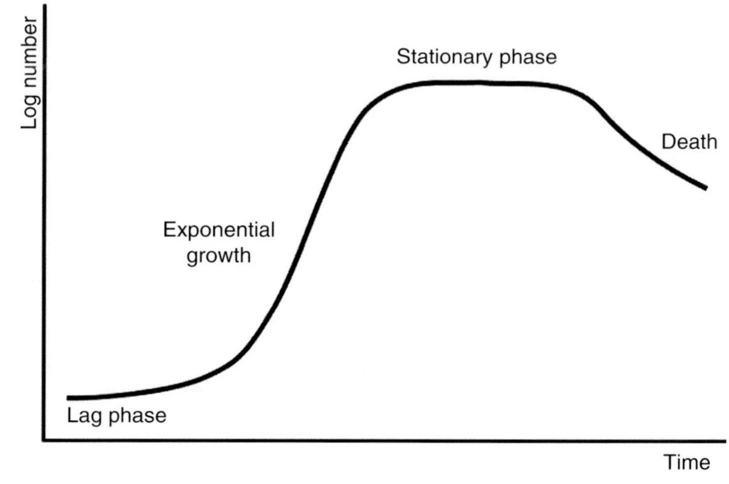

Fig. 31.5 A typical bacterial growth curve.

During the freezing process, the product temperature is lowered and most water in the food is transformed into ice crystals. With decreasing temperature, the liquid phase becomes more and more concentrated. As the volume of ice is about 10% larger than the volume of water, the internal pressure in the food may rise to 10 bar or more, especially during very rapid freezing. Also, during freezing some bacteria are killed, some are in the lag phase, while others may be freeze-damaged (nonlethally injured, sublethally injured). The analytical technique used in microbiological analysis of frozen foods must be modified in order to detect and/or enumerate freeze-damaged cells (Bogh-Sorensen 2000; Jay *et al.* 2005).

Although freezing is a safe preservation method, the food safety issue is still very important. The cold shock response (CSR) could have some implications for frozen foods; often the response is the appearance of cold shock proteins (CSP). The CSR of several microorganisms is being investigated, but comparatively little is known about the response of pathogens to freezing (Kerr 1997). It has been suggested that exposure to stress, for example, freezing, resulting in sublethal injury could render the organism more resistant to the effect of heat, so that heating may not reliably destroy all the bacteria present. Freezing could also result in an increase in the acid tolerance of bacteria, including food-borne pathogens. Another concern regarding the CSR of pathogens is that exposure to low temperatures may have a direct effect on virulence. Kerr (1997) concluded that there is little data in this field, and that further studies of the CSR are clearly required (Bogh-Sorensen 2000).

31.5.2 The impact of freezing on biochemical attributes

As the freezing process converts a large proportion of liquid water into ice, it also concentrates the remaining solution. Enzymes increase the possibility of water being in contact with different substrates. The most common chemical changes that can proceed during freezing and frozen storage are lipid oxidation, enzymatic browning, flavor deterioration, protein denaturation, and degradation of pigments and vitamins. Freezing can give unusual effects on chemical reactions. The temperature and concentration of the reactants in the unfrozen phase (freeze concentration effects) are the main factors responsible for changes in the reaction rates of enzymatic and nonenzymatic reactions during freezing. The formation of ice crystals can produce the release of enclosed contents in food tissues (enzymes and chemical substances) affecting the product during freezing and storage, leading to quality deterioration.

31.5.2.1 Fat and fat decomposition

Fish contain fat, and changes in the fat fraction of fish greatly influence the taste of the frozen product. The amount of fat in the muscle of fish varies from species to species, and great variation may occur even within the same species, for example, in the herring where the content of fat may vary from 1% to 22% in 1 year. The fat content of fish can vary from 0.2% to around 25%.

The fat or lipid content of fish is found in the membranes of cells and as lipid depots just under the skin, in the belly flap, and in the connective tissue. Membrane lipids (primarily phospholipids) make up less than 1% of the total lipid content. Membrane lipids and the fat in lipid depots (primarily triglycerides) consist of long fatty acids. During freezing, the fat decomposes and is converted into substances that taste unpleasant. The changes are due to oxidation or enzymatic degradation.

Fish fat has a particularly high content of polyunsaturated fatty acids, especially susceptible to oxidation, which produces aldehydes and ketones that have a rancid flavor and odor.

Enzymatic and nonenzymatic pathways can initiate lipid oxidation. One of the enzymes that is considered important in lipid oxidation is lipoxygenase, which catalyses the addition of molecular oxygen to a cic-cis-4 pentadiene containing unsaturated fatty acid, releasing a stereo-specific conjugated diene hydroperoxy fatty acid product.

31.5.2.2 Oxidative rancidity

The development of rancidity can still occur quite rapidly even in the frozen state (Ortiz & Bello 1992). The rate of rancidity development once frozen may also be influenced by a number of enzymatic and nonenzymatic events occurring prior to freezing (Haard 1994). In the living muscle, the effects of agents that accelerate lipid oxidation (prooxidants) are balanced by those that inhibit oxidation (antioxidants). Postmortem, however, this balance is disrupted and stimulates the initiation of lipid oxidation. In fresh fillets stored at chill temperature, off-flavor development is less problematic since microbial spoilage will render fish inedible before rancidity has developed to any great extent. However, in the frozen state, where microbial-induced changes are effectively arrested, rancidity is most likely to become the process that limits the shelf life of fatty fish in particular (Harris & Tall 1994).

Oxidation takes place on the surface of the fish, and smaller fish such as mackerel and herring with a relatively large surface go rancid more rapidly than other fish. The cleaning and cutting into filleting of fish increases the total surface area, and the fish therefore become more rancid. Oxidation often takes place very soon after the fish has been killed, especially in the fat layer just under the skin. Moreover, the process is exacerbated by salt and metals such as iron from, for example, the blood of the fish. In codfish, oxidative lipid degradation may give poor flavor, namely, the very characteristic cold-store flavor (due to the substance cis-4-heptenal). Thus, the occurrence of lipid oxidation in frozen foods leads to loss of quality: flavor, appearance, nutritional value, and protein functionality. Decomposition of hydroperoxides of fatty acids to aldehydes and ketones is responsible for the characteristic flavors and aromas known as rancidity.

31.5.2.3 Can oxidative rancidity be avoided?

Oxidative rancidity in frozen fish can be avoided, or at least counteracted, in several ways. First of all, the fish must be frozen as quickly as possible after death, however, allowing time for proper bleeding of the fish in ice water. As mentioned, oxidative rancidity is an oxygen-dependent surface process; glazing and vacuum packing, which reduce the degree of contact with oxygen, are therefore advantageous.

31.5.2.4 Protein changes

In lean fish, protein changes have the greatest influence on the shelf life of the frozen and cold-stored fish. The main causes of freeze-induced damage to proteins are ice formation and recrystallization, dehydration, salt concentration, oxidation, changes in lipid groups, and the release of certain cellular metabolites. Freeze-induced protein denaturation and related functionality losses are commonly observed in frozen fish, meat and poultry. Losses in functional properties of proteins are commonly analyzed by comparing the WHC, viscosity, gelation, emulsification, foaming, and whipping properties. Freeze-induced protein denaturation is often attributed to the formation of ice crystals, dehydration, and the concentration of solutes in the tissue or protein solution. Freezing has an important effect in decreasing the WHC of muscle systems on thawing. This decrease occurs during freezing,

Table 31.5 Enzymes tested as an indicator of quality of frozen fish.

Enzyme	Activity
Myofibrillar ATPase	Drop in activity was not correlated with protein solubility or toughness
Aldolase	Decline in soluble enzyme activity observed for cod and haddock stored at −14°C
Malic enzyme (ME)	Latent form of ME showed a decrease in activity when fish was stored for 5 months at −7°C, but no change at −29°C
Alpha-glycerophosphate dehydrogenase	The test, determination of activity constant from a double reciprocal plot of reaction velocity, was very time consuming
Sarcoplasmic ATPase	Rapid decrease during frozen storage
Cytochrome oxidase	Activity decreased during frozen storage of cod and it was possible to distinguish between frozen storage temperatures −9°C, −20°C, and −40°C
Acid phosphatase	Activity decreased during frozen storage of cod at −30°C
50-Nucleotidase	Activity decreased during frozen storage of cod at −30°C
Phospholipase	Activity increased in the 1st weeks of frozen storage of cod at −30°C, then it declined to reach the original level

Source: From Rehbein (2002).

because water–protein associations are replaced by protein–protein associations or other interactions.

When thawing, plenty of water drains from the fish (drip loss), and the fish becomes less juicy and less suitable for minced fish meat products. Protein changes also affect flavor. Proteins may, in their natural structure, bind undesirable flavors that are released during freezing. Oxidative reactions such as lipid peroxidation are also involved in the deterioration of functional attributes of muscle proteins during frozen storage (Zaritzky 2000).

Protein denaturation in frozen fish should be accompanied by a decrease of enzyme activities. Therefore, a number of enzymes (Table 31.5) have been tested for a regular disappearance of activity during frozen storage. Recently, the activity of cytochrome oxidase, an enzyme located in the inner mitochondrial membrane, was measured in the muscle of fresh and frozen cod (*Gadus morhua*). Activity was enhanced by freezing, and then declined during the frozen storage depending on the temperature (Godiksen & Jessen 2001; Rehbein 2002). The authors conclude that cytochrome b may have potential as an indicator of frozen storage conditions. Release of particle-bound enzymes due to freezing and thawing has been used for differentiation between fresh and frozen fish. Mitochondrial (3-Hydroxyacyl-CoA dehydrogenase [HADH], Glutamate-Oxaloacetate transaminase [GOT], malic enzyme [ME], fumarase) as well as lysosomal enzymes (acid alpha glucosidase, beta-*N*-acetylglucosaminidase) were found to be suitable depending on the fish species.

In frozen gadoid fish, the TMAOase catalyzed the production of formaldehyde (FA), and dimethylamine (DMA) from trimethylamine oxide (TMAO) occurs at temperatures down to −30°C. Both metabolites, which are formed in equimolar amounts from TMAO, can be used as parameters for the deterioration of frozen fish, as the binding of FA to proteins was found to result in texture toughening (Rehbein 2002). During TMAO degradation, small amounts of FA are formed, which apparently bind the proteins and thus make them less able to bind water.

In frozen fish, the degradation of adenosine triphosphate (ATP) and related nucleotides occurs mainly in the temperature range between −5°C and −15°C, and was found to be very slow at lower temperatures. Therefore, the determination of adenosine monophosphate (AMP), inosine monophosphate (IMP), or hypoxanthine has not been applied very much for the quality measurement of frozen fish (Rehbein 2002).

Changes of proteins in fish meat during freezing and cold storage may be also examined and monitored via proteome analysis, a relatively new technique for examination of the protein composition in cells and tissues, such as fish muscle.

31.5.2.5 Can protein changes be avoided?

It is relatively simple to hinder protein changes in frozen fish. Processing the raw material shortly after catching, quick freezing, and cold storage at a low and stable temperature will reduce protein denaturation significantly.

31.5.3 The impact of freezing on sensory attributes

While a number of characteristics affect the overall quality and acceptability of both fresh and frozen meats, tenderness is the major characteristic of eating quality because it determines the ease with which meat can be chewed and swallowed. The tenderness of meat is affected by both chilling/freezing and storage. Under the proper conditions, tenderness is well maintained throughout the chilled/frozen storage life, but improper chilling/freezing can produce severe toughening and meat of poor eating quality (James 2002).

During the freezing and subsequent frozen storage of certain fish species, changes occur that result in the deterioration of the textural quality of the cooked fillet. The textural change has been described as a tendency to express liquid on initial compression in the mouth, with the remaining material being often hard, dry, and fibrous.

The textural changes that occur have been attributed to changes in the myofibrils. Many observations of myofibrillar change have been made; for example, work by Jarenback and Liljemark (1975) showed that during the frozen deterioration of cod (*Gadus morhua*), a decrease in the dimensions of the myosin lattice was observed, coupled with disturbances to the regular hexagonal lattice spacing (based on a comparison of thawed, deteriorated muscle with fresh cod). It was also noted that the myofibrils within a thawed, frozen, deteriorated muscle fiber were pushed closer together, with a commensurate disappearance of the sarcoplasmic reticulum vesicles lying between the myofibrils. Similar effects have also been reported for the *Alaska pollock* muscle that had been subjected to −20°C frozen storage for 2 months. On thawing, the myofibrils within each muscle cell did not recover their original diameter, and also appeared closer together. Further frozen storage (i.e., 12 months at −20°C) rendered the myofibrils even less able to effect a recovery of their original diameters on thawing, and a substantial loss of the order within individual myofibrils was also observed (Hedges & Nielsen 2000).

Although textural change may be significant during the storage of cod and *Alaska pollock*, flavor changes may also occur. One flavor change is often described as the development of cold-store flavor and has been attributed to the development of hept-cis-4-enal, a product of oxidation of specific membrane phospholipids (McGill *et al.* 1977). It has also been suggested that starving cod prior to slaughter reduces the development of off-flavors (Ross & Love 1979; Hedges & Nielsen, 2000).

The appearance of meat at its point of sale is the most important quality attribute governing its purchase. The ratio of fat to lean and the amount of marbled fat are important appearance factors and another is the color of the meat. The changes in the color of the muscle and blood pigments (myoglobin and hemoglobin, respectively) determine the attractiveness of fresh red fish meat, which in turn influences the consumers' acceptance of meat products. Consumers prefer bright red fresh meats, brown or gray-colored cooked meats, and pink cured meats. The color red is more stable at lower temperatures because the rate of oxidation of the pigment decreases. The color of frozen meat varies with the rate of freezing. Those frozen at −34°C to −40°C had the most desirable color and those frozen at −73°C to −87°C tended to be pale (James 2002). At low temperatures, the solubility of oxygen is greater and oxygen-consuming reactions are slowed down. There is a greater penetration of oxygen into the meat and the meat is redder than at high temperatures.

31.5.3.1 How can changes in sensory quality of frozen fish be seen?

The quality of a frozen fish depends on its initial quality and on how it is handled. A white fish such as a cod is cleaned and bled immediately after being killed. This means that the guts are removed and the blood is allowed to drain from the muscles. In an entirely fresh cod, the blood has no effect on quality. However, when the fish is frozen and kept in cold storage, the blood may give the fish a metallic tinge and later contribute to its going rancid. When a lean fish such as cod goes rancid, it will often taste somewhat like wet cardboard (cold-storage flavor). Fatty fish such as salmon and herring do not acquire a cold-storage flavor but will acquire a rancid flavor of cod liver oil if not properly protected in airtight packaging during storage.

31.5.3.2 Differences in the freezing quality of fish populations belonging to the same fish species

Different batches of cold-stored cod can be of very different quality even though the fish were handled identically and stored in the same period of time. This could be due to differences in the composition of the fish, which changes, for example, depending on the time of season and spawning. Researchers believe the reason lies in both hereditary and environmental conditions.

- *Hereditary differences* can occur when a cod in a certain location acts like an individual population, not particularly interested in mixing with cod from another location (belonging to another population). Hereditary differences will be seen between such populations.
- *Environmental differences* can occur when cod from a population stay in two different locations, where conditions differ so that the fish develop differently.

The interplay between the hereditary genes of the individual and its environment defines its "appearance" (size, shape, chemical composition, etc.), in other words it defines the phenotype of the individual fish. A fish product manufactured from several different phenotypes may vary in quality even though the fish were handled identically aboard the fishing boat and at the factory.

If cod is handled with care during and after catching, if it is frozen as quickly as possible and stored at a stable temperature of −30°C (or below), there will be no difference in taste between a thawed and a fresh fish for the first month. After 3 months, rancidity sets in, cold-storage flavor starts to develop, and the taste characteristics of the specific species gradually disappear completely. However, the frozen fish may be suitable for consumption for up to 1 year when kept in cold storage at a low and stable temperature.

Fatty fish, especially large fish like salmon, are more suitable for freezing than cod and at low temperatures may be suitable for consumption for up to 1.5 years (Sørensen *et al.* 1996).

Also here, the taste of the fish becomes more and more neutral; however, off-flavors will only develop after very long storage times, and rancidity occurs only if the fish has not been packed properly.

Numerous consumer studies show that retailed frozen fish taste significantly poorer than fresh fish. As a consumer, you can never be sure of the quality of the fish you buy—there is nothing on the packet to indicate the quality of the fish. Packets are labeled with the packaging date, but say nothing about when the fish was caught, the temperature at which it has been stored, for how long it has been stored as raw material, or if it has been frozen and thawed several times, etc.

The consumer evaluates the eating quality of a fish through the senses—by the look, smell, texture, and taste of the fish—using sensory assessment. As only the consumer can determine what he or she likes, such an evaluation will be subjective. Sensory assessment may also be applied professionally, by individuals who have been trained in the assessment of specific properties of fish. In this way, sensory assessment becomes objective. To handle the quality of fish through the frozen fish chain, the quality index method (QIM) has proven to be a good tool for frozen cod (Warm *et al.* 1998; Jensen *et al.* 2001).

The QIM for frozen fish can be looked upon as a simple profile. Tables 31.6 through 31.9 show the schemes developed for frozen cod from raw material to product.

For frozen fish, the theoretical demerit line (see Chapter 29) has been more difficult to determine than the one for fresh fish because of the extended storage time dependent on the freezing method, temperature of frozen storage, temperature fluctuations, etc. The approach for establishing a quality index for frozen fish (cod) has, therefore, been different from the method used for fresh fish. The decision of which parameters to choose in the QIM schemes for frozen cod was based on parameters for fresh cod showing the largest variation over time. The goal with the development process was to describe each score in a way so that people with little training in sensory assessment of fish understood the description. This required a

Table 31.6 QIM scheme for thawed whole cod.

Quality parameter	Characteristics
Texture	0: Firm texture. Stiff and firm to finger touch 1: Elastic to finger touch. Marks disappear after a few seconds 2: Fairly soft and plastic flesh. Marks do not disappear 3: Very soft. The flesh is easily penetrated on pressure
Remains of guts	0: No remains in the belly 1: Few remains in the belly 2: Many remains in the belly
Shape of fish	0: Normal round shape as freshly caught cod 1: Fairly normal round shape with few marks after freezing 2: Very mechanically damaged during freezing. Very deformed
Marks from fishing tackles/catch handling	0: No marks 1: A few small marks 2: Many marks 3: Very many/big marks
Odor	0: Fresh marine and seaweedy 1: Neutral 2: Slightly sour and metallic 3: Strong sour and metallic. Painty
Appearance	0: Iridescent or opalescent. Bright, shining. No bleaching 1: Slight bleaching 2: Dull and very bleached, no iridescent. Freeze dried
Flesh color in open spaces	0: Open surfaces white and blood in throat cut red 1: Open surfaces gray or slightly yellow 2: Open surfaces yellow or brown 3: Open surfaces very yellow and brown. Milky surfaces as freeze dried

Table 31.7 QIM scheme for thawed fillet.

Quality parameter	Characteristics
Texture	0: Firm and stiff texture. No wateriness 1: Slightly soft, initial wateriness 2: Soft, wateriness noticeable 3: Very soft and pronounced wateriness
Odor	0: Neutral 1: Slightly sour off-odor 2: Very sour off-odor
Color	0: Plain white 1: Grayish 2: Gray, starting yellow maybe slightly red 3: Either yellow or very red. Milky surfaces as freeze dried
Blood stains	0: No stains 1: A single stain (diameter less than 3 mm) 2: Single small stains (1–2 with a diameter under 5 mm) 3: Very discolored from many stains or totally red
Gapping	0: No gapping, coherent 1: Slight gapping 2: Gapping noticeable, disrupted 3: Gapping pronounced, disrupted
Parasites	0: No parasites 1: One parasite 2: More than one parasite

Table 31.8 QIM scheme for cooked fillet from thawed cod.

Quality parameter	Characteristics
Odor	0: Sweet, marine, and seaweedy 1: Loss of odor 2: Neutral 3: Slightly cold-storage odor (cardboard), slightly citric, amine 4: Cold-storage odor (cardboard), citric, strong amine
Color	0: White and opalescent 1: Loss of whiteness 2: Grayish, one small blood stain 3: Slightly yellow, a few small blood stains 4: Light brown. Discolored of blood
Flavor	0: Sweet, marine, and seaweedy 1: Loss of taste, slightly sweet 2: Neutral 3: Slightly cold-storage flavor (cardboard), insipid, slightly as soap, slightly citric, slightly amine 4: Cold-storage flavor (cardboard), dry fish, soap, citric, strong amine
Texture	0: Very succulent, flaky, and coherent also succulent after chewing several times 1: Succulent, flaky, and coherent. Succulent after chewing several times 2: Initial feeling is succulent, watery after first chewing. Dry and tough during following chewings 3: Dry, fibrous, and/or tough. Dry and tough from the first chewing 4: Very dry, fibrous, and/or tough. Very tough from the first chewing

Table 31.9 QIM score—use of product.

Category	**Examples of characteristics**	**Use of product**
High quality: Index 0–4	• Firm texture • Neutral odor • White color • No blood stains • Minimum gapping • No parasites	• Many possibilities for use • Luxury products (loins, tails, etc.)
Medium quality: Quality index 5–12	• A little soft and watery • Somewhat sour odor • Grayish color • Single blood stains • Medium gapping • One parasite	• Breaded products • Several cuttings • Block products
Low quality: Quality index 13–16	• Very soft, very dry • Very sour odor • Yellowish, reddish • Big blood stains • Much gapping • More than one parasite	• Very few possibilities for use • Block products

difference between each score. Ranking the descriptions for each parameter and giving numbers to the ranks in succession from 0 to 4, the indexing method has been established for whole fish, fillets, and cooked fillets. The grading schemes have been evaluated and further developed in a great number of trials by both trained and untrained panel members using frozen cod from the Baltic Sea, Iceland, and Russia. The different parameters have been assessed and the parameters selected to give a picture of the total sensory quality.

A manual containing the total plan for evaluation, explanation of the evaluation terms, and color slides illustrating the different levels of quality of frozen cod and cod fillet has been produced for industrial use.

The three QIM schemes can be used effectively through the chain from fisherman to consumer. It will be possible to check out critical points in a production line by the QIM. When the quality index differs significantly from whole fish to fillet, a decision can be made to change the production conditions. A continuous problem about QIM is the setup of various quality categories. For thawed fillets, three categories were suggested (Warm 2001), as shown in Table 31.9.

An advantage of the QIM is that the limits for quality categories can be set individually depending on the product.

31.5.4 The impact of freezing on physical attributes

Weight may be lost by dehydration or due to physical damage of the fish during the freezing process. Physical damage may be due to damage during freezing, which results in small pieces being broken off; this is likely, for instance, in freezers where the product is fluidized by the cooling air. The other form of physical damage encountered during the freezing process is due to fish adhering to trays or conveyor belts. If the weight loss on releasing fish from trays is excessive, the trays may be sprayed on the underside with water to assist release. Fish frozen in continuous freezers with stainless steel link or mesh belts may stiffer weight losses due to small particles being trapped in the belt. Losses due to physical damage in a freezer should be small and need not be more than about 1% if the freezer and freezing process is suitable for the product (Johnston *et al.* 1994; Campañone *et al.* 2001).

Weight loss due to dehydration in a freezer depends on a number of factors, and the weight losses in air blast freezers give rise to the greatest controversy. Weight loss due to dehydration will depend on the (1) type of freezer, (2) freezing time, (3) type of product (food surface), (4) air velocity, (5) freezer operating conditions, and (6) during storage (room temperature fluctuations) (Gonçalves & Blaha 2010).

The denaturation of fish muscle proteins during frozen storage leads to a decreased water-binding capacity (WBC), and a dry, firm, and tough texture, if the time–temperature profile of storage is unfavorable. The state of water in frozen-thawed fish has been determined as thaw drip (TD), WHC, WBC, or cooking loss (CL). The amount of water released during thawing (TD) can be used as a simple method to get initial information on the properties of the product. WHC was determined in most experiments to describe protein–water interaction during the frozen storage of fish. Some examples are compiled in Table 31.10 showing the suitability for the technique for quality assessment and the shelf life determination of frozen fish (Rehbein 2002).

During freezing, pure water is separated from the system in the form of ice crystals. Solute concentration increases and melting temperature decreases following the thermodynamic equilibrium line. Freezing can also be considered as a dehydration process in which frozen water is removed from the original location in the foodstuff and forms ice crystals. During thawing, water may not be reabsorbed in the original regions, leading to the formation of drip. Factors that affect drip losses are size and location of ice crystals, rate of thawing, the extent of water reabsorption, the status of the tissue before freezing, and the WHC of the tissue (Zaritzky 2000).

Freezing does not stop weight loss. After meat is frozen, sublimation of ice from the surface occurs. If the degree of sublimation is excessive, the surface of the meat becomes dry and spongy, a phenomenon called "freezer burn." Developments in the use of moisture impervious packaging materials have significantly reduced sublimation in frozen meat (James 2002).

Much has yet to be done to correlate the rate of weight loss with differences between storage conditions, but the rate of weight loss has been shown to vary with the following: (1) temperature, (2) temperature fluctuation, (3) humidity, (4) heat transfer, (5) air flow over the product, (6) radiation effects of lighting, (7) the product, (8) shape and size of the product, and (9) type of wrapper (Gonçalves & Blaha 2010).

Table 31.10 Water-holding capacity of frozen and thawed fish.

Fish species and type of product	Storage conditions	Results
Whole cod (*Gadus morhua*)	Different combinations of time and temperature	WHC was strongly negatively correlated with the chemical quality parameters DMA and FA
Fillets of Argentine hake (*Merluccius hubbsi*)	7 weeks at −7°C	WHC was affected by sex, size, and sexual maturity of the fish
Blue crab meat (*Callinectes sapidus*)	32 weeks at −29°C	WHC was improved by cryoprotectants
Single- and double-frozen fillets of cod (*Gadus morhua*)	9 months at −22°C	WHC decreased considerably during storage of double-frozen fillets, but only slightly for single-frozen fillets
Minced prawn flesh (*Penaeus* spp.)	90 days at −12°C	Initial WHC was very high; the slight decline during freezing was diminished by cryoprotectants

Source: From Rehbein (2002).

Most codes of practice only state the temperature for storage. Variations in the other factors that control the rate of dehydration can, therefore, result in cold stores having widely different storage conditions. The rate at which the product loses weight by dehydration can, therefore, vary considerably. There are great differences between the quality of cold stores that may be attributed both to their design and mode of operation as well as to the operating temperature (Johnston *et al.* 1994).

The rate of weight loss within a store can vary considerably with location. Fish stored near fan coolers, where they are subjected to high air velocities, will quickly show signs of dehydration. Fish stored against walls remote from the cooler may be subject to poor air distribution and heat gains from the store walls. This can cause temperature fluctuations in the product, which inevitably results in high dehydration losses (Gonçalves & Blaha 2010).

Storage and transport conditions have a great influence on the quality of frozen foods. Ice recrystallization can be defined as the increase of the average size of the ice crystals. The driving force for this phenomenon is the difference in the surface energy of two adjacent crystals, this energy being proportional to the crystal curvature. Recrystallization reduces the advantages of fast freezing, inducing physicochemical changes that alter product quality. There is a direct relationship between crystal size and the number of faces the crystal has. Small crystals with three or four faces show concave surfaces, and tend to disappear because the limit of the crystal migrates toward the center of the curvature. Six-sided crystals have plane surfaces and are stable, and those with a higher number of faces tend toward growth. Crystal growth occurs at a constant temperature but is accelerated by fluctuations and thermal steps (Zaritzky 2000).

Concepts like dehydration increase of the ionic strength and concentration gradients can be applied to recrystallization. In the course of freezing and storage, each crystal is surrounded by a saline solution of a determined concentration. The increase in ice crystal diameter during recrystallization leads to a redistribution of this solution around the tissue; its interaction with the protein structure contributes to denaturation, which also produces an increase of the exudate released by the tissue after thawing. Temperature fluctuations accelerate the rate of recrystallization. As temperature increases, the small ice crystals melt. Then when the temperature decreases again, since new nuclei cannot be formed, water is converted into ice on the existing crystals, hence increasing their size (Martino & Zaritzky 1987, 1988, 1989; Zaritzky 2000).

The thawing process must be conducted with care if quality and yield are to be preserved. Regardless of the selected procedure, energy must be provided to melt the ice. Thawing must be designed in order to minimize microbial growth, water release, evaporation losses, and deteriorative reactions. Thawing requires longer times than freezing for comparable temperature driving forces. This is because during thawing the heat is transferred through the unfrozen zone of the foodstuff where thermal conductivity is lower than that of the frozen zone. With regard to exudates production, in frozen meats a slow thawing process at low temperatures is sometimes recommended to permit water diffusion in the thawed tissue and its relocation in the fibers (Zaritzky 2000).

31.6 SAFETY AND QUALITY ASSURANCE FOR FROZEN FISH

31.6.1 Frozen supply chain

Fig. 31.6 shows a simplified fish supply chain, going from catch through to the retail cabinet. The main factors that determine frozen fish quality can be grouped into three main areas: prefreezing, the freezing process, and frozen storage/product distribution. It should also be

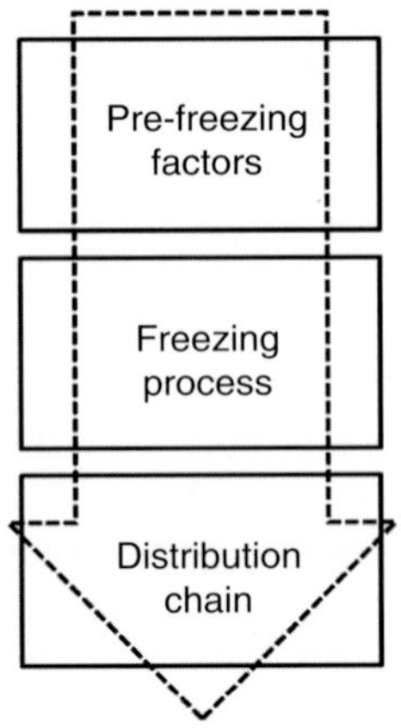

Fig. 31.6 A simplified diagram of key stages in the supply chain for frozen fish products.
Source: Adapted from Hedges and Nielsen (2000).

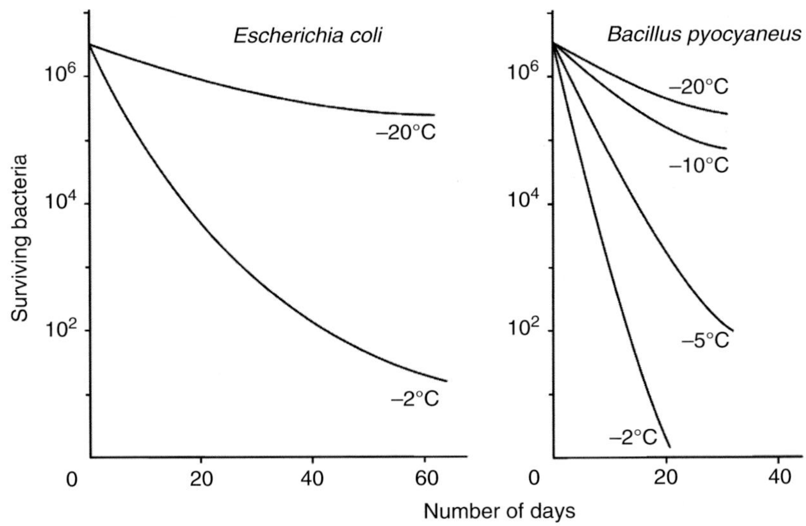

Fig. 31.7 Number of bacteria during storage at different temperatures. The bacteria were frozen to an end temperature of –70°C.
Source: Adapted from Bogh-Sorensen (2000).

borne in mind that one of the least well-defined areas in any supply chain is the consumer handling stage, and it is during this phase that potentially severe thermal abuse may occur. Indeed, thermal abuse may not be confined to the frozen state but may extend to poor cooking of products also (Hedges & Nielsen 2000).

In practice, foods are frozen to be stored for a certain period, often several months. Frozen storage is always included, and it is the combined effect of the freezing (and thawing) process and frozen storage that is of interest. When counting microorganisms in frozen foods during storage at different temperatures, the result is often as shown in Fig. 31.7. Warm storage, that is, temperatures greater than –8°C, results in much larger inactivation of microorganisms than storage at –18°C or less. Freezing and storage at very low temperatures, down to –150°C or even less, seems to result in increasing survival.

During frozen storage, the above-mentioned mechanisms (cell dehydration, membrane damage, etc.) continue to act upon the microorganisms, eventually leading to injury or death of a certain part of the organisms. It is also suggested that the death of microorganisms may be caused by long-time exposure to concentrated solutions, both internal and external. The inevitable temperature fluctuations in the cold chain cause the recrystallization of ice crystals, increasing the salt concentration as well as the damage to microorganisms. The increasing size of the ice crystals during storage reduces the difference between foods frozen very rapidly and foods frozen with a normal freezing rate, and have some influence on the survival of microorganisms (Bogh-Sorensen 2000; Gonçalves & Ribeiro 2009).

31.6.2 Prefreezing factors influencing storage stability

Freezing and cold storage is an efficient method of fish preservation, but it must be emphasized that it does not improve product quality. The final quality depends on the quality of the fish at the time of freezing as well as other factors during handling, prechilling, freezing, cold storage, and distribution. The important requirement is that the fish should at all times be kept in a cool condition before freezing, about 0°C, and the use of ice or other methods of chilling is recommended. Apart from keeping the product chilled, it is also essential to adopt a high standard of hygiene during handling and processing to prevent bacterial contamination and spoilage. In some countries, chemicals are currently used to treat fresh fish in order to assist with such things as color retention and the retention, or even addition, of fluids. The treatment of food with chemicals is usually subject to national and local restrictions and it would be inappropriate to make any general comment on their use in this document (Gonçalves & Blaha 2010).

Clearly, in any fish product supply chain there are many points at which improvements may be made. However, to obtain real benefits the supply chain must be viewed in its entirety. Thus, there may be little to be gained by controlling the quality of raw materials if inappropriate freezing and storage conditions are employed during processing and distribution.

Product preparation has a considerable effect on quality. It is better to eviscerate fish before they are frozen since blood and kidney tissue can be removed. Blood contains iron, a cofactor for TMAOase activity, and kidney remnants may contain proteolytic enzymes that will degrade the fillet structure. Whole fish have longer shelf stability than fillets, which are more stable than minces. This increasing instability, which is more apparent in white fish, probably results from the release of salts and enzymes due to tissue damage, which then leads to more rapid deterioration. Also, the mixing of red and white muscles (red muscle containing more fat and more hem proteins) may result in the dispersion of lipids and enzymes, leading to more deteriorative changes (Gonçalves & Blaha 2010; Gonçalves 2011).

According to Hedges and Nielsen (2000) and Gonçalves and Blaha (2010), many parameters affect the shelf life of fish and the main prefreezing factors are listed as follows:

1. *Fish type*: fish of different types spoil at different rates. In general, it can be stated that larger fish spoil more slowly than small fish, flat fish keep better than round fish (Kim *et al.* 1977), lean fish keep longer than fatty fish (under aerobic storage), and bony fish are edible for longer periods than cartilaginous fish.
2. *Handling*: Rough handling will result in a faster spoilage rate. This is due to the physical damage to the cells of the fish, resulting in easy access for enzymes and spoilage bacteria. Also, rough handling may lead to damage/bruising, which may accelerate deteriorative processes. If fish are kept alive in containers until processing and freezing, microbial spoilage can be avoided. The fish are, however, starved under these conditions

and use up glycogen. Thus, as with catching and handling procedures that cause energy stored to become depleted before death, the pH of the muscle tends to be higher.

3. *Effect of postmortem pH*: The postrigor pH varies between fish species but is generally higher than in meat from warm-blooded animals. The catching method (Botta *et al.* 1987), handling prior to death, time of year, and fishing ground may all influence the final pH of the fish fillet (Love 1975a, b). The ultimate pH of the fish fillet has a direct effect on the WHC of the muscle and on the fish texture. The pH of the muscle influences the shelf life of fish stored on ice, where the shelf life increases with a decreasing pH. The final pH may also influence the properties of the connective tissue, and especially the propensity for fillets to gape (Love and Haq 1970). Gaping is a phenomenon whereby the connective tissue is unable to hold the muscle blocks together.
4. *Prechilling*: The postcatch and postmortem handling of fish is different from that of meat, even though they are both usually chilled. Fish temperature reduction to about 0°C is by far the most important factor for the quality of fish. This should be achieved as rapidly as possible. Traditionally, ice is used in the prechilling of fish. This has several advantages. The first one is that the ice and fish are in good contact, allowing good heat transfer from the fish to the ice. The second is that the melting of the surrounding ice requires a large amount of heat energy to be removed from the fish. The disadvantage with icing is that it can be labor intensive, and for fish in boxes, the contact between the fish and ice may not be very good. Rapid chilling will slow down the rates of enzymatic- (and microbial-) induced changes occurring postmortem. Of these postmortem changes, the earliest ones observed by trained sensory panelists are those that relate to changes in appearance and texture. The most dramatic change is the onset of rigor mortis. The rate of onset and resolution of rigor varies from species to species and is affected by temperature, handling, size, and physical condition of the fish. Indeed, for farmed salmon it has been shown that stress prior to slaughter can influence the speed of rigor development and the textural quality of the fillets. Fish may also be cooled in refrigerated sea water. This allows faster heat removal from the fish and speeds up the chilling process (super chilling). Also, the temperature of the fish can be reduced from −1°C to −2°C, and this may offer advantages in reducing rates of spoilage. An extension to super chilling is to allow the fish to freeze partially. By reducing the temperature of the fish from −2°C to −4°C, fish can be partially frozen. It has been claimed that such partial freezing of fish can extend shelf life. This may offer advantages for preserving fish quality on board trawlers that do not freeze fish at sea (Gonçalves & Blaha 2010).
5. *Addition of cryoprotectants*: compounds that improve the quality and extend the shelf life of frozen foods. A wide variety of cryoprotective compounds are available, and these include sugars, amino acids, polyols, methyl amines, carbohydrates, some proteins, and inorganic salts such as potassium phosphates and ammonium sulfate. The selection of cryoprotectants will depend on whether the application is for a comminuted product, that is, a system into which the cryoprotectants can be intimately mixed, or a whole fish fillet. Probably the most extensive applications of cryoprotectant molecules have been in the stabilization of surimi (e.g., cryoprotectants such as carbohydrates—particularly sugars and polyphosphates—only allowed in some countries) and to minimize the loss of protein functionality properties caused by the freezing and frozen storage processes on surimi. For whole-fillet applications, the most frequently used cryoprotectant to control the WHC is tripolyphosphate. It has been reported that the addition of polyphosphate will improve the texture and color of fish products (Gonçalves & Ribeiro 2008b, 2009; Gonçalves *et al.* 2008).

31.6.3 Postfreezing treatments that protect frozen fish

Deterioration of seafood begins immediately upon harvest, and continues to various degrees depending on storage conditions. The best method of preserving seafood is freezing and storing at low temperatures. If properly frozen, seafood retains quality and flavor. A great problem encountered by producers of both fresh and frozen seafood is the dehydration of the product. Two protective methods are used, usually in combination: glazing and packaging. Good packaging prevents the circulation of air over the surface of the product and protects the moisture in the surface layers of the product (Gonçalves & Blaha 2010).

31.6.3.1 Glazing

Weight loss by dehydration during freezing and storage is directly proportional to the exposed surface area and can be reduced by two methods: covering the surface with packaging material, and surrounding the product with a thin layer of ice. The use of ice-glaze for small and irregularly shaped seafood products, like prawns and shrimp (individually quick frozen [IQF]), may be considered essential when stored without packaging or in pillow packs.

Glazing (or ice-glaze) means applying a coating of ice to the product that has already been frozen. Once frozen, the fish is sprayed with vaporized water, which immediately freezes into a thin layer on the surface of the fish. In this way, the entire surface of the fish is covered in an ice layer hindering air (oxygen) from getting in contact with the fish.

Generally, the glaze is simply water, but ingredients can be added (glycerin, sugar, phosphates, etc.). A glaze provides an excellent barrier to oxidation and freezer burn during frozen storage. Added ingredients include thickeners to ensure a good coating of glaze is applied, antioxidants to assist in preventing fat rancidity and color loss for whole fish, or sometimes just salt to maintain flavor. As soon as seafood is removed from a freezer, they should be glazed or wrapped (unless they have been packaged before freezing) and immediately transferred to a low-temperature store to rapidly refreeze and to preserve taste, smell, and texture as well as to minimize TD loss (Gonçalves & Blaha 2010).

According to Gonçalves and Gindri Junior (2009), a reasonable range of water uptake during the glazing process is between 15% and 20% to guarantee the final frozen shrimp quality. Nevertheless, abuse has been reported with coatings as thick as 25–45% (or up), but the most appropriate way to attain quality assurance is to introduce the standard procedures for ice-glazing, and complying to a regulated glazing content is important to the producers and consumers.

31.6.3.2 Importance of temperature monitoring

There are a number of papers that suggest that temperature fluctuations increase the rate at which fish quality is lost (Gonçalves & Blaha 2010). Indeed, Love (1969) suggested that due to a logarithmic relationship between quality loss and temperature, temperature fluctuation may greatly reduce quality and should therefore be avoided. Also, Scudder (1995) suggested that the degradation of TMAO to FA and DMA may have a maximum rate of around −18°C. Thus, it was recommended that the storage temperature on lean demersal fish should be between −24°C and −30°C. For oily fish, storage temperatures of −50°C to −60°C have been recommended in order to reduce the rate of oxidative changes (Magnussen 1992).

31.6.4 Future trends in frozen seafood

In the future, consumers are going to want more frozen seafood products and that that seafood should have a higher nutritional quality because of improvements in available varieties and in the handling of materials between harvest and processing. The textural quality and the retained nutritional content of frozen seafood after thawing will have improved by the new manufacturing processes, such as pressure shift freezing, that allow the production of smaller ice crystals during freezing. These products will be more stable because the ice crystals will remain small during distribution due to the incorporation of antifreeze proteins and other novel ice actives. Stability will also have improved by the formulation and the understanding of mobility and biochemistry in the unfrozen component. The distribution chain would be better controlled through the use of predictive modeling and through monitoring with time–temperature integrators (Kennedy 2000).

31.7 FINAL CONSIDERATIONS

It is possible to produce high-quality frozen fish if the physical and chemical changes described in this chapter are taken into account during handling immediately after catching, and in connection with the freezing and cold storage of the fish.

Temperature and temperature fluctuations are the most significant factors. It is important to cool the fish as quickly as possible immediately after it has been caught in order to minimize the biochemical and microbiological reactions from the very beginning. Fishing methods and temperature are likewise significant in relation to rigor mortis. In addition, it is essential that fish that are frozen, which will subsequently be thawed and processed, are frozen before rigor sets in, as this will result in the best eating quality and greatest yield.

The fish must be cleaned and bled before freezing, among other things because of the content of compounds in the blood that intensify the degradation processes and because of the very high enzyme activity in the guts.

Thus, a first-class frozen product can be achieved only when using entirely fresh fish and freezing them and storing them at low and stable temperatures. If subjected to poor handling in connection with freezing and cold storage, for example, in terms of a freezing process spanning a whole day and cold-storage temperatures that fluctuate over a longer time, the product will be ruined. Short-term storage of the raw material, gentle processing, quick freezing, short-term storage as a semiprocessed product, and short-term cold storage at temperatures lower than −30°C will result in the best product.

REFERENCES

Bogh-Sorensen, L. (1984) The TTT-PPP concept. In: *Thermal Processing and Quality of Foods* (eds P. Zeuthen, J.C. Cheftel, C. Eriksson, M. Lul, H. Leniger, P. Linko, G. Varela & G. Vos), pp. 511–521. Elsevier Applied Science, Barking, Essex, U.K.

Bogh-Sorensen, L. (2000) Maintaining safety in the cold chain. In: *Managing frozen foods* (ed. C.J. Kennedy), 1st edn., Chapter 2, pp. 5–26, 286pp. CRC Press LLC, Boca Raton, Florida.

Bøknæs, N., Østerberg, C., Nielsen, J. & Dalgaard, P. (2000) Influence of freshness and frozen storage temperature on quality of thawed cod fillets stored in modified atmosphere packaging. *Food Science and Technology*, **3**, 33–37.

Bøknæs, N., Guldager, H.S., Østerberg, C. & Nielsen, J. (2001) Production of high quality frozen cod (*Gadus morhua*) fillets and portions on a freezer trawler. *Journal of Aquatic Food Product Technology*, **10**, 33–47.

Botta, J.R., Bonnell, G. & Squires, B.E. (1987) Effect of method of catch and time of season on sensory quality of fresh raw Atlantic cod (*Gadus morhua*). *Journal of Food Science*, **52**(4), 928–938.

Campañone, L.A., Salvadori, V.O. & Mascheroni, R.H. (2001) Weight loss during freezing and storage of unpackaged foods. *Journal of Food Engineering*, **47**, 69–79.

Cleland, D.J. & Valentas, K.J. (1997) Prediction of freezing time and design of food freezers. In: *Handbook of Food Engineering Practice* (eds K.J. Valentas, E. Rotstein, & R.P. Singh), 1st edn., 736pp. CRC Press LLC, Boca Raton, Florida.

Cortella, G. (2000) Retail display equipment. In: *Managing frozen foods* (ed. C.J. Kennedy), 1st edn., Chapter 12, pp. 233–262, 286pp. Boca Raton, Florida.

Doong, N.F.G. (1998) The effect of freezing and frozen storage on the status of fish tissue. *Diss. Abs. Int. B.*, **48**(8), 2159.

Evans, J. & James, S. (1993) Freezing and meat quality. In: *Food Technology International Europe* (ed. A. Turner), pp. 53–56. Sterling Publications International, London, U.K.

Fagan, J.D., Gormley, T.R. & Mhuircheartaigh, M.U. (2003) Effect of freeze-chilling, in comparison with fresh, chilling and freezing, on some quality parameters of raw whiting, mackerel and salmon portions. *Food Science and Technology*, **36**, 647–655.

FAO. (2010) *The State of World Fisheries and Aquaculture,* 218pp. FAO Fisheries and Aquaculture Department, Rome, Italy.

FAO/WHO. (2009) *Code of Practice for Fish and Fishery Products*, 1st edn., 144pp. Codex Alimentarius Commission, Rome, Italy.

Fellows, P. (2000) Food processing technology–Principles and practice. *Chilling, Part IV Processing by the Removal of Heat*, 2nd edn., Chapter 19, pp. 386–405, 575pp. CRC Press LLC, Boca Raton, Florida.

Godiksen, H. & Jessen, F. (2001) Cytochrome oxidase as an indicator of ice storage and frozen storage. *Journal of Agricultural and Food Chemistry*, **49**, 4488–4493.

Gonçalves, A.A (ed.) (2011) *Tecnologia do pescado: ciência, tecnologia, inovação e legislação,* 608pp. Atheneu, Rio de Janeiro, Brazil.

Gonçalves, A.A. & Blaha, F. (2010) Cold chain in seafood industry. In: *Refrigeration: Theory, Technology and Applications* (ed. M.E. Larsen), 3rd edn., Chapter 7, pp. 287–367. Nova Science Publishers, Inc., Hauppauge, New York, ISBN: 978-1-61668-930-8.

Gonçalves, A.A. & Gindri Junior, C.S.G. (2009) The effect of glaze uptake on storage quality of frozen shrimp. *Journal of Food Engineering*, **90**(2), 285–290.

Gonçalves, A.A. & Ribeiro, J.L.D. (2008a) Optimization of the freezing process of red shrimp (*P. muelleri*) previously treated with phosphates. *International Journal of Refrigeration*, **31**(7), 1134–1144.

Gonçalves, A.A. & Ribeiro, J.L.D. (2008b) Do phosphates improve the seafood quality? Reality and legislation. *Pan-American Journal of Aquatic Sciences*, **3**(3), 237–247.

Gonçalves, A.A. & Ribeiro, J.L.D. (2009) Effects of phosphate treatment on quality of red shrimp (*Pleoticus muelleri*) processed with cryomechanical freezing. *LWT—Food Science and Technology*, **48**(8), 1435–1438.

Gonçalves, A.A., Rech, B.T., Rodrigues, P.M. & Pucci, D.M.T. (2008) Quality evaluation of frozen seafood (*Genypterus brasiliensis, Prionotus punctatus, Pleoticus muelleri* and *Perna perna*) previously treated with phosphates. *Pan-American Journal of Aquatic Sciences*, **3**(3), 248–258.

Haard, N.F. (1992) Biochemical reactions in fish muscle during frozen storage. In: *Seafood Science and Technology* (ed. E.G. Bligh), pp. 176–209. Fishing News Books, Halifax, Nova Scotia, Canada.

Haard, N.F. (1994) Biochemical reactions in fish muscle during frozen storage. In: *Seafood Science and Technology* (ed. E.G. Bligh). Blackwell Scientific, Cambridge, Massachusetts.

Harris, P. & Tall, J. (1994) Rancidity in fish. In: *Rancidity in Foods* (eds J.C. Allen & R.J. Hamilton), pp. 256–272. Blackie Academic and Professional, Glasgow, U.K.

Hedges, N. (2002) Maintaining the quality of frozen fish. Part III—Improving quality within the supply chain. In: *Safety and Quality Issues in Fish Processing* (ed. H.A. Bremmer), Chapter 20, 520pp. Woodhead Publishing Limited and CRC Press LLC, Cambridge, England.

Hedges, N. & Nielsen, J. (2000) The selection and pre-treatment of fish. In: *Managing Frozen Foods* (ed. C.J. Kennedy), 1st edn., Chapter 6, pp. 95–110, 286pp. CRC Press LLC, Boca Raton, Florida.

James, S.J. (2002) New developments in the chilling and freezing of meat. In: *Meat Processing Improving Quality* (eds J. Kerry, J. Kerry, & D. Ledward), 1st edn., 464pp. CRC Press LLC, Boca Raton, Florida.

Jarenback, L & Liljemark, A. (1975) Ultrastructure changes during frozen storage of cod. I. Structure of myofibrils as revealed by freeze etching preparation. *Journal of Food Technology*, **10**, 229–239.

Jay, J.M., Loessner, M.J. & Golden, D.A. (2005) *Modern Food Microbiology*, 7th edn., Food Science Text Series, 790pp. Springer Science+Business Media Inc., New York.

Jessen, F. & Nielsen, J. (2002) God og dårlig frossen fisk—hvorfor er der forskel? *Fisk og Hav*, **54**, 16–25.

Jensen, K.N., Guldager, H.S., Jacobsen, G. & Nielsen, J. (2001) Selection and application of quality indicators to describe quality changes in thawed cod. *Conference Proceeding. IIR International Conference on Rapid Cooling of Food*, 28–30 March, Bristol, U.K.

Jensen, L.H.S., Nielsen, J., Jørgensen, B.M. & Frosch, S. (2010) Cod and rainbow trout as freeze-chilled meal elements. *Journal of the Science of Food and Agriculture*, **90**, 376–384.

Johnston, W.A., Nicholson, F.J., Roger, A. & Stroud, G.D. (1994) Freezing and refrigerated storage in fisheries. *FAO Fisheries Technical Paper No. 340*, 109pp. FAO, Rome, Italy.

Kennedy, C.J. (2000) Future trends in frozen foods. In: *Managing Frozen Foods* (ed. C.J. Kennedy), 1st edn., Chapter 13, pp. 263–278, 286pp. CRC Press LLC, Boca Raton, Florida.

Kerr, K. (1997) The bacterial cold-shock response in food borne pathogens. *Plenary Meeting no. 3*, CT96-1180, Leeds, U.K.

Kim, H.K., Robertson, I. & Love, R.M. (1977) Changes in the muscle of lemon sole (*Pleuronectes microcephalus*) after very long cold storage. *Journal of the Science of Food and Agriculture*, **28**(8), 699–700.

Lee, C.M. (1982) Physical and biochemical changes in fish muscle under various freezing conditions. *Quick Frozen Foods*, **45**(3), 30–32.

Love, R.M. (1969) Time-temperature tolerance. *IFST Proceedings*, **2**(3), 62–67.

Love, R.M. (1975a) Variability in Atlantic cod (*Gadus morhua*) from the Northeast Atlantic: A review of seasonal and environmental influences on various attributes of the flesh. *Journal of the Fisheries Research Board of Canada*, **32**(12), 2333–2342.

Love, R.M. (1975b) The influence of fishing grounds on quality. *Fishing News International*, **14**(4), 16–18.

Love, R.M. (1992) Biochemical dynamics and the quality of fresh and frozen fish. In: *Fish Processing Technology* (ed. G.M. Hall), pp. 1–26. Blackie Academic & Professional, Chapman & Hall, London.

Love, R.M. & Haq, M.A. (1970) Connective tissue of fish. III. The effects of pH on gaping in cod entering rigor mortis at different temperatures. *Journal of Food Technology*, **5**(3), 241–248.

Mackie, I.M. (1993) The effect of freezing on flesh proteins. *Food Reviews International*, **9**, 575–610.

Magnussen, O.M. (1992) *Superfreezing low temperature technology. Scan. Ref.*, **21**(4), 34–41.

Magnussen, O.M., Haugland, A., Hemmingsen, A.K.T., Johansen, S. & Nortvedt, T.S. (2008) Advances in super chilling of food—Process characteristics and product quality. *Trends in Food Science and Technology*, **19/8**, 418–424.

Martino, M.N. & Zaritzky, N.E. (1987) Effects of temperature on recrystallisation of polycrystalline ice. *Sciences des Aliments*, **7**, 147–166.

Martino, M.N. & Zaritzky, N.E. (1988) Ice crystal size modifications during frozen beef storage. *Journal of Food Science*, **53**, 1631–1637, 1649.

Martino, M.N. & Zaritzky, N.E. (1989) Ice recrystallization in a model system and in frozen muscle tissue. *Cryobiology*, **26**, 138–148.

McGill, A.S., Hardy, R. & Gunstone, F.D. (1977) Further analysis of the volatile components of frozen cold stored cod and the influence of these on flavor. *Journal of the Science of Food and Agriculture*, **28**, 200–205.

Olsson, P. (1984) TT-integrators—Some experiments in the freezer chain. In: *Thermal Processing and Quality of Foods* (eds P. Zeuthen, J.C. Cheftel, C. Eriksson, M. Lul, H. Leniger, P. Linko, G. Varela & G. Vos), pp. 782–788. Elsevier Applied Science, Barking, Essex, U.K.

Ortiz, H. & Bello, R. (1992) Composition and stability of fatty fish acids from deboned cachama and sardine meat during frozen storage. *Archivos Latinoamericanos de Nutrición*, **42**(4), 460–466.

Ranken, M.D., Kill, R.C. & Baker, C. (1997) Fish and fish products (Chapter 2) and Food preservation processes (Chapter 15). *Food Industries Manual*, 718pp. Blackie Academic and Professional, London, U.K.

Rehbein, H. (2002) Measuring the shelf-life of frozen fish. Part III—Improving quality within the supply chain. In: *Safety and Quality Issues in Fish Processing* (ed. H.A. Bremmer), Chapter 21, 520pp. Woodhead Publishing Limited and CRC Press LLC, Cambridge, England.

Reid, D. (1998) Freezing preservation of fresh foods: Quality aspects. In: *Food Storage Stability* (eds I.A. Taub & R.P. Singh), 539pp. CRC Press LLC, Boca Raton, Florida.

Ross, D.A. & Love, R.M. (1979) Decrease in the cold store flavour developed by frozen fillets of starved cod (*Gadus morhua*). *Journal of Food Technology*, **14**, 15–122.

Scudder, B. (1995) Icelandic warning on frozen fish. *Seafood International*, **10**(6), 51.

Selman, J.D. (1990) Time–temperature indicators: How they work. *Food Manufacture*, **65**(8), 30–31 and 33–34.

Selman, J.D. & Ballantyne, A. (1988) Time–temperature indicators: Do they work? *Food Manufacture*, **63**(12), 36–38, 49.

Shenouda, S.Y.K. (1980) Theories of protein denaturation during frozen storage of fish flesh. In: *Advances in Food Research* (eds C.O. Chichester, E.M. Mrak & G.F. Steward), Vol. **26**, pp. 275–311. Academic Press, New York.

Sikorsky, Z.E. & Kolakowska, A. (1994) Changes in proteins in frozen stored fish. In: *Seafood Protein* (eds Z.E. Sikorski, B.S. Pan & F. Shahidi), pp. 99–117. Chapman & Hall, Inc., New York.

Sikorsky, Z.E., Olley, J. & Kostuch, S. (1976) Protein changes in frozen fish. *Critical Review in Food Science and Nutrition*, **8**, 97–129.

Sørensen, N.K., Gundersen, B., Nyvold, T.E. & Elvevoll, E. (1996) Evaluation of sensory quality during freezing and frozen storage of whole, gutted Atlantic salmon (*Salmo salar*). In: *Refrigeration and Aquaculture. Proceedings of the Meeting of Commission C2*, Bordeaux, France, 20–22 March, pp. 307–315.

Warm, K. (2001) *Sensory quality criteria for new and traditional fish species of relevance to consumer needs.* PhD thesis, Danish Institute for Fisheries Research, The Royal Veterinary and Agricultural University, Lyngby, Copenhagen, Denmark, pp. 1–55.

Warm, K., Bøknæs, N. & Nielsen, J. (1998) Quality index method for frozen cod. *Journal of Aquatic Food Product Technology*, **7**, 45–59.

Woolfe, M.L. (2000) Temperature monitoring and measurement, Part II. Technologies and processes. In: *Chilled Foods—A Comprehensive Guide* (eds M. Stringer & C. Dennis), 2nd edn., Chapter 5, pp. 100–134, 428pp. CRC Press LLC, Boca Raton, Florida.

Zaritzky, N.E. (2000) Factors affecting the stability of frozen foods. In: *Managing Frozen Foods* (ed. C.J. Kennedy), 1st edn., Chapter 7, pp. 111–135, 286pp. CRC Press LLC, Boca Raton, Florida.

32 Packaging for Chilled and Frozen Seafood

Alex Augusto Gonçalves

Abstract: For the maintenance of quality of the seafood chain, adequate packaging should protect the product, sell the product, and aid in the convenient use of the product. The main technical advantages of current chilled or frozen food packaging materials are discussed: aluminum, paper, semirigid plastics, flexible plastics, or glass. Different modified atmosphere packaging (MAP) materials have advantages and disadvantages: vacuum pack, high O_2 MAP, low O_2 MAP, and controlled atmosphere packaging (CAP). A number of potential antimicrobial agents are used in antimicrobial food packaging systems. External and internal indicators are used in intelligent packaging to give information on storage conditions, package leak, microbial quality, or specific bacteria.

Keywords: Controlled atmosphere packaging (CAP), modified atmosphere packaging (MAP), vacuum packaging, oxygen (O_2)-scavenging packaging, moisture scavenging packaging, antimicrobial packaging, intelligent pack, seafood packaging

32.1 INITIAL CONSIDERATIONS

Nowadays, a large quantity of seafood is still moved to market essentially unpackaged. Whole or gutted fish are iced in fish holds, stored upon landing, and then re-iced in bulk containers to be hauled to local fish markets. At the market, whole, dressed, or filleted fish are displayed and offered for sale (still unpackaged) on a bed of ice. A switch from wooden to waxed or plastic-coated corrugated boxes as the bulk shipping container has been the sole packaging improvement in this marketing scheme, and a folded newspaper remains the most common form of final product "protection" (Gonçalves & Araújo 2010a).

Frozen seafood products have fared somewhat better, judging from a large selection of attractively packaged raw and processed fish products available in the supermarket's frozen food sections. Unfortunately, the appearance of some packages is better than their ability to protect product quality. Thus, consumers complain about off-flavors and rubbery, watery of otherwise attractively packaged products (Lanier 1990; Han *et al.* 2005b; Gonçalves & Araújo 2010a, b).

Obviously, the seafood industry continues to grow despite a lack of adequate packaging. However, future expansion of seafood sales will mean moving products through longer distribution chains to reach inland markets. Many processors understand that a package needs to protect the product but do not realize the important role that packaging plays in selling the product. Thus, packaging will play a key role in the shift from shop to supermarket merchandising of seafood, and will ensure high quality, good sales appeal, and consumer satisfaction (Gonçalves & Araújo 2010a; Gonçalves & Blaha 2010).

Handbook of Meat, Poultry and Seafood Quality, Second Edition. Edited by Leo M. L. Nollet.

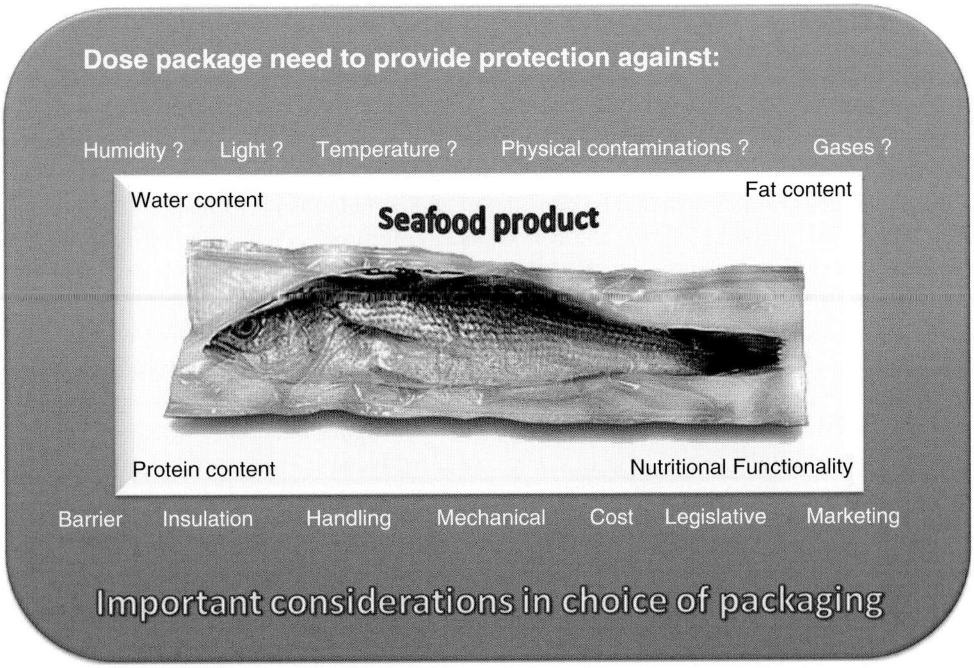

Fig. 32.1 A summary of the main considerations in the choice of seafood packaging.

The consumer is more interested in the product than the package; therefore, a primary function of the package is to provide information about the product. Some consumers, accustomed to evaluating fish quality by smell as much as by sight, may sniff a package in an effort to detect off-odors (flavors). To allow for this inspection, a packaging film of intermediate gas permeability may be the best choice for many seafood products. Its limited permeability, however, allows odors to escape sufficiently to prevent the accumulation of odors that might be objectionable when the package is first opened, regardless of the freshness (Lanier 1990; Han 2005a; Han *et al.* 2005b).

Thus, the aim of any packaging system for fresh muscle foods is to prevent or delay undesirable changes to the appearance, flavor, odor, and texture. Deterioration in these qualities can result in economic losses due to the consumer rejection of the product (Peavey *et al.* 1994; Gill & Gill 2005; Zhou *et al.* 2010).

Any development of novel packaging must address at least four regulatory or food safety criteria: (1) food-contact approval, (2) environmental regulations, (3) a need for labeling (e.g., where the active package may give rise to end-user/consumer confusion), and (4) consideration of the effects of the package on the microbial ecology and safety of the food (see Fig. 32.1). The latter is important to minimize the risk of creating an unnatural "microclimate" within the food package (e.g., an active package that removes O_2 from within the package may also create favorable conditions for the growth of anaerobic pathogenic bacteria).

32.2 DEFINITION, TYPES, AND CHARACTERISTICS

For the maintenance of quality at the seafood chain, adequate packaging should (1) *protect* the product, (2) *sell* the product, and (3) aid in the *convenient use* of the product. Before discussing the many types of packaging materials and machines available for use with seafood, it is important to consider how each of these important functions pertains to the packaging of fishery products (George 2000; Hurme *et al.* 2002; Gill & Gill 2005; Singh & Singh 2005).

Convenience is not only a function of how the product is prepared prior to sale (whole, fillet, breaded portion, smoked slices, etc.), but also of the package construction and function. Packages can protect, help the product sell, and contribute to easy handling and preparation as well (Lanier 1990; Han 2005a).

Handling characteristics should be considered and include the performance characteristics of the packaging materials during the actual packaging process (often called machinability characteristics of the material) as well as the strength and durability of the material when it is in contact with irregular product surfaces and/or when it is subjected to physical abuse in normal distribution. These characteristics, which may be important depending on the type of packaging equipment used, could include tear strength, stretchability, shrinkability, slippage, and heat-sealing or closure-sealing capacities. The toughness of the packaging material is usually a function of cost for a given category of packaging materials, whether clear film, paperboard, metal foils, or rigid plastics. For cost-effectiveness, the materials selected should only be durable enough to provide the level of package protection needed at a particular stage of distribution (Lanier 1990; Hurme *et al.* 2002).

Surface properties of food packaging polymers, such as wet ability, scalability, printability, dye uptake, resistance to glazing, and adhesion to food surfaces or other polymers, are of central importance to food packaging designers and engineers with respect to product shelf life, appearance, and quality control. The most commonly used food packaging polymers are low-density polyethylene (LPDE), high-density polyethylene (HDPE), polypropylene (PP), polytetrafluoroethylene (PTFE), and nylon. Surface properties of these polymers have been extensively studied (Han *et al.* 2005b).

The major function of extended shelf-stable food packaging is barrier protection against the invasion of microorganisms. Beyond this simple barrier function, there has been more research and development regarding the introduction of new purposes to food packaging systems. Among these, significant new functional packaging systems include active packaging, modified atmosphere packaging (MAP), and edible films/coatings (Han 2005a; Rooney 2005; Spencer 2005; Bonilla *et al.* 2010; Zhou *et al.* 2010).

Another significant issue in food packaging is that it should be natural and environmentally friendly. The substitution of artificial chemical ingredients in foods and in packaging materials with natural ingredients is always attractive to consumers. Many ingredients have been substituted with natural components (chemical antioxidants such as butylated hydroxyanisole [BHA], butylated hydroxytoluene [BHT], and tertiary butylhydroquinone [TBHQ]) and have been replaced with tocopherol and ascorbic acid mixtures for food products. This trend will also continue in food packaging system design areas. Designing environmentally friendly packaging systems that are more natural requires, for example, the partial replacement of synthetic packaging materials with biodegradable or edible materials, a consequent decrease in the use of the total amount of materials, and an increase in the amount of recyclable and reusable (refillable) materials (Han 2005a, b).

Biopolymer films have been regarded as potential replacements for synthetic films in food packaging applications in response to a strong marketing trend toward more environmentally friendly materials. However, hydrophilicity is a central limitation that needs to be overcome in order to allow such replacement. To date, most methods of increasing the water barrier ability of biopolymer films depend on the addition of waxes, fatty acids, and lipids (Han *et al.* 2005b; Lacroix & Cooksey 2005; Coma 2008).

Besides the traditional thermal treatments for food preservation, many other new thermal and nonthermal processing technologies have been developed recently (including irradiation, high-pressure processes, pulsed electric fields, UV treatments, antimicrobial packaging, etc.), and some of these processes have been commercially approved by regulatory agencies for food packaging purposes. These new technologies generally require new packaging materials and new design parameters in order for optimum processing efficiency to occur—for example, packages that undergo an irradiation process are required to possess chemical resistance against high energy to prevent polymer degradation, those that undergo UV treatments require UV light transmittable packaging materials, and retortable pouches should resist pressure changes and maintain seal strength. Since these new technologies each possess unique characteristics, packaging materials should be selected with these characteristics in mind. These new packaging materials and/or systems not only need to work technically, but should also be examined scientifically to ensure their safety and lack of toxicity, and be approved by regulatory agencies (Gill & Gill 2005; Han 2005a).

The processing and packaging equipment has new functions that have increased safety, quality, and productivity, and therefore it seems that the development of new packaging functions may go hand in hand with the development of new processes, materials, and equipments. Packages may have new purposes if new functional packaging materials and/or materials containing functional inserts are used. Developing new packaging technologies not only implies new materials but also new packaging design systems (Emblem 2000; George 2000; Han 2005a).

32.3 APPLICATIONS

During recent years, there has been a greatly increased consumer demand for perishable chilled foods that are perceived as being fresh, healthy, and convenient. The major food retailers have satisfied this consumer demand by providing an ever-increasing range of value-added chilled food products. The wide diversity of chilled foods available is accompanied by a huge range of packaging materials and formats that are used to present attractively packaged foods in retail chill cabinets (Day 2000; Gonçalves & Araújo 2010a; Gonçalves & Blaha 2010).

Even when chilled, fishery products are among the most perishable food products known. At least three main factors contribute to this rapid loss of quality: (1) enzyme attack (several digestive and muscle enzymes of fish and shellfish actively break down muscle proteins, which may result in soft, mushy textures), (2) bacterial growth (fishery products are an excellent source of nutrients for bacteria, which convert these nutrients into foul-smelling compound), and (3) fat oxidation (fats and oils of fishery products are unsaturated and tend to break down easily into rancid compounds) (Lanier 1990; Gonçalves & Araújo 2010a; Gonçalves & Blaha 2010).

The action of enzymes can be controlled only by maintaining the product at temperatures as low as possible. Packaging cannot be expected to help much in this respect, except as an insulator during shipment. Low temperature is also essential to minimizing bacterial growth, and packaging can play a key role in its inhibition by controlling the exposed environment (i.e., modified atmosphere—vacuum) (Lanier 1990; Buffo & Holley 2005).

Although frozen fish are not subject to spoilage by bacterial attack, fat oxidation (only in lower velocity), and enzyme activity, moisture migration does contribute to quality loss in many cases. To help control these factors, the maintenance of a low storage temperature with little temperature fluctuation (such as may be caused by cycling of refrigeration equipment) is necessary (Otwell 1997; Gonçalves & Araújo 2010b; Gonçalves & Blaha 2010).

Packaging can play an important role in preserving the quality within frozen foods (i.e., can control fat oxidation and moisture migration very effectively). By selecting the appropriate material and design of package, the food producer can help protect the food product from damage and loss of quality during its distribution from the factory to the consumer. Packaging also makes a key contribution in minimizing quality loss within the food during periods of frozen storage (George 2000).

32.3.1 Requirements

The primary function of food packaging is to protect the food from external hazards. At the same time, the packaging material itself should not affect the food in any way. The packaging material chosen should meet certain minimum technical, legislative, environmental, and commercial requirements. Many of these requirements are intended to ensure that the packaging material provides the product with the necessary protection to ensure both food safety and quality. The capacity of packaging to provide an effective barrier to the ingress of moisture, gases, and contaminants from the environment to the food is essential for preserving both sensory and nutritional characteristics. Similarly, suitable packaging has the ability to provide an effective barrier to the loss of moisture and flavor volatiles from the food to the external environment (George 2000; Pacquit *et al.* 2007).

For chilled products, visibility is important. Clear films with no fogging and dry, fresh-looking products make the best first impression. Consumers like to inspect as much of the product as possible to determine quality. When packaging many fish species, lay fillets with both skin side up and meat side up so the buyer can make a rapid species identification and an adequate quality inspection.

Table 32.1 lists the main requirements of a chilled food package. Depending on the type of food packaged, not all of these requirements will need to be satisfied.

The packaging material must contain the food without leaking, be nontoxic, and have sufficient mechanical strength to protect the food and itself from the stresses of manufacture, storage, distribution, and display. Certain packs require a degree of porosity to allow moisture or gaseous exchange to take place, and packaging materials used in these situations should possess appropriate permeability properties. Alternatively, most modified atmosphere packs require moisture and gases to be retained within the pack and hence the packaging materials used should possess appropriate barrier properties. The specific requirements for modified atmosphere packs are described later. Depending on the type of chilled food product, the packaging material may need to be tolerant of high temperatures experienced during hot filling, in-pack pasteurization, or reheating prior to consumption. The packaging material, particularly with high speed continuous factory operations, may need to be compatible with form–fill–seal machines. The pack closure must have seal integrity but at the same time should be easy to open. There may be a need for reclosure during storage after opening the pack at home. Also, with the increased incidence of malicious contamination, tamperproof or tamper-evident packaging is desirable. The package is the primary means of displaying the contained chilled food and providing product information and point-of-sale advertising. Clarity and printability are two pertinent features that require consideration in the choice of materials (Day 2000; Han 2005a).

With almost any package of chilled seafood, a dry appearance improves the quality image. The product wee page is best controlled by constantly maintaining the product near 0°C or below and using an absorbent pad. "Fog" accumulation, which hides the product from being viewed, can be controlled by coating the underside of the package film with a wetting agent to prevent water droplets from forming. Wet packs of such seafood as oysters and scallops commonly consist of round containers of either metal, plastic, or coated paperboard, with snap-on or crimped lids. The product should be visible through the lid or a side window made of a clear plastic. The rectangular, rigid, heat-sealed containers previously mentioned also

Table 32.1 Main requirements of a chilled food package.

• Contain the product	• Seal integrity
• Be compatible with food	• Prevent microbial contamination
• Nontoxic	• Protect from odors and taints
• Run smoothly on filling lines	• Prevent dirt contamination
• Withstand packaging processes	• Resist insect or rodent infestation
• Handle distribution stresses	• Be cost-effective
• Prevent physical damage	• Have sales appeal
• Possess appropriate gas permeability	• Communicate product information
• Control moisture loss or gain	• Show evidence of tampering
• Protect against light	• Easily openable
• Possess antifog properties	• Be tolerant to operational temperatures

Source: From Day (2000).

make an attractive wet pack. These may be clear or opaque with the product clearly visible through the top film (Lanier 1990).

For frozen products, visibility is also important. Transparent films may be used with frozen fish to heighten the visual impact and quality image. The film should be skintight with no frost accumulation within the package (George 2000).

32.3.2 Materials

The main technical advantages of current chilled food packaging materials are compared in Table 32.2, while the principal types of materials (and their abbreviations) are listed in Table 32.3. For any particular product, a number of materials can generally be used, either as separate components or manufactured as a composite (Day 2000).

Most chilled fish and other raw fishery products are commonly prepackaged on shallow clear or foam plastic trays overwrapped with a transparent plastic film. An absorbent paper pad, covered with plastic to avoid sticking to the product, is wrapped, and a single sheet of transparent film encases the tray and the product. Usually, the film used for this application is polyethylene or polyvinyl chloride (PVC) of the same type used for wrapping red meat cuts (Lanier 1990; Rowe 1990).

These films pose no problem with fishery products, although the product can be smelled through the film. Conversely, some fatty fish may absorb off-odors through such films. Vacuumized bags of a less permeable (but not barrier) film are being evaluated for fresh fish in trays. The overall appearance of the product in trays is reported not to be noticeably different from the traditional overwrap package. Care should be taken, however, to avoid overheating the film in the shrink tunnel, which may cause the foam trays to wrap or break.

A wide variety of metalized plastic films and paperboard products are used in MAP packaging. Films that are employed for gas packaging include polyester terephthalate (PET), crystallized polyester terephthalate (CPET), oriented polypropylene (OPP), PP, HDPE, and occasionally LPDE, PVC, and polyvinylidene chloride (PVDC) composed as monolayers, multilayered, laminated, or metalized film stock, and trays. Among the critical properties of these major packaging resins, the most important characteristic for the purpose of gas packaging is their relative permeability to O_2, carbon dioxide (CO_2), and nitrogen (N_2) or argon. The CO_2 and O_2 gas transmission-rate properties of films are generally well known and can be obtained from the materials specification sheets provided by the manufacturer. The gas transmission rate may also be measured directly by forming

Table 32.2 Comparison of chilled food packaging materials.

Material	Main technical advantages
Aluminum	Impermeability Lightweight Container axial strength Withstands internal pressure
Paper	Great variety of paper grades Ease of decoration Adjunct to all other packaging materials Lightweight
Semirigid plastics	Properties variable with type of plastic Choice of container shape In-house manufacture Lightweight
Flexible plastics	Properties variable by combination Very lightweight containers Tailor-made sizing
Glass	Chemical inertness Impermeability Product visibility Container axial strength Withstands internal vacuum pressure Reuse facility

Table 32.3 Chilled food packaging materials—barrier properties.

Abbreviations	MVTR ($g/m^2/day$)	OTR ($g/m^2/day$)
EVOH (ethylene-vinyl alcohol)	—	—
HDPE (high-density polyethylene)	7–10	1600–2000
HIPS (high impact polystyrene)	—	—
LDPE (low-density polyethylene)	15–20	6500–8500
LLDPE (linear low-density polyethylene)	15–20	6500–8500
MXDE (modified nylon)	—	—
OPP (orientated polypropylene)	5–7	2000–2500
OPS (orientated polystyrene)	—	—
PA (polyamide-nylon)	300–400	50–75
PC (polycarbonate)	—	—
PE (polyethylene)	—	—
PET (polyethylene terephthalate)	15–20	100–150
PETG (modified PET)	—	—
PP (polypropylene)	—	—
PS (polystyrene)	70–150	4500–6000
PVC (polyvinyl chloride)	10–12	3500–4500
PVDC (polyvinylidene chloride)	0.6–1.0	2–4
UPVC (unplasticized polyvinyl chloride)	30–40	150–350

MVTR, moisture vapor transmission rate; OTR, O_2 transmission rate; barrier material of a given thickness (25 μm), commonly used at 25°C and 75%RH or 38°C and 90%RH.
Sources: Adapted from Day (2000) and Emblem (2000).

packages with included gas of known composition and sampling changes over time with a commercially available O_2 and/or CO_2 meter. However, accurate data on N_2 permeabilities are less widely available, and are generally calculated by difference measurement, that is, by measuring that which remains after CO_2 and O_2 have been accounted for. Permeation of argon and other gases is less well known, but some very good data exist (Han *et al.* 2005b; Spencer 2005).

Also, a wide range of commonly available gas packaging machinery was used in these example experiments, and all were successfully and easily modified as above to accommodate argon. A wide range of plastic films and trays were used, and most of these were found to be satisfactory. The key point was to use highly impermeable films and trays. A very large array of tests was conducted to determine the best economical fit of plastic packaging for various product constraints. Generally, excellent results were obtained with highly impermeable (O_2 transmission rate 0–5 cm^3/day/M^2/atm self-powered tray ration heater [SPTRH]) crystallized polyester terephthalate/polyester terephthalate (CPET/PET) trays and either high or medium barrier amorphous polyethylene/polyester terephthalate (APET/PET), ethylene vinyl acetate/polyester terephthalate (EVA/PET), or polypropylene/polyester terephthalate (PP/PET) film combinations (O_2 transmission rate 75–100 cm^3/day/M^2/atm SPTRH) (Spencer 2005).

Creation of activity in packaging materials frequently involves the introduction of additional components into otherwise inactive materials, typically plastics. Besides the intended effect, there may be unintended consequences that must be addressed in order to bring the package to market. These additional considerations have an impact on material properties and the manner in which they must be used either in order to achieve their desired effect or to be processable in the expected manner (Rooney 2000, 2005).

Active components may form homogeneous mixtures with the existing plastic, or may occur in a separate phase. The range of options available has been described in detail (Rooney 2000) and will be discussed later. Some active polymers may be used as polyblends whereas low molecular weight additives may dissolve in the base polymer. Components that must be used as powders may interfere with the optical properties of the plastic as well as the propensity for tearing. Some hard particles can lead to reduced sharpness of slitter blades, and possibly increase the wear in extruder barrels (Han *et al.* 2005a; Lacroix & Cooksey 2005; Coma 2008).

Chemical effects may impact on material properties, where the activity involves substantial chemical change to the active polymer. Such chemical changes may lead to changed compatibility, or the degradation (or cross-linking) of the host polymer if the activity involves free-radical reaction mechanisms (Rooney 2005).

The physical and mechanical properties of packaging materials are affected by the incorporated antimicrobial agents. If the antimicrobial agent is compatible with the packaging materials, a significant amount of the agent may be impregnated into the packaging material without any deterioration of its physical and mechanical integrity (Han & Floros 1997). However, excess antimicrobial agents that are not capable of being blended with packaging materials will decrease physical strength and mechanical integrity (Han & Floros 1997; Han 2003; Cooksey 2005; Han 2005b).

A wide variety of materials have been used for the packaging of frozen foods, including plastics, metals, and card/paperboard. Each of these options has individual advantages and limitations. Retail packages of frozen fishery products may be of the raw types (shrimp, fillets, etc.) or processed/convenience foods (breaded portions, stuffed clams, etc.). Packaging raw seafood is generally simpler in design and more functional (i.e., protecting the product) than eye appealing. Raw products traditionally have been packed by smaller production-oriented seafood firms; the larger marketing-oriented food processing companies have marketed most processed and prepared seafood (Lanier 1990). The characteristics of standard plastic packaging materials in common use for frozen food packaging are given in Table 32.4 (Randell *et al.* 1999; George 2000).

Metal packaging materials can provide highly effective barrier properties for both water vapor and O_2, along with good mechanical strength and ease of handling. The two most common

Table 32.4 Plastic packaging materials for frozen seafood.

Polyethylene (PE); low-density polyethylene (LDPE); High-density polyethylene (HDPE)
Commonly used for the packaging of individually quick frozen (IQF) foods. The bags can be filled on a "form, fill, and seal" basis. The material is relatively easy to seal using standard heat-sealing equipment and good quality printing can be achieved on the surface of the package. PE and LDPE films are relatively inexpensive. HDPE can also tolerate temperatures in excess of 100°C and is often used in "boil-in-the-bag" applications. Both HDPE and LDPE, however, provide relatively poor barrier protection from O_2, although good barrier protection to water vapor
Polyester terephthalate (PET)
These films can withstand high temperatures and are resilient to grease and water vapor. PET and crystallized polyester terephthalate (cPET) form a large and growing sector of packaging materials used in the prepared foods market. The trays are suitable for reheating in both conventional and microwave ovens, with stability at temperatures in excess of 250°C. The materials, however, are relatively expensive and some forms can be "brittle" at freezer temperatures. The material provides good barrier characteristics (semibarrier) to both O_2 and water vapor
Polystyrene (PS)
PS is a general plastic for frozen food applications. It has a high resistance to breakage at freezer temperatures but is relatively expensive and allows relatively high transmission of water vapor and O_2 between the food product and freezer environment. PS is a good thermal insulator and is consequently used as a secondary package material for cases of frozen products
Polyvinyl chloride (PVC)
PVC is generally used for rigid containers. It is cheaper than PS and, when unplasticized, has a much lower permeability to water vapor and reduced permeability to O_2. However, PVC possesses less impact resistance than PS and other plastics and can consequently get damaged during transit
Polyamide (PA)
PA forms a plastic with good strength and molding characteristics. It is suitable for thermoforming laminations, commonly used in "boil-in-the-bag" applications
Polypropylene (PP)
Modified films have found a niche for certain products and are becoming more widely used as a flow-wrap film. As a primary packaging material for frozen foods, PP has relatively poor barrier characteristics to both O_2 and water vapor
Laminates and co-extrusions
It is evident from the characteristics of packaging materials described above that no single synthetic packaging material offers all the desirable features. In such cases, the food producer has to select the package material carefully and tolerate some degree of "trade-off." However, a great deal of development in the field of laminates and co-extrusions aims to combine the desirable characteristics of individual materials to provide a single protective package. By combining films of plastics with different properties, it is possible to optimize the material to gain the specific properties required, for example, the combination of a high water vapor barrier with a high O_2 barrier

Source: Adapted from George (2000).

metals used in food packaging are aluminum and steel. Aluminum's compatibility with foods contributes to its utility as a packaging material. Aluminum foil is used for trays and may also be laminated to plastic films and paper and board to provide specific additional requirements. Aluminum foil is an excellent light and moisture barrier for frozen foods. The applications of aluminum packaging in frozen foods are increasing rapidly. Many prepared foods, for example, ready meals and ready-to-cook products, are commonly enclosed in aluminum trays. Steel trays are also becoming common as food packaging materials and offer the same advantages of barrier properties, mechanical strength, and good heat transfer as aluminum. An important advantage of metal trays and containers over synthetic food packaging materials is the recyclability of metals.

Table 32.5 Relative O_2 and water vapor permeabilities of selected packaging materials.

Package material	Relative O_2 permeability (relative units)	Relative H_2O vapor transmission rate (relative units)
Aluminum (Al)	<50 (barrier)	<10 (barrier)
Ethylene-vinyl acetate (EVOH)	<50 (barrier)	Variable
Modified nylon (MXDE)	<50 (barrier)	10–30 (semibarrier)
Modified polyester (PETG)	50–200 (semibarrier)	30–100 (medium)
Polycarbonate (PC)	200–5 000 (medium barrier)	100–200 (very high)
Polyester (PET)	50–200 (semibarrier)	10–30 (semibarrier)
Polyethylene (PE)		
High-density polyethylene (HDPE)	200–5 000 (medium)	<10 (barrier)
Low-density polyethylene (LDPE)	5 000–10 000 (high)	10–30 (semibarrier)
Polypropylene (PP)	200–5 000 (medium barrier)	10–30 (semibarrier)
Polystyrene (PS)	200–5 000 (medium barrier)	100–200 (high)
Polyvinylidene chloride (PVDC)	<50 (barrier)	<10 (barrier)
Polyvinyl chloride (PVC)		
Unplasticized	50–200 (semibarrier)	Variable
Plasticized	5 000–10 000 (high)	200–300 (very high)

This becomes ever more important as environmental directives drive the food producer toward recyclable and reusable food packaging (Lanier 1990; George 2000; Buffo & Holley 2005).

Paperboard and card are also commonly used for packaging frozen foods. These packaging materials are usually considered in three groups: (1) paper (thickness up to 3 mm), (2) board (thickness between 3 and 11 mm), and (3) fiberboard.

The choice of packaging material for a particular frozen food will depend upon the major deteriorative reactions governing the physical and biochemical changes within the food during frozen storage. These, in turn, have significant bearing on the sensory and nutritional qualities of the food following periods of frozen storage (Otwell 1997; Gonçalves & Araújo 2010b).

From the food product's perspective, the packaging material can have a major role in minimizing the physical and biochemical reactions that lead to sensory and nutritional quality losses. Physical changes to the food product can include color, syneresis, and texture. In terms of color retention in foods, a major factor is the O_2 permeability of the packaging material. Syneresis (drip loss) is also influenced by the choice of packaging material. High O_2 permeability of the package is a major factor that can lead to high drip loss, particularly for high-fat foods. Textural changes within frozen foods can also be influenced by the permeability of the package to both O_2 and water vapor. Table 32.5 shows some relative O_2 and water vapor transmission rates of selected packaging materials (George 2000).

Probably the most popular type of package used today for frozen seafood is the wax- or plastic-coated paperboard carton. When properly sealed, these cartons can provide both a good O_2 and moisture barrier. A protective overwrap of plastic, coated cellophane, or coated paper may be used, and the product may be encased in an inner bag for additional protection. The greatest disadvantage of these packages is the presence of voids when the package is loose-filled. In such cases, frost accumulation occurs from moisture migration within the package, and freezer burn can result. In addition, these cartons do not always provide a good O2 barrier and rancid flavors may develop, but for solid blocks of frozen fillets or other block-frozen seafood, however, such a package can provide good protection and stacks neatly and compactly both in the store and home freezer. For added convenience, fillets are often layer-packed with sheets of plastic film sandwiched between fillets to ease the removal of separate fillets and to facilitate rapid thawing.

Many frozen seafood are also packaged in plain coated cellophane or plastic bags or wrappers. These are some of the least costly packages available and can provide a good moisture vapor–O_2 barrier with the correct film selection and proper sealing (Lanier 1990; George 2000).

Master cartons for frozen seafood that are glazed or packaged in smaller bulk or retail-sized containers may often have many openings to allow the free exchange of cold air with the packaged product. Within the inner bags or cartons, the product may be individually quick frozen (IQF) and loose-filled, or tightly frozen together in either blocks or layer packs. These layer packs should separate the seafood layers by Pliofilm, parchment, or waxed paper; a satisfactory layer pack would ensure the easy removal of each product yet be less expensive than producing IQF products (Lanier 1990).

Higher barrier, shrink-film vacuum packaging certainly offers the most protection to frozen seafood. Whereas a glaze may reduce the likelihood of freezer burn (moisture loss) in loose-filled carton- or wrapper-packaged seafood, rough handling of the package can result in shattering or flaking of the glaze and a lack of protection. Higher barrier vacuum shrink films provide a tougher protective coat.

Foil laminates (metal foil bonded to a plastic or other flexible film) provide excellent barrier/protective properties also. These packaging materials are again most costly, and do not allow a view of the packaged product, but for frozen items they may provide a higher quality image. Additionally, they may be useful in some "cook-in-the-package" applications. The ultimate in convenience packaging for frozen seafood are those packages that may be used directly as containers for cooking the product. One such package is the so-called boil-in-bag pouch, which may be of nylon or a similar transparent, flexible, or foil laminate. These pouches are excellent for such heat-and-eat products as sauces, stews, bisques, and chowders, as well as some chunky-type dishes. For processed seafood of a less fluid form, such as fillets in sauce, stuffed items, and so on, aluminum or the newer oven-proof paperboard trays are popular (Lanier 1990).

32.3.3 Techniques

32.3.3.1 Modified atmosphere packaging (MAP)

To date, modified atmosphere packaging (MAP) appears to be the only technique available that consistently delays microbial growth, ensures freshness, and provides a long shelf life for fresh meat. Nevertheless, shelf-life goals can only be achieved with strict temperature control to achieve maximum microbial inhibition (Day 2000; Buffo & Holley 2005).

Packaging foods in atmospheres containing an inert gas and CO_2 inhibit both oxidative reactions and microbial spoilage. Oxidation of foods is a largely chemical process driven by O_2, which can be blocked by the displacement of O_2 by N_2. Microbial spoilage is caused by the growth of microbes upon foodstuffs, which can be blocked by the microbiocidal and microbiostatic effects of CO_2 (Sivertsvik 2003; Spencer 2005).

The principle of MAP is the replacement of air in the package with a fixed gas mixture. Once the gas mixture is introduced, no further control of the gas composition is exercised, and the composition will inevitably change. MAP is also known as "gas packaging" or "gas exchange packaging," but these are not recommended on packaging labels since consumers often perceive the word gas negatively. Lately, MAP has also been referred to as protective atmosphere packaging or, when used in labeling, "Packaged in a protective atmosphere." MAP should not be confused with controlled atmosphere packaging/storage (CAP/S), where the atmosphere composition is maintained and controlled throughout the storage time. CAP is mostly used in the transport and storage of selected postharvest products (Sivertsvik *et al.* 2002).

Table 32.6 Advantages and disadvantages with modified atmosphere packaging.

Advantages	Disadvantages
• Shelf-life increase by possibly 50–400% • Reduced economic losses due to longer shelf life • Decreased distribution costs, longer distribution distances, and fewer deliveries required • Provides a high quality product • Easier separation of sliced products • Centralized packaging and portion control • Improved presentation—clear view of product and all-around visibility • Little or no need for chemical preservatives • Sealed packages, barriers against product recontamination, and drip from package • Odorless and convenient packages	• Added costs • Temperature control necessary • Different gas formulations for each product type • Special equipment and training required • Product safety to be established • Increased pack volume—adversely affects transport costs and retail display space • Loss of benefits once the pack is opened or leaks • Dissolved CO_2 into the food could lead to pack collapse and increased drip

Source: From Sivertsvik *et al.* (2002).

In order to reduce the concentration of O_2 in contact with foods, the O_2 must be removed by vacuum and/or displaced by an inert gas. The efficiency of this process depends on the design of the packaging machinery, the package (design and materials), and the physical properties of the MAP gas. Removal and exclusion of O_2 by displacing air with MAP gas inhibits the rate of oxidation, which directly slows the rate of degradation of the valuable flavor and color components. To block the oxidation entirely, however, the complete removal of entrained and dissolved O_2 is necessary (Spencer 2005).

Vacuum packaging is also occasionally looked upon as a type of MAP, but this is not generally regarded as MAP since the atmosphere is not altered but only removed from the package. The most apparent advantage of MAP is the achieved increased shelf life, but MAP also has several other advantages and also disadvantages (see Table 32.6). The effectiveness of MAP on extending shelf lives, however, is dependent on several factors: type of food, initial quality of the raw material, gas mixture, storage temperature, hygiene during handling and packaging, gas/product volume ratio, and the barrier properties of the packaging material.

There are four categories of preservative packaging that can be used with raw muscle foods (Table 32.7). These are vacuum packs (VP), high O_2 modified atmosphere packs (high O_2 MAP), low O_2 modified atmosphere packs (low O_2 MAP), and controlled atmosphere packs (CAP) (Day 2000; Buffo & Holley 2005; Gill & Gill 2005).

The three major gases used in the MAP of foods are O_2, N_2, and CO_2. For most food products, different combinations of two or three of these gases are used, chosen to meet the needs of the specific product. Usually for nonrespiring products, where microbial growth is the main spoilage parameter, a 30–60% CO_2 split is used, the remainder being either pure N_2 (for O_2 sensitive foods) or combinations of N_2 and O_2 (Sivertsvik *et al.* 2002; Gill & Gill 2005).

The specific attributes of different raw meat tissues determine which of these various packaging systems is best suited for preservation. MAP is traditionally used to preserve the freshness of fresh produce, meats, and fish by controlling their biochemical metabolism. N_2 flushing, vacuum packaging, and CO_2 injection have been used commercially for many years.

The exception is fish with high dark muscle content, such as active predators like tuna. In those fish, the distinct red color of dark muscle is considered attractive, so the preservation of the red color is desirable. The dark muscle in fillets of such fish discolors, as does other red

Table 32.7 Forms of modified atmosphere packaging.

Type	Packaging gases	Film gas permeability	Atmosphere	Notes
Vacuum pack	Residual air	Low	Anaerobic	Residual O_2 consumed by enzymatic reaction or respiration
High O_2 MAP	O_2, CO_2, and often N_2	Low	Aerobic	O_2 decreases over time
Low O_2 MAP	CO_2, N_2, and residual O_2	Low	Anaerobic	Residual O_2 consumed by enzymatic reaction or respiration
CAP	CO_2, N_2	Impermeable	Anaerobic	Stable atmosphere; O_2 scavenger may be used

Source: From Gill and Gill (2005).

muscle tissue, and discoloration can be retarded or prevented by packaging in VP, CAP, or high O_2 MAP (Oka 1989; Gill & Gill 2005). It should be noted that due to the localization of dark muscle, it may not be visible to the consumer in whole fish and certain fillet cuts. The effects of packaging on the appearance of dark muscle may then have little effect on the acceptability of some forms of fish, even when dark muscle is present.

With fish such as salmon, in which the intense color of the muscle tissue is due to carotenoid pigments assimilated from fish food, any colors imparted by the muscle pigment are immaterial. Carotenoid colors are unaffected by VP, but colors may become lighter during prolonged storage in atmospheres with high CO_2concentrations (Randell *et al.* 1999; Gill & Gill 2005).

Consumer acceptance of whole fish is dependent on the appearance of tissues other than muscle. The appearance of such features as belly flaps, gills, and cornea of the eye has been reported to be adversely affected by high CO_2 atmospheres (Haard 1992; Gill & Gill 2005). However, no deterioration in the appearance of whole salmon was observed with 100% CO_2 by Sivertsvik *et al.* (1999). Low concentrations of carbon monoxide can be used to preserve gill color in anoxic atmospheres, but the same regulatory difficulties that limit its use in red meats exist (Gill & Gill 2005).

To avoid the pack collapse and possible bleaching of muscle that may occur with high concentrations of CO_2, high O_2 MAP is often used with lean fish. Even so, the effects of VP and low O_2 MAP on the appearance of fish can be small, and the appearance of most fish is not greatly enhanced by the inclusion of O_2 in pack atmosphere. There appears to be no use of CAP with fresh fish, perhaps because of the supposed adverse effects of the high concentration of CO_2. However, fatty fish are packaged in low O_2 MAP, with input gases containing up to 80% CO_2, apparently without adverse effects on the appearance of the product (Sivertsvik *et al.* 2002; Gill & Gill 2005).

Argon is another inert gas that has been used in MAP and processing of a variety of foodstuffs. The use of argon improves food quality for a number of reasons. While both N_2 and argon are inert gases, their physical properties differ in significant ways. Since argon displaces and excludes O_2 more efficiently than N_2, it provides better control against oxidation of flavor and color components of foods. In addition, using argon results in the use of less CO_2 to control microbes. As CO_2 is reactive and generally deleterious to product quality, the quality of the product is better with the use of argon rather than N_2 (Spencer 2005).

For a MAP system to work effectively, optimal packaging material with proper gas permeability properties must be selected. The use of MAP systems is attractive to the food industry because there is a fast-growing market for minimally processed fruits and vegetables, nonfrozen chilled meats, ready-to-eat meals, and semiprocessed bulk foods (Day 2000; Han 2005a; Spencer 2005). The type of package used will depend on whether the product is destined for the retail or the catering trade. Popular options include flexible "pillow packs," "bag-in-box," and semirigid tray and lidding film systems (Day 2000).

MAP is used in at least three major packaging areas: retail packages, bulk, and master bags. Retail packages are usually made of barrier packaging materials on a deep drawing (form–fill–seal) or flow-pack machine. Retail packages are also made from preformed trays with barrier lidding film. Bulk packages are larger units meant for transportation and storage, usually as large barrier bags. Master bags are also large MAP units with several nonbarrier retail packages inside. The purpose of master bags is to provide for the transportation and storage of products, and they are usually opened prior to retail sale. The nonbarrier retail packages are then removed from the master bag and sold under non-MAP conditions (Sivertsvik *et al.* 2002).

The packaging gases permeate through the plastic films at different rates according to the type of polymer and the temperature. Packaging materials with sufficient barrier properties need to be used according to the shelf life and storage temperature characteristics of the specific products. Most packaging materials used for MA products consist of several layers in order to achieve low water vapor transmissions, high gas barriers, and mechanical strength to withstand handling during the packaging and distribution process. Typical MA materials consist of laminations of polyester and polyethylene, nylon and polyethylene, polyvinyl dichloride, orientated PP, and many more. The cost of packaging materials is usually directly related to the barrier properties, which force the producer to evaluate the necessary quality of the material versus its cost. Obviously a baguette requiring a shelf life of several months in a 100% CO_2 environment at an ambient temperature will need a better barrier than, for example, a raw fish product stored at 1°C for only 10 days (Sivertsvik *et al.* 2002).

Though changes in the form of myoglobin will occur in fish muscle packaged in VP or low O_2 MAP, this has little effect on the appearance and acceptability to consumers of fish in which there is little dark muscle (Silva & White 1994; Hong *et al.* 1996; Gill & Gill 2005).

32.3.3.2 Vacuum

Vacuum packaging removes most of the available O_2 needed by spoilage bacteria, and also inhibits fat oxidation and rancidity. For such packaging to be effective in inhibiting spoilage, a higher-barrier packaging film is necessary to prevent the leakage of O_2 into the package. The gas permeability of a particular film is determined by its chemical composition as mentioned in Tables 32.3 and 32.5.

In order to maintain a vacuum around the food, high O_2 barrier materials are required. Although the requisite O_2 barrier for VP depends on the type of food packaged, O_2 transmission rates of less than 15 cm3/m2/day1/atm1 are generally required. Also, packaging materials with low water vapor transmission rates must be used. Typical VP materials consist of coextruded or laminated films such as OPP/EVOH/PE, PA/PE, PET/PE, OPP/PVDC/PE, OPP/PVDC/OPP, and PVC/EVOH/PVC (Day 2000).

Unfortunately, removing O_2 from a package may favor the growth of the bacteria, which can produce botulinal toxin; but this eventuality is likely to occur when the product is temperature abused such that it is allowed to warm to a fairly high temperature for the extended period. However, even the slim possibility of botulinal toxin developing is enough to deter most packers from currently considering vacuum packaging of chilled fish in a barrier-type film (Lanier 1990).

Vacuum packaging in permeable films that allow gas exchange with the air does not seem to offer any increased safety risk over conventional meat wraps. Such a package would not by itself, however, increase the shelf life of the product, except by preventing further contamination of the product (i.e., from hands, air, dust, etc.).

Vacuum packaging for frozen fish in O_2-impermeable (high barrier) films is highly recommended, provided the package is removed or opened prior to thawing (George 2000).

32.3.3.3 Active systems

Due to recent outbreaks of contaminations associated with meat products, as well as growing concerns regarding the safety of intermediate moisture foods, active packaging has been greatly developed in recent years (Day 2000; Ahvenainen 2003a, b; Coma 2008).

The term "active packaging" was first applied by Labuza (1987) and may be defined as packaging that performs some desired function other than merely providing a barrier to the external environment. Active packaging should not be confused with "intelligent packaging," which informs or communicates with the consumer regarding the present properties of the food, or records aspects of its history (Hurme *et al.* 2002; Quintavalla & Vicini 2002; Rooney 2002, 2005; Zhou *et al.* 2010).

As mentioned in Quintavalla and Vicini (2002), a packaging material is defined as "a type of packaging that changes the condition of the packaging to extend shelf life or improve safety or sensory properties while maintaining the quality of the food" (Day 2000; Coma 2008).

Active packaging can be defined as performing some desired role in food quality or safety other than to provide an inert barrier to external conditions. Specific reasons for developing active packaging are summarized as follows: (1) removal of an unwanted component, (2) addition of a desired ingredient, (3) package surface antimicrobial activity, and (4) changed physical properties of the package (Hurme *et al.* 2002; Rooney 2002).

The decision to consider active packaging for a food or beverage is commonly based on factors typically involved in any package selection. These considerations include economic advantages, process engineering limitations, convenience in use, environmental impacts, and secondary effects resulting from some other change in processing or packaging. The latter effects may result from the introduction of a new product due to lifestyle changes or the availability of technologies that remove a limitation formerly experienced. Some of the constraints on optimizing the processing, packaging, and distribution of foods and beverages are shown in Fig. 32.2.

Active packaging (AP) is the incorporation of specific compounds into packaging systems that interact with the contents or environment to maintain or extend product quality and shelf life, while intelligent or smart packaging provides for sensing of the food properties or package environment to inform the processor, retailer, and/or consumer of the status of the environment or food (Kerry *et al.* 2006; Zhou *et al.* 2010).

There are many applications of active packaging technologies, several of which have been commercialized and are used in the food industry; these include O_2 scavenging, CO_2-absorbing, moisture-scavenging (desiccation), antimicrobial systems, and others—see Table 32.8 (Day 2000; Hurme *et al.* 2002; Quintavalla & Vicini 2002; Ahvenainen 2003a, b; Han 2005a; Rooney 2005; Coma 2008; Zhou *et al.* 2010).

Active packaging may require some level of adaptation at the stages of plastic extrusion, package fabrication, or when the pack is filled and sealed. If the chemistry causing activity is inhibited by the presence of antioxidants, it may be necessary to control their addition more during extrusion. Some additives would be unstable at extrusion temperatures (200–370°C), and so it is likely that they will be applied to the packaging as internal coatings. It is more likely that adaptation will be required in the package converting or shortly before filling. This occurs with the insertion of sachets or the application of self-adhesive labels to lidding film for thermoformed trays. Triggering of O_2-scavenging systems by means of brief exposure to

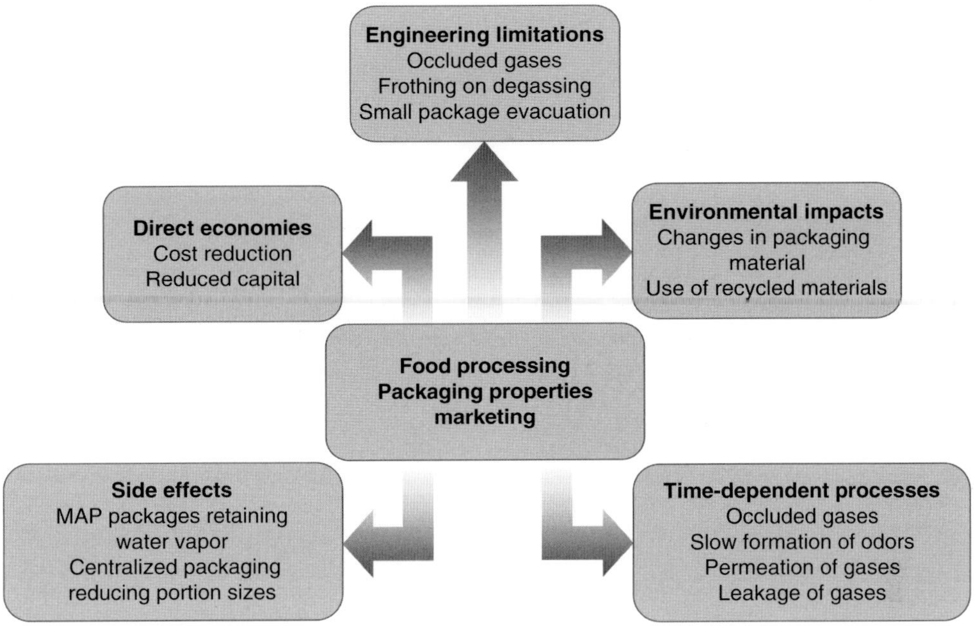

Fig. 32.2 Current constraints on food processing, packaging, and marketing.

UV light may occur at the film stage, with empty packs as with PET bottles, or after package filling (Rooney 2005).

When an edible film and coating contributes to the packaging of a food, the coating performs first as a food component (Lacroix & Cooksey 2005; Bonilla *et al.* 2010; Zhou *et al.* 2010). However, because it is normally made from one or more food constituents it may need protection against microbial activity. Hence, if a mobile antimicrobial agent is incorporated therein, the coating can serve several additional functions. These include providing some self-preservation, helping to reduce the microbial load on the food surface, and providing an outer surface with antimicrobial properties. The latter is potentially important when used with finger foods such as hors d'oeuvres. The requirements of foods that involve edible coatings can be very complex, so the active contribution provided by the additive may not be a major consideration in the formulation of the coating (Sivertsvik 2003; Rooney 2005; Zhou *et al.* 2010).

The use of edible films and coatings as carriers of active substances has been suggested as a promising application of active food packaging. In fact, the use of edible films and coatings for food packaging without active substances is arguably also an application of active food packaging, since the edibility and biodegradability of the films are additional functions not offered by conventional packaging materials (Ahvenainen 2003a, b; Han & Gennadios 2005).

O_2 is responsible for many degradation processes in foods that are commonly delayed with the application of antioxidant agents. The inclusion of strongly flavored antioxidants (essential oils, N-acetylcysteine, glutathione, etc.) in edible films and coatings allow for their encapsulation and reduces their strong aroma. Additionally, oxidation could be effectively reduced by films and coatings through their O_2 barrier effect, which is highly dependent on the water availability. This work reviews the latest studies on the antioxidant effect of edible films and coatings (Coma 2008).

The antioxidant effect of edible films is strongly linked to their O_2 permeability. Films made from proteins and carbohydrates are excellent barriers to O_2, because of their tightly

Table 32.8 Selected examples of active packaging.

Active packaging system	Mechanisms	Actual and potential food applications
O_2 scavengers	1. Iron based 2. Metal/acid 3. Metal (e.g., platinum) catalyst 4. Ascorbate/metallic 5. Enzyme-based beverages	Bread, cakes, biscuits, pizza, pasta, cheese, cured meats and *fish*, coffee, salts snack foods, and dried foods
CO_2 scavengers/emitters	1. Iron oxide/calcium hydroxide 2. Ferrous carbonate/metal halide 3. Calcium oxide/activated charcoal 4. Ascorbate/sodium bicarbonate	Coffee, fresh meats and *fish*, nuts and other snack-food products, and sponge cakes
Ethylene scavengers	1. Potassium permanganate 2. Activated carbon 3. Activated clays/zeolites	Fruit, vegetables, and other horticultural products
Preservative releasers	1. Organic acids 2. Silver zeolite 3. Spice and herb extracts 4. BHA/BHT antioxidants 5. Vitamin E antioxidant	Cereals, meats, *fish*, bread, cheese, snack foods, and fruits and vegetables
Ethanol emitters	1. Encapsulated ethanol	Pizza crusts, cakes, bread, biscuits, *fish*, and bakery products
Moisture absorbers	1. PVA blanket 2. Activated clays and minerals 3. Silica gel	*Fish*, meats, poultry, snack foods, cereals, dried foods, sandwiches, and fruits and vegetables
Flavor/odor adsorber	1. Cellulose triacetate 2. Acetylated paper 3. Citric acid dairy products and fruits 4. Ferrous salt/ascorbate 5. Activated carbon/clays/ zeolites	Fruit juices, fried snack foods, *fish*, cereals, poultry, dairy products, and fruits

packed, ordered hydrogen-bonded network structure (Yang & Paulson 2000). The O_2 permeability of edible materials depends on many factors (temperature, thickness, etc.), and is highly dependent on the relative humidity, as corroborated by Hong and Krochta (2006).

As relative humidity increases, more water molecules interact with the material and the film becomes more plasticized. In these conditions, the mobility and the extensive mass transfer across the film are favored. For this reason, the antioxidant ability of edible films should always be tested under controlled relative humidity (Coma 2008).

In dry conditions (low moisture products), the network structure of the film or coating is tightly packed and its O_2 permeability is very limited. This mechanism by itself may have positive effects on the preservation of the quality, because of the reduced O_2 availability in the coated product. In some cases, the addition of antioxidants can entail further protection due to the enhancement of the O_2 barrier properties of the film. However, in these conditions of reduced mobility the actual chemical activity of the antioxidant agents is masked by the O_2 barrier effect. On the other hand, in wet systems the coating network is plasticized and mass transference is favored. In this context, the O_2 permeability of the film or coating is dramatically increased and the specific activity of antioxidant agents could become more relevant (Coma 2008).

32.3.3.3.1 Oxygen-scavenging systems

O_2-scavenging systems have been commercialized in the form of sachets, packaging films, and closures, with different active ingredients (iron, enzymes, dyes) that removes O_2. An O_2-free environment can prevent food oxidation and rancidity, and the growth of aerobic bacteria and molds (Day 2000; Rooney 2002; Sivertsvik 2003; Han 2005a).

The most successful current approach to O_2 removal is the inclusion of iron-based O_2-scavenging sachets inside the pack. These sachets are capable of reducing the headspace O_2 concentration to less than 0.1%. Despite these advances in the refinement of iron-based systems, the most dramatic advances have occurred in the development of plastics that react with O_2 (Brody *et al.* 2001; Rooney 2002).

32.3.3.3.2 Moisture-scavenging systems

Moisture-scavenging systems have been used for a very long time for packaging dried foods, moisture-sensitive foods, pharmaceuticals, and electronic devices; in these systems, desiccant materials are included in the package in the form of a sachet. Recently, the sachets have contained humectants as well as desiccants to control the humidity inside the package more specifically. Moisture-scavenging systems that are based on desiccation are evolving to control the moisture by maintaining a specific relative humidity inside the package by absorbing or releasing the moisture (Day 2000; Han 2005a).

Excess moisture is a major cause of food spoilage and different humidity absorbers are used to protect dried products from humidity damage. However, these absorbers have a limited effect on seafood products. Several companies manufacture moisture drip absorbent pads, sheets, and blankets for liquid water control in watery foods such as fish. Moisture drip absorber pads or false-bottomed trays are commonly placed under packaged fresh meat, fish, poultry, and prepared fruit to absorb unsightly tissue drip discharge. Larger sheets and blankets are used for the absorption of melted ice from chilled seafood during airfreight transportation (Sivertsvik 2003).

32.3.3.3.3 Antimicrobial packaging

Among many applications such as O_2-scavenging packaging and moisture-control packaging, antimicrobial packaging is one of the most promising innovations of active packaging technologies (Floros *et al.* 1997; Han 2005b), that is, the incorporation of antimicrobial substances in food packaging materials to control undesirable growth of microorganisms on the surface of foods. Antimicrobial packaging is an extremely challenging technology that could extend shelf life and improve food safety in both synthetic polymers and edible films (Day 2000; Brody *et al.* 2001; Han 2003; Zhou *et al.* 2010).

Antimicrobial packaging applications are directly related to food microbial safety and bioterrorism, as well as to shelf-life extension by preventing the growth of spoilage and/or pathogenic microorganisms, and enhancing the safety of packaged products. The growth of spoilage microorganisms reduces the food shelf life, while the growth of pathogenic microorganisms endangers public health. Antimicrobial packaging systems consist of packaging materials, in-package atmospheres, and packaged foods, and are able to kill or inhibit microorganisms that cause food-borne illnesses (Quintavalla & Vicini 2002; Han 2003, 2005a, b; Rooney 2005).

Antimicrobial films can be defined in five basic categories as follows:

1. Incorporation of the antimicrobial substances into a sachet connected to the package from which the bioactive substance is released during further storage. Common packaging materials can be utilized without the use of alternative packaging materials.
2. Direct incorporation of the antimicrobial into the packaging film. When applied in a hot extrusion material, thermoresistance and shearing resistance of the antimicrobial must be considered.
3. Coating of the packaging with a matrix that acts as a carrier for the additive or antimicrobial agent. The substance will not be submitted to high temperature or shearing forces; moreover, it could be applied as the later step. These categories of materials can release the antimicrobial agents onto the surface of the food. The antimicrobial agents may either be released through evaporation in the headspace (volatile substances) or migrate into the food (nonvolatile additives) through diffusion. The system is more efficient than a direct application of the antimicrobial agent onto meat surfaces, because it slows the migration of the agents away from the surface, and thus helps to maintain high concentrations where they are needed.
4. Utilization of inherently antimicrobial polymers exhibiting film-forming properties, such as cationic amino-polysaccharides, or polymers that are chemically modified to produce bioactive properties. Packaging materials with bioactive agents chemically bounded to the polymer can be included in this category and the immobilization of the biocide on the packaging polymer could be used for non-food-grade molecules. For this system, the release of the biocide agent is not required, or is prohibited in the case of non-food-grade biocides. Therefore, legal issues and standards concerning the rate of migration of substances in packaging food products do not limit the development of such bioactive materials. However, the limitation of such a system is the direct contact between packaging and food.
5. Utilization of bioactive edible coatings directly applied onto the foods. The limitation is that the bioactive agent should be approved as a food additive (Sivertsvik 2003; Cooksey 2005; Coma 2008; Zhou *et al.* 2010).

There are various factors to be considered in designing antimicrobial systems. Antimicrobial systems can be constructed by using antimicrobial packaging materials, antimicrobial inserts (such as sachets) to generate antimicrobial atmosphere conditions inside packages, or antimicrobial edible food ingredients in the formulation of foods (Table 32.9).

Since antimicrobial packaging systems are designed to control the growth of microorganisms in packaged foods, the systems essentially consist of packaging materials (or packages), foods, the in-package atmosphere, target microorganisms, and antimicrobial agents. These five elements are related to one another and to the final system design features (Brody *et al.* 2001; Sivertsvik 2003; Han 2005b; Coma 2008).

Most food packaging systems consist of food products, headspace atmosphere, and packaging materials. Any one of these three components of food packaging systems could possess an antimicrobial element to increase antimicrobial efficiency (Brody *et al.* 2001; Sivertsvik 2003; Han 2005b).

The most common chemical antimicrobials used by researchers are the various organic acids. Organic acids are widely used as chemical antimicrobial agents because their efficacy

Table 32.9 Examples of potential antimicrobial agents for antimicrobial food packaging agents.

Classification	Antimicrobial agents	Seafood/seafood products
Organic acids	Acetic acid, benzoic acid, lactic acid, citric acid, malic acid, propionic acid, sorbic acid, succinic acid, tartaric acid, mixture of organic acids	Fish fillets, sausage
Acid salts	Potassium sorbate, sodium benzoate	–
Acid anhydrides	Sorbic anhydride, benzoic anhydride	–
Para benzoic acids	Propyl paraben, methyl paraben, ethyl paraben	–
Alcohol	Ethanol (emitters)	Semi-moist and dried fish products
Bacteriocins	Nisin, pediocin, subtilin, lacticin	Fresh fish, shellfish, sausage
Fatty acids	Laurie acid, palmitoleic acid	–
Fatty acids esters	Glycerol mono-laurate	–
Chelating agents	Ethylenediamine tetraacetic acid (EDTA), citrate, lactoferrin	Fresh fish, shellfish, sausage
Enzymes	Lysozyme, glucose oxidase, lactoperoxidase	Fresh fish, shellfish, sausage
Metals	Silver, copper, zirconium	–
Antioxidants	BHA, BHT, TBHQ, iron salts	Fresh fish fillets
Antibiotic	Natamycin	–
Fungicides	Benomyl, Imazalil, sulfur dioxide	–
Sanitizing gas	Ozone, chlorine dioxide, carbon monoxide, CO_2	Fresh fish fillets
Sanitizers	Cetyl pyridinium chloride, acidified NaCi, triclosan	Fresh fish, shellfish
Polysaccharide	Chitosan	Fresh fish fillets
Phenolics	Catechin, cresol, hydroquinone	–
Plant volatiles/ essential oils	Allyl isothiocyanate, cinnamaldehyde, eugenol, linalool, terpineol, thymol, carvacrol, pinene	Fresh and processed fish (dried fish, nuggets, ground fish meat), shellfish
Plant/spice extracts	Grape seed extract, grapefruit seed extract, hop beta acid, Brassica erucic acid oil, rosemary oil, oregano oil, basil oil, other herb/spice extracts and their oils	Fresh and processed fish (dried fish), shellfish[†]
Probiotics	Lactic acid bacteria (*Lactobacillus reuteri*) to control *E. coli 0157:H7*	–

Sources: From Sivertsvik (2003) and Han (2005b).

is generally well understood and cost-effective. Many organic acids, including fatty acids, are naturally existing chemicals and have been used historically. Currently, most of them are produced by chemical synthesis or chemically modified from natural acids. Organic acids have characteristic sensitivities to microorganisms. Therefore, the correct selection of organic acids is essential to have effective antimicrobial agents. Mixtures of organic acids have a wider antimicrobial spectrum and stronger activity than a single organic acid (Brody *et al.* 2001, Coma 2008; Han 2005b; Zhou *et al.* 2010).

An antimicrobial agent has specific inhibitory activity and mechanisms against each microorganism. Therefore, the selection of antimicrobial agents is dependent on their efficacy against a target microorganism and it is strongly recommended that experiments regarding the efficiency examination of an antimicrobial packaging system should be conducted using real food instead of culture broth or agar media (Han 2005b).

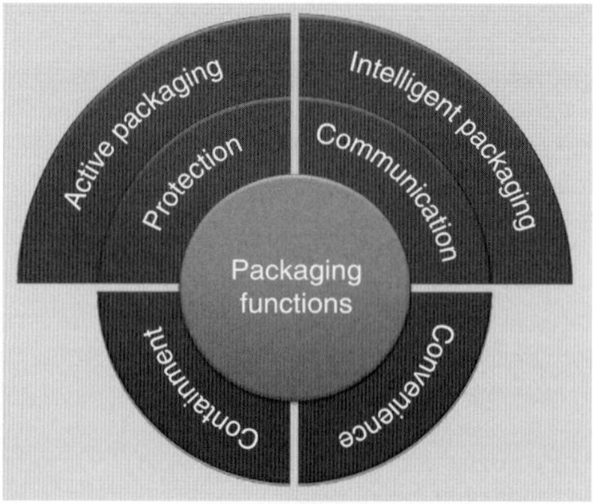

Fig. 32.3 Model of packaging functions.

32.3.3.4 *Intelligent pack*

Intelligent (or smart) packaging has been categorized both as a part of active packaging and as a separate entity, depending on different viewpoints. It contains intelligent functions that have been researched to enhance convenience for food manufacturing and distribution and, increasingly, to improve food security and safety verification (Hurme *et al.* 2002; Han *et al.* 2005a; Zhou *et al.* 2010).

Many intelligent packaging systems use sensors and indicators for a variety of measurements including fluorescence-based O_2 gas detection, temperature monitoring, toxic compounds, freshness through monitoring of specific components, package integrity, and product identification (Zhou *et al.* 2010).

Intelligent packaging refers to a package that has the ability to (1) track the product, (2) sense environmental changes (inside and outside the package), and (3) inform the changes to users (manufactures, retailers, and consumers).

Intelligent packaging devices are capable of providing information about the function and properties of a packaged food and can provide assurances of pack integrity, tamper evidence, product safety, and quality (see Fig. 32.3). Intelligent packaging can also be utilized in applications such as product authenticity, antitheft, and product traceability.

Rodrigues and Han (2003) defined intelligent packaging as having two categories: simple intelligent packaging and interactive or responsive intelligent packaging. In the latter, the packaging contains sensors that notify consumers that the product is impaired, and they may begin to undo the harmful changes that have occurred in the food product (Karel 2000; Han *et al.* 2005a). Such packaging systems contain devices that are capable of sensing and providing information about the functions and properties of the packaged foods (Day 2000; Hurme *et al.* 2002), and/or contain an external or internal indicator for the active product history and quality determination (Han *et al.* 2005a).

These types of devices can be divided into three groups. The first type are the external indicators. These indicators are attached outside the package, and include time–temperature indicators and physical shock indicators. The second type are the internal indicators, which are placed inside the package. They are either placed in the headspace of the package or attached to the lid, for example, O_2 leak indicators, CO_2, microbial, and pathogen indicators (Table 32.10). The third type of devices are the indicators that increase the efficiency of information flow and effective communication between the product and the consumer. These

Table 32.10 Examples of external and internal indicators and their working principles used in intelligent packaging.

Technique	Principle/reagents	Information given	Application
Time–temperature indicators (external)	Mechanical, chemical, enzymatic	Storage conditions	Foods stored under chilled and frozen conditions
O_2 indicators (internal)	Redox dyes, pH dyes, enzymes	Storage conditions Package leak	Foods stored in packages with reduced O_2 concentration
CO_2 indicator (internal)	Chemical	Storage conditions Package leak	Modified or controlled atmosphere food packaging
Microbial growth indicators (internal/external) and freshness indicators	pH dyes, all dyes reacting with certain metabolites	Microbial quality of food (i.e., spoilage)	Perishable foods such as meat, fish, and poultry
Pathogen indicators (internal)	Various chemical and immunochemical methods reacting with toxins	Specific pathogenic bacteria such as *Escherichia coli* 015	Perishable foods such as meat, fish, and poultry

Sources: From Hurme *et al.* (2002) and Han *et al.* (2005a, b).

products include special bar codes that store food product information such as use and consumption date expiration (George 2000; Han *et al.* 2005a; Zhou *et al.* 2010).

The intelligence of a package can be based on the package's ability to give information about the requirements of the product quality like package integrity (leak indicators) and time–temperature history of the product (time–temperature indicators). Intelligent packaging can also give information on product quality directly (see Fig. 32.4) (Smolander 2003).

For these packaging systems to be practical, they should be easy to use, cost-effective, and capable of handling multiple tasks. Color-indicating labels, for example, should have color changes that are irreversible, easy to read, and easily understood by consumers.

Furthermore, convenience has become a priority for most people due to their busy daily work schedules. Customers are seeking quicker and easier means of accessing better quality foods. In addition to convenience of consumption, the convenience at the distribution level and the product quality level are likely to become more important in the future.

Freshness indicators are used to indicate if the product's quality has been impaired due to exposure to unfavorable conditions during storage and transportation. The package is usually equipped with a reversible color-changing device that tells the consumers if the package has undergone deterioration, along with a partial or complete history of the product. Some of the examples described below are internal or external indicators used to monitor elapsed time, time–temperature, humidity, shock abuse, and gas concentration changes (Ahvenainen & Hurme 1997; Smolander 2003; Han *et al.* 2005a).

COX Technologies' "FreshTags®" color-indicating tags consist of a small label attached to the outside of the packaging film. It is used to monitor the freshness of seafood products, and consists of a reagent-containing wick contained within a plastic chip (Millers *et al.* 1999). As the seafood ages, spoils, and generates volatile amines (microbial breakdown products) in the headspace, these are allowed to contact the reagent, causing the wick in the tag to turn bright pink. This indicates that the odor development is showing that the seafood is past the point of useable freshness (Hurme *et al.* 2002; Han *et al.* 2005a; Pacquit *et al.* 2007; Zhou *et al.* 2010).

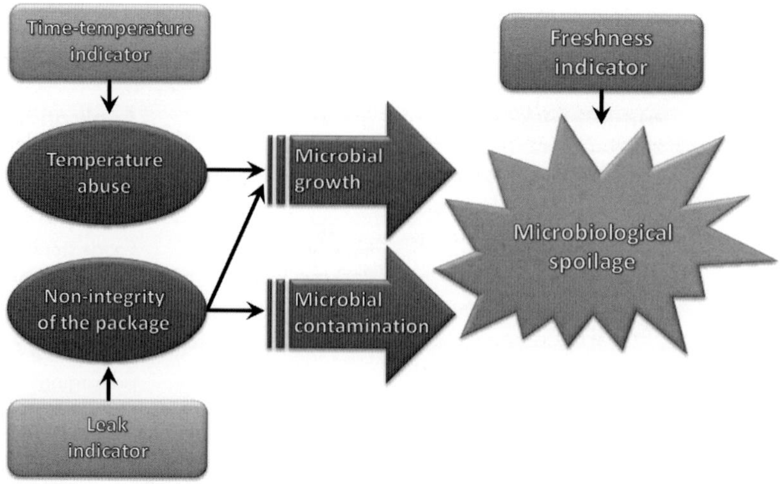

Fig. 32.4 Quality indicators for packaged food products can be either on direct or indirect freshness evaluation.

Food "traceability" is now a legal requirement that establishes a chain of responsibility throughout the whole food chain, and there is consequently great interest from the fisheries industry, retailing industry, consumer's right watchdogs, and food safety controlling bodies in developing accurate, cost-effective, rapid, reliable, noninvasive, and nondestructive methods or devices to evaluate real-time freshness of fish and seafood products. Emphasis is on the ones that would reflect and account for the product's history and its storage conditions from "harvest-to-home." One concept to meet this requirement is that of the development of a simple freshness color indicator in the form of in-package sensor spots that monitor spoilage in fish and seafood products. Volatiles amines, such as trimethylamine (TMA), ammonia (NH_3), and dimethylamine (DMA), contribute to a quantity known as total volatile basic N_2 (TVB-N) and are the characteristic substances responsible for the fishy odor and flavor encountered in fish after having past the initial phase of freshness (Pacquit *et al.* 2007).

Laboratory trials on fresh fish fillets (whiting and cod) showed that the sensor accurately tracks the increase in amine concentrations in the package headspace. The response was also found to correlate to changing microbial populations (total viable count or TVC and *Pseudomonas* spp.) (Hurme *et al.* 2002; Pacquit *et al.* 2007).

In an enclosed food package, as the fish product spoils, a pH increase occurs over time within the headspace, which can be detected with an appropriate pH-indicating sensor. The fundamental characteristic of pH indicator dyes that change color when placed in an acidic or basic environment is the key element of this sensor. The sensor response is independently monitored with a simple, inexpensive reflectance colorimeter that we have developed based on LEDs and a photodiode (Pacquit *et al.* 2007).

The use of pH dyes (e.g., bromothymol blue) as indicators to monitor the formation of CO_2 due to microbial growth is one of the most frequent applications in the food packaging industry. The increase in CO_2 levels can be used to detect microbial contamination in some products due to pH dyes that can react to the presence of this by-product. Other pH dye reagents include xylenol blue, bromocresol purple, cresol red, phenol red, and alizarin. Besides CO_2, other metabolites (such as SO_2, NH_4, volatile amines, and organic acids) have been used as target monitoring molecules for pH-sensitive indicators. Other spoilage detection methods include the incorporation of small analytical tools into the package, such as biosensors for the detection of biogenic amines. The increase in diamines in poultry meat can be detected with a putrescine oxidase reactor combined with an amperometric hydrogen peroxide electrode. The system can also be applied in the detection of histamine from rainbow trout meat and biogenic amines from fish muscles (Ahvenainen 2003a; Smolander 2003; Han *et al.* 2005).

The temperature variations in a food product can lead to changes in product safety and quality. An important aspect of fresh fish distribution and consumption is the effective monitoring of time–temperature conditions that affect both safety and overall quality of fish. Increased attention is focused on the role and the logistics of transport, storage, and handling, and the benefits of taking a supply chain perspective are being appreciated. Since in practice significant deviations from specified conditions often occur, temperature monitoring and recording are prerequisite for any logistics management system that aims on product quality optimization at the consumer's end (Giannakourou *et al.* 2005).

There are two types of temperature indicators: the simple temperature indicators and the time–temperature integrators (TTIs) (Pault 1995; Ahvenainen & Hurme 1997; Woolfe 2000; Taoukis & Labuza 2003; Han *et al.* 2005). Temperature indicators show whether products have been heated above or cooled below a reference (critical) temperature, warning consumers about the potential survival of pathogenic microorganisms and protein denaturation during, for example, freezing or defrosting processes (Pault 1995). To determine the changes in the shelf life of the food product, the second type of indicator, a TTI, should be employed (Pault 1995; Taoukis & Labuza 2003; Gonçalves & Blaha 2010).

These indicators display a continuous temperature-dependent response of the food product. The response is made to chemical, enzymatic, or microbiological changes that should be visible and irreversible, and is temperature dependent (Sivertsvik 2003; Giannakourou *et al.* 2005, Han *et al.* 2005; Nuin *et al.* 2008; Gonçalves & Blaha 2010).

Time–temperature integrators or indicators are tools that can potentially be used to monitor the integrated time–temperature impact on fish quality, offering a cost-effective and user-friendly way to detect any problematic points in the chill chain. TTI are defined as simple, inexpensive devices that can show an easily measurable, time- and temperature-dependent change that cumulatively indicates the time–temperature history of the product from the point of manufacture to the consumer (Selman & Ballantyne 1988; Selman 1990; Woolfe 2000; Hurme *et al.* 2002; Giannakourou *et al.* 2005; Gonçalves & Blaha 2010).

TTIs usually display either the current temperature, or respond to some predefined threshold temperature (e.g., freezing point). TTIs usually utilize a physicochemical mechanism that responds to the integration of the temperature history to which the device has been exposed.

TTIs provide an overall temperature history of the product during distribution. The changes shown by TTI can easily be measured, and have been proven to give consistent responses under the same temperature conditions and when not under the influence of light, humidity, and contaminants. Ideal TTIs are able to adjust the residual shelf life of the food product by assessing the already possible quality deterioration. However, one drawback is that these integrators measure the surface temperature and not the actual temperatures of the product (Ahvenainen & Hurme 1997; Sivertsvik 2003; Taoukis & Labuza 2003; Han *et al.* 2005; Nuin *et al.* 2008).

The application of TTIs has been studied to monitor the time–temperature history of several refrigerated fish (Nuin *et al.* 2008), for example, salmon (Otwell 1997), Mediterranean boque (Taoukis *et al.* 1999), sea bass (Koutsoumanis *et al.* 2002), fresh seafood (Mendoza *et al.* 2004), and sea bream (Giannakourou *et al.* 2005). A systematic approach for fish

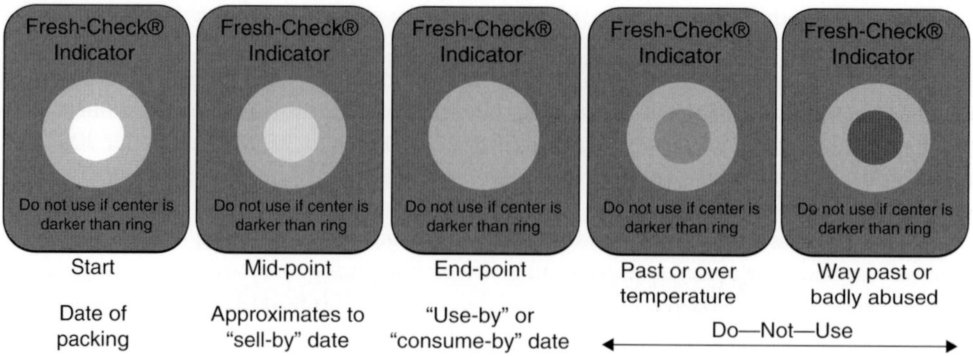

Fig. 32.5 The lifelines "Fresh-Check" indicator—staged examples.

shelf-life modeling and TTI selection in order to plan and apply an effective quality-monitoring scheme for the fish chill chain was developed by Taoukis *et al.* (1999) who modeled the growth of the specific spoilage organisms (SSOs) of the Mediterranean fish boque (*Boops boops*).

Lifelines have developed several indicators all of which show a full time–temperature history. The indicator part consists of polymeric compounds that change color as a result of accumulated temperature exposure. The color change is based on the polymerization of acetylenic monomers, which proceeds faster at higher temperatures leading to a more rapid darkening of the indicator. Developments in Lifelines technology have resulted in the manufacture of the "Fresh-Check®" indicator, designed for consumer use. This device consists of two circles: a small inner circle that contains the polymer, and a printed dark or black outer ring. The inner ring darkens when exposed to time and temperature combinations at a rate predetermined by the durability of the food. The consumer is advised not to consume the product when the inner ring has become darker than the printed outer one—see Fig. 32.5 (Woolfe 2000).

In order to try to link the indicator to microbiological safety as well as quality deterioration, a new development has been the incorporation of a second polymer system in the center ring. If the indicator stays below a pre-set maximum, then the polymer will change color as above, and the change is linked to the durability. If the temperature rises above the maximum, then the second system starts to polymerize, and the center will darken abruptly at the end of a predetermined length of time. Lifelines labels are not physically activated, and once manufactured respond to any temperatures to which they are exposed. Therefore, before use, indicators must, at all times, be stored at −18°C or colder (Woolfe 2000).

The main causes of loss of storage life are fluctuating temperatures and the type of packaging used. Other factors, including type of raw material, prefreezing treatments, and processing conditions, have a great contribution on frozen seafood quality. Time–temperature tolerance (TTT) and product-processing-packaging (PPP) concepts are used to monitor and control the effects of temperature fluctuations on frozen food quality during production, distribution, and storage (Gonçalves & Blaha 2010).

Colored indicators are being developed to (1) show the temperature of food (e.g., liquid crystal coatings that change color with storage temperature), (2) indicate temperature abuse (e.g., wax melts and releases a colored dye when an unacceptable increase in temperature occurs), (3) integrate the time–temperature combination that a food has received after packaging, and (4) to give an indication of the remaining shelf life (see Fig. 32.6).

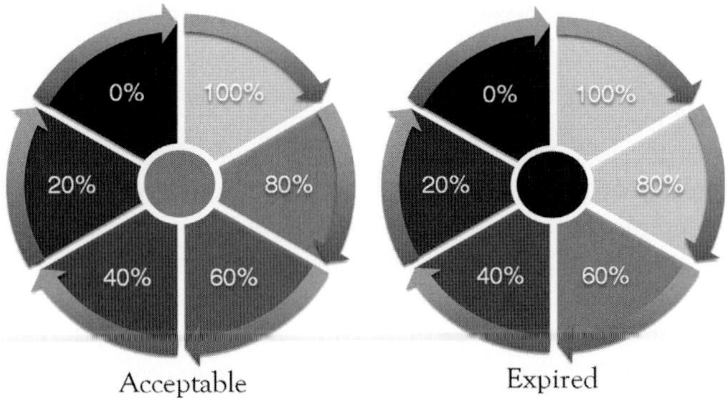

Fig. 32.6 Time–temperature integrator—Indicator based on the color change of the inner circle: the percentages show the equivalence of color with the remaining life.

In the last category, indicators may contain a material that polymerizes as a function of time and temperature to produce a progressive, predictable, and irreversible color change. In another type, a printed label contains diacetylene in the center of a "bull's eye," with the outer ring printed with a stable reference color. The diacetylene gradually darkens in color due to combined time and temperature and when it matches the reference ring the product has no remaining shelf life.

More recently, a bar code system has been developed that is applied to a pack as the product is dispatched. The bar code contains three sections: The first gives information on the product identity, date of manufacture, batch number, etc., to identify each container uniquely; the second identifies the reactivity of a time–temperature indicator; and the third contains the indicator material. When the bar code is scanned by a handheld microcomputer, a display indicates the status and quality of the product with a variety of preprogrammed messages (e.g., "Good," "Don't use," or "Call QC"). A number of microcomputers can be linked via modems to a central control computer, to produce a portable monitoring system that can track individual containers throughout a distribution chain (Fellows 2000; Hurme *et al.* 2002; Gonçalves & Blaha 2010).

Lifelines' Freshness Monitor and Fresh-Check® indicators rely on a solid-state polymerization reaction to give color changes. The indicator consists of a small circle of a polymer surrounded by a printed reference ring. The inside polymer circle darkens if the package has experienced unfavorable temperature exposures (Summers 1992), and the intensity of the color is measured and compared to the reference color scale on the label (Kruijf *et al.* 2002). The faster the temperature increases, the faster the color changes occur in the polymer. Consumers are advised not to consume or purchase the product, regardless of the "use-by" date. These indicators have been used on chilled fresh produce (Järvi-Kääriäinen 2003; Taoukis & Labuza 2003; Han *et al.* 2005).

32.4 MAINTAINING SAFETY IN THE COLD CHAIN

The quality of perishable foodstuffs such as fresh fish can be seriously affected by the temperature control during storage and transportation from processing to consumers. This can be explained by the large impact that temperature and time can have on both microbial and chemical properties of the perishable products (Margeirsson *et al.* 2011).

Fish products, unless appropriately packaged, transported, and stored, will spoil in a relatively short time. Fish shelf life is influenced by a number of factors, such as feeding, initial microbiological quality, season, geographical origin, and handling. Temperature abuse at any stage of the distribution chain usually results in quality loss, decreasing significantly—in the case of extended temperature elevation—the shelf life of fish. Consequently, since temperature largely determines the rate of microbial activity, the main cause of the spoilage of fresh fish products, continuous monitoring of the full time–temperature history of fish would allow for the control of the temperature conditions in the chill chain, which is an issue of central importance (Sivertsvik 2003; Giannakourou *et al.* 2005).

As most changes during the storage of raw muscle foods are undesirable, and changes are slowed by decreasing temperature, it follows that the optimum temperature for storing raw muscle products is the minimum that can be indefinitely maintained without the tissue freezing. That temperature is in practice −1.5 ± 0.5°C. For proper understanding of product storage stability under different circumstances, it is desirable that consideration be given to the storage life at the optimum temperature, as well as at temperatures that might be encountered during storage, distribution, and display. In addition to the preservative effects, the commercial function of the packaging must be considered, but the cost of a packaging is obviously a matter of major commercial concern (Buffo & Holley 2005; Gill & Gill 2005).

Because of the importance of storage and transport temperature, almost all countries in Europe, USA, and many other countries have signed the ATP Agreement (UN 2010) on the international carriage of perishable foodstuffs and on the special equipment to be used for such carriage. Some studies (Giannakourou *et al.* 2005; Margeirsson *et al.* 2008, 2011) have revealed that temperature control in real fish cold chains is quite often far from what is described in the ATP (fish temperature should be as close to 0°C as possible without freezing the products), thereby decreasing product quality, shortening shelf life, and decreasing product value. The temperature regulation in fresh fish distribution chains is actually only a problem for air freight, but not sea freight. This is caused by the fact that more interfaces, where ambient conditions are not well controlled, are found in air logistic chains. Temperature control can in fact be improved in chilled distribution chains for other perishable products such as beef, poultry, and vegetables (Margeirsson *et al.* 2011).

Large quantities of chilled seafood are still shipped loose, packed in ice, in wooden or wax/plastic-coated corrugated boxes. The seafood is usually in direct contact with the ice; however, some delicate products such as fillets may be contained in metal trays or plastic bags immersed in the ice. Although packaging with ice maintains product quality reasonably well, it has several drawbacks: (1) *Product temperature* (melting ice maintains the product temperature at 32°F (0°C)—a more ideal temperature to prolong shelf life would be near 29°F (−1.7°C), just below the freezing point). (2) *Leakage, leaching* (ice packs are leaky and normally must be isolated from other food shipments. Although melting ice may have benefits in terms of "washing" bacteria off the product, leaching of the tissues can also result in significant product weight shrinkage). (3) *Weight of ice* (shipping products on ice increases the shipping weight by 30–50% and is thus more expensive) (Lanier 1990).

Bulk shipping containers that rely on mechanical refrigeration or non-ice refrigerants avoid these problems. The simplest of these bulk shipping containers consist of the same types of inner packs as are used for insulation the product from the ice in iced packs, but they are contained in a master package with no ice. A HDPE deep tray with a flexible film heat-sealed on the top lid is a well-known system introduced. This package is leak- and odor-free and quite strong. The containers are normally made to hold either 10 or 20 lb (4.5–9.1 kg) of seafood and are sized to fit inside a master corrugated carton. Some

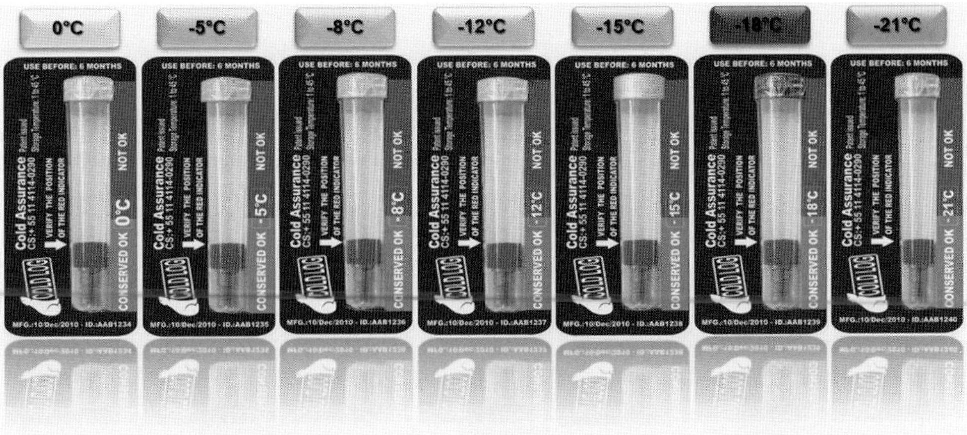

Fig. 32.7 Cold Log® Sensor—Identification to guarantee the cold throughout the cold chain, available in various temperature ranges indicated by different colors.
Source: Courtesy from ColdLog®, www.coldlog.com.br

refrigerants other than ice that may be used in master cartons of chilled fish include CO_2 snow (dry ice at −78.9°C) and the reusable "blue ice" refrigerant packs. Care must be exercised, however, to exhaust CO_2 from the packaging and storage areas to prevent workers from suffocating (Lanier 1990).

The negative impact of unsatisfactory ambient temperature fluctuations during the distribution of perishable products can still be dampened by the thermal insulation of the packaging. Other characteristics of packaging that can influence the quality of the products include cost (including cost related to material disposal, i.e., if the material can be recycled), strength, and space. Here, space includes internal space for cooling both mats and ice in order to maintain low product temperature and space required for storage. Well-insulated expanded polystyrene (EPS) boxes have traditionally been utilized for the export of Icelandic fresh fish products, and these EPS boxes are usually white, manufactured from molded polystyrene beads; up to 98% of the boxes consist of air pores. The air decreases the density and increases the insulation performance but decreases strength and of course increases the required storage volume for the boxes.

Another type of wholesale fresh fish packaging has been receiving increased international attention because of environmental and economic reasons, that is, corrugated plastic (CP) boxes. These boxes are produced from extruded CP (polypropylene) sheets that are 2.0–3.3 mm in thickness. CP boxes can be flat packaged, which can save valuable storage space, but have poor strength; former studies have indicated that the insulation is worse than for EPS boxes (Margeirsson *et al.* 2011).

A new system to avoid loss of frozen foods and increase the quality as well as ensure the integrity of the temperature from the industry to the sale point is the Cold Log® Sensor (see Figs 32.7 and 32.8). The system is equipped with a mechanical device that is activated by freezing the liquid inside the sensor, which enters into merger agreement with the temperature indicated in the model, remaining stable at the same temperature (0°C, −5°C, −8°C, −12°C, −15°C, −18°C, and −21°C). This "spring" is compressed and the position of the plunger on the blue stripe (conserved) indicates that the product was maintained within appropriate temperature conditions.

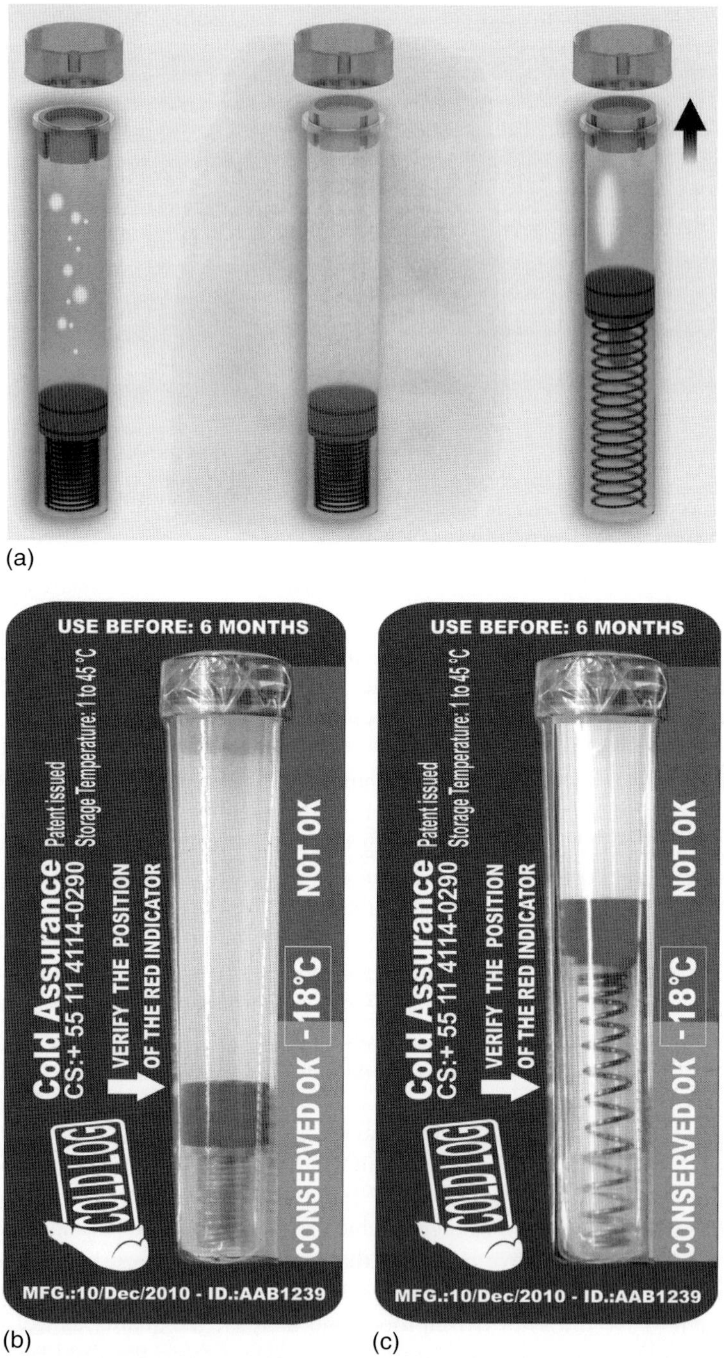

Fig. 32.8 Steps of indication: (a) not activated; (b1) activated; (b2) activated—illustrated along with the blister; (c1) at the position of breaking the cold (changed) inside the blister; and (c2) at the position of breaking the cold—illustrated along with the blister.
Source: Courtesy from ColdLog®, www.coldlog.com.br

Initially, the Cold Log® Sensor is not activated (see Fig. 32.8a) and when the liquid inside gets frozen, it is activated (see Fig. 32.8b). It is recommended to activate the sensor together with the frozen product (food, medicine) that is monitored with Cold Log® Sensor. If there is breakage of the cold chain, spring expels fluid from the sensor tube, being careful to stay the nontoxic fluid within the blister, leaving extended spring irreversibly, and the position of the plunger on the red stripe (changed) indicates that there was a break from the cold (see Fig. 32.8c). Being compact, it offers various forms of application in transport boxes, individual boxes, cold stores and displays, cold rooms, containers, etc.

Considering the importance of the maintenance of the cold chain, Margeirsson *et al.* (2011) investigated the thermal performance of two types of wholesale fresh fish boxes: one made of expanded polystyrene and the other made of CP (polypropylene). They concluded the following: (1) Applying frozen ice packs in fish boxes reduces the impact of temperature abuse on fresh fish fillets. (2) Insulating the performance of expanded polystyrene is significantly better than that of CP. The difference in insulating the performance between the two packaging types is exaggerated when ice packs are used. (3) Insulating properties of the expanded polystyrene boxes makes this type of packaging more suitable for the case of chilled chains with insufficient control. However, since the CP boxes are less insulating, they offer more rapid cooling of the product inside the box. (4) Good agreement was obtained between numerical results and experimental results both for the heat transfer model for the EPS box and the CP box without ice pack. This implies that the models can give valuable information on the temperature distribution inside a thermally loaded freestanding package. The overall absolute error of the numerical model for the EPS box (homogeneous material) was lower (0.4°C) than the corresponding error of the numerical model for the nonhomogeneous CP box (0.7°C).

32.5 QUALITY AND SHELF LIFE FOR PACKED SEAFOOD

The use of packaging to extend the shelf life of foods is not a new phenomenon. The role of packaging, what we call the preservation role, is a fundamental requirement of food packaging. With very few exceptions, the methods by which food is treated to maximize or extend its shelf life rely also on packaging for their success. Without developments in packaging, the food harvesting, processing, and distribution systems would not have developed to their present levels, there would be no easy access to the wide range of foods on offer today, and food wastage due to spoilage, damage, and loss would be high. Packaging, then, is an integral part of the food industry. Apart from the preservation function, packaging has several other important roles to play in delivering safe, wholesome, and attractive foods to the market, and to do this economically and with minimal environmental impact. These roles must be considered at the earliest stages of product development, and not left until the product is ready to launch, when it may be too late to obtain the optimum product/pack combination (Emblem 2000; Gonçalves & Blaha 2010).

During the storage of raw muscle foods at chiller temperatures, various enzyme-mediated and nonenzymic chemical reactions affect the flavor, odor, and texture of tissues. The changes produced by some of the enzymic reactions are desirable. However, most of the changes caused by enzymic activities are undesirable. Therefore, a preservative packaging should ideally inhibit undesirable enzymic activities, but not interfere with, or inhibit, activities that are beneficial. The nonenzymic reactions that affect the organoleptic qualities of raw meats are invariably undesirable, so these should preferably be slowed or prevented by a preservative packaging (Day 1989; Gill & Gill 2005).

Packaging is one of the most important processes to maintain the quality of food products for storage, transportation, and end-use. It prevents quality deterioration and facilitates distribution and marketing. The basic functions of packaging as mentioned earlier are protection, containment, information, and convenience. A good package can not only preserve the food quality but also significantly contribute to a business profit. Beyond the major function of preservation, packaging also has secondary functions—such as selling and sales promotion. However, the main function of food packaging is to achieve preservation and the safe delivery of food products until consumption. During distribution, the quality of the food product can deteriorate biologically and chemically as well as physically. Therefore, food packaging contributes to extending the shelf life and maintaining the quality and the safety of the food products (Day 1989; Emblem 2000; Sivertsvik 2003; Han 2005a; Gill & Gill 2005).

The quality of packaged foods is a combination of attributes that determine their value as human food. These quality factors include visual appearance (freshness, color, defects, and decay), texture (crispness, turgidity, juiciness, toughness, and tissue integrity), flavor (taste and smell), nutritive value (vitamins A and C, minerals, and dietary fiber), and safety (absence of chemical residues and microbial contamination). "Keeping quality" is used more commonly than "shelf life" because of consumer demand for product freshness (Emblem 2000; Singh & Singh 2005).

The quality of the packaged food is directly related to the food and packaging material attributes. Most food products deteriorate in quality due to mass transfer phenomena, such as moisture absorption, O_2 invasion, flavor loss, undesirable odor absorption, and the migration of packaging components into the food. These phenomena can occur between the food product and the atmospheric environment, between the food and the packaging materials, or among the heterogeneous ingredients in the food product itself. Therefore, mass transfer studies on the migration of package components and food ingredients, on the absorption and desorption of volatile ingredients, flavors, and moisture, on gas permeation, and on the reaction kinetics of oxidation and ingredient degradation are essential for food packaging system designs (Han 2005a; Singh & Singh 2005).

Fish flesh contains long chains of polyunsaturated fatty acids with a variety of health benefits; however, polyunsaturated fatty acids are highly susceptible to undergoing oxidation reactions and their composition varies largely between species. Oxidation of fats is one of the most important mechanisms leading to food spoilage, second only to alterations produced by microorganisms. The oxidation of lipids in food leads to a reduction in shelf life due to changes in taste and/or odor, deterioration of the texture and functionality of the muscle, and a reduction in nutritional quality. It is important to have available a good film containing an antioxidant agent for storing fish, as this will increase the lifetime of the product and lead to economic benefits for fishermen and traders, and also allow consumers to enjoy good quality food products for longer (Abreu *et al.* 2010).

Frost accumulation is a common problem found in frozen seafood, and is especially troublesome when less flexible films such as cellophane, which cannot adhere tightly to other than rectangular-shaped products, are used. However, cellophane films are known to handle well an automatic packaging machinery, having fewer machinery problems than many films.

Rupture of the film at any point may cause only a small surface area of the product being exposed to oxidation or freezer burn, as the remainder of the package will cling tightly to the product. These films (i.e., higher vacuum shrink films) are not cheap, however, and some packers are concerned about the safety of such packaging in cases of severe temperature abuse. On consumer package labels, it is wise to include a statement such as "open wrapper before thawing" to ensure proper handling (Lanier 1990).

The development of new food packaging films with the incorporation of antioxidants is expected to improve the shelf life of food and thus increase consumer safety and health. The efficacy of a natural antioxidant derived from barley husks was verified in slowing down lipid hydrolysis and increasing oxidative stability. This method of incorporation of the antioxidant to the film proved highly effective in protecting salmon samples against oxidative processes during frozen storage (Abreu *et al.* 2010).

Textural changes within frozen foods can also be influenced by the permeability of the package to both O_2 and water vapor. Textural properties are closely related to the number and size of ice crystals within the food during frozen storage. It is well known that packages with high permeability to water vapor can ultimately lead to loss of moisture from the product and dehydration at the surface of the food. This is a particular problem if temperatures are fluctuating during frozen storage. Concurrently, loss of moisture from within the package often results in an increase in O_2 tension within the package, with resultant problems of oxidation as mentioned earlier.

Chemical and biochemical reactions are undoubtedly the major cause of deteriorative reactions in frozen foods. Quality deterioration in frozen foods can be associated with enzymic activity, protein denaturation, and lipid oxidation. Enzymic and hydrolytic activities can be minimized by adequate pretreatment of the food prior to freezing, for example, the blanching of fruits and vegetables and the cooking of meats. However, even after adequate pretreatments, it is recommended that the food manufacturer minimizes the effects of slow freezing and recrystallization, to minimize the release of enzymes from the cellular infrastructure. It is known that the denaturation of proteins in frozen foods is caused by ice formation and recrystallization, dehydration, oxidation, and concentration imbalance of salts. Again, fast freezing and prevention of ice recrystallization are among the methods suggested to minimize the effects. Although the role of packaging materials in this respect is less clear, all major food categories undoubtedly benefit from rapid freezing and constant, low frozen storage temperatures. This dictates the production method and food packaging system used by the food manufacturer to some extent, that is, whether to freeze the food product within the package (which may substantially increase the freezing time), or to fill the frozen food product into a precooled package (which may add to production complexity and cost). The decision is dictated as much by the availability of processing equipment and the economics of the operation as the possible effects on food quality (Johnston *et al.* 1994; Harvie 1998; Hedges 2000; Rehbein 2002; Gonçalves & Blaha 2010).

International surveys have revealed that the image of frozen fish is not very positive. Whereas European consumers emphasized the neutral and insipid taste of these products, many US consumers believed that frozen fish is less nutritious and bonier than fresh fish, and has a tough and dry texture, bad smell, and inferior taste (Peavey *et al.* 1994; Rehbein 2002). However, by careful selection of raw material, processing, and storage conditions, frozen fishery products of high quality can be produced. Among them, storage conditions and maintaining the cold chain are the most important factors that the retailer should control to ensure the quality of frozen products. Examples of some problems of image of frozen seafood are referred to in Fig. 32.9 (Gonçalves & Blaha 2010).

Temperature and time of frozen storage have a large influence on the shelf life of frozen stored fish. Temperature should be as low as economically possible, but at least as less as −20°C. A weak point in the cold chain is the temperature of deep freezers in supermarkets,

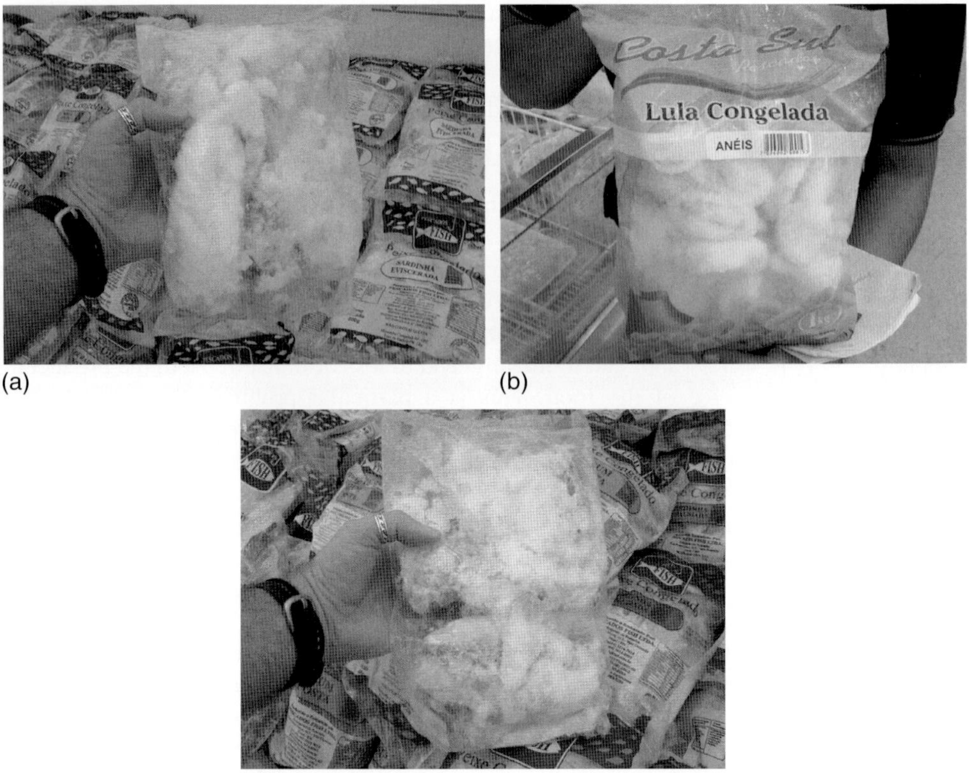

(a) (b) (c)

Fig. 32.9 Frozen seafood selling in an open-top chest cabinet (−18°C) in a Brazilian supermarket: (a) frozen eviscerated sardine (amount of recrystallized ice); (b) frozen squid rings (stuffed with ice); (c) frozen eviscerated sardine (amount of recrystallized ice plus lipid oxidation). The image of frozen seafood is not very positive!

which is often too high and fluctuates over the day. Other conditions, like vacuum packaging or protection from light, may have less influence on deterioration. A trial of storing catfish fillets at −20°C for 11 months did not reveal significant differences in the sensory attributes of fillets, which had either been vacuum packaged or O_2-permeable packaged (Anelich *et al.* 2001; Rehbein 2002).

REFERENCES

Abreu, D.A.P., Losada, P.P., Maroto, J. & Cruz, J.M. (2010) Evaluation of the effectiveness of a new active packaging film containing natural antioxidants (from barley husks) that retard lipid damage in frozen Atlantic salmon (*Salmo salar* L.). *Food Research International*, **43**, 1277–1282.

Ahvenainen, R. (2003a) Active and intelligent packaging. In: *Novel Food Packaging Techniques* (ed. R. Ahvenainen), 1st edn., pp. 4–21. Woodhead Publishing Limited and CRC Press LLC, Boca Raton, Florida.

Ahvenainen, R. (2003b) Introduction. In: *Novel Food Packaging Techniques* (ed. R. Ahvenainen), pp. 1–2. Woodhead Publishing, Cambridge, U.K.

Ahvenainen, R. & Hurme, E. (1997) Active and smart packaging for meeting consumer demands for quality and safety. *Food Additives & Contaminants*, **14** (6/7), 753–763.

Anelich, L.E., Hoffman, L.C. & Swanepoel, M.J. (2001) The quality of frozen African sharptooth catfish (*Claria gariepinus*) fillets under long-term storage conditions. *Journal of the Science of Food and Agriculture*, **81**, 632–639.

Bonilla, J., Atarés, L., Vargas, M. & Chiralt, A. (2010) Edible films and coatings to prevent the detrimental effect of oxygen on food quality: Possibilities and limitations. In: *International Conference on Food Innovation*, pp. 1–4. Universidad Politecnica de Valencia, Valencia, Spain, October 25–29.

Brody, A.L., Strupinsky, E.R. & Kline, L.R. (eds) (2001) *Active Packaging for Food Applications*. CRC Press LCC, Boca Raton, Florida.

Buffo, R.A. & Holley, R.A. (2005) Centralized packaging systems for meats. In: *Innovations in Food Packaging* (ed. J.H. Han), pp. 227–236. Elsevier Academic Press, London, U.K.

Coma, V. (2008) Bioactive packaging technologies for extended shelf life of meat-based products. *Meat Science*, **78**, 90–103.

Cooksey, K. (2005) Effectiveness of antimicrobial food packaging materials. *Food Additives & Contaminants*, **22**, 980–987.

Day, B.P.F. (1989) Extension of shelf-life of chilled foods. *European Food and Drink Review*, **4**, 47–56.

Day, B.P.F. (2000) Chilled food packaging. In: *Chilled Foods—A Comprehensive Guide* (eds M. Stringer & C. Dennis), 2nd edn., pp. 135–150. CRC Press LLC, Boca Raton, Florida, 428pp.

Emblem, A. (2000) Predicting packaging characteristics to improve shelf-life. In: *The Stability and Shelf-Life of Food* (eds D. Kilcast and P. Subramaniam), pp. 145–169. Woodhead Publishing Limited and CRC Press LLC, Cambridge, England.

Fellows, P.J. (2000) Chilling. In: *Food Processing Technology—Principles and Practice* (ed. P.J. Fellows), 2nd edn., pp. 386–405. CRC Press LLC, Boca Raton, Florida, 575pp.

Floros, J.D., Dock, L.L. & Han, J.H. (1997) Active packaging technologies and applications. *Food, Cosmetics and Drug Packaging*, **20** (1), 10–17.

George, M. (2000) Selecting packaging for frozen food products. In: *Managing Frozen Foods* (ed. C.J. Kennedy), 1st edn., pp. 195–211. CRC Press LLC, Boca Raton, Florida.

Giannakourou, M.C., Koutsoumanis, K., Nychas, G.J. & Taoukis, P.S. (2005) Field evaluation of the application of time temperature integrators for monitoring fish quality in the chill chain. *International Journal of Food Microbiology*, **102**, 323–336.

Gill, A.O. & Gill, C.O. (2005) Preservative packaging for fresh meats, poultry, and fin fish. In: *Innovations in Food Packaging* (ed. J.H. Han), pp. 204–226. Elsevier Academic Press, London, U.K.

Gonçalves, A.A. & Araújo, S.M.L. (2010a) El pescado fresco: Situación y fomento del consumo en Brasil. *INFOPESCA Internacional*, **42**, 16–21.

Gonçalves, A.A. & Araújo, S.M.L. (2010b) Pescado congelado: ¿Cómo asegurar la calidad durante la comercialización? *INFOPESCA Internacional*, **43**, 31–34.

Gonçalves, A.A. & Blaha, F. (2010) Cold chain in seafood industry. In: *Refrigeration: Theory, Technology and Applications*. Nova Science Publishers, Inc., Hauppauge, New York (in press—ISBN: 978-1-61668-930-8).

Haard, N.F. (1992) Technological aspects of extending prime quality of seafood: A review. *Journal of Aquatic Food Product Technology*, **1**, 9–27.

Han, J.H. (2003) Antimicrobial food packaging. In: *Novel Food Packaging Techniques* (ed. R. Ahvenainen), Chapter 4, 1st edn., 21pp. Woodhead Publishing Limited and CRC Press LLC, Boca Raton, Florida.

Han, J.H. (2005a) New technologies in food packaging overview. In: *Innovations in Food Packaging* (ed. J.H. Han), pp. 1–11. Elsevier Academic Press, London, U.K.

Han, J.H. (2005b) Antimicrobial packaging systems. In: *Innovations in Food Packaging* (ed. J.H. Han), pp. 80–107. Elsevier Academic Press, London, U.K.

Han, J.H. & Floros, J.D. (1997) Casting antimicrobial packaging films and measuring their physical properties and antimicrobial activity. *Journal of Plastic Film and Sheeting*, **13**, 287–298.

Han, J.H. & Gennadios, A. (2005) Edible films and coatings: A review. In: *Innovations in Food Packaging* (ed. J.H. Han), pp. 239–262. Elsevier Academic Press, London, U.K.

Han, J.H., Ho, C.H.L. & Rodrigues, E.T. (2005a) Intelligent packaging. In: *Innovations in Food Packaging* (ed. J.H. Han), pp. 138–155. Elsevier Academic Press, London, U.K.

Han, J.H., Zhang, Y. & Buffo, R. (2005b) Surface chemistry of food, packaging and biopolymer materials. In: *Innovations in Food Packaging* (ed. J.H. Han), pp. 45–59. Elsevier Academic Press, London, U.K.

Harvie, R. (1998) *Fish for Food*. Seafood Industry Training Organization (Seafood ITO), Wellington, New Zealand.

Hedges, N. (2000) The selection and pre-treatment of fish. In: *Managing Frozen Foods*(ed. C.J. Kennedy), 1st edn., pp. 95–110. CRC Press LLC, Boca Raton, Florida.

Hong, S.I. & Krochta, J.M. (2006) Oxygen barrier performance of whey-protein-coated plastic films as affected by temperature, relative humidity, base film and protein type. *Journal of Food Engineering*, **77**, 739–745.

Hong, L.C., LeBlanc, E.L., Hawrysh, Z.J. & Hardin, R.T. (1996) Quality of Atlantic mackerel (*Scromber scrombus* L.) fillets during modified atmosphere storage. *Journal of Food Science*, **61**, 646–651.

Hurme, E., Sipiläinen-Malm, T. & Ahvenainen, R. (2002) Active and intelligent packaging. In: *Minimal Processing Technologies in the Food Industry* (eds T. Ohlsson and N. Bengtsson), Chapter 5, 1st edn., 37pp. Woodhead Publishing Limited and CRC Press LLC, Boca Raton, Florida.

Järvi-Kääriäinen, T. (2003) Integrating intelligent packaging, store and distribution. In: *Novel Food Packaging Techniques* (ed. R. Ahvenainen), pp. 103–126. Woodhead Publishing, Cambridge, U.K.

Johnston, W.A., Nicholson, F.J., Roger, A. & Stroud, G.D. (1994) Freezing and refrigerated storage in fisheries. *FAO*, FAO Fisheries Technical Paper No. 340, Rome, Italy.

Karel, M. (2000) Tasks of food technology in the 21st century. *Food Technology*, **54** (6), 56–64.

Kerry, J.P., O'Grady, M.N. & Hogan, S.A. (2006) Past, current and potential utilization of active and intelligent packaging systems for meat and muscle-based products: A review. *Meat Science*, **74**, 113–130.

Koutsoumanis, K., Giannakourou, M.C., Taoukis, P.S. & Nychas, G.J. (2002) Application of shelf life decision system (SLDS) to marine cultured fish quality. *International Journal of Food Microbiology*, **73**, 375–382.

Kruijf, N., van Beest, M., Rijk, R. & Sipilainen Malm, T. (2002) Active and intelligent packaging: Applications and regulatory aspects. *Food Additives & Contaminants*, **19** (Suppl.), 144–162.

Labuza, T.P. (1987) Oxygen absorber sachets. *Food Research*, **32**, 276–277.

Lacroix, M. & Cooksey, K. (2005) Edible films and coatings from animal-origin proteins. In: *Innovations in Food Packaging* (ed. J.H. Han), pp. 301–317. Elsevier Academic Press, London, U.K.

Lanier, T.C. (1990) Packaging. In: *The Seafood Industry* (eds R.E. Martin & G.J. Flick), pp. 194–204. An Osprey Book—Van Nostrand Reinhold, New York.

Margeirsson, B., Dornboos, R., Ramirez, A., Mu, W., Xioli, F. & Bogason, S. (2008) D1.10: Flow chart of selected fish supply chains including mapping of temperature and other relevant environmental parameters. Chill on EU project (FP6-016333-2). Technical report. University of Iceland, Reykjavík, Iceland.

Margeirsson, B., Gospavic, R., Pálsson, H., Arason, S. & Popov, V. (2010) Experimental and numerical modelling comparison of thermal performance of expanded polystyrene and corrugated plastic packaging for fresh fish. *International Journal of Refrigeration*, **34** (2), 573–585.

Mendoza, T.F., Welt, B.A., Otwell, S., Teixeira, A.A., Kristonsson, H. & Balaban, M.O. (2004) Kinetic parameter estimation of time–temperature integrators intended for use with packaged fresh seafood. *Journal of Food Science*, **69**, 91–96.

Millers, D.W., Wilkes, J.G. & Conte, E.D. (1999) Food quality indicator device. PCT International Patent Application WO 99/0456.

Nuin, M., Alfaro, B., Cruz, Z., *et al.* (2008) Modelling spoilage of fresh turbot and evaluation of a time–temperature integrator (TTI) label under fluctuating temperature. *International Journal of Food Microbiology*, **127**, 193–199.

Oka, H. (1989) Packaging for freshness and the prevention of discoloration of fish fillets. *Packaging Technology Science*, **2**, 201–213.

Otwell, W.S. (1997) Time, temperature, travel—A quality balancing act. *Seafood International*, September, 57–61.

Pacquit, A., Frisby, J., Diamond, D., *et al.* (2007) Development of a smart packaging for the monitoring of fish spoilage. *Food Chemistry*, **102**, 466–470.

Pault, H. (1995) Brain boxes or simply packed? *Food Processing UK*, **64** (7), 23–26.

Peavey, S., Work, T. & Riley, J. (1994) Consumer attitudes toward fresh and frozen fish. *Journal of Aquatic Food Product Technology*, **3**, 71–87.

Quintavalla, S. & Vicini, L. (2002) Antimicrobial food packaging in meat industry. *Meat Science*, **62**, 373–380.

Randell, K., Hattula, T., Skytta, E., Silvertsvik, M., Bergslien, H. & Ahvenainen, R. (1999) Quality of filleted salmon in various retail packages. *Food Quality*, **22**, 483–497.

Rehbein, H. (2002) Measuring the shelf-life of frozen fish. Part III—Improving quality within the supply chain. In: *Safety and Quality Issues in Fish Processing* (ed. H.A. Bremmer), Chapter 21, 18 pp. Woodhead Publishing Limited and CRC Press LLC, Cambridge, England.

Rodrigues, E.T. & Han, J.H. (2003) Intelligent packaging. In: *Encyclopaedia of Agricultural, Food and Biological Engineering* (ed. D.R. Heldman), pp. 528–535. Marcel Dekker, New York, New York.

Rooney, M.L. (2000) Plastics in active packaging. In: *Materials and Development of Plastics Packaging for the Consumer Market* (eds G.A. Giles & D.R. Bain), pp. 105–129. Sheffield Academic Press, Sheffield, U.K.

Rooney, M.L. (2002) Active packaging: Science and application. In: *Engineering and Food for the 21st Century* (eds J. Welti-Chanes, G.V. Barbosa-Cánovas & J.M. Aguilera), Chapter 32, pp. 1–9. CRC Press LCC, Boca Raton, Florida.

Rooney, M.L. (2005) Introduction to active food packaging technologies. In: *Innovations in Food Packaging* (ed. J.H. Han), pp. 63–79. Elsevier Academic Press, London, U.K.

Rowe, N.J. (1990) Requirements of packaging for seafoods versus those for other types of products. In: *Advances in Fisheries Technology and Biotechnology for Increased Profitability* (eds M.N. Voigt & J. Richard Botta), pp. 215–224. Technomic Publishing Co., Inc., Lancaster, Pennsylvania.

Selman, J.D. (1990) Time–temperature indicators: How they work. *Food Manufacture*, **65** (8), 30–31 and 33–34.

Selman, J.D. & Ballantyne, A. (1988) Time–temperature indicators: Do they work? *Food Manufacture*, **63** (12), 36–38, 49.

Silva, J.L. & White, T.D. (1994) Bacteriological and color changes in modified atmosphere-packaged refrigerated channel catfish. *Journal of Food Protection*, **57**, 715–719.

Singh, R.K. & Singh, N. (2005) Quality of packaged foods. In: *Innovations in Food Packaging* (ed. J.H. Han), pp. 24–44. Elsevier Academic Press, London, U.K.

Sivertsvik, M. (2003) Active packaging in practice: Fish. In: *Novel Food Packaging Techniques* (ed. R. Ahvenainen), pp. 384–400. Woodhead Publishing, Cambridge, U.K.

Sivertsvik, M., Rosnes, J.T., Vorre, A., Randell, K., Ahvenainen, R. & Bergslien, H. (1999) Quality of whole gutted salmon in various bulk packages. *Journal of Food Quality*, **22**, 387–401.

Sivertsvik, M., Rosnes, J.T. & Bergslien, H. (2002) Modified atmosphere packaging. In: *Minimal Processing Technologies in the Food Industry* (eds T. Ohlsson & N. Bengtsson), Chapter 4, 1st edn., 26pp. Woodhead Publishing Limited and CRC Press LLC, Boca Raton, Florida.

Smolander, M. (2003) The use of freshness indicator in packaging. In: *Novel Food Packaging Techniques* (ed. R. Ahvenainen), pp. 127–143. Woodhead Publishing, Cambridge, U.K.

Spencer, K.C. (2005) Modified atmosphere packaging of ready-to-eat foods. In: *Innovations in Food Packaging* (ed. J.H. Han), pp. 185–203. Elsevier Academic Press, London, U.K.

Summers, L. (1992) *Intelligent packaging for quality. Soft Drinks Management International*, May, 32–36.

Taoukis, P.S. & Labuza, T.P. (2003) Time-temperature indicators (TTIs). In: *Novel Food Packaging Techniques* (ed. R. Ahvenainen), pp. 103–126. Woodhead Publishing, Cambridge, U.K.

Taoukis, P.S., Koutsoumanis, K. & Nychas, G.J. (1999) Use of time–temperature integrators and predictive modelling for shelf life control of chilled fish under dynamic storage conditions. *International Journal of Food Microbiology*, **53**, 21–31.

UN—United Nations. (2010) Agreement on the international carriage of perishable foodstuffs and on the special equipment to be used for such carriage (ATP). Economic Commission for Europe—Inland Transport Committee (ECE/TRANS/219). ATP as amended on January 2, 2011. United Nations, New York.

Woolfe, M.L. (2000) Temperature monitoring and measurement. In: *Chilled Foods—A Comprehensive Guide* (eds M. Stringer & C. Dennis), 2nd edn., pp. 100–134. CRC Press LLC, Boca Raton, Florida.

Yang, L. & Paulson, A.T. (2000) Effects of lipids on mechanical and moisture barrier properties of edible gellan film. *Food Research International*, **33**, 571–578.

Zhou, G.H., Xu, X.L. & Liu, Y. (2010) Preservation technologies for fresh meat—A review. *Meat Science*, **86**, 119–128.

Index

Handbook of Meat, Poultry and Seafood Quality, Second Edition. Edited by Leo M. L. Nollet.